T0317704

Automation and Computational Intelligence for Road Maintenance and Management

Automation and Computational Intelligence for Road Maintenance and Management

Advances and Applications

Hamzeh Zakeri
Amirkabir University of Technology
Tehran, Iran

Fereidoon Moghadas Nejad
Amirkabir University of Technology
Tehran, Iran

Amir H. Gandomi
University of Technology Sydney
Ultimo, Australia

This edition first published 2022

Registered Office
John Wiley & Sons, Inc., 111 River Street, Hoboken, NJ 07030, USA

Editorial Office
111 River Street, Hoboken, NJ 07030, USA

For details of our global editorial offices, customer services, and more information about Wiley products visit us at www.wiley.com.

Wiley also publishes its books in a variety of electronic formats and by print-on-demand. Some content that appears in standard print versions of this book may not be available in other formats.

Library of Congress Cataloging-in-Publication Data

Names: Zakeri, Hamzeh, author. | Nejad, Fereidoon Moghadas, author. | Gandomi, Amir Hossein, author.
Title: Automation and computational intelligence for road maintenance and management : advances and applications / Hamzeh Zakeri, Fereidoon Moghadas Nejad, Amir H. Gandomi.
Description: Hoboken, NJ : Wiley, 2022. | Includes bibliographical references and index.
Identifiers: LCCN 2022011360 (print) | LCCN 2022011361 (ebook) | ISBN 9781119800644 (cloth) | ISBN 9781119800651 (adobe pdf) | ISBN 9781119800668 (epub)
Subjects: LCSH: Road construction industry–Automation. | Computational intelligence.
Classification: LCC HD9717.5.R62 Z34 2022 (print) | LCC HD9717.5.R62 (ebook) | DDC 363.12/560285–dc23/eng/20220506
LC record available at https://lccn.loc.gov/2022011360
LC ebook record available at https://lccn.loc.gov/2022011361

Cover Design: Wiley
Cover Image: © Dmitry Kalinovsky/Shutterstock

Set in 9.5/12.5pt STIXTwoText by Straive, Pondicherry, India

Contents

Dedication

This book is dedicated to all researchers
and
intuitive thinkers.

Preface

The content of this book is designed to be used for two purposes. The first application is as a textbook for the introductory courses of pavement management and topics related to automation of evaluation based on image interpretation. The second application is as a reference title for real applications of artificial intelligence methods, new nature-inspired algorithms to solve real and advanced issues of pavement and infrastructure management, and for managers and researchers using analytical information and data analysis in a particular field of expertise.

Rapid advances in computing power and the design of data capture sensors have occurred in recent years, enabling designers and researchers to use to extract new information for developing related technologies. A few examples include image processing techniques, machine vision, AI methods, computational intelligence, ambiguity and solving fuzzy problems with high complexity, optimization based on nature-inspired algorithms and knowledge extraction, an important and increasing role in the development of computational theory, engineering problems, and infrastructure management. Moreover, new computational methods have been developed in other fields that influence the diagnosis, classification and evaluation techniques. Due to the wide range of academic and professional audiences who may use this book, we address these topics as educational content and with the aim of generating ideas from different perspectives on existing theories. That is, the purpose of writing this book is simply to get acquainted with new methods and show some of the applications of these methods in pavement and road management as the main assets of countries. It should be noted that the use of these methods is not limited to road pavement management, civil engineering, and infrastructure management and can be used in other fields, such as astronomy, medical engineering, electrical and electronics, mechanics, industrial engineering, computer engineering, environment studies, forestry, geology, planning, and so on. In this book, it is clearly stated that the combination of different image processing and artificial intelligence methods can be used in the production of new knowledge to solve complex and practical problems simultaneously. This book, comprised of 10 chapters on discrete topics, is recommended to anyone involved in collecting and analyzing big data. Since the publication of this book, science and technology may have been growing exponentially, but the methods discussed in this book are the product of years of research and collection of new material by authors that can generate many ideas for research at the doctoral level and beyond. Gaining expertise for informed and reliable interpretations of the outputs of data collection devices at the network level, such as images and laser data, is one of the topics covered in university courses that do not exist in a single reference and requires searches in different sources. Therefore, this book has been designed and compiled with the aim of collecting these topics in a single reference. The idea of sensor design, digital image processing, and applications is discussed in this book. This version of the book focuses on the acquisition and analysis of digital imaging, while basic information on previous analog sensors and methods is crucial. We have expanded the coverage of lead-based 3D sensing systems, type III fuzzy analysis, and nature-inspired metaheuristics optimization as extensible topics, including image-based multilevel methods and extraction feature methods. Entropy, classification and fragmentation, content-based pattern extraction, deep learning methods, three-dimensional classification of polar support vector, and a meta-innovative algorithm inspired by penguins' behavior are discussed as new topics. Moreover, after studying these methods, the computational power is expected to improve, leading to greater emphasis on techniques. At the end of each chapter, applications of the methods in a specialized field are presented.

The authors would like to say thanks to research colleagues: Dr.Mahsa Payab, Prof. Ahmad Fahimifar, Dr. Behrouz mataei, Prof. MH Fazel Zarandi, Dr.Abolfazl Doostparast Torshizi, Prof. Xin-She Yang, Dr. Sajad Ranjbar, Dr. Seyed Arya Fakhri, Prof.Dr. Serkan Tapkin, Prof. M Saadatseresht, M Hajiali, AA Nik, Prof. A Khodaii, N Taheri, MM Makhmalbaf, N Karimi, SK Azin Sadeghi Dezfooli, A Mehrabi, FZ Motekhases. The authors also would like to say special thanks the Sarah Lemore, Associate Managing Editor, and Viniprammia Premkumar, Content Refinement Specialist for their support and efforts.

In the end, we hope that the topics presented in this book will give readers a new perspective of the innovative ways to solve engineering problems.

Hamzeh Zakeri
Adjunct Researcher/Amirkabir University of Technology (Tehran Polytechnic), Iran

Fereidoon M. Nejad
Professor Amirkabir University of Technology (Tehran Polytechnic), Iran

Amir H. Gandomi
Professor of Data Science at the Faculty of Engineering & Information Technology,
University of Technology Sydney, NSW, Australia

Author Biography

Hamzeh Zakeri

Hamzeh Zakeri/Adjunct Research professor joined Dep. of Civil & Environmental Engineering, Amirkabir University of Technology (Tehran Polytechnic), Iran, Hafez Ave, Tehran, Iran. He received Phd in Civil Engineering, 2016, Iran Amirkabir University of Technology (Tehran Polytechnic), Iran. His research interests include: Automation, Fuzzy type 2, Image Processing, Remote sensing, Machine Learning, Knowledge extraction, Hybrid Meta-heuristic Application in the field of pavement engineering. He has published more than 20 papers in international/national journals and has authored several chapter books.

Fereidoon Moghadas Nejad

Fereidoon Moghadas Nejad holds a Ph.D. degree (1996) from The University of Sydney, Australia, in Civil Engineering. After earning his Ph.D., he was a Postdoctoral Scholar in the Department of Civil Engineering at University of Western Sydney in 1997. In 1998, he joined Dep. of Civil & Environmental Engineering, Amirkabir University of Technology, as an Assistant Professor, and currently is Professor and Head of Transportation Division. Moghadas Nejad's research interests include Pavement Materials Testing, Soil Reinforcement, Pavement Recycling, Nondestructive Testing (i.e. FWD, GPR), Pavement & Railway Management Systems, Pavement Rehabilitation and Maintenance, Computational Models for Asphalt and Concrete Pavements (Thermal modelling, Modelling of Structural Behaviour), Constitutive Models for Pavements (Elastic, Elastoplastic and Viscoelastic Behaviour), Image Processing, Automation, Fuzzy and Numerical Methods in Pavement and Railway Engineering. He has been authored or coauthored more than 190-refereed journal and 200 national and international conference papers, five books and several book chapters, and holds 21 Iran and one US patents.

Amir H. Gandomi

Amir H. Gandomi/Professor of Data Science at the Faculty of Engineering & Information Technology, University of Technology Sydney, NSW 2007, Australia. Amir H. Gandomi is a Professor of Data Science and an ARC DECRA Fellow at the Faculty of Engineering & Information Technology, University of Technology Sydney. Prior to joining UTS, Prof. Gandomi was an Assistant Professor at Stevens Institute of Technology, USA and a distinguished research fellow in BEACON center, Michigan State University, USA. Prof. Gandomi has published over two hundred journal papers and seven books which collectively have been cited 25 000+ times (H-index = 73). He has been named as one of the most influential scientific minds and Highly Cited Researcher (top 1% publications and 0.1% researchers) for five consecutive years, 2017 to 2021. He also ranked 18th in GP bibliography among more than 12 000 researchers. He has served as associate editor, editor, and guest editor in several prestigious journals such as AE of IEEE TBD and IEEE IoTJ. Prof. Gandomi is active in delivering keynotes and invited talks. His research interests are global optimization and (big) data analytics using machine learning and evolutionary computations in particular.

1

Concepts and Foundations Automation and Emerging Technologies

1.1 Introduction

The term "automation" generally refers to a set of automated processes on an input based on expert-inspired thinking. In a broader and more specialized field of management, the term includes automatic data retrieval (DR) and intelligent digital data processing (DP) to understand and quantify the situation. The input data to these systems are a set of numbers that are transmitted online or offline using advanced hardware and even a single image processor. Figure 1.1 shows a simple structure of an automated system for determining road conditions (at Amir Kabir University of Technology, Tehran Polytechnic), which is used for the automation in infrastructure management. Automation encompasses a range of activities from information extraction to robotic navigation, data analysis, solution presentation and knowledge discovery, and knowledge learning and self-learning.

Data on the condition of an infrastructure, such as roads, bridges, tunnels, railroad tracks, or a microscopic image of a bitumen mixture automatically with the machine, are first converted into digital format and stored as input in computer memory or transferred online to the analyzer software. These digital data can be processed or displayed and controlled simultaneously or on a high-resolution monitor. In general, the process of digitization of road infrastructure scans includes all operations of digitization, storage, processing, and display of output through the computer. The program inputs are then transferred to the processor after storage or online through the terminal. After processing the outputs through the same terminal, the data are available and usable. Figure 1.2 shows the chain of automation steps for automating typical processing.

Automated processing and automation have a wide range of applications, such as automatic assessment of road surfaces, automatic assessment of bridges and technical structures, assessment and inspection of tunnels, evaluation of pavement texture roughness, quality control of pavement markings, road safety audit, detection of signs, and classification. For this purpose, advanced devices in-line with modern technology such as remote sensing (RS), use of satellites and other spacecraft, robotics and automatic inspection, and advanced laser equipment are used. The data obtained by the multipurpose usability, such as images, are used for the simultaneous assessment of pavement distress as well as road audits.

This data may be used to isolate and monitor the health of infrastructure or to diagnose damage or other infrastructure characteristics, such as surface drainage, friction, roughness, and slope and arch determination. Images captured by automated systems are utilized as important data to detect various types of failures or to visually evaluate decisions. Figure 1.3 provides examples of several different types of images in infrastructure. Other needs and applications of automation range from robot insights for automation in aerial road imaging to the movement of ligaments on tunnel wall bodies that fall into the automation category. In other words, whenever a machine receives two- or

Automation and Computational Intelligence for Road Maintenance and Management: Advances and Applications,
First Edition. Hamzeh Zakeri, Fereidoon Moghadas Nejad, and Amir H. Gandomi.

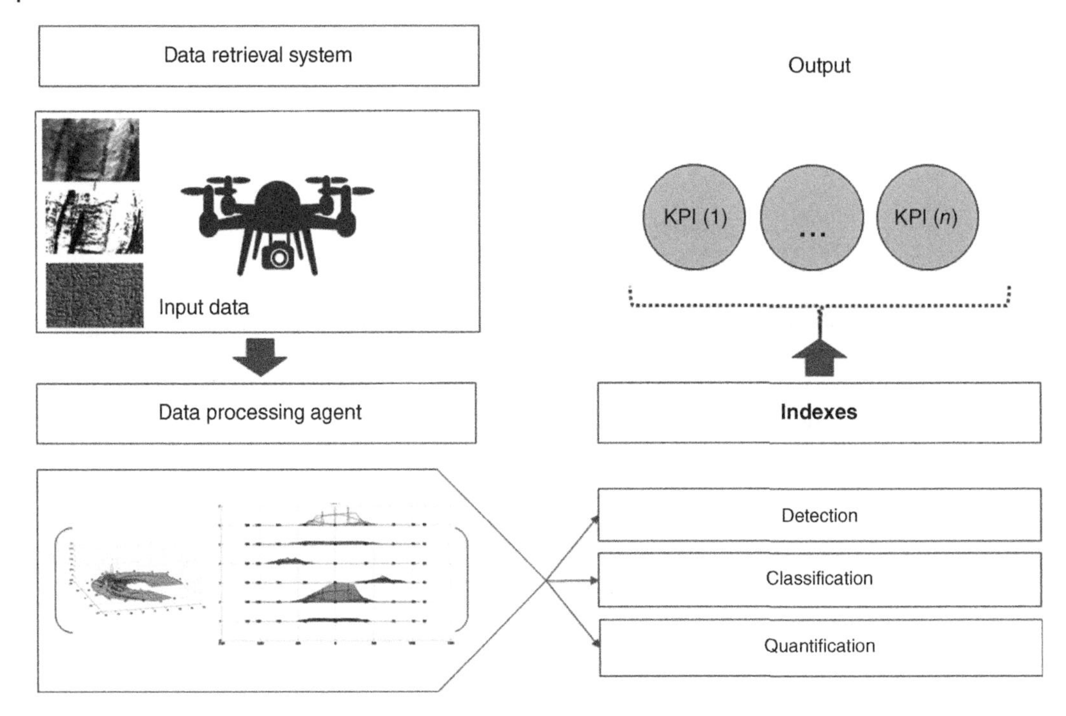

Figure 1.1 An example of a schematic automatic robotic information retrieval system for automation in infrastructure management.

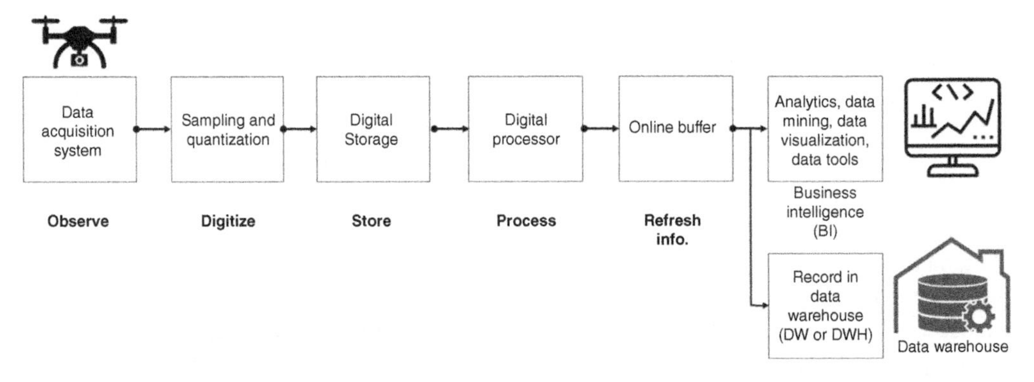

Figure 1.2 The schematic chain of automation steps for automating typical processing.

higher-dimensional data, an image is eventually processed. Although there are many methods and limitations to image processing, in this text, we will consider the following basic classes:

- Structure and framework of automation and key performance indices (KPIs)
- Advanced image processing techniques
- Fuzzy techniques and recent advances
- Automatic detection and its applications in infrastructure
- Feature extraction and fragmentation methods
- Feature prioritization and selection methods

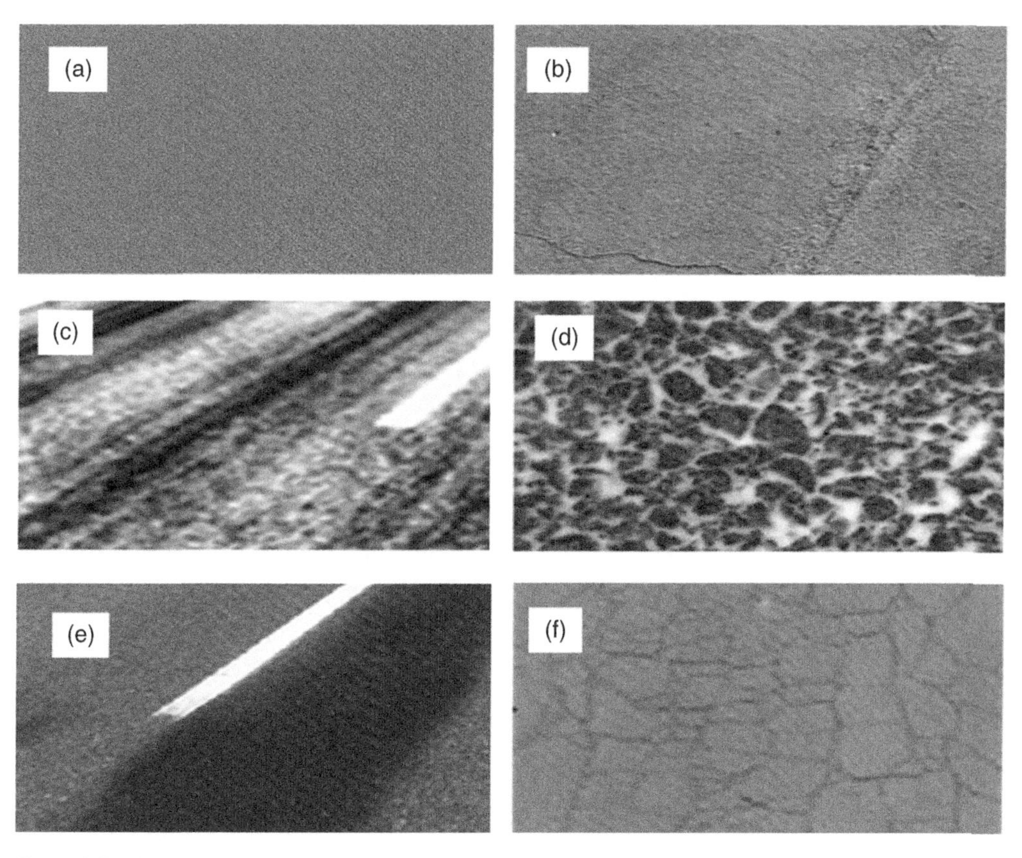

Figure 1.3 Examples of 2D images as an input for the processing step. (a) Pavement without surface damage with coarse texture, (b) pavement with small transverse cracks, (c) pavement with bitumen damage in the path of wheels, (d) pavement with high surface drainage capability, (e) pavement with rutting distress in wheel path, and (f) pavement with alligator-type surface cracking distress (fatigue cracking).

- Classification methods and their applications in infrastructure management
- Models of performance measures and quantification in automation
- Nature-Inspired Optimization Algorithms (NIOAS)

1.2 Structure and Framework of Automation and Key Performance Indexes (KPIs)

Automation is a completely systematic process that requires basic design. If any of the steps are designed incorrectly or the process is not followed correctly, it may lead to the failure of the use of automation in the future. Given the importance of this issue, the elements of infrastructure management and its automation should be considered. The three main components in automation design are the following:

- Include data retrieval (DR)
- Data processing (DP)
- Data and information (DI) interpretation

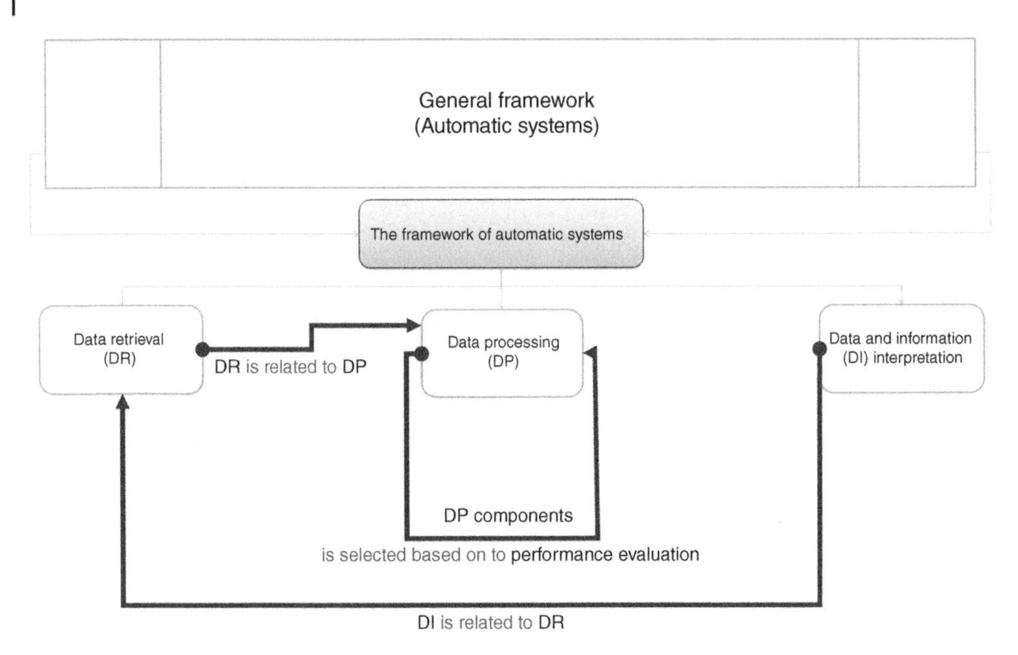

Figure 1.4 General architecture used to design the automation.

Figure 1.4 shows the general architecture used to design the automation. A discussion of the general structure and macroarchitecture of the development of infrastructure models at the network level as well as the main components of the overall system design modules is given in Chapter 2.

The initial selection of indicators and the acceptance of these indicators play an important role in the implementation of automation and its success. As a general principle, the selection of this module directly affects the selection of DR because indicators are ordered according to need. The collection of certain information requires the use of special equipment.

In national and macroautomation systems, the choice of technology depends a lot on the indicators desired by managers and affects the level of management. For this reason, the order of the automation chain is different in practice and requires a top-down design. Before designing any system, it is necessary to have a proper understanding of the types of common indicators in the management of roads and technical buildings. After fully understanding the needs, then the role of emerging technologies and future research of automation largely affects the selection of the method.

1.3 Advanced Image Processing Techniques

In order to evaluate and analyze the images, it is often necessary to extract directional information on the subjects (including cracking, texture, aggregate morphology, morphology of bitumen contents, friction, etc.) in the image. For this reason, multilevel methods are considered as efficient tools due to their ability to decompose information in several levels and the possibility of reconstructing them with the least amount of error.

In this section, various types of single and multilevel methods are introduced, then, using the indicators introduced in Section 1.2, the efficiency of each multilevel method in specialized issues

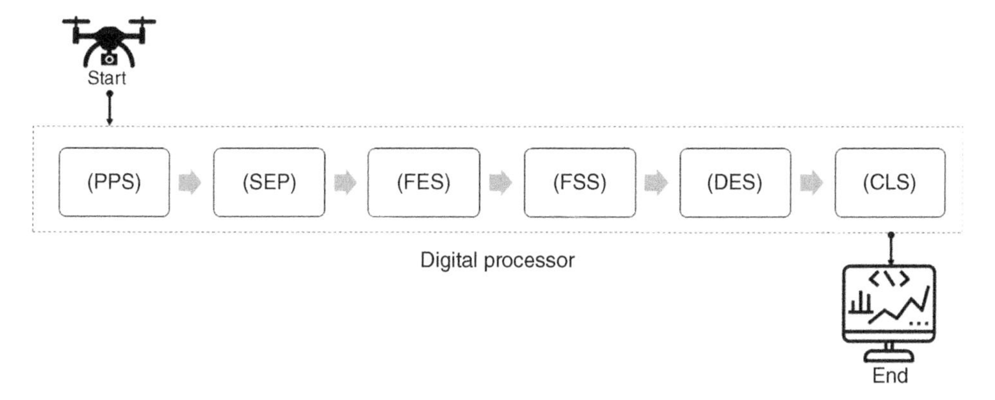

Figure 1.5 General steps of digital processing in automation.

is evaluated. The main characteristics of image quality evaluation, as shown in Figure 1.4, are evaluated for each method and with different filters.

The three main components for any automated system are the following: (i) image capture device, or the image equitization component (IAC); (ii) image analysis system and related algorithms, or the image processing component (IPC); and (iii) interpretation and indexing methods, or the image interpretation component (IIC). To perform analysis on the image, six general steps are required to obtain the result, including preprocessing, segmentation, feature extraction, feature selection, detection, and classification.

One of the most important methods of improving data quality is multilevel analysis that consists of the wavelet approach (WA), curvelet approach (CA), ridgelet approach (RA), and shearlet approach (SA), which has many applications. In this section, the capability of each of these single-level and multilevel methods in noise elimination is clearly presented, then the types of filters and optimal filter selection methods are introduced. Finally, the optimal method is selected from the existing methods as an example for case studies in the field of pavement management according to the capabilities of the method (see Figure 1.5).

The various preprocessing methods presented in Chapter 3 are used to improve the quality of images, with the aim of removing noise and enhancing the image in order to increase the detection power in the separation and detection stages via image processing. Also, different single-level and multilevel methods will be studied in detail. In general, multilevel methods work better than single-level methods. Among the multilevel methods, the complex Shearlet transform (SHT) method is a distinctive method with high capabilities, plays an important role in the image quality improvement stage, and has various filters and optimal analysis, which are widely introduced in Chapter 3. Various indicators, including the peak signal-to-noise ratio (PSNR) index, are used to evaluate the performance of algorithms and methods. This index is higher for the SHT method than other methods. On the other hand, the error index of the Shearlet complex is significantly lower than that of other methods. It should be noted that this method has a high flexibility, and by adjusting the main parameters depending on the type of problem and the nature of the image, practical results can be extracted. Due to the possibility of accessing SHT coefficients in different directions and sections with different scales, this method provides sufficient details and information for an accurate evaluation. Chapter 3 describes practical examples and case studies of the SHT method (Figure 1.6).

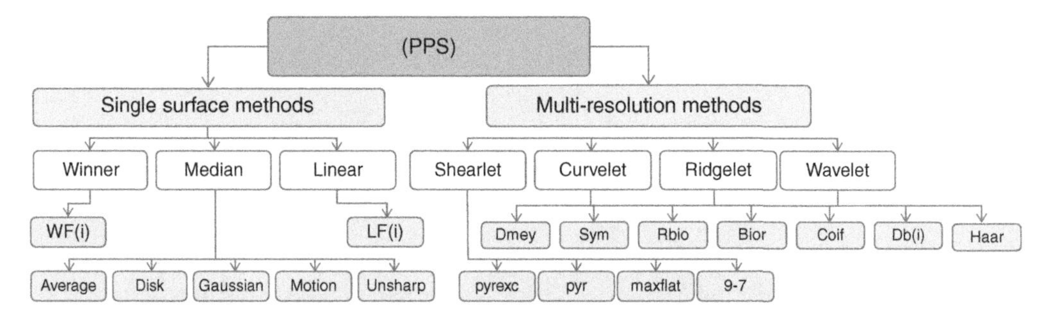

Figure 1.6 Classification of multilevel methods in preprocessing (multiresolution methods).

1.4 Fuzzy and Its Recent Advances

Among the various failures in infrastructure, infrastructure disease symptoms, such as cracking, are the most critical. The different types of pavement surface cracks include longitudinal, transverse, diagonal block, lizard, and track on the road surface that can be visually analyzed and interpreted. The use of image processing tools generates data for analysis that, in many cases, requires fuzzy methods for ambiguity or quantification. In Chapter 4, the concepts of different fuzzy sets are introduced in depth, which can be used as an emerging method to solve ambiguous problems. Fuzzy sets of type 2 and above along with their performance are examined in detail. Given that research on fuzzy set theory has recently entered the field of infrastructure management and image processing, this book summarizes the most important concepts and basic operations methods that can be used to study types 1, 2, and the type 3 fuzzy sets (see Figure 1.7) presented and it is hoped that new ideas will be generated for readers in Chapter 4.

In infrastructure management, especially road paving, we face many descriptive issues, including the severity and extent of damage, which requires the use of advanced methods for analysis. Since the description of this type of distress is usually inaccurate, these indicators are vague (low, medium, and high). By understanding fuzzy laws, fuzzy relations, fuzzy reasoning, and fuzzy facts, a new window opens to solve engineering problems with a degree of ambiguity for decision-making using the fuzzy method.

Fuzzy rules and fuzzy reasoning are essential components of fuzzy inference systems (FISs) that are the most important modeling elements based on the concept of fuzzy set and are widely used in the analysis and modeling of false and ambiguous topics. The following Section 1.4 presents an overview of fuzzy model types, while the latest achievements in the development of type 3 fuzzy modeling are discussed in Chapter 4.

1.5 Automatic Detection and Its Applications in Infrastructure

In the maintenance and management of infrastructure, the most important task and mission of the manager and network engineers are to maintain and improve the efficiency of infrastructure structures, such as roads, bridges, technical buildings, and other structures. However, before starting any treatment, the complication and its initial extent should be determined. Since it is not possible to visually inspect all infrastructures accurately and quickly, human error is one of the most greatest evaluation errors; in this regard, automated systems are used for detection. In Chapter 5, the

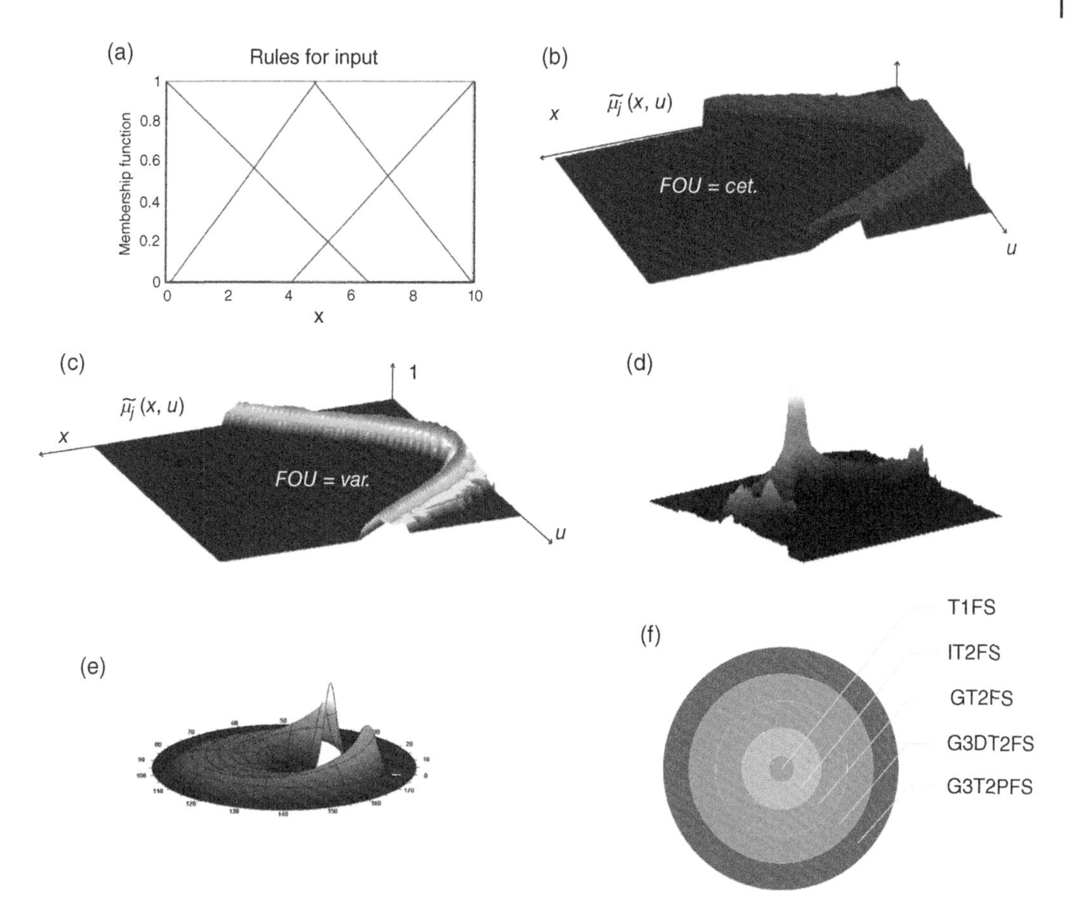

Figure 1.7 Classification of the fuzzy method and its recent advances (a) fuzzy type(T1FS), (b) fuzzy interval type 2(IT2FS), (c) fuzzy general type 2(GT2FS), (d) fuzzy type 3(G3DT2FS), (e) fuzzy type 3 in polar frame(G3DT2PFS), and (f) fuzzy domain.

principles and diagnosis methods as well as the new and effective parameters for diagnosing failure and abnormality are presented. Accurate and fast performance of the diagnostic stage is very effective in later stages of DI, including classification and evaluation.

Despite the advancement of processing methods and tools in detecting infrastructure anomalies, there is still insufficient information in data analysis and interpretation for satisfactory practical applications. Therefore, researchers continue to look for a stronger, more efficient method that is more compatible with harvesting tools. Hypotheses and general rules of diagnostic processing include one of the following statements or a combination of them:

- Abnormal pixels are darker/lighter than the original background.
- The presence of the anomaly has a nonuniform distribution in the gray image histogram.
- Abnormal morphology follows certain patterns.
- Specific anomalies are a set of continuous and continuous veins (such as cracks) or dense patterns (such as holes).
- The characteristics of anomalous patterns can vary but, overall, are a heterogeneous part of the whole set.

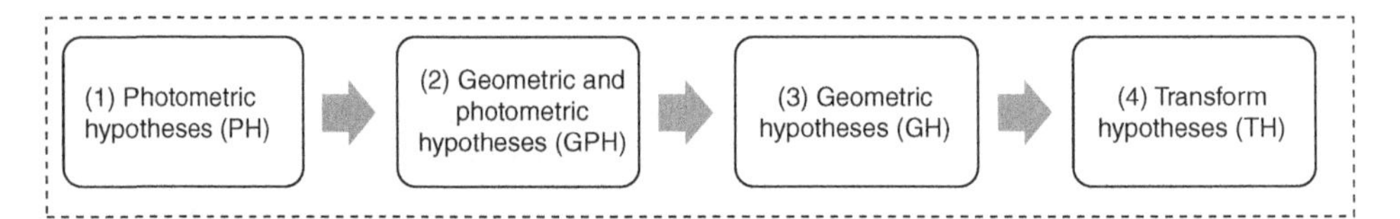

Figure 1.8 The process of the main hypotheses in the stage of diagnosis and segmentation.

- Anomalies usually have vague features around the edge.
- Geometric diagnosis of anomalies is not recommended due to the computational complexity of the diagnosis.
- Analyzing anomalies in transmitted space and applying the properties of the core is simpler and faster than the geometric method.

Each of the methods available for the diagnosis may consider one of the following hypotheses: (i) photometric hypotheses (PH), (ii) geometric and photometric hypotheses (GPH), (iii) geometric hypotheses (GH), and (iv) transform hypotheses (TH) (see Figure 1.8).

1.6 Feature Extraction and Fragmentation Methods

The main purpose of using image processing techniques is to extract meaningful features in order to perform classification and evaluation. The next step after separating the objects (or inconveniences) in the image is to convert the information into properties with the aim of collecting and reducing the amount of data for further processing. Objects separated from the segmentation stage must be converted into a property vector to be classified and constructed so that classification methods can be applied. Extracting features with a greater data transfer capability can increase the speed and efficiency of the method. Different methods for extracting features and their characteristics are examined in Chapter 6. Transformation of information into meaningful and useful features in the category is presented in this section with the aim of summarizing and increasing the speed and accuracy of detection as well as reducing the amount of image data for further processing. Extraction of desirable features to increase the ability to transmit more visual information or increase the amount of information has remained the focus of researchers. Particularly, the results of such extraction in the automatic detection of damaged images have a wide range of applications in infrastructure management (Figures 1.9 and 1.10).

In general, the eight main categories of feature extraction include the following: Low-Level Feature Extraction Methods, Shape Based Feature (SBF), 1D Function-Based Features for Shape Represent, Polygonal-Based Features (PBF), Spatial Interrelation Feature (SIF), Moments Features Extraction (MFE), Scale Space Approaches (SSA) for Feature Extraction, and Shape Transform Features (STF). To evaluate the proposed methods, four case studies in various applications of infrastructure management are presented and discussed with the interpretation of the relevant features.

1.7 Feature Prioritization and Selection Methods

There are several ways to select important and effective features, through which a subset of input variables is identified that can describe the input data more effectively and the negative effects of additional variables and input errors through topics, such as noise. Or this subset can

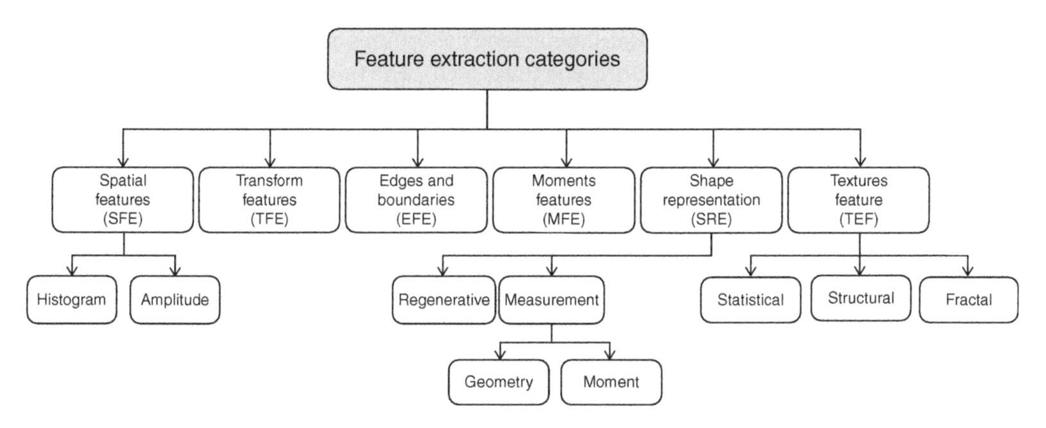

Figure 1.9 Classification of feature extraction methods.

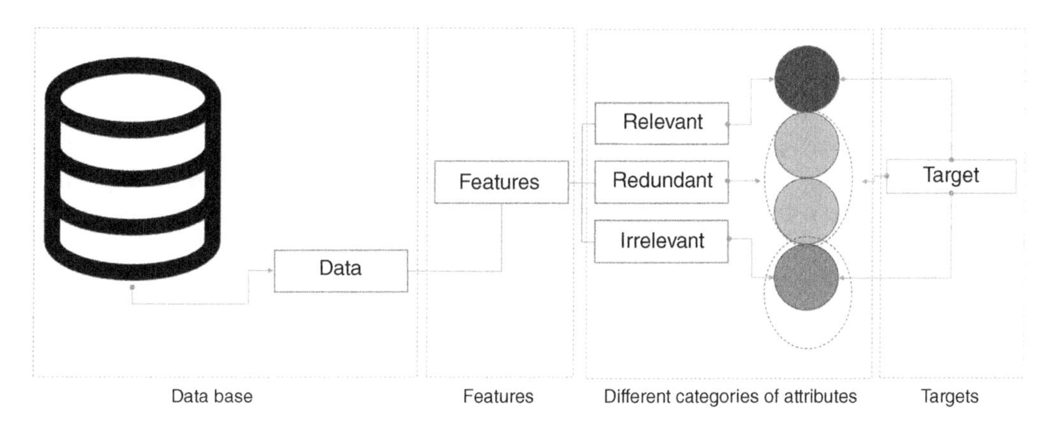

Figure 1.10 General framework of feature selection.

explain additional variables to provide better predictive results with higher speed and less computational complexity. One application of analysis is to influence the properties of a category or the extent to which features overlap to describe a subject. Standard data can contain hundreds of attributes, many of which can be closely related to other variables (for example, when two attributes are perfectly related, only one attribute is sufficient to describe the data and an additional attribute should be removed). However, dependent variables do not contain any useful information about the classification and are, therefore, an extra feature for the classification. This means that the entire content of the information can be obtained by minimizing the number of unique independent features that contain the maximum separating information. By keeping the variables independent, the number of attributes can be reduced, which improves the classifier performance (speed and accuracy). In many applications, variables that are more class-related are retained, while other attributes like noise may reduce classification performance. Therefore, choosing the right features is a prominent and key art form of designers and researchers.

Input attributes are used in attribute selection methods to reduce their number. When determining the criteria for selecting a feature, a method that separates the subset of the best useful features and considers them as a new set should be used.

There are three general types of relationships among the features extracted from the data: related (related), additional (additional), and unrelated (unrelated). The third category includes irrelevant features that do not have significant information about the target and should be removed.

Feature removal methods generally fall into one of four categories

1) Filter methods
2) Wrapper methods
3) Embedded methods
4) Hybrid methods

In filter methods, the operator is used for preprocessing to rank properties, in which high-ranking properties are selected and used for prediction. In this method, other characteristics play an additional role in processing and finally speed and accuracy increase with the optimal feature vector.

In wrapper methods, the criterion for selecting a feature depends on the performance of the predictor, i.e. the predictor is placed on a search algorithm that finds a subset with the highest prediction performance. This method is a kind of optimization with the function of maximizing performance and minimizing the number of features.

Embedded methods involve selecting variables as part of the training process that operate without dividing the data into training and testing sets.

Hybrid methods use a combination of the above three methods. In this chapter, feature selection methods using supervised, unsupervised, and semisupervised learning algorithms are first reviewed, then a new semisupervised method based on the fuzzy method for sample feature vectors is presented. It should be noted that the choice of method depends on the type of issue and quality of the characteristics. The best strategy to select a feature is to develop a method specific to the problem that requires a proper identification of the relationships and an understanding of the concept of features.

1.8 Classification Methods and Its Applications in Infrastructure Management

Classification is a type of prediction and calculation method in which a method is designed to guess the placement of data in a category. If the output is categorized and a class is located, the method is called classification, and if the output is numerical, it is called regression. Descriptive modeling, or clustering, includes assigning each input to a cluster so that similar inputs are placed in the same cluster. Finally, by observing the relationship between similar data, one can discover association rules about the relationship between inputs. By common definition, classification is an important application in data science that predicts the attribution of a variable (goal or class) based on the construction of a model using one or more numerical variables (predictors or attributes). The basis of all variables depends on the categories to which they belong. Such affiliation is calculated by using one of the following: (i) Frequency Table, (ii) Covariance Matrix, (iii) Similarity Functions. Classification is the most commonly used technique for large datasets in the research and practical applications of infrastructure management. It is also prominently applied in a variety of applications, such as detection, feature extraction, feature selection, complication type determination, severity category, scope determination, and evaluation. In this method, data analysis is based on learning-based algorithms that are supervised and consistent with data quality. In these algorithms, the goal is to detect and deduce a relationship that relates the desired variable, qualitatively, to other

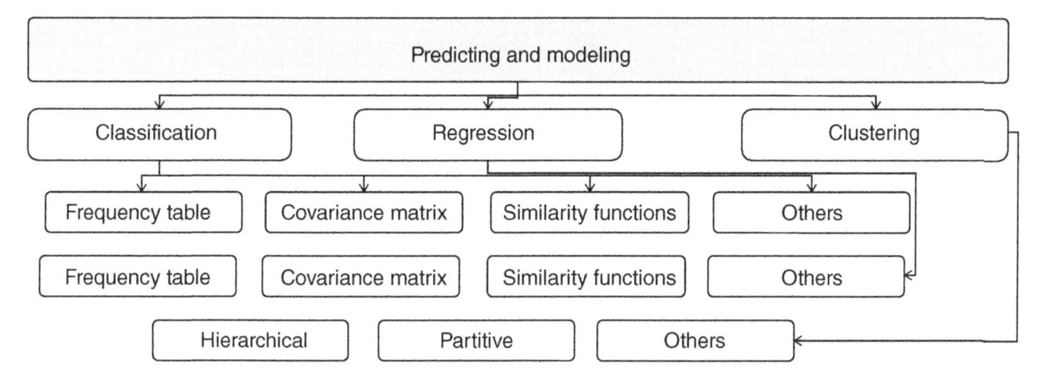

Figure 1.11 The various categories of learning modeling methods and classification.

observed variables that can be predicted. As a result, extracting this relationship is a kind of learning from data behavior that is considered as a branch of data mining. By definition, the classifier algorithm is responsible for classification, and all data are sample observations. It should be noted that classification is used when the desired variable is qualitative, such as determining the severity of a failure (low, medium, and high), type of failure (longitudinal, transverse, oblique, block, and fatigue), and extent of distress (low, medium, high, and very high) in infrastructure management.

The classification method employs various algorithms to obtain useful information, which can be categorized according to Figure 1.11. Particularly, in infrastructure management, this method is used to determine the behavior, i.e. conditions, of roads, bridges, tunnels, and technical buildings. Using these methods to classify is considered a kind of intelligence in management. By classification, a distinction is made between data that are purposefully useful (such as for the presence of cracks in infrastructure/detection of healthy sections) and irrelevant data (as in other types of failures). For example, each section can be divided into two general classes with distress and no failure(yes/no).

Chapter 8 briefly describes the most common classification methods used in infrastructure management, and examples of the application of these methods are presented in Chapter 8 or throughout the book. At the end of this chapter, new hybrid methods are presented as the development of basic methods that can be used to solve various problems. These methods are generally the development of primary methods and are used in combination with other classification methods. In this chapter, the advanced method of classifying the support vector in polar coordinates is clearly presented. Also, the basics of the fuzzy classification method used to classify ambiguous data are given as an extended method with examples.

1.9 Models of Performance Measures and Quantification in Automation

In order to evaluate the overall performance of a method or a classifier, a wide range of criteria with varying degrees of sensitivity have been identified and used to classify methods for subsequent selection and application over the past two decades. Due to the importance of evaluating these methods in infrastructure management, this section is of particular importance. In general, indicators for evaluating the methods are generally complementary, and the adequacy of each indicator alone does not mean the overall guarantee of the method. In general, the results obtained from the evaluation methods depend on the opinion of the analyst, the harvesting instructions, technology

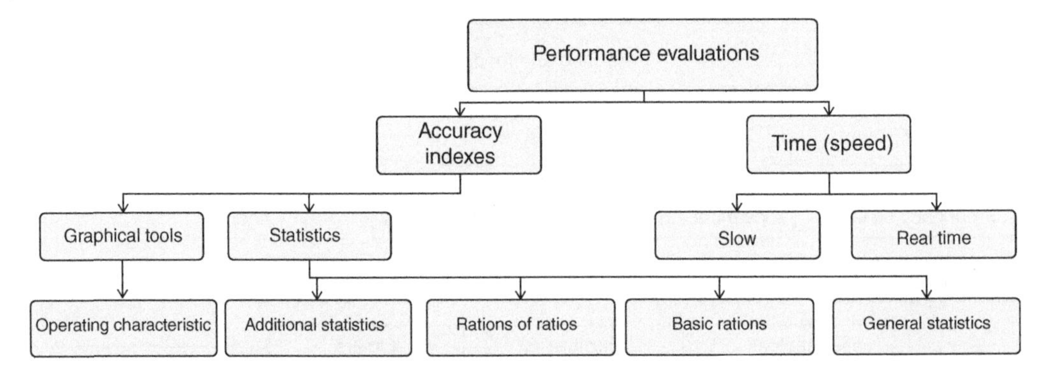

Figure 1.12 Classification of performance measurement indicators.

used, harvesting conditions, analysis speed, and many other dependent factors. For these reasons, different results of evaluating automated methods are expected. In this chapter, performance evaluation methods and indicators are first presented, then the types of general indicators in infrastructure evaluation are summarized. Figure 1.9 shows a general classification of the performance appraisal methods and common indicators. Assessing the true accuracy and performance of automated infrastructure assessment algorithms is critical in choosing the path of analysis and obtaining practical results. Many indices, such as mean square error (MSE), entropy, fuzzy index, signal-to-noise ratio (SNR), PSNR, and mean absolute error (MAE), are used for evaluation. Due to the existence of hidden characteristics, like the number of positive and negative samples as well as categories with different number of samples, the use of the accuracy index alone is not effective for evaluating the overall capability of a method.

In this Chapter 9, evaluation methods are divided into five general categories, including (i) General statistics, (ii) Basic rations, (iii) Rations of ratios, (iv) Additional statistics, and (v) Operating characteristic. Each category is then examined for each diagnosis and classification method. Based on the information extracted from the complexity matrix, indices like accuracy, error, probability of detection, productive index, selectivity, reproduction, negative predicted value (NPV), false positive rate (FPR), false negative rate (FNR), false discovery rate (FDR), false omission rate (FOR), likelihood ratio for positive tests (LRPT), likelihood ratio for negative tests (LRNT), likelihood ratio for positive subjects (LRPS), are likelihood ratio for negative subjects (LRNS), are further defined and calculated. Also, using these indicators, new indicators, such as *F*-measure, balanced accuracy, Matthew's correlation coefficient (MCC), Chisq, χ^2, difference between automatic and manual methods, differentiation index, are used for comparison and evaluation. The overall performance of a method can be evaluated based on the index matrix extracted from the complexity matrix. In Chapter 9, a definition of the classification matrix is provided, followed by an introduction to the database modeling method. Figure 1.12 classifies the types of evaluation indicators, for which the calculation and various examples are presented in Chapter 9. Pertinently, Chapter 9 concludes by describing some general specialized indicators to evaluate infrastructure.

1.10 Nature-Inspired Optimization Algorithms (NIOAS)

NIOAs are inspired by the behavior of natural phenomena to build related models. Some of these behaviors include simulating congestion intelligence, the behavior of biological systems, the behavior of physical and chemical systems, and divisions of miscellaneous behaviors. These models utilize bioinspired and physics- and chemistry-based algorithms. In general, biology-inspired algorithms encompass a class of algorithms based on crowding, swarm, and evolutionary intelligence. In general, NIOAs are an active and important branch of artificial intelligence (AI), which continuously evolve. Thus, it is not unreasonable to expect that, as you read this chapter, several new algorithms will be born.

So far a large number of interesting NIOAs inspired by natural animal behavior, biology, physics, and chemistry have been proposed and applied in many fields, such as engineering and especially in optimization. Some examples include the Genetic Algorithm (GA), Particle Swarm Optimization (PSO) Algorithm, Differential Evolution (DE) Algorithm, Artificial Bee Colony (ABC) Algorithm, Ant Colony Optimization (ACO) Algorithm, Cuckoo Search (CS) Algorithm, Bat Algorithm (BA), Firefly Algorithm (FA), Immune Algorithm (IA), Gray Wolf Optimization (GWO), Gravitational Search Algorithm (GSA), and Harmony Search (HS) Algorithm. Some newer algorithms that have also been proposed are Horse herd optimization algorithm, Mayfly Optimization Algorithm, Chimp Optimization Algorithm, Coronavirus Optimization Algorithm, Water strider algorithm, Newton metaheuristic algorithm, Black Widow Optimization Algorithm, Harris hawks optimization, Sailfish Optimizer, Spider Monkey Optimization, Grasshopper Optimization Algorithm, Fractal Based Algorithm, Bacterial Foraging Inspired Algorithm, Rainfall Optimization Algorithm, Dragonfly algorithm, Sperm Whale Algorithm, Water Wave Optimization, Ant Lion Optimizer, Symbiotic Organisms Search, Egyptian Vulture Optimization Algorithm, Dolphin echolocation, Great Salmon Run, Big Bang-Big Crunch, Flower Pollination Algorithm, Spiral Optimization Algorithm, Galaxy-based Search Algorithm, Japanese Tree Frogs, Termite Colony Optimization, Cuckoo Search, Glowworm Swarm Optimization, Gravitational Search Algorithm, Fast Bacterial Swarming Algorithm, River Formation Dynamics, Imperialistic Competitive Algorithm, Roach Infestation Optimization, Cat Swarm Optimization and krill herd optimization.

In Chapter 10, the original theory and general idea of some NIOAs are summarized, and examines the use of one or more NIOAs and provides a real-world example via a case study of how algorithms respond. Finally, a new algorithm is presented by simulating the behavior of emperor penguins along with theory and case study. The main motivation of this section is to introduce the ability to simulate and model nature-inspired behaviors at the Chapter 10. By reading this chapter, you are expected to be able to design and build your own NIOA to solve optimization problems.

The main idea of the most common NIOPAs is briefly presented based on the levels stated in the framework in Figure 1.13. In Chapter 10, general principles of these methods are presented in a simple way, and it is expected that you will acquire a better understanding of the general concepts and simplified structure of these algorithms. In general, the main steps of each NIOA pattern are the following:

Level-1: Primary population production;
Level-2: Evaluation of individuals by calculating their appropriate values;
Level-3: Creating the next population by changing the original population;
Level-4: Termination criteria; and
Level-5: Choosing the answer that has the highest value of fitness.

The general principles of each of these algorithms are similar, but in step 3 functions, they may have different models for creating the next population or changing position due to different ideas.

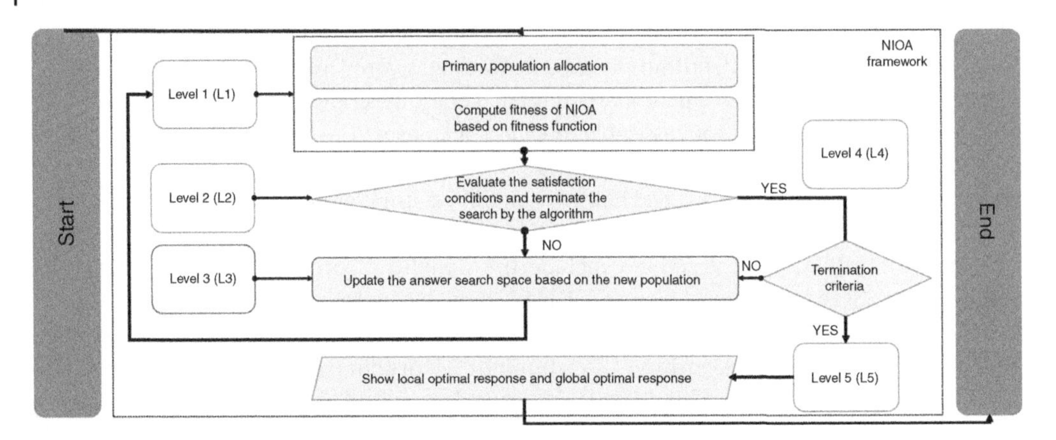

Figure 1.13 General framework and routine process of NIOAs.

1.11 Summary and Conclusion

In this chapter, we first defined automation and data collection methods and provide a summary of this book briefly as well as the importance of each chapter from design to implementation. General categories for the book chapter are presented in Chapter 10.

1.12 Questions and Exercise

1. Define automation and name some examples of its use in infrastructure management. Analyze a system that works intelligently and controls its core components.

2. Include examples of pavement, bridge, and tunnel management systems and specify their smart and automation degrees.

3. The types of key components of an infrastructure management system design are interrelated. Specify the relationship between each.

2

The Structure and Framework of Automation and Key Performance Indices (KPIs)

2.1 Introduction

Regular inspection and maintenance are of particular importance for the proper functioning of important civil infrastructures, such as bridges, pavements, tunnels, railways, and underground structures, and is basically the first and foremost step in planning. Because of the high cost and length of traditional inspection methods, civil infrastructure is not inspected frequently. As such, information about these structures is not up-to-date and usually interferes with proper and integrated management. Also, due to the need for the presence of on-site experts and safety considerations, most traditional and conventional inspection methods require on-site inspection equipment and tools that can disrupt the performance of the desired infrastructure. As an additional risk, inspectors are usually placed in complex and insecure work environments, while accurate and consistent data are challenging to be obtained considering the difficulty of the work. The advent of new imaging technologies, such as drones, unmanned aerial vehicle (UAV), micro-ligaments, 3D laser imaging technologies, sensitive cameras, etc., have led to a change in the field of automated infrastructure inspection. In this chapter, the structure and architecture of automated infrastructure inspection equipment are first examined, then a set of final criteria expected from advanced upgrades is presented. Finally, some emerging technologies that can be effective in increasing the efficiency of the automation evaluation, inspection, and technology development process in this regard are briefly introduced. It is expected that the further development and application of advanced technologies as well as the implementation of complete automation models will be effective in increasing the number of new methods for the automatic quality assessment of infrastructure (Table 2.1).

Recently, the introduction of remote-control systems in the field of infrastructure engineering and evaluation has provided a special opportunity for supervisors and inspectors to obtain the latest information on the state of infrastructure with remote management at the lowest cost and without risk to proper management. This issue can be considered as the biggest infrastructure management change in the current century.

Understanding the macro structure of these technologies and their use in a large network of civil infrastructure to improve the design of macro systems at the country or state level is highly important and emerging as a new specialty. There are several factors that have prevented the rapid and widespread development of infrastructure. These factors include lack of macro-architecture and a coherent and integrated structure for automatic inspection, lack of trust and acceptance of senior managers to the unsystematic results found from infrastructure, and the gap between applied technology and the needs of managers to decide on construction, maintenance, and infrastructure repair. This chapter is designed to highlight the macro view of automated infrastructure assessment

Automation and Computational Intelligence for Road Maintenance and Management: Advances and Applications, First Edition. Hamzeh Zakeri, Fereidoon Moghadas Nejad, and Amir H. Gandomi.

Table 2.1 The future technologies for asphalt pavement distress detection classification and evaluation [11].

	New emerging technology	Level
1	Satellite	R&E&P
2	QUAV	H&R&D&E&P
3	BCI, neuro prosthetics, Neuromorphic	H&R
4	Behavior-based robots (BBR) and Artificial brain	H
5	Swarm robotics (Cloud Robotics)	H&E
6	Hybrid (FPV & UAV and BCI)	H
7	Nanorobotics	H
8	Kirlian photography (Aura)	H

R: Research, D: Development and Diffusion, E: Experiments, P: Prototypes, H: Hypothetical.

and the future direction of developing technological solutions for infrastructure inspection. It is expected that the reader, by the end of this chapter, will be able to explain and design the architecture of new automated systems related to the evaluation of infrastructure with a macro view.

2.2 Macro Plan and Architecture of Automation

2.2.1 Infrastructure Automation

Automation of civil infrastructure assessment is a process to reduce human intervention with infrastructure. In doing so, automated equipment and information technology systems, which have good repeatability, accuracy, and speed, employ specialized and advanced software, and can be connected by non-specialists, should be used.

The science of automation in civil infrastructure management has been able to perform the same evaluation by creating a uniform methodology and by simplifying the indicators, in a shared storage server, to extract key performance indices (KPI) indicators and metrics at the network level. The purpose of this method is to increase accuracy, speed, and improve efficiency, and can greatly improve infrastructure inspection operations.

Infrastructure inspection operations reduce the frequency of repetitive tasks and eliminate the need to work in hazardous environments such as roads and access to hard-to-reach bridges. Automation tools greatly improve the efficiency and speed of inspection operations, while advanced methods, such as image processing and artificial intelligence, allow decisions to be made simultaneously with accurate information.

2.2.2 Importance of Infrastructure Automation Evaluation

Automation is a key component of organizing the management of civil infrastructure, including roads, bridges, tunnels, road safety, and railways, to increase the productivity and digital transformation of an organization. Due to the increasing volume of information and complexity of specialized issues, limited resources, and the need for optimization methods, it is necessary to create automated infrastructure management systems for reducing the size of specialized organizations. The use of new equipment, systems, and methods along with timely updating of infrastructure information can significantly improve the productivity of organizations.

Every specialized road maintenance and rehabilitation organization needs automatic and reliable equipment in accordance with the latest technologies to maintain the network of roads, tunnels, bridges, etc. Automation tools can enhance the equipment and productivity of managers and experts. By eliminating various manual errors and full coverage of status information, it is possible to determine the correct policies. In general, the implementation of automation in infrastructure management is useful for matters that have key indicators and specific objectives, the ability to measure quantity, and tedious and high-volume manual repetition. Such automation can be beneficial for:

- Identifying and isolating pavement sections with distress and determining the type, severity, and extent of damage based on appearance criteria.
- Determining the road pavement status index based on the indices for measuring the surface damage of roads and determining the appropriate option for maintenance.
- Inspecting the level of bridge elements to assess the amount of damage and determine the type, severity, and extent of damage in order to present the general index of bridge health.
- Inspecting the surface condition of the tunnel wall by presenting a numerical index related to the health of the tunnels.

In general, these methods can be used to accurately identify the type, severity, and extent of anomalies and subsequently determine the necessary specialized repetitive tasks based on the location of failure and a set of pre-designed algorithms.

2.3 A General Framework and Design of Automation

In recent years, manual identification and evaluation via visual inspection of the condition of civil infrastructures has gradually given way to automatic methods. Developed countries interested in automation are encouraging researchers to develop new models and methods in this field, while industries are increasingly investing in the implementation of ideas and the use of tools equipped with these methods at the macro level. Processing based on the appearance of infrastructure has an effective and key application in identifying, quantifying, and evaluating the situation at-hand and has recently received a lot of attention from researchers in the field of pavement engineering. Perception and assessment systems of infrastructure health consist of three general parts with well-defined relationships, which are considered when designing a real automation structure. The three components include data retrieval (DR), data processing (DP), and data and information (DI) interpretation. Figure 2.1 shows the macro architecture related to the design of automation structures in infrastructure management.

Based on this architecture, the role of each of the main components in creating automation must be carefully determined. In the DR section, the type of hardware and equipment must be clearly specified. Based on the type of equipment selection and method of use, the inputs of DP and DI sections are determined. In the best case, the total error of the automation system cumulatively depends on the accuracy of each step. In general, DR and DP have more than 90% automation capability. These two parts are closely related, and, thus, their relationship must be properly designed. Each of these sections is, itself, a large subset of components. System designers need to put all the necessary components together properly for an application system to work effectively. The DR method includes a subset of one-dimensional (such as audio), two-dimensional (such as image), three-dimensional (such as lidar), and multidimensional (thermal image) systems. When designing

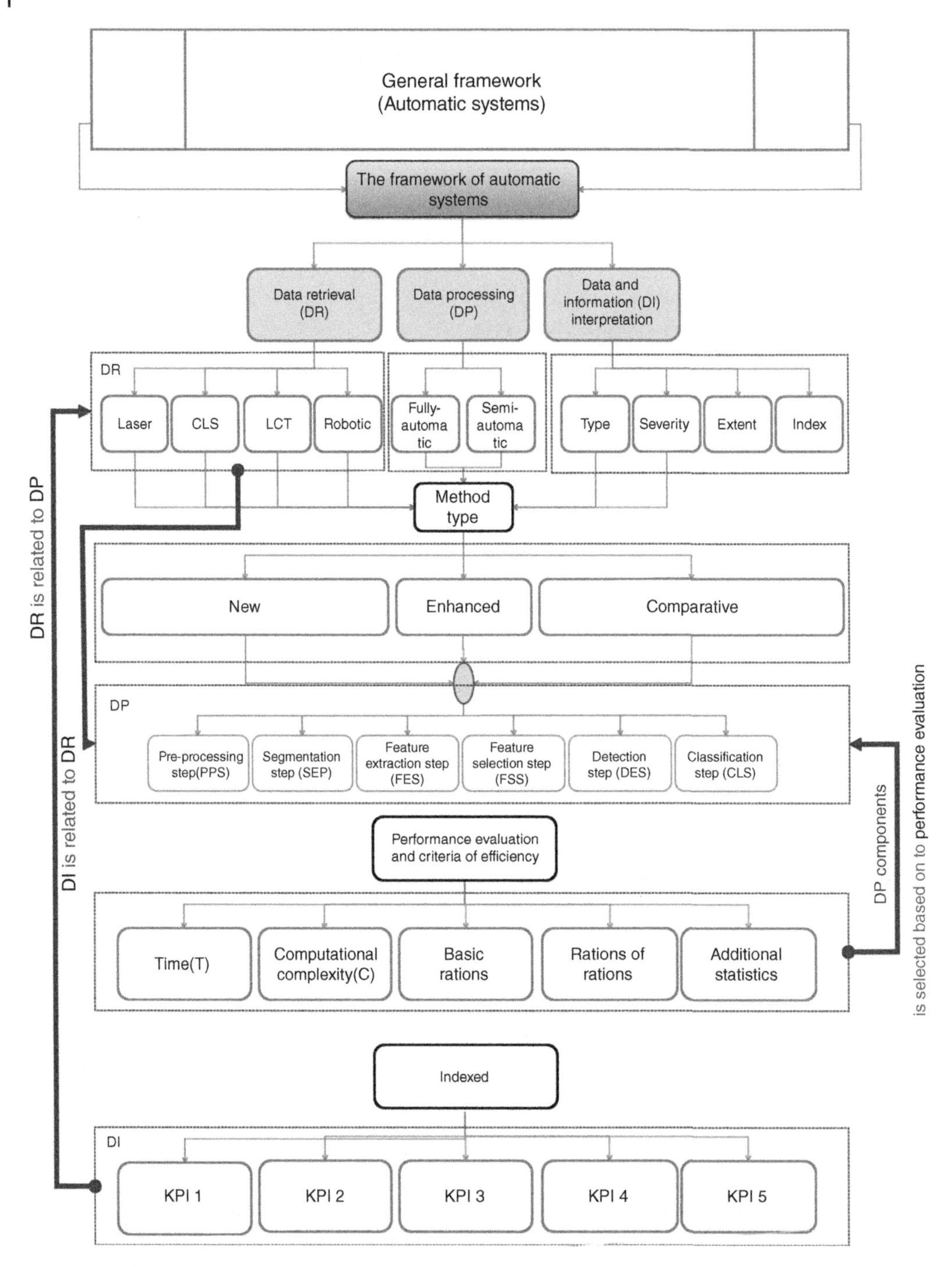

Figure 2.1 The General framework and the macro architecture related to the design of automation structures in infrastructure management.

the DR component, one should always keep an eye on the latest technology, current technological developments that will soon be released, and the research on specialized technologies. For example, over the past five years, the use of UAVs in infrastructure assessment has grown significantly. Therefore, two important parameters in choosing the DR type are current technology and future development.

In the DP section of input data, it should be possible to transfer the maximum amount of available information and to evaluate and categorize topics using new, accurate, and fast methods. For example, if the type of input data is an image, a comprehensive and complete analysis of the image should be performed using the appropriate image processing methods (given in the following chapters). The same output that the expert understands from the image should also be achieved by the module Extract DP. As mentioned earlier, DP and DR are closely related. Recently, various methods for processing input data both in space and in the frequency domain, have been proposed, called multi-level analyses, have many capabilities that will be discussed in detail in the following chapters. This section has different components, such as pre-processing step (PPS), segmentation process (SEP), FES, feature selection step (FSS), distress detection step (DDS), and classification step (CLS), whereby the design of each component is very effective in the success of the DP module. On the other hand, the performance of each of these components can be combined with other methods to yield different results. Various performance indicators are used to measure this section, which are presented in detail in Chapter 9. In general, these criteria are used to evaluate the optimal pattern selection of DR module components. It should be noted that these indicators alone should be assessed for all six components of the DR module, and each should yield the best result. After the optimal selection of each component, i.e. PPS, SEP, FES, FSS, DDS, and CLS, functionality for DR should be selected in the most optimal mode. The latest automation module in infrastructure assessment is the key KPI index, called Data and information (DI) Interpretation.

It should also be mentioned that the initial selection and acceptance of these indicators play an important role in the executive sector. Therefore, the choice of this module directly affects the choice of DR. Depending on the type of KPI, special data may have to be collected, which may require special equipment. In this chapter, due to the fact that the volume of content related to DP is very large and it is necessary to deal with it separately, only the DR and DI module are discussed. The content related to DP is given in detail in the following chapters.

Although the automation process is done from top to bottom in the design and architecture of automated infrastructure assessment systems, its implementation follows the steps below:

1) **Evaluation of KPIs and Selection of Indicators and Protocols**: Based on this, the type and nature of the required data are determined.
2) **Selection of Harvesting Hardware**: According to the type of data required to calculate the key index, sensors and appropriate hardware are selected. Also, the type of platform and how to navigate the equipment (automatic or manual) are designed at this stage.
3) **Processing Methods**: Based on the collected data and the desired indicators, automatic processing methods are used to design six processing components, including PPS, SEP, FES, FSS, DDS, and CLS.

Therefore, in order to design an automated infrastructure evaluation system, the three steps must be performed in the above order. For example, an automated road infrastructure evaluation system that intends to evaluate the pavement surface cracking index completely automatically and with intelligent navigation is based on the architecture shown in Figure 2.2. In this method, the key

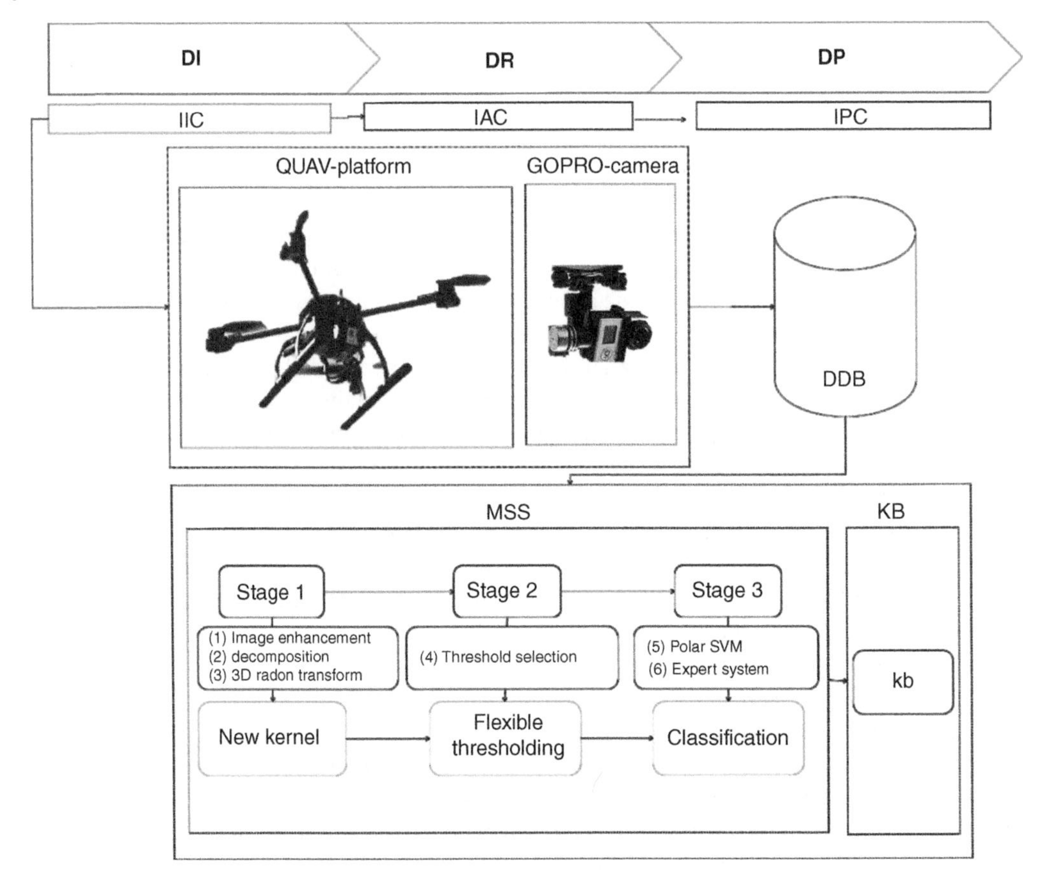

Figure 2.2 Example of designing an automatic system based on the general framework and macro architecture related to the design of automation structures in infrastructure management.

indicator of the cracked surface is determined first, so there is a need for vertical imaging of the surface, which must be completely intelligent according to the set condition. For this purpose, in the second stage, a UAV system with flight planning capability on the desired area is used to calculate the crack index. For this step, image processing methods and six processing steps, including PPS, SEP, feature extraction step (FES), FSS, DDS, and CLS, are used. This chapter presents the common indicators in the evaluation of road infrastructure. Other stages and components related to infrastructure architecture will be examined in the following chapters.

2.4 Infrastructure Condition Index and Its Relationship with Cracking

2.4.1 Road Condition Index

The composition of the effect of pavement failures has been a major problem in pavement management for many years. Combining independent faults and presenting an indicator for the condition of a section of pavement are difficult and complex tasks. The pavement distress index (PDI) is basically an approximate and person-dependent index that is highly dependent on the experience and knowledge of a qualified engineer. Combining the effect of pavement failures in such a way that meaningful information is extracted and can be used at different levels of management is

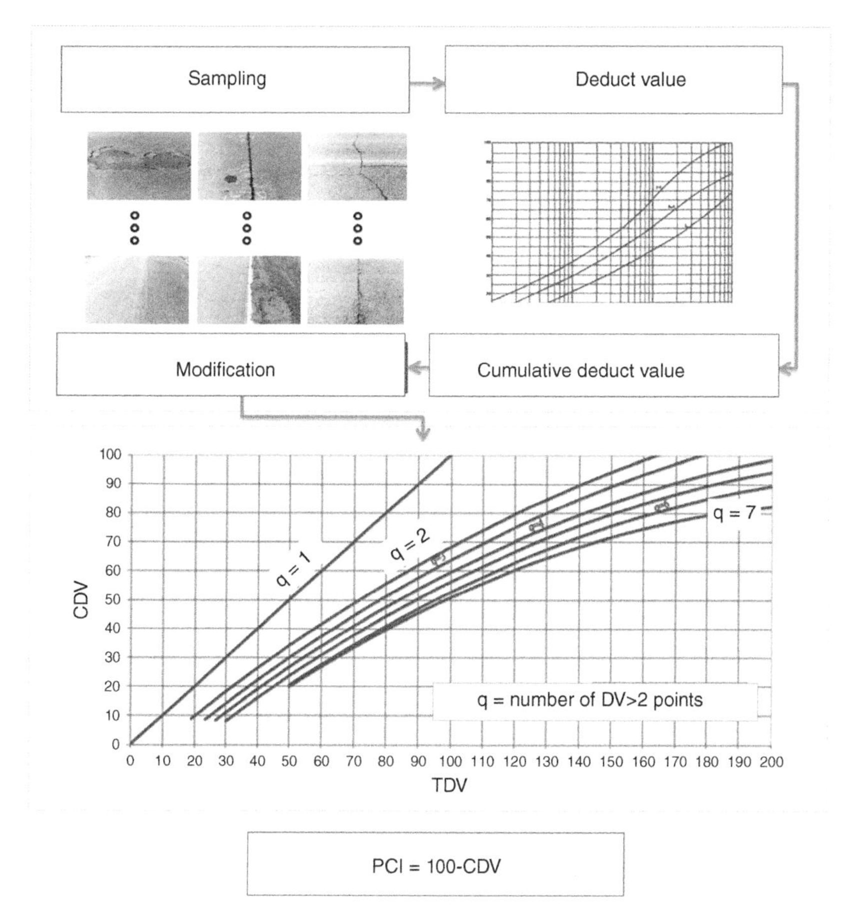

Figure 2.3 General structure of calculating pavement failure index (PCI) based on five main steps.

one of the main purposes of using PDI. Various methods, such as pavement condition rating (PCR), pavement serviceability index (PSI), pavement surface characteristics (PSC), and pavement condition index (PCI) have been proposed so far. Quantification problems are the main obstacle to the comprehensive use of automated harvesting systems. On one hand, there are limited systematic methods to transfer experience from manual to mechanized harvesting. On the other hand, the lack of willingness of organizations to change the structure of the pavement management system has led to more widespread use of manual harvesting than automated methods. Indicators based on three characteristics, including failure type, severity, and extent (density), are used to determine the current condition of the pavement, history of pavement efficiency, and forecast model for the future. Much research has been done to obtain a model to provide an approximate index with acceptable accuracy in order to increase the speed and accuracy of measurements (Figure 2.3).

In general, the independent failure regression method (PRM) and deduct value method (DVM) are applied for PDI formulation. PRM considers failures independently then combines them based on linear or nonlinear relationships, whereas DVM is defined based on the effect of a failure on the pavement performance when it is generated individually on the section. The effect of DVM varies according to the type, severity, and density of failure, and its effect for individual failures is greater than for composite failures.

First PSI index of pavement conditions was presented in 1960 by the Association of State Highway Official (AASHO). This index is independent from a personal and interpretive point-of-view but suffers from proper accuracy in performance simulation. The main equation for flexible paving is as follows:

$$\mathrm{PSI} = 5.03 - 1.91\mathrm{LOG}\left(1 + \overline{\mathrm{SV}}\right) - 1.38\overline{\mathrm{RD}}^{2} - 0.01\sqrt[2]{C + P} \tag{2.1}$$

where

$\overline{\mathrm{SV}}$ is the medium slope change.
$\overline{\mathrm{RD}}$ is the average groove depth.
C is the percentage of cracking.
P is the percentage of patches at the pavement level.

This numerical index of PSI ranges 0–5, where the closer the index is to 5, the better the pavement conditions. The composite PCR index was developed according to the experience of 31 engineers in the harvest experience of 1086 sites, which is based on the PRM method:

$$\begin{aligned} \mathrm{PCR} = {} & 95.5727 - 5.5085(5.0 - \mathrm{R}) - 1.5964(\mathrm{LL1}) - 1.9629(\mathrm{LL2}) - 2.9795(\mathrm{LL3}) \\ & - 0.01630(\mathrm{P}) - 0.07262(\mathrm{B}) - 0.222(\mathrm{A}) - 3.4948(\mathrm{RL1}) - - 7.5269(\mathrm{RL2}) \\ & - 11.2297(\mathrm{RL3}) - 0.03032(\mathrm{L1}) - 0.05484(\mathrm{L2}) - 0.53050(\mathrm{T12}) - 0.69736(\mathrm{T34}) \end{aligned} \tag{2.2}$$

where

LL1 is related to alligator cracking.
R is the current service index.
P refers to patching.
A is rutting.
RL1 is raveling.

In the above equation, RL2 is the groove in the wheel path (No = 0 and detection = 1); the RL3 groove is in a line (No = 0 and detection = 1); L1 is longitudinal cracks of level 1 and 2; L2 is longitudinal crack of level 3 by number; T12 is level 1 and 2 transverse cracks by number; and T34 is level 3 transverse cracks by number. Other similar methods using PRM, such as PSC, PRC, and IRI, have been provided by The Washington State Department of Transportation (WSDOT). The surface condition rating (SCR) method presented by the Georgia Department of Transportation (GDOT) is based on the severity and extent of the distress. Also, the SCR index based on the DVM method is presented. The indicators provided by these methods are very diverse and show different results. However, many influential parameters, such as type, severity and extent, of failures are not considered in these indicators. Moreover, the maintenance control index (MCI) index was presented by Japanese researchers and benefits from its simplicity. This index ranges numerically between 0–10, which is divided into four categories: cracks, potholes and patches, grooves, and roughness. In this method, many failures are not integrated or considered. The MCI value is obtained from the following equations:

$$\begin{aligned} \mathrm{MCI} &= 10 - 1.48C^{0.3} - 0.29D^{0.7} - 0.47R^{0.2} \\ \mathrm{MCI} &= 10 - 1.51C^{0.3} - 0.29D^{0.7} \\ \mathrm{MCI} &= 10 - 2.23C^{0.3} \\ \mathrm{MCI} &= 10 - 1.48C^{0.3} - 1.54D^{0.7} \end{aligned} \tag{2.3}$$

where

D is the wheel groove depth in mm.
C is the crack percentage.
R is the roughness in mm.

The RCI, or Riding Comfort Index, is very similar to PSI but has a different range of 0–10. In this method, a group of experts fills in a form in a descriptive form, which ultimately presents a summary of numerical comments as RCI. The surface distress index (SDI) is a composite index that provides a numerical value based on the level of failure intensity and density for nine types of surface distress.

The use of such indicators for manual or visual harvesting is appropriate, owing to their simplicity and easy understanding by experts. However, the use of these indicators in automatic harvesting tools is difficult and complex, and surface failures are not properly modeled in these indicators. Due to its simplicity, clarity, and logical understanding, the physical PCI model is gaining increased acceptance by organizations compared to other models (DVM), which are expensive and time-consuming. In the method PCI, each distress independently affects the composite index. Therefore, each indicator in this category of models should consider the effect of failure type, intensity, and extent, since the number of types of failure, the degree of severity of failure, and a wide range of densities can produce very different results. It should further be mentioned that several types of failure, three levels of intensity, and a wide range of densities makes problem solving a very heavy regression. The two main steps of such methods include structuring the composition and subsequent adjustment based on on-site testing. In this section, the main and well-known methods based on reducing coefficients are introduced.

As mentioned PCI is a numerical quantity between 0 (for a completely broken pavement) and 100 (for a completely healthy pavement). This index is generally a sign of the condition of the structures and functional conditions of the pavement. The theory of this method is based on the severity of distress and the amount of deduct value (DV). To evaluate sections with this index, another multistage method is used, which includes the following steps: (i) determining and selecting the ratio of the index, dimensions and standards; (ii) identifying the exact type, level of severity, and measurement index; and (iii) adjusting the reduction coefficients. A reduction coefficient of 0 means a small effect of distress, while a reduction coefficient of 100 indicates a very large effect of distress on the functional condition of the pavement.

$$\mathrm{PCI} = 100 - \left[\sum\nolimits_{i=1}^{p} \sum_{j=1}^{m_i} \mathrm{DV}_{ij}\right] \cdot F(\mathrm{TDV}, q) \tag{2.4}$$

where

DV_{ij} is the distress reduction coefficient.

In the above equation, maximum PCI is 100; i is the intensity level of j; i distress type, j distress severity level counter, p is total distress count; m_i is the total number of distress severity level for the ith type of distress; $F(\mathrm{TDV}, q)$ is the adjustment function; and total deduct value (TDV) is the sum of the reduction coefficients calculated from the following equation:

$$\mathrm{TDV} = \left[\sum\nolimits_{i=1}^{p} \sum\nolimits_{j=1}^{m_i} \mathrm{DV}_{ij}\right], m_i = 1 + \left(\frac{9}{98}\right)(100 - \mathrm{HDV}_i) \tag{2.5}$$

The allowable number of reduction coefficients is obtained using the above equation. In the above relation, highest individual deduct value (HDV_i) is the largest value of DV_{ij}. The number of reduction coefficients is reduced to the integer m. The severity of distress has a quantitative and interpretive definition, which still lacks a comprehensive index or method. In most standard protocols, the determination of the intensity level is defined using quantitative (crack width) and qualitative (spalling level) parameters. The type of spalling has a great effect on determining the level of distress intensity, yet no method has been proposed to determine the intensity level of this parameter. Each crack has different levels of severity associated with the same failure. Since all failures are considered independently, the TDV value may be higher than 100 at one point. Corrected deduct value (CDV) diagrams have been used to solve this problem. These coefficients (DV) are based on the opinion and experience of experts. After more than 30 decades of this experimental research, these coefficients cannot be expected to remain constant and unadjusted. It is still not possible to claim that CDV and TDV are accurate criteria or that expertise in determining coefficients has been well-extracted and applied. This becomes even more complicated when the goal is to use automated methods due to systematic errors. Determining the exact type of distress, level of severity, and extent of failure using an automated method remains a difficult task. While attempts have been made to automatically detect the type of distress and the level of compaction, it is only possible to determine the intensity level using simplified definitions. In the standard PCI method, 19 types of distress are considered. This number has been reduced to seven types in the modified PCI (MTC) method, with the aim of increasing the speed of calculation and minimizing the training time.

In order to speed up the evaluation, a number of users tend to reduce the number of distress types. The main goal is to reduce both the computation time and training time of experts for evaluation. The effect of reducing the number of distress on PCI has been previously analyzed, which reveals that the type of distress has a significant effect on the standard PCI index in the modified method. This indicates a numerical difference between 1 and 7. Each organization can decide on the benefits of using a small number of types of distress and deviations from the standard of ASTM6433, depending on the type of prevailing distress patterns. Previously, two modifications have been to the PCI method by various pavement management software designers: (i) combination of reduction coefficients in order to determine the modified reduction coefficient (CDV); and (ii) representation of reduction coefficient curves in standard logarithmic equations. These methods have been applied in several states in the United States and Canada. In the correction method, PDI is a number between 0 and 10, which is expressed as follows:

$$
\begin{aligned}
&\mathrm{PDI} = 10 - \mathrm{CDV}, \quad \mathrm{CDV} = 10^{(0.0012 - 0.3958\mathrm{Log}_{10}(\mathrm{NED}) + 0.9565\mathrm{Log}_{10}(\mathrm{TDV})} \\
&\mathrm{TDV} = \left[\sum\nolimits_{i=1}^{p} \sum_{j=1}^{m_i} \mathrm{DV}_{ij}\right], \quad \mathrm{NED} = \left[\sum\nolimits_{i=1}^{p} \sum_{j=1}^{m_i} \frac{\mathrm{DV}_{ij}}{\mathrm{DV}_{\max}}\right], \\
&\log\left(\mathrm{DV}_{ij}\right) = \mathrm{a}_{ij} + \mathrm{b}_{ij} \times \log\left(\mathrm{PDA}_{ij}\right)
\end{aligned} \tag{2.6}
$$

where

DV_{ij} is the distress reduction coefficient.
a_{ij} and b_{ij} are regression coefficients.

In the first part of Eq. (2.1), CDV is the modified correction factor; number of equivalent distress (NED) is the number of equivalent failure; $j = 1,2,3$ of the counter failure level counter; m_i is the maximum number of types distress; and DV_{ij} is the reduction coefficient for type I distress with severity level j, which is converted to PDI.

In the second part of Eq. (2.2), a_{ij} and b_{ij} are regression coefficients; DV_{ij} is the reduction coefficient for type I distress with intensity level j; and DV_{max} is the maximum reduction coefficient observed in a section and is calculated from the following equation:

$$DV_{max} = Max(DV_{ij}) \tag{2.7}$$

Baladi's Method was proposed in 1991 to formulate the effect of failure. Like other methods, this method is designed by determining the type of distress, level of severity of each distress, and the base index, which is a number between 0 and 100. In this method, based on the definition of the index, the expert is asked to determine the maximum acceptable density for each type and severity of distress. Finally, the reduction coefficient for each other density levels is adjusted for the same type of intensity using linear dimensioning. Status indicators are combined using a user-defined weight factor. This index is used in North Carolina and South Dakota.

In 1991, Chinese researchers proposed the China Method, a multi-step method for formulating a PDI based on the concept of reducing coefficients. In accordance with local experience, various methods for formulating PDI according to a set of distress definitions and measurement methods have been proposed. In general, the China method follows three stages:

1) Standardization of pavement distress, classification, and measurement method
2) Formulation of PDI method
3) Determination of the weight of the function for the composite index

In the first stage, distress is divided into four main categories: cracking, deformation, surface, and potholes. Cracking caused by other cracks is further classified into longitudinal, transverse, block, and alligator cracking. The severity level in this method has less flexibility than that in the PCI method. For example, for longitudinal, transverse, and block cracks, only two levels of low and high intensity are considered. To assessment alligator cracking, three levels of low, medium, and high intensity have been introduced as the standard levels. Moreover, two parameters of crack width and severity have been introduced as criteria, affecting the level of distress intensity due to cracking. However, in this methodology, no index or criterion for determining the level of crack intensification has been presented, and thus, this level is still presented qualitatively. Introducing a quality characteristic in automated systems is a difficult and complex task.

In the second step, the calculation process is weighted step-by-step. In computational form, PDI can be defined as follows:

$$PDI = 10 - CDV, \quad CDV = 10^{(0.0012 - 0.3958 Log_{10}(NED) + 0.9565 Log_{10}(TDV))}$$

$$PDI = \left\{100 - \left\{\sum_{c=1}^{4}\left[\sum_{i=1}^{m_c}\left(\sum_{j=1}^{n_{ic}} DV_{ijc}.w_{ijc}\right).w_{ic}\right].w_c\right\}\right\} \tag{2.8}$$

where

DV_{ijc} is the reduction coefficient for the 1st type of distress.
j is the severity level.

In this equation, 100 is the maximum value of PDI; DV_{ijc} is the reduction coefficient for the 1st type of distress; j is the severity level; and c is the 2nd category of distress; w_{ic} is the combined weight of the composite reduction coefficient value for the 1st type of failure; and the c category of distress after combining with different intensity levels for the type and category of distress. Also, w_{ic} is calculated from the equation as follows:

$$w_{ijc} = \zeta_1 \left\{ \frac{\mathrm{DV}_{ijc}}{\sum_{j=1}^{n_{ic}} \mathrm{DV}_{ijc}} \right\}$$
$$w_{ic} = \zeta_2 \left\{ \frac{\sum_{j=1}^{n_{ic}} \mathrm{DV}_{ijc}.w_{ijc}}{\sum_{i=1}^{m_c} \left(\sum_{j=1}^{n_{ic}} \mathrm{DV}_{ijc}.w_{ijc} \right)} \right\} \tag{2.9}$$

where

DV_{ijc} is the reduction coefficient for the 1st type of distress.

In this equation, w_c is the weight of the combined reduction coefficient for the c of the second category of distress, after combining mc with different intensities for each type of distress in the category, which calculated as:

$$w_c = \zeta_3 \left\{ \frac{\left[\sum_{i=1}^{m_c} \left(\sum_{j=1}^{n_{ic}} \mathrm{DV}_{ijc}.w_{ijc} \right).w_{ic} \right]}{\sum_{c=1}^{4} \left[\sum_{i=1}^{m_c} \left(\sum_{j=1}^{n_{ic}} \mathrm{DV}_{ijc}.w_{ijc} \right).w_{ic} \right]} \right\}. \tag{2.10}$$

where

DV_{ijc} is the reduction coefficient for the 1st type of distress.

The third step is the most important in applying this method, whereby determining the coefficients ζ_1, ζ_2, and ζ_3 is pertinent to obtain valid results. To determine these weights, repetitive methods similar to the Paver method are used. In practice, the main problem is the lack of a sufficient number of samples to calculate the composition coefficient. After drawing the DV against the weights, a diagram with the weighted curve is presented to determine the coefficients, for which the proposed equation is:

$$w = 2.94x - 4.4x^2 + 2.45x^3, \quad \left(r^2 = 0.99\right) \tag{2.11}$$

However, in order to increase the speed, in addition to physically eliminating a number of failures, this method reduces the intensity levels. An innovative and non-repetitive method is used to obtain the index.

Comparing the different methods that work based on the reduction coefficients, it can be understood that the three main criteria of type, severity, and extent are the core of the relationship and can generally be modeled in the following form:

$$\mathrm{PDI} = f(\mathrm{Type}, \mathrm{Severity}, \mathrm{Extent}) \tag{2.12}$$

In different manual methods, these three characteristics are combined and the reduction coefficients generally obtained by experimental methods are used to create a quantitative relationship between descriptive and numerical expressions. However, these methods are unrepeatable, time-consuming, and unrealistic. Weight allocation in these methods is usually done manually or visually. Simultaneously quantifying and automating the interpretation of distress is a difficult task, yet the solution of which can solve many problems related to the unification of perceptions and the realization of indicators. In automated methods, indicators are provided for evaluation, but are approximate and inaccurate. Automatic extraction of reduction coefficients and combining this method with automatic distress and analysis methods, can increase reproducibility, adaptation to real conditions, standardization, and acceptance of using automated systems.

A systematic understanding of these three characteristics and the creation of machine consciousness can be effective in developing a practical index. Most of the activities in the field of automated systems are aimed to diagnose the type of distress. Extent and density of failure can also be quantified using a simple method. The most crucial feature is the determination of the level of distress severity, which is defined differently in the instructions of each method. In automated systems, only the crack width is used to determine the severity of distress. However, most manual harvesting protocols employ another important criterion called spalling for decision-making. According to this relationship, the severity of cracking is a function of the width of the crack and its spalling.

$$\text{Severity} = f(w, s). \tag{2.13}$$

Several methods have been proposed by researchers to determine the PCI, among which the universal cracking indicator (UCI) method is the simplest type of automatic distress indicator. Based on this method, a number is presented as the distress intensity index. For single cracks, such as longitudinal and transverse, the number indicates the ratio of the length and width of cracking in the section and for patterned distress such as block and alligator, the ratio of the distress area to the total section is introduced as the severity of distress. The UCI method provides an index image by dividing the number of bright pixels into image dimensions. Another indicator, such as pavement distress quantification (PDQ), is provided by Zhou, which is based on empirical and qualitative relationships.

$$\text{PDQ} = \text{HAWCP} \times \text{HFEP} \tag{2.14}$$

In the high amplitude wavelet coefficient percentage (HAWCP) index, the percentage of wavelet coefficient is in the high range of the image, and high frequency energy percentage (HFEP) is the percentage of energy in the high frequency. This index is similar to the very basic, simple but fast UCI method. These indicators are not applicable when there are several different types of distress in the section. Generally, the information of such methods cannot be cited, used, or disaggregated into further information about the cause of distress.

In 2014, Tsai, Jiang, and colleagues proposed a new theory of the basic element of crack feature extraction (CFE) cracking based on the multifaceted compositional characteristics of cracking morphology. The multi-level CFE method was developed for accessing crack properties that can be used to classify cracks by linear methods. This classifier with the help of cracking index ($\text{Num}_{\text{cracks}}$) has made it possible to perform classification using a linear kernel separator:

$$\begin{aligned} y* = & -0.0018 \times L_{\text{longi}} + 23.9977 \times \text{Ratio}_{\text{Longi}} - 0.1040 \times \text{Num}_{\text{cracks}} + 0.1209 \\ & \times \text{Num}_{\text{Intersection}} - 0.3955 \times W_{\text{max}} \end{aligned}$$

$$\begin{cases} 1 & \text{if } y^* \geq 6.4 \\ 2 & \text{if } 6.40 > y^* \geq -4.21 \\ 3 & \text{if} -4.21 \geq -19.61 \\ 4 & \text{if } y^* < -19.61 \end{cases} \tag{2.15}$$

where

DV_{ijc} is the reduction coefficient for the 1st type of distress.

L_{longi} is the parameter of longitudinal cracks length (mm); $\text{Ratio}_{\text{longi}}$ is the longitudinal ratio; $\text{Num}_{\text{cracks}}$ is the number of crack lines; $\text{Num}_{\text{intersections}}$ is the number of crack crossings; and

W_{max} is the maximum crack width (mm), which has been proposed to classify the severity of distress.

As can be seen in the study of key indicators, there are different methods for evaluating road procedures, which the selection of each indicator has special requirements and parameters that directly affect the choice of hardware and data collection method. Therefore, selecting the index is the first step in automating infrastructure evaluation.

2.4.2 Bridge Condition Index

The Bridge Health Status Index is often used to assess the condition of structures in various countries, including the United States, Canada, Italy, Japan, and Iran, to prioritize and select maintenance options. In this method, a status index is assigned based on the ratio of current conditions to structural conditions at the beginning of operation. These indicators are often derived from the California bridge health index (BHI) Index, a concept first proposed by the California Department of Transportation to measure the structural performance of a bridge or network of bridges as a number. This type of index evaluates the current condition of a bridge by summing the value of the current condition of all bridge members and comparing it with the value of the total value of the bridge members in optimal condition. The value of each member is proportional to the number of members in the current situation and the economic consequence of the rupture of the member. The cost of member demolition fracture cost (FC) can be considered as a valuation factor that emphasizes the importance of the member in the overall health of the bridge.

The California Index is based on the assumption that the bridge has an initial value at the start of operation, which decreases due to the growth of damage caused by traffic load and environmental impacts. Over time, as the index is repaired, maintained, and rebuilt. The California BHI Index is based on the ratio of the total residual value of the bridge members to the initial value of the total members. The calculation steps are as follows:

1) Extract inspection data at the member level from ASHTO Bridge Management Software (BrM) with the knowledge that parts of a member can be involved in more than one type of situation.
2) Calculate the weighting factor (WF) for each condition via the following:

$$\text{WF} = 1 - ((\text{Condition state Num} - 1)/(\text{Number of condition state} - 1)) \tag{2.16}$$

3) Based on the current state of the member in the first step, calculate the FC for each member in one of two ways:
   - FC member of infrastructure plus FC member user, or
   - The cost of organ replacement in the weight factor (WF) of the member
4) Calculate total element value (TEV) and current element value (CEV) using the following equations:

$$\text{TEV} = \text{TEQ} \times \text{FC}$$

$$\text{CEV} = \sum (Q_i \times \text{WF}_i) \times \text{FC} \tag{2.17}$$

where

TEQ is the total member value.

5) Calculate BHI as a ratio from TEV to CEV, using:

$$\text{BHI} = \frac{\sum(Q_i \times \text{WF}_i) \times \text{FC}}{\text{TEQ} \times \text{FC}} \tag{2.18}$$

Indicators calculated by the weighted average method of the status of individual limbs are the most common types of indices. Their development is based on the status data of structural members, which includes the type, severity, and extent of damage to structural members. Also, some indicators are sufficient for operating data, such as traffic volume and records of the services provided by the bridge. The number of members surveyed and the type of grading system used may vary from country to country.

The British BHI Index is similar to the BHI in California, except that the value of the bridge's members is based on the membership of the bridge as a whole, not on the cost of breaking it. Also, instead of calculating the residual value for bridge members, a simple score based on engineering judgment is used to determine the members' importance factors, as follows:

$$\begin{aligned} \text{BCS} &= \frac{\sum_{I=1}^{N} \text{ECI}_i \times \text{EIF}_i}{\sum_{i=1}^{N} \text{EIF}} \\ \text{ECI} &= \text{ECS} - \text{ECF} \\ \text{BCI} &= 100 - \left(2 \times \left\{(\text{BCS})^2 + (6.5 \times \text{BCS}) - 7.5\right\}\right) \end{aligned} \tag{2.19}$$

where

N is the total number of members.

The bridge condition index (BCI) is a numerical scaling of 0 (worst) to 100 (best condition).

In South Africa, accreditation and prioritization of repair, maintenance, and reconstruction are done using a similar index to the British BCI. This index is calculated based on data obtained from routine structural assessments of the condition of the bridge and the importance factor of the bridge, which depends on the average daily traffic (ADT).

The condition of the structures is assessed based on the degree, extent, and relevancy (DER) of the distress and is assigned by assigning the DER score. The DER rating system identifies and prioritizes faults by assessing their relative importance over the integrity of the bridge structures. It should be noted that the rating is not directly related to the bridge members but, instead, to the defect or failure. Therefore, a member with a defect is given a grade greater than 0 according to the DER rating system.

Each identified failure is assigned a score from 1 (minor) to 4 (severe) based on the degree (severity), extent, and the relationship of the failure. The relationship of the identified failure is related to its overall impact on the integrity of structures, serviceability, and safety of the bridge. Two faults may appear to be identical and of equal magnitude, but their impact on the integrity of the bridge may differ from a general point-of-view. Thus, fault communication helps inspectors obtain information beyond the usual visual ratings by assessing the impact of each fault on the integrity of a bridge structure.

Each failure on the examined member has an IC status index, which is calculated from the following equation:

$$\text{I}_C = 100 \times \left[1 - \frac{(D + E) \times R}{32}\right] \tag{2.20}$$

where

N is the total number of members.

The importance of the bridge depends on its use in the network of communication routes. Therefore, the total BCI is calculated as the sum of all distress status values in all inspected members, weighted by the importance factor of the bridge:

$$\mathrm{BI} = \left[\frac{\mathrm{ADT}_i}{\sum\limits_{i=1}^{n} \mathrm{ADT}_i}\right]$$
$$\mathrm{BI}_i = .\left(\sum\nolimits_{j=1}^{m} I_{Cj}\right)\left[\frac{\mathrm{ADT}_i}{\sum_{i=1}^{n} \mathrm{ADT}_i}\right] \tag{2.21}$$

where BI mean bridge index; ADT is the ADT; i is the number of bridges; and n is the number of bridges in the network; m is the number of members inspected; and j is the number of members in the ith structure.

The Australian BCI is calculated on the basis of a hierarchical framework. The factors of the structural group are assigned to the structure based on the importance of the group. A group of structures (such as columns, foundations, and decks.) consists of a number of members performing similar operations. Bridge condition number (BCN) is calculated based on the average group rating (AGR) obtained for groups of bridge members, using the following equations:

$$\mathrm{BCN} = \sum \mathrm{AGR} \times W_b$$
$$\mathrm{AGR} = \frac{\sum 2 \times \mathrm{ACR} \times E^{0.5}}{\text{Nuber of elemets}} \tag{2.22}$$

where W_b is the coefficient of importance of the structural group; and E is exposure (i.e. environmental).

The Austrian BCI is calculated on the basis of inspection data from bridge members. Each member is assigned five different ratings based on five characteristics as follows:

$$S = \sum\nolimits_{i=1}^{32} \left(G_i \times \sqrt{k_{1i} + k_{2i} + k_{3i}}\right) \tag{2.23}$$

where G_i is the type of distress; k_{1i} is the extent of distress; k_{1i} is the severity of distress; and k_{1i} is the importance of the ith member.

Finland's BCI is based on the average rating of the weighted structural components. In this method, the inputs used to calculate BCI include the cause, location, and effect of the distress on the bearing capacity of the bridge and the urgency of repair. BCI, also known as the repair index, is calculated for a set of identified bridge distress members. During the inspection, the bridge is divided into nine structural sections, the condition of each is evaluated with a rating from 0 (very good) to 4 (very poor). Any distress detected during the inspection is also graded in terms of severity and urgency of repair as follows:

$$\mathrm{KTI} = \max\left(\mathrm{Wt}_i \times C_i \times U_i \times D_i\right) + k \sum \left(\mathrm{Wt}_j \times C_j \times U_j \times D_j\right) \tag{2.24}$$

where KTI is Repair index; G_i is the type of distress; k_{1i} is the extent of distress; k_{1i} is the severity of distress; and k_{1i} is the importance of the ith member.

2.4.3 Tunnel Condition Index

For periodic inspections of underground structures, it is necessary to use indexing methods to select maintenance processes. Most tunnel inspection methods are based on an expert's visual assessment. Automatic cracking assessment of tunnel walls using robotic systems has acquired increasing interest from researchers, in which a valid indicator has been developed to assess the condition of the coating according to the fractal dimensions of the cracks. The D fractal dimension is an indicator to evaluate the rate of tunnel wall cracking. For example, the number of fractures and the order of the patterns in tunnels may be considered as significant indicators for evaluating and selecting the maintenance option. Therefore, selecting the initial index to perform and complete other steps is very important.

2.5 Automation, Emerging Technologies, and Futures Studies

Over the last five years, the growth of technology and use of new methods in engineering sciences has grown significantly. It is predicted that, in the coming years, many capabilities and applications in the field of infrastructure evaluation will be created using these technologies. At present, existing methods do not allow fully automated inspections and simulations of experts' behavior and knowledge. The use of available tools is costly and time-consuming and requires the intervention of the operator in various stages, such as harvesting, preprocessing, processing or summarizing, and analysis.

However, the benefits of this industry have attracted many researchers and investors. It is anticipated that, in the next 10 years, there will be remarkable developments in this field for the evaluation of infrastructure. Multifunction sensors, satellite imaging, drones and robotic assessment, and the like are emerging at a rapid rate. On the other hand, the organizations responsible for the maintenance of roads and infrastructure are trying to move from the traditional to modern situation and, thus, must eventually choose one of the technologies and provide the ground for its growth and development in the future.

Remote sensing (RS) is one of the emerging technologies in this industry, which uses images captured by satellites or drones to assess pavement failure. Due to the quality of the images and high volume of the captured images, satellite images commonly present high error and require high expenses to be updated and improved. In contrast, the use of drone tools can be a good alternative for automated assessment. The quadcopter unmanned aerial vehicle (QUAV) group behavior platform is a good option for intelligent inspection capabilities and robotic routing of the entire network, which creates a new set of intelligent inspections. By building small robots, inspection, evaluation and even repair of pavement will be possible, and intelligent robotic roadblocks can be imagined in the near future. Considering that expert interpretation capabilities are much greater than image processing, knowledge extraction capabilities using brain computer interface (BCI) intellectual communication tools can be used to interpret images and extract specific knowledge. The latter mentioned method, Electroencephalography (EEG) signals are used to interpret the fault images, while the rapid serial visual presentation (RSVP) method is employed for fast image analysis. This branch of science can provide new results of failure assessment and indicators. Moreover, the combination of QUAV, EEG, and BCI methods provides a novel tool for assessing infrastructure failure, while the use of special wave and frequency cameras, such as Kerlian imaging, can create new capabilities in interpreting new data via human visualization.

2.6 Summary and Conclusion

In this chapter, the importance of infrastructure management and its automation was first examined. The position of automation was surveyed as a new function and as a turning point in the field of infrastructure assessment. Then, the general structure and macro architecture of the development of infrastructure models at the network level were presented along with the main components of the general system design modules. Based on the proposed method and protocol, the order and sequence of attention to the topics and components were determined. In macro-level systems, the choice of technology is highly dependent on the indicators desired by managers and affects the level of management. For this reason, the sequence of examining the topics follows a top-down design. A variety of common indicators in the management of roads and technical buildings was subsequently presented. Finally, the role of emerging technologies and futures studies on the method selection were described. In the following chapters, the topics related to each of the components will be presented in more detail.

2.7 Questions

1 For an organization that needs to: semi-automatic, medium cost, and high-speed equipment, define an intelligent infrastructure management system that receives information in a completely intelligent and unmediated manner and does not analyze or evaluate it semi-automatically.

2 For complex maintenance of technical buildings that has bridges out of reach for inspection and needs new information every week, design a routine inspection program that is able to continuously inspect using UAV equipment.

3 Define the types of key components of infrastructure management system design and the involvement of developed technology in the selection of components. Design a system for the next five years for a tunnel maintenance agency.

4 Name three new emerging technologies that can be used in infrastructure asset management, and design a hypothetical plan for their use and development.

Further Reading

1 Barrile, V., et al., *UAV survey of bridges and viaduct: Workflow and application.* in *International Conference on Computational Science and Its Applications.* 2019. Springer. p. 269–284

2 Guerrero, J.A. and Y. Bestaoui, *UAV path planning for structure inspection in windy environments. Journal of Intelligent and Robotic Systems: Theory and Applications*, 2013. **69**(1–4): p. 297–311.

3 Chan, B., et al., *Towards UAV-based bridge inspection systems: A review and an application perspective. Structural Monitoring and Maintenance*, 2015. **2**(3): p. 283–300.

4 Balaguer, C., et al., *Towards fully automated tunnel inspection: A survey and future trends.* 2014 Proceedings of the 31st ISARC, Sydney, Australia, p. 19–33, ISBN 978-0-646-59711-9, ISSN 2413-5844. 2014.

5 He, W., et al., *The structural integrated safety management system in the E'dong Yangtse River Bridge.* 2016. CRC Press/Balkema.

6 Sutter, B., et al., *A semi-autonomous mobile robot for bridge inspection. Automation in Construction,* 2018. **91**: p. 111–119.

7 Chang, P.C., A. Flatau, and S.C. Liu, *Review paper: Health monitoring of civil infrastructure. Structural Health Monitoring,* 2003. **2**(3): p. 257–267.

8 Zakeri, H., F.M. Nejad, and A. Fahimifar, *Rahbin: A quadcopter unmanned aerial vehicle based on a systematic image processing approach toward an automated asphalt pavement inspection. Automation in Construction,* 2016. **72**: p. 211–235.

9 Zakeri, H., et al., *A multi-stage expert system for classification of pavement cracking.* in *Proceedings of the 2013 Joint IFSA World Congress and NAFIPS Annual Meeting, IFSA/NAFIPS 2013,* Edmonton, AB, Canada (24–28 June 2013). IEEE. 2013.

10 Paquis, S., et al., *Multiresolution texture analysis applied to road surface inspection. Proceedings of SPIE - The International Society for Optical Engineering,* 1999. **3652**: p. 242–249.

11 Zakeri, H., F.M. Nejad, and A. Fahimifar, *Image based techniques for crack detection, classification and quantification in asphalt pavement: A review. Archives of Computational Methods in Engineering,* 2017. **24**(4): p. 935–977.

12 Park, S.E., S.-H. Eem, and H. Jeon, *Concrete crack detection and quantification using deep learning and structured light. Construction and Building Materials,* 2020. **252**: p. 119096.

13 Payab, M., R. Abbasina, and M. Khanzadi, *A brief review and a new graph-based image analysis for concrete crack quantification. Archives of Computational Methods in Engineering,* 2019. **26**(2): p. 347–365.

3

Advanced Images Processing Techniques

Introduction

The term "image processing" refers to the two-dimensional image analysis by a digital computer, which in a more general sense refers to the concept of digital processing of any two-dimensional data. Figure 3.1 shows a general process of image processing (at Amir Kabir University of Technology), specifically digital image processing related to the detection of asphalt pavement failure. The image is captured digitally using the relevant hardware and then stored as a matrix of binary digits and used to communicate and control all digitization, storage, processing, and display operations via a computer. Figure 3.2 presents the steps performed in a typical image processing sequence.

The application of image processing has expanded rapidly in recent years. Today, various image-processing algorithms are used to handle many issues related to complex and precise sciences, such as medicine, remote sensing, space and microscopic operations, brain signal analysis, intelligent robots even on other planets, automation. Particularly, images obtained by imaging equipment are used to evaluate and monitor road surface to quantify the condition of road pavement as an important and effective indicator for civil engineers to make the best decisions for maintenance. Some applications based on the science of image processing include the following: identification and evaluation of damage to roads; determining the type, severity, and extent of failures; failure growth rate forecast; quantification of pavement surface friction; evaluation of pavement surface drainage; control of the percentage of materials in the asphalt mix and the distribution of materials; among many others.

The three main components for any automated system are the following: (i) image capture device (image acquisition component, IAC), (ii) image analysis system and related algorithms (image processing component, IPC), and (iii) interpretation and indexing methods (image interpretation component, IIC). This chapter deals with the second part of automated systems, namely image processing. As shown in Figures 3.3 and 3.4, each analysis of an image requires six general steps, which include the following: preprocessing, segmentation, feature extraction, feature selection, detection, and classification.

It should be noted that most research and models based on image processing are structured based off this system. If the hardware part, image processing, and indexing are considered as the main elements of automated intelligent systems, each new model or problem can be designed using the Eq. (3.1):

$$\text{Intelligent automated system} = f(\text{IAC} \cap \text{IPC} \cap \text{IIC}) \tag{3.1}$$

Automation and Computational Intelligence for Road Maintenance and Management: Advances and Applications,
First Edition. Hamzeh Zakeri, Fereidoon Moghadas Nejad, and Amir H. Gandomi.

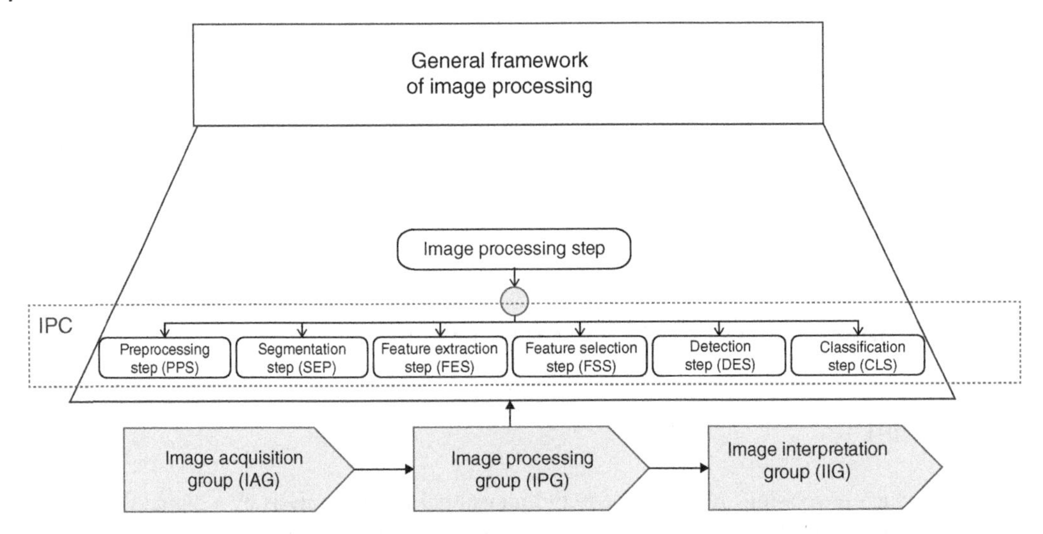

Figure 3.1 A general symbolic framework for digital image processing and image analysis.

In this regard, IAC, IPC, and IIC are the main parameters, which all largely contribute to the success of an automated system based on image processing. Of course, other parameters, such as technology, processing hardware, processing method and its algorithms, data collection conditions, evaluation speed and other components, are also effective, but they do not hold as much prominence as these three components in intelligence.

Among the components of Eq. (3.1), IPC has the greatest impact on intelligent image processing and extraction of key indicators.

In particular, this chapter presents methods related to the IPC component. According to the general framework, three main questions are raised for designing any intelligent automated system: (i) What kind of hardware and sensors are used to capture video information? (ii) Depending on the type of input data, what method should be used for image analysis and processing? (iii) What are the key indicators that will ultimately be used as quantitative indicators for each image? If these three questions are answered correctly, any problem or idea for intelligence can be implemented. In Section 3.2, the six steps of image processing are briefly introduced, and relationships and practical methods are presented.

3.1 Preprocessing (PPS)

Regardless of the quality of the IAG section, most images are taken from subjects in different lighting conditions (day/night), (sun/cloud), and contain subjects and objects that reduce quality. Such factors include random textures, heterogeneity of the surface, nonuniform light, shade, irregularity in the texture of the pavement surface, the presence of various shades, the presence of moisture, water, the effect of tire stretching on the pavement, splashing of oil and accidental elements such as garbage, leaves, dandruff, and even animal carcasses. Therefore, noise or irrelevant objects removal are very important. When using a raw image without an upgrade, choosing a uniform threshold in the segmentation phase is a very challenging problem and cannot be solved using conventional methods. Therefore, it is necessary to design a preprocessing algorithm before processing and feature extraction to achieve good results. At this stage, the main characteristics, such as cracked edge, quality improvement of pavement texture, extraction of aggregate edges, and removal

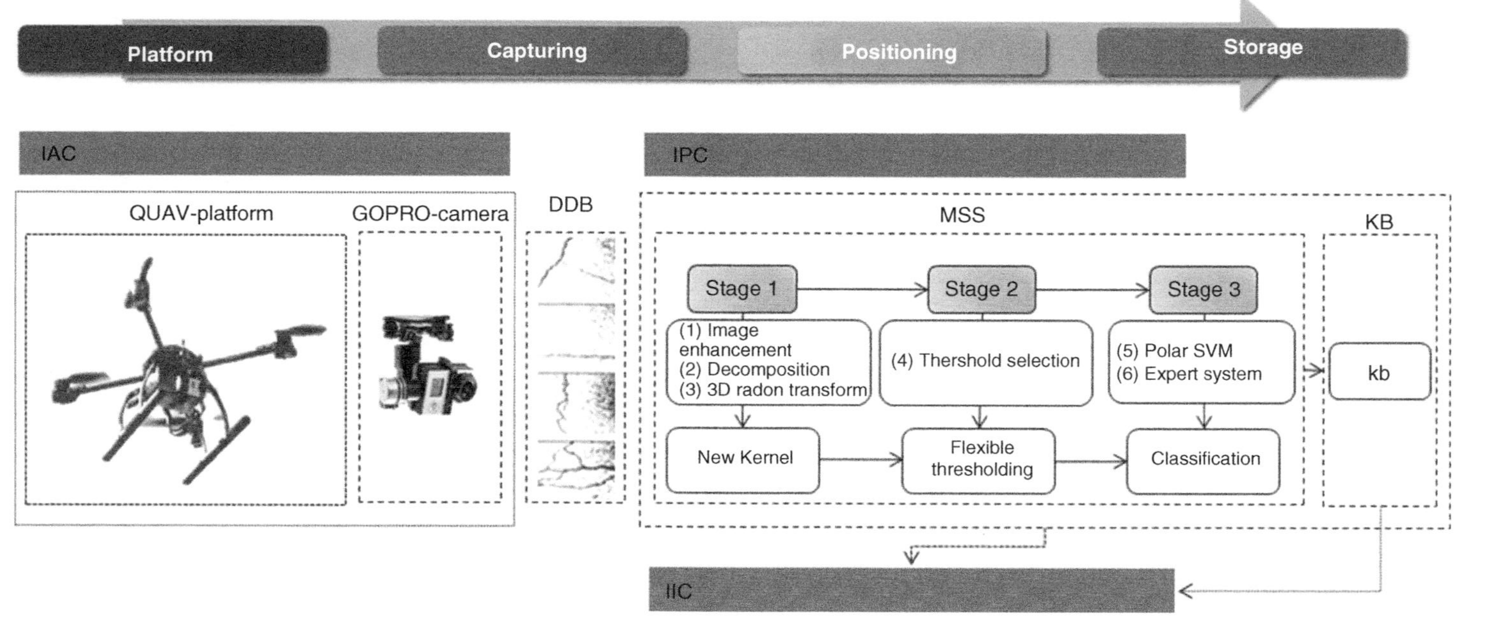

Figure 3.2 The image processing framework process.

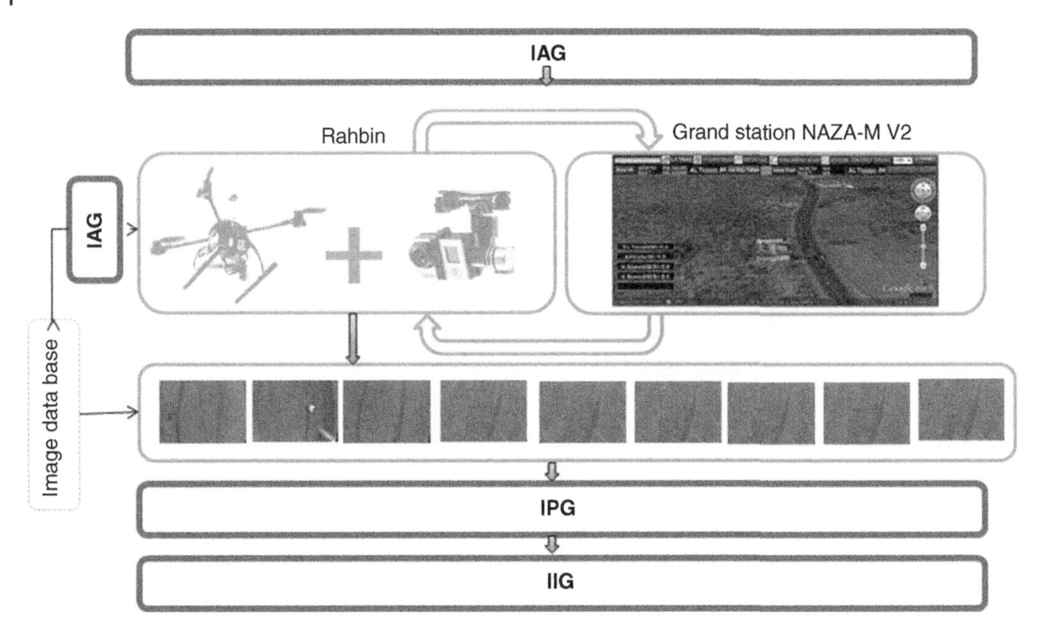

Figure 3.3 An example of an automated robotic system for assessing and inspecting pavement surfaces to detect and classify pavement failure. Source: Google LLC.

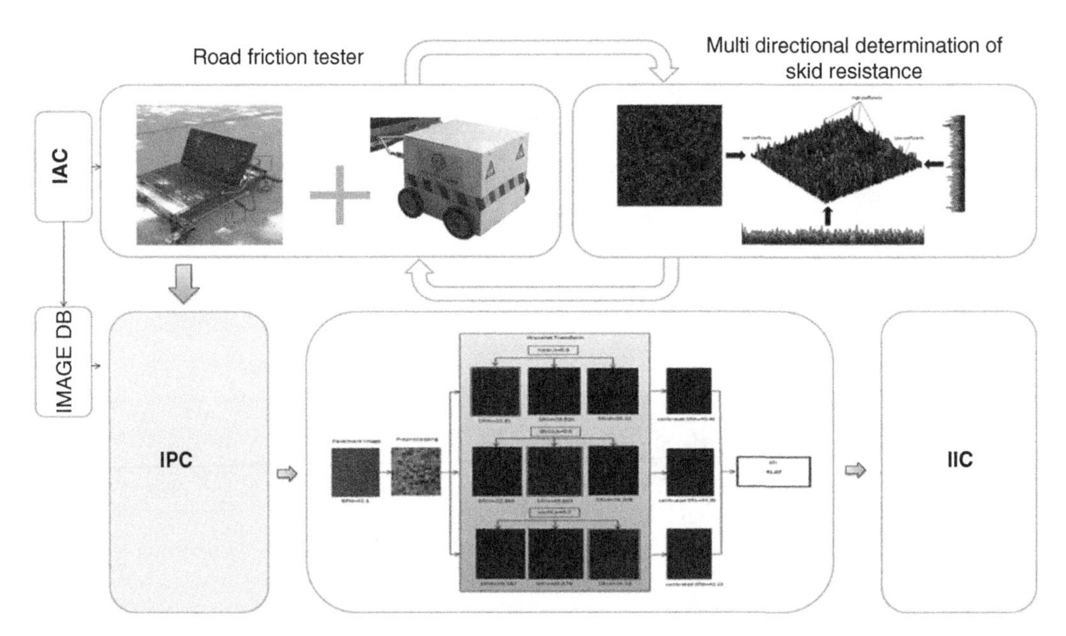

Figure 3.4 An example of an automated system for evaluating friction and surface drainage based on image processing.

of light reflection from the surface of the pavement surface, are upgraded. Image enhancement involves a wide range of methods, including noise reduction, fuzzy edge elimination, filtering, interpolation, magnification, contrast stretching, histogram modeling, conversion operations, and pseudo-staining, as part of the preprocessing methods to increase the accuracy of the next step (feature extraction). Most of these methods have been developed experimentally based on the type of input image and can be effective in obtaining more desirable intelligent results. Some common image enhancement techniques are shown in Figure 3.8. In this section, the image quality evaluation indicators and performance criteria of algorithms in noise elimination are briefly presented.

As an example, to assess the cracking of asphalt pavements, reducing noise and maintaining useful crack information including length, width, direction of rotation, and angle of placement of cracks in the image are important steps in image processing. Uniform energy distribution is a practical method to eliminate noise and correct fault detection. In multilevel methods, the use of high frequency is used to identify and evaluate the characteristics of cracking. Conversely, the use of low frequency is significant in evaluating pavement texture using multilevel methods. In other words, the cracks are entered at high frequency, and the background is placed at low frequency. By using geometry features extracted from the high-frequency band components, various properties, such as fractal, torque, correlation, homogeneity, can be evaluated and interpreted to evaluate cracking, pavement texture, or other damages like potholes and pavement surface bleeding. Subsequently, the method of noise cancellation and quality improvement of subject extraction using multilevel methods is reviewed followed by a discussion on the extraction method of the main characteristics of issues related to pavement and infrastructure.

Quantitative criteria, such as PSNR, IEF, and SSIM, are used to determine image quality. When there is an impulse noise in an image and cracking edges are removed to a certain extent using noise reduction algorithms, or, for example, a large amount of noise is added as pavement texture in the evaluation of pavement texture. To ensure that the enhanced image using such algorithms does not have much effect on the edges and maintains the originality of the subjects, an index called the edge preservation index (EPI) is presented and used to compare different methods.

3.1.1 Edge Preservation Index (EPI)

This index is presented to evaluate "image improvement based on a nonlinear multicriteria method." This index is similar to the numerical correlation index between 0 and 1 and is calculated from the Eq. (3.2):

$$\text{EPI} = \frac{\Gamma\left(\Delta s - \overline{\Delta s},\ \hat{\Delta s} - \overline{\hat{\Delta s}}\right)}{\sqrt{\Gamma\left(\Delta s - \overline{\Delta s},\ \hat{\Delta s} - \overline{\Delta s}\right).\Gamma\left(\Delta s - \overline{\hat{\Delta s}},\ \hat{\Delta s} - \overline{\hat{\Delta s}}\right)}} \tag{3.2}$$

where $\Delta s(i, j)$ is the high-pass filter for $s(i, j)$, which is the standard size of 3×3 for the Laplacian operator. In this index, the amount of similarity and correlation to the original image should be close to one. The closer this index is to zero, the more adverse the algorithm is to the edges.

3.1.2 Edge-Strength Similarity-Based Image Quality Metric (ESSIM)

This index is used to evaluate the range of edges in the image. The accuracy and efficiency of the method can be assessed by evaluating the similarity between the edge patterns. The ESSIM index is defined as follows:

$$\text{ESSIM}(x, xr) = \frac{1}{N}\sum_{i=1}^{N}\frac{2E(x,i)E(xr,i) + C}{(E(x,i))^2 + (E(xr,i))^2 + C} \tag{3.3}$$

where C is used with two purposes to avoid creating ambiguity of the size adjuster.

A different value of C changes the amount of ESSIM. This value is calculated using the relation $C = (BL)^2$, where B is a fixed value, and the value of L is variable. For eight-bit gray images, the edge resistance value is between 0 and 255.

Although the SSIM index has shown good performance in many applications, in matters such as cracking, pavement texture, bitumen, and pavement issues, the obtained quality measurement does not show a correct judgment based on subjective perception and visual information. The QILV index can be used to extract local information and evaluate image quality. This index is based on the assumption that a large amount of structural information of an image is placed in its local variance distribution. For example, the SSIM index calculates the local variances of both images, but the overall index considers only the mean of those values. Therefore, inequality is not considered. Further comparison with local variance properties can allow a more accurate qualitative comparison of the two images.

3.1.3 QILV Index

The local variance of an image is calculated with $\mathrm{Var}(x) = E\{(x - \overline{x})^2\}$, where $\overline{x} = E\{x\}$ is the local average of the image and is calculated based on the weight of adjacent pixels using a Gaussian function. Mean local variance of the relation is $\mu_{V1} = (1/MN) \sum_{i=1}^{M} \sum_{j=1}^{N} \mathrm{Var}(x(i,j))$, which is based on the general standard deviation function. The relation $\sigma_{V1} = \sqrt{((E\{(\mathrm{Var}(x(i,j))\text{-}\mu_{V1})\hat{}2\}))}$ is calculated and estimated based on the Eq. (3.4):

$$\sigma_{Vx} = \sqrt{\frac{1}{NM-1} \sum_{i=1}^{M} \sum_{j=1}^{N} (\mathrm{Var}(x(i,j)) - \mu_{Vx})^2} \tag{3.4}$$

In Eq. (3.4), the covariance between the variances of the two images x and xr is defined based on the following relation:

$$\sigma_{VxVxr} = E\{(\mathrm{Var}(x(i,j)) - \mu_{Vx})(\mathrm{Var}(xr(i,j)) - \mu_{Vxr})\} \tag{3.5}$$

In Eq. (3.5), the value σ ^_VxVxr is obtained from the Eq. (3.6):

$$\sigma_{VxVxr} = \frac{1}{NM-1} \sum_{i=1}^{M} \sum_{j=1}^{N} (\mathrm{Var}(x(i,j)) - \mu_{Vx})(\mathrm{Var}(xr(i,j)) - \mu_{Vxr}) \tag{3.6}$$

Using these definitions, the QILV index between the original image and the enhanced image is calculated as follows:

$$\mathrm{QILV}(x, xr) = \frac{2\mu_{Vx}\mu_{Vxr}}{\mu_{Vx}{}^2 + \mu_{Vxr}{}^2} \frac{2\sigma_{Vx}\sigma_{Vxr}}{\sigma_{Vx}{}^2 + \sigma_{Vxr}{}^2} \frac{2\sigma_{VxVxr}}{\sigma_{Vx}\sigma_{Vxr}} \tag{3.7}$$

There are many similarities between the QILV index and SIMM, the main difference being that QILV is calculated for local variance and SIMM for general statistics.

3.1.4 Structural Content Index (SCI)

The SCI indicator represents the reduction in the amount of noise-depleted image information to the original image. This index is calculated using the Eq. (3.8):

$$\mathrm{SCI}(x, xr) = \sum_{i=1}^{M} \sum_{j=1}^{N} \left(\frac{xr.xr}{x.x}\right) \tag{3.8}$$

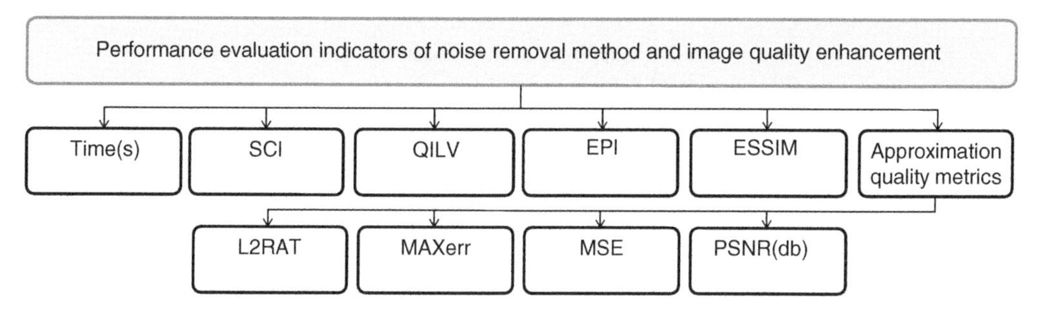

Figure 3.5 Classification of image quality evaluation indicators in order to select image enhancement method, filter type, and analysis level (in multilevel methods).

SCI is a number between 0 and 1, whereby the closer this number is to 1, the more information from the main subjects in the reconstructed image are retained.

3.1.5 Signal-To-Noise Ratio Index (PSNR)

The PSNR index, which represents the maximum signal-to-noise ratio, is based on decibels (db) and is represented by the bit value per pixel. This index is defined based on the following relation:

$$\mathrm{PSNR} = 20 \log_{10} \frac{2^{B}-1}{\sqrt{\mathrm{MSE}}}, \mathrm{MSE} = \frac{\|x - xr\|^{2}}{N}. \tag{3.9}$$

In Eq. (3.9), MSE is the mean squared error squared, and *B* represents the bit ratio for each input image sample. The MSE index represents the absolute value of the *x* approximation square for the image *xr*, which is normalized by the number of elements. The lower the index, the more accurate and less error it indicates. The higher the PSNR, the better the image reconstruction.

3.1.6 Computational time index (CTI)

Computational time is one of the most important indicators to select the best multilevel method, filters, and decomposition and composition steps. The shorter the computational time, the higher the efficiency of the method is for simultaneous computing. This index plays an important role in the classification stage, feature extraction, diagnosis and classification of pavement failure, pavement texture status, pavement bleeding morphology, and other issues related to image-based evaluation.

Figure 3.5 presents the different types of functional indicators for evaluating the method of quality improvement and noise elimination.

3.2 Preprocessing Using Single-Level Methods

In order to evaluate and analyze the images, it is often necessary to extract directional information from the subjects in the image (including cracking, texture, aggregate morphology, morphology of bitumen contents, friction, etc). For this reason, multilevel methods are considered as efficient tools due to their ability to decompose information in several levels and the possibility of reconstructing them with the least amount of error.

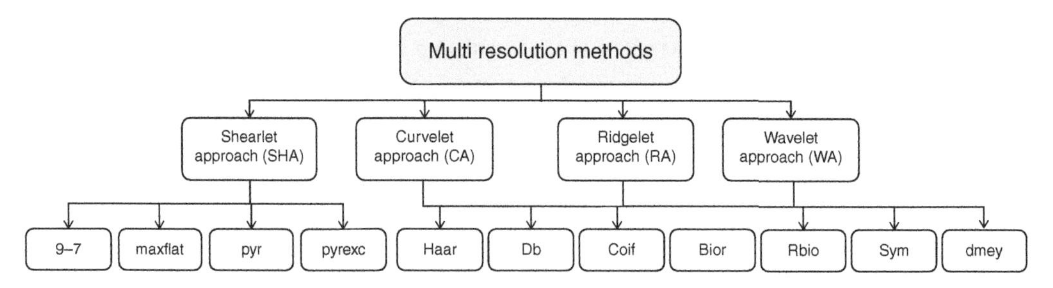

Figure 3.6 Classification of multilevel methods in preprocessing (multiresolution methods).

In this section, various types of single and multilevel methods are first introduced, then the efficiency of each multilevel method in specialized issues is evaluated based on the indicators introduced in the previous section. The main characteristics of image quality evaluation, shown in Figure (3.5), are evaluated for each method and with different filters.

In order to access details of images, one of the most important methods of quality improvement and analysis of pavement images as well as infrastructure is multilevel analysis using wavelet approach (WA), curvelet approach (CA), ridgelet approach (RA), and Shearlet approach (SA). In this section, the capability of each of these single-level and multilevel methods in noise elimination is clearly presented, and then the types of filters and optimal filter selection methods are introduced. Finally, the optimal method is selected from the existing methods as an example for case studies in the field of pavement management according to the capabilities of the method (Figure 3.6).

3.2.1 Single-Level Methods

Single-level methods are commonly used for image enhancement, specifically to blur and reduce noise. Some important applications of these methods include removing small details from the image, bridging small gaps between the cracks, and reducing the effect of noise and background, for example, in pavement texture by blurring with various filters. If these methods are not applied in the preprocessing stage, thresholding on the image to detect, extract, and classify the defect is poorly efficient, while the misdiagnosis of pixels may increase due to the pavement texture. This type of error causes a significant reduction in accuracy in the segmentation, classification, and indexing stages. This section describes the aim to reduce the effect of noise and create connections in linear subjects (e.g. cracking, spall in surface failure, tissue separation, PPP amount) to increase the accuracy in detection and segmentation.

3.2.2 Linear Location Filter (LLF)

Linear spatial filters, also called low-pass filters, are based on the average pixel theory, considering the vicinity of the filter frame as a smoothing response. In this method, the value of each pixel in the image is replaced with the average intensity levels by the filter mask, then the image is created with the property of sharp transitions in the image matrix. The most important feature of this method is to reduce the effect of noise in images with very high noise (due to pavement texture) and also to reduce the effect of edges (i.e. cracked, pavement texture, and aggregate edges) in the X-ray image of an asphalt mixture. Specifically, to reduce the effect of noise caused by the texture of the pavement surface simultaneously save useful information about the objects such as cracking and potholes, the

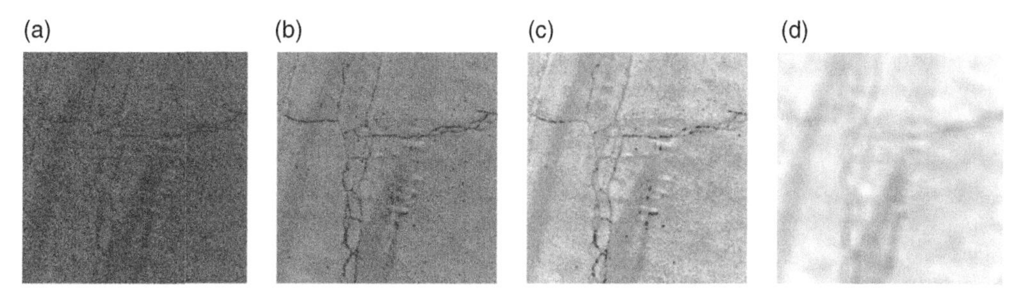

Figure 3.7 The effect of spatial linear filter on noise reduction and image enhancement and application of fixed threshold on the image: (a) image of cracked pavement with fixed threshold; application of linear filter with dimensions (b) [5 5], (c) [10 10], and (d) [50 50].

amount of coarse pavement texture, and PPP percentage play a crucial role. This method blurs the edges and fuzzifies the edges of the main subject in the image.

$$xr = \sum_{s=-a}^{a}\sum_{t=-b}^{b} w(s,t)f(x+s,y+t) \Big/ \sum_{s=-a}^{a}\sum_{t=-b}^{b} w(s,t), \quad (3.10)$$

In Eq. (3.10), the linear spatial filter is an image of size $M \times N$ with a filter of dimensions $m \times n$, where the variables x and y for each pixel in w are smoothed with each pixel in f. The larger the size of the filter, the greater the effect of reducing the noise effect, and the more information about the main subjects (such as cracked edges, surface texture separation, and drainage capacity) is lost.

As shown in Figure 3.7, as the size of the filter increases, the information on the main subjects is lost and the minimum important information (such as cracking, pavement texture, holes and patches, and surface pavement bleeding) also becomes fuzzy. Therefore, it is best to apply this filter in combination with other methods rather than alone. Table 3.1 presents comparisons between the main indicators presented in the previous section and the types of linear filters, from which the optimal linear filter can be selected and applied to the database with crack distress; the results are averages of 10 repetitions.

In Figure 3.7, a linear surface filter with different square dimensions WF1 = [5 5], WF2 = [10 10], WF3 = [15 15], WF4 = [5 10], WF5 = [5 10], and WF6 = [10 15] has been designed and applied on pavement images. The results show that the PSNR index decreased as the filter dimensions increases, and the noise decreased as a result. The average PSNR index is 24.4, and the difference between the minimum and maximum PSNR was 1.54. However, this noise reduction increased both the crack degradation and MSE index, which was determined to be 80 for WF4, with a

Table 3.1 The average computational times over 3 runs to arrive at 100 consequent set.

Filter index	PSNR	MSE	MAXERR	L2RAT	ESSIM	EPI	QILV	SCI	(S) TIME
WF1	25.389	188.017	100.400	0.985	1.000	0.153	0.028	0.985	0.013
WF2	24.133	251.052	122.070	0.980	0.999	0.226	0.001	0.980	0.018
WF3	23.848	268.115	117.164	0.976	0.999	0.241	0.000	0.976	0.018
WF4	24.716	219.509	116.280	0.982	1.000	0.179	0.007	0.982	0.018
WF5	24.399	236.156	113.440	0.980	0.999	0.194	0.003	0.980	0.018
WF6	23.987	259.677	117.540	0.978	0.999	0.233	0.000	0.978	0.019

difference of 80 between the maximum and minimum MSE. It is apparent that the amount of error depends on the pavement texture and the presence of cracks in the surface and that the L2RAT index decreases as the filter size increases. The ESSIM index for the main image is decreased as the filter is increased, and the minimum and maximum values for L2RAT were determined to be 0.976 and 0.98, respectively. The minimum value of QILV index is zero, and its average is 0.007. The average SCI index is 0.98, and the minimum for the linear method is 0.976. The average computational time was estimated to be 0.017 seconds, while the minimum time was 0.013 seconds.

3.2.3 Median Filter

The median filter has shown to be superior to linear filters in removing noise and preserving the effect of edges and objects (including cracks, holes, types of damage, coarse pavement texture, etc.). This method, according to an X matrix and a two-dimensional filter, rotates the created filter based on the created matrix in order to attain a core of 180°. The filtering operation is then performed using convolution. An example of applying a filter with this method is shown in Figure 3.8.

In this method, various filters are used to remove noise and enhance the image, which include averaging, disk, Gaussian, kinetic, logarithmic, and shake filters. Each of these filters is designed so that their size and geometry can be changed accordingly. In this section, a general comparison between the type and size of each of these filters is first provided, then an example in the field of pavement distress detection is given to identify the optimal filter and make comparisons with other methods:

$$xr(n1, n2) = \sum_{s=-\infty}^{\infty} \sum_{t=-\infty}^{\infty} f(s,t)x(n1-k, n2-k) \tag{3.11}$$

Table 3.2 gives a comparison between the main indicators and the types of intermediate filters (small, medium, and large) based on the average of all filters used, including medium, disc, Gaussian, kinetic, logarithmic and vibration filters, and smoothing to select the size. The intermediate optimization was also applied to the database (db1). The results in the table are the averages of ten replications for all filter.

According to this case study, as the filter size increases, the computational time increases and L2RAT decreases. Moreover, it can be seen that the ESSIM index decreased by 0.06%, the QILV index decreased by 88%, and the EPI index decreased by 23%. Moreover, SCI was reduced by

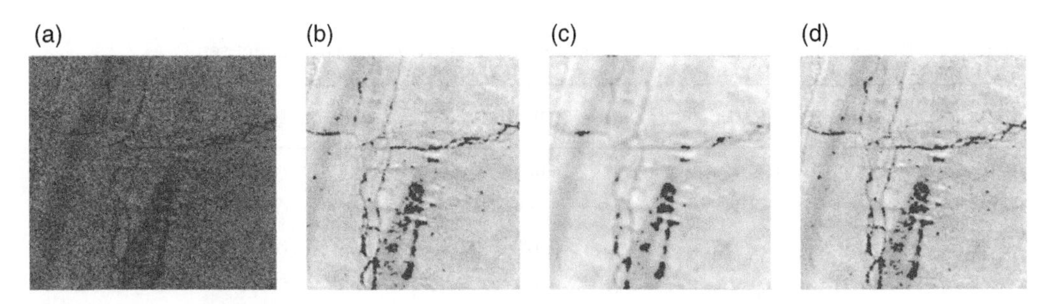

Figure 3.8 The effect of median filter on noise reduction and image enhancement and application of fixed threshold on the image: (a) pavement image with fixed threshold; application of median filter with (b) dimensions [5 5], (c) radius [5], and (d) dimensions [50 50].

Table 3.2 The image smoothing performance evaluation indicators after applying intermediate surface filter in small, medium, and large dimensions.

Filter size	PSNR	MSE	MAXERR	L2RAT	ESSIM	EPI	QILV	SCI	(S) TIME
Small	23.760	361.370	155.080	1.023	1.000	0.387	0.095	1.023	0.014
Medium	23.340	380.980	162.820	1.022	0.999	0.390	0.175	1.022	0.049
Large	23.180	337.900	143.200	0.997	0.999	0.313	0.021	0.997	0.077

Table 3.3 Image enhancement indices after applying median surface filter in different sizes *S*(*i*) and FM filter (*i*).

Filter	Size	PSNR	MSE	MAXERR	L2RAT	ESSIM	EPI	QILV	SCI	TIME
FM1	S1	23.800	268.100	117.200	0.976	0.999	0.240	0.001	0.970	0.075
FM2	S2	23.800	272.900	116.100	0.975	0.990	0.240	0.001	0.970	0.080
FM3	S3	23.900	267.700	117.100	0.976	0.990	0.240	0.001	0.970	0.070
FM4	S4	20.000	656.700	241.300	1.083	1.000	0.510	0.070	1.080	0.008
FM5	S5	20.100	638.500	236.400	1.082	1.000	0.510	0.080	1.080	0.007
Median		22.300	420.800	165.600	1.018	0.990	0.350	0.030	1.010	0.050
Maximum		23.900	656.700	241.300	1.083	1.000	0.510	0.080	1.080	0.080
Minimum		20.000	267.700	116.100	0.975	0.990	0.240	0.001	0.970	0.070

2.66%, indicating edge degradation in linear patterns (e.g. pavement cracking) based on increasing filter size for a 2.5% increase in noise reduction. The computational complexity increased by 81.34% as the filter size increased from small to large. Table 3.3 presents the optimal filters for different shapes, where FM1 to FM5 represent the filters of average, disk, Gaussian, kinetic, and shaking, respectively. Also, the optimal sizes selected with S1 to S5 are designed for sizes 15, 10, 15–30, 10–0, and 0.9, respectively.

The PSNR index for Gaussian filters is higher than that of the other filters, where the average PSNR index is 24.3 and difference between the minimum and maximum PSNR is 3.9. By examining the EPI index, two medium filters and Gaussian filters show the same performance. The maximum error rate in a disk type filter with a size of 10 is less than that of the other filters. The SCI value for the first three filters is 0.97. The MSE for the FM2 filter is lower than the FM4 and FM5 methods, and the average MSE is higher than the FM1 and FM3 filters. According to the study and comparison of different methods of the FM3 filter, 15–30 was selected as the optimal filter in the median method for noise removal, crack connection, and image enhancement. Obviously, the choice of filter size and type depends on the image and quality of the captured data.

3.2.4 Wiener Filter

The Wiener Filter is an adaptive linear filter that uses local variance of the image to eliminate noise. This filter smoothes the image in areas with high variance and performs smoothing with more intensity in areas with low variance. This method often produces better results than linear and median filtering. Preserving the edges and details of cracks and other high-frequency parts of an image is one of the most important features of this method. However, the Wiener filter requires more computational time than linear and median filtering. Based on statistics calculated from a range of each pixel, Wiener considers a pixel-compatible method. The mean and variance around each pixel are calculated using the Eq. (3.12):

$$\mu = \frac{1}{\mathrm{MN}}\sum\nolimits_{n_1,n_2\in\eta} f(n_1,n_2), \sigma^2 = \frac{1}{\mathrm{MN}}\sum\nolimits_{n_1,n_2\in\eta} f^2(n_1,n_2) - \mu^2 \tag{3.12}$$

where η is the local MN neighborhood of each pixel for the image.

Using these parameters, the Wiener filter is calculated from the Eq. (3.13):

$$\mathrm{b}(\mathrm{n}_1,\mathrm{n}_2) = \mu + \frac{\sigma^2 - \mathrm{v}^2}{\sigma^2}(\mathrm{b}(\mathrm{n}_1,\mathrm{n}_2) - \mu) \tag{3.13}$$

In Eq. (3.13), v^2 is the noise variance, which is calculated based on the local mean variance. This method is used as an independent method as well as a hybrid method in multilevel methods.

An example of applying a Wiener filter to a pavement image is shown in Figure 3.9. According to this experiment, the larger the dimensions of the filter, the greater its effect on reducing noise. On the other hand, the adverse effect on crack edges, pavement texture, and cracking patterns will eliminate useful and original image information. This filter is a square for the analysis of cracks, which are heterogeneous and sometimes discrete directional patterns, where the use of directional Wiener filters has a higher efficiency than matched filters to extract cracks. Of course, it

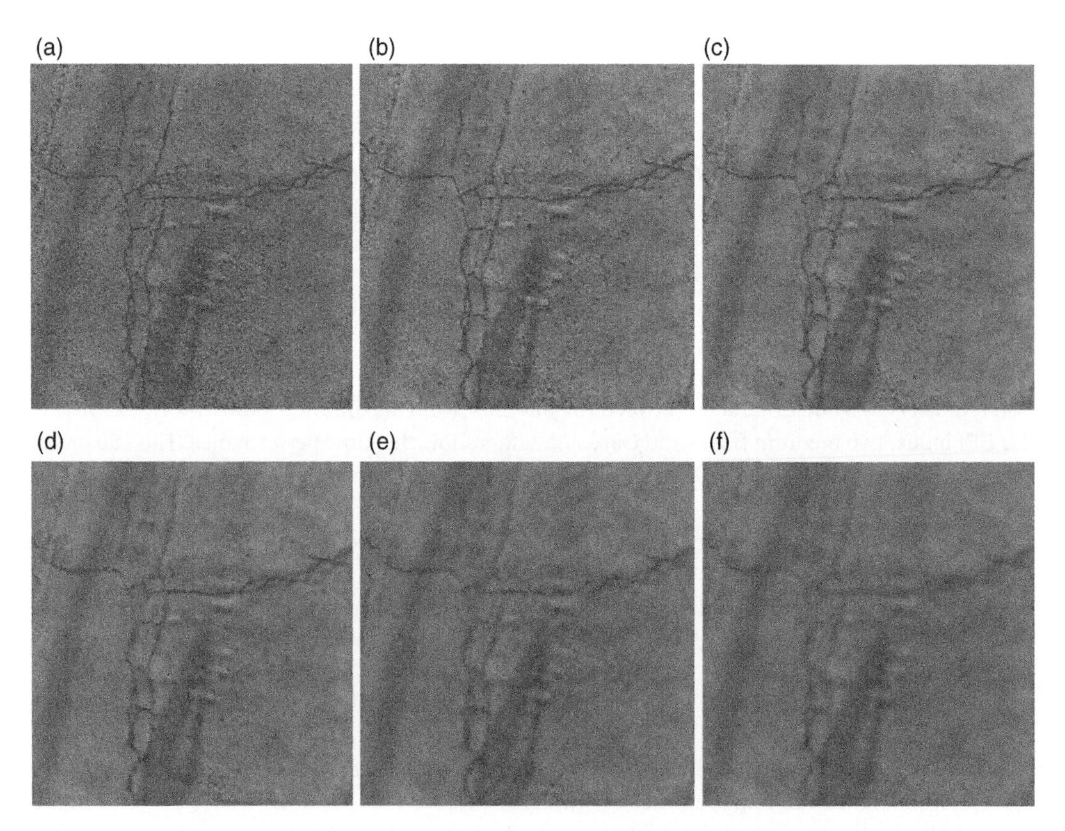

Figure 3.9 The effect of Wiener filter on noise reduction and image enhancement by applying a fixed threshold on the image: applying Wiener filter of size (a) 5, (b) 10, (c) 15, (d) 20, (e) 25, and (f) 30.

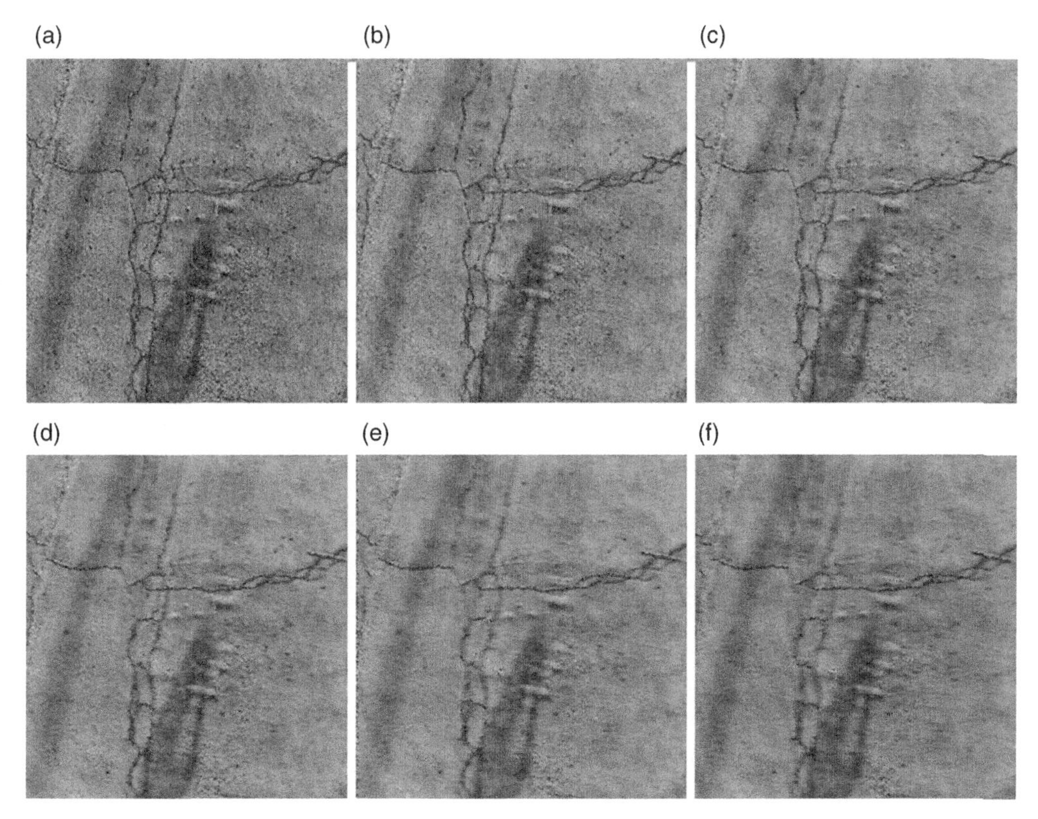

Figure 3.10 The effect of directional Wiener filter on reducing image enhancement noise by applying a fixed threshold on the pavement image with surface defects (such as cracks, spalls): vertical directional Wiener filter of size 10 with coefficient (a) 0.5, (b) 1.5, (c) 2, (d) 2.5, (e) 3, and (f) 3.5.

should be noted that the use of square filters has better results for analyzing the surface texture of pavement. Therefore, the choice of filter pattern depends on the application and type of problem. The use of directional Wiener filters individually is not very efficient and can only detect the predominant direction of cracking. Therefore, determining the direction of dominant cracks is one of the main issues in determining the filter. An example of the use of directional Wiener filters is shown in Figure 3.10, where the extraction of cracks is highly dependent on the direction of cracking. Filters in a certain direction retain only the crack edges in the direction perpendicular to it. Noise in the filter direction is eliminated, and the crack edge information is fuzzified at the same time so that if a fixed threshold is used, the pavement texture and cracking information will be damaged.

As shown in Figure 3.10d–f, by increasing the vertical filter coefficient by 2, 2.5, and 3, the information about the horizontal cracks is preserved and the linear patterns become more recognizable by applying a fixed threshold. Similarly, the larger the filter size, the higher the noise cancellation capability and the higher the PSNR value. In order to select the optimal filter and the appropriate coefficient, a comparison between square, horizontal, and vertical direction filters and the optimal filter is provided in Table 3.4.

In the square method, which applies the filter equally in two directions, the PSNR index is 2% higher than two filters with different dimensions in both horizontal and vertical directions. Also, in the best case, the ESSIM for the square filter is 8% and for and EPI indices 62% greater than the

Table 3.4 Comparison of image smoothing indices after applying Wiener surface filter in different directions, i.e. square, vertical W_v, and horizontal W_h.

Direction	PSNR	MSE	MAXERR	L2RAT	ESSIM	EPI	QILV	SCI	TIME
Square	25.710	174.240	56.440	0.987	1.000	0.363	0.037	0.987	0.193
Vertical	25.270	193.210	65.540	0.986	0.999	0.335	0.014	0.986	0.206
Horizontal	25.100	200.610	63.812	0.985	0.999	0.324	0.008	0.985	0.212
Mean	25.360	189.350	61.931	0.986	0.999	0.341	0.020	0.986	0.204
Maximum	25.710	200.610	65.540	0.987	1.000	0.363	0.037	0.987	0.212
Minimum	25.100	174.240	56.440	0.985	0.999	0.324	0.008	0.985	0.193

Table 3.5 Comparison of square image smoothing filters after applying Wiener surface filter with $\alpha = 1$ for a set of pavement images (case study of pavement failure and pavement texture).

Filter	S	PSNR	MSE	MAXERR	L2RAT	ESSIM	EPI	QILV	SCI	TIME
WF1	5.000	27.460	116.700	45.900	0.990	0.999	0.415	0.240	0.990	0.097
WF2	10.000	25.720	174.200	56.440	0.987	0.999	0.363	0.037	0.987	0.193
WF3	15.000	25.250	194.000	64.770	0.986	0.999	0.335	0.012	0.986	0.208
WF4	20.000	24.970	207.000	67.380	0.985	0.998	0.313	0.006	0.985	0.222
WF5	25.000	24.780	216.200	72.550	0.984	0.998	0.297	0.003	0.984	0.232
WF6	30.000	24.660	222.600	81.990	0.984	0.998	0.286	0.003	0.984	0.244
Mean		25.470	188.500	64.840	0.986	0.991	0.334	0.050	0.986	0.199
Maximum		27.460	222.600	81.990	0.990	0.999	0.415	0.240	0.990	0.244
Minimum		24.660	116.700	45.900	0.984	0.998	0.286	0.003	0.984	0.097

horizontal filter and ESSIM for the square filter is 11% and EPI indices 77% greater than the vertical filter, respectively. This indicator shows that despite improvements in the detection of linear patterns (such as directional-longitudinal and transverse cracks), the use of horizontal and vertical Wiener filters does not detect directional information at other angles. According to this analysis, the Wiener filter with a coefficient of 1 has a higher efficiency than other Wiener filters with coefficients $(\alpha = (m/n)) \neq 1$. Table 3.5 shows the results of different filters with $\alpha = 1$ and of various sizes for different samples, m, n, s.

According to the results of Table 3.5, the larger the size of s, the lower the value of the PSNR index, which reduces the information of the main subjects in the image. The SCI index decreases and the time index increases with increasing S. Also, the WF1 filter exhibited the best results in detecting malfunctions and reducing the amount of noise. It is concluded from this method that although directional methods are more efficient in detecting linear patterns, they may have different characteristics in all directions because these patterns are in fact discontinuous and directional. Therefore, using a single filter in a specific direction only extracts information in the same direction and has little application in real problems. Therefore, it is necessary to use filters that extract directional information in different directions and simultaneously remove or reduce the effect of background noise.

3.3 Preprocessing Using Multilevel (Multiresolution) Methods

3.3.1 Wavelet Method

In this method, a family of library filters are used to analyze single, two-, and three-dimensional waves via the wavelet equation function. Each of these filters can have a different effect on the input image as well as the main subject (such as cracks, textures, holes, spalls).

In signal processing, a wave is $f\epsilon L^2(R)$, where $L^2(R)$ is a measurable spatial vector with a one-dimensional function square that can be multidimensionally expanded using the law of independent linear superposition. This function is

$$f = \sum_i C_i \Psi_i \tag{3.14}$$

In Eq. (3.14), C_i is a coefficient, and $\{\Psi_i\}_{i\epsilon Z}$ is a set of integers independent of $L^2(R)$. If $\{\Psi_i\}_{i\epsilon Z}$ is perpendicular and $i = j$, the result of the internal multiplication $\langle \Psi_i, \Psi_j \rangle$ will be 1; otherwise, it is zero. In this case, the signal amplification can be shown as follows:

$$\mathrm{f} = \sum_{\mathrm{i}} \langle \Psi_\mathrm{i}, \mathrm{f} \rangle \Psi_\mathrm{i}. \tag{3.15}$$

One of the most widely used examples is the Fourier transform with a normal base of e^{-iwx}, as shown in Eq. (3.16):

$$F(\omega) = \int_{-\infty}^{\infty} f(x) e^{-iax} dx. = \int_{-\infty}^{\infty} f(x) e^{-iax} G(x-b) dx. \tag{3.16}$$

where R is a set of real numbers $\omega \in R$; $F(\omega)$ is the instantaneous transfer, which is a function of $f(x)$.

This equation converts waves into frequencies in terms of time and place. In the relation in Eq. (3.16), $G(x)$ is a function of window size, and b is the window transfer for the purpose of time or space. The wavelet function is computed using the Eq. (3.17):

$$W_{(a,b)} = |a|^{-0.5} \int_{-\infty}^{+\infty} f(x).\overline{\Psi\left(\frac{x-b}{a}\right)} dx. \quad a, b \in R, a \neq 0. \tag{3.17}$$

The two-dimensional wavelet function is generated using the Eq. (3.18):

$$\begin{aligned} \Phi(x,y) &= \varphi(x)\varphi(y), \\ \psi^H(x,y) &= \varphi(x)\Psi(y), \\ \psi^V(x,y) &= \Psi(x)\varphi(y), \\ \psi^D(x,y) &= \Psi(x)\Psi(y) \end{aligned} \tag{3.18}$$

In these relations, ψ^H, ψ^V, and $\psi^D$are WT components for the horizontal, vertical, and diagonal directions, respectively; and Φ is a two-dimensional function with approximate information from the image that is converted to other dimensions:

$$\begin{aligned} \Phi(x,y) &= 2^{-k}\varphi(2^{-k}x-m)\varphi(2^{-k}y-n), \\ \Psi^H_{kmn}(x,y) &= 2^{-k}\varphi(2^{-k}x-m)\Psi(2^{-k}y-n), \\ \Psi^V_{kmn}(x,y) &= 2^{-k}\varphi(2^{-k}x-m)\varphi(2^{-k}y-n), \\ \Psi^D_{kmn}(x,y) &= 2^{-k}\Psi(2^{-k}x-m)\Psi(2^{-k}y-n). \end{aligned} \tag{3.19}$$

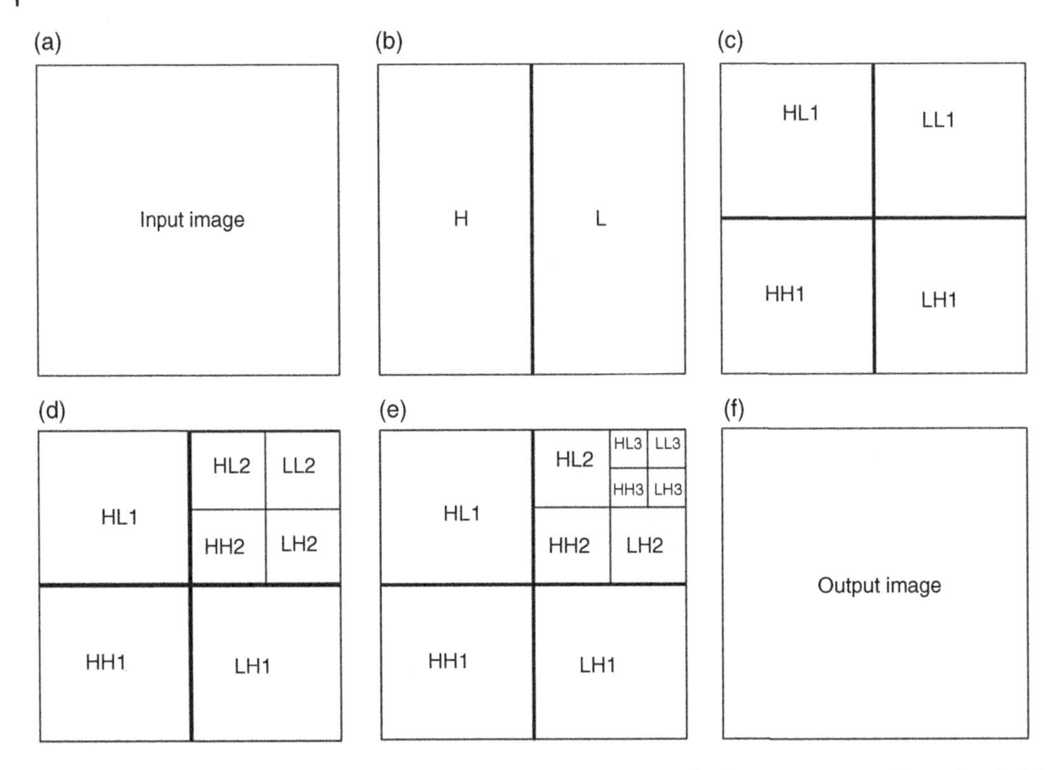

Figure 3.11 The general structure of the wavelet method in the analysis of coefficients at different levels: (a) main image, (b) separation into low and high-frequency bands, (c) first level of decomposition, (d) second band of decomposition, (e) third band of decomposition, and (f) combination of coefficients and reconstruction of the image.

Based on the approximation and analysis of the data, the above calculations (3.19) can be performed using wavelet coefficients. An example of a wavelet structure for an image is shown in Figure 3.11, D_1^H D_1^V, and D_1^D are the details of the first level obtained for ψ^H, ψ^V, and ψ^D. The details and A2 are D_2^H, D_2^V, and D_2^D approximately below the second level. The upper left rows in Figure 3.11 mean that the decomposition can continue until the desired subband frequency is reached.

In order to reconstruct an image, $f_{k-1}(x, y)$ can be calculated in $v_k$space in three states, namely w_k^l, $w_k^h$and w_k^d, using the following equation:

$$\begin{aligned} f_{k-1}(x,y) = 2^{-k}\sum_{m,n} LL_k(m,n)\varphi(2^{-k}x-m)\varphi(2^{-k}y-n) \\ + 2^{-k}\sum_{m,n} HL_k(m,n)\varphi(2^{-k}x-m)\Psi(2^{-k}y-n) \\ + 2^{-k}\sum_{m,n} LH_k(m,n)\Psi(2^{-k}x-m)\varphi(2^{-k}y-n) \\ + 2^{-k}\sum_{m,n} HH_k(m,n)\Psi(2^{-k}x-m)\Psi(2^{-k}y-n). \end{aligned} \tag{3.20}$$

where LL_K, HL_K, LH_K, and HH_K are the normalized images in the spatial environment V_K, w_k^l w_k^h, and w_k^d, which are, respectively, calculated using the following equations:

$$\begin{aligned} LL_k(i,j) &= \sum_{m,n} l(m-2i)l(n-2j)LL_{k-1}(m,n), \\ HL_k(i,j) &= \sum_{m,n} h(m-2i)l(n-2j)LL_{k-1}(m,n), \\ LH_k(i,j) &= \sum_{m,n} l(m-2i)h(n-2j)LL_{k-1}(m,n), \\ HH_k(i,j) &= \sum_{m,n} h(m-2i)h(n-2j)LL_{k-1}(m,n). \end{aligned} \tag{3.21}$$

Then, A_k can be calculated as follows:

$$\begin{aligned} A_k &= 2^{-\frac{3k}{k-1}} \sum_n \varphi(2^{-k}x-l), \varphi(2^{-k+1}x-n) f(x), \varphi(2^{-k+1}x-n) \\ &= 2^{-k}\sqrt{2}\varphi(2^{-k}x-l), \varphi(2^{-k+1}x-n) A_{K-1}. \\ \varphi(2^{-k}x-l) &= 2^{-k+1} \sum_n \varphi(2^{-k}x-l), \varphi(2^{-k+1}x-n)\varphi(2^{-k+1}x-n). \end{aligned} \tag{3.22}$$

In Equation (3.22), $A_{k-1} = 2^{-(k-1)/2} \sum_n \langle f(x), \varphi(2^{-k+1}x-n) \rangle$. If the equation $2^{-k}(x-l)$ is replaced by t/2, then $\langle \varphi(2^{-k}x-l), \varphi(2^{-k+1}x-n) \rangle = \langle \varphi(0.5t), \varphi(t-(n-2l)) \rangle$, where l(n) is calculated as follows:

$$\begin{aligned} l(n) &= \sqrt{2}\langle \varphi(0.5t), \varphi(t-(t-n)) \rangle = \sqrt{2}\int_R \varphi(0.5t)\varphi(t-n)dt. \\ h(n) &= \sqrt{2}\langle \Psi(0.5t), \varphi(t-(t-n)) \rangle = \sqrt{2}\int_R \Psi(0.5t)\varphi(t-n)dt. \end{aligned} \tag{3.23}$$

In the above equations, $A_n^K$and φ_n^K refer to the normalized image in space $V_K$and W_K, which is calculated using equation (3.22). The parameter h (n) is a ¼ symmetric filter of ln, and in the relation of h and l, the one-dimensional wave decomposed in LL is the main image. The image reconstruction algorithm is shown in 3.24

$$\begin{aligned} LL_{k-1}(i,j) = &\sum_{m,n} LL_k(m,n)l(i-2m)l(j-2m) + \sum_{m,n} HL_k(m,n)h(i-2m)l(j-2n) \\ &+ \sum_{m,n} LH_k(m,n)l(i-2m)h(j-2n) + \sum_{m.n} HH_k(m,n)h(i-2m)h(j-2m). \end{aligned} \tag{3.24}$$

An example of image decomposition/composition processes using the wavelet method is shown in Figure 3.12.

Noise removal in the wavelet method is done using a shrinkage or thresholding methodology, which is based on maintaining high-amplitude coefficients and eliminating and smoothing low-amplitude coefficients. Cracking and failure can be observed in coefficients with high amplitude and frequency. In this method, high-amplitude coefficients are maintained, and low-amplitude coefficients are reduced to zero.

Filters and the family filters of wavelet transform are summarized in Table 3.6. As shown in Figure 3.13, as the db number increases, the PSNR increases and MSE decreases. In order to evaluate the efficiency of filters used in the wavelet method, from the set of each filter, the optimal

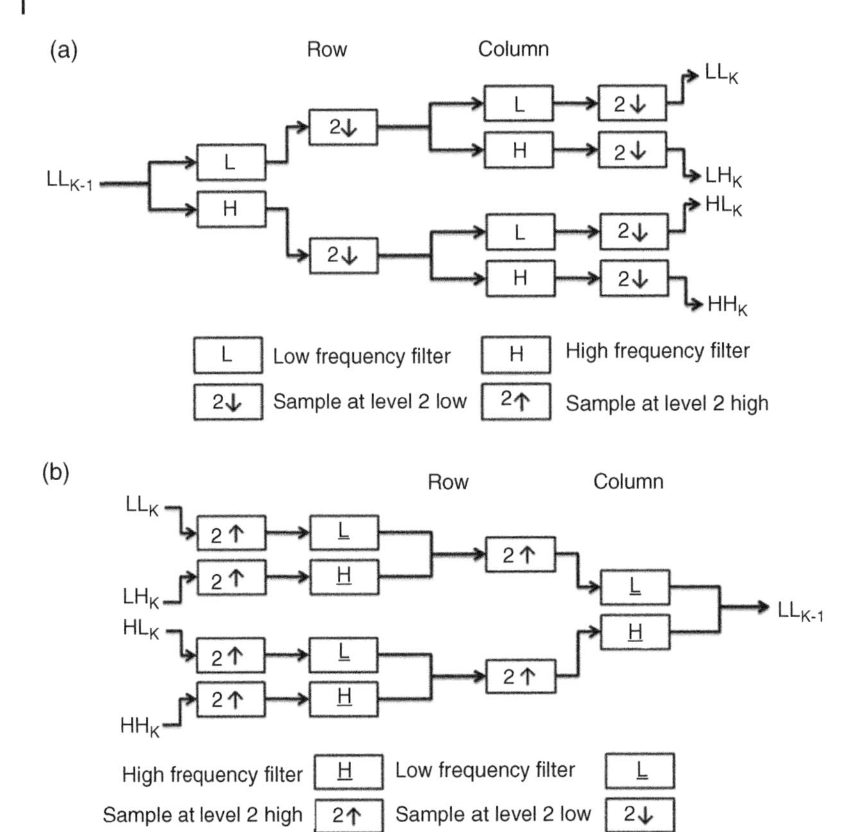

Figure 3.12 An example of the image (a) decomposition and (b) reconstruction processes using wavelet coefficients.

Table 3.6 The family filters of wavelet transform.

Filter	Symbol	Filter family
Daubechies	"db"	"db1" or "haar", "db2", ..., "db10", ..., "db45"
Coiflets	"coif"	"coif1", ..., "coif5"
Symlets	"sym"	"sym2", ..., "sym8", ..., "sym45"
Discrete Meyer	"dmey"	"dmey"
Biorthogonal	"bior"	"bior1.1", "bior1.3", "bior1.5"
		"bior2.2", "bior2.4", "bior2.6", "bior2.8"
		"bior3.1", "bior3.3", "bior3.5", "bior3.7"
		"bior3.9", "bior4.4", "bior5.5", "bior6.8"

option for the case study of pavement was selected, and the types of db filters in Table 3.7 were compared. Finally, the best filter was selected from these filters to smooth the image and remove noise. An example of each filter at different levels for the wavelet method is shown in Figure 3.13. Based on the results of other researches, the db filter has provided acceptable results at level 3 and

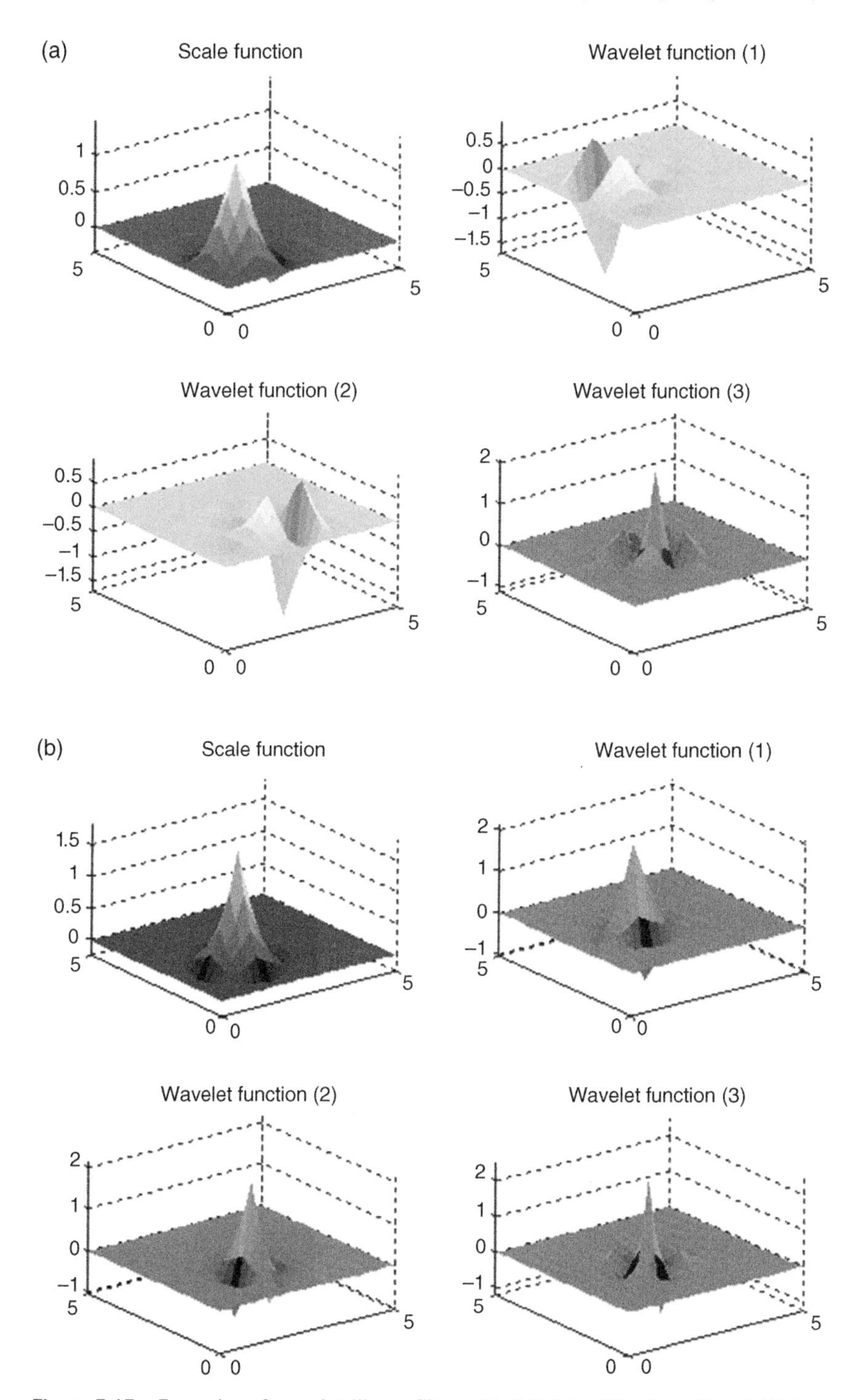

Figure 3.13 Examples of wavelet library filters: (a) db2, (b) coif1, (c) sym2, and (d) dmey.

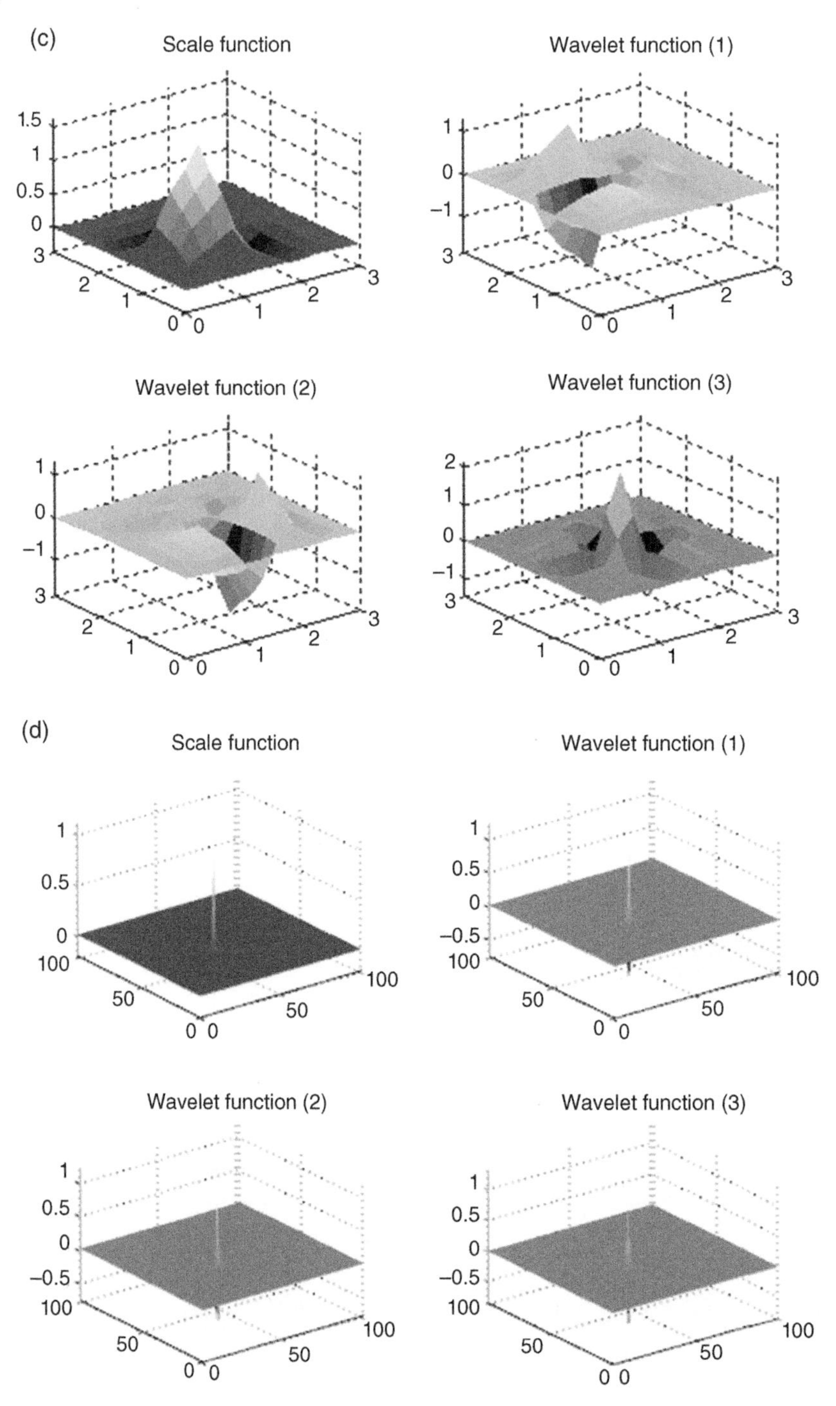

Figure 3.13 (Continued)

Table 3.7 Comparison of db(l) family wavelet filters in smoothing and noise reduction.

Filter	PSNR	MSE	MAXERR	L2RAT	ESSIM	EPI	QILV	SCI	TIME
"db1"	25.188	196.923	59.705	0.987	0.998	0.360	0.025	0.987	0.117
"Db2"	25.383	188.265	58.359	0.987	0.998	0.372	0.028	0.987	0.113
"db5"	25.519	182.450	57.687	0.987	0.998	0.387	0.031	0.987	0.117
"db10"	25.565	180.553	60.762	0.987	0.998	0.391	0.029	0.987	0.126
"db15"	25.569	180.372	58.283	0.987	0.998	0.394	0.027	0.987	0.154
"db30"	25.560	180.732	60.158	0.987	0.998	0.395	0.024	0.987	0.199

for the analysis of the pavement image and its texture in the preprocessing stage with coefficient filter 2. As shown in Table 3.7, the analysis speed of the db2 method at level 3 is higher than other methods, while the PSNR difference of this method is 25.3, indicating good performance for preprocessing and removing pavement background. On the other hand, the wavelet coefficients at level 3, L2RAT, are lower than those of other db coefficients.

Based on the analysis of the filters in the wavelet library and their application to pavement texture images, a coefficient was selected for each wavelet family followed by a comparison between the types of filters. The results of the indicators are summarized in Table 3.8. As the level increases in the wavelet method, the amount of noise decreases, the PSNR index increases, the cracking information and details decrease, and the computational time increases. Therefore, according to the studies, level 3 has been set as a standard for increasing speed and maintaining information. The evaluation of the types of wavelet family filters at level 3 and the introduced indicators are presented in Table 3.8.

The table reveals that the higher the filter coefficient, the higher the value of the PSNR index, which is selected in return for the increase in computational time.

Also, information about the crack edge, crack width, and spall is more accurately estimated by reducing the filter coefficient and reducing the window size. It can be seen that the WF2 filter has the best results in preprocessing and noise reduction. Although the directional methods are more

Table 3.8 Comparison of family wavelet filters in image enhancement and noise reduction.

Filter	PSNR	MSE	MAXERR	L2RAT	ESSIM	EPI	QILV	SCI	TIME
"db2"	25.129	199.600	54.731	0.986	0.998	0.386	0.029	0.986	0.089
"coif3"	25.560	180.740	60.944	0.987	0.998	0.386	0.035	0.987	0.124
"sym8"	25.530	182.020	58.548	0.987	0.998	0.390	0.033	0.987	0.116
"dmey"	25.630	177.860	57.503	0.987	0.998	0.395	0.035	0.987	0.258
"bior5.5"	25.053	203.150	60.046	0.987	0.998	0.363	0.012	0.987	0.117
"rbio3.5"	25.097	201.100	63.868	0.988	0.998	0.349	0.073	0.988	0.119
Mean	25.333	190.750	59.273	0.987	0.998	0.378	0.036	0.987	0.137
Maximum	25.630	203.150	63.868	0.988	0.998	0.395	0.073	0.988	0.258
Minimum	25.053	177.860	54.731	0.986	0.998	0.349	0.012	0.986	0.089

efficient in detecting cracks, patterned linear subjects (such as cracks) may grow in all directions and have different thicknesses since they are discontinuous and directional. Therefore, using a single filter in one direction only extracts information in the same direction and does not apply to real problems. As such, it is necessary to use filters that extract crack direction information in different directions and simultaneously remove or reduce the effect of background noise. This is also very important for pavement texture analysis because, in addition to the edges of pavement texture, the angles of the pavement texture are also changing.

The difference between the maximum and minimum PSNR index was determined to be about 2% of the PSNR_max value. The average computational time of the optimal filters was 0.138 s for the 512 × 512 image. The value of changes in the SCI index for this category of filters was a maximum of 015%. The difference between the MSE index and MSE max for this set was 14.3%, which reduced to 2.28% for the three filters (coif3, sym8, and dmey). On the other hand, the maximum difference of MAXERR for the whole set was 14.3%, which reduced to 3.93% by limiting the set of filters to only three (coif3, sym8, and bior5.5). The structure of crack edges, according to the QILV index, compared to the maximum of this set was equal to 83.3%, which was reduced to 5.95% by selecting coif3, sym8, and demy filters. Also, by selecting coif3, rbio3.5, and bior5.5, the maximum time difference was less than 5.1%. According to this analysis, the coif3 filter has a higher efficiency than other wavelet family filters for extracting damage as well as noise elimination and smoothing. It should be noted that the difference between this filter and other filters in the wavelet family is not very noticeable, which all display a similar efficiency in the analysis of pavement texture. Based on this analysis, the coif3 filter in the third level was selected as a representative of the wavelet method and will be evaluated further. Figure 3.14 shows the coif3 filter and the application of the wavelet method with this filter for a pavement image in level 3.

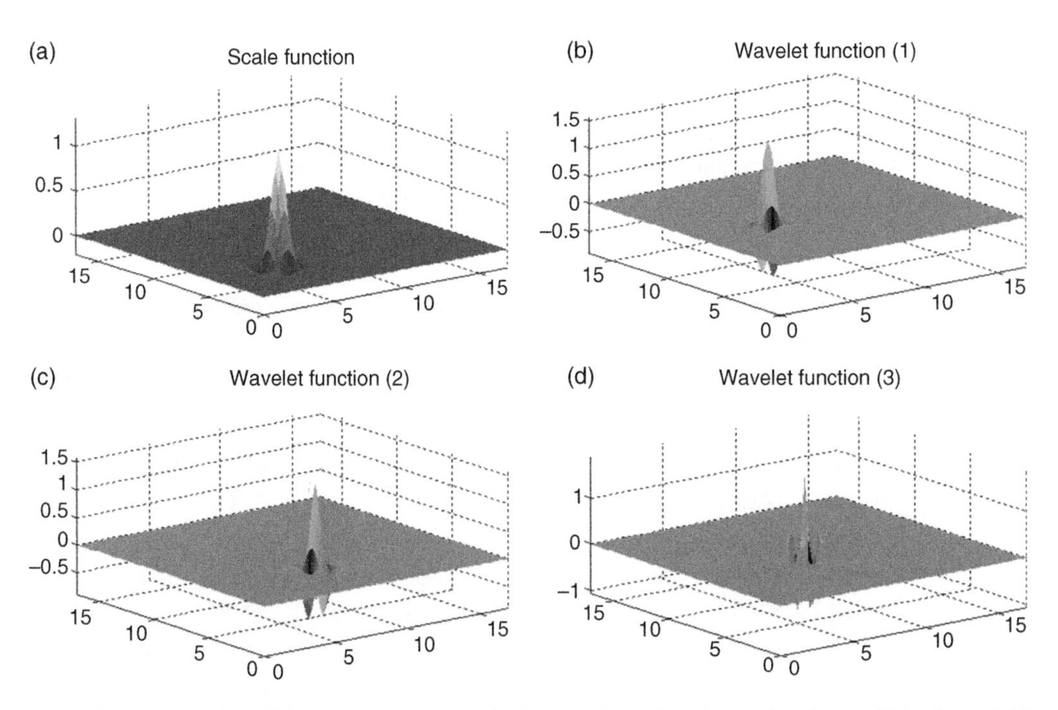

Figure 3.14 View of coif3 from the wavelet family library: (a) size function estimation an (b) horizontal, (c) vertical, and (d) diagonal directions.

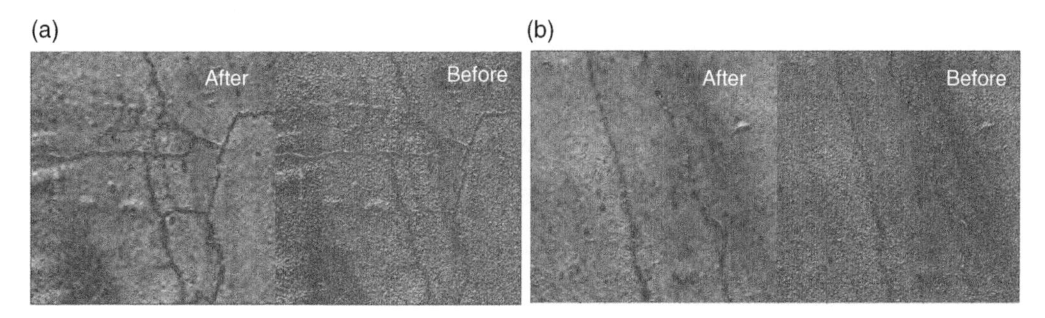

Figure 3.15 Examples of applying coif3 function in wavelet method at a fixed threshold: (a) noisy image of pavement texture containing fatigue cracking, and (b) noisy image of pavement with longitudinal cracking.

As shown in Figure 3.15a, the dominant linear patterns are identified in a specific direction more than the other directions, and Figure 3.15b reveals that the background is modified, and noise is reduced by preserving crack information.

3.3.2 Ridgelet Transform

Ridgelet transform is one of several methods for analyzing images belonging to the wavelet family, which uses Fourier transform, wavelet, and Radon to analyze different bands of the image (from small to large_ and is separated using thresholding.

The patterns obtained from an image due to the polar rotation around the center are called radon transitions. Per definition, this conversion creates a new pattern of meaning by stacking at different angles. By thresholding on this pattern, the most important use of this method in medical engineering is for three-dimensional imaging using a coat scanner. This method is also applied in image processing to detect the edge, determine the line parameters, and detect the direction. The parameter θ, or the transfer angle, is the angle between the principal and the transmitted components. The radon transfer $R_\theta(x')$ is a function of f (x, y) and is calculated from the linear sum f (x, y) along the axis y' based on the following relations:

$$R_\theta(x') = \int_{-\infty}^{+\infty} f(x,y)dy',$$
$$R(p,\tau) = \int_{-\infty}^{+\infty} f(x, px + \tau)dx \tag{3.25}$$

In Equation 3.29, R (p, τ) is the radon transfer for the two-dimensional function f (x, y), which is calculated from the sum of the values of f along the line slope. The location of the line is determined using the slope parameter p and the output τ.

$$R(p,\tau) = \int_{-\infty}^{+\infty}\int_{-\infty}^{+\infty} f(x,y)\delta(y - px - \tau)dxdy. \tag{3.26}$$

The rotational components and the polar environment of the equation are:

$$\begin{bmatrix} x \\ y \end{bmatrix} = \begin{bmatrix} \cos\theta - \sin\theta & \\ \sin\theta & \cos\theta \end{bmatrix}\begin{bmatrix} x' \\ y' \end{bmatrix}. \tag{3.27}$$

Radon transfer is obtained using the function f (x, y), which is calculated by the two following equations:

$$R_\theta(x') = \int_{-\infty}^{+\infty} f(x'\cos\theta - y'\sin\theta, x'\sin\theta + y'\cos\theta)dy'$$
$$R(\rho,\theta) = \int_{-\infty}^{+\infty}\int_{-\infty}^{+\infty} f(x,y)\delta(\rho - x\cos\theta - y\sin\theta)dxdy \tag{3.28}$$

In Equation (3.28), δ is a function of the Dirac delta, whereby the larger the size of an image in one direction with the active pixels within the specified threshold, the higher the radon value. In the radon method, the angle varies from 0 to 180. Radon transform is important because of its linear role in integrating image intensity in all directions, in which a line in the image is converted to a maximum or minimum point in radon. It is clear that linear patterns (such as pavement cracks) often have a specific direction; therefore, when placed in the dominant direction, the crack intensities add up and are converted to a max point in the radon domain using radon transform. In other words, if there is a linear algorithm, there is a maximum point in the radon medium. For a more accurate analysis, the third dimension, which is the value of the integral for each (ρ, θ), can be used. An example of this conversion for a linear pattern is shown in Figure 3.16.

According to the definition of ridgelet, the ridgelet transfer (RIT) of the function *f*(*x*) in the environment R^2 for a patterned image is defined as follows:

$$RIT_f(a,b,\theta) = \int_{R^2} \psi_{a,b,\theta}(x)f(x)dx, \tag{3.29}$$

where $\psi_{a,b,\theta}(x)$ for a two-dimensional image is defined by the one-dimensional wavelet function ψ(x) as follows:

$$\psi_{a,b,\theta}(x) = \frac{1}{\sqrt{a}}\psi\left(\frac{(x_1\cos\theta + x_2\sin\theta - b)}{a}\right) \tag{3.30}$$

An example of an RIT located in the θ direction and fixed along the line $x_1\cos\theta + x_2\sin\theta = \text{const}$) is shown in Figure 3.17. Based on the ridgelet definition and comparison with wavelet, it can be concluded that both methods are similar. It is pertinent to note that the only difference is in the display of WT with point parameters, while RIT is defined using line parameters b, θ. In the image, lines and cracks can be modeled using the RT theory; thus, WT is related to RIT conversion using radon transfer. According to the definition, RIT in a one-dimensional environment is defined as follows:

$$RIT_f(a,b,\theta) = \int_{R^2} \psi_{a,b}(t)R_f(\theta,t)dt, \tag{3.31}$$

Also, according to Equation 3.31, the Fourier transform can be used instead of the wavelet transform; this relationship is transformed as follows:

$$F_f(\xi\cos\theta, \xi\sin\theta) = \int_{R^2} e^{-j\xi t}R_f(\theta,t)dt, \tag{3.32}$$

This relation represents the projection-slice theorem, which is commonly used in many methods for reconstructing an image from its transition image. The relationship between these three conversions is shown in Figure 3.17.

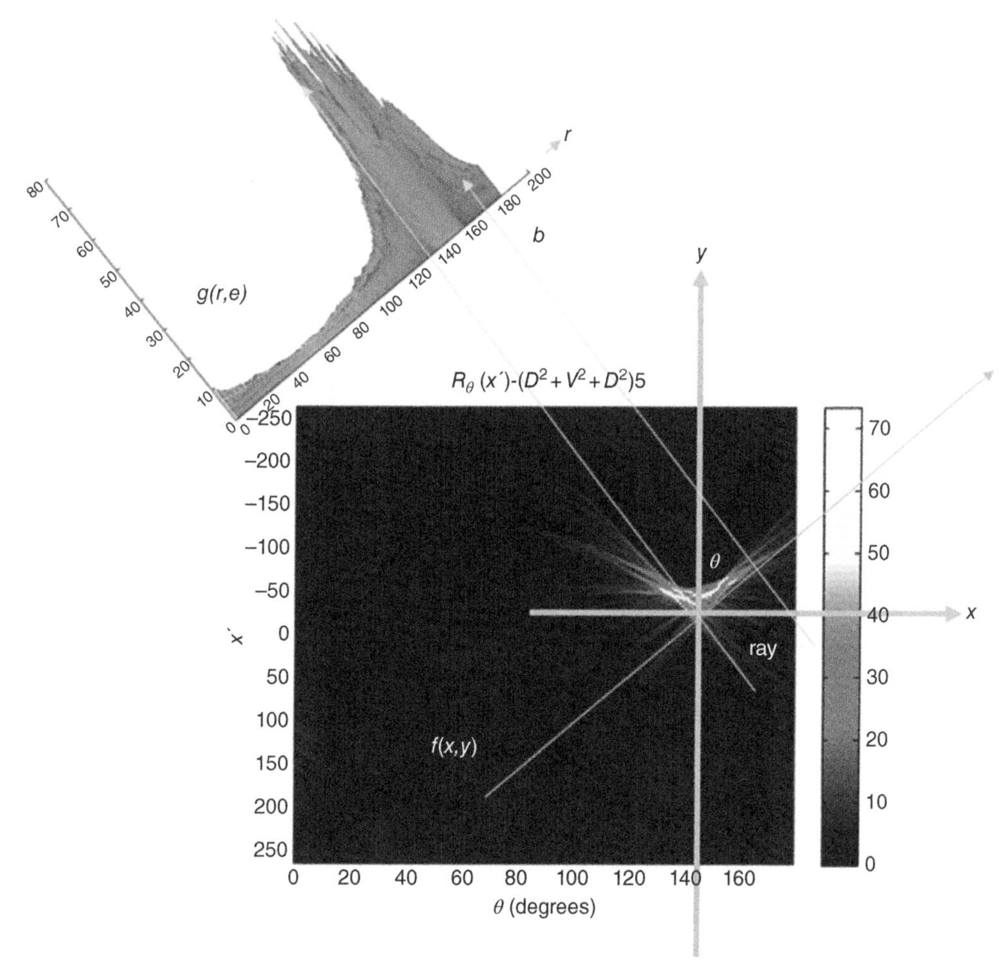

Figure 3.16 Principles of radon transform capability in discrete linear patterns and radon coefficient accumulation.

The RIT is applied from one-dimensional wavelet transform with RT fragments, while two-dimensional Fourier is applied from one-dimensional Fourier transform then RT is applied to it. The main structure of this method for image analysis is shown in Figure 3.18.

Noise reduction and image enhancement are based on the ridgelet method using thresholding. The basis of this method is based on maintaining high-amplitude coefficients and eliminating and modifying low-amplitude coefficients. In calculations related to pavement assessment (cracking and types of damage), the main issues are in the coefficients with high amplitude and frequency. In this method, high-amplitude coefficients are kept, while low-amplitude coefficients are reduced to zero. This method works similar to the wavelet method.

Based on the extracted results from the analysis of filters in the wavelet library from the previous section (Table 3.1), the selected filters were applied to the database. By applying this set on the pavement cracking images, suggested indicators were obtained for each wavelet family and are presented in Table 3.9.

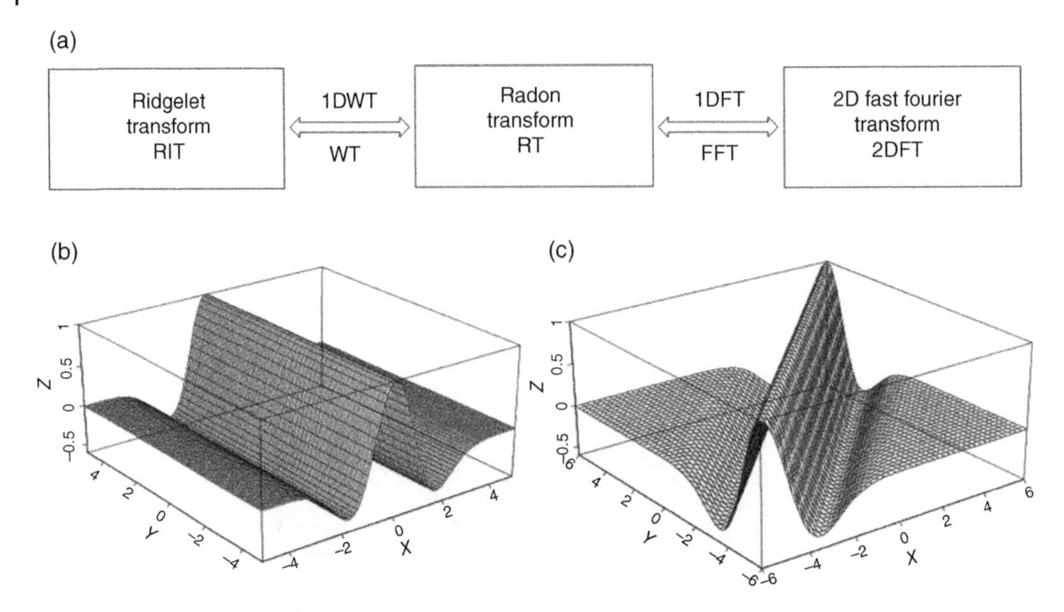

Figure 3.17 (a) Relationship between radon and Fourier transform functions in ridgelet; (b) Example of wavelet filter $\psi_{a,\, b}(t)$; and (c) Example of ridgelet filter $\psi_{a,\, b,\, \theta}(x)$.

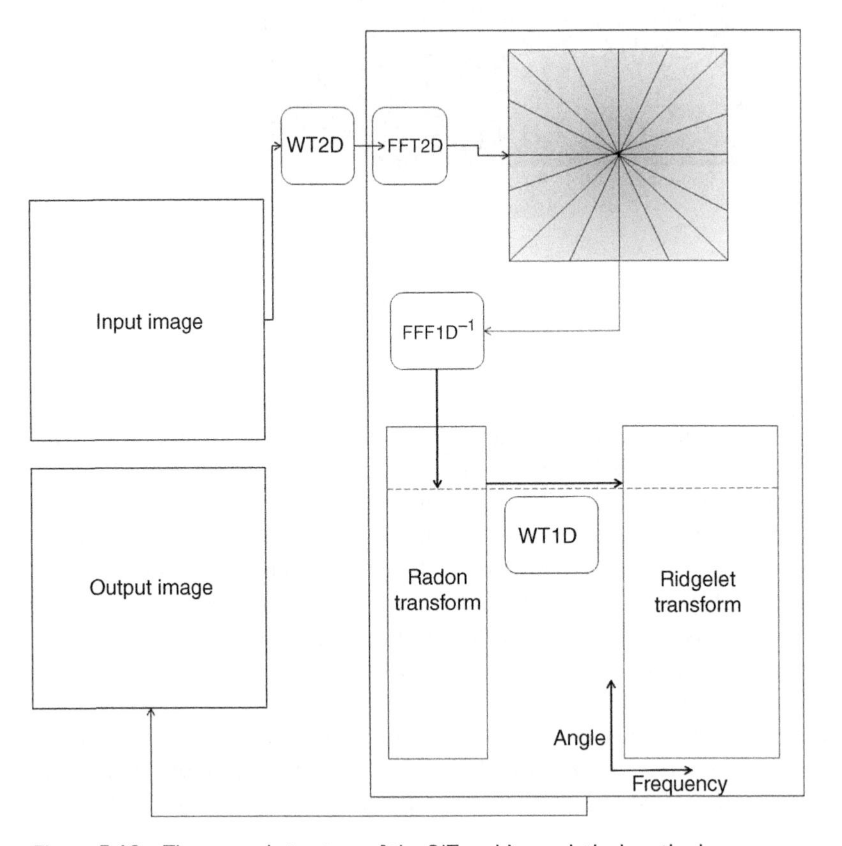

Figure 3.18 The general structure of the RIT and its analytical method.

Table 3.9 Comparison of family ridgelet filters in image enhancement and noise reduction.

Filters	PSNR	MSE	MAXERR	L2RAT	ESSIM	EPI	QILV	SCI	TIME
"db2"	26.984	130.222	41.155	0.987	0.998	0.550	0.226	0.987	0.086
"coif3"	27.072	127.612	40.385	0.988	0.998	0.551	0.243	0.988	0.091
"sym8"	27.071	127.642	46.531	0.988	0.998	0.552	0.244	0.988	0.090
"dmey"	27.130	125.928	44.932	0.988	0.998	0.552	0.257	0.988	0.160
"bior5.5"	26.981	130.310	46.789	0.987	0.998	0.552	0.234	0.987	0.089
"rbio3.5"	27.013	129.367	45.667	0.988	0.998	0.557	0.280	0.988	0.089
Mean	27.042	128.513	44.243	0.988	0.998	0.552	0.247	0.988	0.101
Maximum	27.130	130.310	46.789	0.988	0.998	0.557	0.280	0.988	0.160
Minimum	26.981	125.928	40.385	0.987	0.998	0.550	0.226	0.987	0.086

As the level increases in the Ridgelet method, similar to the wavelet method, the amount of noise decreases, the PSNR index increases, the cracking information and details decrease, and the computational time increases. The evaluation of different types of filters similar to the wavelet method are presented in Table 3.9.

From the results obtained by the Ridgelet method, the db2 filter exhibited the fastest computational time, which was 46% faster than the demy filter. Also, the MSE index shows a difference of 3.4% for these two filters. The highest MAXERR index was obtained for the Symlets filter and the lowest for Coiflets, with a maximum difference of 13.7%. The indices related to the structure of cracks and spalls for EPI in the db filter were less than the other filters, with a maximum difference of 1.2%. Also, r the rbio3.5 filter showed the maximum QILV index, while the db filter had the lowest.

The computational time ratio of db/demy was determined to be 0.53 in s, which indicates the faster speed of this method compared to other filters. It should be noted that bior and rbio filters are very similar to the db filter in terms of computational time. Based on the analysis performed and the results obtained, the demy filter achieved the best results from the Ridgelet structure than the other methods. An example of the images analyzed by this method for the selected filters is shown in Figure 3.19.

Figure 3.20 displays the performance of the main RIT parameters, including the number and size of windows, in extracting the cracking properties and pavement texture to evaluate the amount of image noise reduction based on the resizing of different windows. As the threshold value increased, the PSNR index decreased, and the MSE index increased. MAXERR decreased at first and then increased.

These results indicate that the optimum threshold is in the initial zone of about 10. The L2RAT index exhibits a decreasing trend, and structural indicators show a rapid decrease of the index as the threshold increases. Computation time was evaluated for windows with different dimensions. The graphs in Figure 3.21 reveal that as the window size increased, the computational time increased to less than 2%, and the maximum computational time for thresholds was less than 10 s.

By comparing the reconstructed results and images, it can be concluded that the demey filter, at threshold of 10, can detect cracks well and decrease the cracking structure based on structural

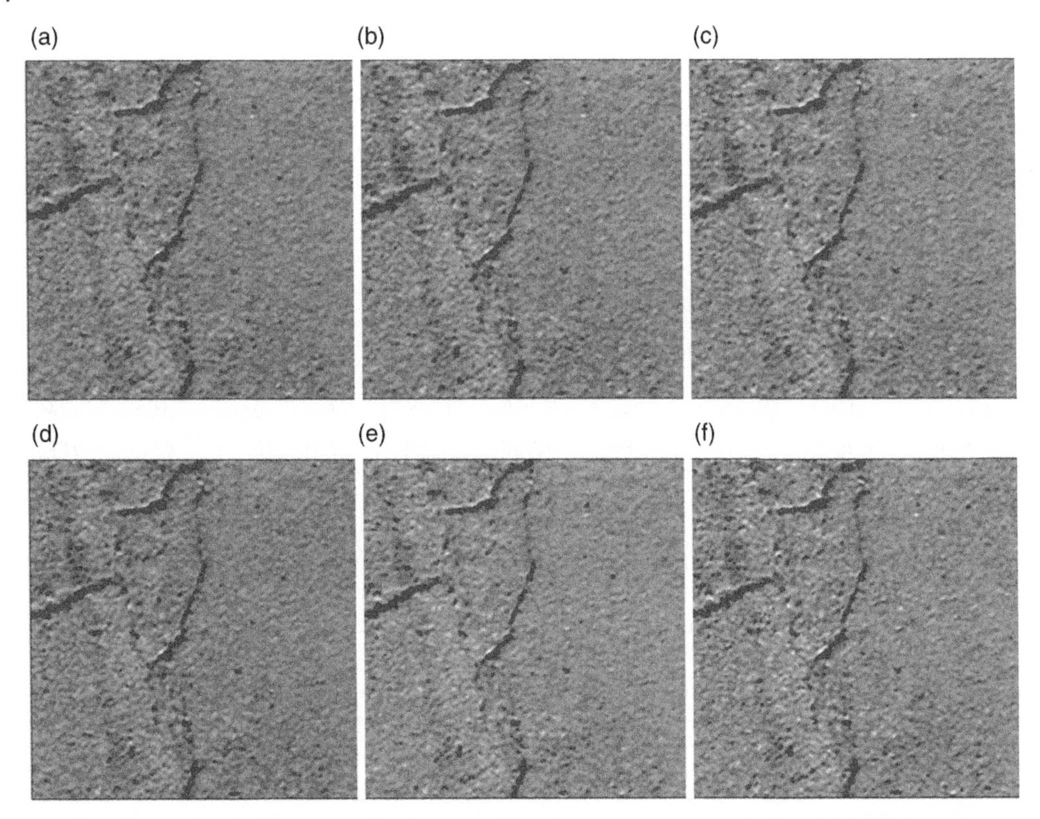

Figure 3.19 Investigation of the effect of filters of ridgelet method based on wavelet library and application of fixed threshold on pavement cracking image: application of (a) db3 filter, (b) coif3 filter, (c) sym8 filter, (d) dmey filter, (e) bior5.5 filter, and (f) rbio3.5 filter.

indicators by increasing the threshold, resulting in decreased computational time. In the threshold range of 60, due to image errors, pavement crack detection, or evaluation of coarse pavement texture, images became blurred. As a result, the computational time increased unrealistically, and a significant computational error entered the analysis. As the threshold increased again, the computational time decreased once more. Also, as the window size increased, the computation time increased. Therefore, the window size of 3×3 is considered optimal to select the type of RIT filter.

3.3.3 Curvelet Transform

The curvelet transform (CT) method is one of the developed RIT methods in the field of preprocessing (image enhancement noise reduction and separation) of 2D and 3D signals. In this method, the reconstruction of crack edges and pavement texture is performed more accurately when using RIT in small windows of the image. As a result, the main characteristics can be obtained by the subjects (such as crack length and width, pavement texture, drainage rate, image morphology, aggregates). Moreover, adjusting the CT parameters preserves the maximum image information. Discrete CT is designed using wavelet filter banks in different face sizes, which includes the following steps:

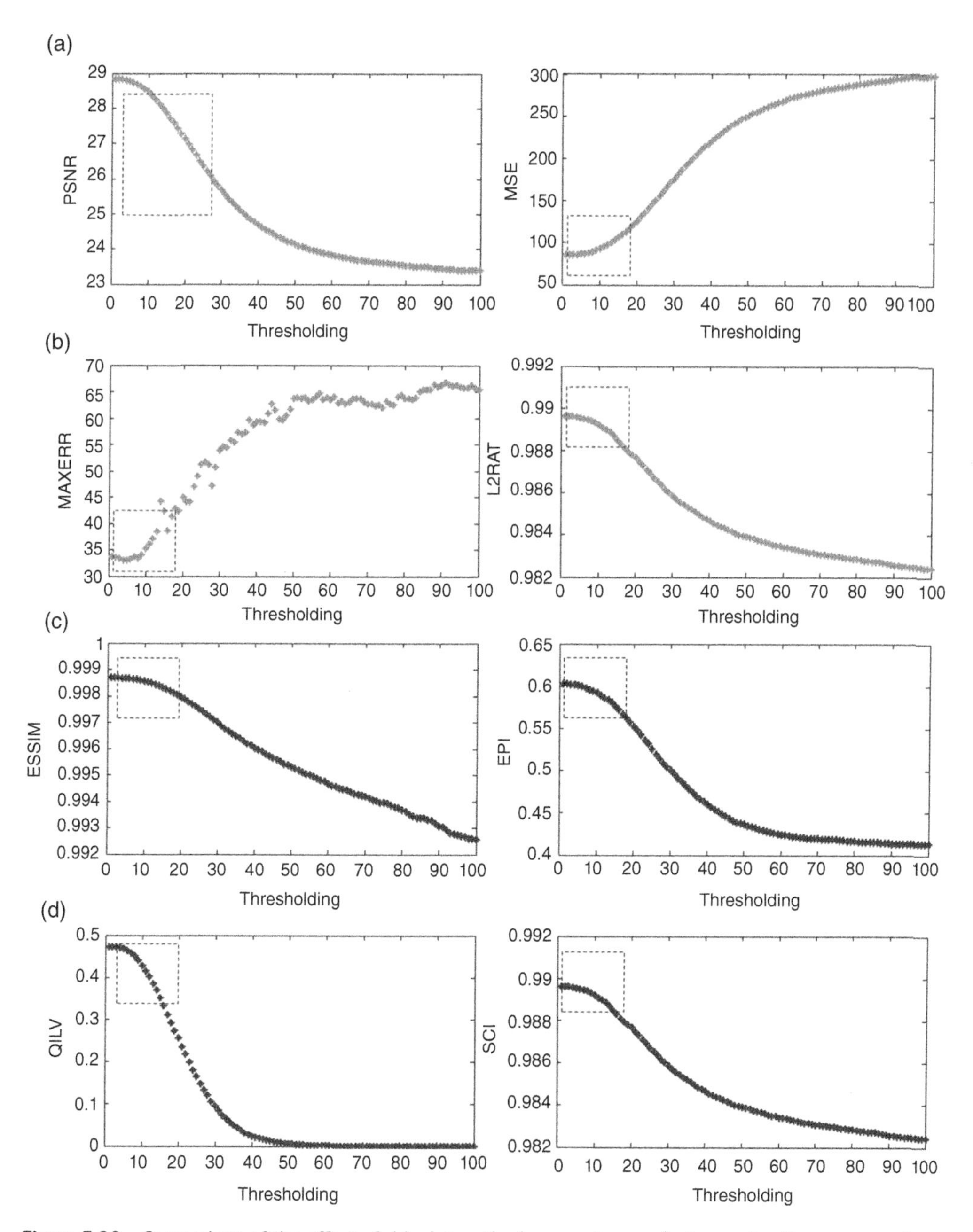

Figure 3.20 Comparison of the effect of ridgelet method parameters on feature extraction, pavement cracking, and its effect on reducing image noise with window size of 3 × 3: (a) PSNR, (b) MAXERR, (c) ESSIM, and (d) QILV.

1) **Fragmentation into subbands**: The purpose of this section is to parse the x image into the details of each frequency band as follows:

$$x \rightarrow (P_0 x, \Delta_1 x, \Delta_2 x, \ldots), \Delta_s x \rightarrow (w_Q \Delta_s x)_{Q \in Q_s} \tag{3.33}$$

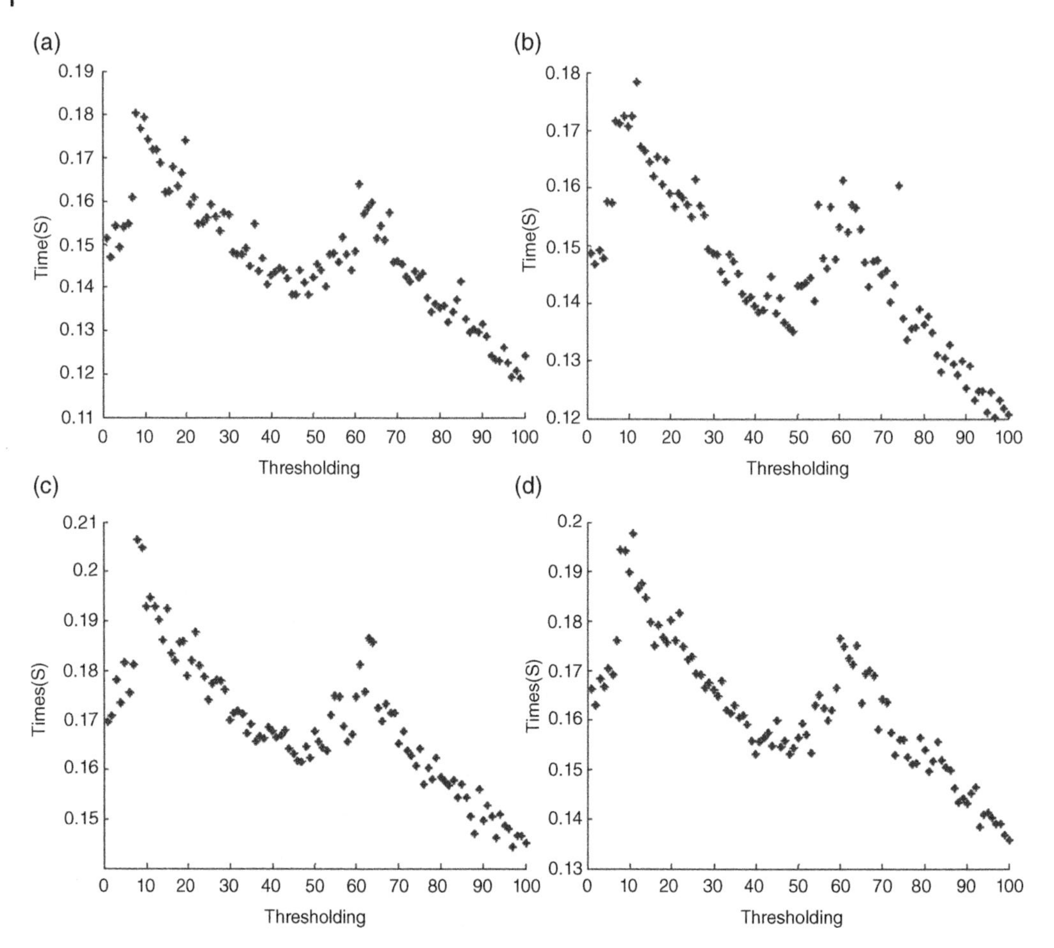

Figure 3.21 Comparison of the effect of filter size and threshold on computational time in the ridgelet method: (a) size 3 × 3, (b) size 5 × 5, (c) size 10 × 10, and (d) size 20 × 20.

2) **Smoothing the pieces in the subbands**: Each image is divided into smaller pieces. This ratio is 2^{-s}.
3) **Normalize**: Each smoothed square image is normalized and resized. This ratio is 2^{-s}.
4) **Window analysis using RIT**: Each image in the subbands is analyzed, and its coefficients are calculated using the RIT described in the previous section. The results are then combined and reconstructed using the relation below:

$$\mathrm{xr}(X, Y) = c_J(X, Y) + \sum_{j=1}^{J} w_j(X, Y). \tag{3.34}$$

In Eq. (3.34), $c_J$is an example of the large and small texture details of image x, and w_j is the detail of the image x for the window size of 2^{-j}. Also, $j = 1$ is used for fine-textured pavement and noise removal.

Figure 3.22 displays the structure of the CT method, in which the RIT function is applied separately in each window. However, one disadvantage of this method is that it has a much longer computational time than other multilevel methods.

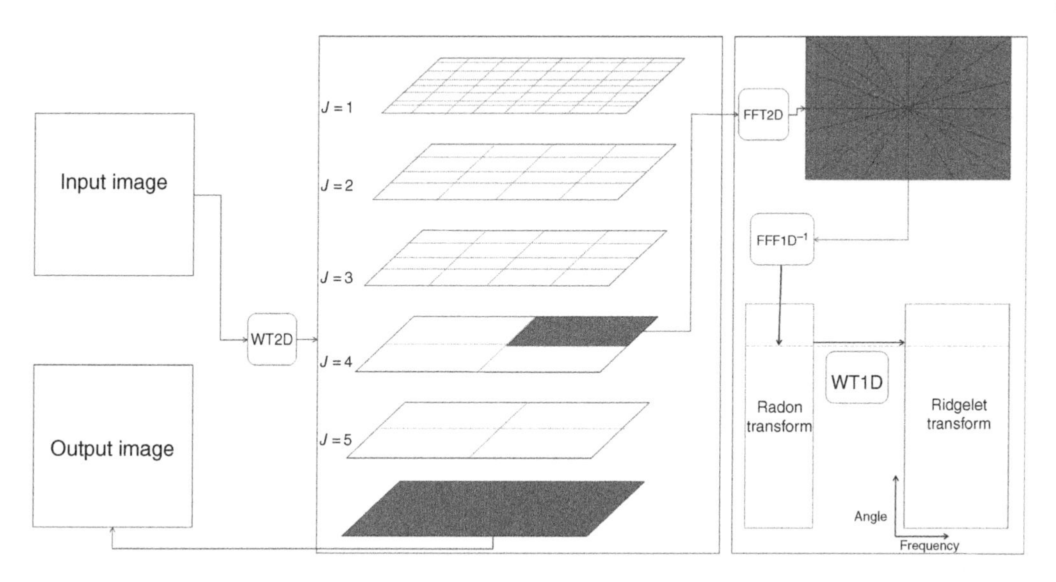

Figure 3.22 The structure of curvelet transform (CT) and the relationship with ridgelet transform (RIT).

Regarding the optimal filter for RIT discussed in the previous section, computational time is important in addition to the structural features. In the CT method, despite the high noise indices as well as the preservation of cracking properties in the faulty pavement image, the computational time due to window analysis is several times that of other multilevel methods. Although this method has high accuracy in noise removal, crack quality enhancement, and optimal image smoothing, it suffers from high computational complexity. An example of the CT method to reduce noise, improve crack coefficients, and reduce the background effect on the image of pavement is shown in Figure 3.23. Using the Harr filter, CT coefficients were extracted; the difference between the original image and the enhanced image and its effect on background and noise removal are presented in Figure 3.23(c). As it turns out, the CT coefficients separate the noise details and then are adjusted by applying a threshold. As a result, the cracking details are preserved for the high CT

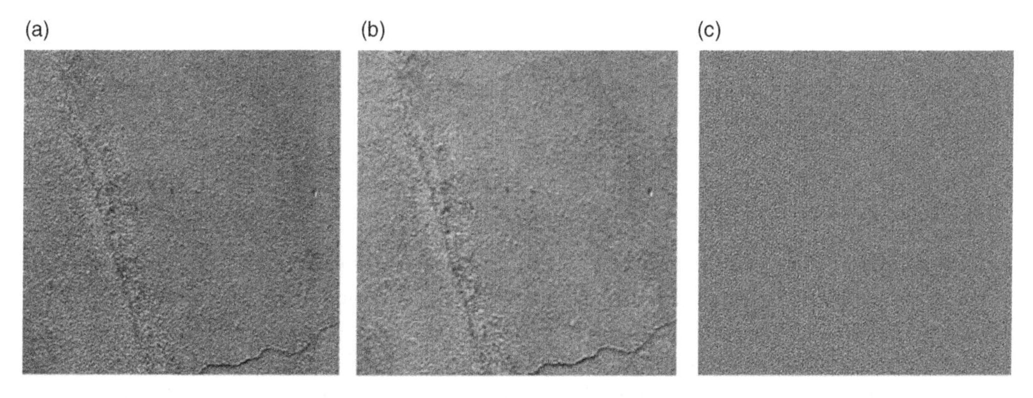

Figure 3.23 An example of the application of CT method to reduce noise, improve crack coefficients, and reduce background effect: (a) original image, (b) coefficients extracted for Harr filter (db1) by Coroll method, and (c) difference between the original image and enhanced reconstructed image, showing its effect on removing background and noise.

Table 3.10 Comparison of family curvelet filters in image enhancement and noise reduction.

Filters	PSNR	MSE	MAXERR	L2RAT	ESSIM	EPI	QILV	SCI	TIME
"db2"	28.3604	94.85	42.73	0.9907	0.9994	0.5000	0.1569	0.99	247.79
"coif3"	28.3640	94.77	42.77	0.9907	0.9994	0.5001	0.1572	0.99	248.24
"sym8"	28.3641	94.77	42.80	0.9907	0.9994	0.5001	0.1572	0.99	248.26
"dmey"	28.3647	94.76	42.79	0.9907	0.9994	0.5002	0.1572	0.99	248.59
"bior5.5"	28.3589	94.88	42.78	0.9907	0.9994	0.5000	0.1568	0.99	247.83
"rbio3.5"	28.2433	97.44	43.50	0.9906	0.9994	0.4965	0.1487	0.99	248.09
Mean	28.3426	95.25	42.89	0.9907	0.9994	0.4995	0.1557	0.99	248.13
Maximum	28.3647	97.44	43.50	0.9907	0.9994	0.5002	0.1572	0.99	248.59
Minimum	28.2433	94.76	42.73	0.9906	0.9994	0.4965	0.1487	0.99	247.79

coefficients, while the low coefficients are corrected. The CT method was analyzed by calculating the evaluation indicators for a set of images, which are presented in Table 3.10. One of the most important features of this method is the possibility of modeling and detecting the main issues (pavement cracking, coarse pavement texture, texture morphology, etc.) in a discrete manner.

The evaluation results on the database are shown in Table 3.10 in which noise removal operations were performed with a slight difference for CT filters. The maximum PSNR index for the filter was determined to be 28.36 with a minimum index of 0.4%; therefore, it can be concluded that all filters are able to remove noise. On the other hand, the CT method showed a lower MSE rate than that of the other methods. The minimum index is 94.75, which is significantly lower than other methods. Also, the MAXERR and L2RAT indices for selected filters did not change significantly, with changes less than 0.01%. Structural indicators, including EPI, QILV, QILV, and SCI, do not show significant differences for the proposed filters. The difference in the EPI index is less than 1% and less than 6% for QILV. The computational time for different filters is longer and much more complex compared to that of other multilevel methods. Specifically, the difference in computational time is less than 1%, and the computational time is much higher than other methods.

Figure 3.24 presents examples of a cracked pavement image for comparison, which was processed using different filters and the CT method. Based on the analysis performed by the CT method, it can be seen that the reconstructed images are not significantly different from the original image in terms of appearance, the cracks in all filters are well preserved, and the image smoothing and noise removal are almost similar to each other.

3.3.4 Decompaction and Reconstruction Images Using Shearlet Transform (SHT)

Shearlet transform (SHT) has been designed as a new multiresolution method in applications related to the extraction of geometric properties of multidimensional signals. In 2007, irregular SHTs were studied, and a framework for constructing a variety of discrete multidimensional-dimensional systems with the ability to detect directions affected by continuous conversion was proposed. Then, in 2008, researchers introduced an integrated Fourier operator based on a square format of SHT. An evaluation of the method showed that it is well capable of reconstruction and combining two-dimensional and three-dimensional data types. In 2010, the Discrete Shearlet Transform (DST) was developed by adding supplementary base elements to the SHT system to

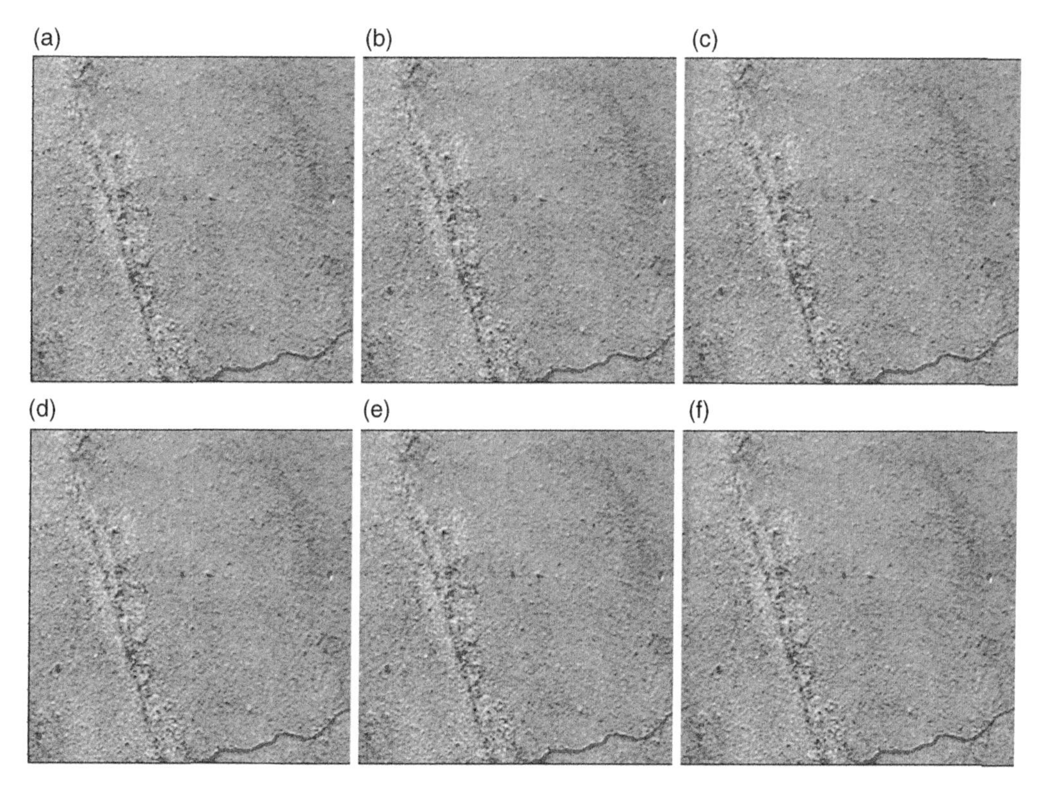

Figure 3.24 Comparison of the effect of CT method filters based on wavelet library and application of fixed threshold on pavement cracking image: application of (a) db3 filter, (b) coif3 filter, (c) sym8 filter, (d) dmey filter, (e) bior5.5 filter, and (f) filter rbio3.5.

obtain the true values of each feature matrix. In 2015, researchers used a discrete deformation design method based on the Laplacian hybrid method with specific shear filters to isolate, evaluate, and classify surface damages.

3.3.5 Discrete Shearlet Transform (DST)

Shearlet functions are constructed on the basis of nonisotropic nature (direction-dependent information), including shear, transfer, and scale. This function is made based on the converter $\psi \in L^2(R^2)$ in the following forms:

$$\psi_{a,s,t} = \psi(S_s A_a(. - t)), \tag{3.35}$$

where $S_s = \begin{pmatrix} 1 & s \\ 0 & 1 \end{pmatrix}$, $A_a = \begin{pmatrix} a & s \\ 0 & a^\alpha \end{pmatrix}$, parameters $a \in R^+$ and $s \in R$ and $t \in (R^2)$, and $\alpha \in [0,1]$ indicates the degree of nonisotropy. When $\alpha = 1$, the A_a is completely nonisotropic, and when $\alpha = 0$, the matrix is an anisotropic scaling matrix. The most common way to build a shearlet generator is presented below as follows:

$$\psi(x_1, x_2) = \psi_{1D}(x_1)\varphi_{1D}(x_2), \tag{3.36}$$

where ψ_{1D} is a wavelet function, and φ_{1D} is a function similar to the Gaussian method. In these relationships, shear functions play a torsional role to represent a change of direction.

Using the $\psi \in L^2(R^2)$ generator (R ^ 2) and the set of parameters $\Gamma \subset R^+ \times R \times R^2$, the Shearlet-based wave decomposition function is defined as $f \in L^2(R^2)$:

$$SH_{\psi,\Gamma}(f) = \left(\langle f, \psi_{a,s,t}\rangle\right)_{a,s,t\in\Gamma}. \tag{3.37}$$

In Equation 3.37, S^{-1} is obtained using a simple calculation with a positive and self-connected operator. According to wavelet theory, the function $f \in L^2(R^2)$ is called a shearlet when it satisfies the following condition:

$$\int_{R^2} \frac{|\hat{\psi}(\xi_1, \xi_2)|^2}{\xi_1{}^2} d\xi_1 d\xi_2 < \infty, \tag{3.38}$$

In Equation 3.38, $\hat{\psi}$ is the Fourier transform function ψ. Therefore, basically, any function supported in $f \in L^2(R^2)$ using the Fourier function can be implemented using the SHT function, and in the case of the wavelet function, integral convergence means the inverse conversion function of the Shearlet transfer. One of the most important limitations of the first-generation Shearlet, which is constructed with a one-dimensional wavelet and the Fourier function, is that the Shearlet drastically shifts to a horizontal axis by increasing the shear parameter k. However, this elongation occurs gradually. If this shear shape is completely in the horizontal direction, k tends to infinity. This process causes the filter to become thinner in the horizontal direction, creating different forms in the implementation and use of SHT in applied problems. One way to solve this problem is to divide the frequency medium into four conical sections (C1 to C4) according to Figure 3.25. Then, the generator function ψ has the Fourier value C1∪C3 and C2∪C4 for the vertical direction using the second generator in the horizontal direction. Assuming the function ψ as SHT, the first generation can be obtained:

$$\hat{\psi}(\xi_1, \xi_2) = \hat{\psi_1}(\xi_1)\hat{\psi_2}(\xi_2/\xi_1), \tag{3.39}$$

where ψ_1 is defined as a discrete wavelet function, and ψ_2 is an impact function, $\psi(\xi_1, \xi_2) = \tilde{\psi}(\xi_2, \xi_1)$. In order to analyze the low frequency, the square segment is cut as a cross, and the

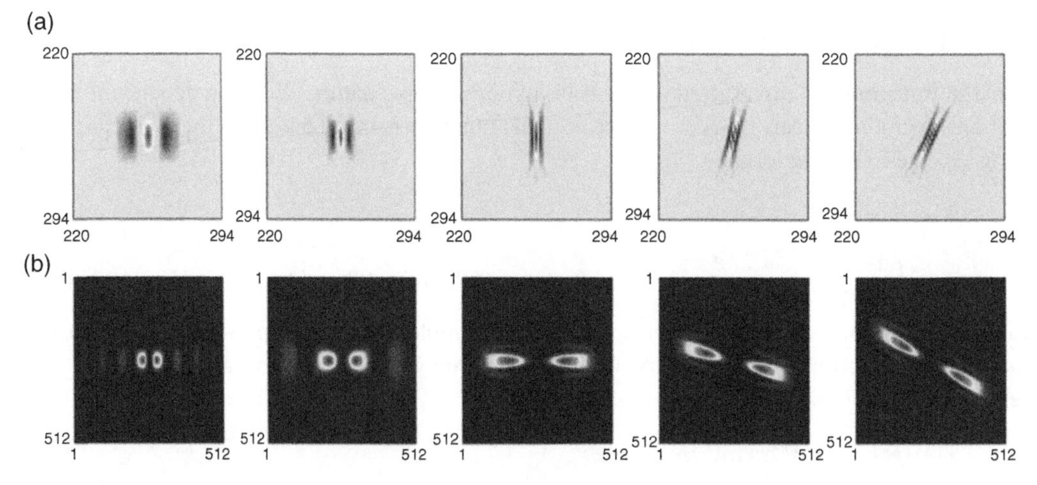

Figure 3.25 The effect of using Shearlet transform parameters: (a) the effect of size and shear parameter on a Shearlet generator in time mode, and (b) the frequency domain (response).

parameter size $j \geq 0$ and the transmission function are applied to the value of the functions. Based on the definitions provided, the first generation Shearlet Discrete Cone System is defined as follows:

Definition 1. Cone-shaped discrete Shearlet method – assuming $f \in L^2(R^2)$ based on Equation 3.38, the frequency of the horizontal cone, and definition $\tilde{\psi} \in L^2(R^2)$, the main matrices of the size parameter for a vertical cone are defined as follows:

$$S_k = \begin{pmatrix} 1 & k \\ 0 & 1 \end{pmatrix}, \tilde{A}_j = \begin{pmatrix} 2^{j/2} & 0 \\ 0 & 2^j \end{pmatrix}, , A_j = \begin{pmatrix} 2^j & 0 \\ 0 & 2^{j/2} \end{pmatrix} \tag{3.40}$$

As a result, the function of a discrete conical Shearlet $SH(\phi, \psi, \tilde{\psi})$ is defined as follows:

$$\begin{aligned} SH(\phi, \psi, \tilde{\psi}) &= \Phi(\phi)\Psi(\psi) \cup \tilde{\Psi}(\tilde{\psi}), \\ \Phi(\phi) &= \{\phi_m = \phi(. - m) : m \in Z^2\}, \\ \Psi(\psi) &= \left\{\psi_{j,k,m} = 2^{\frac{j3}{4}}\psi(S_k A_j . - m) : j \in N, |k| < 2^{\frac{j}{2}}, m \in Z^2\right\}, \\ \tilde{\Psi}(\tilde{\psi}) &= \left\{\tilde{\psi}_{j,k,m} = 2^{\frac{j3}{4}}\tilde{\psi}\left(S_k{}^T\tilde{A}_j . - m\right) : j \in N, |k| < 2^{\frac{j}{2}}, m \in Z^2\right\} \end{aligned} \tag{3.41}$$

Various functions can be used to make SHT frames based on $SH(\phi, \psi, \tilde{\psi})$. By selecting the size function and transferring the sets $\Psi(\psi)$ and $\tilde{\Psi}(\tilde{\psi})$ to the frequency cones, a frame of size $A = B = 1$ is generated using the first-generation SHT.

Definition 2. Framework based for $L^2(R^2)$ – according to $\psi \in L^2(R^2)$, which is the first-generation Shearlet, the size function $\phi \in L^2(R^2)$ can be chosen using the following equation:

$$|\hat{\phi}(\xi)|^2 + \sum_{j \geq 0} \sum_{|k| \leq \lceil 2^{j/2} \rceil} |\hat{\psi}(S^T_{-k}A_{-j}\xi)|^2 1_C + \sum_{j \geq 0} \sum_{|k| \leq \lceil 2^{j/2} \rceil} \left|\hat{\tilde{\psi}}\left(S^T_{-k}\tilde{A}_{-j}\xi\right)\right|^2 1_{\tilde{C}} = 1, \tag{3.42}$$

For each $\xi \in R^2$, C is a horizontal frequency cone and $\tilde{C}$ is a vertical frequency cone, where $C = \left\{(\xi_1, \xi_2) \in R^2 : \left|\frac{\xi_2}{\xi_1}\right| \leq 1\right\} : C = \frac{R^2}{C}$. In these relations, $\Phi(\phi), \Psi(\psi), \tilde{\Psi}(\tilde{\psi}) \subset L^2(R^2)$ and $P_C\Psi(\psi)$transfer $\Psi(\psi)$on C and also P_C̃ ($\tilde{\Psi}$The transfer $\tilde{\Psi}(\tilde{\psi})$ is on $\tilde{C}$. Based on this, the coefficients and details can be combined as follows:

$$\Phi(\phi) \cup P_C\Psi(\psi) \cup P_{\tilde{C}}\tilde{\Psi}(\tilde{\psi}) \tag{3.43}$$

The above formula is a framework for the state $L^2(R^2)$ with range Shearlet structure is expected to be faster than that of the wavelet. In fact, the coefficients are calculated faster in the presence of nonisotropic patterns (such as cracking, pavement texture, pits, and bitumen bleeding) using the Shearlet transform.

3.3.6 Shearlet Decompaction and Reconstruction

Decomposition and composition of Shearlet transform can be conducted in one of two ways: multilevel and directional. In the Non-Subsampled Shearlet Transform (NSST) method, different frequencies are categorized and analyzed using six steps. In the first stage, low and high frequencies are decomposed. In the second stage, the two-dimensional signals are decomposed into different dimensions; then in stages 3 to 5, different subbands are extracted in different directions (Figure 3.26). Figure 3.27 displays the three stages of decomposition of a two-dimensional image,

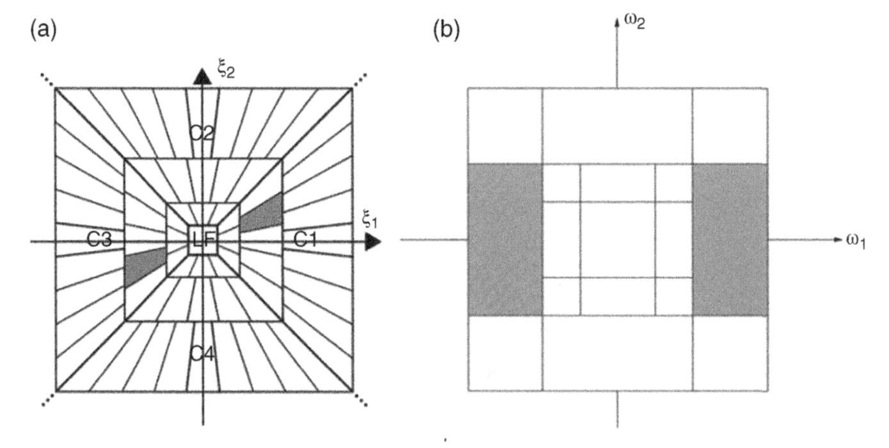

Figure 3.26 The frequency band windowing extracted from the (a) shearlet and (b) wavelet methods, where the dark regions represent the frequency of the small part of the function. The division into four parts and the low-frequency square are shown in (a).

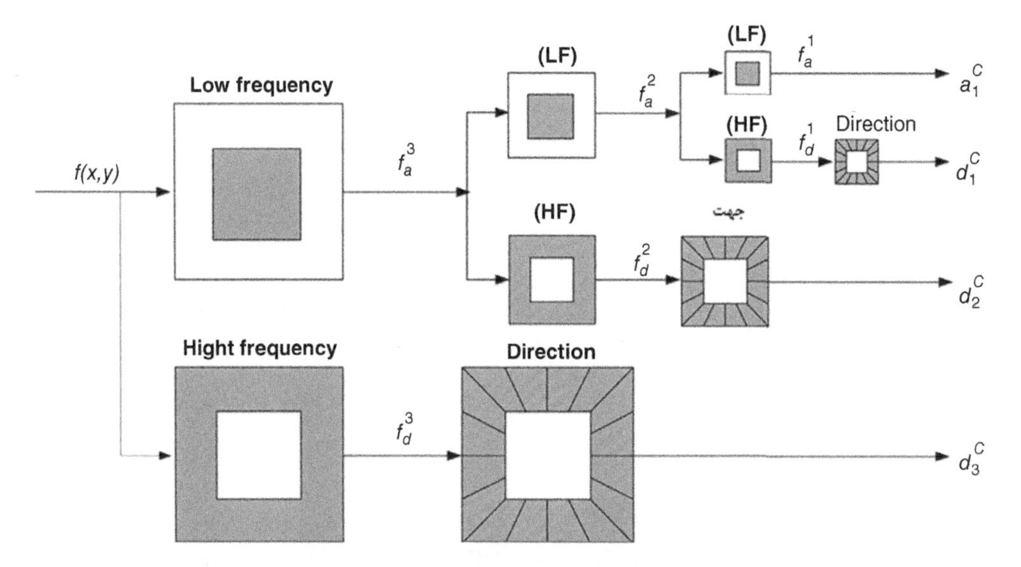

Figure 3.27 An example of three levels of SHT decomposition and directional extraction, at three levels for low and high frequency using the NSST method.

where a_1^c is the low-frequency shearlet coefficient and d_j^c is the high-frequency shearlet coefficient for j = 1,2,3,..., L. Directional filters are applied to the image details at each level. In each level, d_j^c is shown as a set of coefficients. For the high-frequency band, the texture becomes smaller at each step.

An example of image analysis using the NSST method for the two-level sample image shown in Figure 3.27, in which the direction vector is $\vec{n} = (n_1, n_2) = (4, 6)$ (Figure 3.28).

In the first level of decomposition, the agents are divided into four subband directions, and in the second level of converters, there are six directional subbands. The direction of the subbands varies at each level, which can be created using the Shearlet converter.

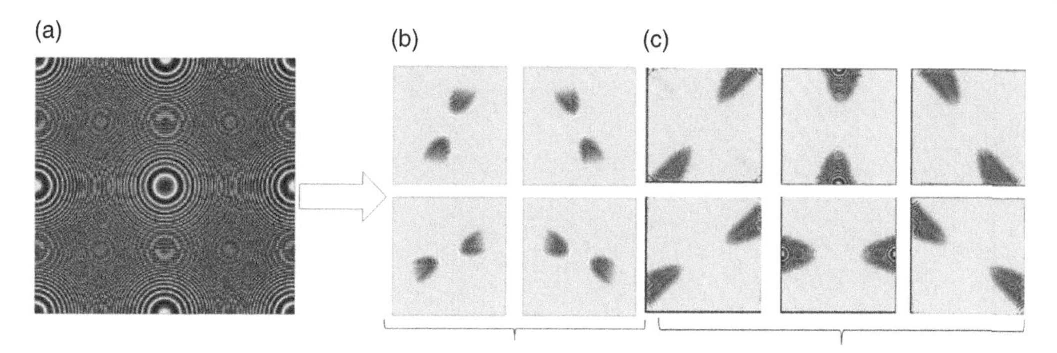

Figure 3.28 (a) Filters of three levels of SHT decomposition and directional extraction at three levels for (b) low-frequency using the NSST method and (c) high frequency using the NSST method.

3.3.7 Shearlet and Wavelet Comparison

Figure 3.29 shows the difference in the amplitude of the coefficients based on the shearlet and wavelet methods. Based on the coefficients, the diameter values in the size matrix are equal to $A_j = diag(2^{-j}, 2^{-j})$, which has the dimensions of isotropic and identical filters. In other words, in both horizontal and vertical directions, the size of the filter is symmetrical. Despite wavelet sensitivity to direction, this method does not perform well in asymmetric discrete environments (such as pavement cracks or asymmetrically shaped pavement potholes). Compared to other multilevel methods, the shearlet method has the following three distinguishing features in surface surveying:

1 – Properties of multilevel analysis related to time-frequency-local wavelet shows good performance for analyzing two-dimensional and three-dimensional surfaces due to its nonisotropic nature in transmitting geometric properties in the direction of surfaces.

2 – High directional sensitivity, compared to discrete two-dimensional wavelet methods, 2D DWT (horizontal, directional, and vertical) of the shearlet method has no limit on the number of directions and provides the ability to analyze surface texture and various patterns.

3 – Allows the analysis of large data with larger dimensions in a tight environment without significant error and provides better performance than other multiresolution methods.

These three features can be clearly seen in Figure 3.29, which shows that shuttle windows are heterogeneously variable in size, and the directional sensitivity and computational speed change with size and continuity. In the wavelet method, the analysis of images is done in a square with different dimensions, then the curve is estimated as points in this range.

As the dimensions get smaller, the number of nonzero points increases exponentially, which makes the wavelet method less efficient at displaying discontinuous objects (such as cracks in the real environment or discrete pavement texture). Despite this limitation, the SHT method can create a good approximation of curved patterns with the help of a set of coefficients. On the other hand, due to its ability to construct nonisotropic models, the SHT method is highly capable of analyzing and combining complex geometric patterns in an image. On the other hand, due to the immutability of shearlet, it is possible to recombine in different dimensions after filtering images by deleting the least effective data. This feature has made it possible to improve quality while retaining the most useful visual information. An example of combining coefficients with different methods is shown in Figure 3.30, from which it is apparent that the Shearlet method has better results for nonisotropic image reconstruction.

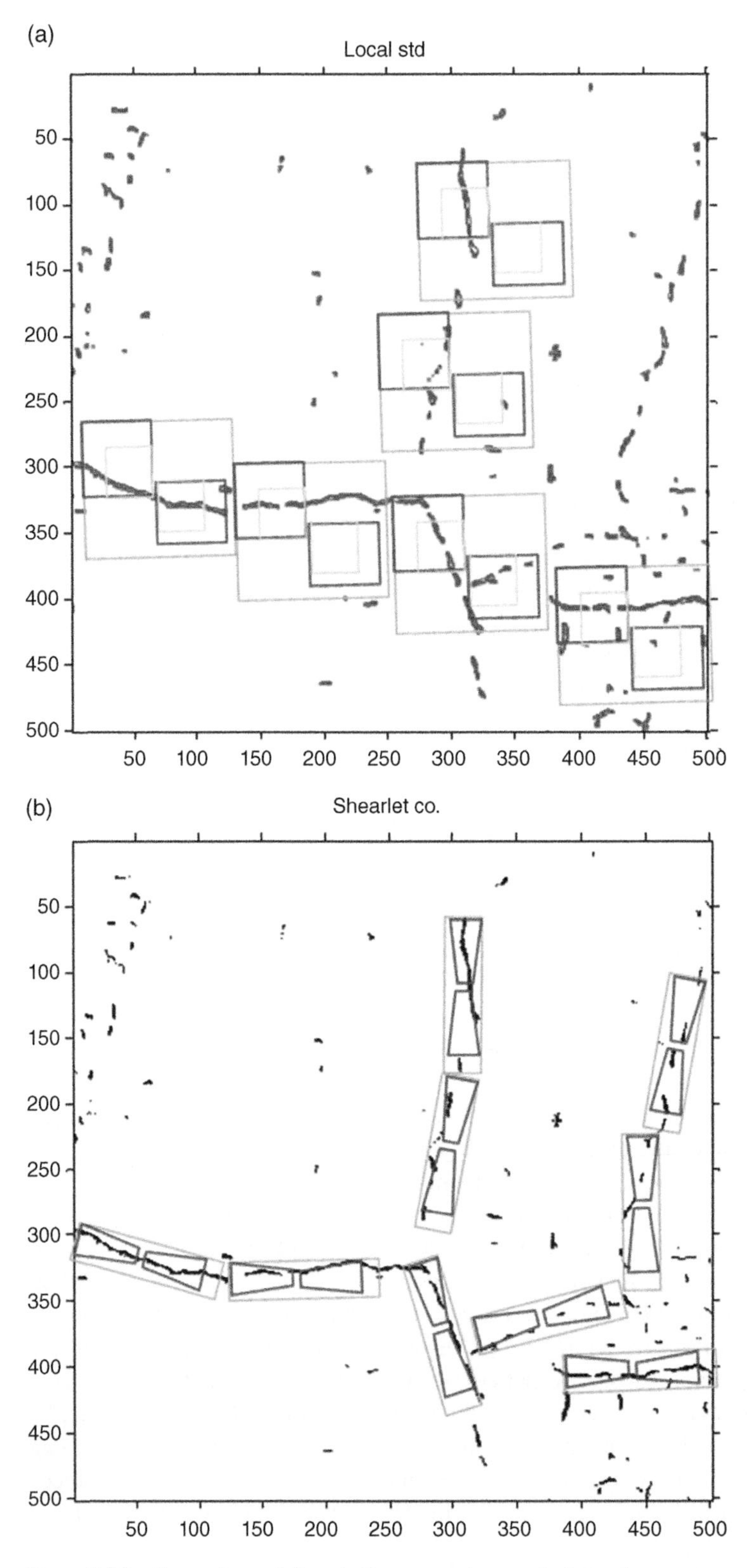

Figure 3.29 Comparison of discrete linear curvature analysis (such as pavement cracking) using two methods: (a) two-dimensional discrete wavelet and (b) two-dimensional shearlet.

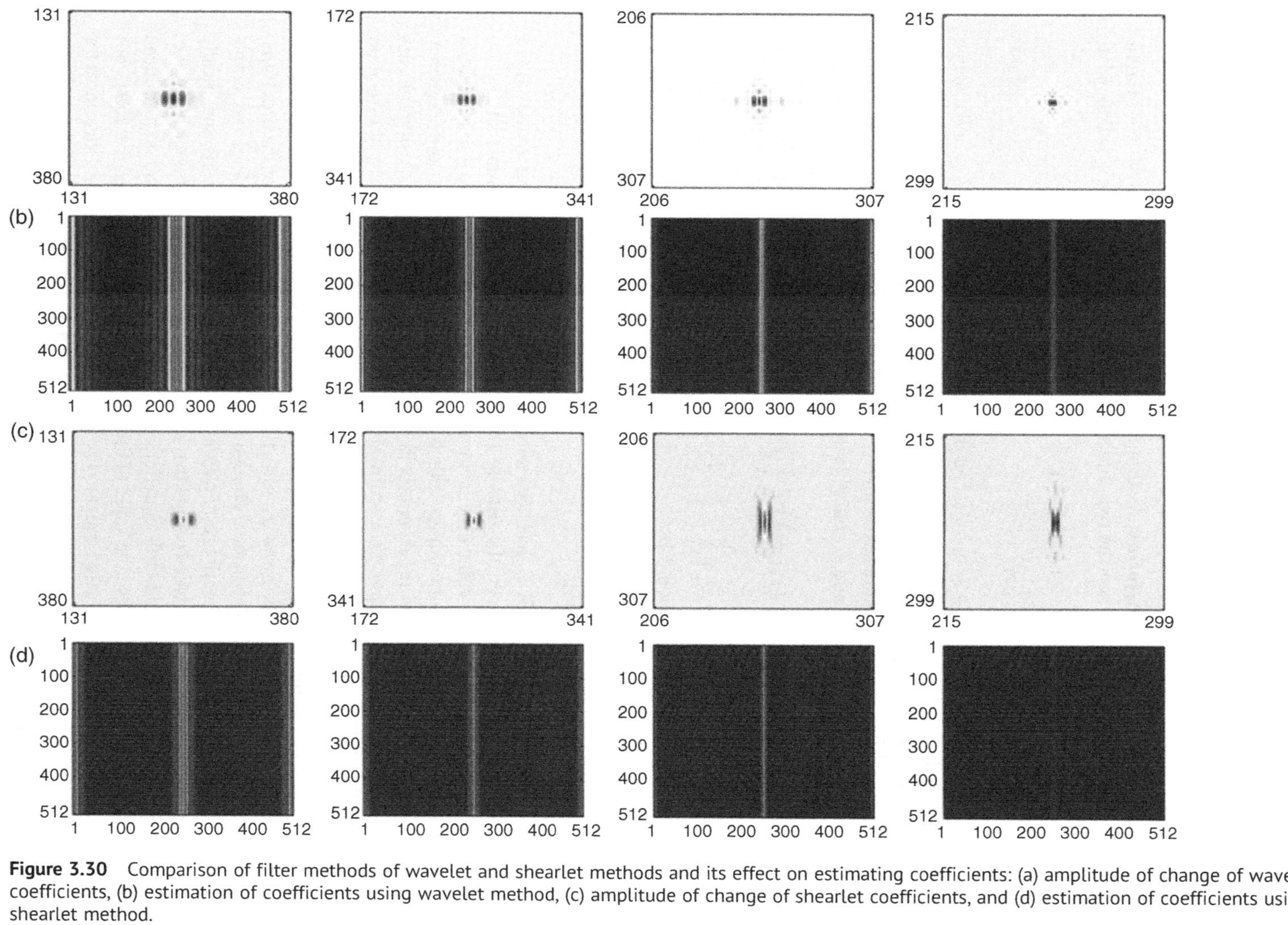

Figure 3.30 Comparison of filter methods of wavelet and shearlet methods and its effect on estimating coefficients: (a) amplitude of change of wavelet coefficients, (b) estimation of coefficients using wavelet method, (c) amplitude of change of shearlet coefficients, and (d) estimation of coefficients using shearlet method.

The following definitions describe the characteristics of the wavelet and shearlet methods:

One-dimensional wavelet transfer: Assuming $\psi \in L^2(R)$ for wavelet analysis with multilevel analysis, then the one-dimensional wavelet function for ψ is defined as follows:

$$f \rightarrow (W_{\psi}f)(j, m) = \langle f, \psi_{j,m} \rangle \tag{3.44}$$

where $f \in L^2(R)$and $j \in Z$, $m \in Z$, and the wavelength coefficients $\psi_{j,\ m}$are derived from Equation 3.44.

Two-dimensional wavelet transforms: Assuming $\phi \in L^2(R)$ for wavelet analysis with multi-resolution analysis and $\psi \in L^2(R)$, then the two-dimensional wavelet function for ψ is defined as follows:

$$f \rightarrow (W_{\psi,\phi}f)(k, j, m) = \langle f, \psi^{(k)}{}_{j,m} \rangle \tag{3.45}$$

where $k \in \{1, 2, 3\}$, $f \in L^2(R^2)$, and $j \in Z$, $m \in Z^2$, and the wavelength coefficients $\psi_{j,\ m}$ of the above theory is extracted.

Shearlet transform: Assuming $\psi, \widetilde{\psi} \in L^2(R^2)$ for the production of shearlet with multilevel analysis and $\phi \in L^2(R^2)$ as a function of scale, then $SH(\phi, \psi, \widetilde{\psi})$ is a framework for $L^2(R^2)$, which is defined two-dimensionally for ψ and ϕ and ψ using the following equation:

$$f \rightarrow \left(SH_{\phi,\psi,\widetilde{\psi}}f\right)\left(m', (k, j, m), \left(\widetilde{k}, \widetilde{j}, \widetilde{m}\right)\right) = \left(\langle f, \phi_{m'} \rangle, \langle f, \psi_{j,k,m} \rangle, \left\langle f, \widetilde{\psi}_{\widetilde{j,k,\widetilde{m}}} \right\rangle\right) \tag{3.46}$$

where $m, m', \ \widetilde{m} \in Z^2$, and $j, \widetilde{j} \in N$, k, $k, \widetilde{k} \in \left\{-\left\lceil 2^{\frac{j}{2}} \right\rceil, \ldots, \left\lceil 2^{\frac{j}{2}} \right\rceil\right\}$. The functions $\phi_{m'}$, $\psi_{j,\ k,\ m}$, and $\widetilde{\psi}_{\widetilde{j,k,m}}$are calculated based on Equation 3.19.

3.3.8 Complex Shearlet Transform

Shearlet Discrete Transformation (SDT) has been successfully applied in analyzing image data in the frequency-time domain. However, the cells in this transform, despite its simple shape, are termed complex cells due to their nonlinear behavior and have a definite fixed size that is suitable for spatial indexing. This property of complex cells is used to change the rotation and frequency (modeled by the S_k shear matrix and the A_j matrix in the SHT shell). Mixed values, using the capabilities of sin and cos functions, and the nonlinear frequency ξ are combined by the relation $\sqrt{\sin(\xi.)^2 + \cos(\xi.)^2}$. The wavelet and the shearlet functions are based on a certain value of the Fourier transform, while in the definition of the first type of shearlet, the response parameters show the acceptable size of functions $\psi_1$and ψ_2. Also, the effect of size and amount of shearlet transform shear can be observed by a proportional change in the frequency level. A general and simple method for the fixed wavelet and shearlet responses is to use the sin and cos relations, specifically the Fourier mode structure that is created using the relation $e^{i\xi} = \cos(\xi.) + i\sin(\xi.)$ for each ψ wavelet and shearlet function. According to this equation, each part can be changed and is introduced with ψ^*. Also, in this relation, the absolute value of the Fourier transforms ψ and ψ^* are equal, but the value of the cos waves in ψ^* is created equal to the value of the sin wave in ψ, and vice versa for sin. This feature can be done using the Hilbert Transform (TH) properties, which is defined by the following equations.

Definition of Hilbert Transform (HT): Assumingf $\in L^2(R)$, the HT function is defined as follows:

$$(H_,f)(x) = \frac{1}{\pi}\int_{R} \frac{f(t)}{t-x} dt, F(H_,f)(\xi) = -i \ \text{sgn} (\xi)\hat{f}(\xi) \tag{3.47}$$

where i represents the imaginary part, sgn represents the sign function, and the value of the function for the case $x = t$ is zero. Also, the value of the integral is equal to the value of the Cauchy principal value, and the relation can be written as a multiplication of $-\frac{1}{\pi}.*f$; therefore, the value of the function is the same as that stated in the Fourier convolution theory (FCT). Hilbert transform provides a linear and continuous conversion in $L^2(R)$ to create a positive correlation between conversion and corrosion:

$$\begin{aligned} Hf(.-m) &= (Hf)(.-m), \\ Hf(a.) &= (Hf)(a.) \end{aligned} \tag{3.48}$$

The properties of the Hilbert transform are used to make connections between the even and odd parts of the wavelet and shearlet transform. This conversion creates a connection between the two sin and cos waves. Accordingly, if $f \in L^1(R) \cap L^2(R)$ and f are real parts, then

$$\begin{aligned} \left(F^{-1}\hat{f}\right)(x) &= \frac{1}{\pi}\int_0^{\infty} \text{Re}\left(\hat{f}(\xi)\right)\cos(\xi x) - \text{Im}\left(\hat{f}(\xi)\right)\sin(\xi x) d\xi, \\ \left(F^{-1}\hat{Hf}\right)(x) &= \frac{1}{\pi}\int_0^{\infty} \text{Re}\left(\hat{f}(\xi)\right)\sin(\xi x) + \text{Im}\left(\hat{f}(\xi)\right)\cos(\xi x) d\xi \end{aligned} \tag{3.49}$$

The above relationships indicate that Hilbert transform works well on size and transmission. With this feature, the following definition is provided for the complex wavelet, also known as the ψ complex wave:

$$\psi^c = \psi + iH\psi \tag{3.50}$$

This wave contains real and imaginary parts, including the amplitude $|\psi^c|$, wherein the phase $\text{Re}\left(\frac{\psi^c}{|\psi^c|}\right)$ is defined as follows:

$$\psi(x) = |\psi^c| \ \text{Re}\left(\frac{\psi^c(x)}{|\psi^c(x)|}\right) \tag{3.51}$$

One of the important features of this transform is the existence of only positive frequencies, for each $\xi < 0$, $\psi^c(\xi) = 0$.

Definition of Positive Frequency of Complex Wave: One of the important characteristics of complex wave is the absence of negative frequency. Considering $\psi \in L^1(R) \cap L^2(R)$, then the relation for each $\xi \in R$ is:

$$\hat{\psi^c}(\xi) = F(\psi + iH\psi)(\xi) = \begin{cases} 2\hat{\psi}(\xi) & \text{if } \xi > 0 \\ 0 & \text{else} \end{cases} \tag{3.52}$$

Definition of One-dimensional Complex Discrete Wavelet Transform: Assuming $\psi \in L^2(R)$, the real value of the wavelet function is extracted in multilevel analysis, and the value of the wavelet complex is created using the following equation:

$$\left\{\psi^{c}_{j,m} = \psi^{c}_{j,m} + i(H\,\psi)_{j,m} \quad : (j, m) \in Z \times Z\right\} \tag{3.53}$$

In this relation, the Hilbert transform H and functions $\psi^{c}_{j,\,m}$and $\psi^{c}_{j,\,m}$ are defined based on the positive frequency theory of the complex wave; hence, the definition is

$$f \mapsto \left(W^{c}_{\psi} f\right)(j, m) = \left\langle f, \psi^{c}_{j,m}\right\rangle \tag{3.54}$$

where $\psi \in L^2(R)$and $m \in Z^2$. Based on the definition of the complex wavelet, the next step is to develop the Hilbert transform for two-dimensional problems and build the complex Shearlet model. The Fourier method is the most commonly used method for Hilbert transform in higher dimensions, based on Eq. (3.47), which is called partial Hilbert transform (PHT).

Definition of Partial Hilbert Transform (PHT): Assuming $\xi \in R^2 \setminus \{0\}$ and $\psi \in L^2(R^2)$, then the PHT of f in the direction $\xi_0$is calculated using the following equation:

$$F\left(H_{\xi_0} f\right)(\xi) = -i\ \mathrm{sgn}\left(\langle \xi, \xi_0\rangle\right)\hat{f}(\xi) \tag{3.55}$$

where sgn represents the sign function(signum function). To generate the real part of the shearlet transform, $\psi \in L^2(R^2)$, the shearlet complex function is considered as follows:

$$\psi^{c} = \psi + iH_{\xi_0}\psi \tag{3.56}$$

where $\xi_0 \in R^2 \setminus \{0\}$ (value of ξ_0 is considered equal to $(1, 0)^T$ or $(0, 1)^T$). This allows the phase/amplitude decomposition for one-dimensional cases. Fourier transform for complex wavelet using HT removes the negative part, and this property for PHT is also defined as $\langle \xi_0, \xi_0\rangle < 0$.

Two-dimensional complex waves containing all frequencies: Assuming $\xi_0 \in R^2 \setminus \{0\}$ and $\psi \in L^1(R^2) \cap L^2(R^2)$, based on Equation 3.55, then the complex wavelet for each $\xi \in R$ is

$$\hat{\psi^{c}}(\xi) = F\left(\psi + iH_{\xi_0}\psi\right)(\xi) = \begin{cases} 2\hat{\psi}(\xi) & \text{if } \langle \xi, \xi_0\rangle > 0 \\ 0 & \text{else} \end{cases} \tag{3.57}$$

Complex waves related to linear invertible mappings: Assuming $\xi_0 \in R^2 \setminus \{0\}$ and $A : R^n \rightarrow R^n$, then for every $f \in L^2(R^2)$, all PHTs are linked together using the following relation:

$$A\left(H_{\xi_0} f\right) = H_{A\xi_0}(Af) \tag{3.58}$$

In particular, because

$$\mathrm{sgn}\left(\langle \xi, \lambda\xi_0\rangle\right) = \mathrm{sgn}\left(\langle \xi, \xi_0\rangle\right) \tag{3.59}$$

for all coefficients $\lambda > 0$, this relation becomes

$$A\left(H_{A\xi_0} f\right) = H_{\xi_0}(Af) \tag{3.60}$$

where $\xi_0$of the unit vector A is a positive unit function, and therefore,

$$S_k \begin{pmatrix} 1 \\ 0 \end{pmatrix} = \begin{pmatrix} 1 & k \\ 0 & 1 \end{pmatrix} \begin{pmatrix} 1 \\ 0 \end{pmatrix} = \begin{pmatrix} 1 \\ 0 \end{pmatrix} \tag{3.61}$$

where S_k is the shear matrix that is calculated based on Equation 3.39. The PHT response, denoted as $H_{(1,0)^T}$, is related to the S_k shear operators for all $k \in Z$. Therefore, it can be proven that PHT affects all shearlet transforms, including transmission, scaling, and shearing, and can create complex discrete wavelet transforms (Figure 3.31).

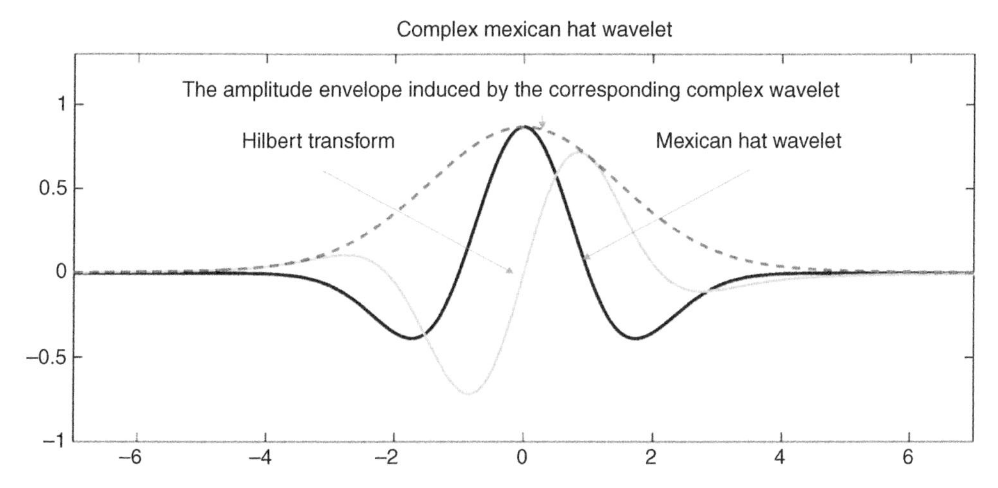

Figure 3.31 Mexican wavelet display, Hilbert transform, mixed wavelet. The Mexican wavelet is symmetric if the Fourier transform has a real value (with cos part). Using the Hilbert transform on the Mexican wavelet, the output has only a sine wave and an odd part.

Partial conical complex discrete shearlet transform (DC_CSHT): Assuming $\psi,\ \hat{\psi} \in L^2(R^2)$, the real part of $\phi \in L^2(R^2)$ is a function of scaling, then based on the relation $SH(\phi, \psi, \hat{\psi})$, which is a frame in the environment $L^2(R^2)$, and using Equation 3.42, DC_CSHT is defined as follows:

$$\begin{aligned}
&SH^c(\phi, \psi, \widetilde{\psi}) = \Phi(\phi) \cup \Psi(\psi) \cup \widetilde{\Psi}(\widetilde{\psi}),\\
&SH^c(\phi, \psi, \widetilde{\psi}) = \Phi(\phi) \cup \Psi^c(\psi) \cup \widetilde{\Psi}^c(\widetilde{\psi}),\\
&\Phi(\phi) = \{\phi_m = \phi(. - m) : m \in Z^2\},\\
&\Psi^c(\psi) = \left\{\psi^c_{j,k,m} = \psi_{j,k,m} + i\left(H_{(1,0)^T}\psi\right)_{j,k,m} : j \in N_0, |k| < 2^{\frac{j}{2}}, m \in Z^2\right\},\\
&\widetilde{\Psi}^c(\widetilde{\psi}) = \left\{\widetilde{\psi}^c_{j,k,m} = \widetilde{\psi}_{j,k,m} + i(H_{(1,0),^T}\widetilde{\psi})_{j,k,m} : j \in N_0, |k| < 2^{\frac{j}{2}}, m \in Z^2\right\}
\end{aligned} \tag{3.62}$$

The definition of functions $(H_{(1,0),^T}\widetilde{\psi})_{j,k,m}$ and $\left(H_{(1,0)^T}\psi\right)_{j,k,m}$ are presented in (3.41). According to these relations, for $f \in L^2(R^2)$, the DC_CSHT conversion is defined as follows:

$$f \rightarrow \left(SH^c_{\phi,\psi,\widetilde{\psi}}f\right)\left(m', (k, j, m), (\widetilde{k}, \widetilde{j}, \widetilde{m})\right) = \left(\langle f, \phi_{m'}\rangle, \left\langle f, \psi^c_{j,k,m}\right\rangle, \left\langle f, \widetilde{\psi}^c_{\widetilde{j,k,m}}\right\rangle\right) \tag{3.63}$$

where $m, m',\ \widetilde{m} \in Z^2$ and $j, \widetilde{j} \in N_0$, $k, \widetilde{k} \in \left\{-\left\lceil 2^{\frac{j}{2}}\right\rceil, \ldots, \left\lceil 2^{\frac{j}{2}}\right\rceil\right\}$. Functions $\phi_{m'}$, $\psi^c_{j,\ k,\ m}$, and $\widetilde{\psi}^c_{\widetilde{j},\widetilde{k},\widetilde{m}}$ are defined in (3.62). Due to the stability of the Fourier transform under the Hilbert transform, $\phi \in L^2(R^2)$ satisfies the complementary conditions of the shearlet given in relation $H_{\xi_0}\psi$. Based on these relationships, a new type of shearlet can be combined with the first type of shearlet and implemented in the form of a conical transformation according to Hilbert theory, the second-generation Shearlet in the form $SH\left(\phi, H_{(1,0)^T\psi}, H_{(0,1)^T\widetilde{\psi}}\right)$ is provided. The pair of and odd part for symmetric shearlet transform can be converted using the Hilbert transform, which is a practical method for extracting edges and shells in complex shearlet transforms.

3.3.9 Complex Shearlet Transform for Image Enhancement

One of the applications of the shearlet method is to reduce the amount of noise using thresholding similar to that applied in the wavelet method. In this method, the image shearlet f is calculated using the shearlet transform, and by thresholding, part of the information related to the image noise is separated. The image with noise is modeled as a complex image and the noise as y = f + n. By thresholding and filtering coefficients below the threshold, image noise can be reduced, and the SNR index can be increased. Other methods, such as partial differential equations (PDE), can be used to reduce noise and minimize the change and division of the equations. The basis of the complex shearlet method is to minimize the amount of image energy using the following equation:

$$\tilde{E}\left(f;\lambda,\tilde{f}\right) = \frac{\lambda}{2}\int_{\Omega}\left(f-\tilde{f}\right)^{2}dx\,dy + P(f) \tag{3.64}$$

where the function f is amplified in the domain $\Omega \in R^2$ and $\tilde{f}$ is the image. The first sentence of 3.64 indicates the degree of similarity between the two images, $\tilde{f}$ and f, and the second sentence refers to the amount of fine $(f, \nabla f)$ that is used to control the answer. Equation 3.64 describes the relationship between decomposing and regulating filters. The following equation is then applied to determine the amount of total variation (TV), which includes minimizing the following Eqs. (3.65 and 3.66):

$$\int_{\Omega}\emptyset(\|\nabla f\|)dx\,dy + \frac{\lambda}{2}\int_{\Omega}\left(f-\tilde{f}\right)^{2}dx\,dy \tag{3.65}$$

where $\emptyset \in C^2R$ is an even function, and the rate of change of the penalty function is f, which is defined for the one-dimensional wavelet function at level 1, as follows:

$$TV(f) = \int_{\Omega}\emptyset(\|\nabla f\|)dA \tag{3.66}$$

where $\nabla f = (\partial u/(\partial x\_1), \partial u/(\partial x\_2))$, and || is the value of the standard distance function. To convert SHT, Eq. (3.67) is rewritten a:

$$\left(\frac{\partial f}{\partial t}\right) = \nabla.\left(\frac{\emptyset'(\|\nabla P_s(f)\|)}{(\|\nabla P_s(f)\|)}\nabla P_s(f)\right) - \lambda_{x,y}\left(f-\tilde{f}\right) \tag{3.67}$$

The boundary conditions are $\frac{\partial f}{\partial n} = 0$ for $\partial\Omega$, the initial conditions are $\frac{\partial f}{\partial n} = f(x,y,0) = \tilde{f}(x,y)$, and $x, y \epsilon \Omega$ and x, $y\epsilon\Omega$. The parameter $\lambda_{x,\,y}$ is a special variable that is calculated based on the size of the local variance, and its value changes after each step or with time. The shearlet domain, Eq. (3.68), uses the boundary conditions $\frac{\partial f}{\partial n} = 0$ and $\partial\Omega$ as well as the Newman boundary conditions, $f\,f(x, y, 0) = f_0(x, y)$ for x, $y\epsilon\Omega$ as follows:

$$\left(\frac{\partial f}{\partial t}\right) = \nabla.(\rho(\|\nabla P_s(f)\|))\nabla P_s(f) \tag{3.68}$$

An example of a two-dimensional image output using this method is shown in Figure 3.32. The complex method provides another way to reduce noise, using Eq. (3.69):

$$f = \frac{1}{n}\sum_{k=1}^{n} f_1(k) f_2^*(k) = |f|e^{i\psi} \tag{3.69}$$

Figure 3.32 An example of the results extracted from the noise reduction method using SHT: (a) original image and (b) noise image, SNR = 11.11 dB. (c) The decomposition method with 53 repetitions, SNR = 14.78 dB. (d) SHT with six repetitions, SNR = 16.15 dB. (e) SHT TV method with two repetitions, SNR = 16.29 dB. (f) TV method with 113 repetitions, SNR = 14.52 dB.

where the dimensions of the image and its phase quality depend on the amplitude of the correlation coefficient and are calculated as follows:

$$\rho = \frac{E[\, f_1 f_2^*]}{\sqrt{E\left[|\, f_1|^2 E|\, f_2^*|^2\right]}} = |\rho| e^{j\theta} \tag{3.70}$$

In this regard, ρ is the degree of convergence, and θ is the phase of the complex correlation coefficient. By selecting the appropriate phase noise model, reducing the shearlet coefficient leads to noise elimination and improving the image quality for the next stage of analysis. Figure 3.33 presents a comparison of the images obtained by selecting the appropriate phase and the phase estimated using this method.

3.3.10 Low and High frequencies of Complex Shearlet Transform for Image Denoising

In large images, the components of low frequency and energy in the image are used to enhance the image. Due to the presence of noise in the image and the lack of continuity in issues, such as cracking, the amount of noise must be reduced and the energy distribution must be adjusted before analyzing and performing the next steps. Using background modulation via linear interpolation, the low-frequency component is a practical method for preprocessing and noise reduction. For this

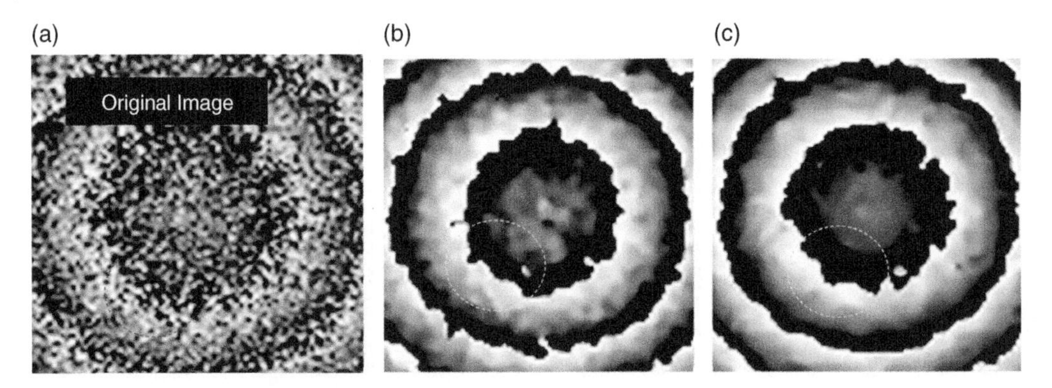

Figure 3.33 Comparison of phase filtration method to reduce noise using: (a) image noise with $|\rho| = 0.5$ (number of residual noises = 14 119), (b) wavelet method (number of residual noises = 80), and (c) shearlet method (number of remaining noise = 20).

purpose, the low-frequency component of the image is divided into small blocks that do not overlap, where the size of each block depends on the size of the original image. If the size of each image is $N \times N$, then the size of each block is $n \times n$, which will be $n = N/16$. Using the processing of each of these cells, the image is calculated as follows:

$$\begin{aligned} f(x,y) &= f(i+u,j+v) = (1-u)(1-v)f(i,j) + u(1-v)f(i+1,j) \\ &\quad + (1-u)vf(i,j+1) + uvf(i+1,j+1)f(i,j) \\ &= \text{mean}(S(i,j) + \varepsilon(\text{mean}(S(i,j)) - \text{mean}(f(x,y)) \end{aligned} \tag{3.71}$$

where $S(i, j)$ is the block (i, j) of the image, and the parameter ε is calculated from Eq. (3.72):

$$\varepsilon(x) = \begin{cases} 0, x \geq 0 \\ -1 \times a|x|, x < 0 \end{cases} \tag{3.72}$$

where a is a constant parameter with a default value of 0.5. In general, to apply this method in the detection and evaluation of pavement cracking, the value of the gray pixel (pavement cracking is usually darker than the background) is less than the average value of the image. Figure 3.34 provides an example of the analysis for the low-frequency band of an image with cracking. The shearlet coefficients contain the high-frequency details of the image with a small amount of cracking and indicate small changes in edges and noise. In pavement images with small dimensions, a lot of noise and hair edges are common (more than 80% of the image's information). In Fourier transform, cracking has higher shearlet coefficients than noise and image edges. Therefore, by selecting an optimum threshold, a good classification of cracks in the pavement image can be done with this method. As shown in the example in Figure 3.34, two shearlet coefficients in five levels are considered to separate cracks and the background. From the first to the fourth surface cut, the cutting direction 10, 10, 6, and 6, and the processed coefficients for sizes 3–4 are calculated using Eq. (3.73):

$$S'_{x,y}(l,j) = S_{x,y}(l,j).\varepsilon\left(g_{x,y}(l,j) - T_1\right), g_{x,y}(l,j) = \frac{|S_{x,y}(l,j)|}{\max\left(|S_{x,y}(l,j)|\right)} \tag{3.73}$$

where $l = 1, 2, 3, ...6$, which indicates direction, and $j = 3{,}4$ indicates the size. The value of T_1 for a large number of experiments, as determined by trial-and-error, is equal to 0.25. The first and second sizes of high-frequency image contain information about key pavement features, such as edges,

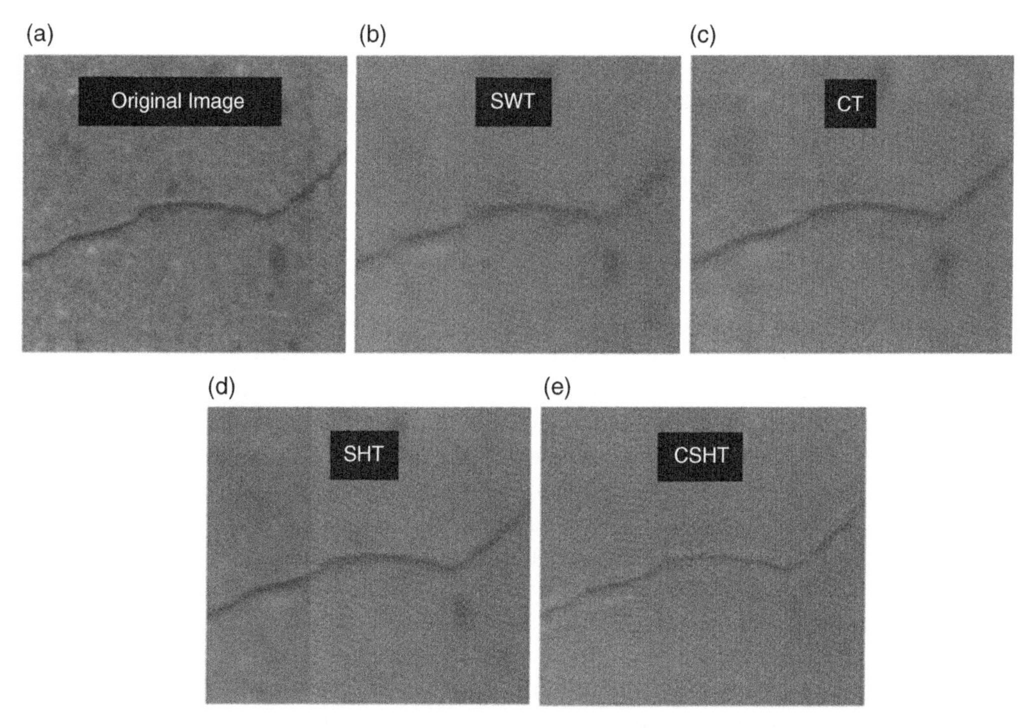

Figure 3.34 Thresholding on coefficients and evaluation of noise removal methods and image quality enhancement: (a) main image, (b) result of SWT method, (c) CT method, (d) SHT method, and (e) SHT method modified in different frequencies and sizes with fixed threshold (CSHT).

cracks, and coarse textures. In the next steps, the goal is to keep track of information and while eliminating noise. Also, for cracking, which is a nonisotropic and directional subject, the value of component l can be used as a functional feature to introduce the characteristics of cracking and pavement texture. Hence, the coefficients obtained from the first and second steps are calculated based on Eq. (3.74):

$$\begin{aligned} S'_{x,y}(L(l),j) &= S_{x,y}(L(l),j).\varepsilon\left[\left(S_{x,y}(L(l),j).\varepsilon\left(g_{x,y}(l,j)-T_2\right)-T_1\right]\right. \\ \mathrm{S}'_{\mathrm{x,y}}(\mathrm{L(1),j}) &= \mathrm{S}_{\mathrm{x,y}}(\mathrm{L(1),j}).\varepsilon\left[\left(\mathrm{S}_{\mathrm{x,y}}(\mathrm{L(1),j}).\varepsilon\left(\mathrm{g}_{\mathrm{x,y}}(1,3)-\mathrm{T}_2\right)-\mathrm{T}_1\right]\right. \end{aligned} \tag{3.74}$$

where $l = 1, 2, 3, \ldots 6$, which indicates the direction, and $j = 1, 2$ indicates the size. The value of T_2 a large number of experiments, as determined by trial-and-error, is equal to 0.65. Figure 3.34 presents an example of the analysis performed using the parameters $T\_1 = 0.25$, $T\_2 = 0.65$, and $a = 0.5$ for the image of pavement with distress due to cracking. In this analysis, 5 SHT sizes were used in the Laplace in frame. For sizes 1–4, the number of shear directions is 10, 10, 6, and 6, respectively. From the appearance of the images gained by different methods, it can be concluded that in the SWT method, the output: is m coarser in different directions. Comparatively, the curvelet method has a Gibbs-type complication, while the other image enhancement methods removed important cracking information and complicated crack detection and the pavement texture analysis. In the SHT method, in addition to image preprocessing and noise reduction, important information can be obtained from the main objects of the pavement image, such as crack geometry, pavement texture, and morphology of pavement distress.

The real and imaginary components of the image are separated to eliminate noise and improve the quality of the image. Since the imaginary part contains information about noise, by examining and filtering the imaginary part of the image and using the Eq. (3.75) (complex diffusion), noise can be removed from the image:

$$I_t = \mathrm{div}(c(\mathrm{Im}(I))\nabla I), C = e^{\left(\frac{-x^2}{\rho^2}\right)}.\rho, c(\mathrm{Im}(I)) = \left[\frac{e^{i\theta}}{1 + \left[\left(\frac{\mathrm{Im}(I)}{k\theta}\right)\right]^2}\right] \tag{3.75}$$

where k is the threshold parameter, and $\theta \ll 1$.

The main advantage of the shearlet method over wavelet is the access to directional, multilevel information. Three different algorithms in the shearlet domain were designed based on shearlet theory to separate the image into coefficients. Figure 3.35 presents these three algorithms, namely DSST1, DSST2, and DSST3, can be used to eliminate noise and improve the quality of the pavement image. SHT coefficients are extracted based on the determination of input parameters, and then are calculated based on each algorithm. It should be noted that the choice of shear, size, and direction parameters are important in the final determination of coefficients. Noise reduction, crack quality improvement, and evaluation of coarse pavement texture are performed on the SHT method by adjusting Dst and threshold methods. The basis of this method is similar to the RIT method, particularly to maintain high-amplitude coefficients and eliminate and modify low-amplitude coefficients. As with other methods, cracking and failure occur at high-amplitude frequency coefficients. In this method, high-amplitude coefficients are kept, and low-amplitude coefficients are reduced to zero.

Increasing the level reduces the amount of noise and, thus, increases the PSNR index, decreases the information and details of cracking and coarse paving texture, and increases the computational time. An evaluation of three Myer filters, including "7-9," "maxflat," "pyr," and "pyrex," and the introduced indicators for the methods (Figure 3.35) are presented in Table 3.11.

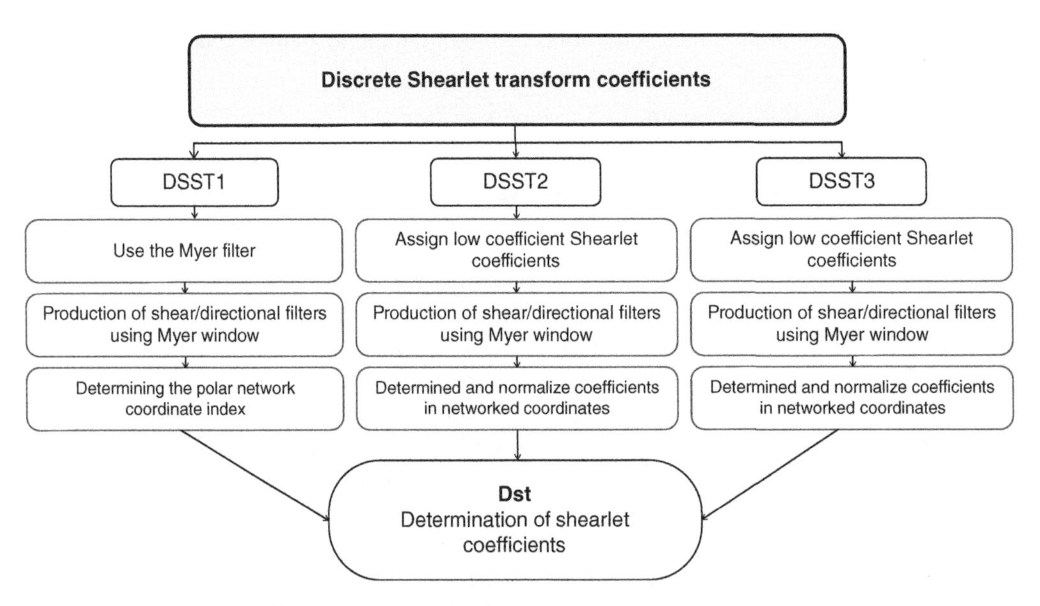

Figure 3.35 Shearlet coefficient extraction algorithms.

Table 3.11 Comparison of Discrete Shearlet transform coefficients.

	PSNR	MSE	MAXERR	L2RAT	ESSIM	EPI	QILV	SCI	TIME
Method 1	nsst-dec1e								
"9-7"	28.980	82.170	37.740	0.991	0.999	0.514	0.469	0.991	0.144
"maxflat"	29.440	73.940	35.680	0.992	1.000	0.524	0.544	0.992	0.451
"pyr"	28.950	82.820	35.060	0.991	0.999	0.517	0.439	0.991	0.104
"pyrex"	28.960	82.710	35.040	0.991	0.999	0.517	0.444	0.991	0.121
Method 2	nsst-dec1								
"9-7"	28.980	82.280	37.720	0.991	0.999	0.514	0.468	0.991	0.143
"maxflat"	29.440	73.900	35.750	0.992	1.000	0.525	0.544	0.992	0.458
"pyr"	28.940	82.980	35.030	0.991	0.999	0.517	0.438	0.991	0.103
"pyrex"	28.950	82.740	35.980	0.991	0.999	0.517	0.444	0.991	0.121
Method 3	nsst-dec2								
"9-7"	29.435	74.060	33.270	0.991	0.999	0.530	0.495	0.991	0.528
"maxflat"	29.931	66.070	32.870	0.992	1.000	0.539	0.570	0.992	0.826
"pyr"	29.630	70.800	33.550	0.991	0.999	0.538	0.500	0.991	0.472
"pyrex"	29.620	70.970	33.720	0.991	0.999	0.537	0.504	0.991	0.496

The method DSST1 was obtained using one-dimensional samples 7-9. The DSST2 method is based on a one-dimensional filter, resulting from the removal of 4 moments, and the third method is based on a one-dimensional filter resulting from the removal of two moments. The pyrex filter is similar to the pyr filter, except that it has been replaced by two high-frequency filters. The output of these methods is four filters based on the wavelet filter library, which produces high-pass and low-pass filters. In order to evaluate the filters, conv2(h0, g0) + conv2(g1, h1) = 1 was used. The analysis of DSST1 for the filters in the SHT filter bank for samples 7–9 is shown in Figure 3.36. By applying a normalization coefficient of 0.5 for h1 and g0, the same relation becomes a direct function, which is a rule that also applies to other filters. The SHT coefficient is created at different levels based on the convolution of the low-pass filter obtained by this method and the shear filter. Various experiments have shown that the simulated coefficient can be modeled with generalized Gaussian distributions so that these thresholds should be at risk of about 5%. To estimate the signal variance in each local subband, the neighbor coefficient in the square window was used to determine the maximum probability. The variance was estimated using the Monte Carlo method for several pavement images with normal noise, and then these estimates were averaged. Threshold parameters were selected based on the following relationship:

$$\tau_{i,j} = \left(\frac{\sigma^2_{\epsilon_{i,j}}}{\sigma^2_{i,j,n}} \right) \tag{3.76}$$

where $\sigma^2_{i,j,n}$ is the variance of the nth factor and it-h is subband shear in the size j and noise variance $\sigma^2_{\epsilon_{i,j}}$ in the shear direction is i. A comparison of filters and past research experience shows that SHT coefficients can be obtained using the Gaussian distribution with the help of optimal bias. In order to determine the variance in each subband, the adjacent coefficients in each window and the maximum neighborhood value are used.

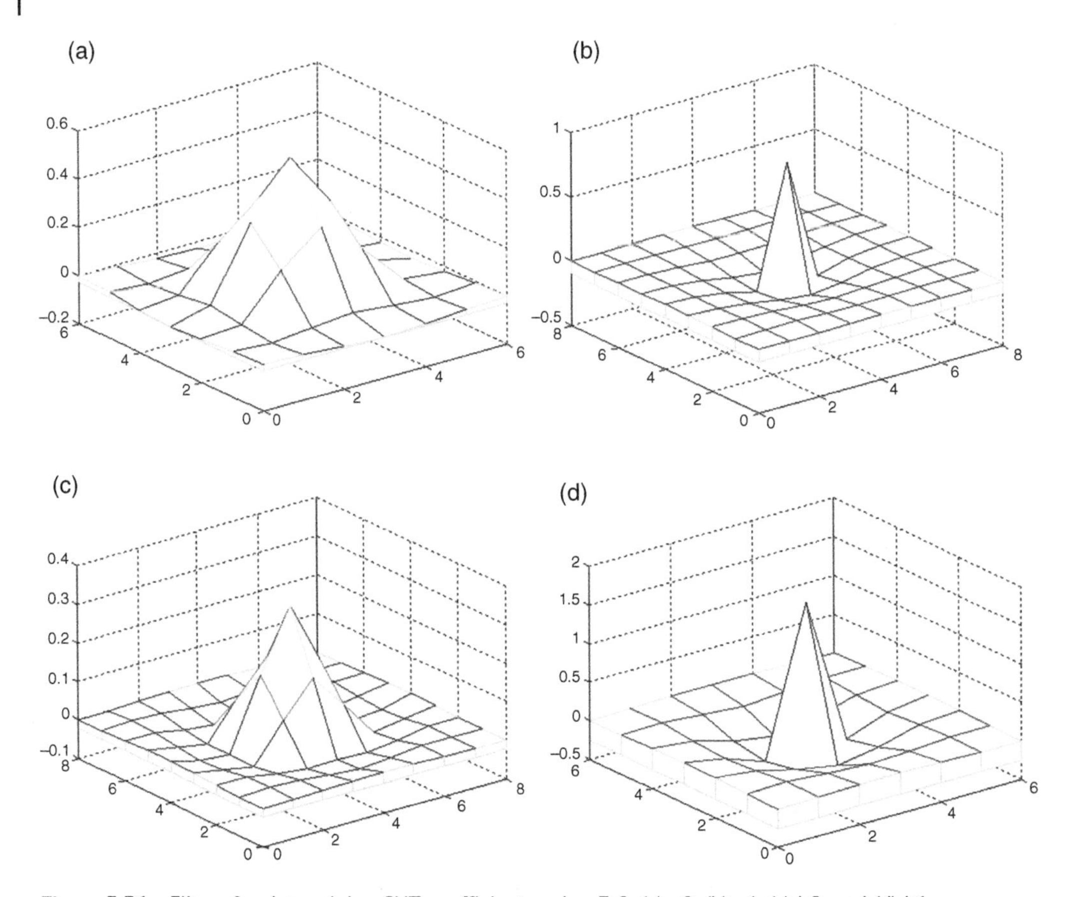

Figure 3.36 Filters for determining SHT coefficients using 7-9: (a) g0, (b) g1, (c) h0, and (d) h1.

In order to compare the effect of different parameters in the noise removal algorithm and improve the quality of the pavement image using the SHT method, a study was performed for a set of pavement image databases. Initially, the effect of the sigma parameter on each of the main characteristic of the method, including its evaluation, selection rules, filters, and thresholds, was investigated (Figures 3.37–3.39).

Figures 3.40–3.42 present the change diagrams of the main characteristics, including PSNR, MSE, MaxErrt, L2RAT, ESSIM, EPI, QILV, and SIC time, for the three different methods, where shear parameters and T scalars threshold = [0 1 1], shear decomposition parameters = [1 2 2], shear parameters show resizing = [1 2 2], and deviation of different standards was from 1 to 100.

In the first method, the PSNR index could not be obtained for the mentioned filters according to a standard deviation of more than 50, and a significant effect on noise reduction was not observed. On the other hand, the error rate and MSE exceeded 50 and destroyed the information needed to evaluate the main characteristics of the structure. The turning point in the structural indices of cracking was between 30 and 50. By selecting a turning point of 30, the results show reduced noise, cracks, coarse pavement and spall, and crack edges and improved image quality to correct faulty detection.

The results obtained for methods 2 and 3 are given in Figures 3.41 and 3.42, respectively. By changing the value of parameters related to noise, examination of the nine indicators reveal a slight difference among the different filters between the methods, whereby changes in sigma has a greater effect than the filter type. According to this analysis, with increasing sigma, the amount of PSNR,

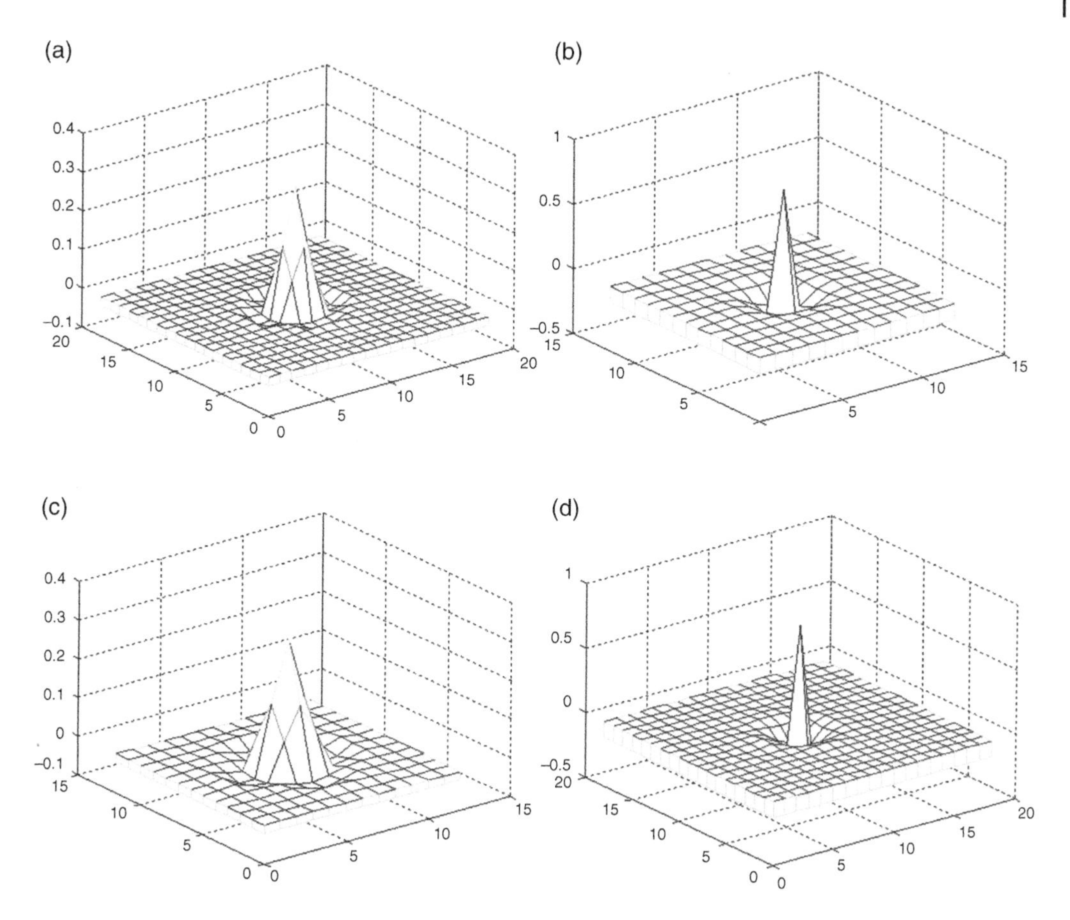

Figure 3.37 Filters for determining SHT coefficients using Maxflat filters: (a) g0, (b) g1, (c) h0, and (d) h1.

ESSIM, L2RAT, QILV, EPI, and SCI indices decreased and other indices, such as MSE and MaxErr, increased. Although the main structure of cracking was lost, the amount of noise reduction increased. Figure 3.40 shows the change in image quality and the preservation of failure characteristics and coarse pavement texture based on the altered parameters.

Then, the optimal method was selected by comparing the parameters and computational speed as well as the ability of models to enhance images. Computational time is another influential factor that affects method selection. The evaluation results of the important characteristics are shown in Table 3.11 for the three methods, in which the threshold T scalars = [0 1 1], the shear parameters of the decomposition = [1 2 2], and the shear parameters of the scaling = [1 2 2]. A comparison was performed on a set of 100 pavement images, and the average of computational time of different filters was determined.

The maximum difference in the PSNR index between each method DSST1 and DSST2, method DSST2 and DSST3, and DSST1 and DSST3 was determined to be 1.67%, 1.71%, and 1.66%, respectively. The average of this index for method 1 is greater than method 2 and smaller than method 3, while the average computational time of method 3 is 2.8 times that of method 1.

According to this analysis, method 1 is fastest in performing calculations. The minimum difference ΔMSE for the maxflat filter between methods 1 and 2 is 0.04 and between methods DSST1 and DSST3 is 7.8.

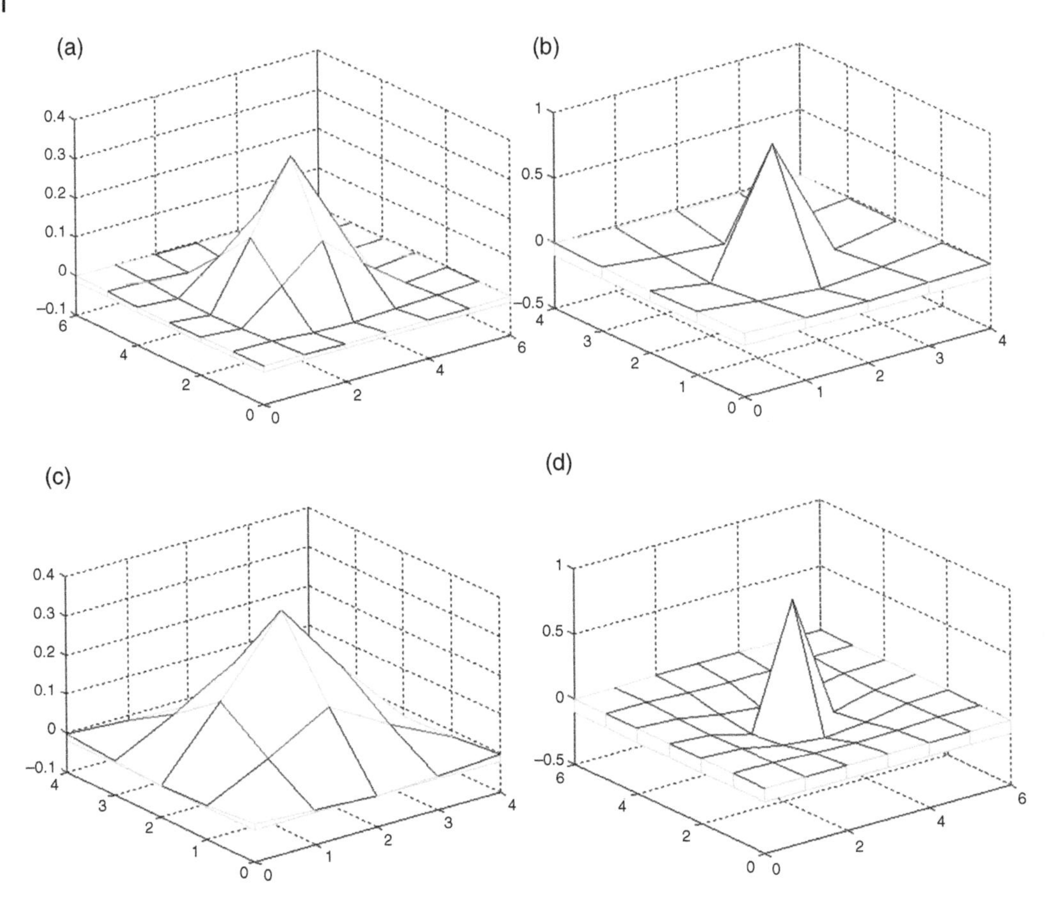

Figure 3.38 Filters for determining SHT coefficients using pyr filters: (a) g0, (b) g1, (c) h0, and (d) h1.

This comparison shows that the difference between methods 1 and 2 is less and method DSST3 has less MSE index. However, as mentioned, the computational time in method 3 is considerably larger compared to 2, which has less computational complexity in the identification, evaluation, and classification of faults and coarse pavement texture. Also, $\Delta MSE_{(1,\ 3)}$ is equal to 2.2 and for $\Delta MSE_{(1,\ 2)}$is 1.7. The L2RAT index is very similar for the three methods with a minimal difference. Also, the ESSIM index for methods DSST1 and DSST2 is 3% smaller than that of method DSST3. Other structural features for cracking and retention of original characteristics are similar for methods 1 and 2. In addition, the EPI index is 0.51, the QILV index is 0.47, and the average SCI is 0.99 for methods DSST1 and DSST2.

Based on the above evaluation, the method 1, nsst-dec1e, with a pyr filter is proposed as the optimal method due to the minimum computational time and cutting structure [1 2 2], based on the 30-σ variable for pavement image analysis. Figure 3.43 shows that the structure selected for cracking assessment is complex with crack distress.

In order to compare the effect of effective threshold selection in reducing noise and image enhancement, the three bands, namely high-pass, mid-pass, and low-pass, were evaluated separately. At different thresholds, the effect of each band on the main characteristics of the image was extracted and compared. The average result of each of these indicators was evaluated for 100 pavement images, and its average for the three thresholds is shown in Figure 3.43. As the

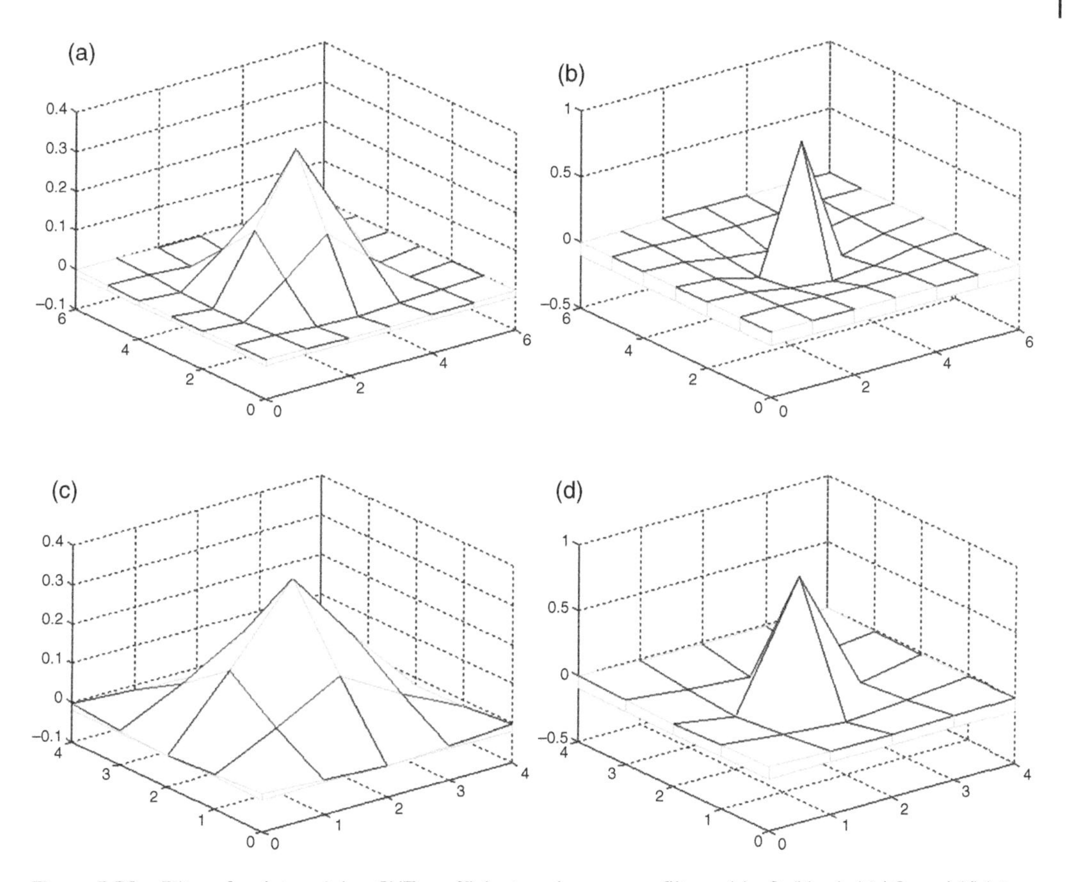

Figure 3.39 Filters for determining SHT coefficients using pyrexe filters: (a) g0, (b) g1, (c) h0, and (d) h1.

diagrams show, the mid-pass band exhibits more sensitivity to the threshold changes. As the median threshold increased, the PSNR index decreased, and the error rate increased. It should be noted that the change in the high-pass and low-pass threshold had little effect on the main characteristics of the image. Also, the structural characteristics of cracks generally decreased when the third threshold level increased. Since the effect of the middle band is far more important than the thresholds of high-pass and low-pass coefficients, by selecting a threshold close to zero for the middle band, the maximum structural indicators and error-related indicators can be minimized.

3.4 General Comparison of Single/Multilevel Methods and Selection of Methods for Noise Removal and Image Enhancement

In this section, a comparison of the methods is presented in order to select the optimal method. Based on all analyses, one-level and multilevel methods and filters with higher efficiency were selected. Table 3.12 summarizes the optimal methods and filters and average results for 100 images. Based on the analysis of the extracted results, it can be concluded that the PSNR index for the SHT method is higher than the other methods. On the other hand, the shearlet error index is significantly lower than other multi-resolution methods, and the SHT method has the highest error value which is 35.6. On the other hand, the SHT has the maximum L2RAT and structural indicators show the highest

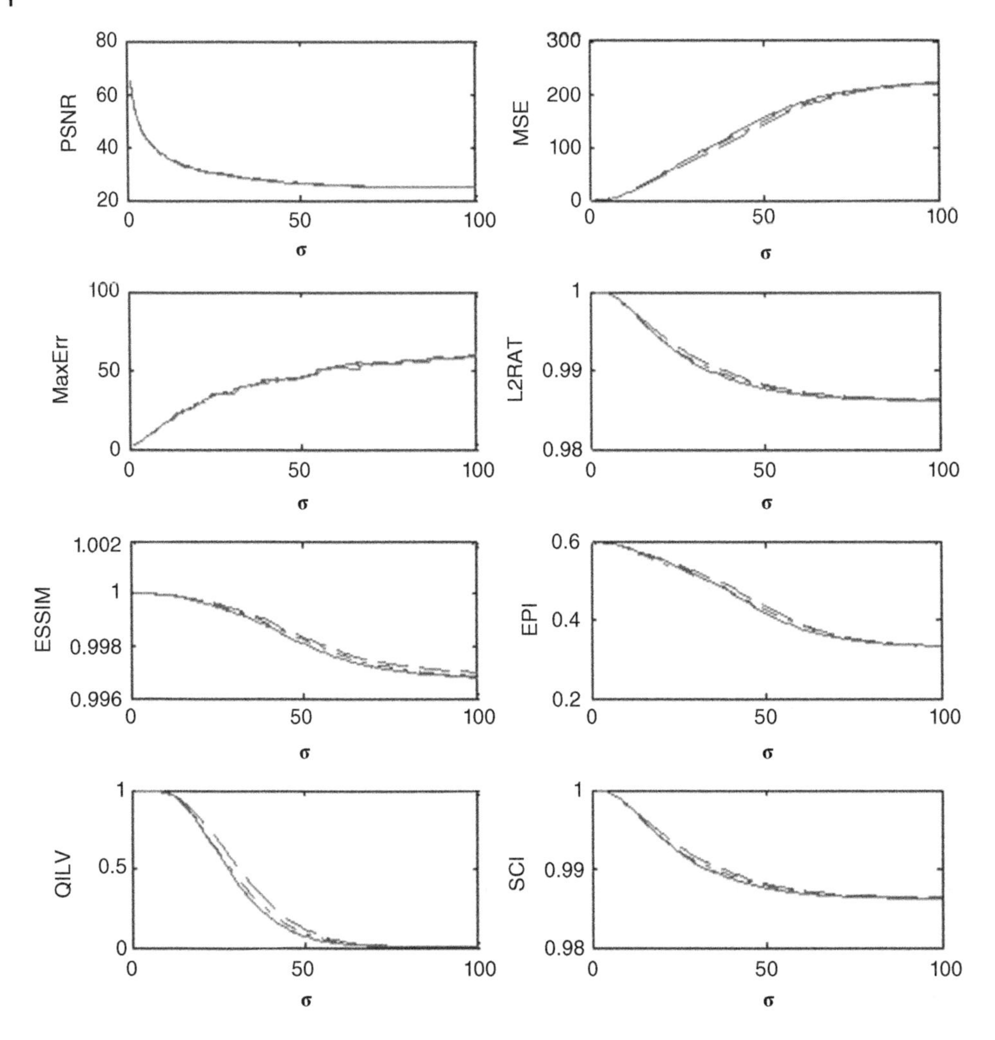

Figure 3.40 Comparison of the altered pavement image characteristics caused by changing the standard deviation of sigma noise for the first method, nsst_dec1e.

efficiency with the Shearlet method. Therefore, the SHT method, using the pyr filter, can provide a multilevel analysis with the appropriate accuracy and speed. It should be noted that this method is flexible and, by adjusting the main parameters, can achieve acceptable and practical results. Due to the possibility of accessing SHT coefficients in different directions and sections on different scales, this method can provide sufficient details and information in order to correctly assess the distress and texture details and subsequently determine the type and orientation of pavement distress.

3.5 Application of Preprocessing

3.5.1 Pavement Surface Drainage Condition Assessment

As an example, a database containing images of pavement surface after drainage was collected by an invented device (patent number "US10197486B2"). These images were taken at 2-seconds time

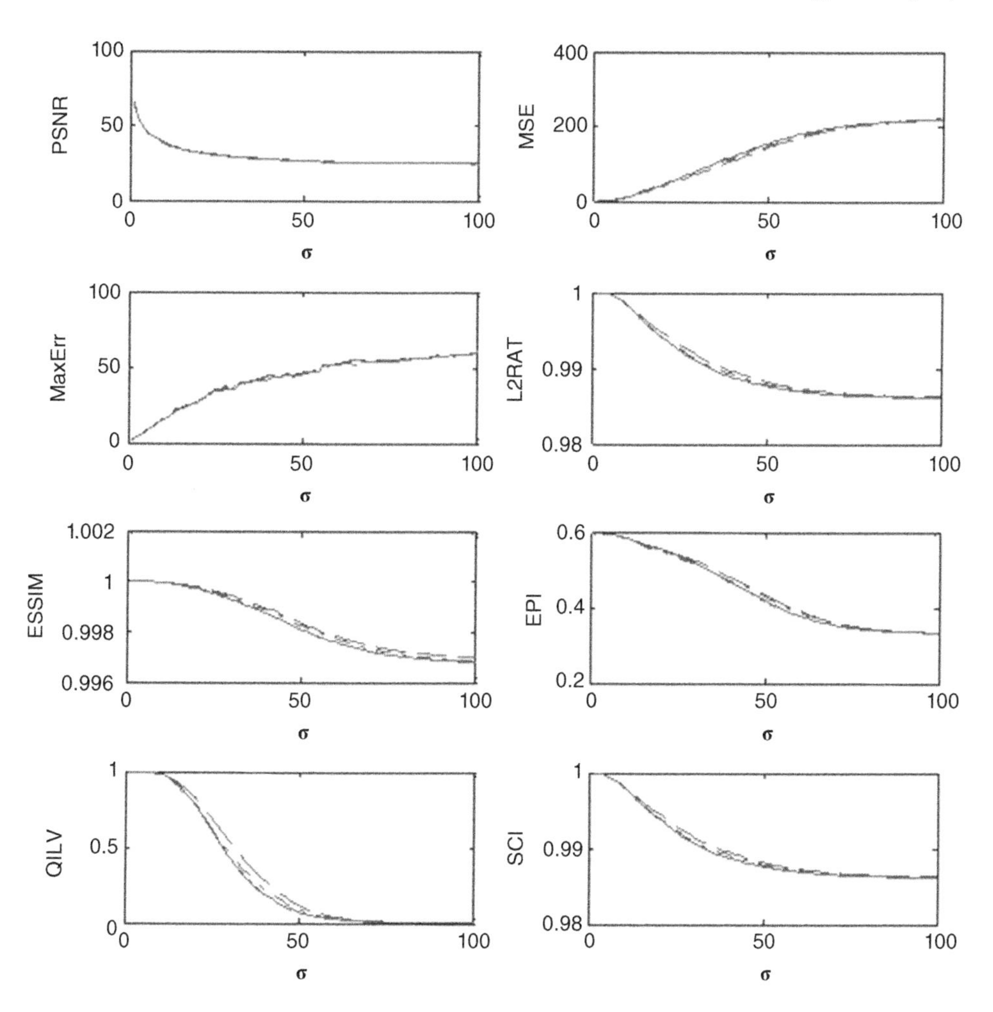

Figure 3.41 Comparison of the altered pavement image characteristics caused by changing the standard deviation of sigma noise for the second method, nsst_dec2.

intervals from the surface saturation stage (at time zero = absolute saturation) to the end of the initial drainage (no surface drainage of the pavement). To enhance these images, a white liquid was used to create a tint contrast and reduce the effect of noise in the image. Alternatively, a suitable preprocessing stage can be performed to eliminate the need for the white liquid. Figure 3.44 presents the images of the pavement surface during drainage using a device (patent number "US10197486B2") (Figure 3.4).

A modified SHT converter with advanced parameters and a new filter called "Mpyrexc" based on the pyrexc filter were developed for pavement texture analysis. This method is useful for images that are distorted or have patterns that change over time (such as pavements with changing texture of). Table 3.13 displays the modified filter coefficients and shear parameters using the SHT method, where h and g are the low-pass and high-pass decomposition filters, respectively.

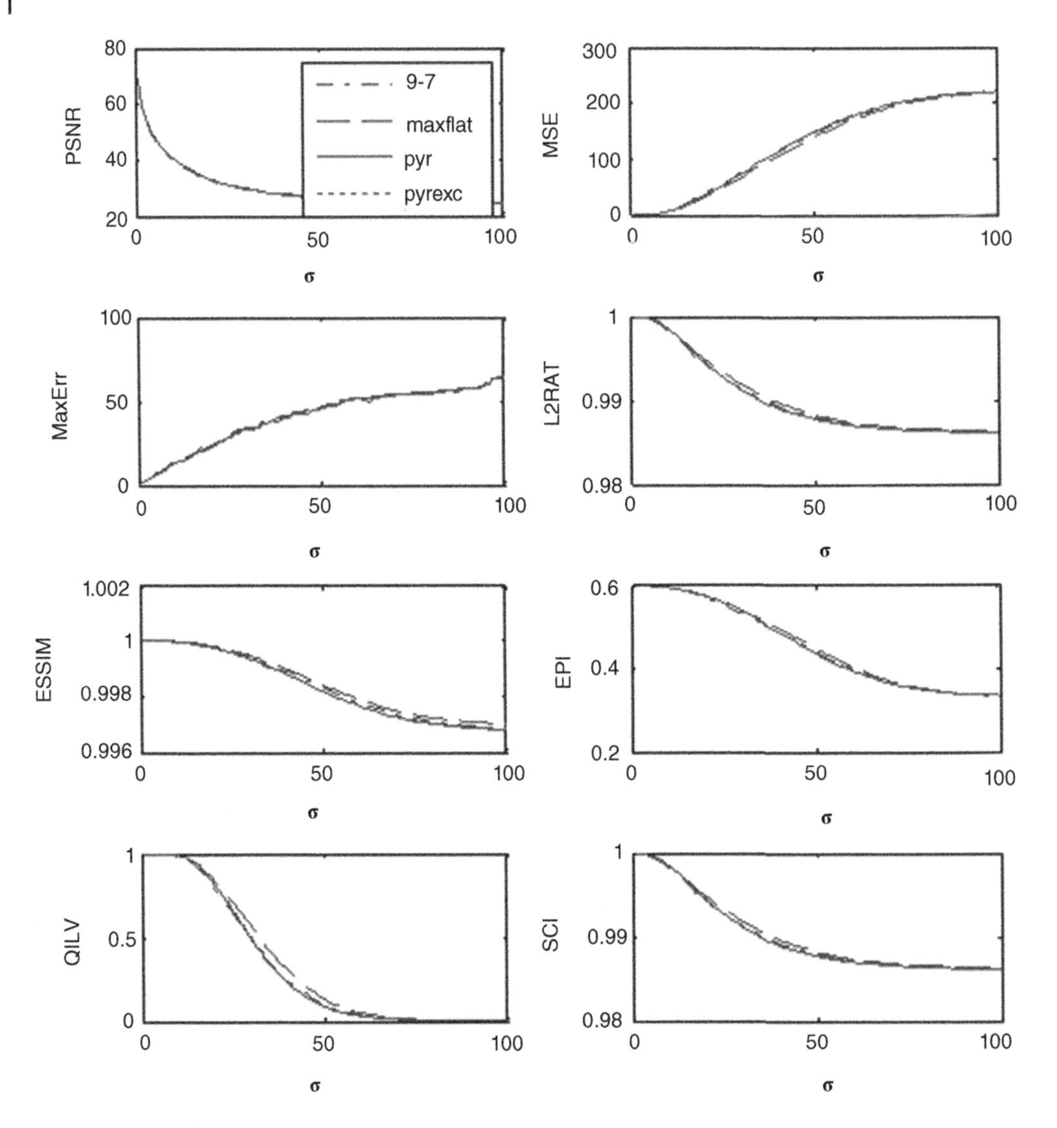

Figure 3.42 Comparison of the altered pavement image characteristics caused by changing the standard deviation of sigma noise for the third method, nsst_dec3.

Table 3.14 shows the results of the modified conversion and normal SHT conversion using different filters and key parameters for comparing and evaluating filter performance, including time, MSE, MAE, PSNR, and SNR. The results show that modified filters have better results for specific problems. Although multilevel methods with wavelet library function filters can achieve good results for solving some problems compared to older methods, innovative filters should be designed and used instead of wavelet library family filters to extract superior results. An example of these filters is shown in Figure 3.45.

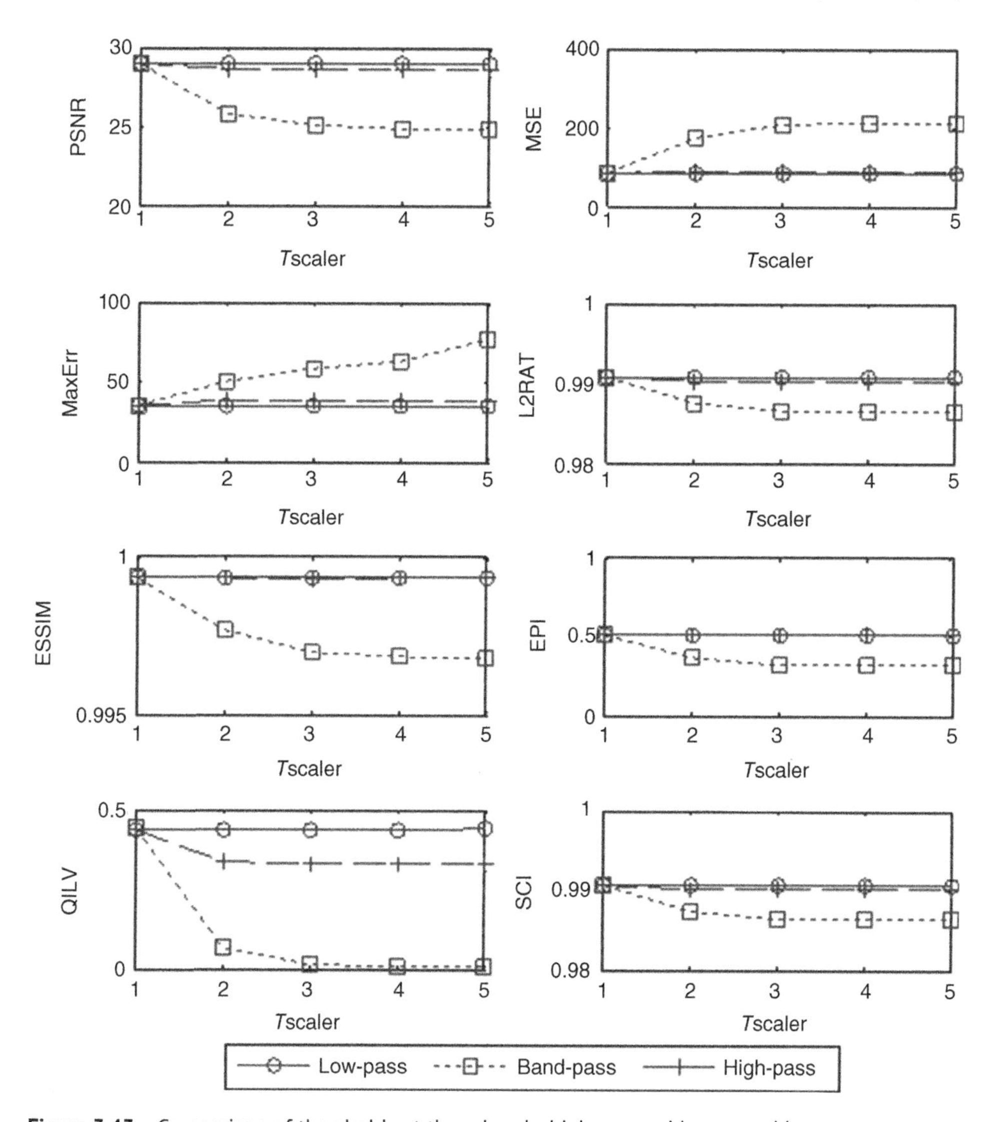

Figure 3.43 Comparison of thresholds at three bands: high-pass, mid-pass, and low-pass.

Table 3.12 General comparison of image enhancement and noise reduction methods.

Method	PSNR	MSE	MAXERR	L2RAT	ESSIM	EPI	QILV	SCI	TIME	Filter
LLF	25.389	188.017	100.400	0.985	1.000	0.153	0.028	0.985	0.013	WF1
MF	23.855	267.670	117.114	0.976	0.999	0.241	0.001	0.976	0.074	FM2
WF	27.459	116.729	45.902	0.990	1.000	0.416	0.240	0.990	0.097	WF2
WT	25.560	180.743	60.944	0.987	0.998	0.386	0.035	0.987	0.124	coif 3
RIT	27.130	125.928	44.932	0.988	0.998	0.552	0.257	0.988	0.160	demy
CT	28.243	97.443	43.499	0.991	0.999	0.497	0.149	0.991	248.000	rbio3.5
SHT	29.442	73.940	35.683	0.992	1.000	0.524	0.544	0.992	0.451	pyr

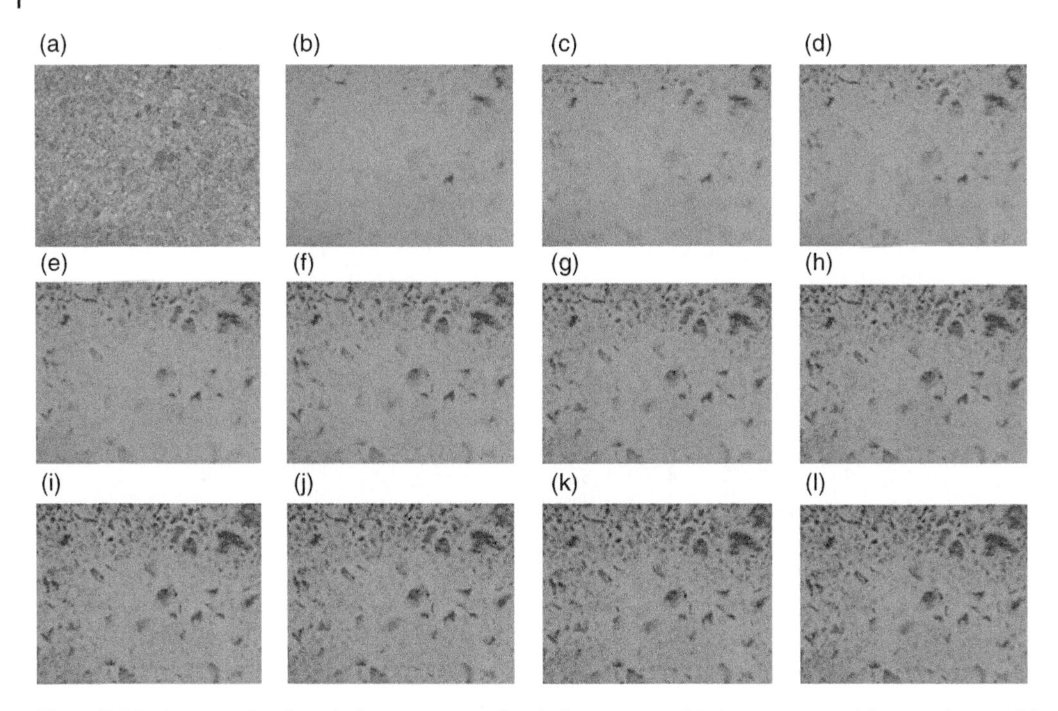

Figure 3.44 Images of surface drainage process of asphalt pavement: (a) dry pavement without moisture with saturation degree 0, and (b–l) saturated pavement with different degrees, with drainage process at 2-seconds intervals.

Table 3.13 Modified filter coefficients and shear parameters for image enhancement in pavement texture evaluation.

9Parameters		Values
Modified SHT "mpyrexc" coefficients	h0	[−0.0032 −0.0129 −0.019; −0.013 0.062 0.150; −0.0194 0.150 0.3406];
	h1	[−0.0032 −0.0129 −0.019; −0.0129 −0.0625 −0.099; −0.0194 −0.0991 0.840];
	g0	[−0.00016 −0.00100 −0.002 −0.003; −0.00100 −0.005 −0.011 −0.015; −0.0025 −0.0119 0.067 0.154; −0.0033 −0.015 0.154 0.332];
	g1	[0.00016 0.0010 0.0025 0.0033; 0.001005 −0.0012 −0.0139 −0.0234; 0.002513 −0.013 −0.067 −0.1024; 0.00335 −0.023 −0.1024 0.8486];
Shear decomposition parameter	[2 1 3 3]	
Shear resize parameter	[8 32 48 16]	

Table 3.14 General comparison of the modified conversion and normal SHT conversion using different filters.

Filter	Time (s)	Psnr (db)	SNR (db)	MSE(-14)	MAE(-27)
9-7	0.048	312.128	296.8768	4.89	3.98
Maxflat	0.064	306.777	291.5258	9.02	13.7
Pyr	0.0504	316.9911	301.7399	2.63	1.30
Pyrexc	0.0469	317.0163	301.7651	2.62	1.29
Mpyrexc	0.0222	317.6468	302.3956	2.40	1.12

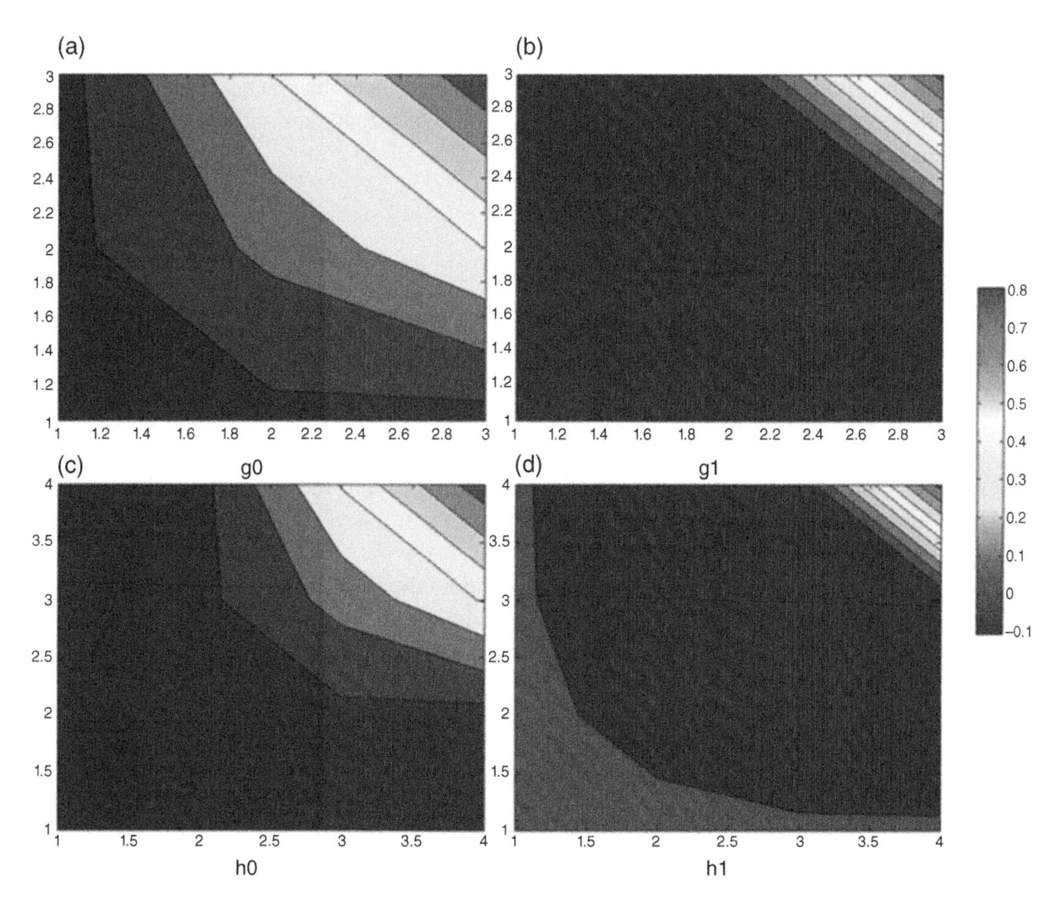

Figure 3.45 An example of innovative filters designed in the SHT method to evaluate pavement surface drainage through image processing and image enhancement.

3.6 Summary and Conclusion

This chapter describes the various methods to improve the quality of images, with the aim of eliminating noise and enhancing the image in order to increase the recognition power in the isolation and detection stage using image processing. A detailed discussion on various single-level and multilevel methods is also presented. Based on a general comparison, multilevel methods perform better than single-level methods. Among the multilevel methods, the complex SHT method was identified as the superior method in the image quality improvement stage, for which the filters and optimal analysis methods in this approach were reviewed and introduced. By comparing and evaluating the obtained results, it can be concluded that the PSNR index for the SHT method is higher than that of the other methods. On the other hand, the error index of the shearlet complex method is significantly lower than other methods and the PSNR is highest for the SHT method, which has the lowest MAXERR of 35.6. Additionally, the SHT has the maximum L2RAT and the highest efficiency based on the structural indicators. Therefore, the SHT method, using the pyr filter, can achieve the appropriate accuracy and speed for multilevel analysis. It should be noted that this method is highly flexible, and by adjusting the main parameters, can produce practical results depending on the type of problem and nature of the image. Given the possibility of accessing SHT coefficients in different directions and sections with different scales, this method provides sufficient details and information for accurate evaluation. According to the practical example of using the SHT method with an innovative filter, given at the end of the chapter, the use of special innovative filters is recommended in multilevel methods.

3.7 Questions and Exercises

1 Give an example of three imaging-based.

2 Hardware for assessing pavement condition. Prepare a database of 100 images and apply a variety of single-level filters to eliminate noise and enhance the image. Compare the evaluation indicators presented in Section 3.1 for these methods and introduce the optimal single-level method.

3 Name the types of multilevel image preprocessing methods. Implement different multilevel methods on the image database in Question 1 with different wavelet family filters. Compare the results with key indicators and determine the optimal multilevel method.

4 With the parameters from Eq. (3.13), use the Wiener filter to determine the coarse pavement texture and crack pavement failure and evaluate its performance using the indicators in Section 3.1. Can the use of a fixed index be true for all conditions? What is the alternative solution for better results?

5 What is the difference between the complex shearlet transform and shearlet transform? Analyze a two-dimensional image of the pavement using the complex shearlet transform method and then reconstruct the image. What is the reconstruction error index? What is the reason for this error? Design an error reduction algorithm that does not destroy key image information, such as pavement texture or crack damage.

6 Using Shearlet coefficient extraction algorithms, calculate the directional and shear coefficients of pavement texture in three modes: fine, medium, and large. Choose a fixed threshold and calculate the sum of the pixels greater than the threshold. Apply different wavelet filters on it and then compare the results again.

7 Explain the types of partial Hilbert transform (PHT) properties for developing Shearlet coefficient extraction algorithms. Prove each of these properties, and give an example for each on a two-dimensional wave.

8 Compare discrete linear curvature analysis (such as pavement cracking) using two methods: (a) two-dimensional discrete wavelet and (b) two-dimensional section on 10 pavement images. Then, calculate the average of the results obtained for each method.

9 Compare thresholds at three bands: high-pass, mid-pass, and low-pass with SHT. For a pavement image database, with cracks and breakdowns, calculate key indicators and plot diagrams for each of the three bands.

10 Consider real and imaginary components of the image to eliminate noise and improve image quality for a paving image bank containing 100 images (referring to the database in Q1). As mentioned in Chapter 3, Section 3.4.9, the imaginary part contains information about noise. Calculate the main Khai index to evaluate the image. By examining and filtering the imaginary part of the image and using complex propagation, remove the noise from the image and recalculate the indicators.

4

Fuzzy and Its Recent Advances

4.1 Introduction

As mentioned in Chapter 1, there are different types of road surface failures that can be visually analyzed and interpreted. The use of image processing tools generates data for analysis, which in many cases, and requires the use of fuzzy methods to remove ambiguity or quantification. This chapter introduces the concepts of various fuzzy sets that will be required in the following Chapters 6, 7, and 8. Type-2 and -3 fuzzy sets as well as their operations will be discussed in Sections 4.1.2 and 4.1.6. Since research on fuzzy set theory only emerged recently within the last 50 years, a thorough understanding is still lacking. Therefore, this book provides a summary of the basic concepts and operations that are applicable to the study of Type-1, -2, and new on -3 fuzzy sets.

We also summarize the definition of fuzzy variables and explain how this linguistic is to be used in various fuzzy rules, as a powerful method for modeling nonquantitative ambiguous descriptive and computable quantitative parameters. In infrastructure management, especially road paving, we encounter many descriptive issues, including the severity and extent of damages. The description of this type of failure is usually inaccurate, vague, and unclear. As such, these indicators themselves are ambiguous and descriptive (low, medium, and high). Quantitative modulation of these types of nonnumerical parameters requires special quantified methods. In fact, in the real world, we deal with many similar and inaccurate issues. By understanding fuzzy rules, fuzzy relations, fuzzy reasoning, and fuzzy facts, a new window opens to solve engineering problems with a degree of ambiguity for decision-making using the fuzzy method.

Fuzzy rules and fuzzy reasoning are essential components of fuzzy inference systems (FIS) that are the most important modeling elements based on fuzzy set concept and are widely used in the analysis and modeling of inaccurate and ambiguous subjects. The following is a brief overview of the types of fuzzy models followed by a discussion of the latest achievements in the development of Type-3 fuzzy modeling.

4.1.1 Type-1 Fuzzy Set Theory

For decades, researchers have offered interesting applications of fuzzy set theory in a variety of disciplines. The main purpose of this section is to provide an introduction to Type 1 fuzzy topics and does not intend to address its theoretical foundations. You can refer to the reference to read more about Type 1 fuzzy sets.

Fuzzy sets are an efficient tool for quantitative modeling of words or sentences in natural or artificial language that quantitatively describes different designs of fuzzy reasoning by interpreting fuzzy rules as fuzzy relations. The fuzzy inference method is based on the concept of the combined

Automation and Computational Intelligence for Road Maintenance and Management: Advances and Applications, First Edition. Hamzeh Zakeri, Fereidoon Moghadas Nejad, and Amir H. Gandomi.

law of inference to draw conclusions from a set of fuzzy laws and known facts. Fuzzy rules and fuzzy reasoning are essential components of FIS and are the most important part of ambiguous problem analysis.

In general, the basic structure of a Type 1 FIS consists of three keys: the "rule of thumb or rule base," which includes a bank of fuzzy rules. A "database" (or glossary), which defines the membership functions (MFs) (simply and geometrically) used in fuzzy rules, and a "reasoning mechanism," which models the method of inference. This model leads to a logical result (quantitative and numerical) based on the rules created from the data. It can be said that a FIS is an operator that converts from qualitative input space to quantitative output space.

A fuzzy set is the degree to which a member belongs to a set. According to this definition, the characteristic function of a fuzzy set can be values between 0 and 1, which indicates the degree of membership of an element in a set [1].

The "fuzzification" is a general concept of fuzzy set theory that enables a comprehensive method for developing exact memberships in fuzzy mathematical expressions. This method converts the mapping of the function f to fuzzy domain.

The "if–then" rule takes the form if x is A, then y is B, where A and B are qualitative linguistic values defined by fuzzy sets in the X and Y discourse world. Frequently "x is A" is called "antecedent" and "y is B" is called the "consequence." There are many examples of ambiguous if–then rules in pavement engineering and road management, for instance:

- If maximum radon transform (RT) is high, then pavement distress severity is extensive [2].
- If entropy of polypropylene fibers (PPF's) image is high, then absorption rate of asphalt is high [3].
- If the number of peaks of RT is high, then pavement distress is extensive [2].
- The temperature is high, then rutting the asphalt pavement is high.
- If the PPF's additive is less, then durability of asphalt mixture is less [3].
- If the friction of the pavement surface is high, then the hydroplaning is less [4, 5].

The most important fuzzy inference methods that are widely used in solving Type 1 fuzzy problems are the following: The Mamdani, Sugeno, and Tsukamoto (TSK). These methods have different results due to the different structure of the FIS and consequently their fuzzy rules, and the final answer is sometimes different [6, 7].

Ambiguity is a very important characteristic feature of data. Today, most decision-making procedures involve retrieving and evaluating evidence, most of which are inaccurate, incomplete, noisy, and complex. As an efficient tool, Type 1 fuzzy sets (T1 FS) have been used for a long time in various disciplines such as qualitative subject analysis [2], pattern recognition [2, 8, 9], medical issues, extensive engineering problems, behavioral, classification, multicriteria decision-making [10–16], fuzzy filtering [14], fuzzy cluster, fuzzy segmentation, and thresholding [17, 18], and has shown good results [10–13, 19–21], etc. For more information about the first type of fuzzy, refer to the Refs. [1, 22–24].

4.1.2 Type-2 Fuzzy Set Theory

The T1 FS method has been used to solve some problems, but its use is exclusively recommended for problems with a degree of ambiguity. The T1 FSs have good application for many cases; however, it is not entirely satisfactory in more uncertain environments [9]. T1 FSs give a single value from [0, 1] to each member of the primary domain as a membership value, but membership of an object to the universe of discourse is uncertain itself. This property has led the researchers to propose interval

Type-2 fuzzy sets (IT2 FSs), whose membership functions are an interval instead of a single value. Similar to Type-1 fuzzy systems, Type-2 fuzzy logic systems consist of a set of rules in the knowledge bank. In most applications, the knowledge used to make these rules is ambiguous and basically requires a way to resolve this uncertainty. The existence of such uncertainty leads to the production of laws whose consequences are uncertain and need to be clarified in some way to the final result. In Type-2 fuzzy systems, at least one or both of the "if–then" membership functions are of Type-2 fuzzy.

IT2 FSs are a unique form of general Type-2 fuzzy sets (GT2 FSs), that secondary membership values for all members of the primary domain are 1 [9, 25]. This property reduces computational efforts needed for the analysis of GT2 FSs, while remarkably enhancing imprecision capabilities of traditional T1 FSs (see Figure 4.2).

Most fuzzy methods form a base for the concept of T1 FSs' need to complete a process called defuzzification [7, 25–27] in order to mix environment uncertainty and to give a crisp value, which denotes the entire fuzziness of the environment. In contrast, T2 FSs need an additional step for representing a single value as the characteristic of the vagueness. This step is called type-reduction, which converts a T2 FS into its T1 counterpart. Type-reduction is a necessary step in different computational intelligence fields, including fuzzy systems, clustering, and classification [25, 28].

The general framework of T2 FS is shown in Figure 4.1, which is similar T1FLS. It can be seen that the main difference between T1 FS and T2 FS is the defuzzification step.

4.1.3 α-Plane Representation of General Type-2 Fuzzy Sets

By definition, an α-plane of a GT2 FS $\tilde{A}$ is the union of the whole primary memberships of $\tilde{A}$ whose secondary marks are greater than or equal to $\alpha(0 \le \alpha \le 1)$. α-Plane of $\tilde{A}$ is symbolized by $\tilde{A}_\alpha$ [9]

$$A_\alpha = \int_{\forall x \in X} \int_{\forall u \in J_x} \{(x, u) \mid f_x(u) \ge \alpha\} \tag{4.1}$$

An α-cut of the secondary membership functions (SMFs) is represented by $S_A(x \mid \alpha)$:

$$S_A(x \mid \alpha) = [S_L(x \mid \alpha), S_R(x \mid \alpha)] \tag{4.2}$$

Consequently, $\tilde{A}_\alpha$ can be a formation of all α-cuts for all its SMFs:

$$A_\alpha = \int_{\forall x \in X} S_A(x \mid \alpha)/x = \int_{\forall x \in X} \left(\int_{\forall u \in [S_L(x|\alpha), S_R(x|\alpha)]} u \right) /x \tag{4.3}$$

By means of α-planes, footprint of uncertainty (FOU) is defined. The FOU is equal with the lowest α-plane i.e.

$$\text{FOU}(A) = A_0 \tag{4.4}$$

Individually, α-plane is limited from Eq. (4.4) by its upper membership function, $\bar{\mu}_A(x \mid \alpha)$, and lower membership function, $\underline{\mu}_A(x \mid \alpha)$. The upper and lower membership functions [9] of a plane $\tilde{A}_\alpha$ can be defined in relation of α-cuts as follows:

$$\begin{aligned} \bar{\mu}_A(x \mid \alpha) &= \int_{\forall x \in X} S_R(x \mid \alpha) \\ \underline{\mu}_A(x \mid \alpha) &= \int_{\forall x \in X} S_L(x \mid \alpha) \end{aligned} \tag{4.5}$$

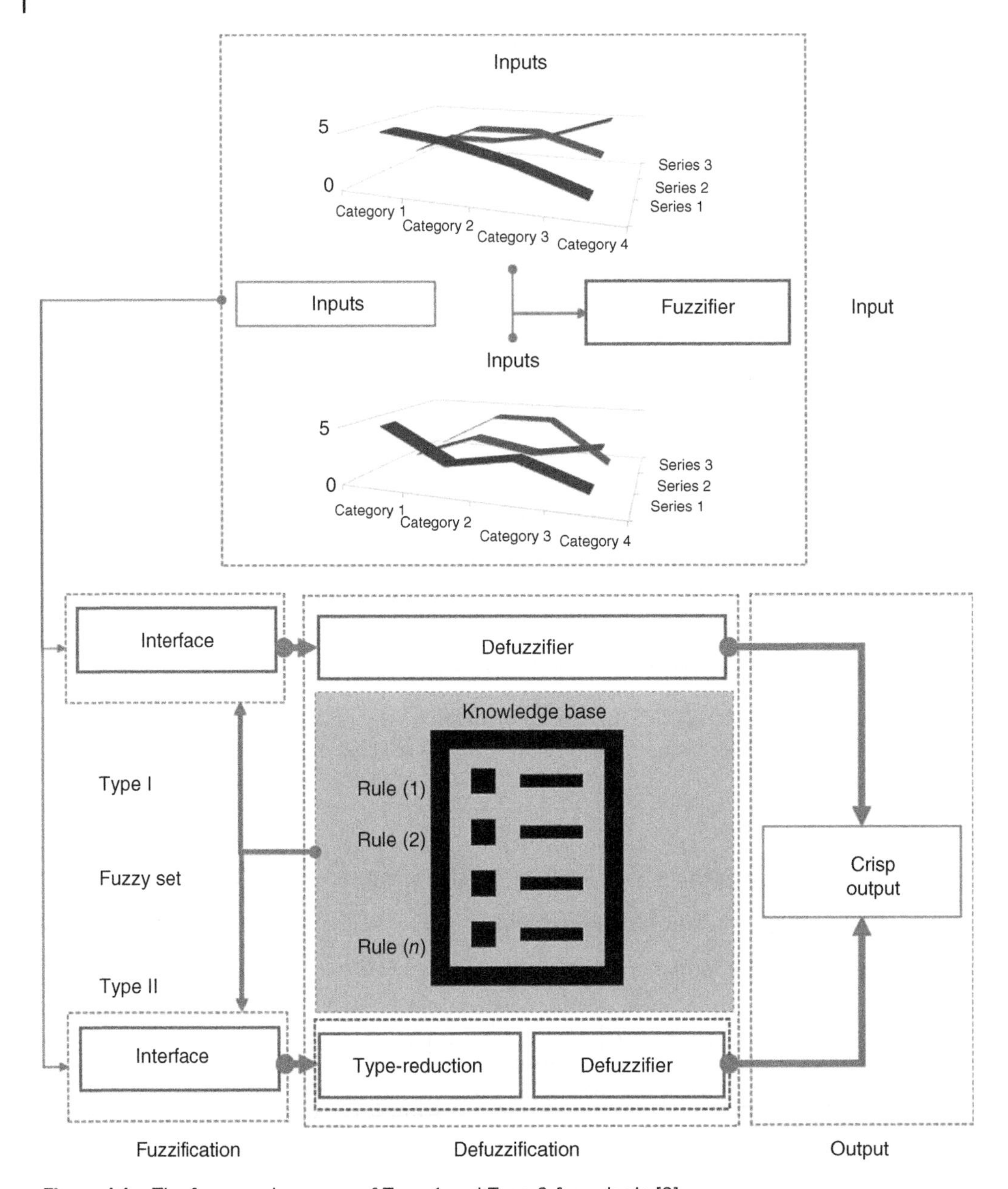

Figure 4.1 The framework process of Type-1 and Type-2 fuzzy logic [9].

Each α-plane is an interval T2 FS with the centroid $C_{\tilde{A}}(x)$. The centroid of a GT2 FS $\tilde{A}$, $C_{\tilde{A}}(x)$, is an arrangement of all its α-planes, i.e.:

$$C_A(x) = \bigcup_{\alpha \in [0,1]} \alpha / C_{A\alpha}(x) \tag{4.6}$$

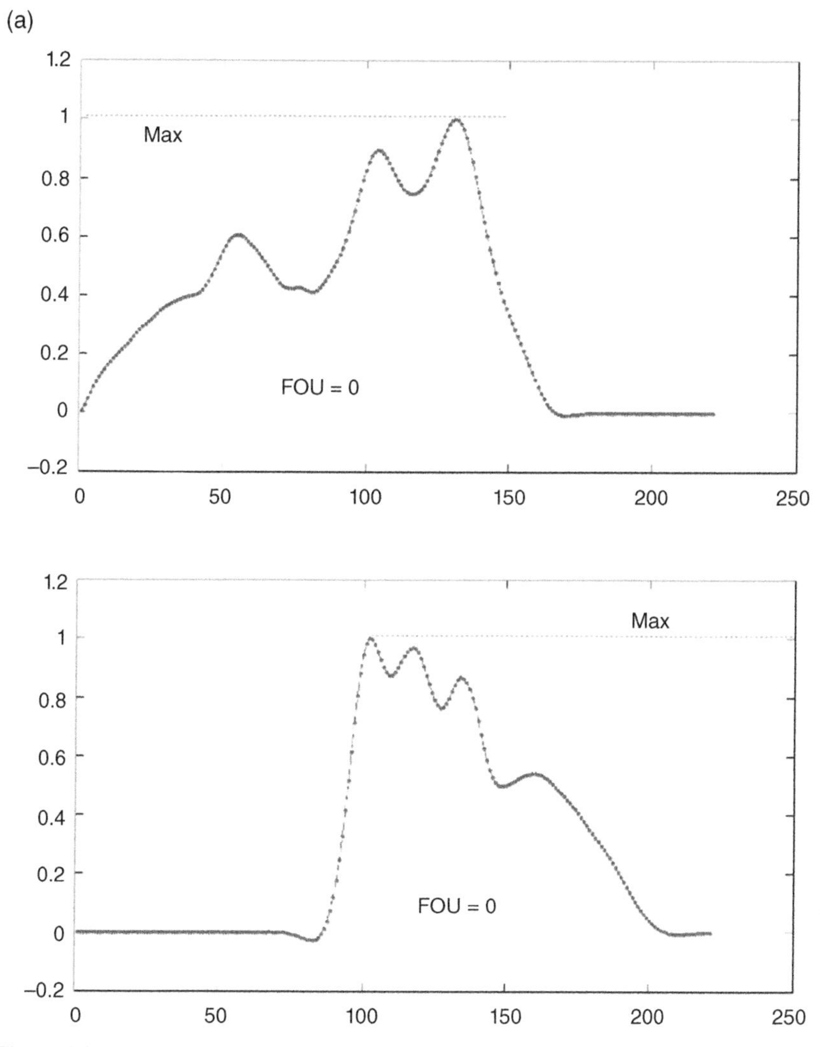

Figure 4.2 The secondary MF of (a) Type-1 fuzzy set, (b) interval Type-2 fuzzy set, and (c) general Type-2 fuzzy set.

Each α-plane, $C_{\tilde{A}\alpha}(x)$, has a lower and an upper bound so that centroid of a GT2 FS$\tilde{A}$ can be defined as

$$C_A(x) = \bigcup_{\alpha \in [0,1]} \alpha / [c_l(A \mid \alpha), c_r(A \mid \alpha)] \tag{4.7}$$

4.1.4 Type-Reduction

Creation of T1 FS from T2 FS and nonfuzzy numerical foundation from T1 FSs is completed with the type reduction (TR) operator [6, 9, 25, 29]. Type-reduction is further recognized by the acronym (COS) in the fuzzy set. This model is defined as follows:

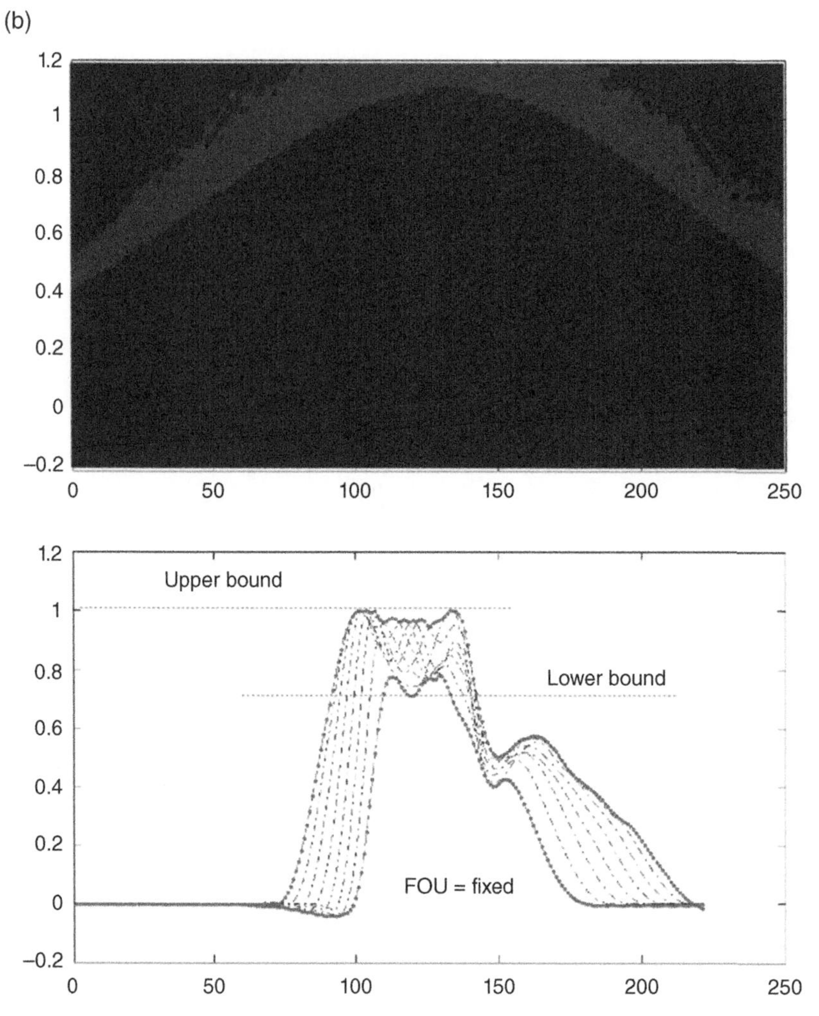

Figure 4.2 (Continued)

$$Y_{\cos}(x) = [y_l, y_r] = \int_{y^l \in [y_l^1, y_r^1]} \cdots \int_{y^l \in [y_l^M, y_r^M]} \int_{f^l \in [f_l^1, f_r^1]} \cdots \int_{f^m \in [f^m, f^{-m}]} 1 \Big/ \left(\frac{\sum_{i=1}^{M} f^i y^i}{\sum_{i=1}^{M} f^i} \right) \tag{4.8}$$

$$C_{\tilde{G}^i} = \int_{\theta_1 \in J_{y_1}} \cdots \int_{\theta_N \in J_{y_N}} 1 \Big/ \left(\frac{\sum_{i=1}^{N} y^i \theta^i}{\sum_{i=1}^{N} \theta^i} \right) = [y_l^i, y_r^i] \tag{4.9}$$

$$y_l = \frac{\sum_{i=1}^{M} f_l^i y_l^i}{\sum_{i=1}^{M} f_l^i}, y_r = \frac{\sum_{i=1}^{M} f_r^i y_l^i}{\sum_{i=1}^{M} f_r^i} \tag{4.10}$$

This method [9] is modeled using two endpoints, the center of the distance y_l and y_r corresponds to the center of the result of the set, T2 FSs, and is equivalent to $C_{\tilde{G}^i}$.

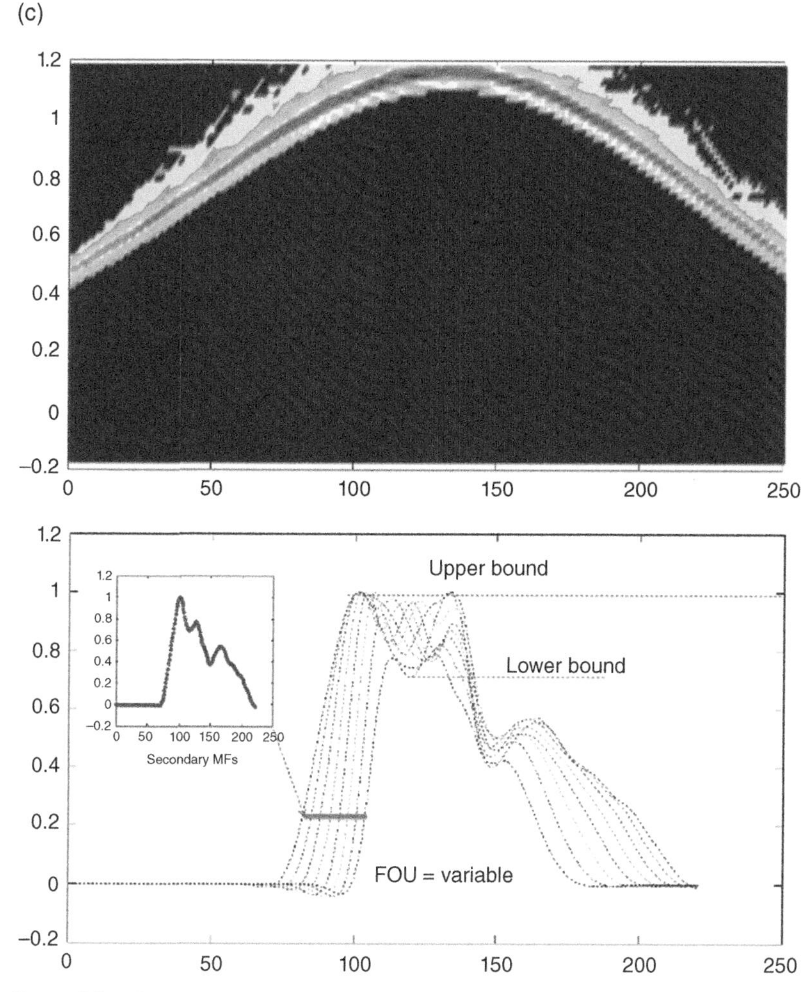

Figure 4.2 (Continued)

4.1.5 Defuzzification

According to the definition, defuzzification is the process of converting fuzzy sets to a definite number. This step works exactly the opposite of the fuzzy method.

There are two methods or steps for this part, including defuzzification to a point (DP) and multiplier defuzzification to a set (DS). The DS is generally used in the TR phase. DS criteria include a Kernel function, proper transfer, and proper order for a fuzzy set. The criteria for DP are core selection criteria, fixed dimensions, uniformity, *t*-conorm criteria, transmission, scaling, and continuity. The DP is divided into three categories: maximum methods defuzzification (MMD), distributed methods defuzzification (DMD), and area methods defuzzification (AMD).

In MMD, an element from the core of the fuzzy set is selected as the fuzzy generator function, the simplicity of the method being its most important feature. The MMD method performs the opacification operation by converting the MF to a probabilistic distribution and calculating the expected value. The most important feature of this method is the application of connectivity. In the AMD method, the amount of inertia is determined by dividing the regions containing the MF into equal pieces. For the MMD model, two submodels are introduced:

1) Random choice of maxima (RCM):

$$\text{Prob}(D(A) = x_0) = \begin{cases} \dfrac{1}{|\text{core}(A)|} x_0 \epsilon \text{core}(A) \\ 0 \text{ otherwise} \end{cases} \tag{4.11}$$

2) First of maxima, last of maxima, middle of maxima (FLMM):

$$\begin{cases} \text{FOM}(A) = \text{minimum core}(A) \\ \text{LOM}(A) = \text{maximum core}(A) \end{cases} \tag{4.12}$$

There are also six methods for the defuzzification model:

1) Generalized-level set defuzzification (GLSD):

$$\text{GLSD}(A, \gamma) = \frac{\sum_{i=1}^{N} c_i m_i \gamma^i}{\sum_{i=1}^{N} c_i \gamma^i} \tag{4.13}$$

In the model, $\gamma \epsilon (0, \infty)$ is the system reliability coefficient, and N is the α-cut number. In this equation, $c_i = | Aaj |$, and $m_i$is the mean α-cut for i.

2) Indexed center of gravity (ICG):

In this equation, α is used as a threshold. If it is below the value of α, it is 0; otherwise, it is considered equal to MF itself:

$$\text{ICOG}(A) = \frac{\sum_{\gamma \epsilon A_\alpha} x.A(x)}{\sum_{\gamma \epsilon A_\alpha} A(x)} \tag{4.14}$$

3) Fuzzy mean (combines aggregation and defuzzification) (CAD):

$$\text{FM}(A) = \frac{\sum_{i=1}^{N_A} \alpha_i . a_i}{\sum_{i=1}^{N_A} \alpha_i} \tag{4.15}$$

Here, N_A is the output fuzzy set number, α_i is the degree of inference law, and a_i is the numerical value for the fuzzy output set (α_i = MOM (A_i)).

4) Centre of gravity (COG):

$$\text{ICOG}(A) = \frac{\sum_{X \min}^{X \max} x.A(x)}{\sum_{X \min}^{X \max} A(x)} \tag{4.16}$$

5) Mean of maxima (MEOM):

$$\text{MEOM}(A) = \frac{\sum_{x\text{core}(A)} X}{|\text{core}(A)|} \tag{4.17}$$

6) Basic defuzzification distribution (BADD):

$$\text{BADD}(A, \gamma) = \sum_{X \min}^{X \max} x.A \tag{4.18}$$

In this relation, $\gamma \epsilon$ [0, ∞) is adjusted by the problem conditions. For AMD, the TR relation is calculated using the following equation:

$$\left| \sum_{X = \inf(x)}^{\mathrm{COA}(A)} A(X) - \sum_{X = \mathrm{COA}(A)}^{\sup(X)} A(X) \right| \tag{4.19}$$

where COA(A) is the minimum value.

The α-cut and α-mean means have been introduced for the DS model instrument. Based on the previous step, which decreases the rank, the set $Y\cos(x)$ is obtained. The average y_l and y_r are used for the purpose of defuzzification. The output of IT2 FLS can be estimated based on the following simple equation:

$$y(x) = \frac{y_l + y_r}{2}$$

$$C = \int_{(\rho,\theta)_1 \epsilon J_{x_1}} \cdots \int_{(\rho,\theta)_N \epsilon J_{x_N}} \left[f_{(\alpha,\vartheta)_1}((\rho,\theta)_1) \widetilde{\ } \cdots \widetilde{\ } f_{(\alpha,\vartheta)_N}((\rho,\theta)_N) \right] / \left(\left(\sum_{i=1}^{N} (\alpha,\vartheta)_i (\rho,\theta)_i \right) / \left(\sum_{i=1}^{N} (\rho,\theta)_i \right) \right) \tag{4.20}$$

4.1.6 Type-3 Fuzzy Logic Sets

Very recently, notable developments have been made from Type 1 fuzzy logic systems (T1FLSs) to generalized Type-2 fuzzy logic sets (GT2FLSs). Particularly, many researches have focused on the concept of Type-2 fuzzy sets (T2 FSs) and its components. The general Type-2 fuzzy set (GT2FS) is a T2 FS [9] with Type-1-blur weight or a secondary membership value between [0, 1].

Because of the high complex structure and computational time of general Type-2 FLSs, it is considered as an unbefitting method for real-world applications. The general Type-2 fuzzy sets are more detailed than interval Type-2 sets and T1FLS because of their ability to model vagueness based on the FOU. In addition, the secondary grades in GT2FLS are fuzzy themselves and transfer more information than the IT2 FS; the main structure of GT2FLS is defined by its FOU, lower bound membership function (LMF), and upper bound membership function (UMF). The real-world applications of GT2FLS are a challenging issue especially with the increase in the number of rules. One of the applications of this method is the use of the fuzzy method in resolving the ambiguity of road surface failures, including the detection of crack edges and potholes. All operations can also be used in GT2FS but for each of their α-level T2 FS.

In order to moderate the complexity of GT2FL, several tricks have been proposed, such as α-planes and z-slices. The automatic generation of fuzzy membership function (FMF) is too precious to handle the uncertainties, truly. Many procedures based on Type 2 fuzzy membership function (T2FMF) have been recommended; however, automatic T2 FMF generation approaches have been seldom discussed in the literature and are a relatively infrequent issue [9, 30, 31].

This chapter presents a different method to automatically produce a general Type-2 fuzzy logic system in the frame of polar coordinates. The main purpose of this chapter is to make automatically produce GT2FLS and represent the uncertainty of the vertical-slice and horizontal-slice based on a 3D FOU in the polar domain [9, 19].

The rest of this chapter is organized as follows: Section 4.2.1 provides the background of general Type-2 fuzzy sets. Section 4.3 describes the automatic membership function generators and three important methods. Section 4.2 presents ambiguity modeling in the fuzzy methods and background of general Type-2 fuzzy sets in Section 4.2.1. In Section 4.3, the theory of automatic methods for MF generation is presented. Section 4.4 shows steps and components of general 3D Type-2 fuzzy logic systems (G3DT2 FL). Section 4.7 presents application of G3DT2FLS in pattern recognition. Section 4.2.1 presents some applications of G3DT2FLS. Final section discusses some important issues about the complexity of general 3D Type-2 polar fuzzy logic systems (G3DT2PFLS) and provides the conclusion.

4.2 Ambiguity Modeling in the Fuzzy Methods

4.2.1 Background of General Type-2 Fuzzy Sets

Conditional on the level of uncertainty, FLs can be T1FLS, interval Type-2 fuzzy logic systems (IT2LFs), or GT2FLS, which needs to be handled.

T1FLS sets A, denoted by $\mu_A(x)$, where $x \in X$, is represented by FS $A = \{x, \mu_A(x) \vee x \in X\}$, where $\mu_A(x, u) \in [0, 1]$. In this instance, Figure 4.1a illustrations an example T1 FSs. The defuzzification procedure can be completed using the centroid, center-of-sums, or heights. In IT2LFs, Ã is defined by $\underline{\mu_{\hat{A}}(x)}$ and $\overline{\mu_{\hat{A}}(x)}$, as represented by the lower and upper MFs of $\mu_{\hat{A}}(x)$, and is represented by $\hat{A} = \int_{\forall x^l \in \left[\underline{\mu_{\hat{A}}(x)}, \overline{\mu_{\hat{A}}(x)}\right]} \left(\frac{1}{x^l}\right)$, where $x \in X$, l is the xth antecedent and lth rule. Figure 4.1b exemplifies an instance of IT2FS. The rules for IT2LFs are matching to those of T1FLS, and the type-reducer of an IT2LFs can be used in many approaches. In conclusion, the defuzzification process is finalized based on the mean rule to find a crisp value.

A GT2FS involves several units: fuzzifier, inference (rules), type-reducer, and defuzzifier. In GT2FS, $\widetilde{A}$ is subject to alternatives with Cartesian product space $X \times [0, 1]$ and is defined by $\mu_{\hat{A}}(x), \hat{A}: X \times [0, 1] \rightarrow [0, 1]$. The main variable $x \in X$ can be showed by $\mu_{\hat{A}}(x)$ and is classically stated as follows:

$$A = \int_{\forall x \in X} \frac{\mu_A(x)}{x} \tag{4.21}$$

$$\mu_{\tilde{A}}(x) = \mu_{\tilde{A}}(u \mid x) = \int_{\forall u \in U} f_x(u) \tag{4.22}$$

where $u \in U = [0, 1]$ is the secondary variable; $\mu_{\hat{A}}(x)$ is the secondary MF at x, which is a T1FS; and $f_x(u)$ is the secondary membership grade at (x, u).

Instances of secondary MF of $\widetilde{A}$ are shown in Figure 4.2a. GT2FS Ã defines a universe of discourse X using MF $\mu_{\tilde{A}}(x', u)$, where $x \epsilon X$and $u \in J_x$:

$$A = \int_{x \in X} \int_{u \in J_x} \mu_A(x', u)/(x, u), J_x \subseteq [0, 1] \tag{4.23}$$

where the parameters of set $\{x, u\}$ are primary and secondary variables; and J_x represents the primary membership function of x.

The operator $\int .$ denotes union over all possible values of x and u, and $\mu_{\tilde{A}}(x, u) \in [0, 1]$.

Mendel et al. newly presented a new standard description for $u \in J_x$ in terms of the set $\left\{\left(u \subseteq [0, 1] \mid \mu_{\tilde{A}}(x, u) > 0\right)\right\}$ using new illustration methods, such as vertical slice, wavy slice, z-slices, and α-planes, and more revisions are being made on the theoretical facets of GT2FS. The vertical slice can be considered as $x = x'$, $\mu_{\tilde{A}}(x', u)$ for $x' \in X$ and $\forall u \in J_{x'} \subseteq [0, 1]$:

$$\mu_{\widetilde{A}}(x', u) = \mu_{\widetilde{A}}(x') = \int_{u \in J_x} f_{x'}(u)/(u), J_{x'} \subseteq [0, 1] \tag{4.24}$$

Here, $f_{x'}(u)$ represents the amplitude of the secondary MF and $f_{x'}(u) \in [0, 1]$. Assuming that the domain of x is discretized using N samples, GT2FS $\widetilde{A}$ can be denoted as a configuration of all its vertical slices:

$$\widetilde{A} = \sum_{i=1}^{N} \left[\left(\int_{u \in J_x} f_{x'}(u)/(u), \right) \right] / x_i \tag{4.25}$$

The wavy slice representation can be defined by embedded FS, $\widetilde{A}_e$, which is defined in a separate universe of discourse with N elements, $\widetilde{A}_e$ as follows:

$$\widetilde{A}_e = \sum_{i=1}^{N} [f_{x'}(\lambda_i)/(\lambda_i)]/x_i \tag{4.26}$$

where $f_{x'}(\lambda_i)$ $\lambda_i \in J_x \subseteq U = [0, 1]$. The $f_{x'}(\lambda_i)$ has N elements, as it contains one element from λ_1, λ_2,..., λ_i, namely, u_1, u_2, ..., u_i, each with secondary grades, namely $f_{x'}(\lambda_1)$, $f_{x'}(\lambda_2)$, ...$f_{x'}(\lambda_i)$ [9].

The GT2FS $\widetilde{A}$ is defined as a union of all of its n surrounded T2 FS, and denoted as follows:

$$\widetilde{A} = C_{j=1}^{n} \left[\sum_{i=1}^{N} [f_{x'}(\lambda_i)/(\lambda_i)]/x_i \right]_j \tag{4.27}$$

IT2FSs are a different form of GT2FSs, whose secondary membership value for the whole values of the primary domain and primary membership space stays 1. Hence, the third dimension of IT2FSs discloses no extra data and is discounted. This property has made such FSs prevalent since they need less computational efforts and have lower computational difficulty than GT2FSs. Based on (4.21), an IT2 FS can be specified as (4.28):

$$\widetilde{A} = \int_{x \in X} \int_{u \in J_x} 1/(x, u), J_x \subseteq [0, 1] \tag{4.28}$$

An α-plane $\widetilde{A}_\alpha$ of GT2FLs $\widetilde{A}$ can be represented as the union of all primary memberships of $\widetilde{A}$ with $\mu_{\widetilde{A}}(x', u) = \mu_{\widetilde{A}} \geq \alpha$ and can be stated as follows:

$$\widetilde{A}_\alpha = \int_{\forall x \in X \forall} \int_{u \in J_x} \{(x, u) \mid f_x(u) \geq \alpha\}, \alpha \subseteq [0, 1] \tag{4.29}$$

An α – cut $\widetilde{A}_\alpha$ of secondary memberships of $\widetilde{A}$ with $\mu_{\widetilde{A}}(x) = \mu_{\widetilde{A}} \geq \alpha$ can be denoted as $S_{\widetilde{A}}(x \mid \alpha)$ and defined as follows:

$$S_A(x \mid \alpha) = \left[S_A^l(x \mid \alpha), S_A^r(x \mid \alpha) \right] \tag{4.30}$$

Hence, α – plane is a composition of all α – cut of all SMFs and is stated as follows:

$$\widetilde{A}_\alpha = \int_{\forall x \in X} S_A(x \mid \alpha) = \int_{\forall x \in X} \left(\int_{\forall u \in \left[S_A^l(x|\alpha), S_A^r(x|\alpha) \right]} u, \right) / x, \quad \alpha \subseteq [0, 1] \tag{4.31}$$

Based on $\widetilde{A}_\alpha$, the FOU can be stated as follows:

$$\mathrm{FOU}\left(\widetilde{A}\right) = \widetilde{A}_{(\alpha = 0)} = \int_{\forall x \in X} S_A(x \mid (\alpha = 0)) = \int_{\forall x \in X} \left(\int_{\forall u \in \left[S_A^l(x|\alpha = 0), S_A^r(x|\alpha = 0) \right]} u, \right) / x, \alpha \subseteq [0, 1] \tag{4.32}$$

The edges of each slice can be represented as follows:

$$\begin{aligned}\mu_{\widetilde{A}}(x \mid \alpha)_l &= \mu_{\widetilde{A}}(c')_l = \int_{\forall x \in X} S_A^l(x \mid \alpha), \alpha \subseteq [0,1] \\ \mu_{\widetilde{A}}(x \mid \alpha)_r &= \mu_{\widetilde{A}}(c')_r = \int_{\forall x \in X} S_A^r(x \mid \alpha), \alpha \subseteq [0,1]\end{aligned} \tag{4.33}$$

The upper and lower MFs $\left\{\mu_{\widetilde{A}}(x \mid \alpha)_l, \mu_{\widetilde{A}}(x \mid \alpha)_r\right\}$ are enclosed by $\alpha -$ plane. A special IT2LFs ($\alpha -$ level T2FS) is designed by increasing the $\widetilde{A}_\alpha$ to the level α:

$$\begin{aligned}&L_A(x', u) = \alpha/\widetilde{A}_\alpha = \alpha/\int_{\forall x \in X} S_A(x \mid \alpha) = \alpha/\left(\int_{\forall x \in X}\left(\int_{\forall u \in \left[S_A^l(x|\alpha), S_A^r(x|\alpha)\right]} u,\right)/x\right) \\ &\alpha \subseteq [0,1] \\ &\forall x \in X, \forall u \in J_x\end{aligned} \tag{4.34}$$

Then, GT2 FLS $\widetilde{A}$ can be formed as a summation of all of its discrete $\alpha -$ levelT2FSs:

$$\widetilde{A} = \underset{\alpha \in [0,1]}{\mathrm{C}} \alpha/\widetilde{A}_\alpha = \left(\int_{\forall x \in X} S_{\widetilde{A}}(x \mid \alpha) = \int_{\forall x \in X}\left(\int_{\forall u \in \left[S_A^l(x|\alpha), S_A^r(x|\alpha)\right]} u,\right)/x\right) \tag{4.35}$$

where $\int .$ denotes the union operation and computes the $\max\left(\mu_{\widetilde{A}}(x \mid \alpha)_l\right) \wedge \max\left(\mu_{\widetilde{A}}(x \mid \alpha)_r\right)$ for $\alpha -$ planes.

The *z*-slices are a comparable technique to α-planes as a depiction procedure of GT2 FLSs. Using this idea, Wagner and Hagras established the theoretical structure of *z*-slice created on GT2 FLSs to demonstrate its characteristics. An illustration of GT2 FLS is illustrated in Figure 4.3a,b.

$$\widetilde{A}_z = \int_{\forall x \in X}\int_{\forall y \in [0,1]} z/(x, y), \alpha \subseteq [0,1] \tag{4.36}$$

The union of *z*-slices can be seen in Figure 4.2 and is defined as follows:

$$\widetilde{A} = \underset{\alpha \in [0,1]}{\mathrm{C}} \widetilde{A}_z = \left(\int_{\forall x \in X}\int_{\forall y \in [0,1]} z/(x, y)\right) \alpha \subseteq [0,1] \tag{4.37}$$

The interpretation of a GT2FLS can be summarized into two main actions, specifically meet and join, as distinct below:

$$\begin{aligned}\mu_{\widetilde{A}}(x) \mathrm{C} \mu_{\widetilde{B}}(x) &= \int_{u \in J_x^u}\int_{u \in J_x^w} f_x(u) * g_x(w)/(u \vee w) \\ \mu_{\widetilde{A}}(x) \mathrm{E} \mu_{\widetilde{B}}(x) &= \int_{u \in J_x^u}\int_{u \in J_x^w} f_x(u) * g_x(w)/(u \wedge w)\end{aligned} \tag{4.38}$$

The sample knowledge-based (KB) rules for GT2FLS maintain the same format as previous instances of FLS, T1, and IT2, with a difference in notation, as shown in (4.39):

$$\begin{aligned}&\text{Rule 1 : If } X_1 \text{ is } (f')_{11} \text{ and } X_2 \text{ is } (f')_{12} \text{ and...and } X_m \text{ is } (f')_{1m}.\text{Then } Y \text{ is } (g')_1 \\ &\text{Rule 2 : If } X_1 \text{ is } (f')_{21} \text{ and } X_2 \text{ is } (f')_{22} \text{ and...and } X_m \text{ is } (f')_{2m}.\text{Then } Y \text{ is } (g')_2 \\ &\text{Rule } n \text{ : If } X_1 \text{ is } (f')_{n1} \text{ and } X_2 \text{ is } (f')_{n2} \text{ and...and } X_m \text{ is } (f')_{nm}.\text{Then } Y \text{ is } (g')_n\end{aligned} \tag{4.39}$$

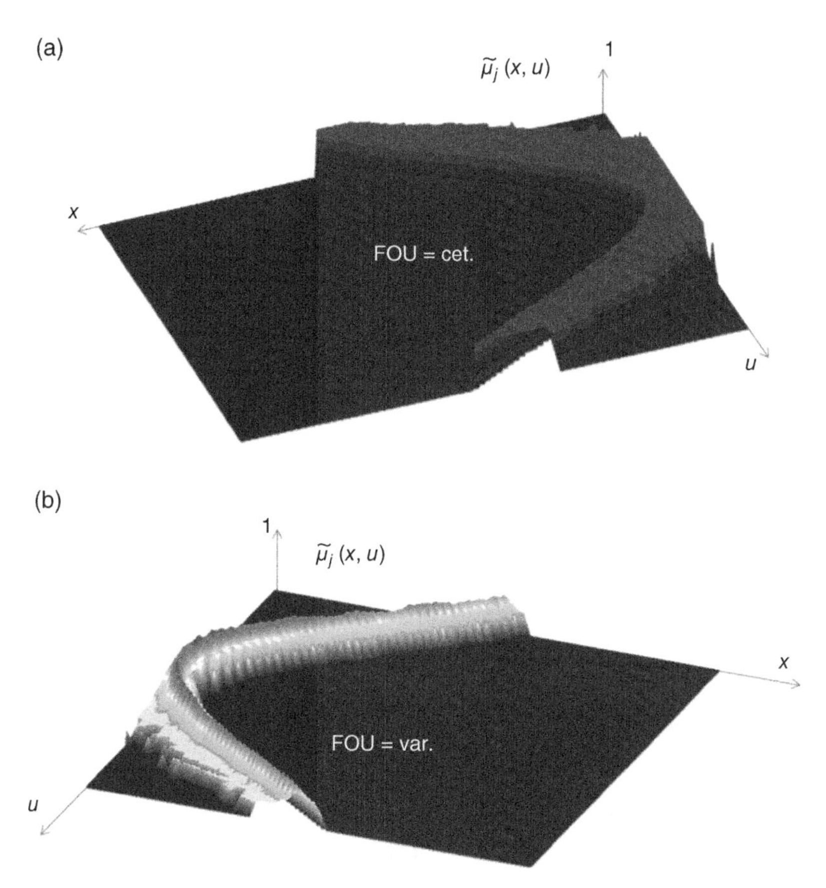

Figure 4.3 3D Type-2 fuzzy logic (T2FL): (a) interval Type-2 fuzzy logic (IT2FL) and (b) general Type-2 fuzzy logic (GT2FL).

Where X_p denotes the pth Type 2 polar fuzzy membership (T2PFM) antecedent; Y denotes the GT2FLS consequence; and $(f')_{nm}$ is the n,mth consequence of the GT2FLS fuzzy set of Rule n?

The fuzzy inference from a GT2FLS is best labeled by (4.40), where $\underline{f}^l(x')$ and $\overline{f}^l(x')$ represent the firing sets definite by (4.41) and (4.42), and b is an interval defined by (4.42).

$$\mu_{\widetilde{B}}(y) = \int_{b\epsilon} \left[\begin{matrix} \left[\left[\underline{f}^{1*}\underline{\mu}_{G_N}(y)\right]\cdots\left[\underline{f}^{N_{M}*}\underline{\mu}_{G_N}(y)\right]\right], \\ \left[\left[\overline{f}^{1*}\overline{\mu}_{G_N}(y)\right]\cdots\left[\overline{f}^{N*}\overline{\mu}_{G_N}\right]\right] \end{matrix} \right](1/b), y\epsilon Y \tag{4.40}$$

$$\left[\underline{f}^l(x'), \overline{f}^l(x')\right] = \left[\left[\underline{\mu}_{F_1^l}(x_1')\widetilde{*}\ldots\widetilde{*}\underline{\mu}_{F^l p_1}(x_P')\right], \left[\overline{\mu}_{F_1^l}(x_1')\widetilde{*}\ldots\widetilde{*}\overline{\mu}_{F^l p_1}(x_P')\right]\right] \tag{4.41}$$

$$\left[\underline{b}^l(y), \overline{b}^l(y)\right] = \left[\underline{f}^l\widetilde{*}\underline{\mu}_{G^l}(y), \overline{f}^l\widetilde{*}\overline{\mu}_{G^l}(y)\right] \tag{4.42}$$

Based on the idea of type-reduction, Doostparast et al. provided a wide-ranging assessment on numerous type-reduction and defuzzification approaches for general Type-2 fuzzy sets and systems. The type-reduction approaches can be separated into four categories: exact method (EM), uncertainty method (UM), approximate method (AM), and geometrical method (GM), in which G is

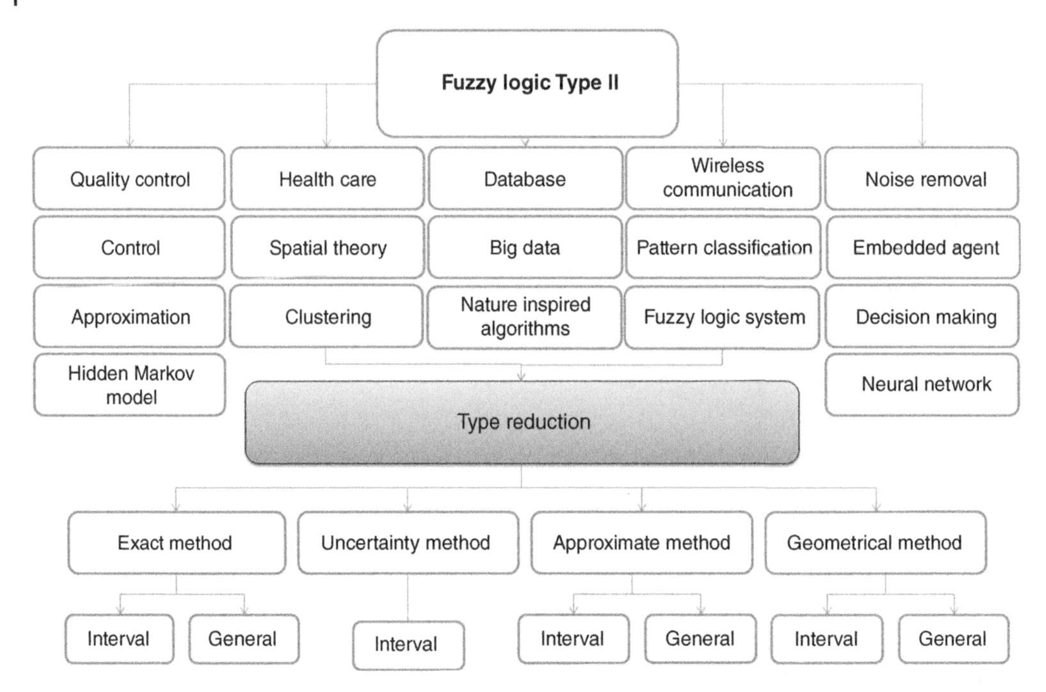

Figure 4.4 Type reduction in T2 fuzzy logic. *Source:* Torshizi et al. [25]/with permission of Elsevier.

general and I is interval. According to this study, only 37.5% of the current type-reduction procedures is completely designed for GT2FLS and the residual works on IT2 FLSs (Figure 4.4).

The centroid is one of the prevalent methods used for type-reduction of a GT2FLS, where θ_N is a combination associated with the secondary degree ω:

$$\omega = f_{x_1}(\theta_1)\tilde{*}\ldots\tilde{*}f_{x_N}(\theta_N).$$
$$C = \int_{\theta_1 \in J_{x_1}} \cdots \int_{\theta_N \in J_{x_N}} [f_{x_1}(\theta_1)\tilde{*}\ldots\tilde{*}f_{x_N}(\theta_N)] / \left(\left(\sum_{i=1}^{N} x_i\theta_i\right) / \left(\sum_{i=1}^{N} \theta_i\right)\right) \tag{4.43}$$

4.3 Theory of Automatic Methods for MF Generation

4.3.1 Automatic Procedure to Generate a 3D Membership Function

Many methods are used under the title of Artificial Intelligence to automate the T1 FMS generation. The heuristic-based design process automatically makes the IT2 FMF using heuristic T1 FMS. The IT2 fuzzy c-means (FCM)-based method uses the derived formulas of IT2 FMS in the IT2 FCM algorithm. The heuristic T1 FMF $\mu_i(x)$ for feature i is calculated by modifying the parameters of the function, which are regularly provided by experts or according to the maximum entropy principle (MEP). This converts the UMF $\bar{\mu}_i(x)$ of the IT2 FMF. The LMF $\underline{\mu}_i(x)$ is acquired by scaling the UMF by a factor between 0 and 1, which is also determined by an expert or based on optimization methods [8, 9, 19]. The FOU of the heuristic method for feature i can be expressed as follows:

$$\underset{\forall x \in X}{\mathrm{C}} \left[\underline{\mu}_i(x), \bar{\mu}_i(x)\right] = \underset{\forall x \in X}{\mathrm{C}} \left[\underline{\mu}_i(x), \alpha_k \cdot \bar{\mu}_i(x)\right],$$
$$0 < \alpha_k < 1 \text{ and } k \geq 1 \tag{4.44}$$

FOU is expressed as follows:

$$\text{FOU} = \underset{\forall x \in X}{\text{C}} \left[\underline{\mu}(x), \bar{\mu}(x)\right] = \underset{\forall x \in X}{\text{C}} \left[\min\left\{\underline{\mu}_i(x_i)\right\}, \min\left\{\alpha_k.\bar{\mu}_i(x_i)\right\}\right], \\ 0 < \alpha_k < 1 \text{ and } k \geq 1 \tag{4.45}$$

where $\underline{\mu}(x)$ and $\bar{\mu}(x)$ are the minimum LMF and UMF among all LMFs and UMFs for all features, respectively.

The selection of α_k related to the flexibility of MF describes the uncertainties in circumstances in which the information is blurred. These calculation methods – the extension of α_k – include drawing the manifold IT2LFs 3D GT2FLS.

$$\left[\underline{\mu}_{\tilde{A}}(x), \bar{\mu}_{\tilde{A}}(x)\right] = \begin{bmatrix} \sup\left\{\left(u \mid u\epsilon[0,1], \mu_{\tilde{A}}(x,u) > 0\right)\right\}, \\ \inf\left\{\left(u \mid u\epsilon[0,1], \mu_{\tilde{A}}(x,u) > 0\right)\right\} \end{bmatrix} \tag{4.46}$$

Parameter optimization and fine-tuning are significant subjects in automatically defining the MF. There are a wide range of meta-heuristic family procedures that can be used, such as the simulated annealing (SA) algorithm, genetic algorithm (GA), ant colony optimization (ACO), particle swarm optimization (PSO), bee colony, among others, which are discussed in detail in Chapter 10. The FOU is extracted as follows:

$$\begin{aligned} \text{FOU} &= \left[\underline{\mu}(x), \bar{\mu}(x)\right] \cong \left[f^L(x), f^U(x)\right] \\ &= \left[\min{}_i\{f_i^L(x_i)\}, \min\left\{\alpha_k.f_i^U(x_i)\right\}\right], \\ i &\geq 1, 0 < \alpha_k < 1 \text{ and } k \geq 1 \end{aligned} \tag{4.47}$$

where f^L is the $\underline{\mu}(x)$, f^U is the $\bar{\mu}(x)$, i is the feature number ($i \geq 1$), k is the α number for $k \geq 1$.

Mendel and John presented the theory of the domain of uncertainty (DOU) for a T2FS, $\tilde{A}$, as the of $\tilde{A}$:

$$\begin{aligned} \text{DOU}\left(\tilde{A}\right) &= \underset{x \in X}{\text{C}}(J_x) = \left\{((x,u))\epsilon X \times [0,1] \mid \mu_{\tilde{A}}(x,u) > 0\right\}, \\ J_x &= \left\{(x,u) : u \in \left[\underline{\mu}_{\tilde{A}}(x), \bar{\mu}(x)\right]\right\} = \left\{u \in [0,1], \mu_{\tilde{A}}(x,u) > 0\right\} \end{aligned} \tag{4.48}$$

where J_x is the "uncertainty" in "DOU" expressed by the SMF.

4.4 Steps and Components of General 3D Type-2 Fuzzy Logic Systems (G3DT2 FL)

This section presents some important theories that are used to develop the polar FIS.

4.4.1 General 3D Type-2 Fuzzy Logic Systems (G3DT2 FL)

Based on the DOU description, the uncertainty of 3D MFs can be modeled using a 3D shape, which is called the 3D FOU. The general 3D Type-2 fuzzy set, represented by $\tilde{\tilde{A}}$, is defined by a 3D Type-2

MF $\mu_{\tilde{\tilde{A}}}(x, y, u, v)$, where $x \in X$, $y \in Y$ $u \in Jx \subseteq [0, 1]$, $v \in Jy \subseteq [0, 1]$, $0 \le \mu_{\tilde{\tilde{A}}}(x, u) \le 1$, $0 \le \mu_{\tilde{\tilde{A}}}(y, v) \le 1$, and can be represented by:

$$\begin{aligned}
&\widetilde{A} = f\left(\widetilde{A}_x, \widetilde{A}_y\right) = f\left(\int_{x \in X}\int_{u \in J_x} \mu_{\widetilde{A}_x}(x, u)/(x, u), \int_{y \in Y}\int_{v \in J_y} \mu_{\widetilde{A}_y}(y, v)/(y, v)\right),\\
&J_x \subseteq [0, 1] \text{ and } J_y \subseteq [0, 1],\\
&\widetilde{A}_x \subseteq \left\{\left((x, u), \mu_{\widetilde{A}_x}(x, u) \mid \forall x \epsilon X, \forall u \epsilon J_x \subseteq [0, 1]\right)\right\},\\
&\widetilde{A}_y = \left\{\left((y, v), \mu_{\widetilde{A}_y}(y, v) \mid \forall y \epsilon Y, \forall v \epsilon J_y \subseteq [0, 1]\right)\right\}
\end{aligned} \tag{4.49}$$

where $\int\int$ means the union over all acceptable input variables (x, y) and(u, v).

A secondary MF is a vertical slice of $\left\{\mu_{\widetilde{A}_y}(y), \mu_{\widetilde{A}_x}(x)\right\}$.There are μ $\mu_{\widetilde{A}_x}(x = x', u)$, for $x \in X$ and $\forall u \in Jx \subseteq [0, 1]$, and $\mu_{\widetilde{A}_y}(y = y', v)$, for $y \in Y$ and $\forall v \in Jy \subseteq [0, 1]$, as defined by

$$\begin{aligned}
\mu_{\widetilde{A}}(x', u) &= \mu_{\widetilde{A}}(x') = \int_{u \in J_x} f_{x'}(u)/(u), J_{x'} \subseteq [0, 1],\\
\mu_{\widetilde{A}}(y', v) &= \mu_{\widetilde{A}}(y') = \int_{u \in J_y} f_{y'}(v)/(v), J_{y'} \subseteq [0, 1],\\
\left\{\widetilde{A}_x, \widetilde{A}_y\right\} &= \left\{\int_{x \in X} \frac{\mu_{\widetilde{A}_x}(x)}{x}, \int_{y \in Y} \frac{\mu_{\widetilde{A}_y}(y)}{y},\right\}\\
&= \left\{\int_{x \in X}\int_{u \in J_x \in [0,1]} \mu_{\widetilde{A}_x}(x, u)/(x, u), \int_{x \in X}\int_{u \in J_x \in [0,1]} \mu_{\widetilde{A}_x}(x, u)/(x, u), \int_{y \in Y}\int_{v \in J_y \in [0,1]} \mu_{\widetilde{A}_y}(y, v)/(y, v)\right\}\\
&= \left\{\int_{x \in X}\left[\int_{u \in J_x \in [0, 1]} f_{\widetilde{u}}(u)/(u)\right], \int_{y \in Y}\left[\int_{v \in J_y \in [0, 1]} f_{\widetilde{v}}(v)/(v)\right]\right\}
\end{aligned} \tag{4.50}$$

The exemplification of a G3DT2 MF is illustrated in Figure 4.5.

The IT2PFM algorithm can automatically adjust the vagueness of MF. In this section, the IT2FMF procedure is presented, and then IT2PFM is defined. Cubic smoothing spline (CSS) is used to moderate the 3DMFs in the upper and lower MFs. The fitting objective is to minimize the equation as follows:

$$\mathrm{FM} = p\sum\nolimits_{y=1}^{m}\sum\nolimits_{x=1}^{n}\left((f(x, y) - s(x, y))^2\right) + (1 - p)\int\int\left(D^2 S(x, y)\right)^2 \mathrm{d}x\mathrm{d}y \tag{4.51}$$

where the equation consists of two parts: compactness and smoothness.

Equation (4.51) regulates the scales around data (with $p = 1$) and is a severely smooth spline (with $p = 0$). For $p = 0$, f is the least-squares conventional line fit to the data, and for $p = 1$, f is neutral. As p changes from 0 to 1, the profile of the smoothing spline is transformed. The parameter p regulates the smoothness form of 3D MF. The RT controls the general shape of MFs. The representation $E_\alpha(p) \equiv E_\alpha(p_l, p_u)$ is used to describe the entropy of the 3D fuzzy event $\left\{\widetilde{A}_{x,}, \widetilde{A}_y\right\}$, which can be distinct as follows:

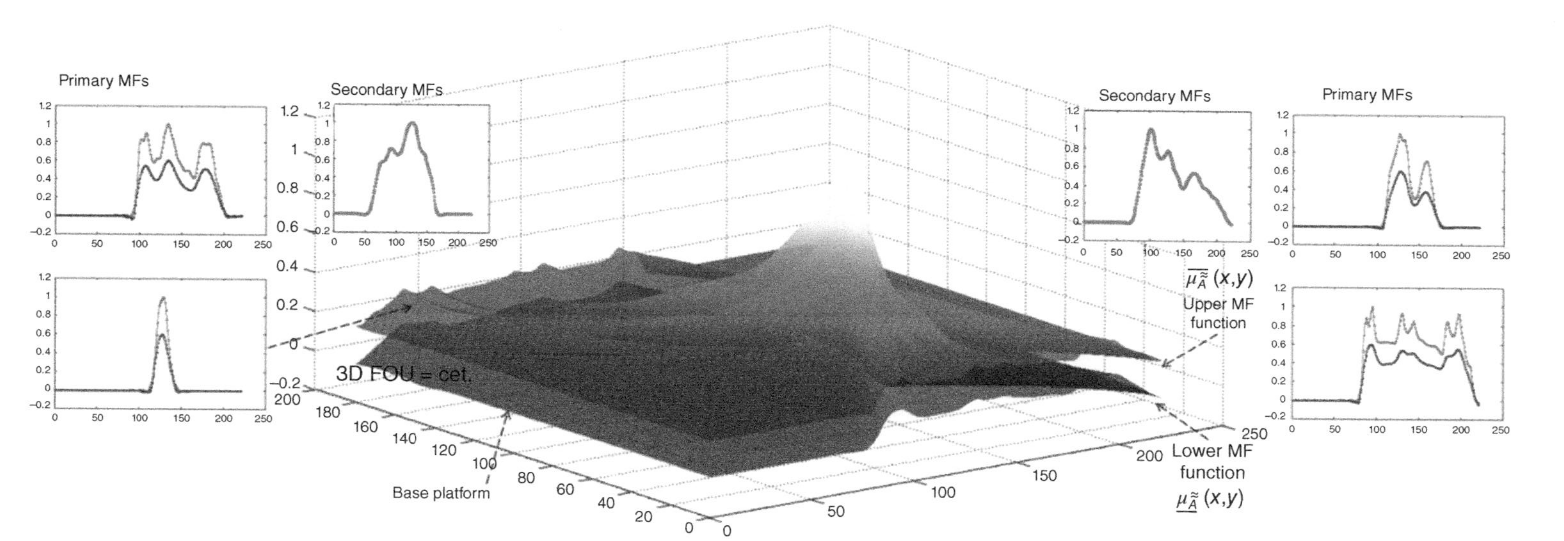

Figure 4.5 The interval 3D Type-2 membership function.

$$
\begin{aligned}
E(A) &= -P(A)\log(P(A)) - (1-P(A))\log(1-P(A)) \\
E\left(\left\{\widetilde{A}_x, \widetilde{A}_y\right\}\right) &= -P\left(\left\{\widetilde{A}_x, \widetilde{A}_y\right\}\right)\log\left(P\left(\left\{\widetilde{A}_x, \widetilde{A}_y\right\}\right)\right) \\
&\quad -\left(1-P\left(\left\{\widetilde{A}_x, \widetilde{A}_y\right\}\right)\right)\log\left(1-P\left(\left\{\widetilde{A}_x, \widetilde{A}_y\right\}\right)\right) \\
&= -P\left(\left\{\int_{x\in X}\left[\int_{u\in J_x\in[0,1]}\frac{f_{\tilde{u}}(u)}{u}\right], \int_{y\in Y}\left[\int_{v\in J_y\in[0,1]}\frac{f_{\tilde{v}}(v)}{v}\right]\right\}\right) \\
&\quad * \log\left(P\left(\left\{\int_{x\in X}\left[\int_{u\in J_x\in[0,1]}\frac{f_{\tilde{u}}(u)}{u}\right], \int_{y\in Y}\left[\int_{v\in J_y\in[0,1]}\frac{f_{\tilde{v}}(v)}{v}\right]\right\}\right)\right) \\
&\quad -\left(1-P\left(\left\{\int_{x\in X}\left[\int_{u\in J_x\in[0,1]}\frac{f_{\tilde{u}}(u)}{u}\right], \int_{y\in Y}\left[\int_{v\in J_y\in[0,1]}\frac{f_{\tilde{v}}(v)}{v}\right]\right\}\right)\right) \\
&\quad * \log\left(1-P\left(\left\{\int_{x\in X}\left[\int_{u\in J_x\in[0,1]}\frac{f_{\tilde{u}}(u)}{u}\right], \int_{y\in Y}\left[\int_{v\in J_y\in[0,1]}\frac{f_{\tilde{v}}(v)}{v}\right]\right\}\right)\right)
\end{aligned}
\tag{4.52}
$$

Vagueness can be denoted by an index, which is designed from the variances of $\left|E\left(\widetilde{A}_{x,}\right)_l - E\left(\widetilde{A}_{x,}\right)_u\right|$. Based on equation, $(x, u) \in X \times [0, 1] \vee \mu\,\tilde{}A(x, u) > 0$, it is suggested to use the DOU for this explanation. The RT can be written as follows:

$$
P(\rho, \theta) = R(\rho, \theta)\{F(x, y)\} = \int_{-\infty}^{+\infty}\int_{-\infty}^{+\infty} f(x, y)\delta(\rho - x\cos\theta - y\sin\theta)\,\mathrm{d}x\mathrm{d}y \tag{4.53}
$$

where $f(x, y)$ represents 2D data; $P(\rho, \theta)$ is the RT of $f(x, y)$; θ represents the line direction; and ρ is the distance away from the origin of coordinates.

The set requirement be in [0,1]. The standardization form 3DMF is gained by dividing every point by max three-dimensional radon transform (3DRT) $f(x, y)P(\rho, \theta)f(x, y)\rho$. The RT is useful to make MF and possesses properties (linearity, periodicity, semisymmetry, translation, rotation, and scaling) that are useful for IT2 and GT2 FLs [8, 19].

$$
f(x, y)P(\rho, \theta)f(x, y)\rho
$$

The RT is applied to generate MF and possesses properties (linearity, periodicity, semisymmetry, translation, rotation, and scaling) that are beneficial for IT2 and GT2 FLs.

$$
\begin{aligned}
&R_{f+g}(\rho, \theta) = R_f(\rho, \theta) + R_g(\rho, \theta), \\
&R_f(\rho, \theta) = R_f(\rho, \theta + 2\pi k), \forall k \in Z, \\
&R_f(\rho, \theta) \to R_f(-\rho, \theta \pm \pi), \\
&R_f(\rho, \theta) \to R_f(\rho - x_0\cos(\theta) - y_0\sin(\theta), \theta), \\
&R_f(\rho, \theta) \to \left(\frac{1}{\alpha}\right) R_f(\alpha\rho, \theta)
\end{aligned}
\tag{4.54}
$$

Based on the properties stated in (4.54), RT is linear and RT $f(x, y)$ is periodic in the variable θ with period 2π. A translation of $f(x, y)$ by a vector $\vec{u} = (x_0, y_0)$ implies a shift in the variable ρ of $R_f(\rho, \theta)$ by a distance $d = x_0 \cos\theta + y_0 \sin\theta$ that is equal to the length of the projection of $\vec{u}$ onto the line $x \cos\theta + y \sin\theta = \rho$ (4.43). Also, based on (4.44), a rotation of $f(x, y)$ by an angle θ_0 implies a circular shift in the variable θ of $R_f(\rho, \theta)$ by distance θ_0. Scaling $f(x, y)$ by a factor α results in scaling the variable ρ and the amplitude of the factor α and $1/\alpha$, respectively. The normalization form 3DMF is obtained by dividing every point by maximum 3DRT:

$$\mathrm{RTMF}_{(i,j)} = \left(\left[v_{(3\mathrm{DRT})}{}^{h}_{(i,j)}\right] \Big/ \max\left[v_{(3\mathrm{DRT})}{}^{h}_{(i,j)}\right]\right)^{\frac{1}{h}} \tag{4.55}$$

$$\begin{aligned} \mu_{(i,j)} &= \left(\left[v_{(3\mathrm{DRT})}{}^{h}_{(i,j)}\right] \Big/ \max\left[v_{(3\mathrm{DRT})}{}^{h}_{(i,j)}\right]\right)^{\frac{1}{h}} + H, \\ \mathrm{GR}_{(i,j)} &= \frac{1}{(MN)^h} \sum\nolimits_{j=1}^{N} \sum\nolimits_{i=1}^{M} \left(\left[v_{(3\mathrm{DRT})}{}^{h}_{(i,j)}\right] \Big/ \max\left[v_{(3\mathrm{DRT})}{}^{h}_{(i,j)}\right]\right)^{\frac{1}{h}} \end{aligned} \tag{4.56}$$

where M and N denote the size of the 3DMF platform; H is platform height $H\epsilon[0,1]$, $h\epsilon(1,\infty)$; $v_{(3\mathrm{DRT})}{}^{h}_{(i,j)}$ are the 3DRT values in the position i and j.

The maximum GR is equal to 1. To extend the fuzzy membership to Type-2 fuzzy sets, ultrafuzziness is set to 0. The amount of ultrafuzziness will increase by raising the uncertainty bound (see Figure 4.6) [8, 9, 19].

The maximal ultrafuzziness, equal to 1, is worth complete vagueness.

$$\begin{aligned} \text{Ultrafuzziness} &= \{\text{FOU and/or DOU}\} \\ &= \left\{ \begin{matrix} \min_i\{ f_i^L(x_i)\},\ \min\{\alpha_k . f_i^U(x_i)\}, \\ \left\{\left((x,u)\epsilon X \times [0,1] \mid \mu_{\tilde{A}}(x,u) > 0\right)\right\} \end{matrix} \right\}, i \geq 1, 0 < \alpha_k < 1 \text{ and } k \geq 1 \end{aligned} \tag{4.57}$$

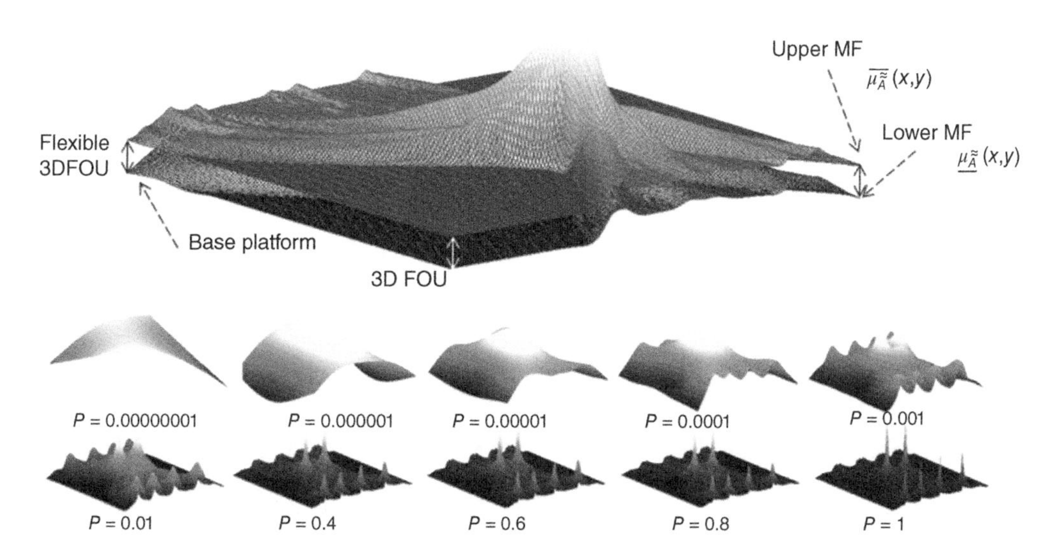

Figure 4.6 The structure of 3DT2MF.

Based on fuzziness for a set $A \in r_n(x)$:

$$
\begin{aligned}
H_k a(A) &= (2/n^k) d(A, A^{\text{near}}) \\
d(A, A^{\text{near}}) &= \left(\sum_{i=1}^{n} |\mu_i - \mu_{A^{\text{near}},i}|^q \right)^{1/q} \\
H_{ka}(q, A) &= \left(2/\left(n^{\frac{1}{q}}\right) \right) \left(\sum_{i=1}^{n} |\mu_i - \mu_{A^{\text{near}},i}|^q \right)^{1/q} \\
\tilde{\gamma}\left(\tilde{A}\right) &= \frac{1}{MN} \sum_{g=0}^{L-1} \left[\mu_U\left(g_{ij}\right) - \mu_L\left(g_{ij}\right) \right] \times h(g), \\
\tilde{\gamma}(A) &= \left(\frac{1}{(MN)^{\frac{1}{q}}} \right) \left[\sum_{j=1}^{M-1} \sum_{i=1}^{N-1} \left| \mu_{u(i,j)} - \mu_{L(i,j)} \right|^q \right]^{1/q}
\end{aligned}
\tag{4.58}
$$

where $k \in R^+$, d is a distance, and A^{near} is the crisp set close to A; $d(A, A \mid \mid \text{near})$ can be linear or quadratic; and $H_{ka}(A)$ is calculated by q-norms.

Based on the Kaufmann theory, the smallest, extreme, equivalent, and compact ultrafuzziness are satisfied for a three-dimensional case, such as 3DMF, the fuzziness is defined as follows:

$$
M = p \sum_{y=1}^{m} \sum_{x=1}^{n} (f(r, \theta) - s(r, \theta))^2 + (1 - p) \int\int \left(D^2 S(r, \theta)\right)^2 \mathrm{d}r\mathrm{d}\theta \tag{4.59}
$$

where $M \times N$ is subset $\mathrm{A} \subseteq \mathrm{X}$ with L RT value; $\mathrm{r} \in [0, -1]$ represents the histogram $h(\mathrm{RT})$; and $\mu_{A\left(r_{i,j}\right)}$ is the membership function.

Similar to the 3DMF presented before, CSS is used to make the polar MF:

Next, the 3D PMF(r,θ) principals for the automatic MF generation process are described in Algorithm 4.1 [9]:

This method has less computational load and complexity than the histogram-based IT2 FMF method. In addition, by reducing the dimensions and compatibility of the histogram smoothing process, the computational time is reduced and the pattern identity data (ID) is interpreted as

Algorithm 4.1: Approximating UMF and LMF in the polar frame

Approximating UMF and LMF in the polar frame, they are ordered in the following way:

1) Approximate the three-dimensional surface (3DS) using RT to generate the 3D surface from 2D data and construct the 3D surface.
2) Transform data to the polar coordination (3D Polar), transfer 2D data to the polar domain to make data uniform in multiscale.
3) Use the polar histogram generator (PHG) to generate the polar histogram of the multi-scale data.
4) Adjusting the smoother fitting parameters (SF), apply the SF parameter to obtain the approximate parameter value (p).
5) Use the polar smooth generator (PSG) to generate the histogram of the overall polar surface.
6) Use PSG to fit the upper and lower histogram values.
7) Extract the PFMF, and normalize the height of the upper PSG and LMF using the lower PSG.

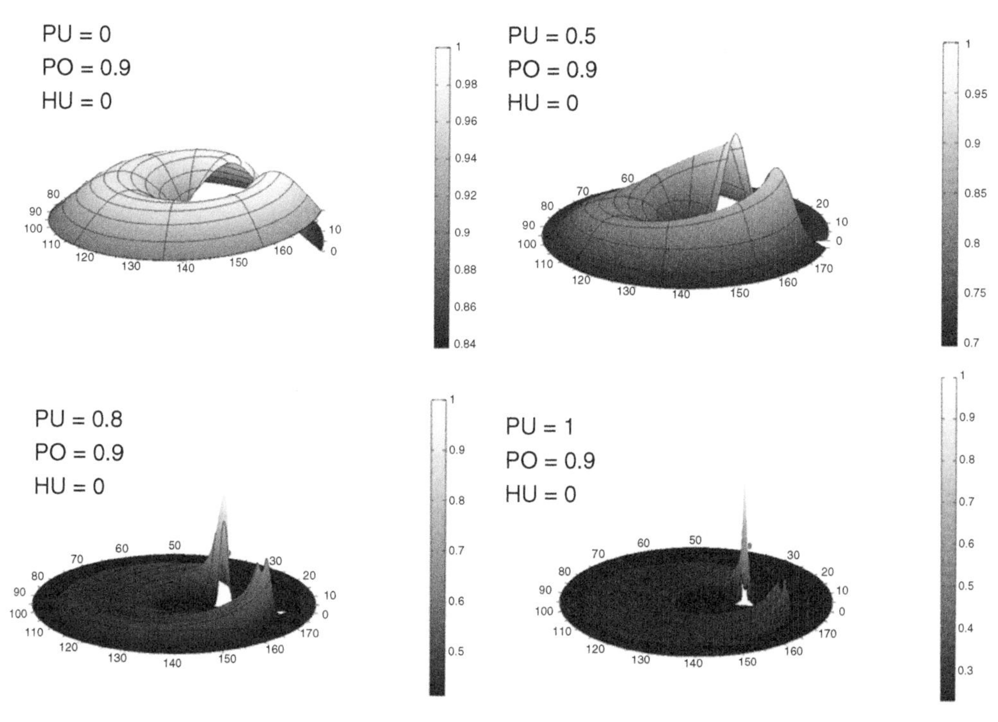

Figure 4.7 A sample of polar membership functions: As p moves from 0 to 1, the smoothing spline changes from one extreme to the other.

an effective new feature. Figure 4.7 illustrated an example that is created based on proposed algorithm. According to the vague distance between the UMF and LMF of 3DPFOU, the polar UMF and LMF surface is defined. Figure 4.7 displays the GT2 PMF removal, where the 3DPFOU is created by the shaded area bordered by 3D UMF and 3D LMF.

As clarified in Figure 4.8, 3DPFOU can be efficiently useful to regulate the ambiguity and can be stated as follows:

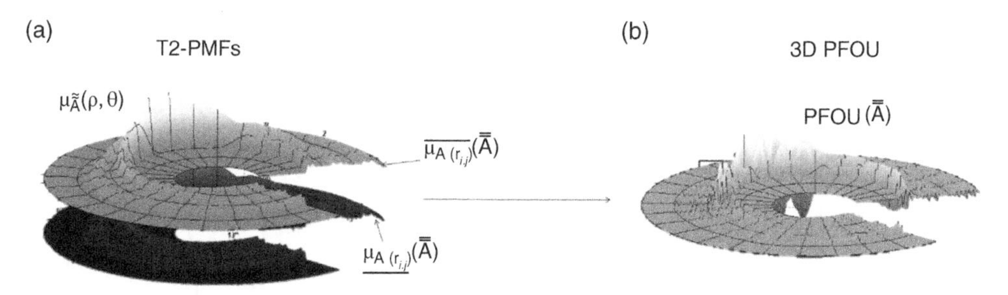

Figure 4.8 (a) 3DPFOU created by polar upper bound and lower bound and (b) FOU for 3D polar fuzzy sets (3DPFOU).

$$\begin{aligned}\text{PFOU}\left(\overline{\overline{A}}\right) &= \int \text{PFOU}\left(\overline{\overline{A}}\right)_{(\theta,r)} \\ &= \sum_{\theta=1}^{N}\left\{\int\int_{0}^{r}\left(\left(\max\left\{\overline{\mu_{A(r_{i,j})}}\left(\overline{\overline{A}}\right)\right\}\right)-\left(\min\left\{\underline{\mu_{A(r_{i,j})}}\left(\overline{\overline{A}}\right)\right\}\right)\right) dr.d\theta\right\}, \\ 1 &< N < 360 \\ \text{PDOU}\left(\widetilde{A}\right) &= \underset{x\in X}{\text{C}}(J_x) = \left\{\left((x,u)\epsilon X \times [0,1] \mid \mu_{\widetilde{A}}(r,\theta) > 0\right)\right\}\end{aligned} \tag{4.60}$$

The PDOU(A) and J_x are defined above. In the FOU, the word "uncertainty" in "DOU" denotes the ambiguity that is stated by the secondary membership in the polar frame.

4.5 General 3D Type-2 Polar Fuzzy Method

4.5.1 Automatic MF Generator

To find the automatic 3DFOU, the 3DT2 FMS must to be considered. The 3D UMF and LMF are formed by selection of right general fuzzy (GF) sets. A degree of ultrafuzziness is calculated according to the 3DMF fuzzy set [8]:

$$\text{GR}_{(i,j)} = \frac{1}{2(2\pi\rho)^h}\int_0^{2\pi}\int_0^{\rho}\left(\left[v_{(3\text{DMF})}{}^{h}_{(\rho,\theta)}\right]/\max\left[v_{(3\text{DRT})}{}^{h}_{(\rho,\theta)}\right]\right)^{\frac{1}{h}} d\rho d\theta \tag{4.61}$$

where 2π denotes the size of the three-dimensional polar membership function (3DPMF) platform, H is the controllable leveling of polar platform $H\epsilon[0, 1]$, $H\epsilon(1, \infty)$ and $v_{(3\text{DRT})}{}^{h}_{(\rho,\theta)}$ are the 3DMF values in the position ρ, θ shows the resulting upper and lower 3DPMF functions, p values used for fitting and showing the 3D Type-2 FMF are acquired using the suggested method.

The shaded area among the $\underline{\text{PFOU}}(\acute{A})$ and $\overline{\text{PFOU}}(\acute{A})$ specifies the 3D FOU. A degree of ultrafuzziness $\widetilde{\gamma}$ for a polar 3D FMF, polar 3D FMF, and the membership function $\mu_{\widetilde{P}(\rho,\theta)}$ is defined based on the equation as follows:

$$\widetilde{\gamma}(P) = \left(\frac{1}{(\text{Area})^{\frac{1}{q}}}\right)\left[\int_0^{2\pi}\int_0^{r}\left|\mu_U\left(g_{(\rho,\theta)}\right)-\mu_L\left(g_{(\rho,\theta)}\right)\right|^q\right]^{\frac{1}{q}} = 0, for\ q \in [1, \infty) \tag{4.62}$$

The $d(P, P||\text{near})$ and linear or quadratic $H_{ka}(P)$ can be determined by q-norms like 3D FMF based on

$$\begin{aligned} d(P, P^{\text{near}}) &= \left\{\left(\sum_{i=1}^{n}\left|\begin{array}{l}\left(\left(\left[v_{(3\text{DMF})}{}^{h}_{(\rho,\theta)}\right]/\max\left[v_{(3\text{DRT})}{}^{h}_{(\rho,\theta)}\right]\right)^{\frac{1}{h}}+H\right) \\ -\left(\left(\left[v_{(3\text{DMF})}{}^{h}_{(\rho,\theta)}\right]/\max\left[v_{(3\text{DRT})}{}^{h}_{(\rho,\theta)}\right]\right)^{\frac{1}{h}}+H\right)_{\text{near}}\end{array}\right|^q\right)^{1/q}\right\}, \\ H_{ka}(q,P) &= \left(\frac{2}{n^{\frac{1}{q}}}\right)\left(\sum_{i=1}^{n}\left|\begin{array}{l}\left(\left(\left[v_{(3\text{DMF})}{}^{h}_{(\rho,\theta)}\right]/\max\left[v_{(3\text{DRT})}{}^{h}_{(\rho,\theta)}\right]\right)^{\frac{1}{h}}+H\right)_i \\ -\left(\left(\left[v_{(3\text{DMF})}{}^{h}_{(\rho,\theta)}\right]/\max\left[v_{(3\text{DRT})}{}^{h}_{(\rho,\theta)}\right]\right)^{\frac{1}{h}}+H\right)_p\end{array}\right|^q\right)^{1/q}\end{aligned} \tag{4.63}$$

A new index that continues in the polar domain, based on (4.59), for a special case can be presented as follows:

$$\tilde{\gamma}\left(\tilde{P}\right) = \frac{1}{4\pi\rho^2}\int_0^{2\pi}\int_0^{r}\left(\overline{\mu_{A(r_{i,j})}}\left(\overline{\overline{A}}\right)\left(g_{(\rho,\theta)}\right) - \underline{\mu_{A(r_{i,j})}}\left(\overline{\overline{A}}\right)\left(g_{(\rho,\theta)}\right)\right) d\rho d\theta$$
$$\cong \left(\frac{1}{4\pi\rho^2}\int_0^{2\pi}\int_0^{r}\left(\begin{array}{l}\min_i\{f_i^L(x_i)\}\left(g_{(\rho,\theta)}\right)\\ -\max_i\left\{\alpha_k.f_i^U(x_i)\left(g_{(\rho,\theta)}\right)\right\}\end{array}\right)\right)d\rho d\theta, \tag{4.64}$$
$$i \geq 1, 0 < \alpha_k < 1 \text{ and } k \geq 1 \cong \text{PFOU}\left(\overline{\overline{A}}\right)$$

4.5.2 A Measure of Ultrafuzziness

A measure of ultrafuzziness $\tilde{\gamma}$ for a polar 3D FMF, and the membership function $\mu_{\tilde{P}(\rho,\theta)}$ was developed as follows [8, 9, 19]:

$$\tilde{\gamma}(P) = \left(\frac{1}{(\text{Area})^{\frac{1}{q}}}\right)\left[\int_0^{2\pi}\int_0^{r}\left|\mu_U\left(g_{(\rho,\theta)}\right) - \mu_L\left(g_{(\rho,\theta)}\right)\right|^q\right]^{1/q} \tag{4.65}$$

The ambiguity in the polar space is measured based on FOU in 3D polar fuzzy sets (3D PFOU), according to the following properties (see Figures 4.9 and 4.10):

a) Minimum, maximum, equal, and reduced ultrafuzziness are evaluated for the polar method. The polar index is qualified for a measure of ultrafuzziness in the 3D polar domain with the conditions presented in reference [9].
   - IF $\mu_{(\rho,\theta)}$ is considered a Type-1 polar fuzzy set, then $\mu_{u(\rho,\theta)} = \mu_{L(\rho,\theta)}$ and $\tilde{\gamma}(P) = 0$.
   - IF $|\mu_{u(\rho,\theta)} - \mu_{L(\rho,\theta)}| = 1$ (high ambiguity), then $\tilde{\gamma}\left(\tilde{P}\right) = 1$.

b) $\tilde{\gamma}\left(\tilde{P}\right) = \tilde{\gamma}\left(\overline{\tilde{P}}\right)$, where $\left(\tilde{P}\right)$ is the Type-2 fuzzy set in the polar frame, and its complement set can be determined by $1\text{-}\mu_{u(\rho,\theta)}$ and $1\text{-}\mu_{L(\rho,\theta)}$; therefore, the complement set is defined as follows:

$$\overline{\tilde{P}} = \left\{x, \mu_{u(\rho,\theta)}, \mu_{u(\rho,\theta)} \;\middle|\; \begin{array}{c} 1-\left(\dfrac{v_{(3\text{DPRT})}{}^{h}_{(\rho,\theta)}}{\max\left[v_{(3\text{DPRT})}{}^{h}_{(\rho,\theta)}\right]}\right)^{\frac{1}{h}} + H_U \\ , \\ 1-\left(\dfrac{v_{(3\text{DPRT})}{}^{h}_{(\rho,\theta)}}{\max\left[v_{(3\text{DPRT})}{}^{h}_{(\rho,\theta)}\right]}\right)^{\frac{1}{h}} + H_L \end{array}\right\} \tag{4.66}$$

where $H_U = H_L = 0$ for the complement set, and the ultrafuzziness $\tilde{\gamma}$ is equal to 0.

$$\tilde{\gamma}\left(\overline{\tilde{P}}\right) = \left(\frac{1}{(\text{Area})^{\frac{1}{q}}}\right)\left[\int_0^{2\pi}\int_0^{r}\left|1-\mu_U\left(g_{(\rho,\theta)}\right) - 1 + \mu_L\left(g_{(\rho,\theta)}\right)\right|^q\right]^{\frac{1}{q}}$$
$$= \tilde{\gamma}\left(\tilde{P}\right) \text{for } q \in [1, \infty) \tag{4.67}$$

c) IF $3\text{DPFOU}_{(\rho,\theta)} < 3\text{DPFOU}_{(d,c)}$, then $\tilde{\gamma}\left(\tilde{P}_{(\rho,\theta)}\right) < \tilde{\gamma}\left(\tilde{P}_{(d,c)}\right)$.

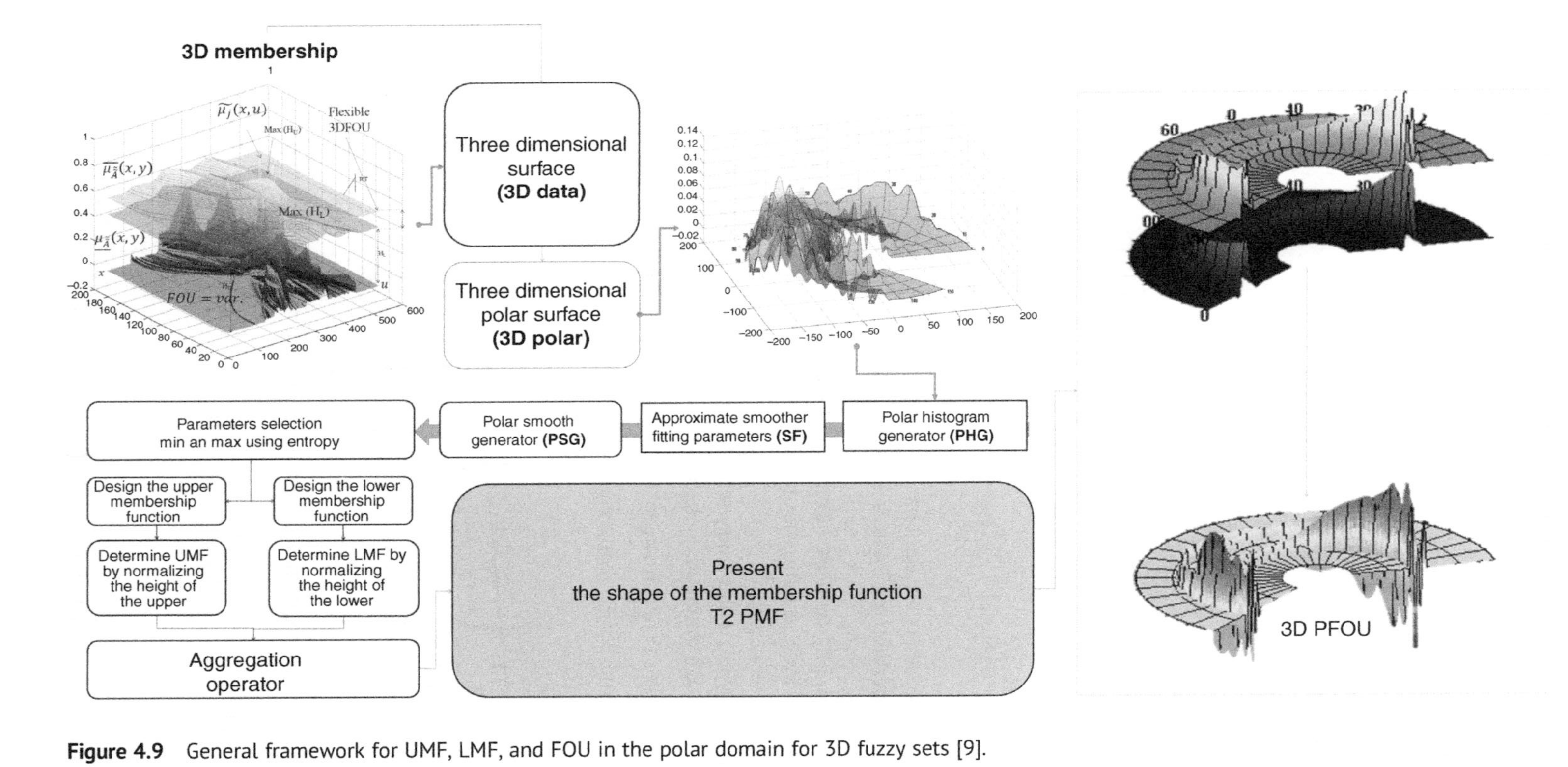

Figure 4.9 General framework for UMF, LMF, and FOU in the polar domain for 3D fuzzy sets [9].

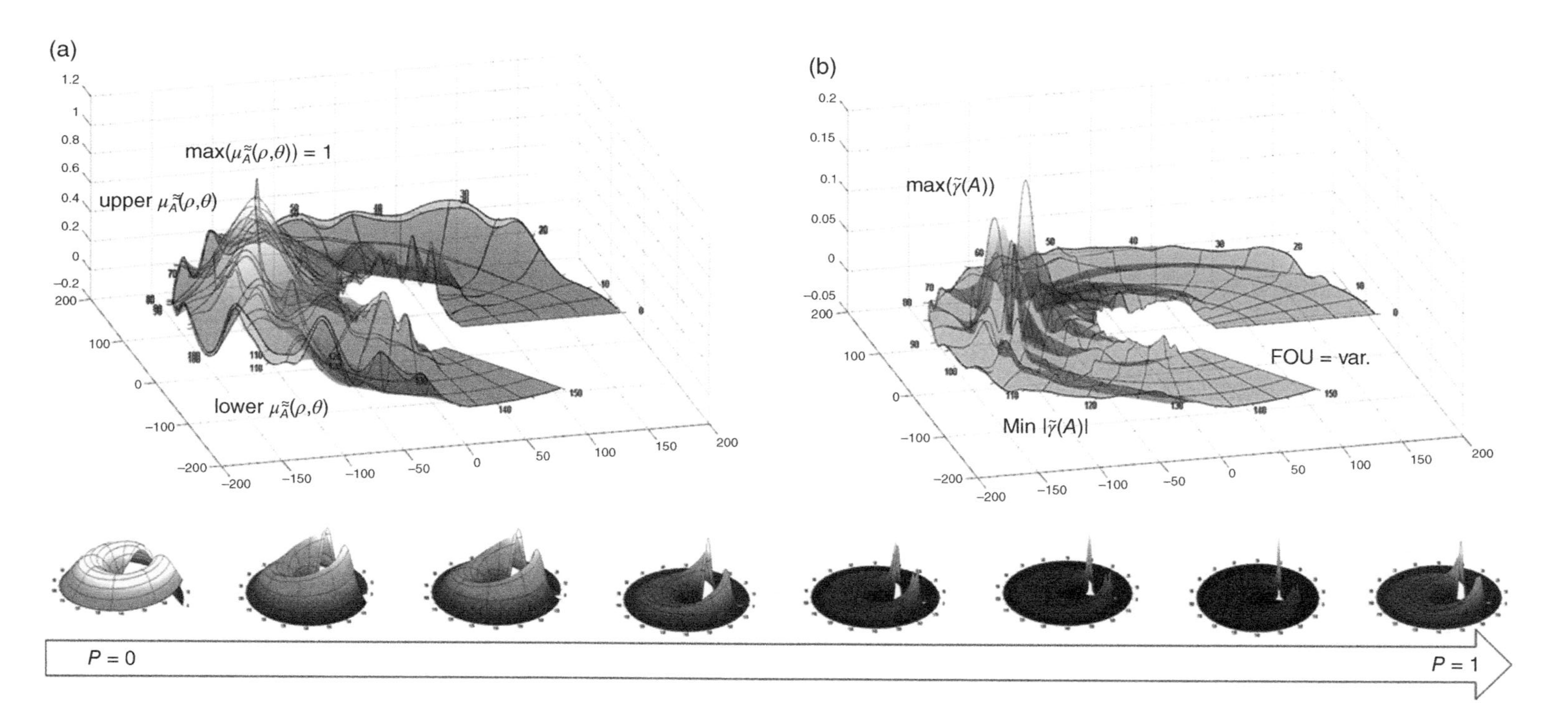

Figure 4.10 Examples of (a) $\underline{\mu_{A(r_{ij})}}\left(\acute{A}\right)$ and (b) $\mu_{A(\bar{r}_{ij})}\left(\acute{A}\right)$, PDOU$\left(\widetilde{A}\right)$, and PFOU$\left(\acute{A}\right)$ in the polar frame for 3D fuzzy sets (3D PFOU) [9].

4.5.3 Theoretic Operations of 3D Type-2 Fuzzy Sets in the Polar Frame

Consider two 3DT2 FS, $\tilde{\tilde{A}}$ and $\tilde{\tilde{B}}$, in a universe $X = f(\rho, \theta)$ and $Y = f(\rho, \theta)$. Let $\mu_{\tilde{\tilde{A}}}(\rho, \theta)$ and $\mu_{\tilde{\tilde{B}}}(\rho, \theta)$ be the membership grades (fuzzy sets in $J\rho \subseteq [0; 1]$ and $J\theta \subseteq [0; 1]$) of these two sets, which represent each (ρ, θ), as $\mu_{\tilde{\tilde{A}}} = \left\{ \int_{x \in X} \left[\int_{u \in J_x \in [0,1]} f_{\tilde{u}}(u)/(u) \right], \int_{y \in Y} \left[\int_{v \in J_y \in [0,1]} f_{\tilde{v}}(v)/(v) \right] \right\}$ and $\mu_{\tilde{\tilde{B}}} = \left\{ \int_{x \in X} \left[\int_{u \in J_x \in [0,1]} g_{\tilde{u}}(u)/(u) \right], \int_{y \in Y} \left[\int_{v \in J_y \in [0,1]} g_{\tilde{v}}(v)/(v) \right] \right\}$. Also, u; $w \in J_x$ shows the primary memberships of x, and $\int_{u \in J_x \in [0,1]} f_{\tilde{u}}(u)/(u)$ indicates the secondary memberships of x. The main principles, like union, intersection, and complement of 3D polar Type-2 fuzzy sets $\tilde{\tilde{A}}$ and $\tilde{\tilde{B}}$, are defined as follows:

1) **Similarity**: Two 3D polar Type-2 fuzzy sets $\mu_{\tilde{\tilde{A}}}(\rho, \theta)$ and $\mu_{\tilde{\tilde{B}}}(\rho, \theta)$ are similar if and only if the sets have the same universe $f(\rho, \theta)$ in the polar domain, and $\mu_{\tilde{\tilde{A}}}(\rho, \theta) = \mu_{\tilde{\tilde{B}}}(\rho, \theta)$ for all $x \in X$ and $y \in Y$. That is, $\mu_{\tilde{\tilde{A}}}(\rho, \theta) - H = \mu_{\tilde{\tilde{B}}}(\rho, \theta) - H$. For example, in Figure 4.9, all MFs are similar. Based on the definition, H is a high platform in which [0, 1], $h\epsilon(1, \infty)$ and $\mu_{\tilde{\tilde{A}}}(\rho, \theta) - H$ is the original MF value in position ρ, θ. $\mu_{\tilde{\tilde{A}}}(\rho, \theta)$ and $\mu_{\tilde{\tilde{N}}}(\rho, \theta)$ are similar if and only if $\left\{ \text{PFOU}(\acute{A})_{(\theta,r)} \right\} = \left\{ \text{PFOU}(\acute{N})_{(\theta,r)} \right\}$.
2) **Correspondence of Polar Fuzzy Sets**: Polar fuzzy set $\mu_{\tilde{\tilde{A}}}(\rho, \theta)$ is a correspondence of polar fuzzy set $\mu_{\tilde{\tilde{B}}}(\rho, \theta)$ if and only if $\mu_{\tilde{\tilde{A}}}(\rho, \theta) \leq \mu_{\tilde{\tilde{B}}}(\rho, \theta)$ for all $x \in X$ and $\mu_{\tilde{\tilde{A}}}(\rho, \theta)$-$\mu_{\tilde{\tilde{B}}}(\rho, \theta) \neq 0$. That is, $\mu_{\tilde{\tilde{A}}}(\rho, \theta) \subset \mu_{\tilde{\tilde{B}}}(\rho, \theta)$. $\mu_{\tilde{\tilde{A}}}(\rho, \theta)$ and $\mu_{\tilde{\tilde{N}}}(\rho, \theta)$ are in correspondence if and only if $\left\{ \text{PFOU}(\acute{A})_{(\theta,r)} \right\} \propto (\eta)_{(\theta,r)} \left\{ \text{PFOU}(\acute{N})_{(\theta,r)} \right\}$.
3) **(NOT)-Complement of a Polar Fuzzy Set (PFS)**: The (NOT)-complement of two PFSs simply refers to a polar set containing the entire domain without the elements of that polar set. For PFSs, the complement of the set O is O, and the membership degrees differ. Let O denote the complement of set O. Then, for all $x \in X$, $\mu_O(x) = 1 - \mu_O(x)$, which means that $O \cap O = \emptyset$ and $O \cup O$ = Domain ($2\pi\rho 1$). As a consequence, we have the following result: (NOT)-complement is defined as $\left\{ \text{PFOU}(\acute{A})_{(\theta,r)} \right\} = -\left\{ \text{PFOU}(\acute{A})_{(\theta,r)} \right\} + cet.$
4) **(OR)-T-norms of PFS**: The union is the max polar set in both sets at the intersection that is known by t-norms. Several methods have been proposed for the t-norms calculation in Type-1 and Type-2 fuzzy sets. Using Zadeh's Extension Principle, the membership grades for the union of Type-2 fuzzy sets $\acute{A}$ and $\acute{B}$ can be defined as follows:Several patterns of 3DPMFs = $\left\{ \mu_{\tilde{\tilde{A}}}(\rho, \theta), \mu_{\tilde{\tilde{B}}}(\rho, \theta) \right\}$ and its correspondence of Type-2 polar sets can be represented as follows:

$$\overline{\overline{A}} \cup \overline{\overline{B}} = \mu_{\overline{\overline{A}} \cup \overline{\overline{B}}}(\rho, \theta) = \mu_{\overline{\overline{A}}}(\rho, \theta) \sqcup \mu_{\overline{\overline{B}}}(\rho, \theta) = \int_u \int_w f_x(u) * g_x(w)/(u \hat{e}\, w) \tag{4.68}$$

where $\cap$ represents the max t-conorm; $*$ represents a t-norm; and $\sqcap$ is the join operation.

5) **(OR)-*T*-Conorms of PFS**: *T*-conorms refers to the intersection of the min polar sets in both sets. Like *T*-norms, several methods are proposed for *t*-conorms calculation in Type I and Type II. All of these *t*-norms can be developed for the polar form. If P1 and P2 are two polar fuzzy sets, then,

$$\overline{\overline{A}} \cap \overline{\overline{B}} = \mu_{\underset{A \cup B}{\approx}}(\rho, \theta) = \mu_{\underset{A}{\approx}}(\rho, \theta) \sqcap \mu_{\underset{B}{\approx}}(\rho, \theta) = \int_u \int_w f_x(u) * g_x(w)/(u \wedge w) \tag{4.69}$$

where $\cap$ represents the minimum *t*-conorm; $*$ represents a *t*-norm; and $\sqcap$ is the join operation.

Examples of three-dimensional fuzzy sets in polar coordinates are shown in the Figure 4.11.

4.5.4 Representation of Fuzzy 3D Polar Rules

The sample shape of the polar KB in the generalized 3D Type-2 FLS is the typical format of Type-1 FLS. Nevertheless, it is supposed that the antecedents and consequent sets are denoted by G3DT2FS. So for a Type-2 polar FLS with p inputs $x1 \in X1, \ldots, xp \in XP$ and one output $y \in Y$, single input single output (SISO) and multiple input single output (MISO), if we assume there are M rules, the kth rule in the generalized Type-2 polar FLS is expressed as follows:

$$\text{If } X_1 \text{ is } (P\acute{A})_{11} \text{ and...and } X_m \text{ is } (P\acute{A})_{1m}, \text{Then } Y \text{ is } (P\acute{C})_{11} \text{ and } (P\acute{C})_{12} \text{ and...and } (P\acute{C})_{1r} \tag{4.70}$$

As mentioned earlier, there are several ways to combine rules, and using each will not necessarily have the same answer. The choice of method is done using the nature of the problem and the amount of minimum error and maximum speed. In Figure 4.11, the response extracted at different angles is calculated, and its value is quantified based on various methods, including weighted average. The maximum amount and area under the polar membership function are then calculated. Assuming a knowledge bank in polar coordinates and membership functions, different methods are compared in Figure 4.12. As seen, the combined polar function can have completely different results quantitatively and formally.

4.5.5 ϑ-Slice and α – Planes

A ϑ-slice for the general 3D polar Type-2 (G3DPT2), $\tilde{A}$, symbolized by $\tilde{\tilde{A}}_\vartheta$, is the union of all primary membership functions of $\tilde{\tilde{A}}$ whose secondary grades are equal or greater than 1 $(0 \le \alpha \le 1)$ in slice $\vartheta(0 \le \vartheta \le 360)$. According to the definitions an α – plane $\tilde{A}_\alpha$ of GT2FS, $\tilde{A}$ can be indicated as the union of all primary memberships of $\tilde{A}$ with $\mu_{\tilde{A}}(x', u) = \mu_{\tilde{A}} \ge \alpha$ for slice ϑ, as follows:

$$\tilde{\tilde{A}}_\alpha(\vartheta) = \int_{\forall x \in X} \int_{\forall u \in J_x} \{(x, u) \mid f_x(u, \vartheta) \ge \alpha \text{ and } \vartheta\}, \alpha \subseteq [0, 1], \vartheta \subseteq [0, 360] \tag{4.71}$$

An α – cut $\tilde{\tilde{A}}_{\alpha,\vartheta}$ of secondary memberships of $\tilde{A}$ with $\mu_{\underset{A}{\approx}}(x) = \mu_{\underset{A}{\approx}} \ge \alpha$ in ϑ-slice can be symbolized as $S_{\underset{A}{\approx}}(x \mid (\alpha, \vartheta))$ and defined as follows:

$$S_{\underset{A}{\approx}}(x \mid (\alpha, \vartheta)) = \left[S^l_{\underset{A}{\approx}}(x \mid (\alpha, \vartheta)), S^r_{\underset{A}{\approx}}(x \mid \alpha(\alpha, \vartheta)) \right] \alpha \subseteq [0, 1], \vartheta \subseteq [0, 360] \tag{4.72}$$

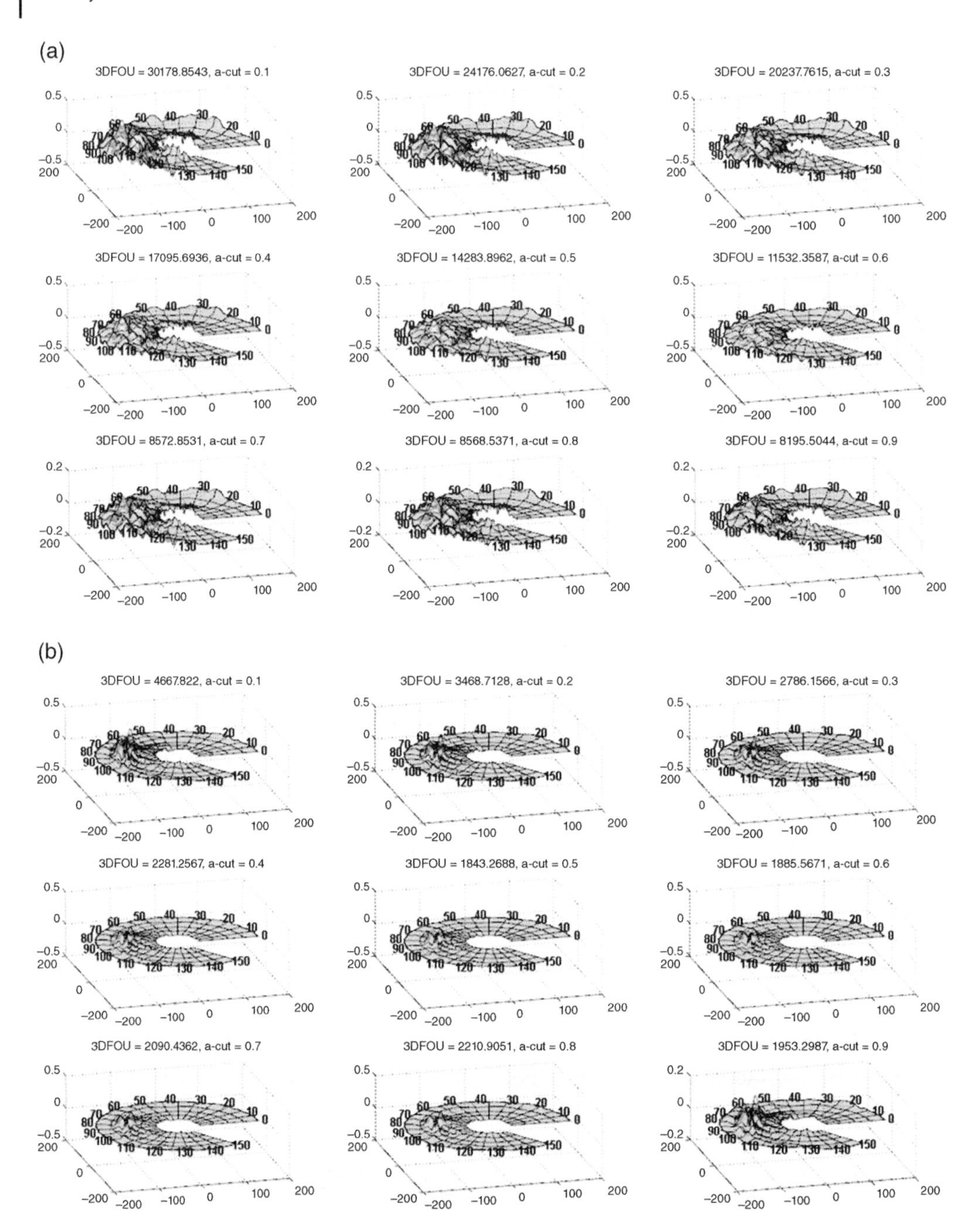

Figure 4.11 Examples of 3D fuzzy sets of (a) $\underline{\mu_{A(r_{ij})}}\left(\acute{A}\right)$ and (b) $\mu_{A(\bar{r}_{ij})}\left(\acute{A}\right)$, PDOU$\left(\widetilde{A}\right)$ in the polar frame (3D PFOU).

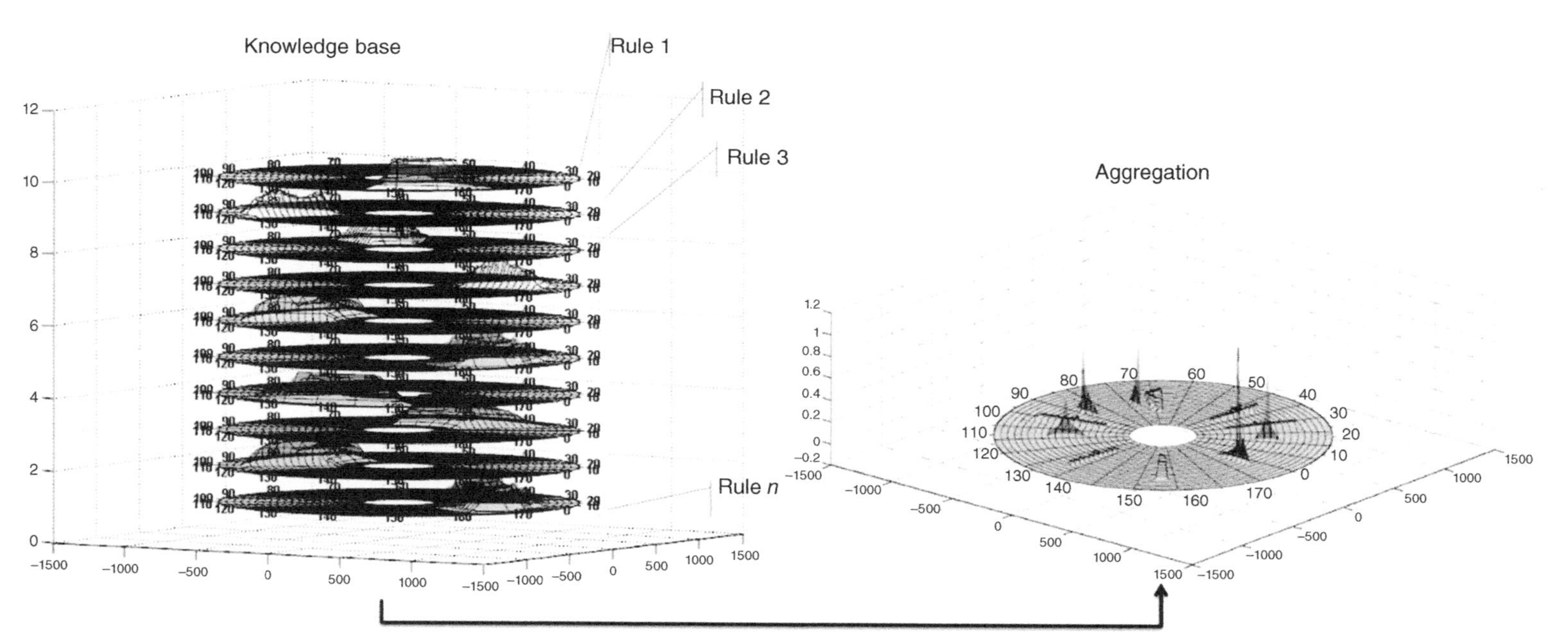

Figure 4.12 Representation of sample fuzzy 3D polar rules from a polar KB.

where an α-plane is an arrangement of all α-cut of all SMFs in ϑ-slice and is expressed as follows:

$$\tilde{\tilde{A}}_{\alpha}(\vartheta) = \int_{\forall x \in X} S_{\tilde{\tilde{A}}}(x \mid (\alpha, \vartheta)) = \int_{\forall x \in X} \left(\int_{\forall u \in \left[S^{l}_{\tilde{A}}(x|(\alpha,\vartheta)), S^{r}_{\tilde{A}}(x|(\alpha,\vartheta)) \right]} u, \right) / x, \alpha \subseteq [0,1], \vartheta \subseteq [0,360],$$

$$\text{3DPFOU}\left(\tilde{\tilde{A}}_{\alpha}(\vartheta)\right) = \tilde{\tilde{A}}_{(\alpha=0)}(\vartheta) = \int_{\forall x \in X} S_{\tilde{\tilde{A}}}(x \mid (\alpha = 0, \vartheta)) = \int_{\forall x \in X} \left(\int_{\forall u \in \left[S^{l}_{\tilde{A}}\left(x|(\alpha = 0,\vartheta)\right), S^{r}_{\tilde{A}}(x|(\alpha = 0,\vartheta)) \right]} u, \right) / x,$$
$$\alpha \subseteq [0,1], \vartheta \subseteq [0,360] \tag{4.73}$$

Based on this theory, examples of three-dimensional polar fuzzy function operators are shown in the Figure 4.12. The edges of each slice can be denoted as follows:

$$\mu_{\tilde{\tilde{A}}}(x \mid (\alpha, \vartheta))_l = \mu_{\tilde{\tilde{A}}}(c')_l = \int_{\forall x \in X} S^{l}_{\tilde{\tilde{A}}}(x \mid (\alpha, \vartheta)), \mu_{\tilde{\tilde{A}}}(x \mid (\alpha, \vartheta))_r = \mu_{\tilde{\tilde{A}}}(c')_r = \int_{\forall x \in X} S^{r}_{\tilde{\tilde{A}}}(x \mid (\alpha, \vartheta)),$$
$$\alpha \subseteq [0,1], \vartheta \subseteq [0,360] \tag{4.74}$$

The polar upper and lower MFs $\left\{ \mu_{\tilde{\tilde{A}}}(x \mid (\alpha, \vartheta))_l, \mu_{\tilde{\tilde{A}}}(x \mid (\alpha, \vartheta))_r \right\}$ are enclosed by $\alpha -$ plane in slice $\vartheta(0 \leq \vartheta \leq 360)$. A different polar IT2 FS ($\alpha -$ level PT2FS) is formed by increasing $\tilde{\tilde{A}}_{\alpha}(\vartheta)$ to the level α in slice ϑ. Also, the 3D PGT2 FS, $\tilde{\tilde{A}}_{\alpha}(\vartheta)$, is formed as a collection of all of its separate α-level P T2 FSs in ϑ-slice $(0 \leq \vartheta \leq 360)$:

$$L_{\tilde{\tilde{A}}}(x', u) = \frac{\alpha_{\vartheta}}{\tilde{\tilde{A}}_{\alpha}(\vartheta)} = \alpha_{\vartheta} / \int_{\forall x \in X} S_{\tilde{\tilde{A}}_{\alpha}}(x \mid (\alpha, \vartheta)) = \alpha_{\vartheta} / \left(\int_{\forall x \in X} \left(\int_{\forall u \in \left[S_{\tilde{\tilde{A}}_{\alpha(\vartheta)}}{}^{l}(x|(\alpha,\vartheta)), S_{\tilde{\tilde{A}}_{\alpha(\vartheta)}}{}^{r}(x|(\alpha,\vartheta)) \right]} u, \right) / x \right),$$

$$\tilde{\tilde{A}} = C_{\alpha \in [0,1]} \left[\left(\frac{\alpha_{\vartheta}}{\tilde{\tilde{A}}_{\alpha}(\vartheta)} \right) \right] = \int_{\forall x \in X} S_{\tilde{\tilde{A}}_{\alpha}}(x \mid (\alpha, \vartheta)) = \int_{\forall x \in X} \left(\int_{\forall u \in \left[S_{A}{}^{l}(x|(\alpha,\vartheta)), S_{A}{}^{r}(x|(\alpha,\vartheta)) \right]} u, \right) / x \tag{4.75}$$

where $\wedge$ denotes the union operation, which computes the $\max\left(\mu_{\tilde{\tilde{A}}}(x \mid (\alpha, \vartheta))_l \right) \wedge \max\left(\mu_{\tilde{\tilde{A}}}(x \mid (\alpha, \vartheta))_r \right) \alpha -$ planes in slice ϑ.

The ϑ-slices for polar MF are similar to α-planes and z-slices. The ϑ-slice based GT2 FLSs in the polar frame are developed in this work. A graphic view of polar GT2 FS is illustrated by its ϑ-slices in Figure 4.13.

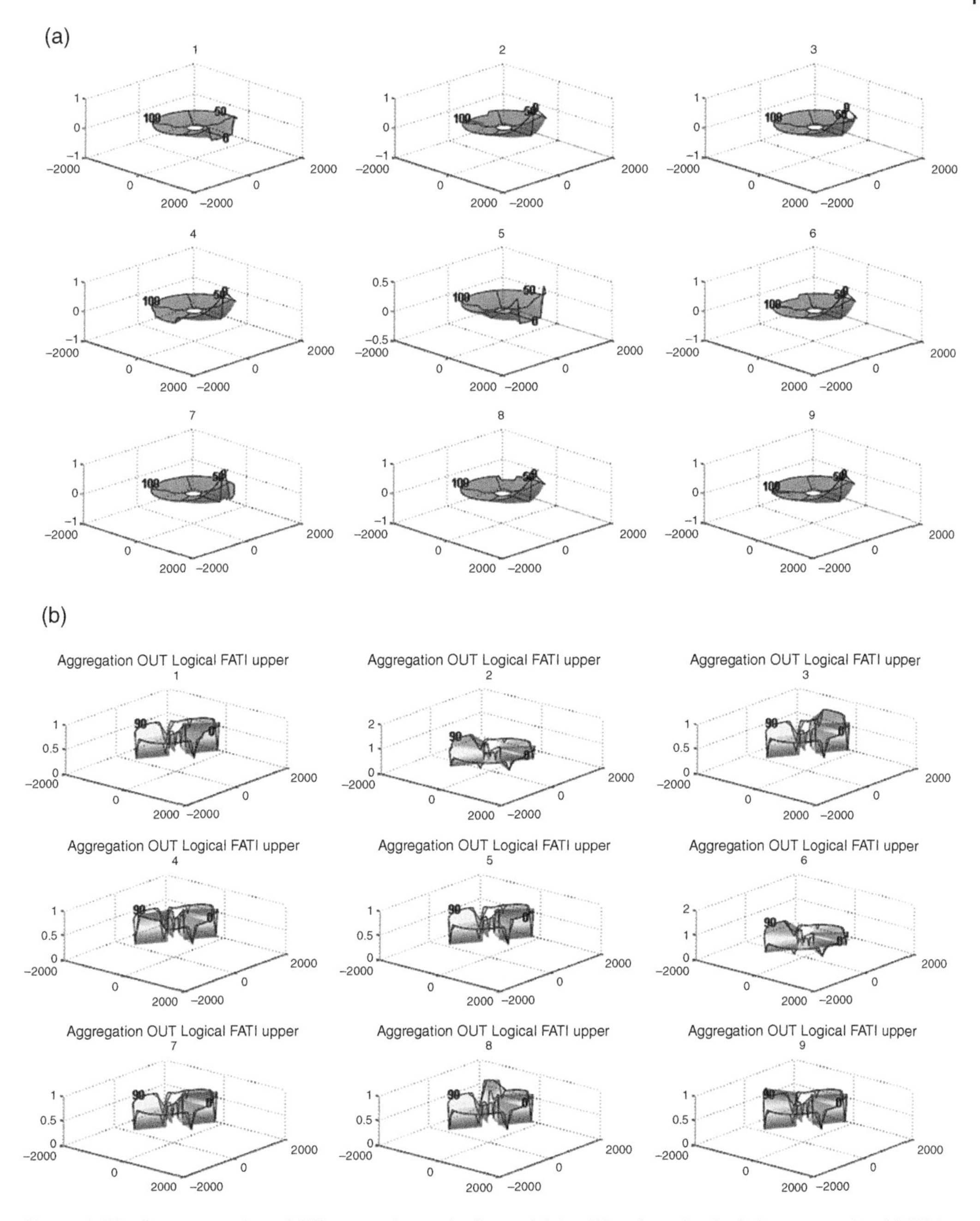

Figure 4.13 Representation of differences in results in combining KB polar rules for inference engine (a) SISO, (b) MISO.

$$\begin{aligned}
\tilde{\tilde{A}}_z(\vartheta) &= \int_{\forall r \in R \forall} \int_{\vartheta \in [0,360]} z/(r,\vartheta),\\
\tilde{\tilde{A}} &= C_{\alpha\in[0,1]} \tilde{\tilde{A}}_z(\vartheta) = \left(\int_{\forall r \in X} \int_{\forall \in [0,1]} z/(r,\vartheta) \right) \quad \alpha \subseteq [0,1]\\
\tilde{\tilde{A}} \cup \tilde{\tilde{B}} \leftrightarrow \mu_{\tilde{A} \cup \tilde{B}}(\rho,\theta) &= \mu_{\tilde{A}}(\rho,\theta) \sqcup \mu_{\tilde{B}}(\rho,\theta) = \int_{u(\rho,\theta) w} \int_{(\rho,\theta)} f_{(\rho,\theta)}(u) * g_{(\rho,\theta)}(w)/(u\hat{e}\, w),\\
\bar{\bar{A}} \cap \bar{\bar{B}} \leftrightarrow \mu_{\tilde{A} \cup \tilde{B}}(\rho,\theta) &= \mu_{\tilde{A}}(\rho,\theta) \sqcap \mu_{\tilde{B}}(\rho,\theta) = \int_{u(\rho,\theta) w} \int_{(\rho,\theta)} f_{(\rho,\theta)}(u) * g_{(\rho,\theta)}(w)/(u \wedge w)
\end{aligned} \tag{4.76}$$

The rules for the polar general Type-2 fuzzy logic systems (PGT2FLS) have the same format as the previous instances of FLS, T1, and IT2 with a difference in notation.

$$\begin{aligned}
&\text{Rule}\, n: \text{If}\, X_{n1(\rho,\theta)}\, \text{is}\, (P\acute{A})_{n1}(\rho,\theta)\, \text{and}\, X_2(\rho,\theta)\, \text{is}\, \left(\acute{f}\right)_{n2}(\rho,\theta)\, \text{and...and}\, X_m(\rho,\theta)\, \text{is}\\
&\left(\acute{f}\right)_{nm}(\rho,\theta)\, \text{Then}\, Y\, \text{is}\, (\acute{g})_n(\rho,\theta)
\end{aligned} \tag{4.77}$$

where $\left(\acute{f}\right)_{nm}(\rho,\theta)$ denotes the ith T2PFM antecedent; $\left(\acute{g}\right)_n(\rho,\theta)$ denotes the general polar Type 2 fuzzy system (GPT2FS) consequence; and $\left(\acute{g}\right)_n(\rho,\theta)$ is the nth consequence of GPT2FS fuzzy set of Rule n.

The fuzzy inference from the GT2FLS in the polar frame is defined as follows:

$$\begin{aligned}
\mu_{\tilde{B}}(\rho,\theta) &= \int\int_{b\epsilon} \begin{bmatrix} \left[\left[\underline{f}^{1*}\underline{\mu}_{G_N}(\rho,\theta)\right]\cdots\left[\underline{f}^{N_M*}\underline{\mu}_{G_N}(\rho,\theta)\right]\right], \\ \left[\left[\bar{f}^{1*}\bar{\mu}_{G_N}(\rho,\theta)\right]\cdots\left[\bar{f}^{N*}\bar{\mu}_{G_N}(\rho,\theta)\right]\right] \end{bmatrix} (1/b), (\rho,\theta)\epsilon\mathcal{R},\\
&\left[\underline{f}^l(\rho,\theta), \bar{f}^l(\rho,\theta)\right]\\
&= \left[\left[\underline{\mu}_{F^l_1}(\rho,\theta)\tilde{*}...\tilde{*}\underline{\mu}_{F^l p_1}(\rho,\theta)\right], \left[\bar{\mu}_{F^l_1}(\rho,\theta)\tilde{*}...\tilde{*}\bar{\mu}_{F^l p_1}(\rho,\theta)\right]\right],\\
&\left[\underline{b}^l(\rho,\theta), \bar{b}^l(\rho,\theta)\right] = \left[\underline{f}^l\tilde{*}\underline{\mu}_{G^l}(\rho,\theta), \bar{f}^l\tilde{*}\bar{\mu}_{G^l}(\rho,\theta)\right]
\end{aligned} \tag{4.78}$$

As shown, the centroid is one of the dominant procedures that are recommended for type-reduction of G3DPT2FS, where θ_N is a combination associated to the secondary degree ω in the position (ρ, θ):

$$\omega_{(\rho,\theta)} = f_{(\alpha,\vartheta)}(\theta_1)\tilde{*}...\tilde{*}f_{(\alpha,\vartheta)}(\theta_N), C = \int_{(\rho,\theta)_1 \epsilon J_{x_1}} \cdots \int_{(\rho,\theta)_N \epsilon J_{x_N}} \left(\frac{\left[f_{(\alpha,\vartheta)_1}((\rho,\theta)_1)\tilde{*}...\tilde{*}f_{(\alpha,\vartheta)_N}((\rho,\theta)_N)\right]}{\left(\left(\sum_{i=1}^{N}(\alpha,\vartheta)_i(\rho,\theta)_i\right)/\left(\sum_{I=1}^{N}(\rho,\theta)_i\right)\right)} \right) \tag{4.79}$$

4.6 Computational Performance (CP)

The CP of the G3DT2FLS in the polar border and its interpretation processes were related with the discrete system (DS) and geometric system (GS). Five basic examples were developed for MISO. G3DT2 FLS, holding 2, 4, 6, 8, and 10 rules, respectively. Then each case was learned to display values over a series of 50 cycles for 3DPMF with repetition three times to moderate

Table 4.1 The average computational times.

Number of rules		Mean		Mean		Mean		Mean
Method presented in [9]	Discrete	>1.6		>1.6		>1.6		>1.6
	Geometric	>6		>6		>6		>6
CI	($S = 1$)	0.23	SI	0.25	NNI	0.22	BI	0.23
	($S = 0.5$)	0.68		0.57		0.48		0.51
	($S = 0.1$)	10.84		5.18		6.99		7.42
Mean		3.9		2.0		2.6		2.7

SI, spline Interpolation; CI, bi-cubic Interpolation; BI, bilinear Interpolation; NNI, nearest neighbor interpolation, the results for 3 runs to at 100 consequent sets; S, ϑ-slice.

errors in $\vartheta -$ slice= 1, 0.5, and 0.1 with diverse mesh scales and interpolation approaches, which involved [9]:

- Spline interpolation (SI)
- Bilinear interpolation (BI)
- Bicubic interpolation (CI)
- Nearest neighbor interpolation (NNI) are compared.

The regular time for trial tests is stated for 100 consequent sets over the three runs in Table 4.1. The SI has a higher performance in comparison with bilinear interpolation, bicubic interpolation, and nearest neighbor interpolation, with a speed rate of 1.55–3.04 times higher than the discrete method and 20.60–40.37 times higher than the geometric method. Normally, these outcomes display that a G3DT2 FL in the polar frame is an effective technique for pattern recognition and handling ambiguity in 3D with high speed. The general 3D Type-2 polar fuzzy logic systems (G3DPT2 FLs) outperformed both geometric and discrete methods, exhibiting a higher flexibility of ϑ-slice > 0 [9].

The G3DPT2 FLs directed an approximately linear increase in computation time as the $\vartheta -$ slices and number of rules increased. When $n = 3$ for the G3DPT2 FLs, the SI interpolation reduced the computation time by over 67% and 98% compared to the discrete and geometric method, respectively. For G3DPT2 FLs, the CI interpolation reduced the computation time by over 36% and 95% compared to the respective methods [9].

4.7 Application of G3DT2FLS in Pattern Recognition

4.7.1 Examples of the Application of Fuzzy Methods in Infrastructure Management

In this section, two applications of this method in road pavement management are briefly presented. The first application is the automatic detection of the most important road pavement failure, namely cracking, using image processing. In the second application, image processing is used to detect polymer in bitumen using image processing and the proposed theory.

In this section, the application of a new method is introduced for automatically generating a membership function in a Type-2 fuzzy logic context in polar coordinates. The purpose of this method is to provide an example of the application of the automated modified GT2FLS structure for fuzzy logic methods (FLS), which are capable of modeling nonindependent structures to enhance the ability to model uncertainty in descriptive rules based on FOU 3D. An automatic generator of membership functions can be extracted using the trained information gain methods and optimal membership functions. General Type-2 fuzzy logic systems (G3DT2FLS) have been proposed as a new solution in cracking pattern analysis. The purpose of this section is to perform experiments related to the automatic MF generator of general Type-2 polar fuzzy membership (GT2PFM) and to provide an ambiguity measurement index, namely logical operators in polar frames. By constructing a FMF in the polar context, it is possible to develop the applied models. For this purpose, several experiments were performed to select a FMF based on the information gain.

The choice of the optimal membership function is related to the information gain. To evaluate the ambiguity and uncertainty in the membership function, the entropy index was used in this study. The higher the probability of occurrence in one direction, the more uniform the information gain (IG) information gain. CSS is used for smoothing based on the polar coordinate membership function presented in the auto-generating algorithm. In this method, by setting the p parameter, the value of 3DPMF (r, θ) is estimated in three dimensions. By minimizing the estimated page, the amount of error can be controlled. According to a seven-step model, the p parameter controls the estimation between two points, whereby the closer p is to 1, the less ambiguity the function has. In this section, based on the algorithm presented in Figure 4.14. Different entropies are calculated for different α-cuts, and the maximum entropy value is considered to be α-cut (_opt). If the value is α_opt, the entropy committee will have values up to 1. The implication of this is that Type-2 fuzzy is downgraded to Type-1, and in practice, the amount of ambiguity is reduced. In the method provided, entropies will have values up to 1, which means that Type-2 fuzzy has been downgraded to Type-1, and in practice, the amount of ambiguity is reduced.

Each point with a higher degree of membership is associated with pavement cracking in a particular direction, and this feature can be used to classify the type of crack. If the position of membership functions of $\overline{\mu_{A(r_{i,j})}}(\acute{A})$, and $\underline{\mu_{A(r_{i,j})}}(\acute{A})$ is larger than the threshold, the pattern can be used as a feature for the classification of the type of failure and are used to determine the location and the direction of linear cracks. If the MF is larger than the threshold, the profile is used as a feature to classify and determine the main characteristics of the pavement. Due to the ambiguity in the elements with a lower degree of membership, choosing a threshold with a suitable substrate reduces the ambiguity and increases the accuracy in selecting the maximum points. An example of these membership functions is shown in Figure 4.15.

The relation μ_(i, j) is set using the variable H, where the higher the H value, the higher the threshold and the lower the crack detection rate. The combination of two obscure parts, namely the membership function and dynamic filter screen, reduces errors and increases accuracy in detecting cracks. Figure 4.15 provides an example of a dynamic threshold that corresponds to the membership function. In order to smooth out CSS with the control parameter p, the value of 3DPMF (r, θ) is estimated in three dimensions. Also, by minimizing the estimated plate, the error value is minimized, and the closer p is to 1, the threshold becomes harder. The experiment was conducted on the db1 images, consisting of 260 sections, which were analyzed and evaluated. Table 4.2 shows the results for the maximum value up to radon transform maximum value (RTMV) for failure with different intensities, revealing that the amount of RTMV increases as the width of cracks and spalls of Type-1 and Type-2 increase. By applying the adaptive filtering method, the image related to the crack is separated. This method provides a new indicator for evaluating crack fragmentation.

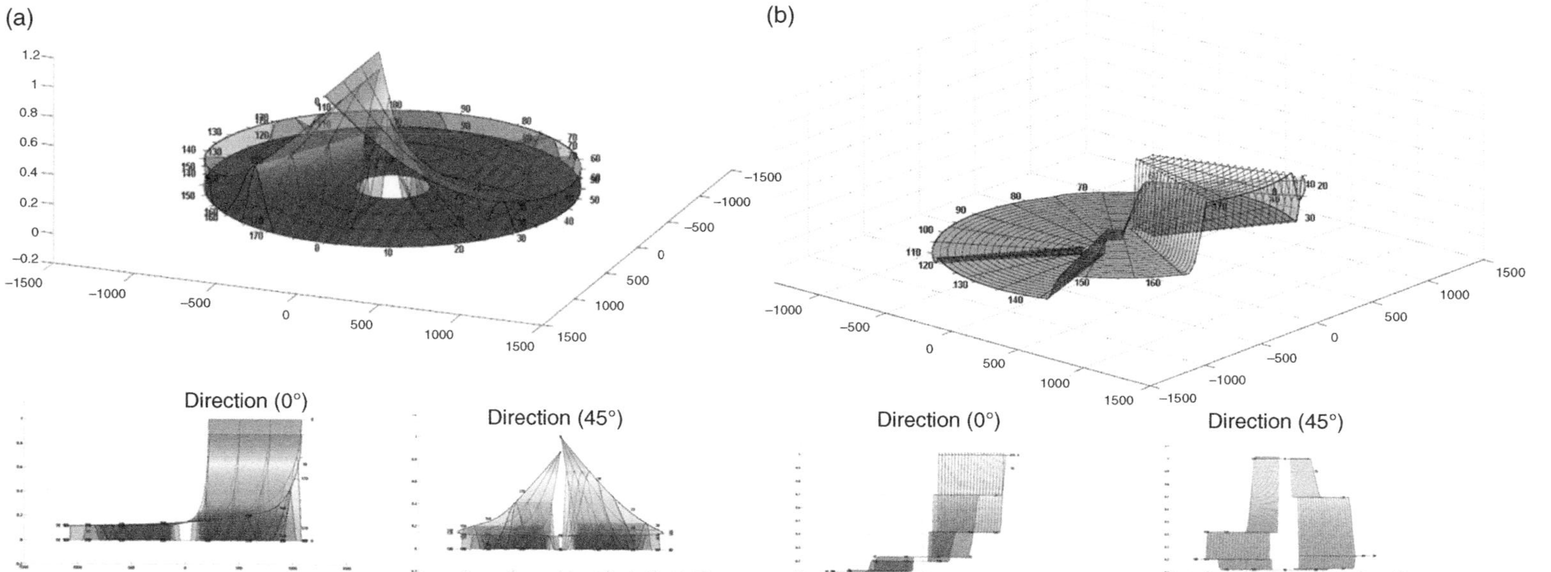

Figure 4.14 Representation of 3D view of T2 FSs in polar frame: (a) GT2 FS represented by α-planes and (b) GT2 FS represented by ϑ-slices [9].

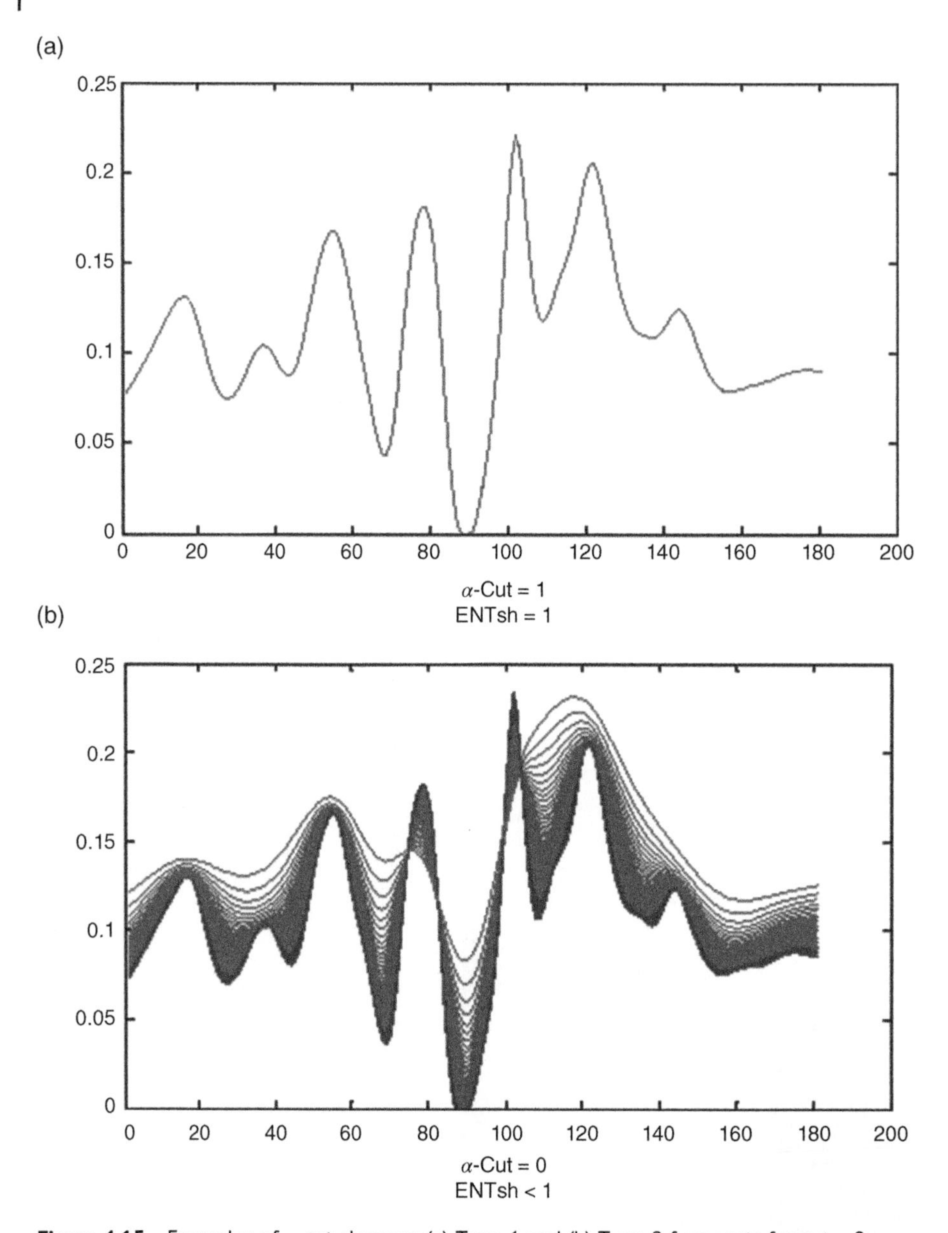

Figure 4.15 Examples of α-cut changes: (a) Type-1 and (b) Type-2 fuzzy sets for cut = 0.

To classify pavement distress, five rules that are extracted based on theory are presented. Therefore, the KB of fuzzy system is constructed based on the rules shown in Table 4.2. The zone and some consequences for five classes are presented below [2].

If the maximum firing rule is positioned in zones L, T, and D and is single, then the number of peaks increases by one, and the position of peaks is used for classification. This process continues until no peak remains. The parameter ($\Delta\theta$) is considered as a control rule, where $\Delta\theta_1$ is the difference between the two-line slope or angle between two polar support vector machines (PSVMs) in the polar domain.

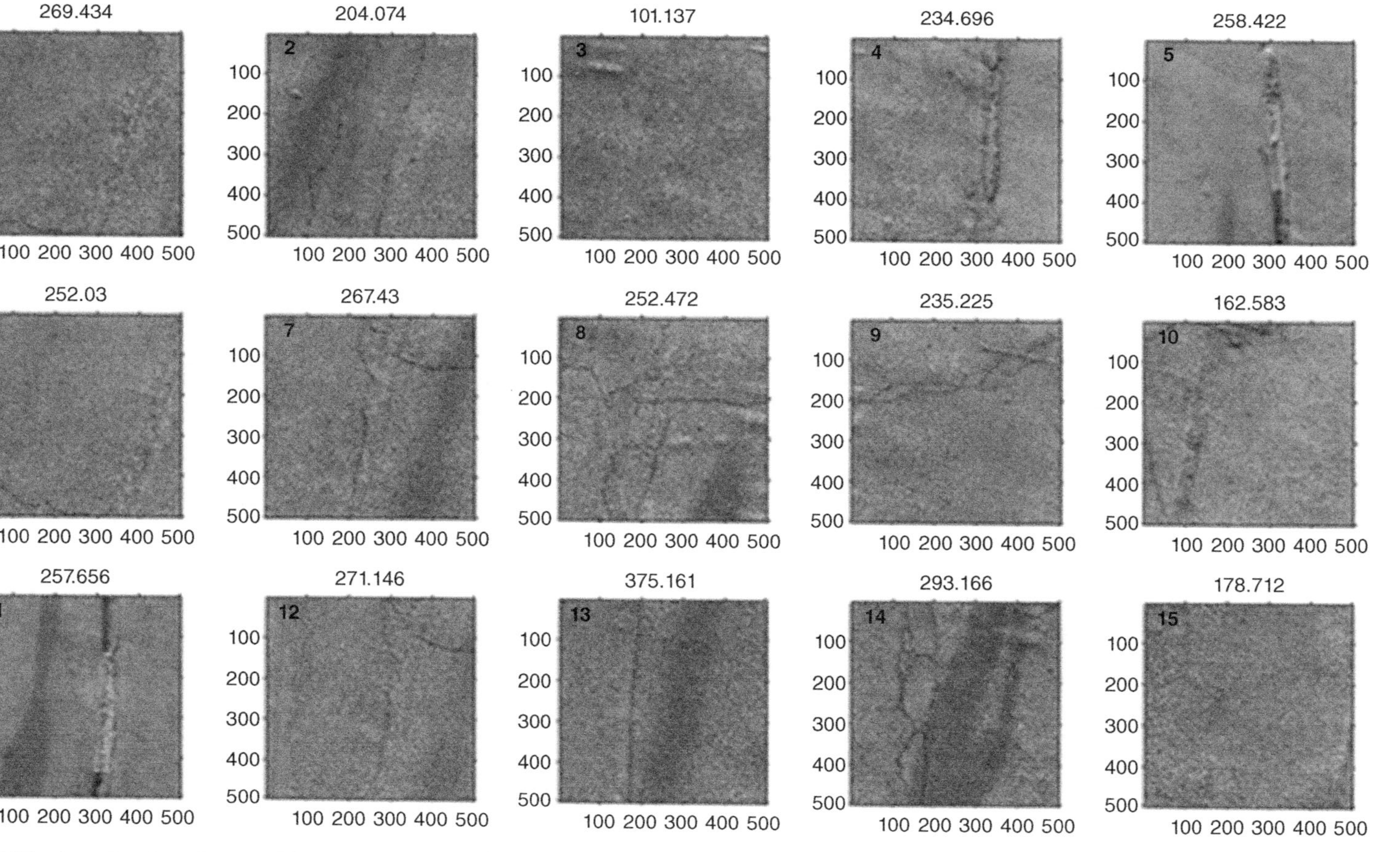

Figure 4.16 Investigation of types of failures with different intensities in order to extract the highest probability of recurrence of $\mu\_(i, j)$.

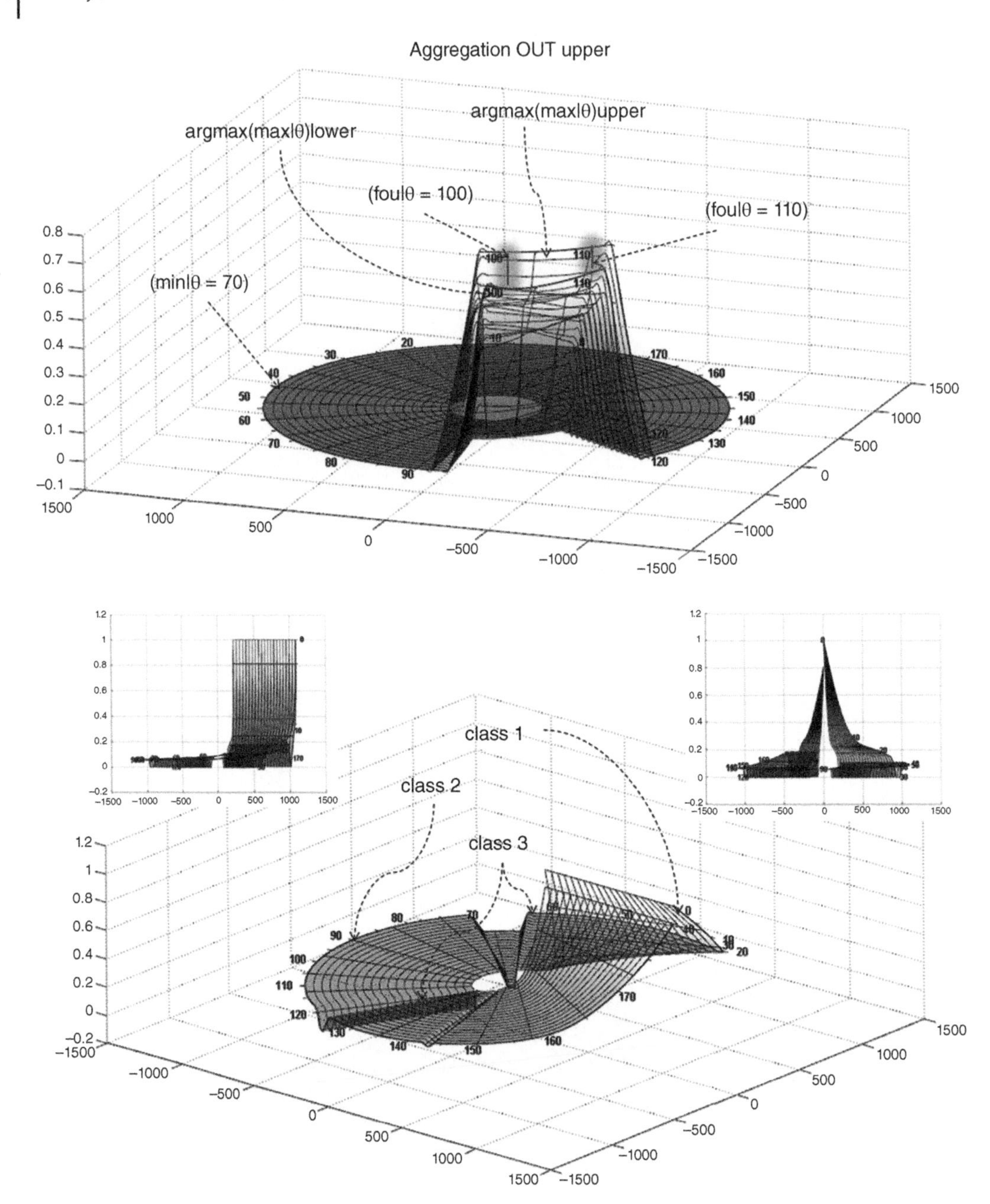

Figure 4.17 Representation of max firing rule of 3DPT2 from Kb for pavement crack detection and classification.

Despite its complexity, the developed Type-3 fuzzy method offers a variety of artificial intelligent (AI) indicators for analysis, evaluation, and classification. The 3D view of T2 FSs in polar frame: Fuzzy rule base of the described fuzzy controller designed for pattern recognition of pavement cracking distress (Figures 4.16–4.18).

Table 4.2 The average computational times over 3 runs to arrive at 100 consequent sets (in seconds).

Rule	IF	And/Or	Then
Rule 1	SC	($\Delta\theta_5$ or$\Delta\theta_4$) and (DD)	(LC)
Rule 2	SC	($\Delta\theta_3$ or$\Delta\theta_1$) and (DD)	(TC)
Rule 3	SC	($\Delta\theta_2$) and (DD)	(DC)
Rule 4	MC	($\Delta\theta_5$ and$\Delta\theta_4$) and (DD)	(BC)
Rule 5	MC	($\Delta\theta_1$ or$\Delta\theta_3$) and ($\Delta\theta_2$and/or$\Delta\theta_5$ and/or$\Delta\theta_4$a) and (DD)	(AC)

AC, aligátor cracking; BC, block cracking; DC, diagonal cracking; DD, distress detected; LC, long cracking; MC, multicrack; ND, no defect; SC, single crack; TC, transver cracking.

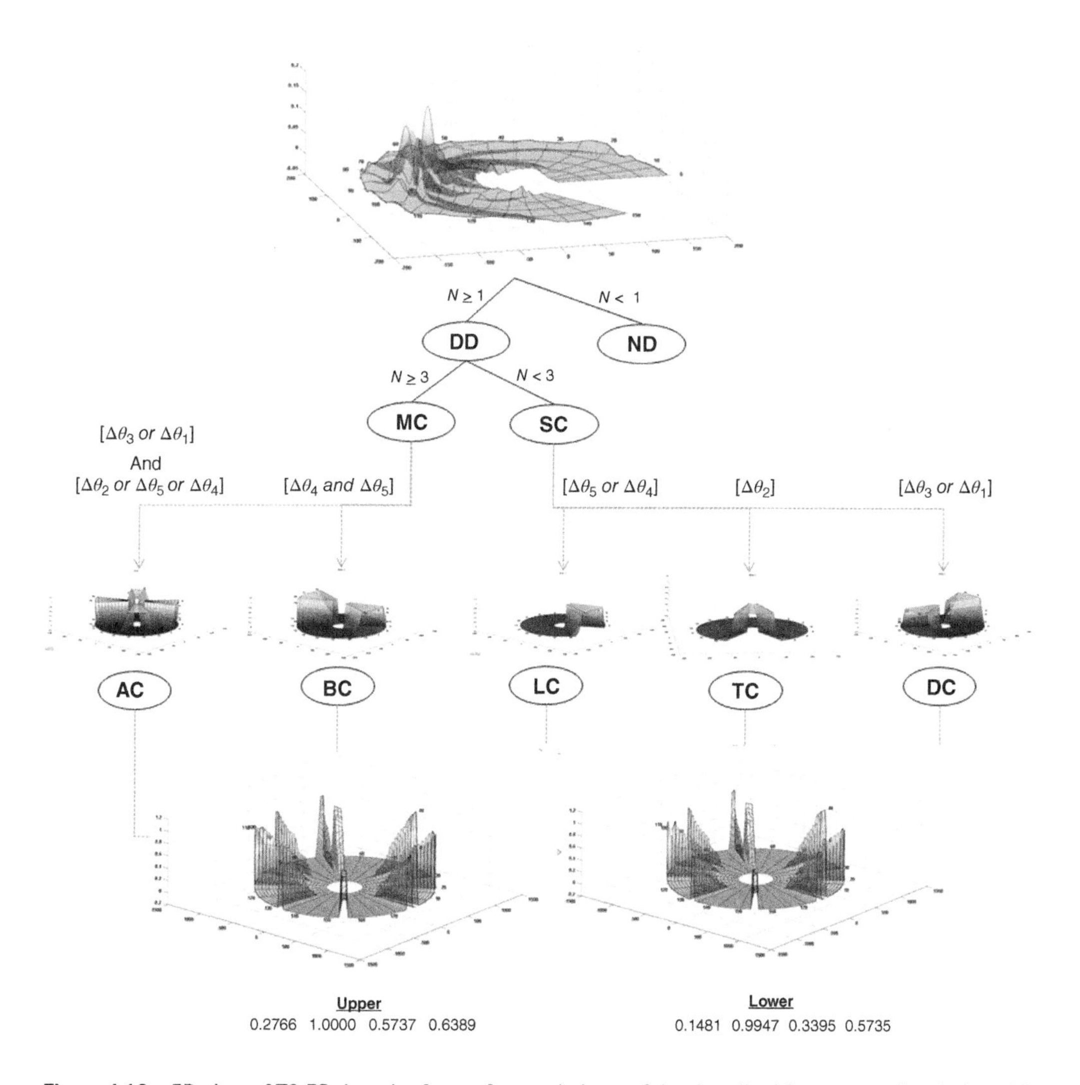

Figure 4.18 3D view of T2 FSs in polar frame: fuzzy rule base of the described fuzzy controller designed for pattern recognition of pavement cracking distress.

Another application of the 3DPT2 method in intelligent calculations is the interpretation of bitumen and the amount and dispersion of additives to the mixture. The PPF absorption is determined in an area that can be computerized using image processing and AI methods [3].

Two-phase systems can be considered for the distribution of polymer within the bitumen on the microscale: (i) polymer-rich phase and (ii) asphaltene-rich phase. In images, the polymer phase appears lighter than the bitumen phase. A new classification method between the PMBs and bitumen phase is proposed based on watershed segmentation and RT (Figure 4.19).

Automatic interpretation of polypropylene modified bitumen (PMB) based on image processing is another new application of this method. The main purpose is to provide a new automatic technique for the interpretation of PMB properties. An image-based system using several statistical criteria was developed based on fuzzy segmentation (FCM) and RT for the approximation of polypropylene features, which include the high amplitude radon percentage (HARP), high energy radon percentage, standard deviation, momentum, and a set of morphological features. The vector of features is selected by applying the flexible threshold on 3DRT.

Uneven boundaries PPF(s), such as 1%, 3%, and 5%, are transformed into high amplitude coefficients, while 13%, 11%, and 9% are transformed into lower amplitude coefficients. Therefore, counting the coefficients that are larger than the threshold proves a good index for quantifying the properties of PPF(s). To further demonstrate the advantages of the proposed method, the RT is shown in 3D by expanding the RT value as a component. After this step, 3D RT is binarized with an adoptive threshold hyperplane surface as follows:

$$D_1(x', \theta) = \begin{cases} 1 \text{ if } \mathrm{NRT}_{(x',\theta)} \geq C_{\mathrm{thu}(x',\theta)} \\ 0 \text{ if } \mathrm{NRT}_{(x',\theta)} < C_{\mathrm{thl}(x',\theta)} \end{cases} \tag{4.80}$$

where $\mathrm{NRT}_{(x',\theta)}$ is normalized radon transform; $D_1(x', \theta)$ is the binarized RT; $C_{\mathrm{thu}(x',\theta)}$ is the adaptive upper threshold of RT; and $C_{\mathrm{thl}(x',\theta)}$ is the adaptive lower threshold.

We used CSS to generate the two-thresholding surface as the 3D RTs. The approximation of the $C_{\mathrm{thu}(x',\theta)}$ and $C_{\mathrm{thl}(x',\theta)}$ can be extracted from CSS of the 3D RT $f(x', \theta)$.

PPF (%) content has an influence on the mixture, HARP, which increases to some degree with the reduction of PPF (%) in different images. Also, as PPF (%) increases, there are fewer changes to HARP and the effect of PPF decreases. As a conclusion, a connection can be found between the PPF and HARP. It is pertinent to note that the relations are shown in Figure 4.18. Apply only for PPF of 5–13%. Specifically, Δ(HARP) decreases less noticeably when the PPF rate is increased from 5 to 13 (%), and even less remarkably for PPF 1–3 (%) [3].

4.8 Summary and Conclusion

In this chapter, the theoretical framework for G3DPT2 FLS, as an extension of a Type-2 FLs, is presented. The automatic MF Generator (GT2 PFM), new geometric operators in polar frame, inference consisting of fuzzy 3D polar rules, and antecedents/consequents ϑ-slice and α-planes are presented. The indices of compactness and smoothness were used for tuning the 3DMFS. Also, several desirable properties of the technique to generate the MFs are described. A new index of ultrafuzziness, according to the 3DMF fuzzy set, is also provided. For G3DPT2 FLs, the novel 3D FOU and DOU are defined. The minimum, maximum, equal, and reduced ultrafuzziness were evaluated in the polar domain. Additionally, the set theoretic operations of 3D Type-2 fuzzy sets in the polar

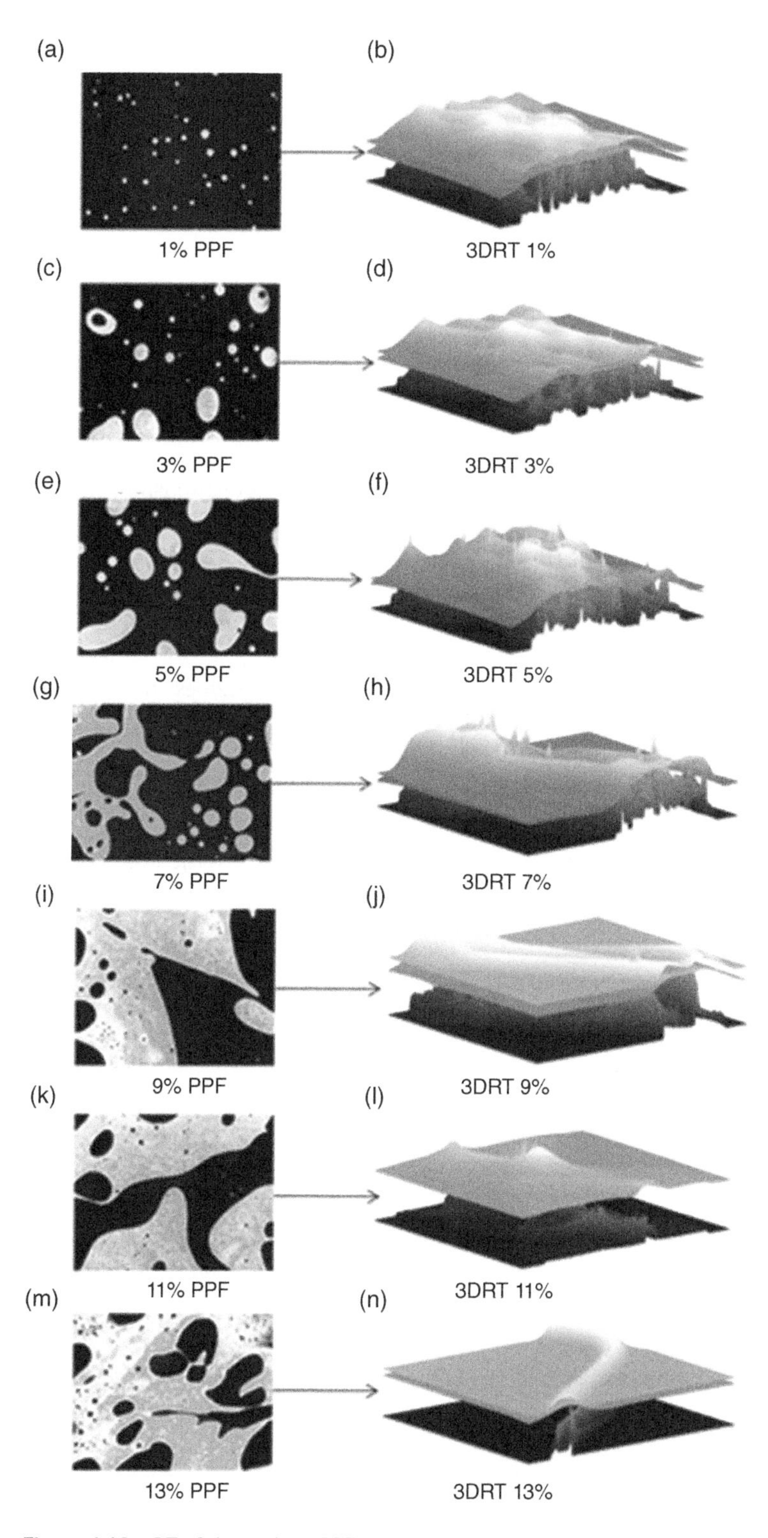

Figure 4.19 RT of the various PPF(s) percentages based on the proposed thresholding method, (a) PPF 1%, (b) 3DRT 1%, (c) PPF 3%, (d) 3DRT 3%, (e) PPF 5%, (f) 3DRT 5%, (g) PPF 7%, (h) 3DRT 9%, (i) PPF 9%, (j) 3DRT 9%, (k) PPF 11%, (l) 3DRT11%, (m) PPF 13%, (n) 3DRT 13%. *Source:* Tapkin et al. [3]/Springer Nature.

frame are discussed. The 3DGT2PFS was developed for upper and lower bounds, which were automatically extracted. In conclusion, it is now possible to quantify the uncertainty of 3D MFs in the polar frame. Due to this flexibility, in ϑ-slice, α-planes, and 3DPFOU, G3DPT2 FLs is a promising alternative for the computation of general Type-2 FLSs.

Also, two applications of the proposed method in pavement management and automation are presented in detail.

4.9 Questions and Exercises

1 Give three examples of the application of fuzzy topics related to pavement engineering, road pavement management, and infrastructure.

2 Explain the steps of generating a 3D membership function and prove its relationships in two- and three-dimensions.

3 Define a measure of ultrafuzziness and prove its equation. Provide a new relationship for fuzziness and ambiguity based on minimizing the degree of ambiguity using an AI method.

4 PROOF OF Kaufmann Conditions (four parts): (i) for minimum ultrafuzziness or $\tilde{\gamma}(\mathrm{A}) = 0$, one; (ii) for maximum ultrafuzziness or $\tilde{\gamma}\left(\tilde{\mathrm{A}}\right) = 1$; (iii) for equal ultrafuzziness or $\tilde{\gamma}\left(\tilde{\mathrm{A}}\right) = \tilde{\gamma}\left(\tilde{\tilde{\mathrm{A}}}\right)$; and (iv) for reduced ultrafuzziness or $\tilde{\gamma}(\bar{A}\_(i,j)) < \gamma\ (\tilde{\gamma}(\bar{A}\_(d,c))$.

5 PROOF OF Similarity Operators in Polar Frame: Consider $\overline{\mu_{A(r_{i,j})}}(\acute{A})$ to be the upper bound (surface) and $\underline{\mu_{A(r_{i,j})}}(\acute{A})$ to be the lower bound (surface) of 3D polar MFs.

6 PROOF OF Correspondence of Fuzzy Sets in Polar Frame: Consider $\overline{\mu_{A(r_{i,j})}}(\acute{A})$ to be the upper bound (surface) and $\underline{\mu_{A(r_{i,j})}}(\acute{A})$ to be the lower bound (surface) of 3D polar MFs.

7 PROOF OF (NOT)-Complement Operators in Polar Frame: $\left\{\mathrm{PFOU}\left(\acute{A}\right)_{(\theta,r)}\right\} = \left\{\mathrm{PFOU}\left(\acute{A}\right)_{(\theta,r)}\right\}$.

Further Reading

1 Zadeh, L.A., G.J. Klir, and B. Yuan, *Fuzzy sets, fuzzy logic, and fuzzy systems: Selected papers*. Vol. **6**. 1996. World Scientific.

2 Zakeri, H., F.M. Nejad, and A. Fahimifar, *Rahbin: A quadcopter unmanned aerial vehicle based on a systematic image processing approach toward an automated asphalt pavement inspection. Automation in Construction*, 2016. **72**: p. 211–235.

3 Tapkın, S., H. Zakeri, A. Topal, F. M. Nejad, A. Khodaii, and B. Şengöz, A brief review and a new automatic method for interpretation of polypropylene modified bitumen based on fuzzy radon transform and watershed segmentation. *Archives of Computational Methods in Engineering*, 2020. **27**(3): p. 773–803.

4 Mataei, B., H. Zakeri, and F.M. Nejad, An overview of multiresolution analysis for nondestructive evaluation of pavement surface drainage. *Archives of Computational Methods in Engineering*, 2019. **26**(1): p. 143–161.
5 Mataei, B., F.M. Nejad, H. Zakeri, and A.H. Gandomi, *Computational intelligence for modeling of pavement surface characteristics*. in *New materials in civil engineering*, Samui, P., N. R. Iyer, Butterworth-Heinemann. 2020, 65–77, https://doi.org/10.1016/B978-0-12-818961-0.00002-8.
6 Castillo, O. *Introduction to Type-2 Fuzzy Logic Control. In: Type-2 Fuzzy Logic in Intelligent Control Applications*. Studies in Fuzziness and Soft Computing, Vol. **272.** 2012. Springer, Berlin, Heidelberg. https://doi.org/10.1007/978-3-642-24663-0_1
7 Castillo, O., et al., A comparative study of type-1 fuzzy logic systems, interval type-2 fuzzy logic systems and generalized type-2 fuzzy logic systems in control problems. *Information Sciences*, 2016. **354**: p. 257–274.
8 Zakeri, H., et al. A new automatic MF generator (AMFG) for general 3D type-ii fuzzy in the polar frame. in 2014 IEEE Conference on Norbert Wiener in the 21st Century (21CW). 2014. IEEE.
9 Zakeri, H., F.M. Nejad, and A. Fahimifar, General 3-D type-II fuzzy logic systems in the polar frame: concept and practice. *IEEE Transactions on Fuzzy Systems*, 2017. **27**(4): p. 621–634.
10 Jayaraman, S., M. Ramachandran, R. Patan, M. Daneshmand, and A. H. Gandomi, Fuzzy deep neural learning based on Goodman and Kruskal's Gamma for search engine optimization. *IEEE Transactions on Big Data*, 2020. **8**(1).
11 Krishankumar, R., et al., Solving renewable energy source selection problems using a q-rung orthopair fuzzy-based integrated decision-making approach. *Journal of Cleaner Production*, 2021. **279**: p. 123329.
12 Krishankumar, R., K. S. Ravichandran, P. Liu, S. Kar, and A. H. Gandomi, A decision framework under probabilistic hesitant fuzzy environment with probability estimation for multi-criteria decision making. *Neural Computing and Applications*, 2021: p. 1–17, https://doi.org/10.1007/s00521-020-05595-y.
13 Krishankumar, R., et al. *Selection of apt renewable energy source for smart cities using generalized orthopair fuzzy information*. in *2020 IEEE Symposium Series on Computational Intelligence (SSCI)*. 2020. IEEE.
14 Liu, S., S. Wang, X. Liu, A. H. Gandomi, M. Daneshmand, K. Muhammad, and V. H. C. de Albuquerque, Human memory update strategy: a multi-layer template update mechanism for remote visual monitoring. *IEEE Transactions on Multimedia*, 2021. 2188–2198.
15 Rani, P., A. R. Mishra, R. Krishankumar, K. S. Ravichandran, and A. H. Gandomi, A new Pythagorean fuzzy based decision framework for assessing healthcare waste treatment. *IEEE Transactions on Engineering Management*, 2020. 1–15.
16 Sivagami, R., R. Krishankumar, V. Sangeetha, K. S. Ravichandran, S. Kar, and A. H. Gandomi, Assessment of cloud vendors using interval-valued probabilistic linguistic information and unknown weights. *International Journal of Intelligent Systems*, 2021. **36**(8), 3813–3851.
17 Jawahar, C., P.K. Biswas, and A. Ray, Investigations on fuzzy thresholding based on fuzzy clustering. *Pattern Recognition*, 1997. **30**(10): p. 1605–1613.
18 Aja-Fernández, S., A.H. Curiale, and G. Vegas-Sánchez-Ferrero, A local fuzzy thresholding methodology for multiregion image segmentation. *Knowledge-Based Systems*, 2015. **83**: p. 1–12.
19 Zarandi, M.F., F.M. Nejad, and H. Zakeri, *A type-2 fuzzy model based on three dimensional membership functions for smart thresholding in control systems*. in *Fuzzy controllers-recent advances in theory and applications*, Iqbal, S., N. Boumela, National University of Science and Technology: Russia, 2012: p. 85–118.

20 Krishankumar, R., et al., Interval-valued probabilistic hesitant fuzzy set-based framework for group decision-making with unknown weight information. *Neural Computing and Applications*, 2021. **33** (7): p. 2445–2457.
21 Abdulgader, M.M. *Bio inspired evolutionary fuzzy system for data classification*. 2019. University of Toledo.
22 Zadeh, L.A. and R.A. Aliev. *Fuzzy logic theory and applications: part I and part II*. 2018. World Scientific Publishing.
23 Terano, T., K. Asai, and M. Sugeno. *Fuzzy systems theory and its applications*. 1992. Academic Press Professional, Inc.
24 Klir, G. and B. Yuan. *Fuzzy sets and fuzzy logic*. Vol. **4**. 1995. Prentice Hall, New Jersey.
25 Torshizi, A.D., M.H.F. Zarandi, and H. Zakeri, On type-reduction of type-2 fuzzy sets: A review. *Applied Soft Computing*, 2015. **27**: p. 614–627.
26 Tahayori, H., A. Sadeghian, and A. Visconti. Operations on type-2 fuzzy sets based on the set of pseudo-highest intersection points of convex fuzzy sets. in 2010 Annual Meeting of the North American Fuzzy Information Processing Society. 2010. IEEE.
27 Liu, F., An efficient centroid type-reduction strategy for general type-2 fuzzy logic system. *Information Sciences*, 2008. **178**(9): p. 2224–2236.
28 Nejad, F.M. and H. Zakeri. The hybrid method and its application to smart pavement management. *Metaheuristics in Water Geotechnical and Transport Engineering*, 2012. **439**, 439–484.
29 Karnik, N.N. and J.M. Mendel. Operations on type-2 fuzzy sets. *Fuzzy Sets and Systems*, 2001. **122**(2): p. 327–348.
30 Zhai, D. and J.M. Mendel. Computing the centroid of a general type-2 fuzzy set by means of the centroid-flow algorithm. *IEEE Transactions on Fuzzy Systems*, 2010. **19**(3): p. 401–422.
31 Wagner, C. and H. Hagras. Toward general type-2 fuzzy logic systems based on z-slices. *IEEE Transactions on Fuzzy Systems*, 2010. **18**(4): p. 637–660.

5

Automatic Detection and Its Applications in Infrastructure

5.1 Introduction

In infrastructure maintenance and management, the most important task and mission of the network manager and engineers are to maintain and improve the efficiency of infrastructure structures, such as roads, bridges, technical buildings, and structures. Before starting any treatment, the distress and its initial extent must be determined. Infrastructure distress occurs due to various causes, including aging, overload, repeated loading and fatigue of materials, poor performance, weather conditions and extreme heat or cold, and frequency of temperature changes, or a combination of all these factors. Visual defects are signs of an abnormality in the infrastructure, and clinical signs can lead to practical and effective solutions. Since it is not possible to visually inspect all infrastructures accurately and quickly, human errors are the main cause of evaluation errors, and therefore, automatic systems are used for detection. In this chapter, the basic principles and working methods of diagnosis and new effective parameters in diagnosing failure or anomaly are presented.

Various hypotheses and ideas with different structures have been presented in recent decades for processing, evaluating, and diagnosing based on the principles of image processing, which has resulted in diagnoses of acceptable speed and accuracy. Accurate and fast performance of the diagnostic step optimizes the next steps, including classification and evaluation. This method has grown significantly in recent years, and advances in science are expected to revolutionize the emergence of advanced technologies, including intelligent infrastructure maintenance robots, remote control, detection with vehicles using the network and internet of things (IoT), the use of satellite imagery in detection, and tracking anomalous growth in the network.

Despite improvements in the processing methods and tools for detecting anomalies in infrastructures, there is still not enough intelligence for data analysis and interpretation for satisfactory practical applications. As such, researchers continue to look for a stronger, more efficient, and compatible method with harvesting tools. General hypotheses and rules in processing for diagnosis belong to one of the following rules or a combination of them:

- The anomalous pixels are darker/lighter than the original background.
- The presence of an anomaly has a nonuniform distribution in the gray image histogram.
- Anomalous morphology follows certain patterns.
- Specific anomalies are a collection of continuous and continuous veins (such as cracks) or dense intertwined patterns (such as holes).
- The features within anomalous patterns can vary, but at a holistic level, they are a heterogeneous part of the whole set.

Automation and Computational Intelligence for Road Maintenance and Management: Advances and Applications, First Edition. Hamzeh Zakeri, Fereidoon Moghadas Nejad, and Amir H. Gandomi.

- Anomalies usually have vague features around the edge.
- Geometric diagnosis of anomalies is not recommended due to the computational complexity of the diagnosis.
- Analyzing anomalies in transmitted space and applying kernel properties is easier and faster than the geometric method.
- Using a set of features is more practical and faster than geometric analysis for diagnosis.

Figure 5.1 shows examples of these anomalies. As can be seen, a fast and accurate detection of images displaying some kind of texture heterogeneity can be achieved with skim images and do not need to access accurate image details.

Each method that is available for diagnosis can be considered categorized as one of the following: (i) photometric hypotheses (PH), (ii) geometric and photometric hypotheses (GPH), (iii) geometric hypotheses (GH), and (iv) transform hypotheses (TH). A brief introduction to each of these methods is subsequently provided.

5.1.1 Photometric Hypotheses (PH)

According to this method, two main ideas are used to identify and evaluate heterogeneity or damage: (i) darker pixels being more than the average pixel value of the original texture, and (ii) histogram distribution properties have characteristics that recognize heterogeneity specifically and independently. Although these hypotheses are ideal and simple, their application to real problems is very time-consuming and difficult, and they work in reverse in many real passages and examples of such methods and hypotheses. For example, in some sections, the cracks are lighter than the pavement surface, or on a crack-covered surface, the pavement surface and crack histogram are completely integrated, and no specific, independent characteristics can be provided for

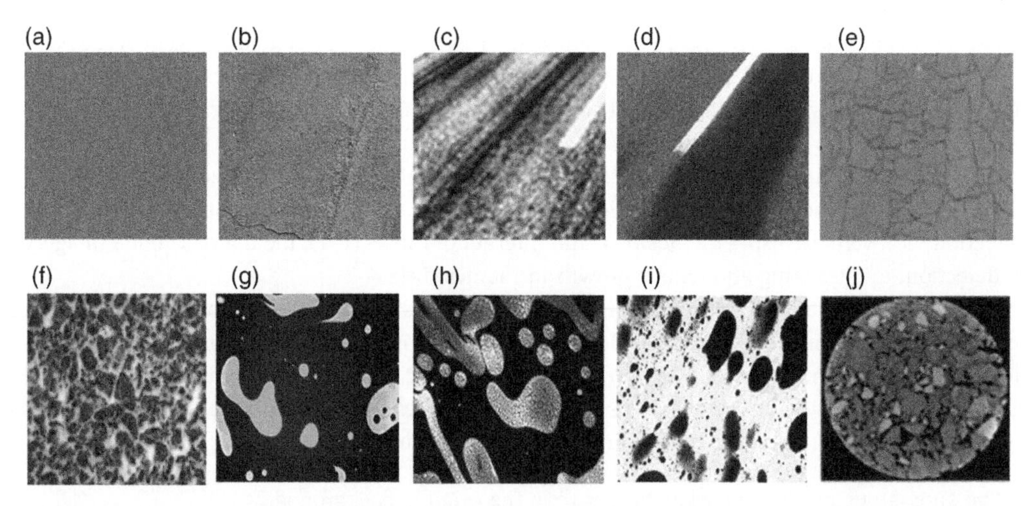

Figure 5.1 Existence of anomalous types in the images taken from the surface of the infrastructure or elements of materials used in its construction. (a) Surface of a healthy pavement without anomalies; (b) Road surface with low-intensity cracks; (c) High-intensity bleeding on pavement surface; (d) Uniform bitumen on the pavement surface and low friction on the car wheel path; (e) Lizard skin cracks or fatigue on the pavement surface; (f) Heterogeneity and uniformity of pavement texture; (g) Percentage of percentage of polypropylene bitumen (PPB) equal to 3%; (h) Percent PPB equal to 5%; (i) Percent PPB equal to 7% in the modified asphalt mixture; (j) Granulation of asphalt mixture and investigation of uniform distribution of aggregates and detection of heterogeneity or anomaly.

segregation and classification. These methods generally work using a fixed threshold. Type of material, image capture speed, type of failure, angle and degree of lighting, presence of various subjects (shadow, rubber effect, potholes and other pavement failures, etc.), harvest time (season, night and day, hour, air temperature, etc.) are factors that may disrupt the results and create various specifications. Choosing an optimal image enhancement, noise reduction, and enhancement of the appropriate image quality can greatly eliminate the effects of these factors (see Chapter 3).

5.1.2 Geometric and Photometric Hypotheses (GPH)

Based on this hypothesis, main ideas have been used in various methods: (i) the abnormal morphology follows certain patterns; (ii) specific anomalies are a set of continuous and continuous veins (such as cracks) or dense intertwined patterns (such as holes). In this category of hypotheses, the real conditions of failure are not considered, and identification and evaluation by type, are ignored. For example, according to these definitions, cracking should be a narrow and continuous pattern, while in many images and in real conditions, cracking is accompanied by other failures or fractures at the edge, which is itself the cause of the error. On the other hand, according to fracture, a set of continuous parts with different orientations is irregular or regular, but in the presence of other failures, this hypothesis is not able to distinguish between cracks and other failures. According to another hypothesis in this category, the crack area can be of different widths or, for example, remain in the bleeding part of a healthy surface, which can cause a diagnosis error. A clear boundary for the determination and separation of composite faults has not yet been proposed at the diagnostic stage. Therefore, this method cannot be used with a simple threshold alone, and it is necessary to use predictable thresholding methods. As a result, simplifying hypotheses creates an error that affects the diagnosis phase.

5.1.3 Geometric Hypotheses (GH)

According to this hypothesis, there may be photometric or geometric cracks in the margin of the crack. Based on the hypothesis that the anomaly pixels are darker/lighter than the original background, only the possibility of cracking based on the presence of an anomaly with an uneven distribution in the gray image histogram is considered in this method, which has the aforementioned limitations. Determining an anomaly based on this hypothesis is a problem that requires a new method for diagnosis. Defects can be discrete in the image, so the anomalous section can be a set of inhomogeneities that make it difficult to diagnose.

5.1.4 Transform Hypotheses (TH)

Based on this hypothesis, abnormalities can be analyzed and interpreted in other environments, creating more regular, analyzable patterns. Although the use of this method increases speed and accuracy, its transmission is a cause of error and ambiguity in the analysis. In this type of data, we generally encounter a degree of uncertainty, where choosing a fixed threshold cannot yield satisfactory results. Dynamic intelligent or adaptation thresholds for separating analytical models, in addition to being time-consuming, cannot be developed for application systems. On the other hand, collecting a sufficient amount of training data to obtain a suitable result for this method is necessary. Using the knowledge and abstract of extracted rules and benefiting from the rules in fuzzy domain can be a good approach to solve the problem.

The purpose of extracting image features is to perform rapid anomaly detection and isolation operations. The conversion of information into features with the aim of summarizing and reducing the amount of image data for further processing is presented in this section. Extracting features with more data transfer capability can increase speed and efficiency by reducing search and categorization space. Various methods have been proposed for the automatic detection of anomalies in infrastructure, including road pavements, bridges, and technical structures. In this section, effective features in automatic detection of images with anomalies are presented, and examples of its application in infrastructure management are given. In general, this section introduces various characteristics in the space of wavelet transform, Shearlet transform (SHT) in real and imaginary space, edge detection by the mixed Shearlet method, and momentum family characteristics, including Central_Moments_q, Central_Moments_*q*, Hu_Moments, Bamieh_ Moments, Zernike_ Moments, energy, Contrast, correlation, Homogeneity, entropy, Local_range, deviation from local standard (Local_STD), and fractal characteristics in order to detect images with abnormalities. After selecting the appropriate features vector using the feature selection methods presented in Chapter 7, the features are categorized, and the optimal attributes are selected based on special methods. Finally, the diagnostic process is completed using the classification methods presented in Chapter 8.

5.2 The Framework for Automatic Detection of Abnormalities in Infrastructure Images

This section provides a general framework for identifying the types of failures in infrastructure management, including road pavement management, technical building management, tunnel walls, dam walls, and building walls. All of these infrastructures use a similar structure to detect nonuniformity. Since an accurate analysis of all levels is not possible, it is optimal that only images with abnormalities are first isolated, and then only the same database of new images is used to perform a detailed analysis using advanced algorithms. The following figure shows the general framework for automatic detection of images with abnormalities (Figure 5.2).

As shown in the framework, image capturing is performed continuously, while only images that are detected as abnormal based on the fault detection method are stored in the image database. Images that are isolated in this way are also selected to perform the decomposition and classification steps. In Sections 5.2.1–5.2.20, various diagnostic methods based on feature extraction and the effect of each in detecting images containing anomalies are presented.

5.2.1 Wavelet Method

In this method, a family of library filters are used to analyze single two- and three-dimensional waves via the wavelet equation function. Each of these filters can have a different effect on the input image as well as the main subject (such as cracks, textures, holes, spalls).

5.2.2 High Amplitude Wavelet Coefficient Percentage (HAWCP)

Irregular patterns and shapes, such as failures, are transmitted to high-amplitude coefficients, and regular patterns like noise are transmitted to low-amplitude, low-frequency subsurface coefficients in the violet medium. In order to determine an index, the coefficients can be counted in the W medium and at high-frequency levels that have a range greater than the T-threshold. An index for the image can be obtained using the normalization function of all pixels. In order to avoid

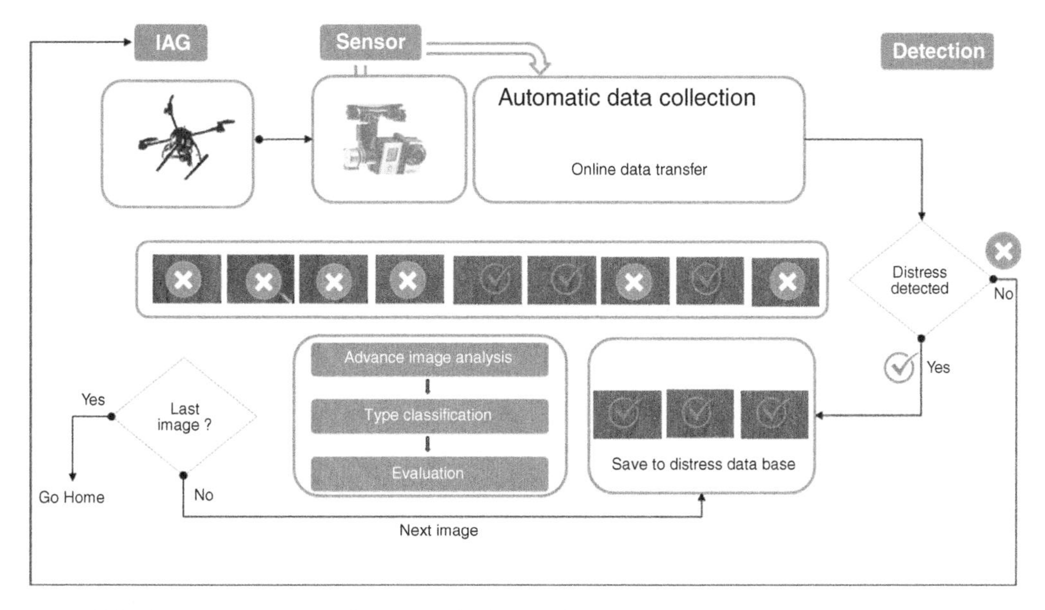

Figure 5.2 A general framework for automatic distress detection.

the calculation of HAWCP in each subsurface and the need to maintain all the information, the wavelet module is proposed using a combination of horizontal, vertical, and diagonal details in the violet environment and is calculated using Eq. (5.1):

$$M_k(p,q) = \sqrt{\mathrm{HL}_{\mathrm{K}}^2(p,q) + \mathrm{LH}_{\mathrm{K}}^2(p,q) + \mathrm{HH}_{\mathrm{K}}^2(p,q)} \tag{5.1}$$

where HLK, LHK, and HHK are the wavelet coefficients below the horizontal, vertical, and oblique at the point (p, q) and at the K-th level, respectively. As mentioned in Section 5.2.1, the wavelet coefficient (WC) in the first stage has the highest frequency component and is a good criterion for determining anomalies in the infrastructure. In this section, HAWCP is used in the first level to determine the failure. The modulus w at the first level is initially calculated and converted to binary form using Eq. (5.2):

$$D_1(p,q) = \begin{cases} 1 \text{ if } & M_1(p,q) \geq C_{\mathrm{th}} \\ 0 \text{ if } & M_1(p,q) < C_{\mathrm{th}} \end{cases} \tag{5.2}$$

where $D1$ (p, q) is the binary wavelet modulus in the first level and C_{th} is the threshold for the wavelet modulus, which is presented based on the wavelet thresholding method (Donoho— method) in the form of Eq. (5.3):

$$C_{\mathrm{th}} = \sqrt{2\sigma^2 \log(n)} \tag{5.3}$$

In these relationships, the amount of noise is based on the standard deviation of wavelet coefficients and n. The size of each tube contains wavelet coefficients that can be calculated based on the length and width below the surface in the first tube. In this work, the average standard for all images in db was calculated, and the image threshold was about 44.99; threshold was set at

44.99 based on the evaluation of pavement images. As a result, HAWCP can be calculated as follows:

$$\text{HAWCP} = \sum_{p=0}^{W/2} \sum_{p=0}^{l/2} \frac{D_1(p,q)}{(.25 * w * l)} \tag{5.4}$$

In this regard, l and w are the length and width of the sample image, respectively. The size of the sample in the first subsurface is equal to half the dimensions of the image. The HAWCP index has a value between (0 and 1), where a value of 0 indicates a suitable pavement level without anomalies, and a value of 100 indicates a completely broken pavement level. The HAWCP coefficient is similar to the unified crack index (UCI) uniform failure index calculation method. This method was used to evaluate the pavement anomaly caused by automatic cracking based on the UCI index. This index can be used to measure the extent of failure and to determine failure, isolation, and evaluation.

In order to avoid the calculation of HAWCP in each subband and the need to maintain all the information, the wavelet module using a combination of horizontal, vertical, and diagonal details in the wavelet domain is a more suitable indicator to increase speed and accuracy.

The higher the severity of an anomaly in the image, the higher the HAWCP coefficients at different levels and the higher the overall HAWCP index. On the other hand, it can be concluded that with the spread of anomalies, cracks change state and deform into the next type of failure. For example, a block crack changes into a type of fatigue crack after passing the level of severe failure. This property can be used to predict the spread of failure.

5.2.3 High-Frequency Wavelet Energy Percentage (HFWEP)

One way to measure the severity of a failure based on wavelet coefficients is to calculate the energy of the coefficients. Failures are transmitted to high-amplitude violet coefficients in the high-frequency subband, whereby high-amplitude coefficients have more energy than low-amplitude coefficients. Therefore, the more severe the anomaly, the more energy it will have as a result. The high frequencey energy precentage (HFEP) index for each subband is calculated using the sum of squares of all the coefficients in each subband and then is normalized using the total energy for the sample section. The energy in the K-level and HFEP_{H_k} can be calculated using Eq. (5.5):

$$\text{HFEP}_{H_k} = \sum_{p=0}^{W/2^k} \sum_{q=0}^{L/2^k} \text{HL}_k^2(p,q) \Big/ \sum_{m=0}^{W} \sum_{n=0}^{L} I^2(m,n) \tag{5.5}$$

where $I(m, n)$ is the pixel value of $p(m, n)$ in the main image. The length and width k of the subband are calculated as $w/2$ and $l/2$. In order to consider the high frequency of each subband and reduce the calculation time in this section, the percentage of energy with low frequency is used. The HFEP index can be calculated by subtracting the percentage of low-frequency energy from 1. Here, HFEP is the sum of the energy in all the high-frequency details that is calculated using Eq. (5.6):

$$\text{HFEP} = \frac{\sum_{k=0}^{K} \sum_{p=0}^{\frac{W}{2^k}} \sum_{q=0}^{\frac{L}{2^k}} \left[\text{HL}_K^2(p,q) + \text{LH}_K^2(p,q) + \text{HH}_K^2(p,q)\right]}{\sum_{m=0}^{W} \sum_{n=0}^{L} I^2(m,n)} \tag{5.6}$$

In this equation, the image k is decomposed. When the images are decomposed into a set of subbands, the approximate energy and energy details are calculated according to Eq. (5.7).

$$\sum_{m=0}^{W}\sum_{n=0}^{L} I^2(m,n) = \sum_{k=0}^{K}\sum_{p=0}^{2^k}\sum_{q=0}^{2^k}\left[\mathrm{HL}_K^2(p,q) + \mathrm{LH}_K^2(p,q) + \mathrm{HH}_K^2(p,q)\right] + \sum_{p=0}^{\frac{W}{2^k}}\sum_{q=0}^{\frac{L}{2^k}} LL_k^2(p,q)$$

$$\mathrm{HFEP} = 1 - \left\langle \frac{\left[\sum_{p=0}^{\frac{W}{2^k}}\sum_{q=0}^{\frac{L}{2^k}} LL_k^2(p,q)\right]}{\left[\sum_{m=0}^{W}\sum_{n=0}^{L} I^2(m,n)\right]} \right\rangle \tag{5.7}$$

The wavelet coefficient is below its approximate level. The implication of this relationship is that the total energy of all subbands, details, and approximations must be equal to the energy of the original image. No energy is reduced by wavelet experimentation.

5.2.4 Wavelet Standard Deviation (WSTD)

The histogram scattering distribution of wavelet coefficients can be used to determine the type, severity, and extent of abnormality, where the wider a histogram, the higher the heterogeneity index. Different statistical parameters, such as mean, range, STD, skewness, and Kurtosis, are used to describe and analyze histograms. Wavelet coefficient curves can be considered symmetric, which proves the nonuse of skewness for curves and shows that it is not a suitable feature for describing histograms. Calculation of Kurtosis is time-consuming and is not a suitable criterion for determining image turbulence simultaneously. The two characteristics of STD and range are unique in different histograms. The range index is calculated by subtracting the minimum value from the maximum wavelet coefficient. The maximum and minimum coefficients are affected by different conditions, such as light intensity and color contrast. In this case, using histograms, the STD property can be used to analyze infrastructure damage. The STD index shows the difference of each value from the mean of the values. The probability of coefficients $c(i)$ is shown using $p(i)$. The average can be calculated using Eq. (5.8):

$$\mu = \sum_{i=-\infty}^{+\infty} \langle c(i)p(i) \rangle \tag{5.8}$$

The standard deviation σ is calculated by distributing a set of data via Eq. (5.9):

$$\sigma = \sqrt{\sum_{i=-\infty}^{+\infty} (c(i) - \mu)^2 p(i)} \tag{5.9}$$

For a set of cracked images, the standard deviation of the wavelet coefficients under the horizontal, vertical, and diagonal bands in the first level is calculated. The highest STD in D, H, and V is considered the standard image deviation in this section. In this experiment, a minimum standard deviation of 2.19, maximum of 30.072, and STD ratio of maximum to minimum of 13.73 were calculated. The STD coefficient of healthy pavement samples (no failure) or with very low failure is less than 5. The superficial pavement damage categories with closely related STDs have similar HFEPs. According to STD, the first group includes failures, such as low-intensity cracking, compaction, and transverse cracking. In the second group, failures pertain to rutting, holes, cracks, and compound cracks, which are slightly different in HFEP. In the third group, both fatigue and block cracking are considered as surface distress. The STD index for fatigue among pavement cracking is still the highest value, where HAWCP, HFEP, and STD are considered for severe damage. Therefore, the STD index is a suitable criterion for distress diagnosis and isolation.

5.2.5 Moments of Wavelet

There are important properties for a set of data, one of which is momentum. The momentum is split into two types: central momentum and second momentum. The general equation for the qth moment of the central momentum coefficient (CMC) is calculated using Eq. (5.10):

$$cm_q = \sum_{i=-\infty}^{+\infty} [c(i) - \mu]^q \cdot p(i) \tag{5.10}$$

When q is equal to 1, $cm1$ must be equal to zero. When q is equal to 2, $cm2$ is equal to the square of the STD. For q equal to 3, the mean is close to zero. When q equals 4, $cm4$ is a good measure of detection and separation. The $cm4$ index is not used in this task due to the time-consuming nature of the calculations and the importance of time. The $cm4$ index is a very good measure for the future, as processor speeds increase, and can be a good alternative to calculations related to fault detection compared to the CMC central momentum, the qth moment of the centralized momentum mq is calculated using Eq. (5.11):

$$m_q = \sum_{i=-\infty}^{+\infty} |c(i)|^q . p(i) \tag{5.11}$$

In this equation, if $q = 1$, the moment value is equal to zero, and when $q = 2$, it is equal to the standard deviation. When q is equal to 1, we have the first moment in $m1$, which is a very good criterion for detecting and isolating faulty images. The second momentum is not a good measure because it is very similar to HFEP. Both of these criteria are considered square coefficients. The HFEP index is a more appropriate criterion because the image is normalized by total energy. This factor reduces the effects of light and weather conditions. For example, when the light intensity is high, the contrast is very high and energy in the high-frequency band is increased. The total energy of the sample increases with this method. Therefore, with normalization, the HFEP index remains constant. The dual momentum is normalized by the total number of pixels, which is not affected by changing weather conditions.

5.2.6 High Amplitude Shearlet Coefficient Percentage (HASHCP)

In SHT, pavement surface failures are transferred to Shearlet coefficients and are entered at high frequencies of these coefficients. Nonuniformity in Shearlet coefficients is generally due to the presence of defects, such as cracking or texture heterogeneity, in asphalt pavement. Meanwhile, noise and coefficients of healthy pavement (without cracking damage or with homogeneous texture) have coefficients with low amplitude and high frequency. In this method, the percentage of coefficients above the high-frequency threshold can be used as an indicator for the detection and identification of paved or abnormal pavement image. The HASHCP index is a useful index for detecting unusual images by normalizing the image size. Due to the importance of computational time in the detection stage and in order to preserve information in different sections and perform separate calculations in each subband, the Shearlet module is used, which is a combination of directional and shear details of the subbands in real and imaginary Shearlet environment. This module is calculated based on Eq. (5.12):

$$M_k(x,y) = \left(\text{Imag}\left(\prod_{i=1}^{k} \text{coeffs}(x,y,k)\right) \times \text{real}\left(\prod_{i=1}^{k} \text{coeffs}(x,y,k)\right)\right) \tag{5.12}$$

In Eq. (5.12), the Shearlet coefficients are converted into real and imaginary parts in the direction i and in the place (x, y) at the level k. Due to the fact that computational speed is more important in the detection stage, the frequency components in the third level provide sufficient information to identify and detect the normal image. According to this issue, the details and coefficients of the first level are used. In the third stage, the Shearlet module is used with thresholding for the initial segmentation of cracking using Eq. (5.13):

$$\text{SH}_{bk}(x,y) = \begin{cases} 1 & \text{if } M_k(x,y) \geq \text{Th}_{\text{Adaptive}} \\ 0 & \text{if } M_k(x,y) < \text{Th}_{\text{Adaptive}} \end{cases} \tag{5.13}$$

where $\text{SH}_{bk}(x, y)$ is the component with abnormal coefficients in the pavement image.

$\text{Th}_{\text{Adaptive}}$ is the adaptive threshold based on the mean and median.

The efficiency of this method depends on the number and size of local windows and image segmentation. The smaller the size of this segment, the more carefully the threshold is selected and the shorter the computational time. The adjustment parameters of Shearlet method, $W = \min([x, y])/7$ and $G = \min([x, y])/20$, size of each cone 2 and at level 2, have been evaluated for different sections.

Various adaptive methods, such as adaptive threshold selection using the principle of Stein's Unbiased Risk Estimate, Heuristic variant of the first option, Donoho, Threshold is $\sqrt{(2 * \log(\text{length(X)})}$, and Minimax thresholding, have been proposed in the past. Choosing an optimal method to eliminate noise largely affects the index as well as the separation of unusual images. In order to investigate the effect of threshold selection on the index, results of different adaptive thresholding methods on the sample image are shown in Figure 5.3. As can be seen, different results are obtained using different methods. The heuristic variant of the first option method and the minimax thresholding show more accurate results than the other methods. As shown in Figure 5.3, the use of different thresholding methods in the extraction of data related to abnormal images can lead to the presentation of a wide range of HASHCP indices. Choosing the right threshold plays an important role in the performance of this indicator.

Stein and Threshold is $\sqrt{(2 * \log(\text{length(X)}))}$ methods estimate the abnormality (such as cracking) above the actual value and have an edge error and higher coefficients than the heuristic method. In contrast, the median–mean method, by choosing the threshold 0.001, shows a lower estimate of crack width. The minimax thresholding and the heuristic variant of the first option methods provide an acceptable approximation of cracks and have good performance in extracting failure. After comparing the HASHCP indices for the thresholds, the adaptive heuristic method was selected (Figure 5.4):

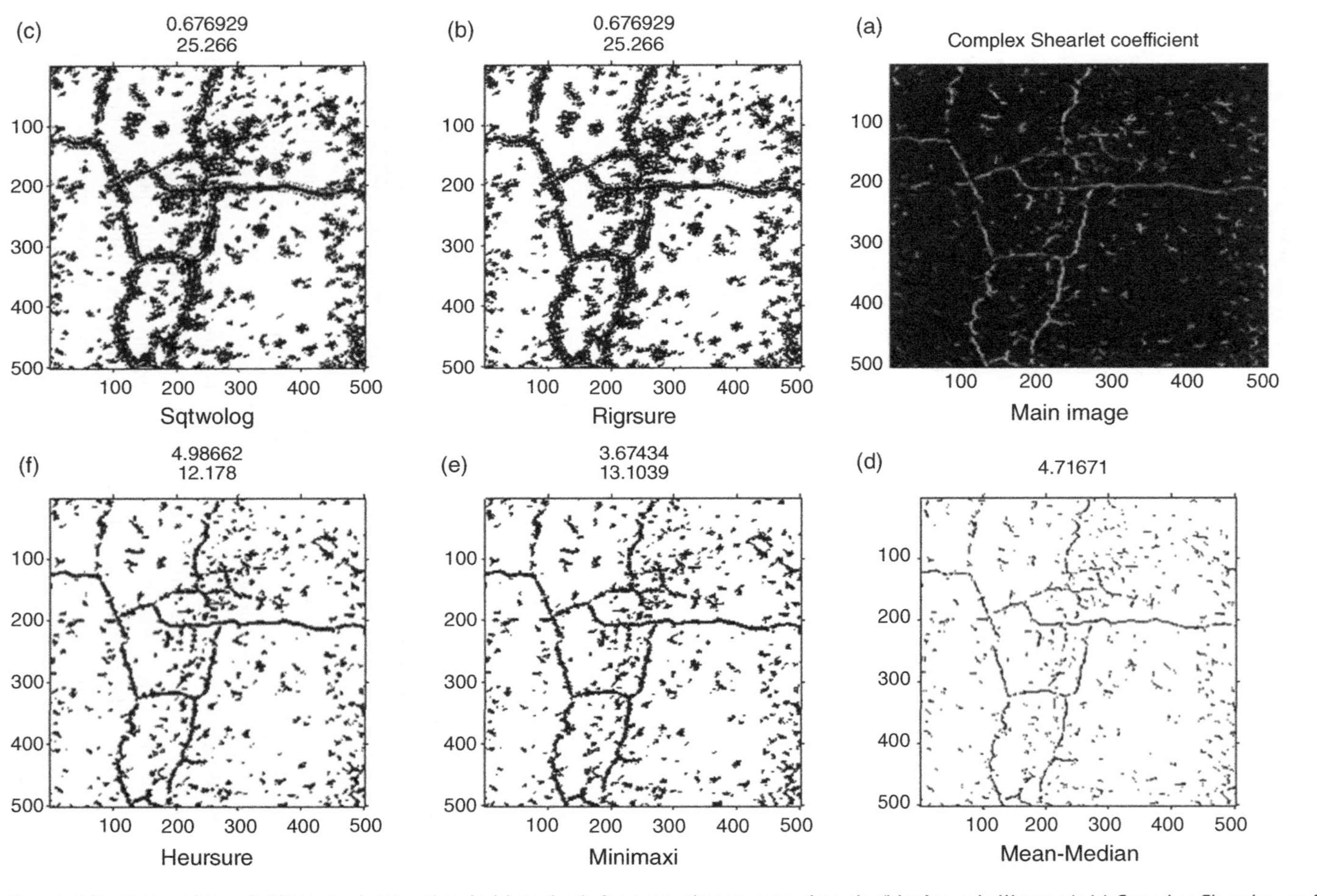

Figure 5.3 Comparison of different adaptive threshold methods for extracting patterned cracks (blocks and alligators): (a) Complex Shearlet coefficients, (b) Stein's method, (c) Threshold is $\sqrt{(2^* \log(\text{length}(X))}$, (d) Medium–Mean method, (e) Minimax thresholding, and (f) Heuristic variant of the first option.

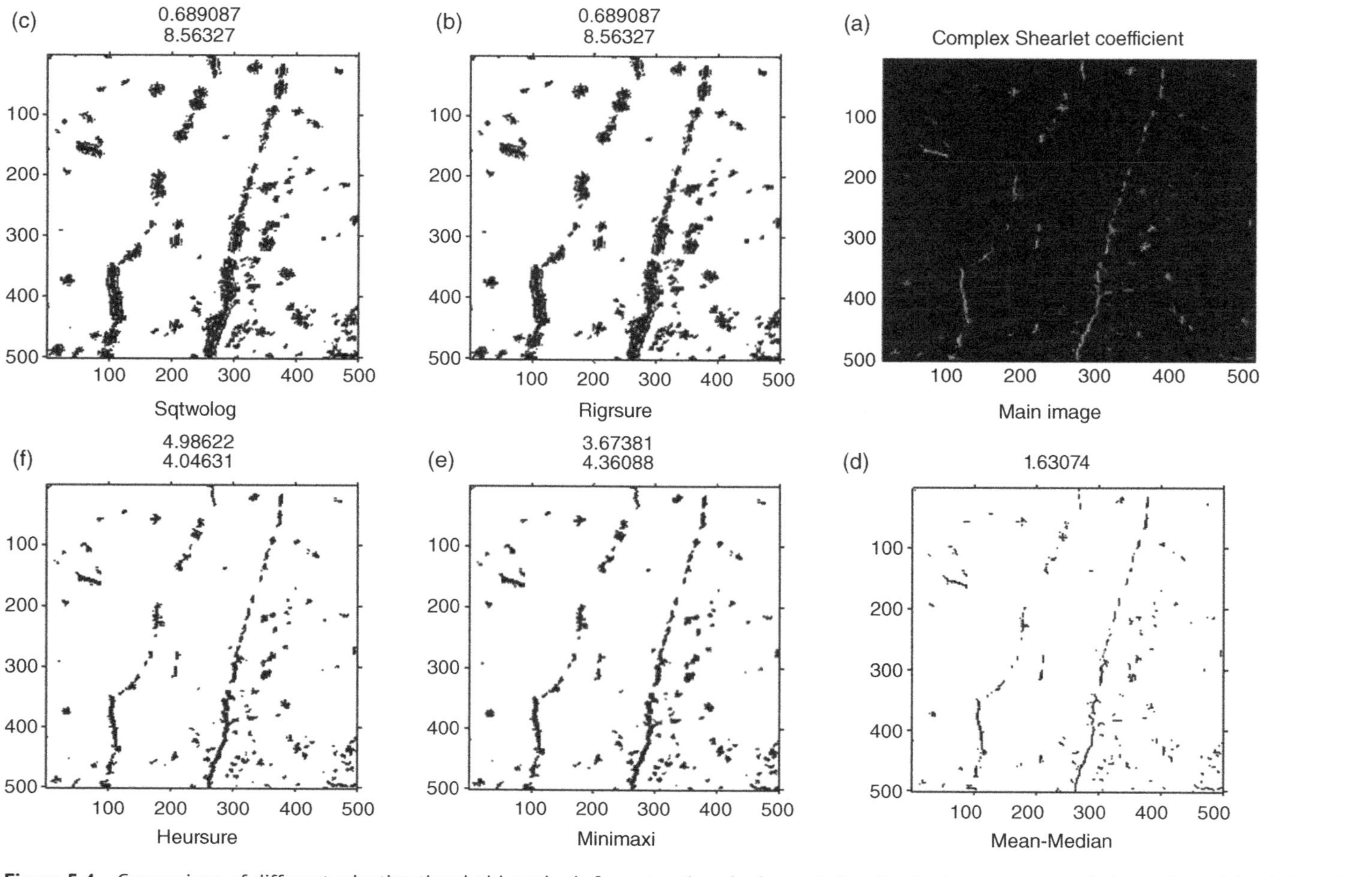

Figure 5.4 Comparison of different adaptive threshold methods for extracting single crack (longitudinal, transverse, and diagonal cracks): (a) Complex Shearlet coefficients, (b) Stein's method, (c) Threshold is $\sqrt{(2^{*}\log(\text{length}(X))}$, (d) Medium–Mean method, (e) Minimax thresholding, and (f) Heuristic variant of the first option.

$$
\begin{aligned}
&\mathrm{Th}_{\mathrm{Adaptive}} = \begin{cases} \mathrm{th} = \sqrt{(2 * \log(n))} & \textit{if } \alpha < \beta \\ \mathrm{th} = \min\left(\sqrt{(2 * \log(n))}, r\right) & \textit{if } \alpha \geq \beta \end{cases}, \\
&\alpha = \frac{\mathrm{norm}(x).^{2} - n}{n}, \\
&\beta = \frac{\left(\dfrac{\log(n)}{\log(2)}\right)^{1.5}}{\mathrm{sqrt}(n)}, \\
&r = \mathrm{sqrt}(\mathrm{sort}(\mathrm{abs}(x))^{2}(\min(\mathrm{risks}))), \\
&\mathrm{risks} = \frac{n - (2 * (1:n)) + (\mathrm{cumsum}(S) + (n-1: -1:0).* S)}{n} \\
&S = \mathrm{sort}\left(\mathrm{abs}(x)^{2}\right)
\end{aligned} \tag{5.14}
$$

After applying the threshold, the anomalous index is calculated by the mixed Chert method and is defined as follows for an image that has a crack (patterned or linear):

$$
\mathrm{HASHCP} = \sum_{X=1}^{W} \sum_{Y=1}^{l} \mathrm{SH}_{b\bar{k}}(x, y)/w \times l \tag{5.15}
$$

Based on this relationship, the HASHCP∈ index [0,100] is 0 for healthy pavement (image without damage), and the closer it gets to 100, the higher the failure rate. The HASHCP index is similar to the HAWCP index, except that in the wavelet-based method, the directional information is parsed only in the horizontal, vertical, and diagonal directions, while in this method, different directions are considered. HAWCP also calculates the threshold which is $\sqrt{(2 * \log(\mathrm{length}(X)))}$ relation from the fixed threshold. This index is used as a criterion for diagnosing and evaluating the extent of failure. The algorithm for calculating the HASHCP index is shown in Figure 5.5.

This index has a higher value than the HAWCP index, which indicates that directional cracks and lateral branches are extracted better than the wavelet method by the Shearlet method. In order to evaluate the performance of the proposed index, a set of pavement images, were used to evaluate the algorithm (examples of which are shown in Figure 4.16). The existing algorithm was applied to the image, and the HASHCP index was calculated for each image. The results show that with the increasing failure rate of this index, the extent and width of cracking also increased, for which this index shows a greater value; therefore, the HASHCP index is a practical feature for detection of abnormal sections.

Table 5.1 shows the HASHCP index for different images and types of distress. As can be seen, this index varied with increasing severity of distress, extent, and type. Particularly, the value ranges between 0 and 100, where the higher this number, the lower the pavement efficiency in service.

This comparison was performed for 100 images in six categories, with a maximum $\mathrm{HASHCP}_{\mathrm{max}}$, average $\mathrm{HASHCP}_{\mathrm{ave.}}$ HASHCP_(ave.), and minimum $\mathrm{HASHCP}_{\mathrm{min}}$ index for summary purposes. The number of samples was approximately equal, the number of samples with failure was 67, and the number of healthy samples was 33. The ratio of the number of healthy samples to the total sample is 33%, and the sample with a failure is 67%. Among the broken images, there are three levels of intensity and magnitude. According to the results of $\mathrm{HASHCP}_{\mathrm{ave.}}$, the sections with more failure than healthy the average is two times higher than healthy sections. This ratio of($\mathrm{HASHCP}_{\mathrm{max}}/\mathrm{HASHCP}_{\mathrm{ave.}}$) is changed for different types of cracks. Area cracks (fatigue and block) have the highest value, while linear cracks (longitudinal, transverse, and oblique) show the lowest value of the proposed index. For failure with low intensity/extent level, the ratio of $\mathrm{HASHCP}_{\mathrm{min}}$ index for healthy section to failure is 34% and, for medium and high levels, is 52% and 53%, respectively. These ratios indicate that for the

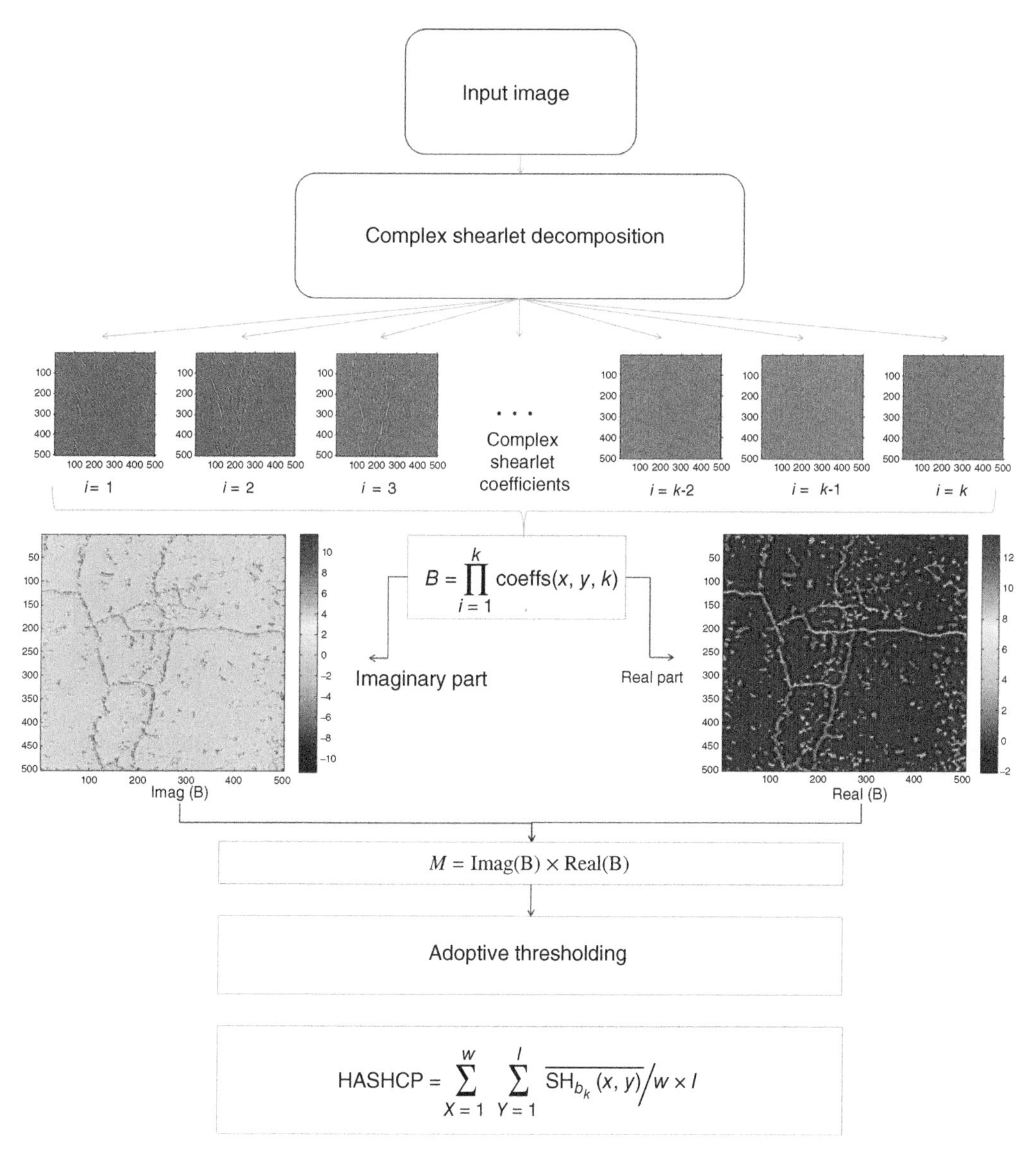

Figure 5.5 Algorithm of complex Shearlet coefficients for calculating the HASHCP index.

Table 5.1 Comparison of the effect of type and severity of distress on HASHCP index.

	Log.	Tr.	Di.	Bl.	Al.	Co.	Nd.
$\text{HASHCP}_{\text{min}}$	6.66	6.53	4.63	8.14	9.15	7.37	2.17
$\text{HASHCP}_{\text{ave.}}$	9.91	8.33	6.63	10.92	13.94	10.23	4.85
$\text{HASHCP}_{\text{max}}$	14.39	13.49	9.84	16.77	16.27	13.15	6.93
Number of samples	14.00	11.00	10.00	9.00	12.00	11.00	33.00

intensity and extent of medium and high failure, the difference between the indicators is greater and the separation of faulty sections from healthy sections can be performed with less risk. It should be noted that the difference between these indicators is to introduce the extent of failure. In the $HASHCP_{min}$ category, the indices are close to the index of healthy sections, and even in some samples, healthy sections have a higher index than the sample with distress. Figure 5.6 shows the ranges obtained for each type of distress.

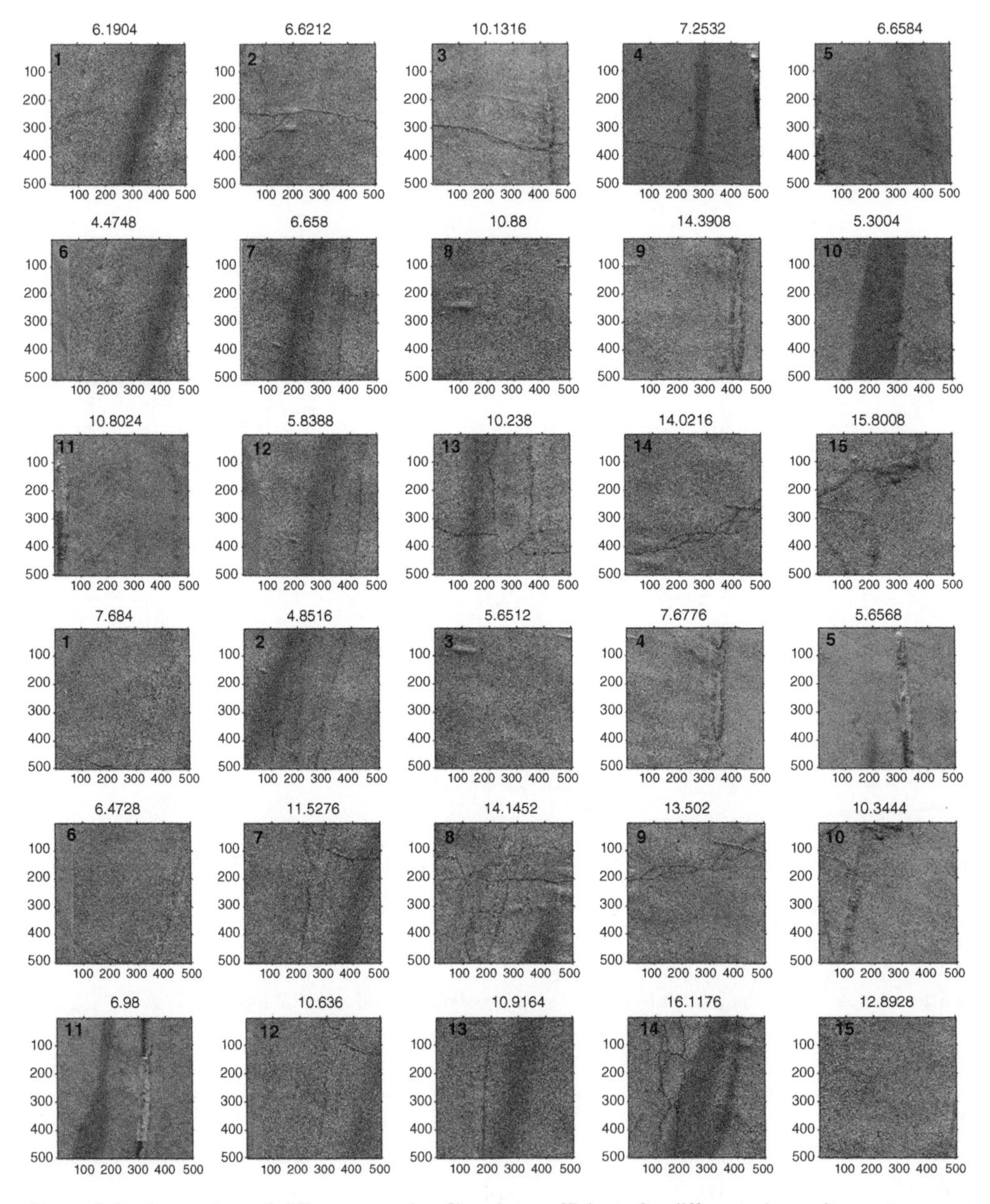

Figure 5.6 Comparison of different complex Shearlet coefficients for different abnormity on pavement images.

According to the classification, as a general rule, samples with irregularities exhibit a higher index than healthy samples. On average, samples with an index higher than 4.85 have a distress. Samples of healthy images have an index below this value. If the 4.85 index is selected as the threshold, a number of healthy images may also be selected and supplemented. This has no effect on the final performance and only increases the computational time (Figure 5.7).

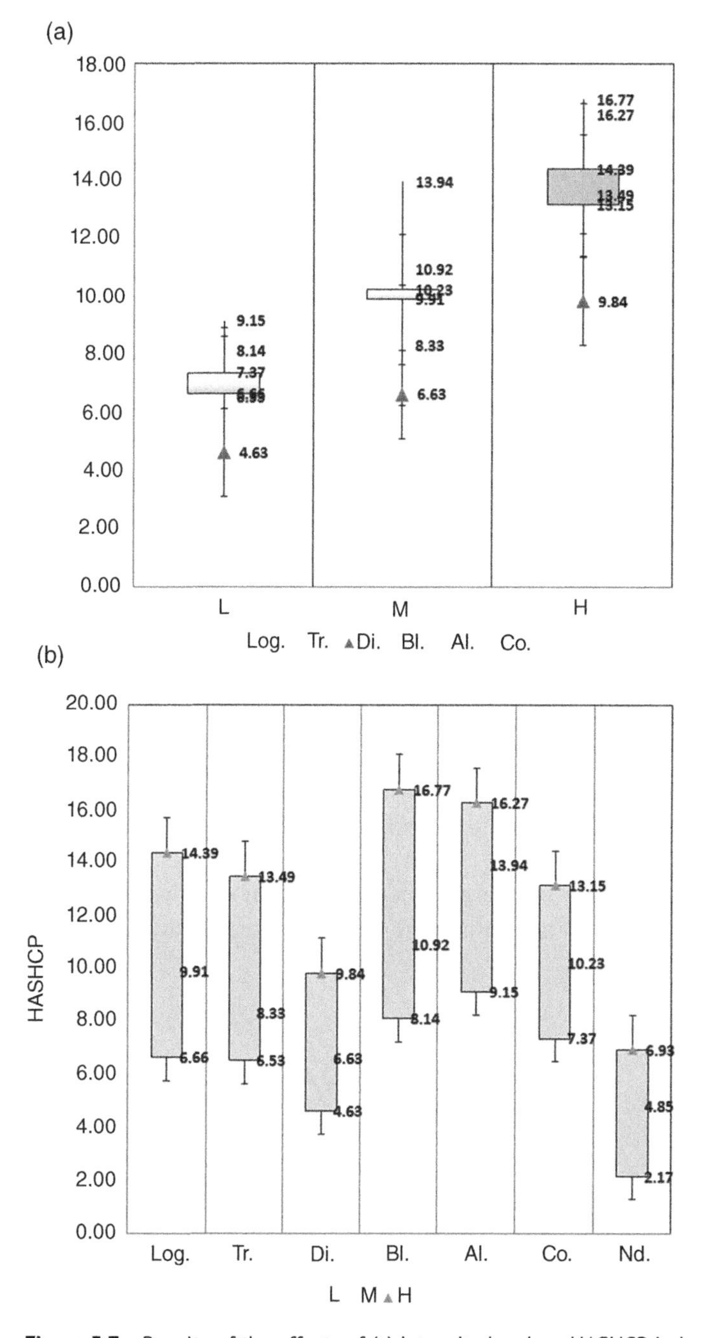

Figure 5.7 Results of the effects of (a) intensity level on HASHCP index and (b) failure type on HASHCP.

5.2.7 High-Frequency Shearlet Energy Percentage (HFSHEP)

One of the effective methods in measuring the severity of failure is to calculate the energy of complex coefficients. This method is similar to the method of determining the crack intensity in the wavelet, except that there are different directions and the possibility of further coverage of the quantification of the orientation and width of the crack. The cracks are transferred to high-amplitude SHT coefficients, so these sections have more energy than the nonfailure sections. The more severe the damage, the more energy is used. The HFSHEP index is obtained by summing the squares of all coefficients in different directions and then normalizing the sum with the original image. The energy of the details in the horizontal section and at the k-level is calculated from Eq. (5.16):

$$\mathrm{HFSHEP}_{\mathrm{sh}_k} = \left(\sum_{k=1}^{K}\sum_{X=1}^{w}\sum_{Y=1}^{l}\left(\mathrm{SH_{d1\mathit{k}}}^2(x,y) + \mathrm{SH_{d2\mathit{k}}}^2(x,y) + \mathrm{SH_{d3\mathit{k}}}^2(x,y) + \dots + \mathrm{SH_{dn\mathit{k}}}^2(x,y)\right)_H\right) \Bigg/ \left(\left(\sum_{X=1}^{w}\sum_{Y-1}^{l}\left(\mathrm{SH}_{a1k}{}^2(x,y)\right)\right)_L\right) + \left(\sum_{k=1}^{K}\sum_{X=1}^{w}\sum_{Y=1}^{l}\left(\mathrm{SH_{d1\mathit{k}}}^2(x,y) + \mathrm{SH_{d2\mathit{k}}}^2(x,y) + \mathrm{SH_{d3\mathit{k}}}^2(x,y) + \dots + \mathrm{SH_{dn\mathit{k}}}^2(x,y)\right)\right)_H \tag{5.16}$$

where HFSHEP is High-Frequency Shearlet Energy Percentage, $f(x, y)$ is the pixel intensities in the coordinates (x, y) for the original image; SH_{bk} is the high-amplitude coefficients; and $a1$ represents the Shearlet coefficients below the low-frequency approximated band at its corresponding point. Due to the fact that in different sections, the coefficients are different and must be added together and then normalized, the computational complexity is higher than the low-frequency band calculation. Given that the value of the $\mathrm{HFSHEP}_{\mathrm{sh}_k}$ index is equal to the sum of the energies of all high-frequency details, the total energy of the image is a constant value, and low energy at the desired level can be used instead of high-frequency coefficients to reduce computational time. Given that the energy level for each image is a constant value, the equation (5.16) is converted as follows. Compared to Equation (5.16), Equation (5.17) increases the computational time 30-fold and is, therefore, a good feature for detecting and separating faulty pavement sections. The method and algorithm for calculating the index are presented in Figure 5.9. According to this method and with the help of the fixed law of energy level 1, for each image, a low-frequency band is used to calculate the index. The low-frequency of image related to noise reduction and used to enhanced images. After normalizing the cross section at low frequency as well as normalizing the high-frequency image, the energy ratio and HFSHEP index are calculated based on the percentage. An example of the outputs obtained by this method, low and high frequency, is shown in Figure 5.8.

$$\mathrm{HFSHEP}_{\mathrm{sh}_k} = \frac{\left(1 - \left(\left(\sum_{X=1}^{w}\sum_{Y=1}^{l}\left(\mathrm{SH}_{a1k}{}^2(x,y)\right)\right)_L\right)\right)}{\left(\left(\sum_{X=1}^{W}\sum_{Y=1}^{l}\left(\mathrm{SH}_{a1k}{}^2(x,y)\right)\right)_L + \left(\sum_{k=1}^{K}\sum_{X=1}^{W}\sum_{Y=1}^{l}\left(\mathrm{SH_{d1\mathit{k}}}^2(x,y) + \mathrm{SH_{d2\mathit{k}}}^2(x,y) + \mathrm{SH_{d3\mathit{k}}}^2(x,y) + \dots + \mathrm{SH_{dn\mathit{k}}}^2(x,y)\right)\right)_H\right)} \tag{5.17}$$

According to this equation, the higher the frequency of a section, the greater the failure in that section (Figure 5.9). On the other hand, the lower the number of points containing low frequency,

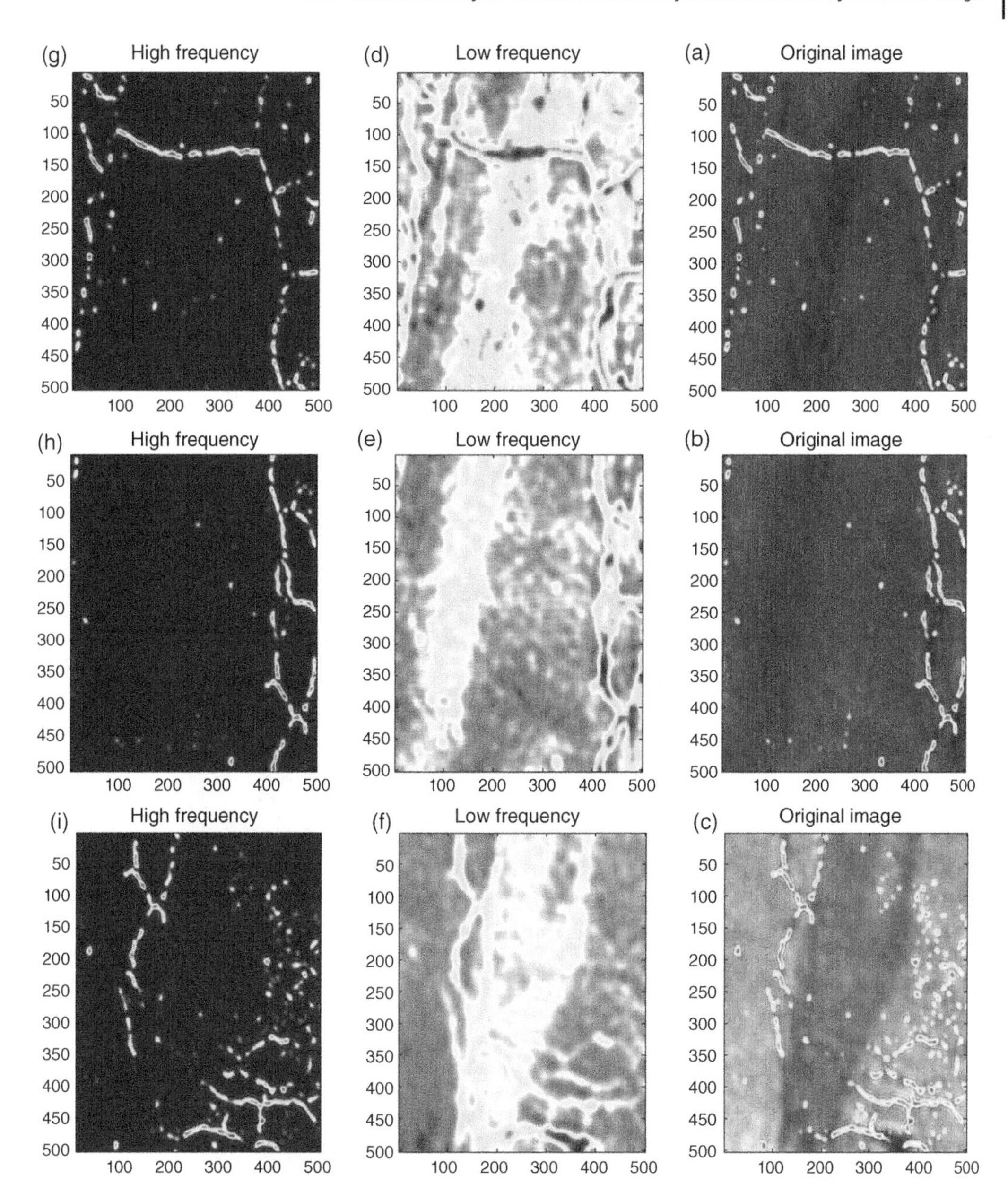

Figure 5.8 Examples of low- and high-frequency separation using Shearlet method: (a–c) Main images; (d–f) Low-frequency (LF); and (g–i) High-frequency (HF).

the higher the number of points with high frequency. As a result, the higher the breakdown energy, and finally the higher the high frequency energy index.

For the set of images in Figure 5.10, the calculated HFSHEP index is shown on each image. As can be seen, the sections with a failure of this index have a higher value than those with no or little failure. The higher the failure rate in the sections, the higher the energy, and thus, in the high-frequency band, the HFSHEP index increases.

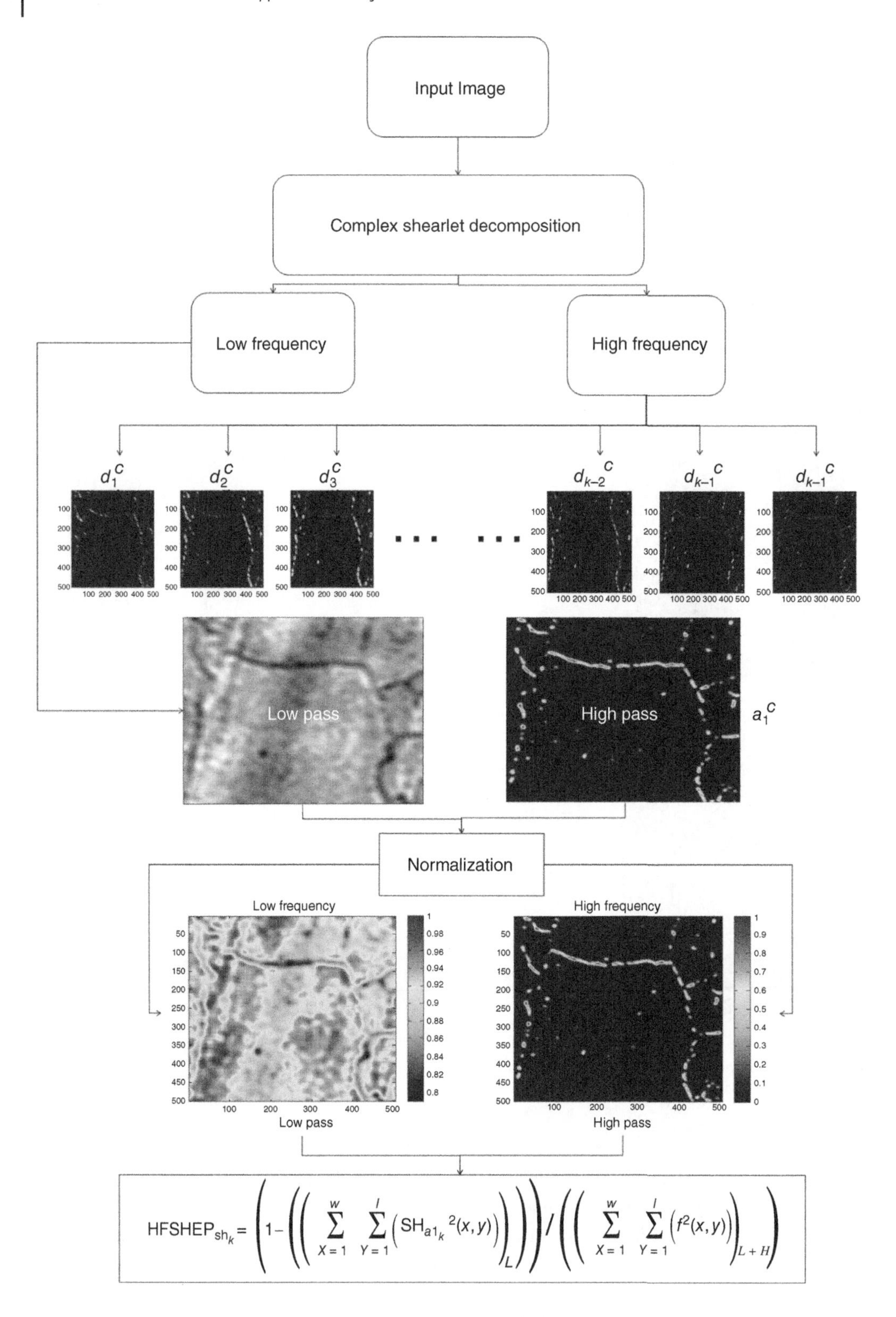

Figure 5.9 The complex Shearlet algorithm to calculate the HFSHEP index.

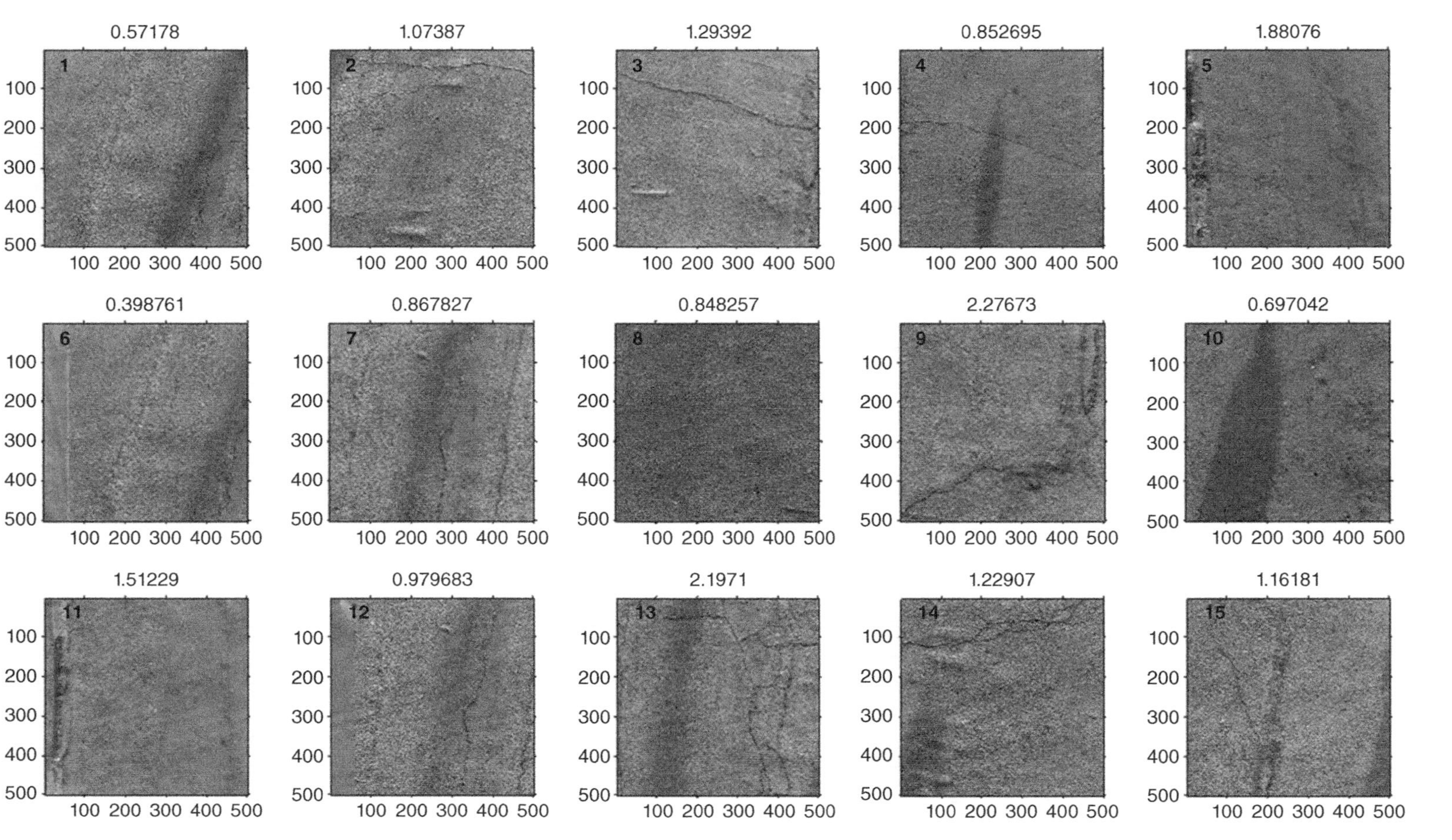

Figure 5.10 Comparison of different complex Shearlet coefficients for different distress on pavement images.

Table 5.2 Comparison of the effect of type and severity of distress on HSHEP index.

	Log.	Tr.	Di.	Bl.	Al.	Co.	Nd.
$HSHEP_{min}$	0.630	0.475	0.385	0.858	1.008	0.826	0.399
$HSHEP_{ave.}$	1.257	0.980	0.738	1.228	1.842	1.070	0.785
$HSHEP_{max}$	2.360	2.248	1.701	2.197	2.658	1.881	1.506
Number of samples	14	11	10	9	12	11	33

Similar to the $HASHCP_{max}$ index, this review was performed for 100 images in six categories. The maximum $HASHCP_{max}$ index, average $HSHEP_{ave.}$, and minimum $HASHCP_{min}$ are given in Table 5.2. The percentage of sections with damage to healthy samples is 67%, and the percentage of sections with no damage is 33%. Three levels of intensity and extent are considered in all sections. Based on the results of HSHEP, sections with failure displayed a ratio of 2.54, 2.34, and 1.76 for low, medium, and high failure intensity, respectively. As can be seen, this ratio is greater for low-intensity sections. Compared to the HASHCP index, the new index is more efficient in detecting sections at low and medium intensity levels and is, therefore, effective for diagnosis and isolation, at these intensity levels.

In areas where the pavement is healthy, other types of damage, such as grooves, baldness, may be present, and the HSHEP index could go up to a maximum of 1.5. In this case, the image may be in the category of defective images, and other indicators, such as HASHCP, are a good criterion for detection. The HSHEP index is directly related to the failure intensity and puts different failures in the same range with the same intensity level. For example, cracks (longitudinal, transverse, block, and fatigue) can be at a high intensity level, at the same level, or within shorter intervals (36% of the maximum value) from the low and medium level (60% and 63% of the maximum value). One of the capabilities of this method is to differentiate the failure intensity for single cracks (longitudinal and transverse) from other cracks. This index separates single-line failures by more than four times the difference in intensity level. It should be noted that this ratio is more than 2.2 for breakdowns, block, fatigue, and various cracks. For high-intensity sections, the HFSHEP ratio is 6.9 for healthy sections.

Unlike the HASHP index, which is detected and isolated with less risk in the severity and extent of moderate and high failure, the HFSHEP index detects the severity of low and moderate failures with the least risk. Also, the severity of the failure shows a good relationship with this index. Figure 5.11 shows the ranges obtained for each type of damage.

5.2.8 Fractal Index

One of the practical indicators in assessing the damage of infrastructures is a UCI section that represents the ratio of the damaged surface to the total surface. In this method, the number of clear pixels and the size of the cross section is defined as a simple and practical indicator in the automatic assessment of damage. One of the most important advantages of this method is its simplicity and high speed of calculation. However, the UCI does not have the ability to detect and index the type, severity, or extent of failure and does not provide crack information to the user. Cracking is an example of a natural fractal event with properties related to fractal geometry. If each crack is divided by the number r of the smaller component, then the following relation is given for the number of N small cracks:

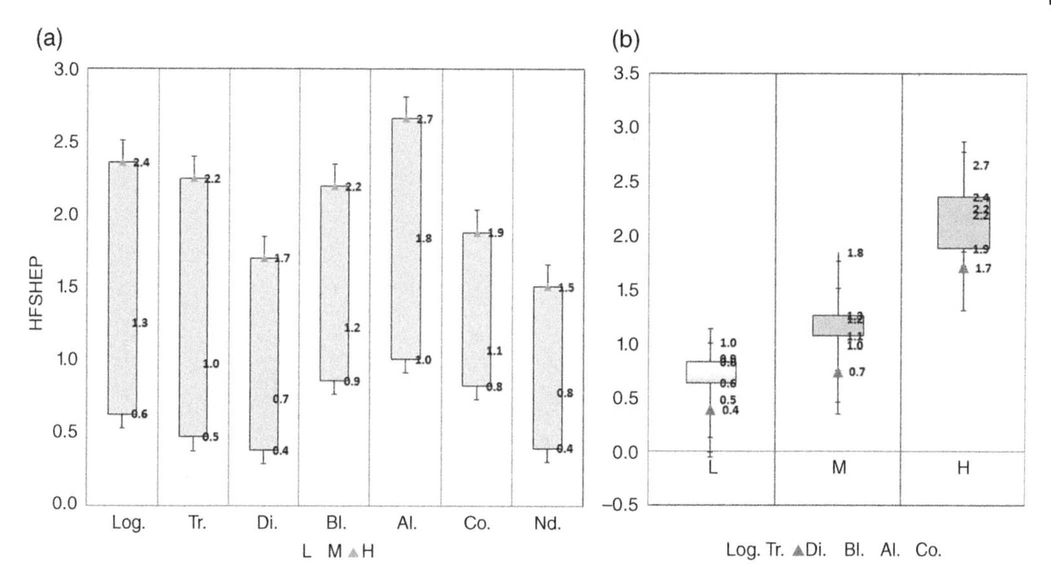

Figure 5.11 Results of HASHCP index in samples with different type and intensity: (a) effect of failure type on the HASHCP index, and (b) effect of intensity level on the HASHCP index.

$$N = r^{-\mathrm{FD}} \rightarrow \mathrm{FD} = -\frac{\log N}{\log r}, D = 1, 2, 3, ..., n \tag{5.18}$$

Accordingly, there is a relationship between the minimum size used and the activated surface for counting crack parts. This relationship can be seen by plotting the ratio of logarithmic changes of $\log N$ to $\log r$. The slope of the line is calculated using the regression method, which represents the fractal dimension of cracking. The fractal dimension, as the most important parameter of this method, is calculated using different methods. The grid counting method, or Minkowski or Minkowski–Bouligand dimension, considers the fractal dimension of a set of points in Euclidean space and the set of points distributed in a section. Then, the number of active points in this set is calculated and based on the fractal relationship, and the fractal dimension is calculated using Eq. (5.19):

$$fD_{\mathrm{box-count}} = \lim_{\varepsilon \to 0} \frac{\log N(\varepsilon)}{\log(1/\varepsilon)} \tag{5.19}$$

where $N(\varepsilon)$ is the number of components of the faulty section. When this function tends to the pixels of the image (i.e. the number of grids is maximum and the grid size is equal to the pixel size), the calculated index is equal to UCI. The fractal dimension is calculated using this index, which is a number between 1 and 2 depending on the type of crack (Figure 5.12).

In general, if the fractal dimension is greater than 1, the analyzed pattern, cracks, or linear subjects are considered to be two-dimensional. When the index is larger than 1, the more similar the fractal dimension will be to the fatigue cracking pattern. Linear cracks are in the range of 1–1.5, while nonlinear cracks (such as fatigue and blockage) are in the range of 1.5–2. Accordingly, linear cracks with spalls have a higher $fD_{\mathrm{box-count}}$index than nonspall cracks. The fractal dimension is a numerical scale used to quantify the amount of pavement space filling without failure by a specific pattern to describe the extent of damage to the infrastructure.

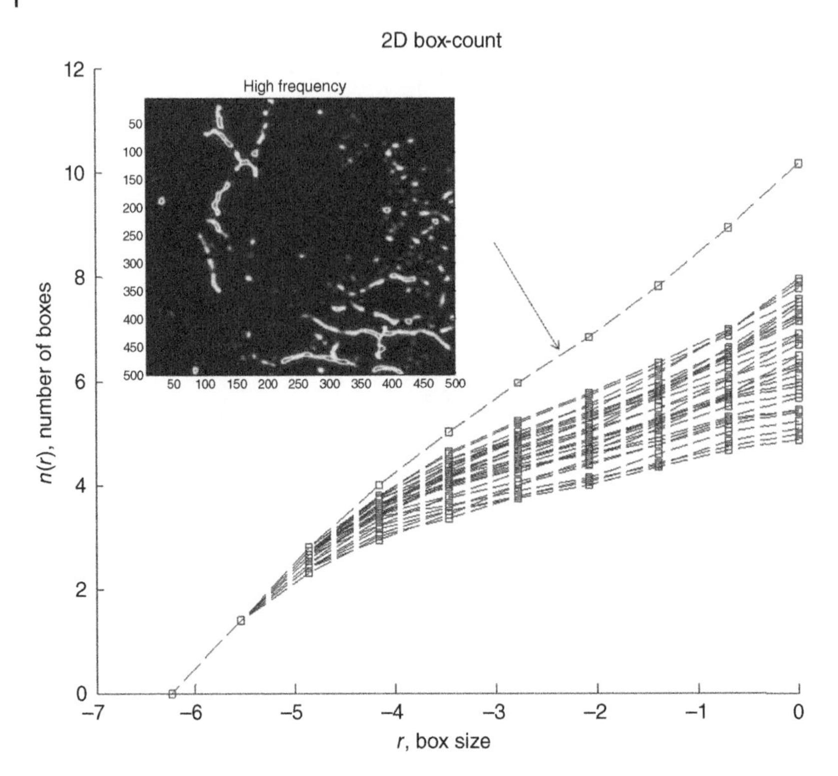

Figure 5.12 Fractal dimension changes $fD_{box-count}$ with size and number of parts.

Each direction and detail of SH has a dimension of the fractal index, which is used to determine the type of damage. The higher the fractal dimension of an injury, the more similar it is to the type of fatigue crack with a severe spall, and the more extensive the linear damage with a larger spall.

In the last component of the high-frequency reconstructed image, the combination of details shows a higher fractal index. This result is due to the cohesion in the heterogeneous and nonisotropic environment of cracking. As shown in Figure 5.13, the real part of the information has a larger dimension than the imaginary part. On the other hand, the combination of two parts, real and imaginary, has a higher dimension than both parts, which is equal to 1.54 at the end of the graph for a low-intensity fatigue crack image:

$$\begin{aligned} &fD_{box-count}(\text{complex}) > fD_{box-count}(\text{real}) > fD_{box-count}(\text{img}),\\ &fD_{final} = \{fD_{box-count}(\text{complex}), fD_{box-count}(\text{real}), fD_{box-count}(\text{img})\},\\ &\{fD_{box-count}(\text{complex}) \cong fD_{box-count}(\text{real})\} > \{fD_{box-count}(\text{img})\} \end{aligned} \tag{5.20}$$

In the above relations, in order to correctly assess the type and severity of damage, the maximum fractal index obtained from the two main parts of the image as well as its mixed composition are used. One of the main applications of this method is to determine the direction of the mold inherently similar to the principle of damage. In this method, using the ratio of fractal dimension to final fractal dimension *fD_final*, the details of the cross section with mold damage are identified, from which cracking is extracted in order to determine the direction and main dimension. This method is also used to detect damage through thresholding.

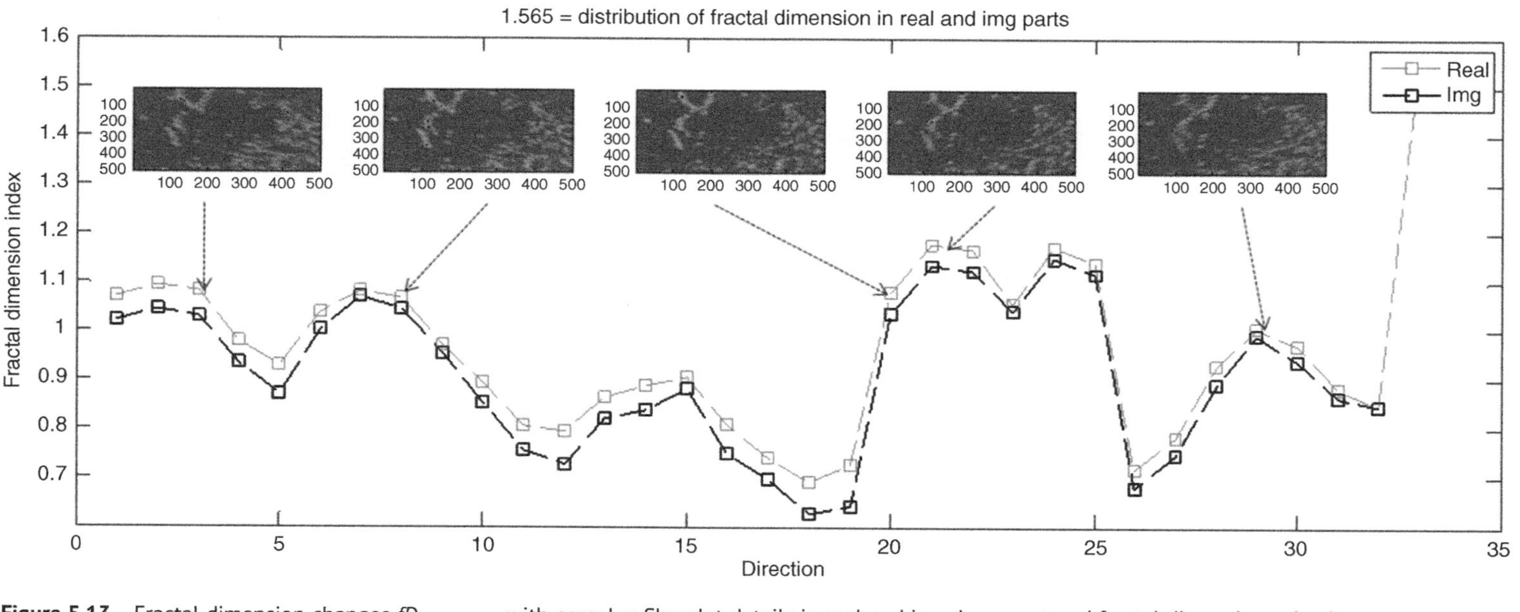

Figure 5.13 Fractal dimension changes $fD_{\text{box-count}}$ with complex Shearlet details in real and imaginary part and fractal dimension selection.

As shown in Figure 5.14, the fractal dimension for an image with fatigue damage is approximately equal in the real and absolute parts and smaller than that in the imaginary part. In general, it can be concluded that the real and absolute parts represent the fractal structure of the whole crack, and the imaginary part is related to the general structure and skeleton of the damage and main branches; therefore, the difference between these two fractal indices can be used to evaluate the characteristics of spall branches and branching cracks.

For the set of images in Figure 5.15, the indices $fD_{box-count}(\text{img})$ and $fD_{box-count}(\text{complex}) \cong fD_{box-count}(\text{real})$ are calculated and shown on each image. As can be seen, the sections with damage of this index show a higher value than the sections with healthy or low damage. The higher the amount of spall points in the sections, the higher the index in the real part, and the fractal index in the imaginary part is a constant value.

For the sample image database, $fD_{box-count}(\text{img})$, $fD_{box-count}(\text{real})\text{HSHEP}_{ave.}$, and $fD_{box-count}(\text{abs})$ are given. The percentage of sections with damage to healthy samples is 67%, while the percentage of healthy samples is 33%. The evaluation results and fractal indices for the three sections are given in Table 5.3. Based on the results, fD_(box-count) (abs) has been extracted for sections with low, medium, and high failure rates of 1.219, 1.156, and 1115, respectively, while the respective minimum values are 1.42, 1.48, and 1.57. Also, the difference between the high and low levels for the average is the highest and equal to 16%. The diagram shows the limits and how to change this index for cracking damage. As shown in the diagrams in Figure 5.16, in the mixed image section, the numerical fractal index is between 1.28 and 1.5 for healthy pavement and between 1.42 and 1.6 for pavement with fatigue and block damage.

The results of the index presented for the Shearlet image analyzed in the real section are almost similar. The difference between the two coefficients is less than 1%, and the index for the mixed image is 1% higher than the real image, but the results in the imaginary part show a greater difference. As can be seen from the diagrams of these three sections, the overall numerical fractal index is between 0 and 1, which increases with increasing level of failure as well as the severity of failure (related to failure sources). In the imaginary part, this index is between 1 and 1.3. As the level of severity and extent of damage increased, a growth of 17% was achieved for the low level of intensity compared to the maximum index, while medium- and high-levels exhibited growths of 13% and 6%, respectively.

The complex index has a greater ability to isolate the three blocks of damage, fatigue, and compound, and the other three failures at the medium and high-intensity levels have a more fractal dimension and are more detectable. Although the real part is not much different from the imaginary mode, the three linear failures have less overlap due to the input of background information. Therefore, this method shows more efficiency than the first case. In the imaginary part, due to the presence of background and pavement texture, the amount of fractal dimension decreases to less than 1.3. Therefore, the fractal dimension for these three methods was compared, and the efficiency of the real part was evaluated more than other imaginary part. It should be noted that all three methods are highly sensitive to damage, spall, and extent of failure, whereby the fractal dimension increases with increasing similarity (Figure 5.16).

5.2.9 Moments of Complex Shearlet

Moments are practical indicators used to describe image properties. Thus far, two main categories of moments have been presented to define the moment index. In this section, using the image of complex Shearlet coefficients, different moments are evaluated to quantify the characteristics of pavement damage. Each moment can provide an indicator to introduce the characteristics of

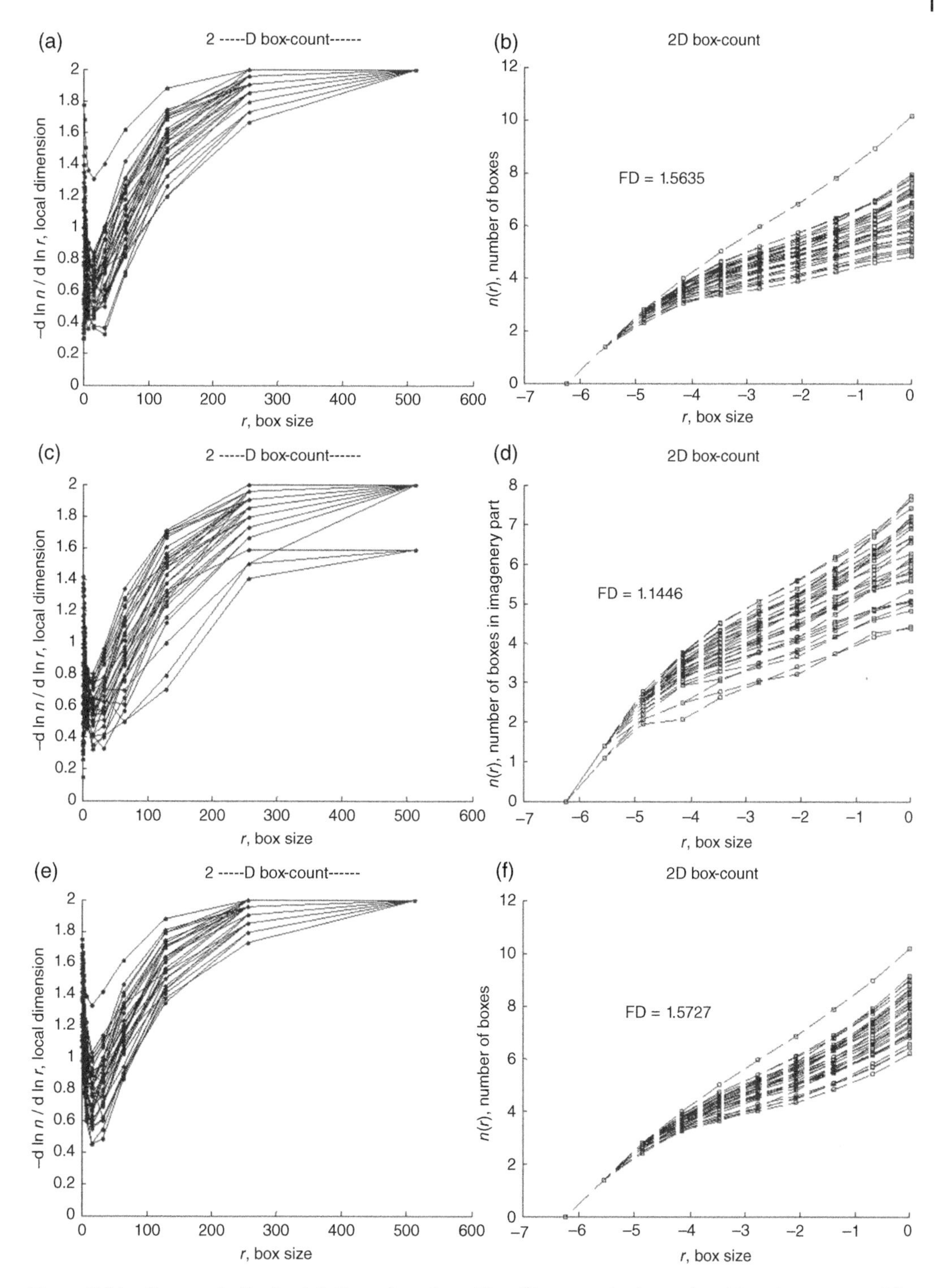

Figure 5.14 Changes in the fractal dimension of cracking $fD_{box-count}$ in the real and imaginary parts of the complex Shearlet: (a,b) real part; (c,d) imaginary part; and (e,f) original image.

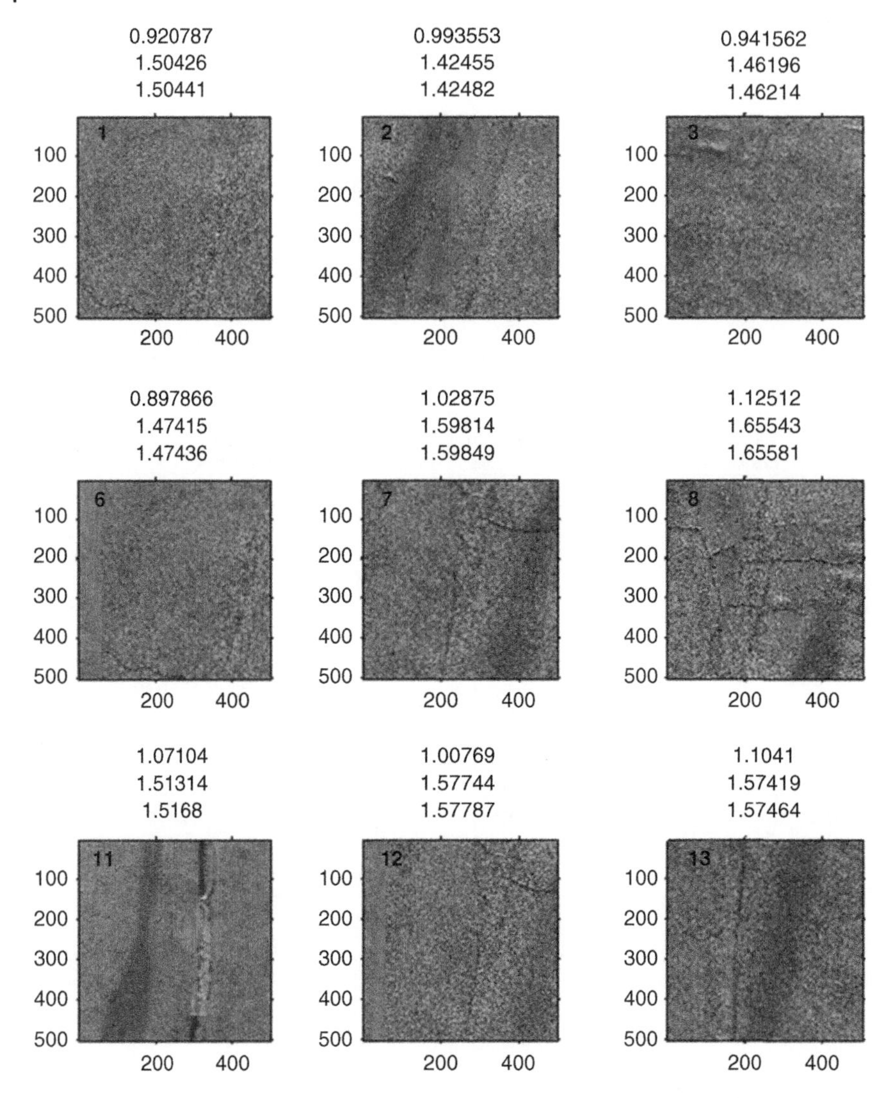

Figure 5.15 Several examples of fractal dimension results for pavement surface cracks using Shearlet results in $fD_{box-count}$(real) and imaginary $fD_{box-count}$(img).

Table 5.3 The effect of type and severity of surface damage of infrastructure on complex fractal index.

		Log.	Tr.	Di.	Bl.	Al.	Co.	Nd.
$f\mathrm{Dmin}_l$	Fdc	1.479	1.474	1.423	1.519	1.561	1.514	1.281
fDava_m		1.567	1.534	1.485	1.587	1.648	1.580	1.425
fDmax_h		1.662	1.646	1.572	1.692	1.685	1.641	1.518
fDmin_l	Fdr	1.479	1.473	1.423	1.519	1.560	1.514	1.281
fDava_m		1.567	1.533	1.484	1.587	1.647	1.579	1.424
$f\mathrm{Dmax}_h$		1.661	1.646	1.571	1.692	1.684	1.641	1.514
$f\mathrm{Dmin}_l$	Fdi	0.984	0.898	0.897	1.004	1.078	0.958	0.793
$f\mathrm{Dava}_m$		1.039	1.020	0.978	1.070	1.121	1.060	0.961
$f\mathrm{Dmax}_h$		1.143	1.156	1.104	1.160	1.170	1.138	1.071

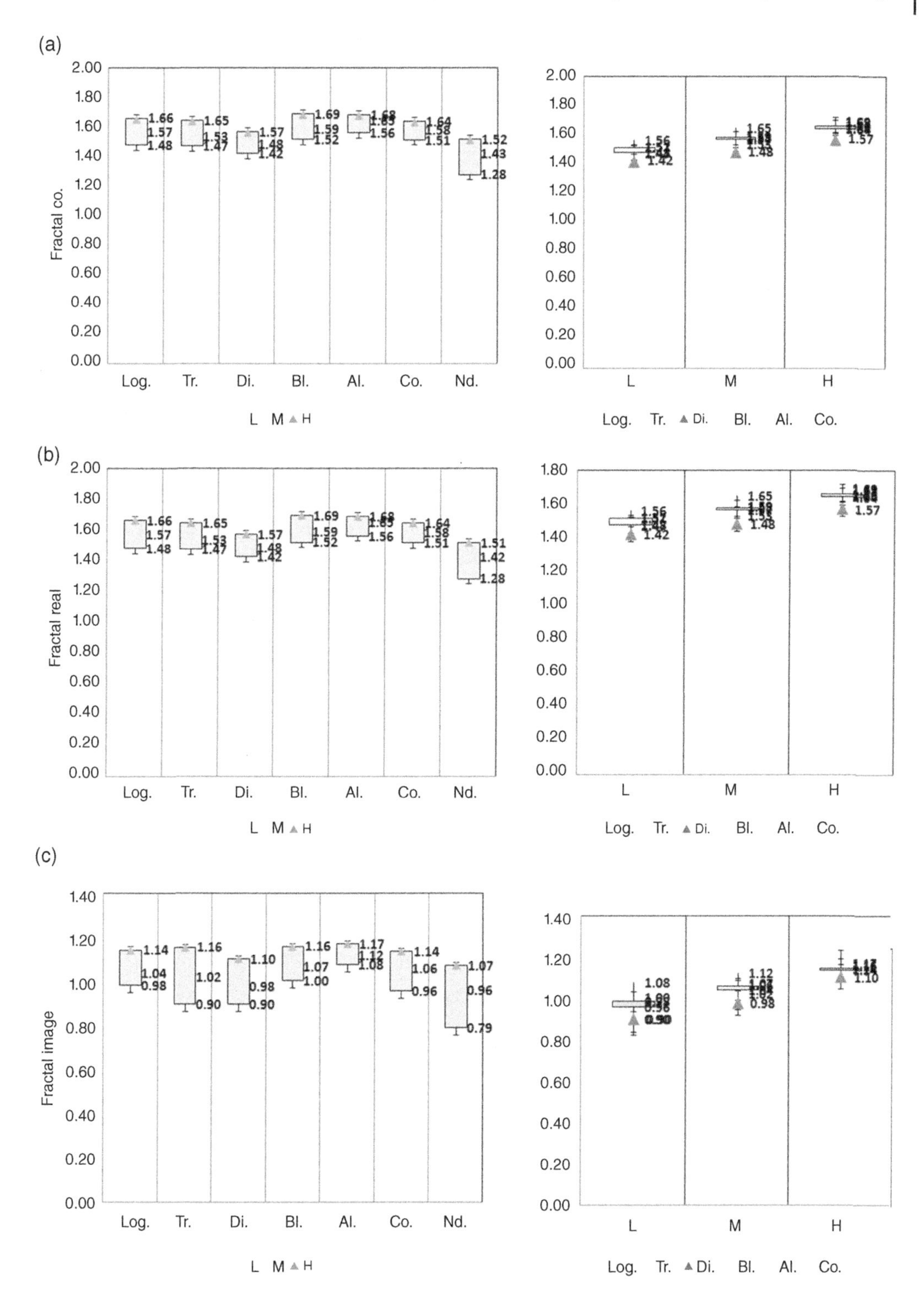

Figure 5.16 Results of fractal dimension of cracking $fD_{box-count}$ in the real and imaginary parts of the complex Shearlet: (a) Fractal co.; (b) real part; and (c) imaginary part.

the damage. For example, at different levels, different characteristics can be used, such as extracted and indexed area, average, energy, and centrality. In the following sections, the main points for diagnosing and separating pavement damage are introduced.

5.2.10 Central Moments *q*

The central moment q is defined by Eq. (5.21):

$$cm_q = \sum_{i=-\infty}^{+\infty} [c(i) - \mu]^q p(i) \quad (5.21)$$

In Eq. (5.21), if $q = 1$, the moment value is equal to zero, and when $q = 2$, it is equal to the standard deviation. When $q = 3$, the value $(\overset{\mathrm{cm}_3}{\rightarrow} 0)$ tends to zero, and when $q = 4$, the computational time increases dramatically despite good performance in detecting and isolating damage. If the value is $\mu = 0$, this relation becomes the central moment relation q. In this case, $q = 1$ is a good criterion for detecting and isolating failure. If this value is positive, it indicates a malfunction in the image. Given that $q = 2$ is very similar to the HFSHEP criterion, the difference is that this index is not normalized if the HFSHEP index is normalized with the total energy of the image. In contrast, light, shadow, and environmental conditions have less effect on the results than the HFSHEP index. The higher the moment, the longer the computation time, and the slower the detection of failure. In the stage of determining the sections with failure, levels 1–4 are sufficient to detect distress.

For example, in assessing the cracking of asphalt pavement, three criteria of type, intensity, and extent are important. In addition to the pattern properties, the direction, rotation, and size are used to determine these properties. This feature is modeled using the relation of translation, rotation, and scaling (TRS) parameters:

$$x' = sR.x + t, R = \begin{pmatrix} \cos\alpha & -\sin\alpha \\ \sin\alpha & \cos\alpha \end{pmatrix} \quad (5.22)$$

where t is the transfer vector; s is the size factor; R is the rotation matrix; and α is the angle of rotation. Due to the change in angle and height, imaging should be used in a way that the results are independent of angle, rotation, and direction. Regardless of size, direction, and angle, fracture failures are severe and extensive and are indexed based on translation, rotation, and scaling (TSR)-based moments:

$$\begin{aligned} cm_{pq} &= \sum_{k=0}^{p} \sum_{j=0}^{q} \binom{p}{k} \binom{q}{j} (-1)^{k+j} x_c^k y_c^j m_{p-k,q-j} \\ x_c &= \frac{m10}{m00, y_c} = \frac{m01}{m00} \end{aligned} \quad (5.23)$$

This relation (5.22) is used to quickly calculate the central moment for a geometric moment. It is also obtained with independent size using the normalization of each moment, where each moment is fixed by the experimental normalization factor assigned to the image set. Moments with a lower order are more resistant to noise and are also calculated with less speed and complexity. The cm_{pq} is normalized using μ_{00}:

$$\begin{aligned} s_{pq} &= \mu_{pq}/\mu_{00}^{w}, w = \frac{(p+q)}{2} + 1 \\ \mu'_{pq} &= s^{p+q+2} \mu_{pq} \end{aligned} \quad (5.24)$$

The C_{pq} complex moment for the pavement image with the function $f(x, y)$ is calculated using Eq. (5.25):

$$\begin{aligned} c_{pq} &= \iint_{-\infty}^{+\infty} (x+iy)^p (x-iy)^q f(x,y) \mathrm{d}x\mathrm{d}y = \int_0^{\infty}\int_0^{2\pi} r^{p+q+1} e^{i(p-q)\theta} f(r,\theta)\theta \mathrm{d}r\mathrm{d}\theta \\ x &= r\cos\theta, y = r\sin\theta, r = \sqrt{x^2+y^2}, \theta = \arctan\left(\frac{y}{x}\right) \\ c'_{pq} &= e^{-i(p-q)\alpha}.c_{pq} \end{aligned} \tag{5.25}$$

In Eq. (5.25), assuming a rotation of α, the relative moment value of $e^{-i(p-q)\alpha}$. Assuming $p = q$, these two values will be equal.

5.2.11 Hu Moments

An independent moment for rotation was proposed by Hu et al. in 1962. In this section, the results of Hu's independent algebraic theory, based on seven independent parameters in the two-dimensional plane and around the crack, are used as an example in the assessment and management of infrastructure. According to this theory, Hu's weekly moments are calculated based on Eq. (5.26):

$$\begin{aligned} &\varnothing_1 = m20 + m02, \\ &\varnothing_2 = (m20 + m02)^2 + 4m11^2, \\ &\varnothing_3 = (m30 - 3m12)^2 + (3m21 - 3m03)^2, \\ &\varnothing_4 = (m30 + m12)^2 + (m21 - m03)^2, \\ &\varnothing_5 = (m30 + 3m12)(m30 + m12)\left((m30 + m12)^2 - 3(m21 + m03)^2\right) \\ &\quad + (3m21 - m03)(m21 + m03)\left(3(m30 + m12)^2 - (m21 + m03)^2\right), \\ &\varnothing_6 = (m20 - \mathrm{m}02)\left((m30 - m12)^2 - (m21 - m03)^2\right) + 4m11(m30 - m12)(m21 - m03), \\ &\varnothing_7 = (3m21 - m03)(m30 + m12)\left((m30 + m12)^2 - 3(m21 + m03)^2\right) - (3m21 + m03)^2 \\ &\quad - (m30 - 3m12)(m21 + m03)\left(3(m30 + m12)^2 - (m21 + m03)^2\right) \end{aligned} \tag{5.26}$$

For example, for the moment 1 $\varnothing_1$, the second time, the moment after rotation, change of angle, and change of size using angle α are determined using relations (5.27):

$$\begin{aligned} \mu'_{20} &= \cos^2\alpha.\mu_{20} + \sin^2\alpha.\mu_{02} - \sin 2\alpha.\mu_{11}, \\ \mu'_{02} &= \sin^2\alpha.\mu_{20} + \cos^2\alpha.\mu_{02} + \sin 2\alpha.\mu_{11}, \\ \mu'_{11} &= \frac{1}{2}\sin 2\alpha.(\mu_{20} - \mu_{02}) + \cos 2\alpha\mu_{11}, \\ \varnothing'_1 &= m'20 + m'02 = \left(\sin^2\alpha + \cos^2\alpha\right)(\mu 20 + \mu 02) = \varnothing_1 \end{aligned} \tag{5.27}$$

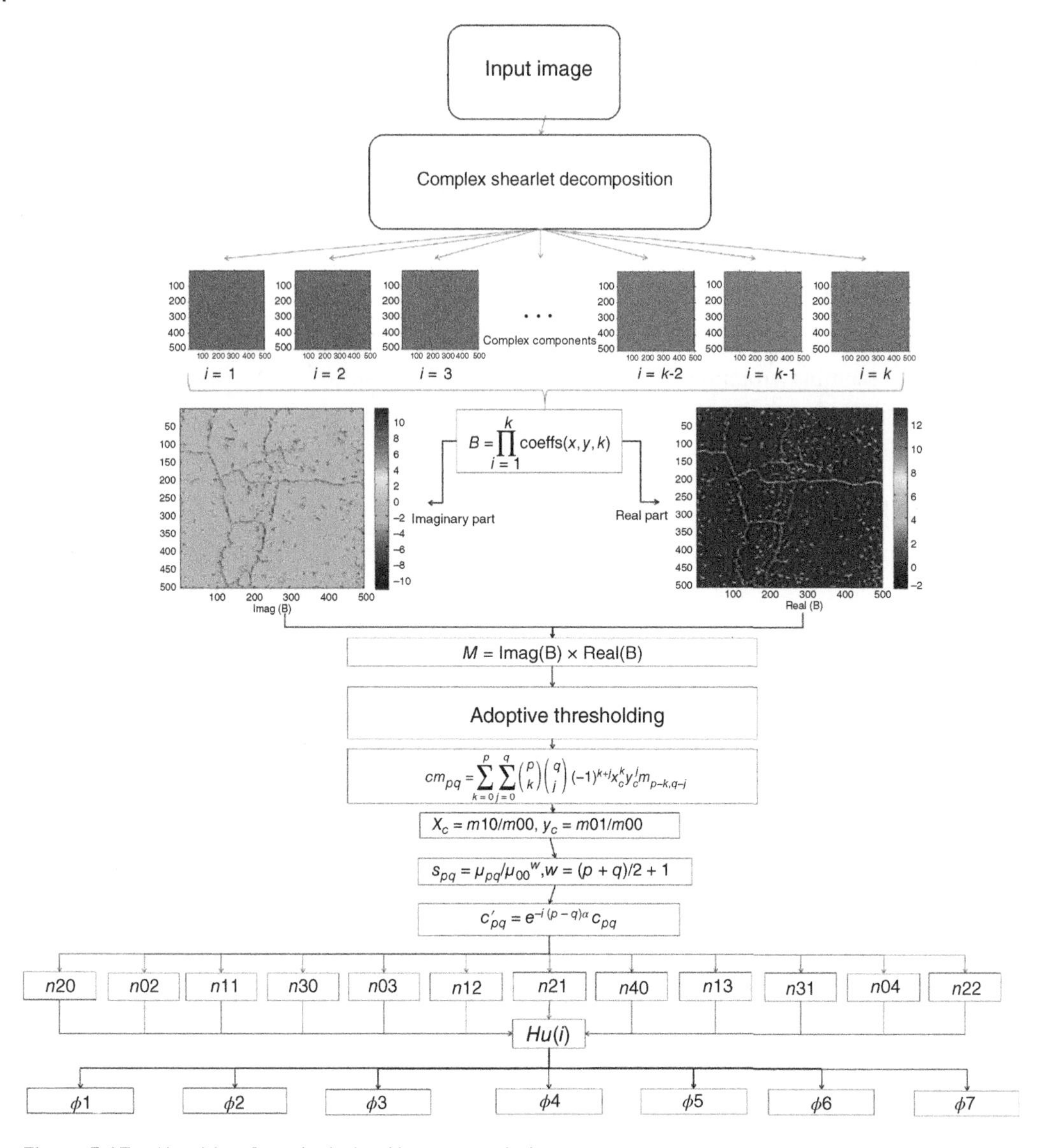

Figure 5.17 Algorithm for calculating Hu moment index.

Similarly, considering $\cos 2\alpha = \cos^2\alpha - \sin^2\alpha$, $\varnothing_2 = \varnothing_2'$, and other Hu moments, this relationship can be proved (Figure 5.17).

In the figure above, the algorithm for calculating the Hu (i) indices at different levels is given for the crack image of the pavement complex. Based on these relationships, the seven indicators are calculated for a set of sample images, as shown in the Figure 5.18.

In the table of results, the calculation of moment $\varnothing_1$using the Hu (1) relations is summarized for the detection and separation for 100 images. As can be seen from the indicators, the value $\varnothing_1$increases with increasing intensity and the amount of failure. This indicator shows a higher number for area failures. The lower the isotropic amount of damage, the higher the index value 1 $\varnothing$ 1. For example, this index shows a lower number for linear injuries. According to the analysis on the image bank, the moment ratio for an image with a defect to a healthy image is a number greater than 1 (Table 5.4).

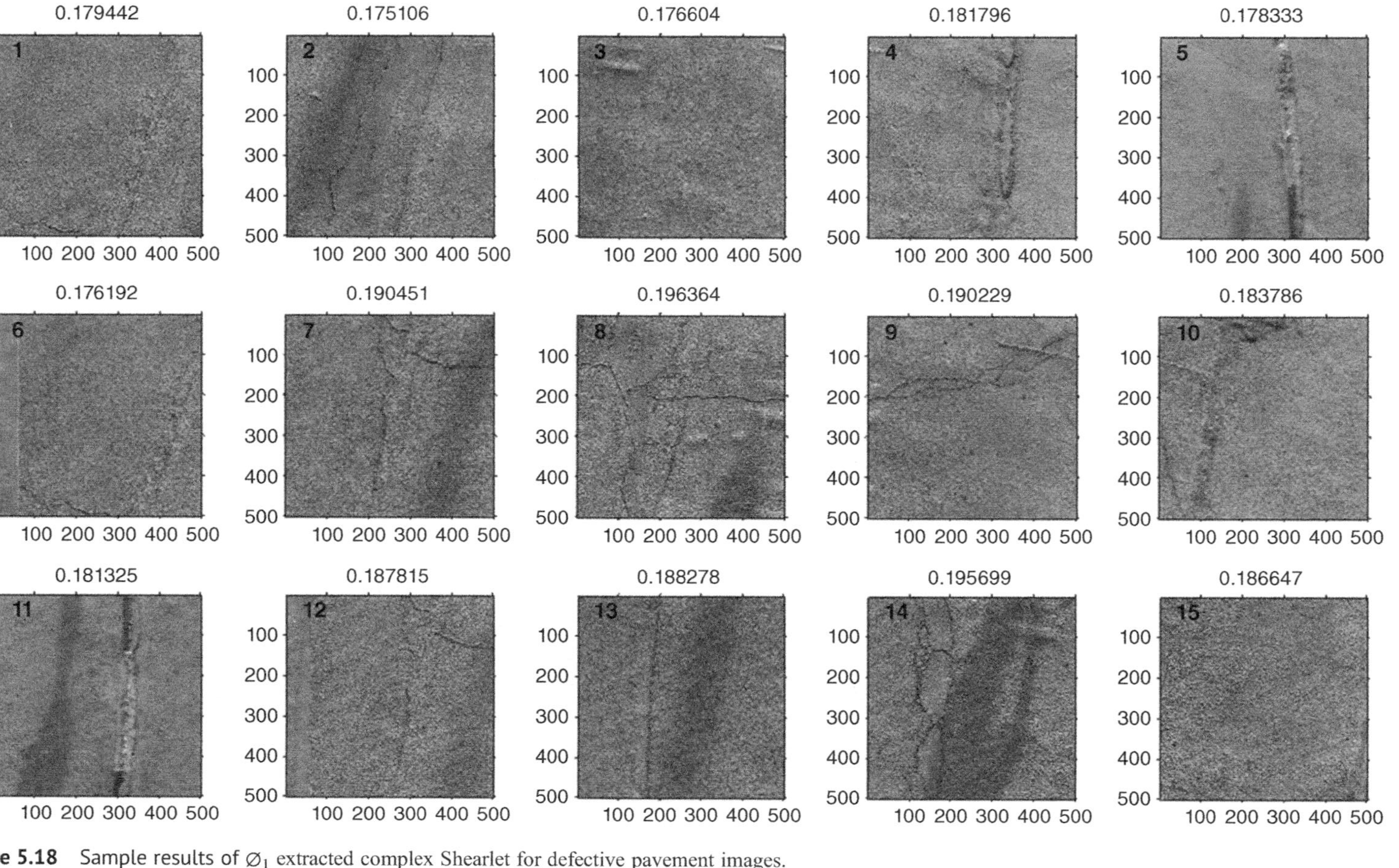

Figure 5.18 Sample results of $\varnothing_1$ extracted complex Shearlet for defective pavement images.

Table 5.4 Investigation of the effect of type and severity of failure on the first moment of Hu $\varnothing_1$.

	Log.	Tr.	Di.	Bl.	Al.	Co.	Nd.
$\varnothing 1min$	0.177	0.175	0.175	0.179	0.186	0.175	0.171
$\varnothing 1ave.$	0.185	0.180	0.178	0.186	0.193	0.185	0.175
$Q(1)_{max}$	0.193	0.190	0.183	0.194	0.198	0.196	0.181
Average	0.185	0.182	0.179	0.186	0.193	0.185	0.176
Thresholding				0.17			

The ratio of damaged sections to healthy sections is {1.087, 1.104, 1.093} for low, medium, and high intensity and extent, respectively. The percentage difference of the indicators is 6–8% for different intensities. The threshold for that can be used in this method is 0.17. Sections with a value greater than this threshold are placed in the failure category. It is also possible to use this method for three different levels to classify the type of injury based on severity (Figure 5.19).

The value of the index $\varnothing_2$ is presented in the table based on the type and severity of pavement damage for the images. As shown in Table 5.5, the distress ratio to the healthy cross section is a positive number greater than zero. As the severity and extent of infrastructure damage increase, this ratio also increases. For low, medium, and high intensity, this index is equal to 5.3, 14.3, and 38.5, respectively. The higher this value, the greater the distinction between healthy and damaged sections.

As shown in Figure 5.20, the value of this index for complex damage indicates a higher value than other damages. Thus, this index can be an effective indicator to determine the diversity of types of distress. On the other hand, due to the severity of the damage and the extent of this growth index, it shows seven times the intensity. Based on the presented relations, all indices from $\varnothing_3$ to $\varnothing_7$ were calculated for the set of images and are shown for comparison and evaluation (Figure 5.20).

Index $\varnothing_3$ gives the maximum value for block cracking, which is a number close to zero for healthy sections. This ratio is four to five times that of healthy cross-sections for medium and large intensities and areas. Also, the index $\varnothing_4$ has the value 21 times in low-intensity cracking and 7 times in severe damage. This index is a good criterion for detecting low-intensity failures. The highest indicators are related to severe block cracks and complex cracks. Index $\varnothing_4$ is similar to the ratio of separation of damaged sections from healthy sections to 12, 21, and 53, experimentally. The higher the intensity and extent, the higher the index. This index is zero for healthy sections and increased by damaged sections. Block cracks have the highest index, while alligator and complex cracks have the lowest index. The absolute value of index $\varnothing_6$ for cracking is greater than zero, which increases with the greater extent of injury. This index has a higher value for block cracks than other types of cracks. The ratio of sections with damage to healthy sections is a minimum of 2.11 and maximum of 4.6. The absolute index $\varnothing_7$ has a higher value for sections with cracks than for other sections, whereby the lower the failure value, the closer the index is to zero. The sensitivity of this index to block cracking is higher than other cracks.

The maximum value of the Hu moment was not very effective for pavement with low cracking, while it performed the separation and detection steps well only for medium- and high-intensity damage. The ratio of coefficients for medium to healthy index is 2.5 times, and for high-intensity failure, this index is more than 25 times that the cross-section with medium intensity failure and low coefficient. According to the analysis of each moment, $\varnothing_1$ to $\varnothing_7$, the capability of these methods in diagnosis was proven, moment has been used as an effective feature in the feature vector (Figure 5.21).

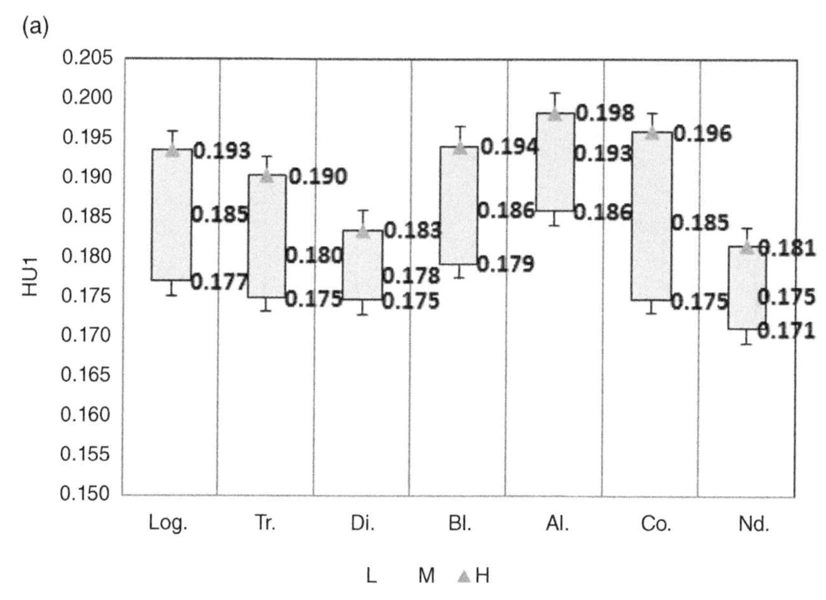

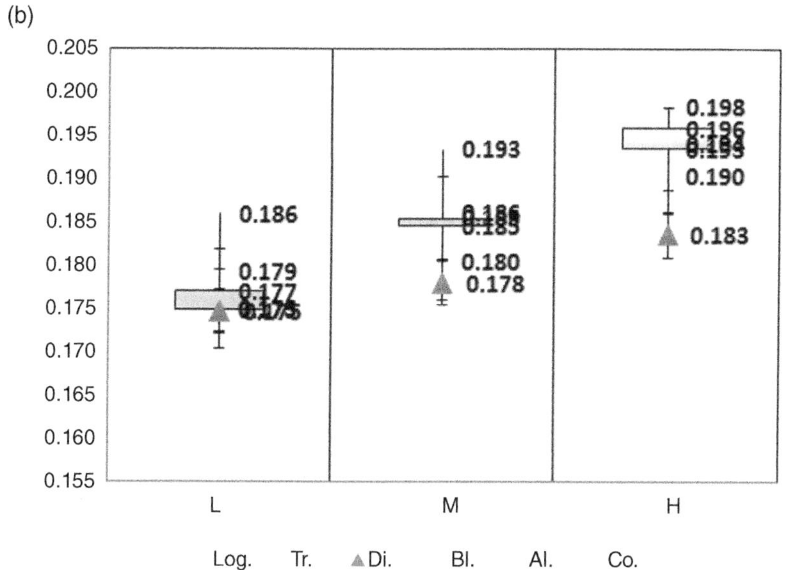

Figure 5.19 Results of index $\varnothing_1$ in samples with different type and intensity: (a) effect of intensity level on the index $\varnothing_1$, and (b) effect of type of distress on $\varnothing_1$.

Table 5.5 Investigation of the effect of type and severity of failure on the first moment of Hu $\varnothing_2$.

	Log.	Tr.	Di.	Bl.	Al.	Co.	Nd.
$\varnothing 2_{min}$	0.452	0.417	0.464	5.322	1.036	0.957	0.066
$\varnothing 2_{ave.}$	7.924	11.541	6.826	14.044	14.319	13.868	4.948
$Q(2)_{max}$	22.731	35.808	24.620	22.982	33.178	38.554	15.935
Average	10.369	15.922	10.637	14.116	16.178	17.793	6.983

(a)

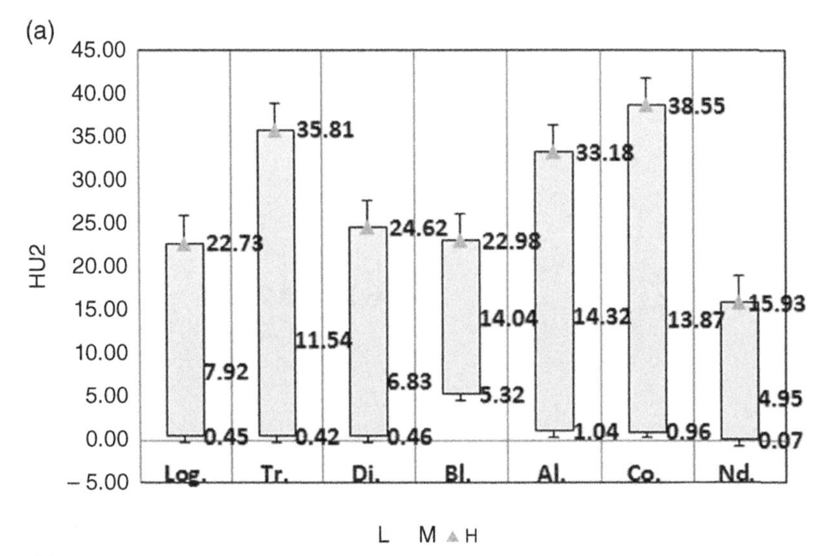

(b)

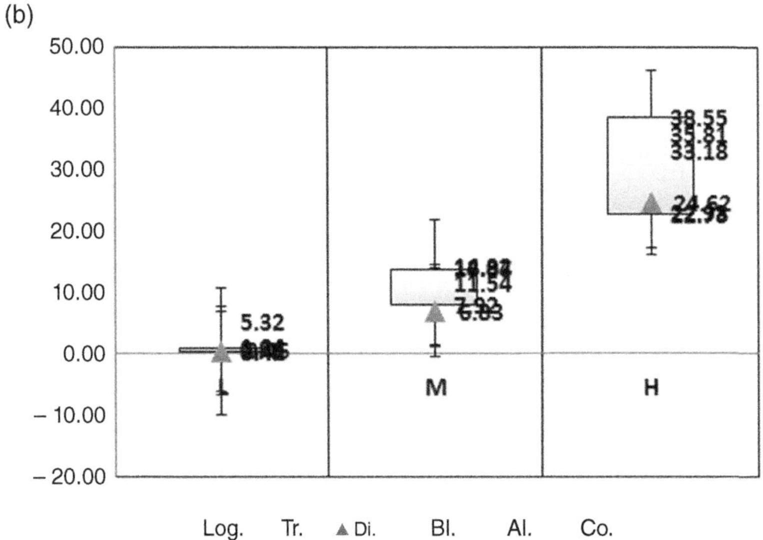

Figure 5.20 Results of index $\varnothing_2$ in samples with different types and intensities: (a) effect of intensity level on the index $\varnothing_2$, and (b) effect of type of distress on $\varnothing_2$.

5.2.12 Bamieh Moments

The Bamieh moment algebraic independent was introduced by Bamieh et al. in 1986 for diagnosis and evaluation of cracks based on visual characteristics. This feature enables the ability to model and detect crack failures more quickly than other moments. Bamieh is independent, and moments are extracted based on central tensor moments using Eq. (5.28):

$$\mathrm{BMI}(i,j,k) = \int_{-\infty}^{+\infty}\int_{-\infty}^{+\infty} x^i x^j x^k \ldots f(x,y)\mathrm{d}x\mathrm{d}y \tag{5.28}$$

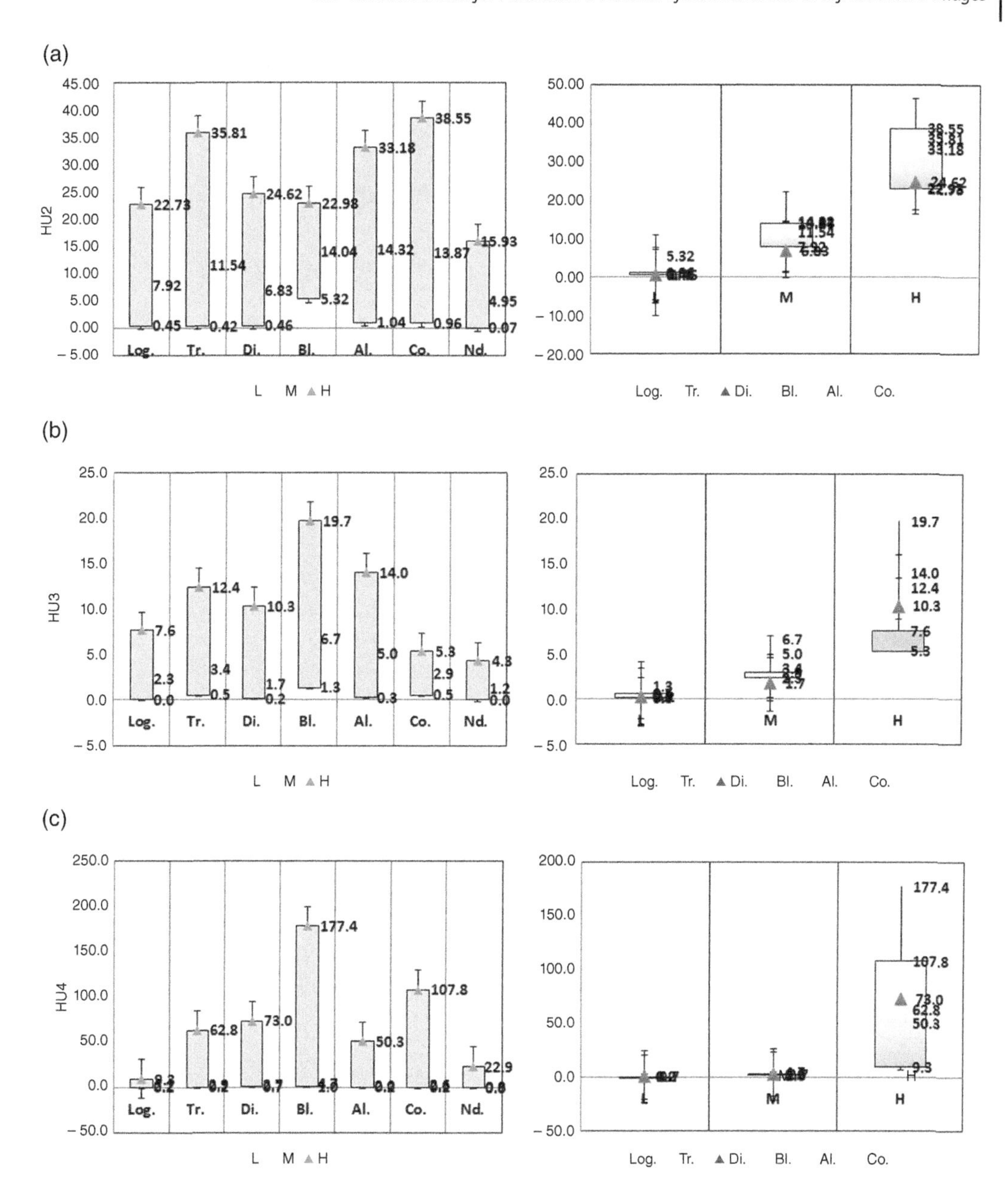

Figure 5.21 Results of index $\varnothing_{2-4}$ in samples with different types and intensities: (a) effect of intensity level on the index $\varnothing_2$, (b) effect of type of distress on $\varnothing_3$, and (c) effect of type of distress on $\varnothing_4$.

Based on Eq. (5.28), the central moment equation is calculated from Eq. (5.29):

$$n_{pq} = \iint xc^p yc^q f(xc, yc)\mathrm{d}x.\mathrm{d}y \; cm_{pq} = \sum_{k=0}^{p}\sum_{j=0}^{q}\binom{p}{k}\binom{q}{j}(-1)^{k+j}x_c^k y_c^j m_{p-k,q-j}$$
$$x_c = \frac{m10}{m00, y_c} = \frac{m01}{m00} \tag{5.29}$$

where *yc* and *xc* are the components; and $f(xc, yc)$ is the function of the cross-sectional image. The set of numbers, n_{pq}, is a single representation of the main elements of the damaged image function. The sum of $p + q$ yields the order of the moment:

$$\begin{aligned} B_1 &= (n20 * n02) - n11^2 \\ B_2 &= (n03 * n03 - n12 * n21)^2 - 4 * \left(n03 * n12 - n21^2\right) * \left(n21 * n30 - n12^2\right) \\ B_3 &= (n40 * n04) - (4 * n31 * n13) + 3 * \left(n22^2\right) \\ B_4 &= (n40 * n04 * n22) - 2 * (n31 * n13 * n22) - \left(n40 * n13^2\right) - \left(n04 * n31^2\right) - n22^2 \end{aligned} \tag{5.30}$$

Any algebraic expression or weighted two-dimensional function of order d for tensor A is a linear combination of expressions constructed by multiplying the principal tensors. The algorithm for calculating these properties is shown in Figure 5.22. Based on this algorithm, the image of mixed Shearlet coefficients is calculated from the image, then the binary image is extracted using steps and used as the input of okra functions. Using the relations, the bays B_1 to B_4 are extracted as properties by the Bamieh method and placed in the property vector for detection. As an example of the application results, the okra algorithm is shown on the images in Figure 5.23, for B_1 to B_4.

This section discusses the evaluation of the effect of existing damage caused by cracking in sections using Bamieh spikes. The Bamieh indices for the experimental image bank, including 100 images, were calculated, and the results are summarized in Table 5.6 for B_2 based on the type and intensity level (Figures 5.23 and 5.24).

As shown in Table 5.6, the minimum second Bamieh value is slightly different from the healthy state. On the other hand, for linear failures, such as longitudinal, transverse, and oblique, the average value of the second Bamieh moment is equal to the maximum state of this index at the maximum value; therefore, this moment is a good indicator for detecting and isolating pavement failures. In sections where the moment is greater than 0.2, the image is defective and stored as an analyzable section in the defective database (Figure 5.25).

Based on this case comparison, it was found that the ratio of the second Bamieh index to detect the damage of infrastructure to the healthy cross section $(B_2)_d/(B_2)_{nd}$ increased to 43. As the rate of disruption and irregularity increased, this index also showed growth. In general, this index has a higher ratio for complex and block failures. On the other hand, the highest value in all sections with failure is more than 0.48, and the average of these sections is higher than 0.2.

For index B_1, the average pavement damage is more than 0.77. The highest average failure rate at three levels is related to fatigue cracking with 0.93. In general, the ratio of cross section to damage of healthy cross section is close to 1.2. As shown in Figure 5.26a, the differences between failure sections and healthy sections are close, except in fatigue cracking. Therefore, this index seems to be suitable for diagnosing fatigue, block and complex cracks. Regarding index B_3, longitudinal cracks have the highest value. The difference of this index for the separation of anomalous sections with a low-intensity level for this type of failure is 1.5 to 2.5 times the average value of this index for healthy sections or low failure; therefore, a threshold of 0.25 is considered for detection. Index B_4 is very similar to index B_3 in terms of the efficiency of detection and isolation. According to the results, the ratio of index B_3 for failure to B_3 of healthy section, in failure at a high-intensity level, is higher than average- and low-intensity levels. This ratio increased from 1.4 for low severity distress to 2.8 for high severity; therefore, the separation of sections with high failure intensity levels can be done more accurately using B_4. The average threshold of 0.07 was selected for separation.

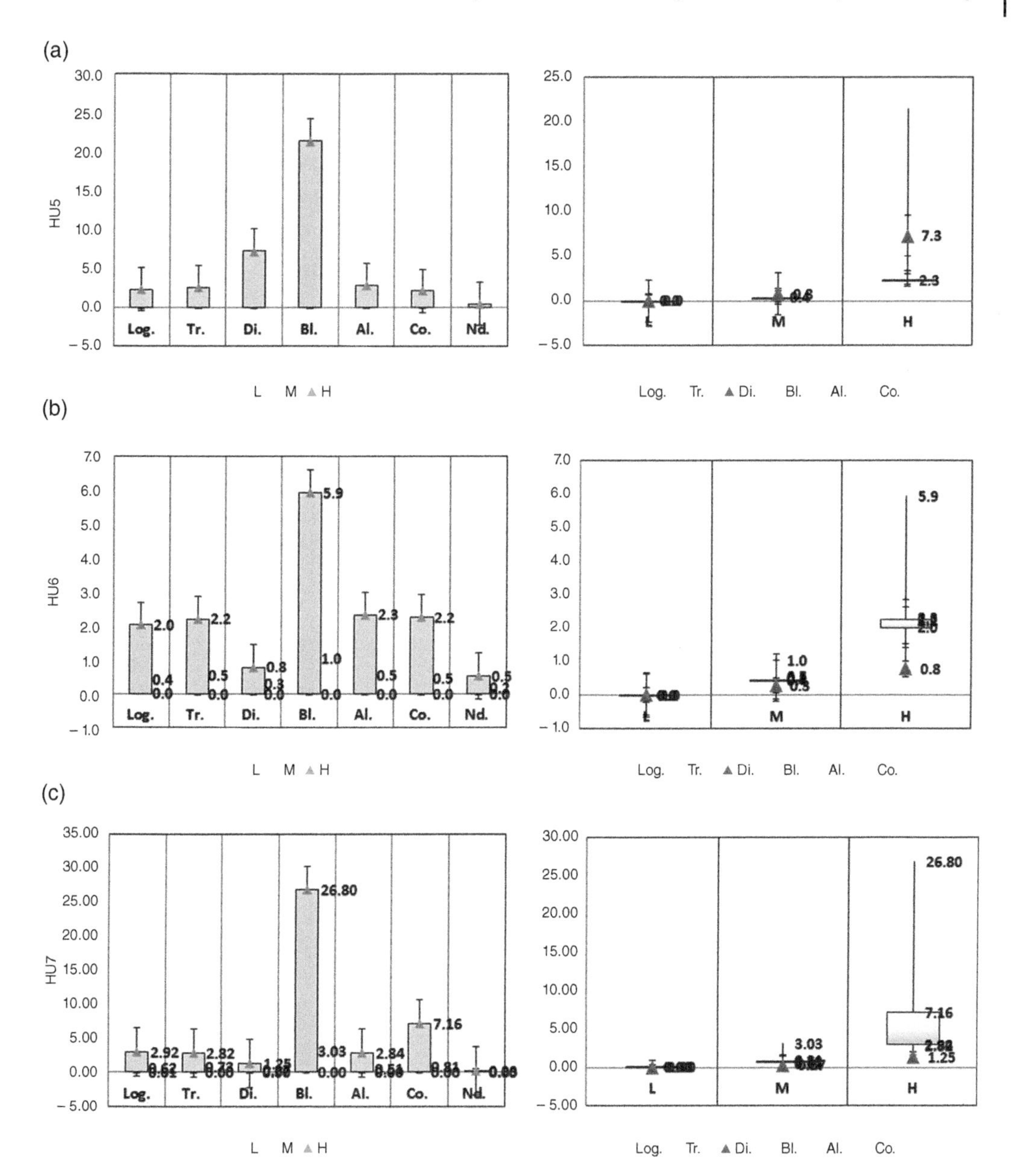

Figure 5.22 Results of index $\varnothing_{5-7}$ in samples with different types and intensities: (a) the effect of intensity level on the index $\varnothing_5$, (b) effect of type of distress on $\varnothing_6$, and (c) effect of type of distress on $\varnothing_7$.

5.2.13 Zernike Moments

The Zernike moment (ZM) shows the number of cracks in different directions and different rotations, with the same index. Also, the value of moments 2 and 3 is equal to the Hu moment in the geometric display mode. This moment can be calculated up to eight times. If so, can that be explained in more detail. The value of ZM for order n is calculated based on the number of iterations β using Eq. (5.31):

Table 5.6 Comparison of the effect of type and severity of distress on the first Bamieh of Hu B_2.

	Log.	Tr.	Di.	Bl.	Al.	Co.	Nd.
$B_{2\min}$	0.000	0.000	0.000	0.000	0.000	0.000	0.000
$B_{2\text{ave.}}$	0.438	0.673	0.382	2.456	0.409	0.819	0.105
$B_{2\max.}$	2.374	2.800	3.130	20.980	2.750	7.071	0.480
Average	0.937	1.158	1.171	7.812	1.053	2.630	0.195

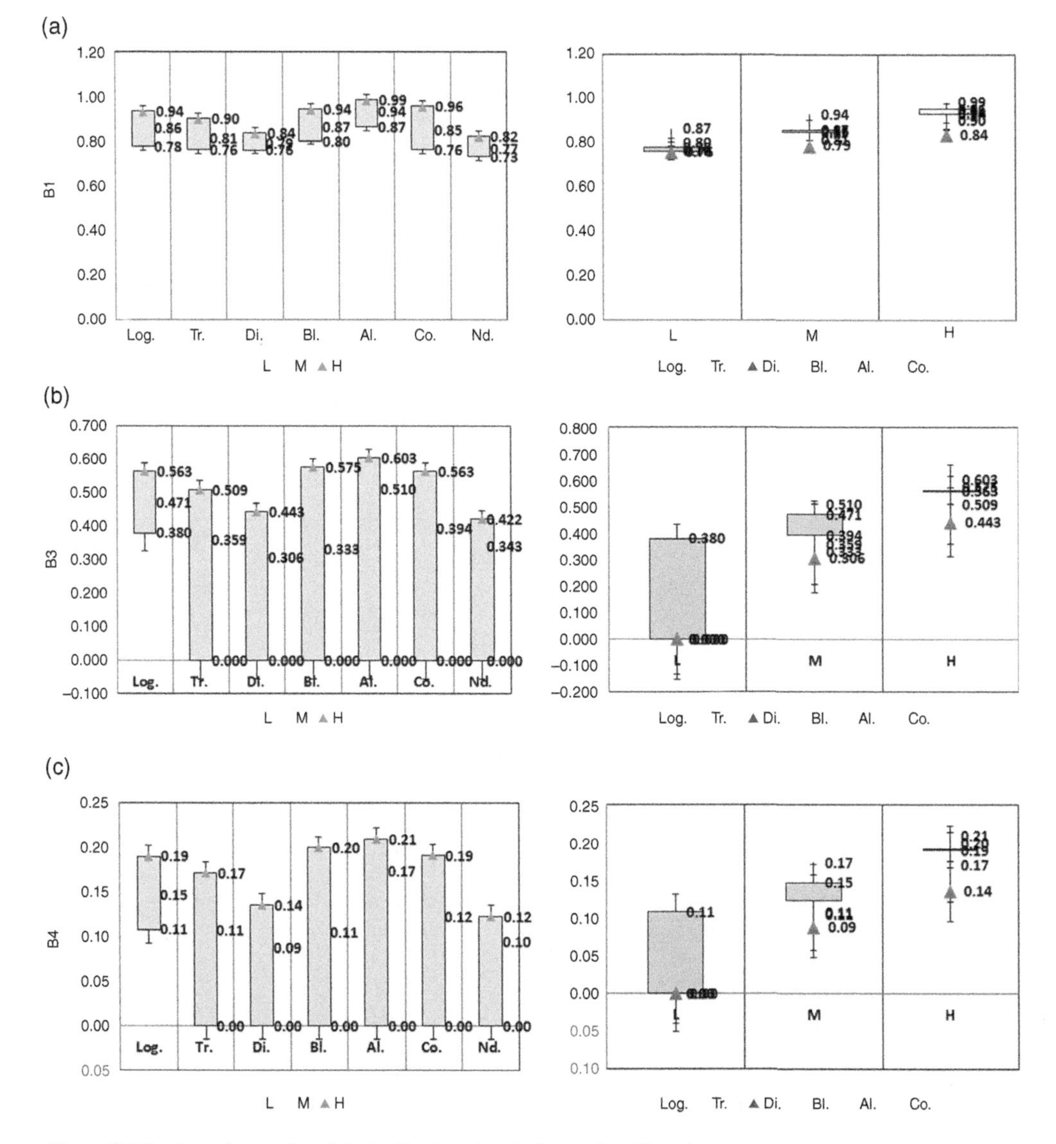

Figure 5.23 Sample results of B_1 to B_4 via extracted complex Shearlet.

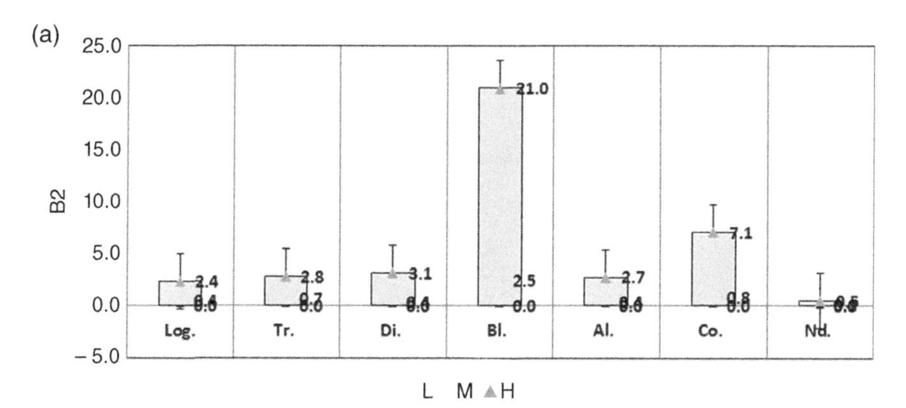

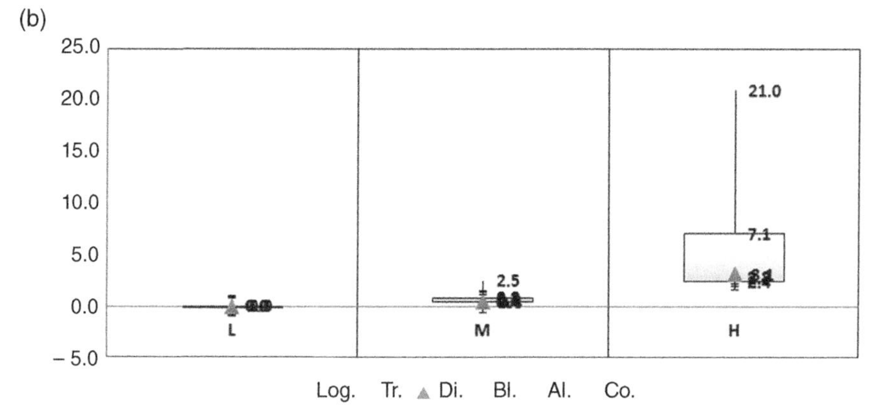

Figure 5.24 Index results in samples based on type and severity of damage to different infrastructures: (a) effect of type of failure and (b) severity level on B_2.

$$A_{n\beta} = \frac{n+1}{\pi}\int_0^{2\pi}\int_0^1 V^*_{n\beta}(r,\varphi)f(r,\varphi)r\mathrm{d}r\mathrm{d}\varphi, n = 0,1,2,3,\ldots,\beta = -n, -n+2,..n \tag{5.31}$$

In this regard, the difference between $n - |\beta|$ is always an even number. The value of $V^*_{n\beta}$ is a multiplication, and the Zernike index of polynomials is defined as follows:

$$V^*_{n\beta}(r,\varphi) = V_{n\beta}(r)e^{i\beta\varphi} \tag{5.32}$$

The behavior of the Zernike moment in rotation is similar to that of the mixed moment. If the image is rotated by θ, the magnitude of ZM changes as follows (Figure 5.27):

$$A'_{n\beta} = A_{n\beta}(r)e^{-i\beta\theta} \tag{5.33}$$

The independent moment value is calculated using size $|A_{n\beta}|$. Using the size $A_{n\beta}$ removes some image information that calculates the results in mixed moments with different variations. By normalizing the principal moments, the phase is calculated using Eq. (5.34):

$$\varnothing = \left(\frac{1}{\beta_r}\right)\arctan\left[\frac{Im\left(A_{m_r\beta_r}\right)}{Re\left(A_{m_r\beta_r}\right)}\right], A_{m\beta} = A_{m\beta}e^{-i\beta\varnothing} \tag{5.34}$$

This relationship shows that the amount of moment is independent of rotation, transmission, and size. An example of how these properties affect the image of cracking is shown in Figure 5.28.

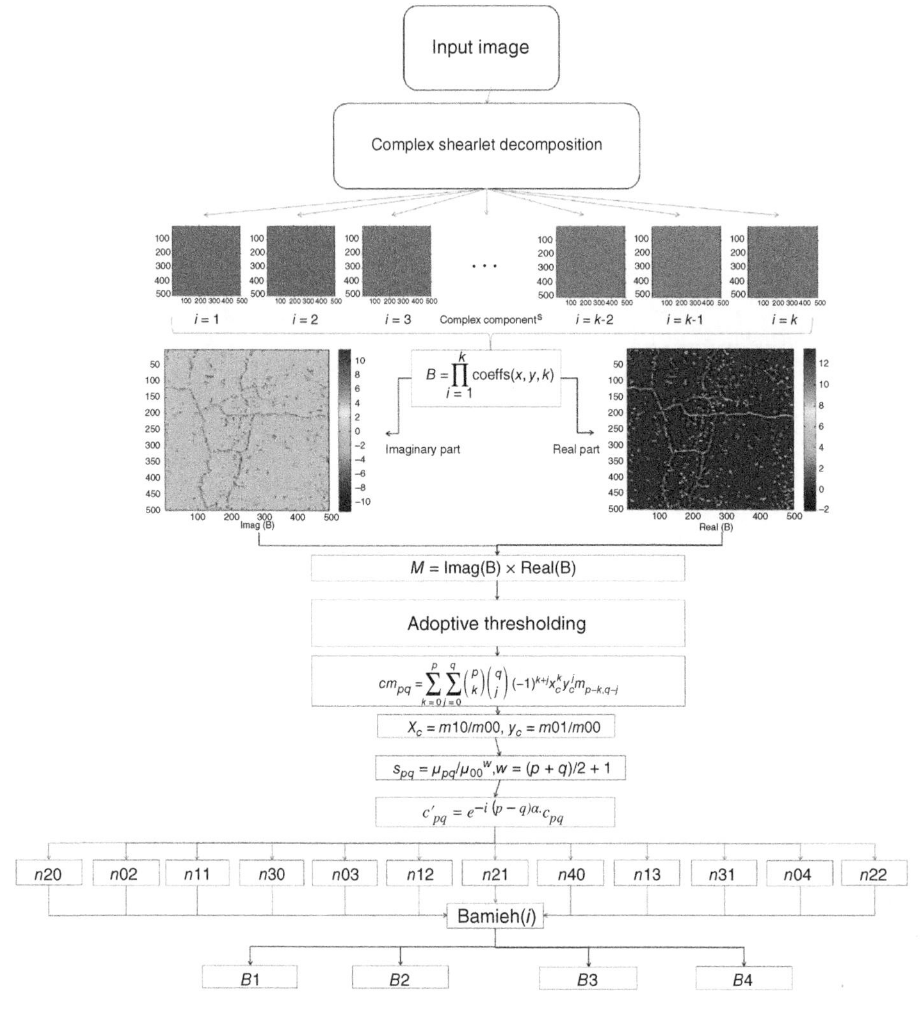

Figure 5.25 Algorithm for calculating Bamieh moment index.

$$
\begin{aligned}
M1 &= ((2 * (n20 + n02)) - 1) * \frac{3}{\pi},\\
M2 &= \left((n20 - n02)^2 + 4 * n11^2\right) * \frac{9}{\pi^2},\\
M3 &= ((n30 - 3 * n12)^2 + (n30 - 3 * n12)^2) * \frac{16}{\pi^2},\\
M4 &= ((n03 + n21)^2 + (n30 + n12)^2) * \frac{144}{\pi^2},\\
M5 &= (((n03 - 3 * n21) * (n03 + n21) * ((n03 + n21)^2 - 3 * (n30 + n12)^2))\\
&\quad - \left((n30 - 3 * n12) * (n30 + n12) * ((n30 + n12)^2 - 3 * (n03 + n21)^2)\right)) * \frac{13824}{\pi^4},\\
M6 &= (864/\pi^3) * (((n02 - n20) * ((n30 + n12)^2 - (n03 + n21)^2))\\
&\quad + (4 * n11 * (n03 + n21) * (n30 + n12)))
\end{aligned}
\tag{5.35}
$$

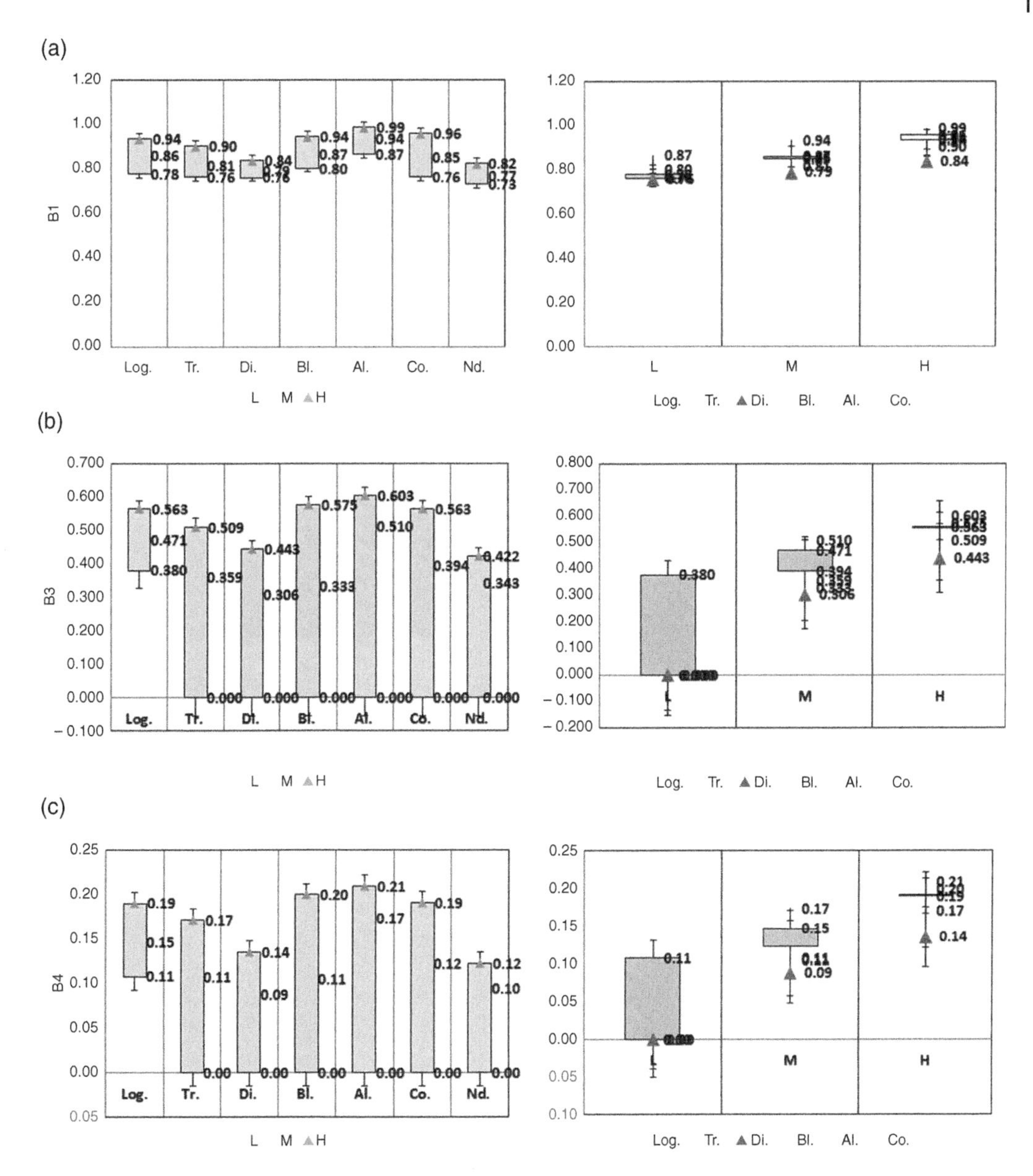

Figure 5.26 Results of index B_l in samples with different types and intensities: (a) effect of intensity level on the index B_1, (b) effect of type of distress on B_3, and (c) effect of type of distress on B_4.

In this relation, $n30$ is calculated using equation $A_{n\beta}$. A total of six image properties were extracted using this moment and used as input to the feature matrix for selection in the next step. In this section, using Zernike theory, the types of pavement failures in different levels of analysis and the main indicators, including their size and phase, are shown in the image. The results indicate that the value of $A_{m\beta}$ is constant for both numerical states. Also, $\varnothing$ is the value of the variable that represents the dominant angle of crack failure. In Figure 5.29, the set of Zernike properties is obtained for Figure C {0.0015 0.0007 0.0060 0.0938 8.9369 0.4478}, and for Figure D, which is a

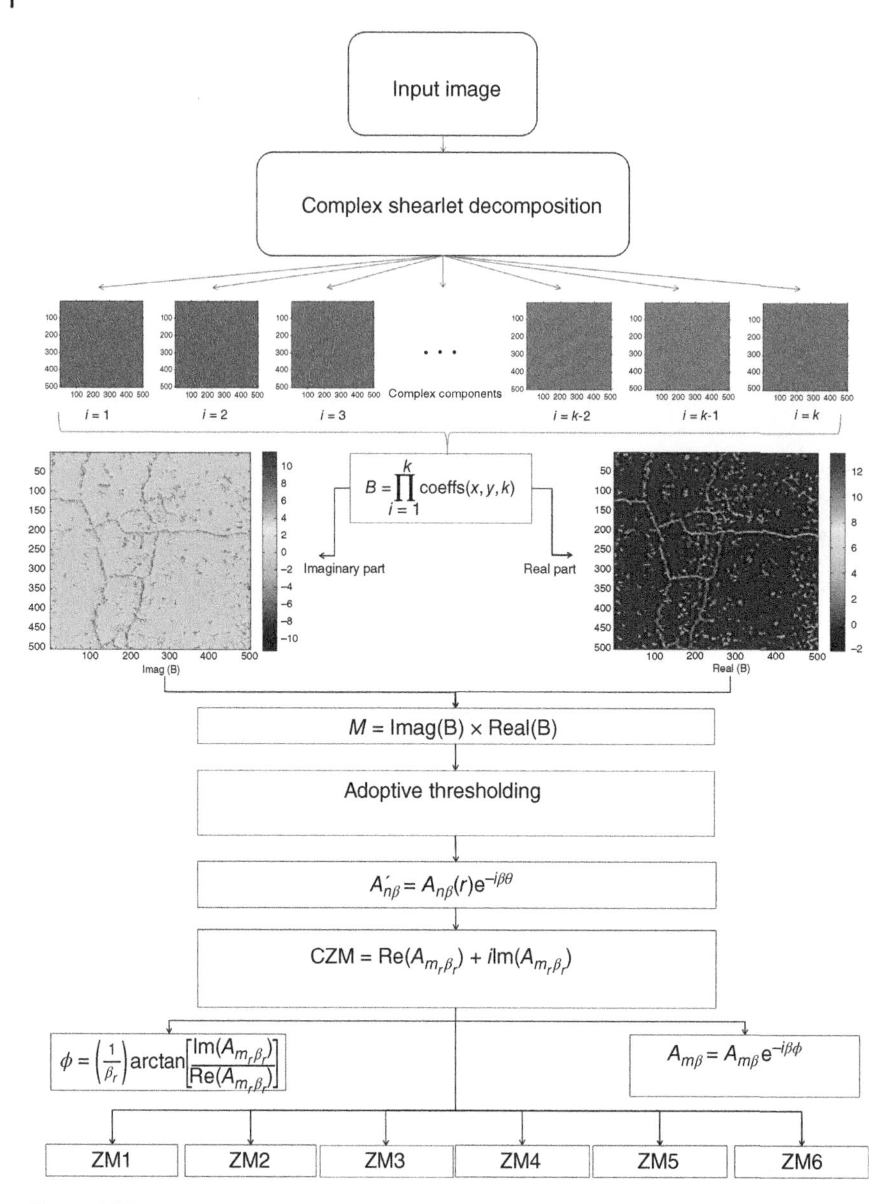

Figure 5.27 Algorithm for calculating Zernike moment index.

90° rotation of Figure C, the same value is obtained. Another example of crack failure is given in Figure 5.30 for comparison.

As shown in Figure 5.28b, each failure has a specific Zernike moment equation. For example, this equation is computed for a mixed order Zernike of order 2. Figure 5.29 shows an example of mixed Zernike functions to calculate the Zernike moments for a set of images.

As shown, the size of the Zernike in each rank is constant by rotating and resizing. The phase change in these functions can be used to determine the direction of failure placement (Table 5.7).

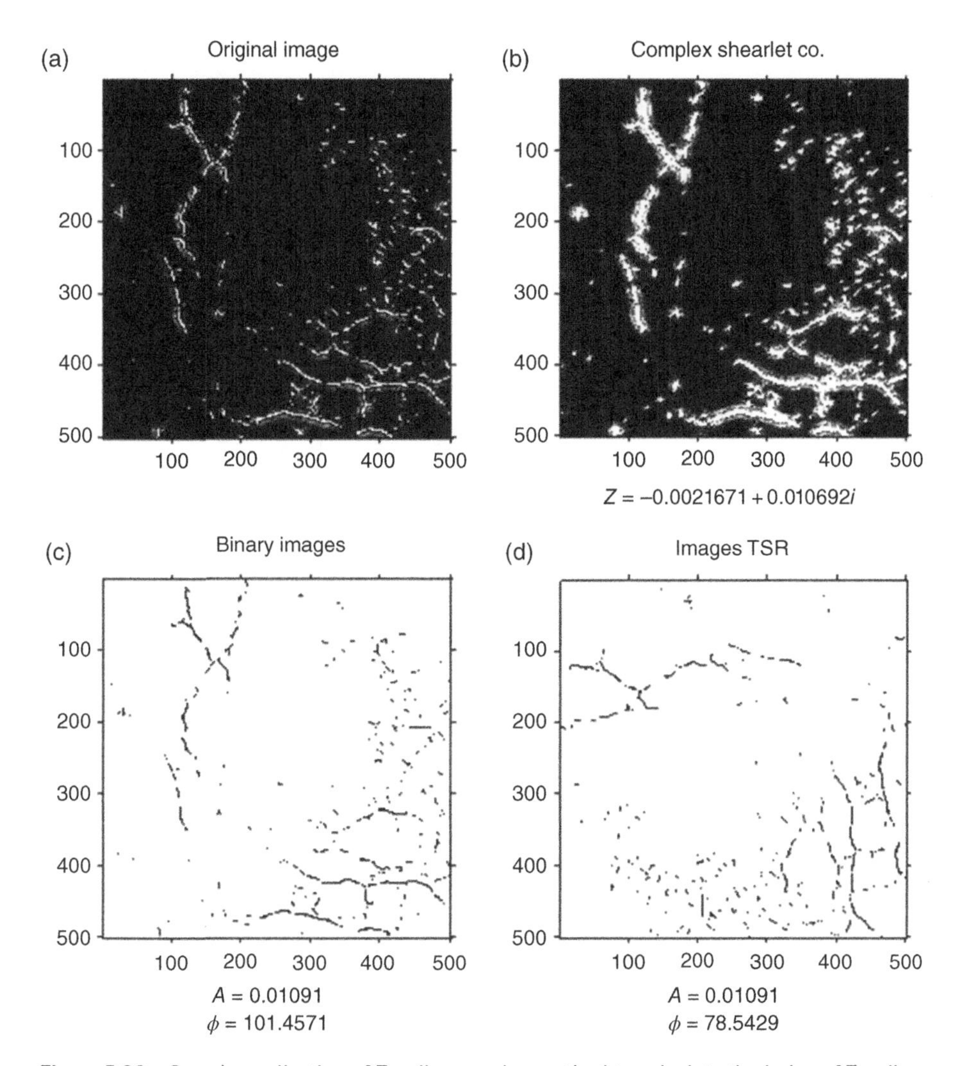

Figure 5.28 Sample application of Zernike complex method to calculate the index of Zernike moments and compare the results with 90° rotation mode to leave fatigue: (a) main image, (b) mixed Shearlet coefficients, (c) determination of size and phase of Shearlet coefficients, and (d) size and phase of Shearlet coefficients with rotation 90°.

As shown in Figure 5.30, for a set of images with different failures, the value of moment $Z1$ is higher value for the sections with failure. This index grows as the amount and type of failure increase, whereby the higher the breakdown order, the lower the number index. As shown in Figure 5.30a, block, fatigue, and mixed failures show a higher Zernike index of $Z1$. The closer this value is to zero, the more it indicates the order in the development of the fault on its axis of symmetry. The maximum value of the index equal to −0.59 is related to fatigue failure, which has a range equal to 0.024, from the highest to the lowest value. Also, the lowest value for healthy cross section is −0.628. The ratio of healthy to maximum cross section is 93%. Also, the value of the $Z1$ index is equal to 98% for the least amount of failure. According to these results, the higher the $Z1$

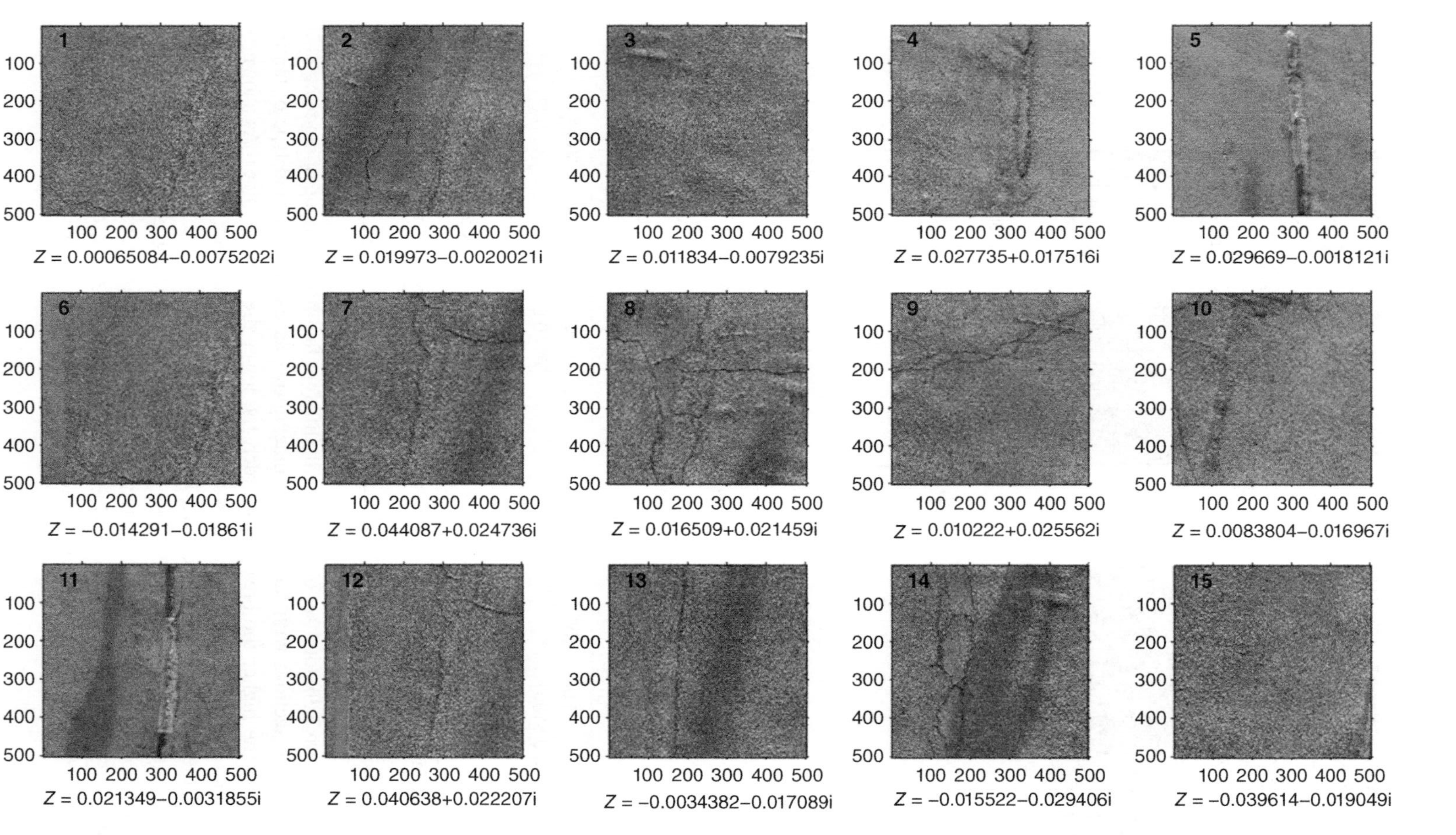

Figure 5.29 An example of the results of the extracted Zernike complex functions for each section of order 2, in order to calculate the Zernike moment $Z(1)-Z(6)$.

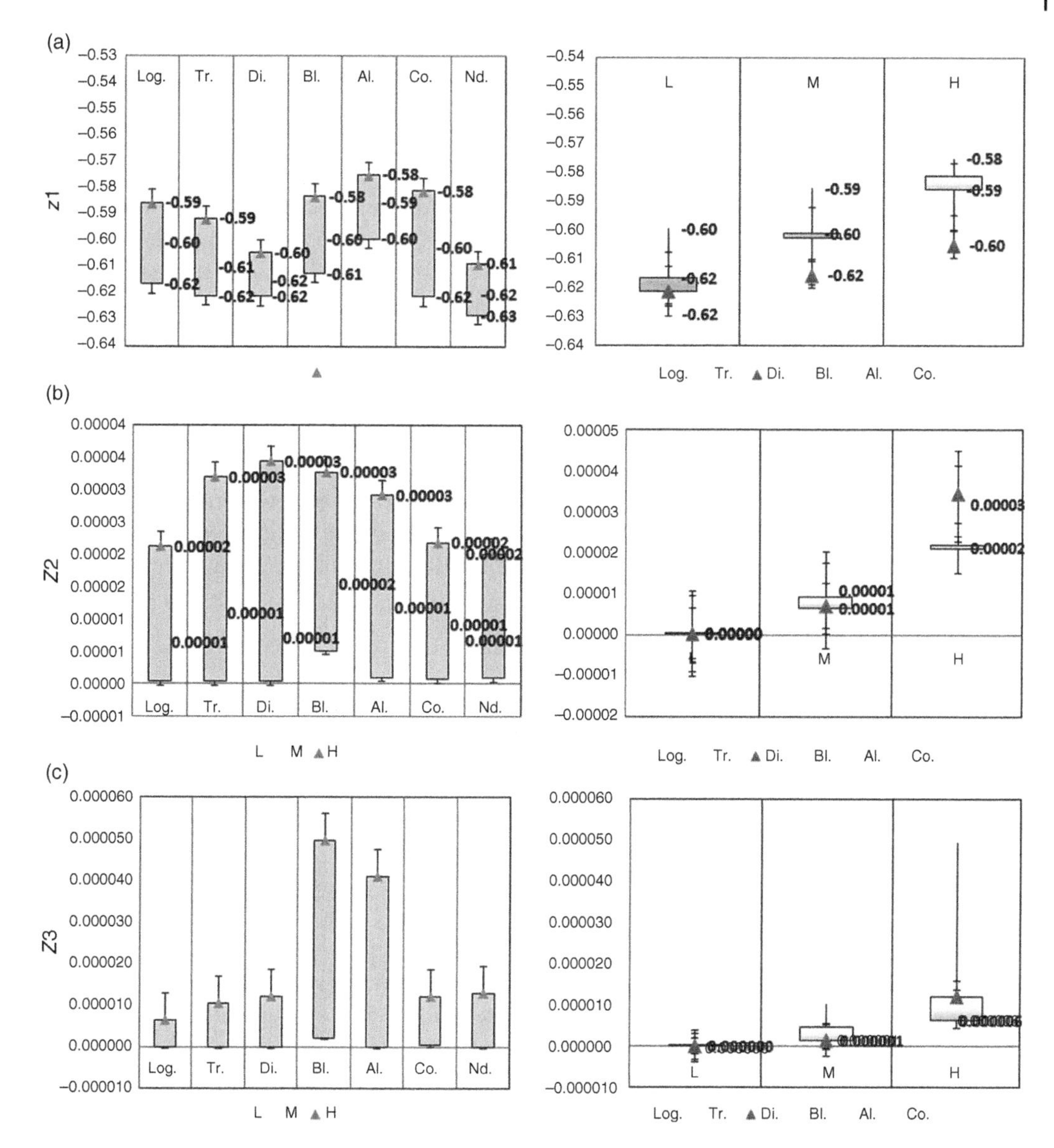

Figure 5.30 Results of index Z_i in samples with different types and intensities: (a) effect of intensity level on the index $Z(1)$, (b) effect of type of distress on $Z(2)$, and (c) effect of type of distress on $Z(3)$.

Table 5.7 Comparison of the effect of type and severity of distress on the first Zarnik of $Z(1)$.

	Log.	**Tr.**	**Di.**	**Bl.**	**Al.**	**Co.**	**Nd.**
$Z_{1\text{min}}$	−0.617	−0.621	−0.621	−0.613	−0.599	−0.621	−0.628
$Z_{1\text{ave.}}$	−0.601	−0.610	−0.615	−0.599	−0.585	−0.603	−0.620
$Z_{1\text{max}}$	−0.586	−0.592	−0.605	−0.583	−0.575	−0.581	−0.609
Samples	14	11	10	9	12	11	33

index in a section, the greater the probability of failure, where the type of failure is more similar to area failures (block, fatigue, and composite). For $Z2$, the minimum index is related to block failures, which is better identified than other failures. The higher the value of $Z2$, the higher the probability of failure. The index value $Z2_d/Z2_n$ of failure of healthy sections is more than five times maximum and minimum 1.7 times.

For the $Z3$ moment, this index has grown significantly for area cracks (block, fatigue) and for linear and composite cracks. The $Z4$ index is a measure of workmanship in the separation of faulty sections, where the higher the index, the more likely it is that cracks will occur at that point. Based on the experiment, a threshold of 0.0004 was obtained for cross-section failure. This threshold varied from five to seven times in sections with failure. The $Z5$ moment can be either positive or negative for failures and is positive for healthy sections. Also, if there is a failure, this index is increased, whereby the higher this value, the more likely it is that there is a failure in the section. Moment $Z6$ is an indicator that has high performance in detecting transverse, oblique, and compound failures. This index can be either positive or negative depending on the direction and severity of the failure. In general, severe damage can be better distinguished than moderate and low damage. As shown in (c) of Figures 5.30 and 5.31, the range of change in the index increases as the severity of the failure decreases. This increase in the change interval reduces the power of the $Z6$ index at the detection stage.

5.2.14 Statistic of Complex Shearlet

Figure 5.32 shows the main statistical characteristics for evaluating the Shearlet complex coefficients. Each feature is used to detect and isolate a faulty section. Using the results of faulty isolation section and to propose an adaptive threshold, the main statistical characteristics are extracted.

The following section introduces each of these characteristics and their relationship with failure with the help of mixed Shearlet coefficients.

5.2.15 Contrast of Complex Shearlet

The sum of the differences between each Shearlet coefficient and its adjacent coefficient is called the contrast index. This index can be a number between 0 and 1 for an image without cracks up to the cross-sectional area. The greater the crack width, the higher the Shearlet coefficient, and the greater the contrast at the crack edge points as well as the crack center. This feature is used after normalization to detect the appearance of apparent damage to infrastructures.

$$S1 = \arg\left(\max \left(\sum\nolimits_{i,j} |i-j|^2 p_{\mathrm{sh}}(i,j) \right) \right) \tag{5.36}$$

where $S1$ is the coefficient of conflict coefficient and $p_{\mathrm{sh}}(i,j)$ is the value of the Shearlet coefficient in the coordinates (i,j). The value of the coefficients is normalized to the cross-sectional size $a \times b$ and is an index between 0 and the maximum normalized Shearlet coefficients, which is related to the severity of the damage (e.g. crack width). The higher this index, the higher the severity of the injury and the more fatigue and complex the becomes. Characteristic $S1$ with increasing spall rate is known as one of the effective factors in assessing the severity of failure and can be effective in detection algorithms. Figure 5.33 shows an example of cracking failures and Shearlet coefficients to calculate the contrast of the image. As it turns out, the cracks are entered in the mixed chert coefficients at a high frequency of the image, and the highest contrast is related to this part. Using the size of Shearlet coefficients |.|, the highest contrast represents this index. The lower the

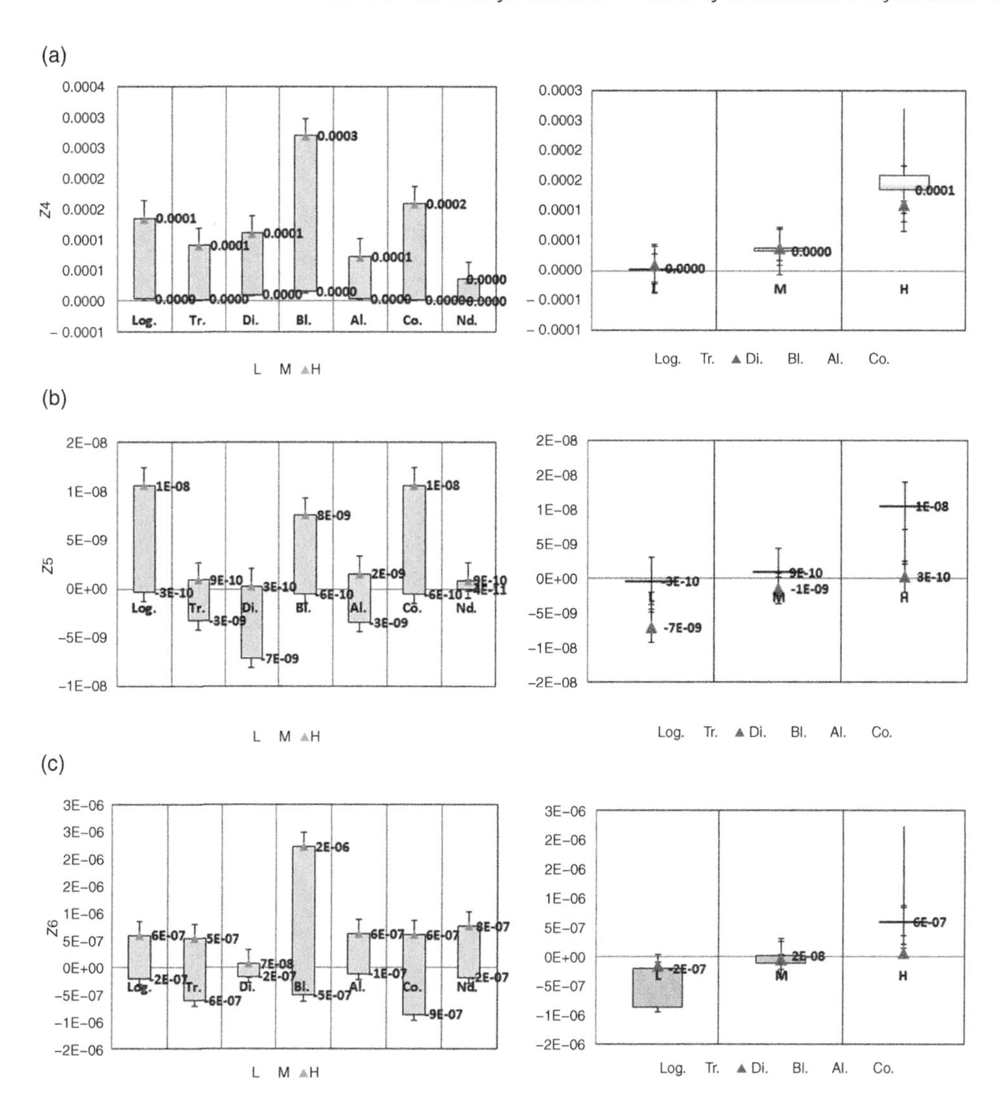

Figure 5.31 Results of index Z_l in samples with different types and intensities: (a) effect of intensity level on the index $Z(4)$, (b) effect of type of distress on $Z(5)$, and (c) effect of type of distress on $Z(6)$.

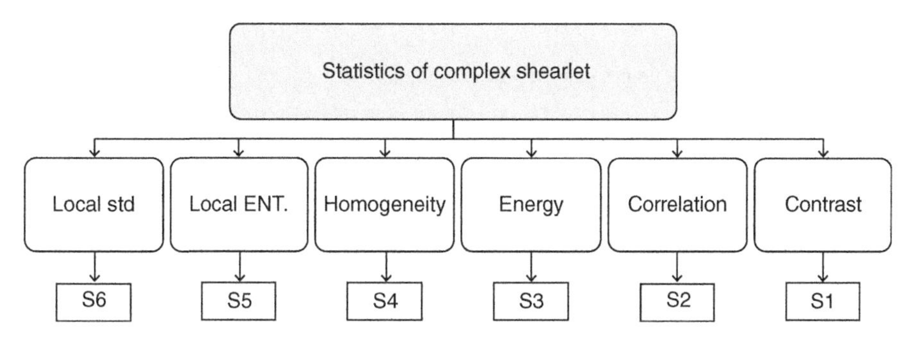

Figure 5.32 Classification of statistical characteristics used for the Shearlet complex coefficient.

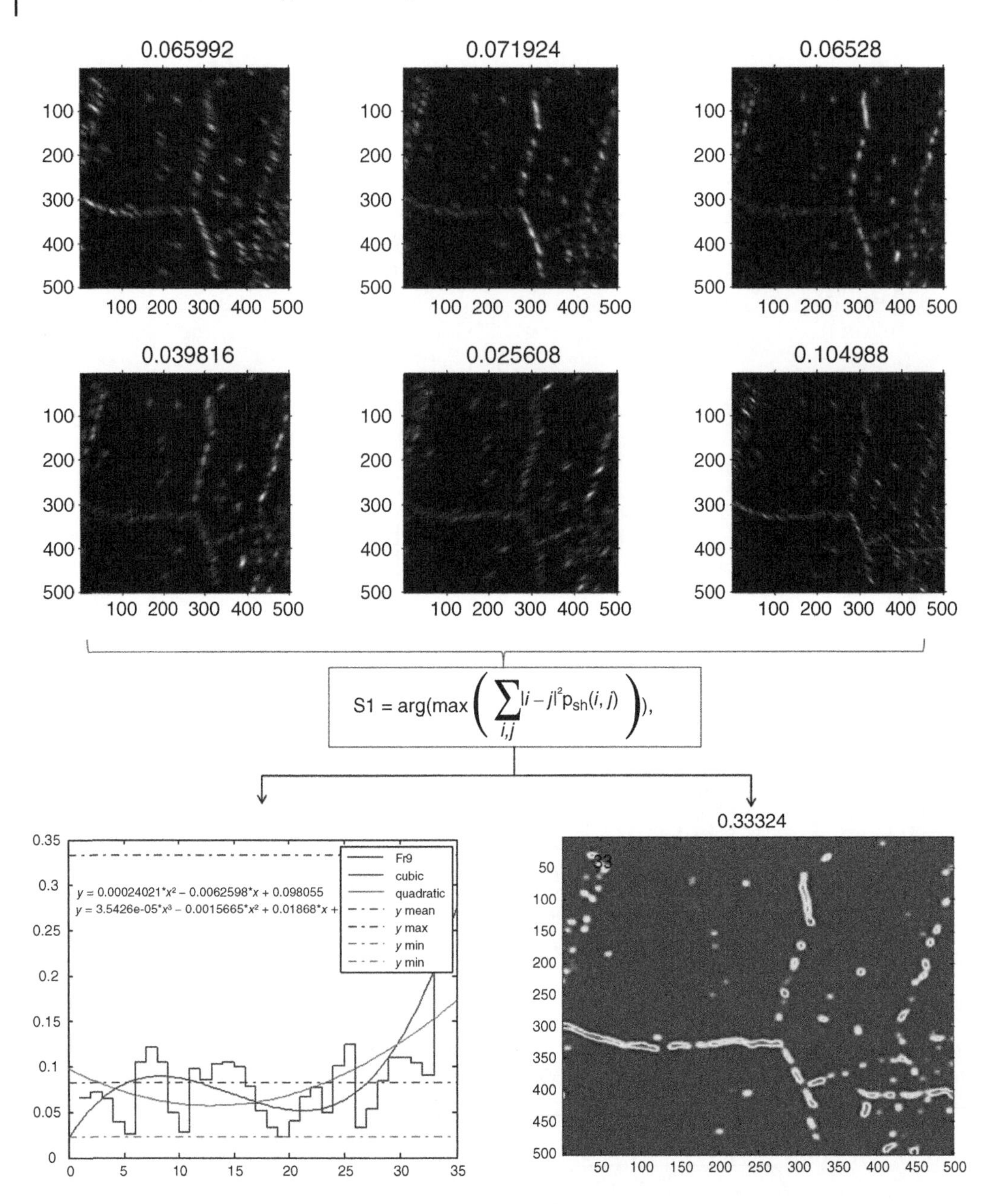

Figure 5.33 Algorithm and sample of results obtained for S1 index.

coefficients, the lower the contrast, and the more similar $S1$ is to a healthy cross section with an index of 0. The size of the $S1$ index is an indication of the presence of cracks in the image as well as the level of intensity and extent of cracks.

The real and imaginary parts of Shearlet coefficients have different $S1$ indices. The maximum value of this index is in the real part for a high-frequency coefficient. In order to extract a practical index in this research, the size of Shearlet coefficients has been varied. The value of $S1$ indicates a larger value of the complex coefficients, which reduces the computational time and increases the accuracy of the diagnosis. An example of the results of the $S1$ index for a set of images of different conditions is shown in Figure 5.34. As it is known, this index increases with increasing intensity and extent of distress and decreases for sections with less failure.

5.2.16 Correlation of Complex Shearlet

The degree of correlation of a Shearlet coefficient to its neighborhood is a practical indicator for fault detection. The higher the correlation, the closer the absolute value of the index is to 1. This numerical index is in the range of 1–1. If the coefficient is positively or negatively correlated with the impression, the value of the $S2$ index is 1 or −1, respectively. For a healthy image without cracks, this value does not exist. The implication is that there is no degree of correlation between the components:

$$S2 = \arg\left(\max \left(\sum\nolimits_{i,j} \frac{(i-\mu i)(j-\mu j) p_{\mathrm{sh}}(i,j)}{\sigma_i \sigma_j} \right) \right), S2 \in [-1\ 1] \quad (5.37)$$

where $S2$ is the correlation coefficient and $p_{\mathrm{sh}}(i,j)$ is the value of the Shearlet coefficient in the coordinates (i,j), (Figures 5.35–5.38).

5.2.17 Uniformity of Complex Shearlet

Sections have nonuniform image failures that can be assessed using the uniformity index. As a hypothesis, healthy sections have a higher uniformity index than sections with severe and large cracks, so this index has good efficiency in diagnosis and separation. In general, this index is calculated using Eq. (5.38):

$$S3 = \arg\left(\max \left(\sum\nolimits_{i,j} p_{\mathrm{sh}}(i,j)^2 \right) \right), S3 \in [0\ 1] \quad (5.38)$$

where $S3$ is the uniformity of complex Shearlet and $p_{\mathrm{sh}}(i,j)$ is the value of the Shearlet coefficient in the coordinates (i,j). This equation is the normalized image energy, which is always between 0 and 1 for a numerical image. The sum of energies of the image frequencies, i.e. the upper and lower bands, is equal to 1. The lower the failure rate in a section, the lower the value of the $S3$ index. This index is related to the surface energy of the cross-sectional damage.

5.2.18 Homogeneity of Complex Shearlet

Defective sections are generally known as nonhomogeneous sections with anomalies. As a hypothesis, the sections that display a smooth and uniform surface without damage are isotropic and have a higher homogeneity index. This study found that as the amount of cracking in the image increases, the degree of homogeneity decreases and eventually the homogeneity index decreases. The numerical homogeneity index is between 0 and 1 and is calculated from Eq. (5.39):

$$S4 = \arg\left(\max \left(\sum\nolimits_{i,j} \frac{p_{\mathrm{sh}}(i,j)}{1 + |i-j|} \right) \right), \quad S4 \in [0\ 1] \quad (5.39)$$

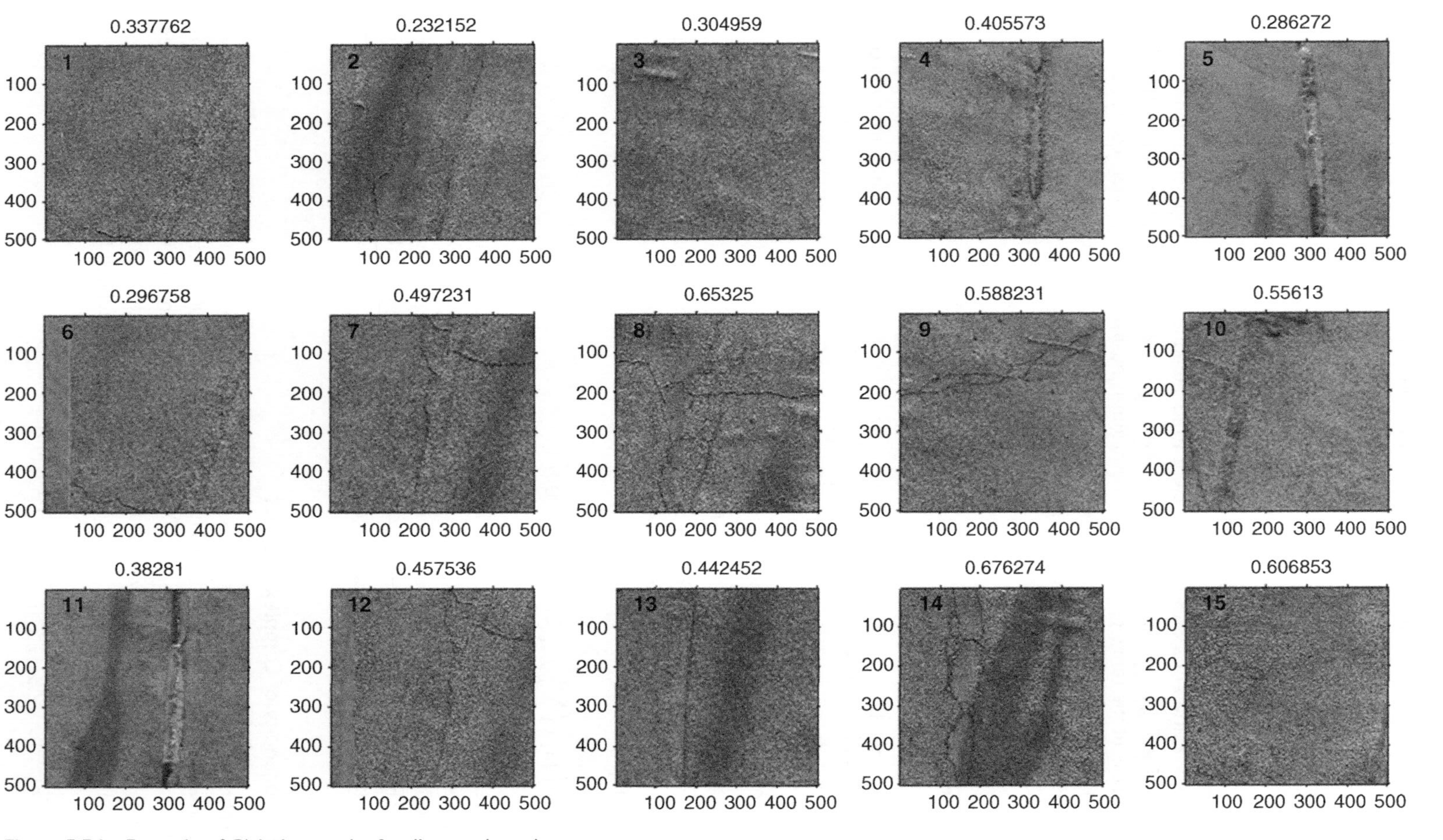

Figure 5.34 Example of *S*1 index results for distress detection.

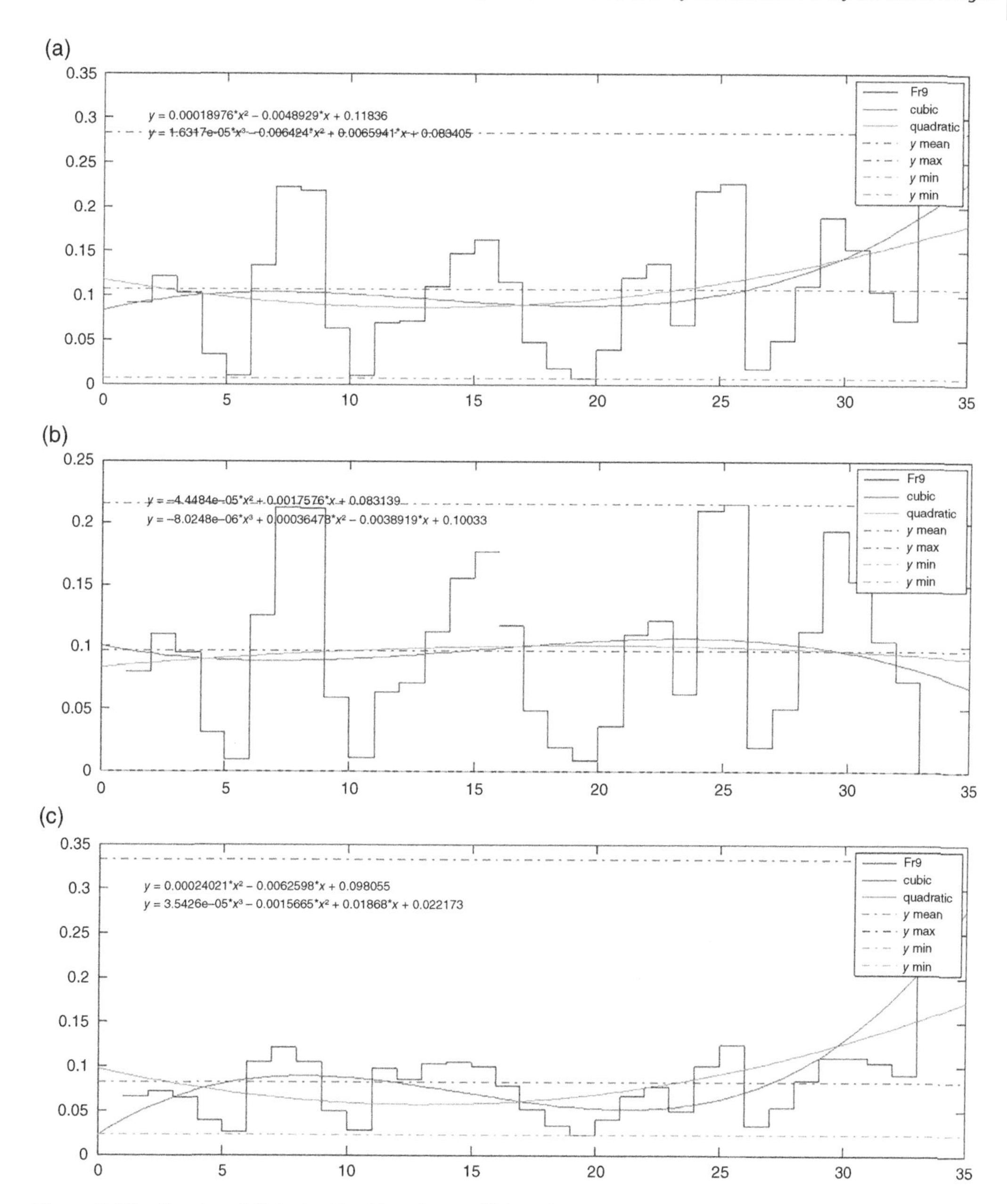

Figure 5.35 Results of *S*1 index for (a) real part, (b) imaginary part, and (c) Shearlet coefficients.

where *S*4 is the homogeneity of complex Shearlet and $p_{sh}(i, j)$ is the value of the Shearlet coefficient in the coordinates (*i*,*j*). Figure 5.39 shows an example of the results obtained using Eq. (5.39) on the data bank images for the index *S*4.

5.2.19 Entropy of Complex Shearlet

The local entropy index is a statistical measure that is calculated using Eq. (5.40) for a 9×9 square range and is normalized to the size of the image:

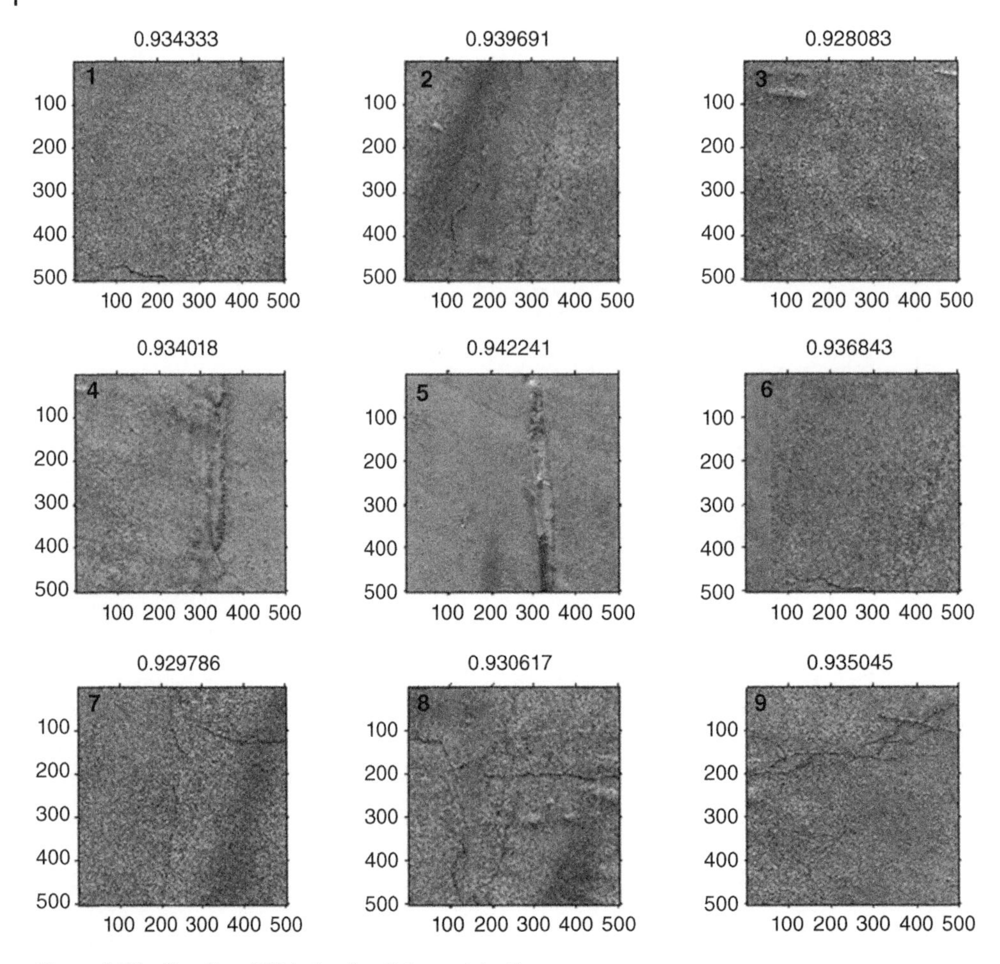

Figure 5.36 Results of $S2$ index for distress detection.

$$S5 = \arg\left(\max \frac{\left(\sum_{i,j} p_{sh}(i,j) \log_2^{p_{sh}(i,j)}\right)}{m \times n}\right), \quad S5 \in [0\ 1] \tag{5.40}$$

where $S5$ is the entropy of complex Shearlet and $p_{sh}(i,j)$ is the value of the Shearlet coefficient in the coordinates (i,j). Figure 5.40 presents an example of the results using Eq. (5.40) on the data bank images for the index $S5$. Similar to the $S3$ index, this index also uses the size of mixed Shearlet coefficients in order to determine the local entropy index. The higher the magnitude and severity of the failure, the greater the value of $S5$. The maximum value of this index is 1 for a completely damaged section and 0 for a healthy section. As shown in Figure 5.40, the value of the $S5$ index is higher than that of other sections that are healthy or have no damage. Therefore, this index depends on the type of injury and its severity and extent, and with increasing or decreasing each of these features, its value changes (Figure 5.41).

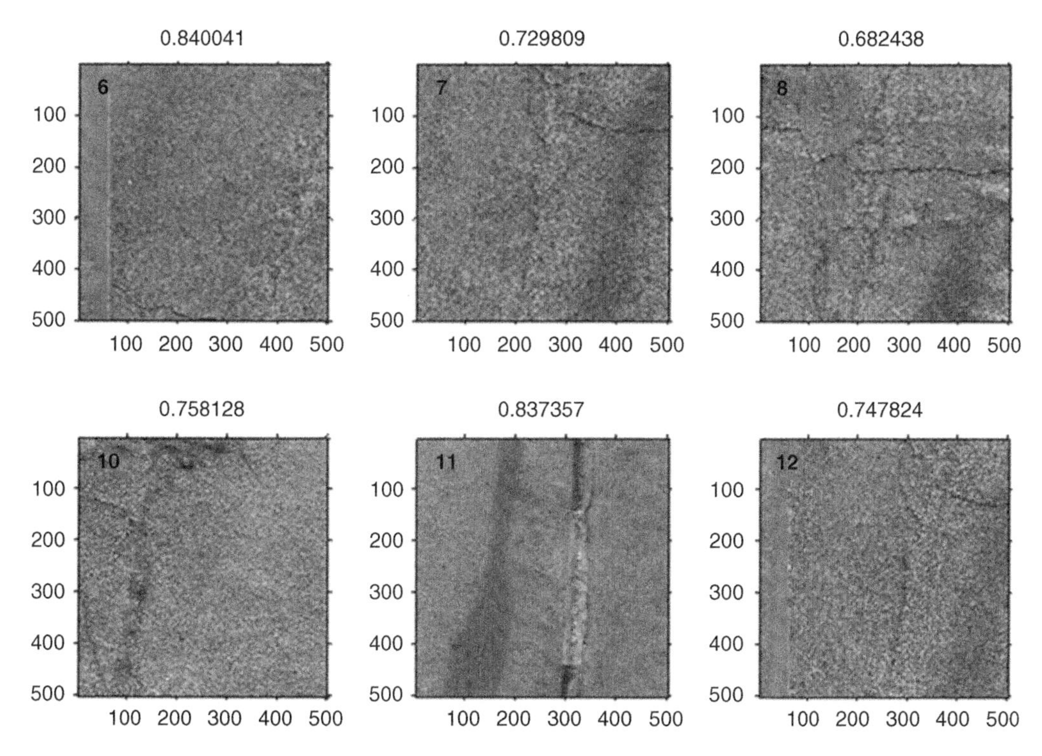

Figure 5.37 Results of $S3$ index for distress detection.

5.2.20 Local Standard Deviation of Complex Shearlet Index (F_Local_STD)

The local standard deviation entropy index is calculated using Eq. (5.41) for Shearlet complex coefficients and is normalized to the image size:

$$S6 = \arg\left(\max\frac{\left(\sum_{i,j} p_{\text{sh}}(i,j) > 0\right)}{m \times n}\right), S6 \in [0\ 1] \tag{5.41}$$

where $S6$ is the local standard deviation of complex Shearlet index; and $p_{\text{sh}}(i,j)$ is the value of the Shearlet coefficient in the coordinates (i,j). This index is similar to the UCI index. The higher the magnitude and severity of the damage, the greater the value of $S6$ (Figure 5.42). The maximum value of this index is equal to 1 for a section with complete damage and 0 for a section without distress, similar to the $S5$ index. Therefore, this index also depends on the damage and its severity and extent, and with the increase or decrease of each of these characteristics, its value changes.

Based on the results extracted from the set of 100 images used to build the database, the characteristic is that $S1$ for low-intensity failure is 3.7 times that for the healthy cross section and nearly two times that for the medium and severe cross sections. The mean threshold value is 0.025 based on the mean of the coefficients of $S1$.

Selecting this index will separate and detect the maximum number of faulty sections, but a number of healthy sections will also be included in the feature vector bank. Characteristic $S2$ is very small for all samples, for which the ratio of the cross section with failure to a healthy cross section is less than one and increases slightly with increasing failure rate. The value of this index

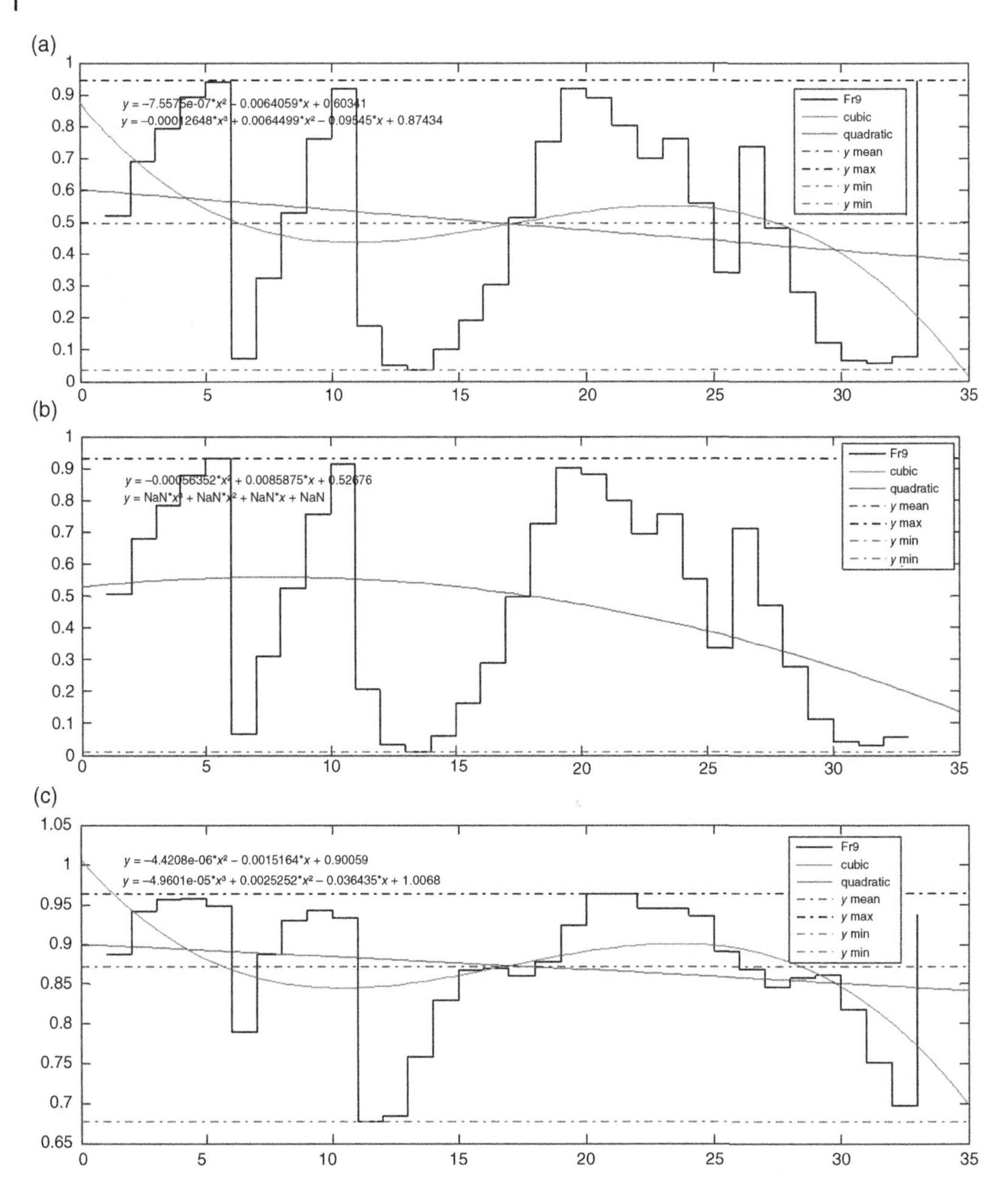

Figure 5.38 Results of *S*2 index for (a) real part, (b) imaginary part, and (c) Shearlet coefficients.

for low, medium, and high failure is equal to 3% and 1%. As shown in Figure 5.43c, the slam section has a higher *S*3 than that of the damaged section. Area failures, such as composite, block, and fatigue, have a higher *S*3 value than other linear cracks. Similarly, the *S*4 index has a higher value than that for the sections with failure. The cross section with fatigue, compound, and block failure has a lower *S*4 index than that for linear failure. This relationship is shown in Figure 5.44a.

The *S*5 index is a number between 0 and 1 for healthy sections or low failure, where the maximum numerical index ranges between 0.127 and 0.328. On average, for block failures and

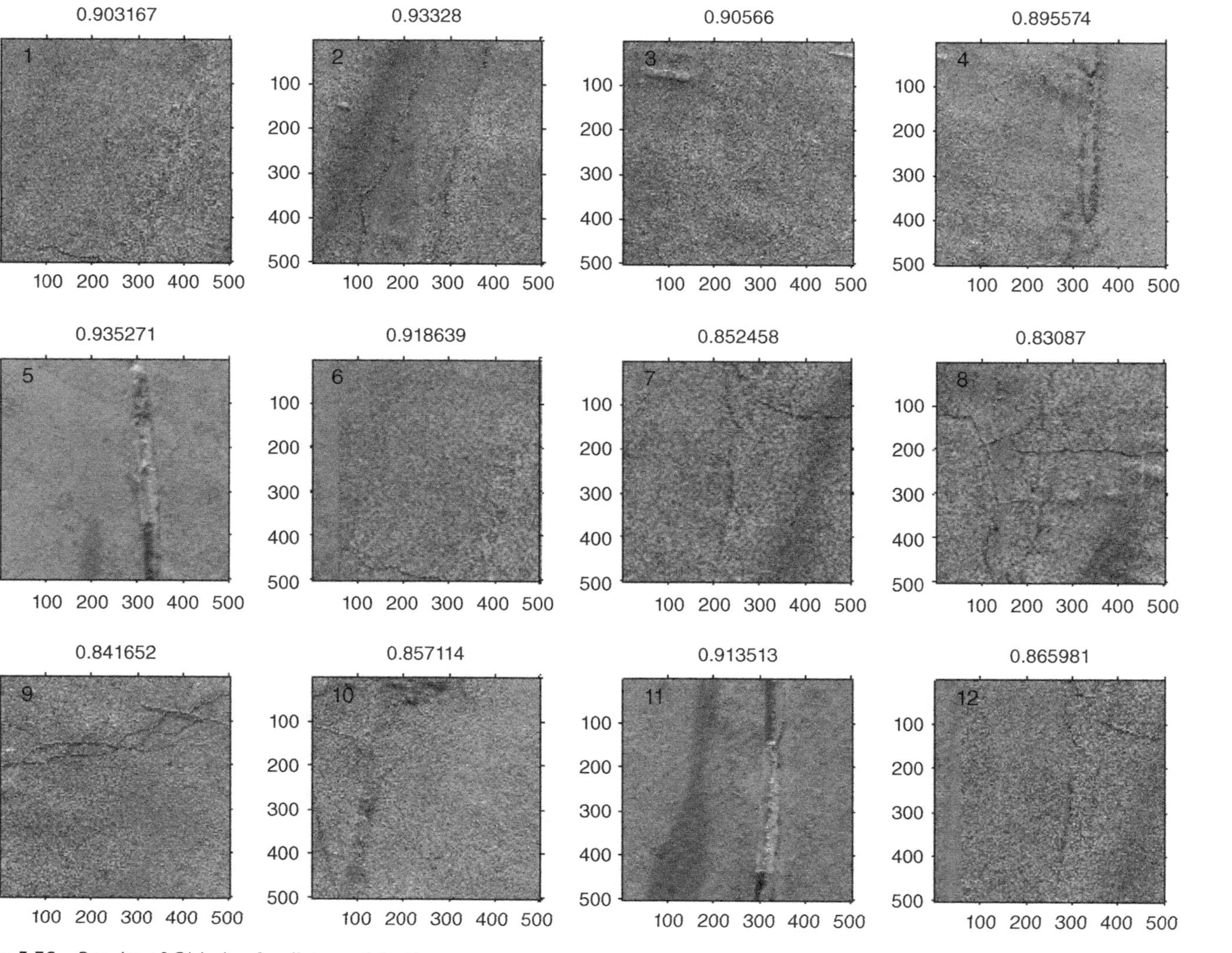

Figure 5.39 Results of $S4$ index for distress detection.

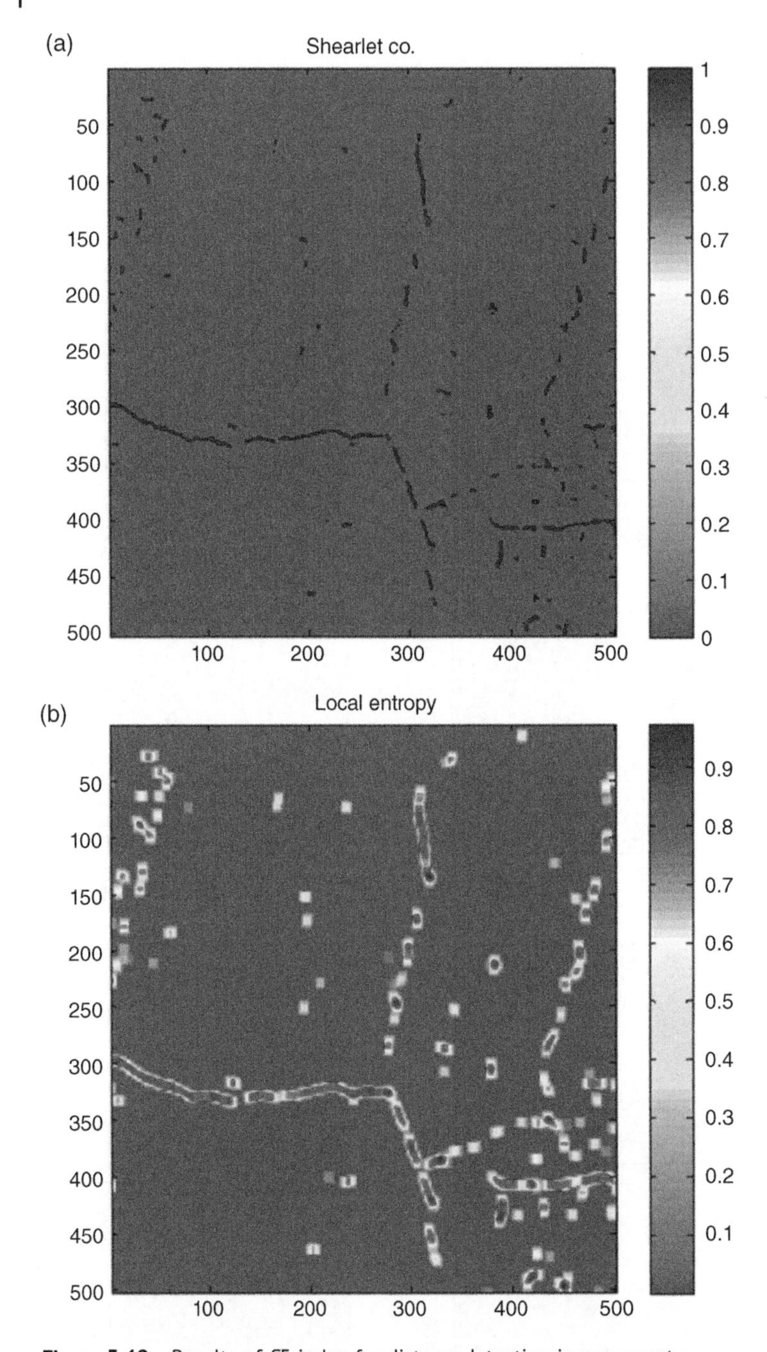

Figure 5.40 Results of *S*5 index for distress detection in pavement.

fatigue, the index is higher than the maximum value of *S*5 for healthy sections. This ratio is three times that for healthy sections, and two times that for medium surface distress. Also, the *S*6 index acts similarly to *S*5 and shows a higher value for damage. This index is also greater than the maximum *S*6 index for healthy cross sections with distress and block failure. Also, linear cracking has a lower index than that for superficial distress (Figure 5.45).

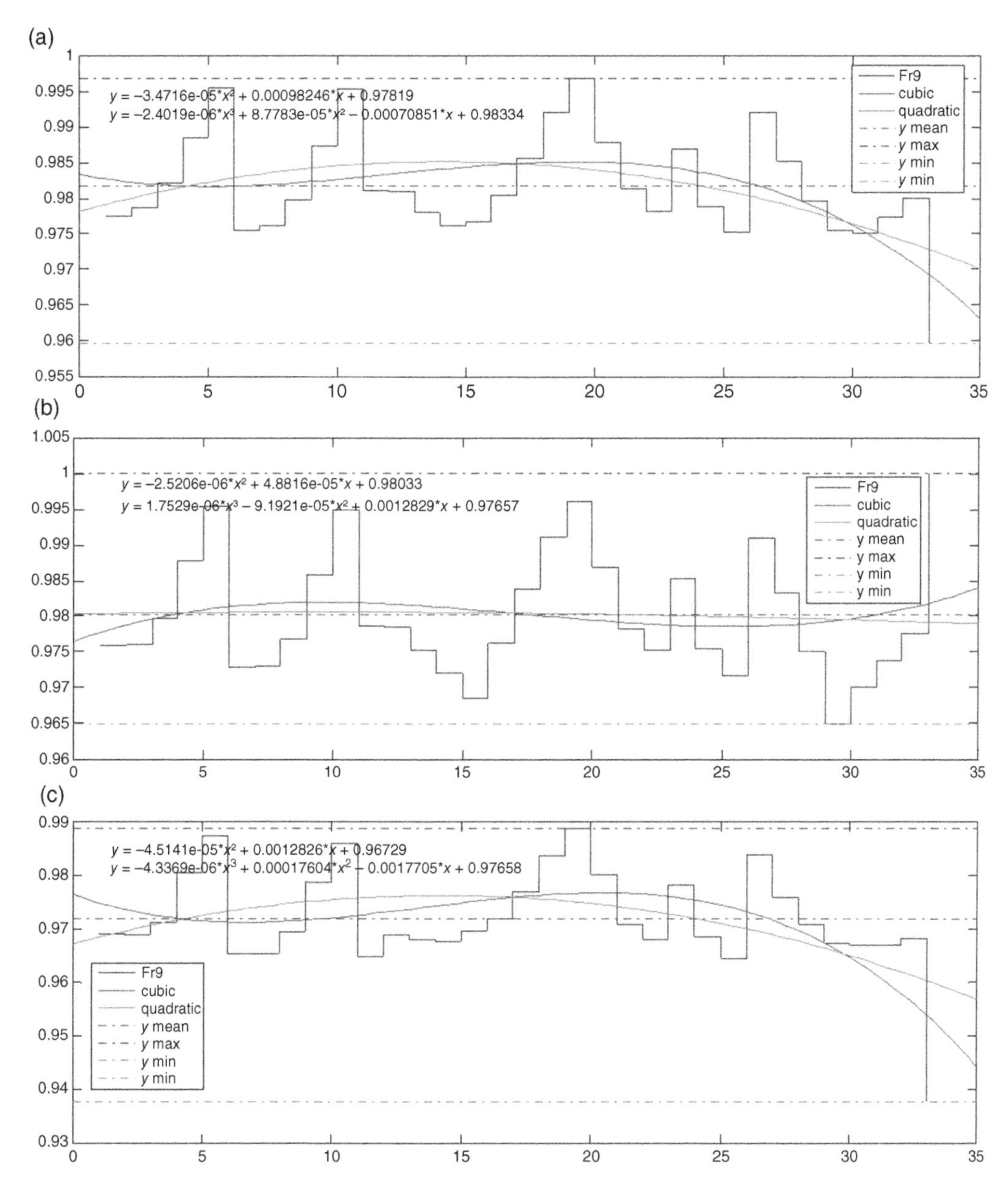

Figure 5.41 Results of the *S*4 index for (a) real part, (b) imaginary part, and (c) Shearlet coefficients.

5.3 Summary and Conclusion

In this chapter, the purpose of extracting image features was to perform rapid anomaly detection and isolation operations. The conversion of information into meaningful features was presented with the aim of summarizing and increasing the speed and accuracy of detection as well as reducing the amount of image data for further processing. Extracting features with more data transfer capability can increase both speed and efficiency by reducing the search and categorization space.

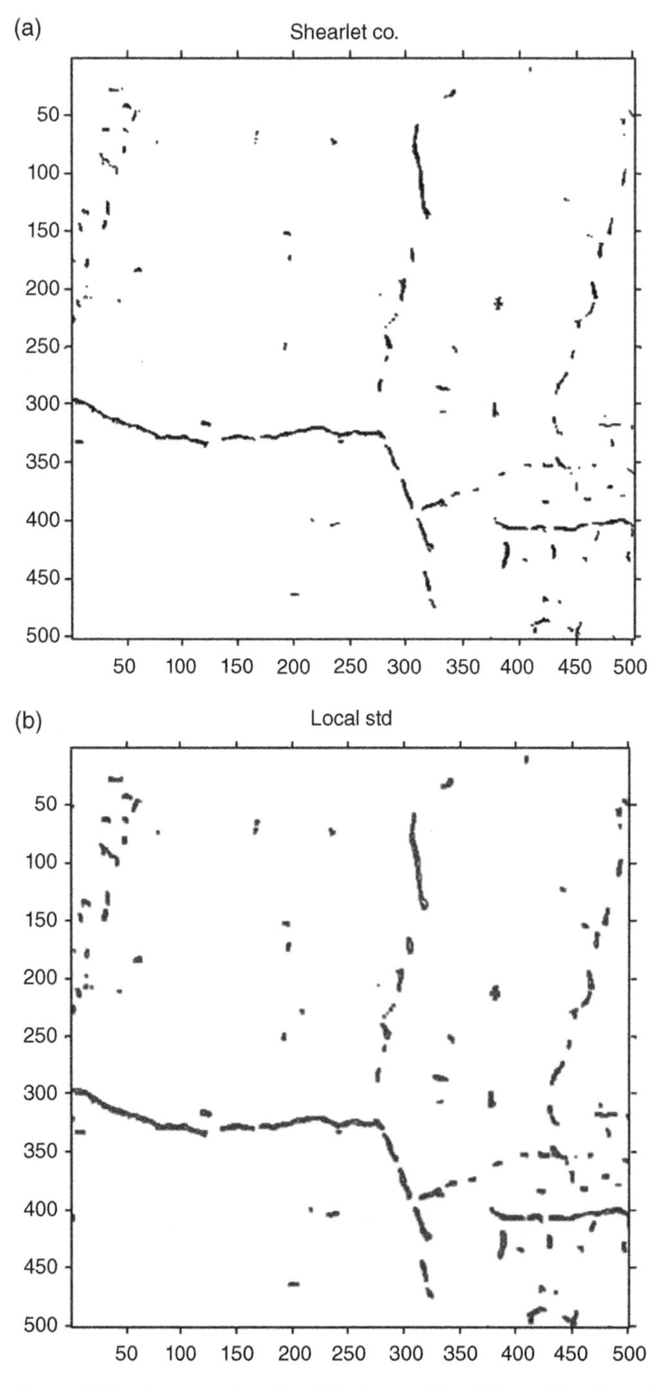

Figure 5.42 An example of the *S6* index and its ability to identify and detect: (a) Shearlet coefficients, and (b) local standard deviation.

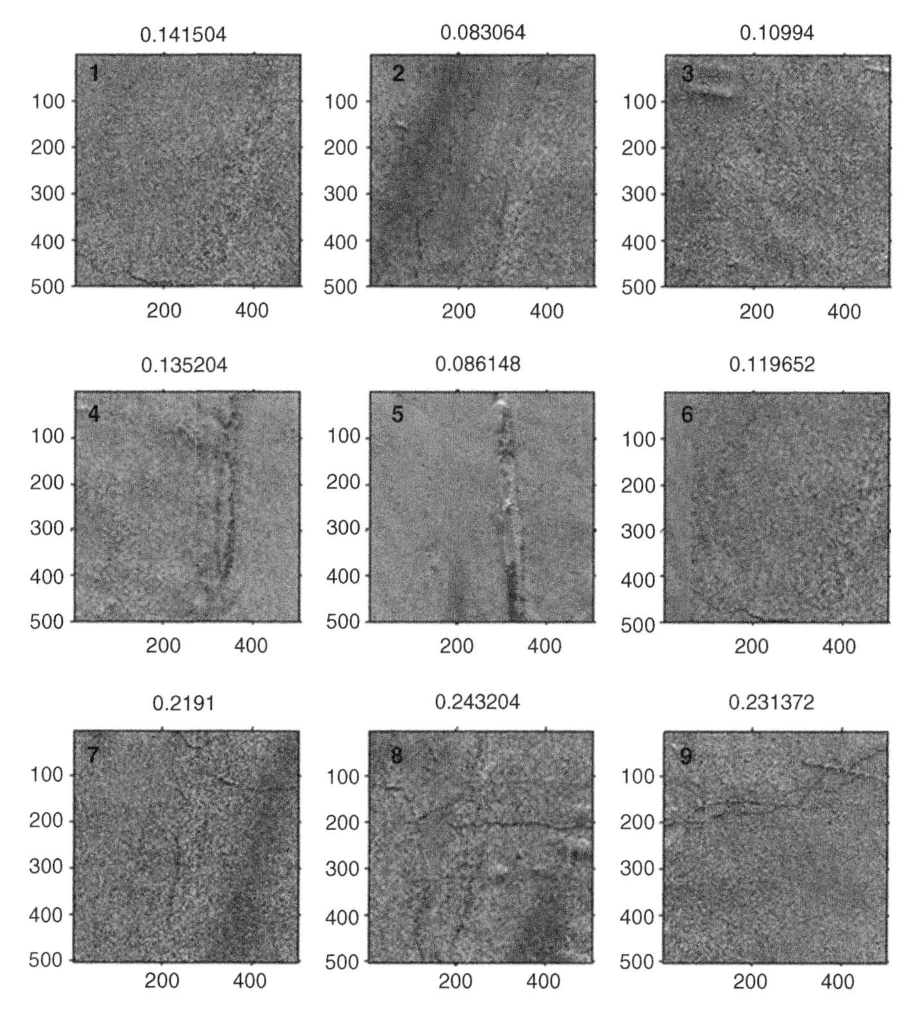

Figure 5.43 Results of *S6* index for distress detection.

This chapter introduced various methods for automatic detection of anomalies in infrastructure, including road paving, that can be further developed for other infrastructure, such as tunnel walls, structures, dams, silos, and power plant walls. In addition, effective features in the automatic detection of images with anomalies are presented along with examples of their application in infrastructure management. In general, various features were presented in wavelet transform space, Shearlet, in real and imaginary space, edge detection by a combined method, statistical family features such as Central_Moments_*q*, Central_Moments_*q*, Hu_Moments, Bamieh_ Moments, Zernike_ Moments, energy contrast, correlation, homogeneity, entropy, local range, deviation from local standard (Local_STD), and fractal properties to distinguish images with anomalies. After selecting the appropriate attribute vector using the selection methods discussed in the next chapter, the attributes can be classified, and the optimal attributes can be selected based on intelligent methods. Finally, the diagnostic process is completed in by detection process and performed using the classification methods.

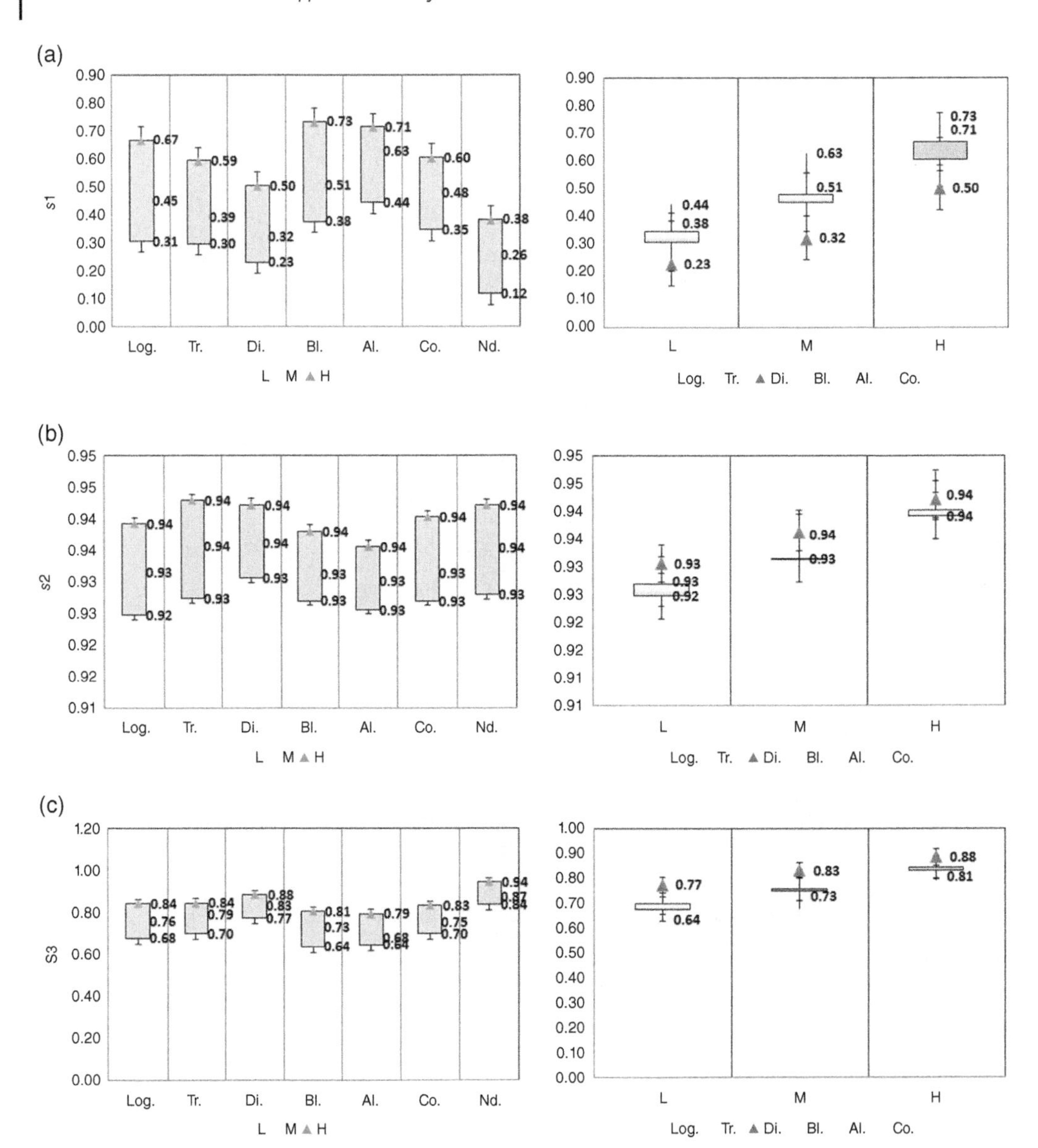

Figure 5.44 Results of index S_l in different types and intensities in terms of the statistics of complex Shearlet: (a) effect of complex Shearlet on index *S*(1), (b) effect of type of distress complex Shearlet on *S*(2), and (c) effect of complex Shearlet of distress on *S*(3).

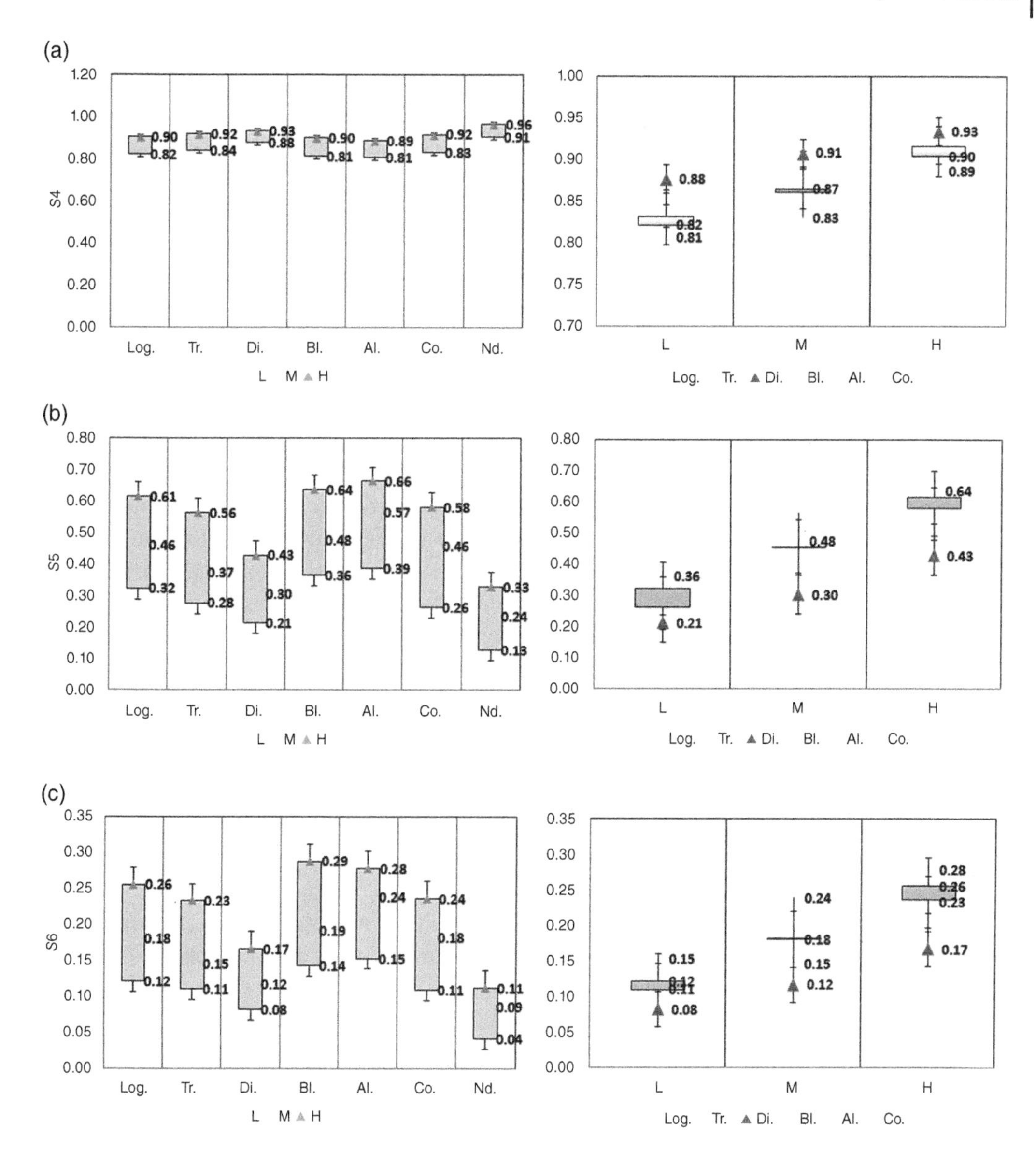

Figure 5.45 Results of index S_l in samples with different types and intensities in terms of statistics of complex Shearlet: (a) effect of complex Shearlet on index *S*(4), (b) effect of type of distress complex Shearlet on *S*(5), and (c) effect of complex Shearlet of distress on *S*(6).

5.4 Questions and Exercises

1 What are the general hypotheses and rules in diagnostic processing to detect anomalies? What is the most important hypothesis used in algorithms in this regard? Can new relationships and hypotheses be developed regarding the diagnosis of anomalies? Give an example of a type of infrastructure anomaly and collect a database of more than 100 images.

2 Explain the steps of generating a 3D membership function and prove its relationships in two and three dimensions. What are the general hypotheses and rules in diagnostic processing to detect anomalies? What is the most important hypothesis used in algorithms in this regard? Can new relationships and hypotheses be developed regarding the diagnosis of anomalies? Give an example of a type of infrastructure anomaly and collect a database of more than 100 images for it.

3 Prove the wavelet module relation. Is it possible to design another module consisting of horizontal, and vertical components so that the details of infrastructure anomalies can be better identified? Calculate evaluation components, such as peak signal-to-noise ratio (PSNR), for the new module and compare with the previous method. Is the fixed threshold for images with anomalies PSNR and the same detection accuracy? Use a dynamic threshold method for detection. What effect does the use of a dynamic threshold have on diagnosis and accuracy?

4 What does the HAWCP equation have to do with the HFEP equation? Are these equations computable separately? How does the rate of diagnosis and evaluation change if used independently (diagonal, vertical, and horizontal details)? Perform a sensitivity analysis to select the optimal band and level on the database in question one. What is the best subband based on optimal time and accuracy?

5 Calculate the value of cm_q for the image data base of Q1. Prove relationships cm_q for wavelet and Shearlet. What is the relationship between standard deviation and the types of infrastructure cracking? Assuming that a crack has a fixed length, what changes as the index crack width increases in different bands? How does this relationship change for cracks of constant width and variable length? Calculate the equation of the severity and extent of the crack for the different types of cracks.

6 Prove the relationship between the main indicators of Shearlet coefficient and the type of cracking and severity. Calculate the Shearlet coefficients converted into real and imaginary independently for the Q1 database $M_k(x, y)$. What is the difference between indices calculated on the basis of a fixed threshold $\mathrm{SH}_{bk}(x, y)$ and indices calculated on the basis of a dynamic threshold $\mathrm{Th}_{\mathrm{Adaptive}}$? Is it possible to improve the obtained module based on separate thresholds based on direction ($\mathrm{HFSHEP}_{\mathrm{sh}_k}$)? Provide an optimal method to increase the possibility of anomaly diagnosis.

7 What is the relationship between the main indicators of the fractal index and the type of single and patterned cracking?

$$N = r^{-\mathrm{FD}} \rightarrow \mathrm{FD} = -\frac{\log N}{\log r}, D = 1, 2, 3, \ldots, n, fD_{\mathrm{box-count}} = \lim_{\varepsilon \to 0} \frac{\log N(\varepsilon)}{\log(1/\varepsilon)}$$

As the number of cracking branches increases, how does the intensity change, and at what rate does the fractal index increase? Calculate this index independently for the Q1 database. What is the difference between indices calculated on the basis of a fixed threshold and indices calculated on the basis of a dynamic threshold? Is the fractal index related to the crack principle? Are the cracking patterns reproducible and similar? Why is crack detection based on the fractal index less likely than crack detection based on other indicators? How does the fractal index relate to the severity of cracking?

8 After calculating the Shearlet coefficients, find the relationship between the main indices $c'_{pq} = e^{-i(p-q)\alpha}$.What is c_{pq} for single cracking and pattern?As the severity and extent of cracking increase, how does this index change based on the angle? Can this index be used in combination with other indicators, such as fractal index, for diagnosis? Why is crack detection based on this indicator more likely than crack detection based on other indicators? What is the computational speed and complexity of this index compared to the wavelet index?: Identify the other criteria, such as accuracy, sensitivity, accuracy, translation, rotation, and scaling (TRS) parameters.

Further Reading

1 Abdel-Qader, I., S. Yohali, O. Abudayyeh, and S. Yehia, *Segmentation of thermal images for non-destructive evaluation of bridge decks. NDT and E International*, 2008. **41**(5): p. 395–405.
2 Abouhamad, M., T. Dawood, A. Jabri, M. Alsharqawi, and T. Zayed, *Corrosiveness mapping of bridge decks using image-based analysis of GPR data. Automation in Construction*, 2017. **80**: p. 104–117.
3 Acosta, J.A., J.L. Figueroa, and R.L. Mullen, *Algorithms for pavement distress classification by video image analysis. Transportation Research Record*, 1995(1505): p. 27–38.
4 Adhikari, R.S., O. Moselhi, and A. Bagchi, *Image-based retrieval of concrete crack properties for bridge inspection. Automation in Construction*, 2014. **39**: p. 180–194.
5 Bergeson, W. and S.L. Ernst, *Tunnel Operations. Maintenance, Inspection and Evaluation (TOMIE) Manual*, 2011 Federal Highway Administration.
6 Adu-Gyamfi, Y.O., T. Tienaah, N. O. Attoh-Okine, and C. Kambhamettu, et al., *Functional evaluation of pavement condition using a complete vision system. Journal of Transportation Engineering*, 2014. **140**(9), 04014040.
7 Agnisarman, S. and S. Lopes, *A survey of automation-enabled human-in-the-loop systems for infrastructure visual inspection. Automation in Construction*, 2019. **97**: p. 52–76.
8 Ahmed, M., C.T. Haas, and R. Haas, *Toward low-cost 3D automatic pavement distress surveying: The close range photogrammetry approach. Canadian Journal of Civil Engineering*, 2011. **38**(12): p. 1301–1313.
9 Asakura, T. and Y. Kojima, *Tunnel maintenance in Japan. Tunnelling and Underground Space Technology*, 2003. **18**(2–3): p. 161–169.
10 Mataei, B., H. Zakeri, M. Zahedi, and F.M. Nejad, *Pavement friction and skid resistance measurement methods: A literature review. Open Journal of Civil Engineering*, 2016. **6**(4): p. 537–565.
11 Bao, Y., Z. Tang, H. Li, and Y. Zhang, *Computer vision and deep learning-based data anomaly detection method for structural health monitoring. Structural Health Monitoring*, 2018. **18**(2): p. 401–421.
12 Brandes, H.G. and J.G. Hirata, *An automated image analysis procedure to evaluate compacted asphalt sections. International Journal of Pavement Engineering*, 2009. **10**(2): p. 87–100.

13 Burrow, M.P.N., H.T. Evdorides, and M.S. Snaith, *Road marking assessment using digital image analysis. Proceedings of the Institution of Civil Engineers: Transport*, 2000. **141**(2): p. 107–112.

14 Burrow, M.P.N., H.T. Evdorides, and M.S. Snaith, *Segmentation algorithms for road marking digital image analysis. Proceedings of the Institution of Civil Engineers: Transport*, 2003. **156**(1): p. 17–28.

15 Ceylan, H., M.B. Bayrak, and K. Gopalakrishnan, *Neural networks applications in pavement engineering: A recent survey. International Journal of Pavement Research and Technology*, 2014. **7**(6): p. 434–444.

16 Chambon, S. and J.M. Moliard, *Automatic road pavement assessment with image processing: Review and comparison. International Journal of Geophysics*, 2011. **2011**, Article ID 989354, 20 pages.

17 Chang, K., J. Chang, and J. Liu, *Detection of pavement distresses using 3D laser scanning technology. Computing in Civil Engineering*, 2005: p. 1–11.

18 Chang, P.C., A. Flatau, and S.C. Liu, *Review paper: Health monitoring of civil infrastructure. Structural Health Monitoring*, 2003. **2**(3): p. 257–267.

19 Chen, G., R. McDaniel, M. Brower, and D. Pommerenke, *Crack detectability and durability of coaxial cable sensors in reinforced concrete bridge applications. Transportation Research Record*, 2010. **2172**(1): p. 151–156.

20 Chen, G., H. Mu, D. Pommerenke, and J.L. Drewniak, *Damage detection of reinforced concrete beams with novel distributed crack/strain sensors. Structural Health Monitoring*, 2004. **3**(3): p. 225–243.

21 Cheng, H.D. and M. Miyojim, *Automatic pavement distress detection system. Information Sciences*, 1998. **108**(1–4): p. 219–240.

22 Cheng, H.D. and M. Miyojim, *Novel system for automatic pavement distress detection. Journal of Computing in Civil Engineering*, 1998. **12**(3): p. 145–152.

23 Cheng, H.D., X.J. Shi, and C. Glazier, *Real-time image thresholding based on sample space reduction and interpolation approach. Journal of Computing in Civil Engineering*, 2003. **17**(4): p. 264–272.

24 Cheng, Y.M. and S.S. Leu, *Constraint-based clustering model for determining contract packages of bridge maintenance inspection. Automation in Construction*, 2008. **17**(6): p. 682–690.

25 Chou, J., W.A. O'Neill, and H. Cheng, *Pavement distress evaluation using fuzzy logic and moment invariants. Transportation Research Record*, 1995(1505): p. 39–46.

26 Chu, X.M., R.B. Wang, J.W. Chu, and C. Wang, *Study of asphalt pavement surface distress image segmentation. Zhongguo Gonglu Xuebao/China Journal of Highway and Transport*, 2003. **16**(3): p. 11.

27 Chu, X.M., X.P. Yan, and Z. Mao, *Automatic classify method of traffic sign. Journal of Traffic and Transportation Engineering*, 2006. **6**(4): p. 91–95.

28 Cord, A. and S. Chambon, *Automatic road defect detection by textural pattern recognition based on AdaBoost. Computer-Aided Civil and Infrastructure Engineering*, 2012. **27**(4): p. 244–259.

29 DeVault, J.E., *Robotic system for underwater inspection of bridge piers. IEEE Instrumentation & Measurement Magazine*, 2000. **3**(3): p. 32–37.

30 Dinh, K., N. Gucunski, and T. Zayed, *Automated visualization of concrete bridge deck condition from GPR data. NDT & E International*, 2019. **102**: p. 120–128.

31 Dorafshan, S. and M. Maguire, *Bridge inspection: Human performance, unmanned aerial systems and automation. Journal of Civil Structural Health Monitoring*, 2018. **8**(3): p. 443–476.

32 Dorafshan, S., R.J. Thomas, and M. Maguire, *Comparison of deep convolutional neural networks and edge detectors for image-based crack detection in concrete. Construction and Building Materials*, 2018. **186**: p. 1031–1045.

33 Dung, C.V. and L.D. Anh, *Autonomous concrete crack detection using deep fully convolutional neural network. Automation in Construction*, 2019. **99**: p. 52–58.

34 Duran, O., K. Althoefer, and L.D. Seneviratne, *Automated pipe defect detection and categorization using camera/laser-based profiler and artificial neural network. IEEE Transactions on Automation Science and Engineering*, 2007. **4**(1): p. 118–126.

35 Ellenberg, A., A. Kontsos, F. Moon, and I. Bartoli, *Bridge deck delamination identification from unmanned aerial vehicle infrared imagery. Automation in Construction*, 2016. **72**: p. 155–165.

36 Gavilán, M., D. Balcones, O. Marcos, D.F. Llorca, M.A. Sotelo, I. Parra, M. Ocaña, P. Aliseda, P. Yarza, and A. Amírola, *Adaptive road crack detection system by pavement classification. Sensors*, 2011. **11**(10): p. 9628–9657.

37 German, S., I. Brilakis, and R. DesRoches, *Rapid entropy-based detection and properties measurement of concrete spalling with machine vision for post-earthquake safety assessments. Advanced Engineering Informatics*, 2012. **26**(4): p. 846–858.

38 Golparvar-Fard, M., V. Balali, and J.M. De La Garza, *Segmentation and recognition of highway assets using image-based 3D point clouds and semantic Texton Forests. Journal of Computing in Civil Engineering*, 2015. **29**(1), 04014023.

39 Guan, H., J. Li, Y. Yu, M. Chapman, and C. Wang, *Automated road information extraction from mobile laser scanning data. IEEE Transactions on Intelligent Transportation Systems*, 2015. **16**(1): p. 194–205.

40 Guan, H., J. Li, Y. Yu, M. Chapman, H. Wang, C. Wang, and R. Zhai, *Iterative tensor voting for pavement crack extraction using mobile laser scanning data. IEEE Transactions on Geoscience and Remote Sensing*, 2014. **53**(3): p. 1527–1537.

41 Guo, W., L. Soibelman, and J.H. Garrett, *Visual pattern recognition supporting defect reporting and condition assessment of wastewater collection systems. Journal of Computing in Civil Engineering*, 2009. **23**(3): p. 160–169.

42 Hallermann, N. and G. Morgenthal. *Visual Inspection Strategies for Large Bridges Using Unmanned Aerial Vehicles (UAV)*. 2014. Taylor and Francis: Balkema.

43 Ham, Y., K.K. Han, J.J. Lin, and M. Golparvar-Fard, *Visual monitoring of civil infrastructure systems via camera-equipped unmanned aerial vehicles (UAVs): A review of related works. Visualization in Engineering*, 2016. **4**(1), 1–8.

44 He, S.H., X.M. Zhao, J. Ma, Y. Zhao, H.S. Song, H.X. Song, L. Cheng, Z.Y. Yuan, F.W. Huang, J. Zhang, B. Tian, L.Y. Wang, and X.Z. Qi, *Review of highway bridge inspection and condition assessment. Zhongguo Gonglu Xuebao/China Journal of Highway and Transport*, 2017. **30**(11): p. 63–80.

45 Hou, Z., K.C. Wang, and W. Gong. *Experimentation of 3D pavement imaging through stereovision*. in *International Conference on Transportation Engineering*. 2007.

46 Hu, Y., C.X. Zhao, and H.N. Wang, *Automatic pavement crack detection using texture and shape descriptors. IETE Technical Review (Institution of Electronics and Telecommunication Engineers, India)*, 2010. **27**(5): p. 398–405.

47 Huang, Y. and B. Xu, *Automatic inspection of pavement cracking distress. Journal of Electronic Imaging*, 2006. **15**(1): p. 013017–013017-6.

48 Hüthwohl, P. and I. Brilakis, *Detecting healthy concrete surfaces. Advanced Engineering Informatics*, 2018. **37**: p. 150–162.

49 Hüthwohl, P., R. Lu, and I. Brilakis, *Multi-classifier for reinforced concrete bridge defects. Automation in Construction*, 2019. **105**: p. 102824.

50 Iyer, S. and S.K. Sinha, *A robust approach for automatic detection and segmentation of cracks in underground pipeline images. Image and Vision Computing*, 2005. **23**(10): p. 921–933.

51 Iyer, S. and S.K. Sinha, *Automated condition assessment of buried sewer pipes based on digital imaging techniques. Journal of the Indian Institute of Science*, 2005. **85**(5): p. 235–252.

52 Iyer, S. and S.K. Sinha, *Segmentation of pipe images for crack detection in buried sewers. Computer-Aided Civil and Infrastructure Engineering*, 2006. **21**(6): p. 395–410.

53 Jahanshahi, M.R., J.S. Kelly, S.F. Masri, and G.S. Sukhatme, *A survey and evaluation of promising approaches for automatic image-based defect detection of bridge structures. Structure and Infrastructure Engineering*, 2009. **5**(6): p. 455–486.

54 Jahanshahi, M.R. and S.F. Masri, *Adaptive vision-based crack detection using 3D scene reconstruction for condition assessment of structures. Automation in Construction*, 2012. **22**: p. 567–576.

55 Jahanshahi, M.R. and S.F. Masri, *A new methodology for non-contact accurate crack width measurement through photogrammetry for automated structural safety evaluation. Smart Materials and Structures*, 2013. **22**(3), 035019.

56 Jahanshahi, M.R., S.F. Masri, C.W. Padgett, and G.S. Sukhatme, *An innovative methodology for detection and quantification of cracks through incorporation of depth perception. Machine Vision and Applications*, 2013. **24**(2): p. 227–241.

57 Jahanshahi, M.R., S.F. Masri, and G.S. Sukhatme, *Multi-image stitching and scene reconstruction for evaluating defect evolution in structures. Structural Health Monitoring*, 2011. **10**(6): p. 643–657.

58 Jiang, J., H. Liu, H. Ye, and F. Feng, *Crack enhancement algorithm based on improved EM. Journal of Information and Computational Science*, 2015. **12**(3): p. 1037–1043.

59 Kaseko, M.S., Z.P. Lo, and S.G. Ritchie, *Comparison of traditional and neural classifiers for pavement-crack detection. Journal of Transportation Engineering*, 1994. **120**(4): p. 552–569.

60 Kaseko, M.S. and S.G. Ritchie, *A neural network-based methodology for pavement crack detection and classification. Transportation Research Part C*, 1993. **1**(4): p. 275–291.

61 Kim, H., E. Ahn, M. Shin, and S.-H. Sim, *Crack and noncrack classification from concrete surface images using machine learning. Structural Health Monitoring*, 2019. **18**(3): p. 725–738.

62 Kim, Y.S. and C.T. Haas, *A man-machine balanced rapid object model for automation of pavement crack sealing and maintenance. Canadian Journal of Civil Engineering*, 2002. **29**(3): p. 459–474.

63 Kirschke, K.R. and S.A. Velinsky, *Histogram-based approach for automated pavement-crack sensing. Journal of Transportation Engineering*, 1992. **118**(5): p. 700–710.

64 Kitada, T., Y. Maekawa, I. Nakamura, and Y. Horie, *A bridge management system for elevated steel highways. Computer-Aided Civil and Infrastructure Engineering*, 2000. **15**(2): p. 147–157.

65 Koch, C. and I. Brilakis, *Pothole detection in asphalt pavement images. Advanced Engineering Informatics*, 2011. **25**(3): p. 507–515.

66 Koch, C., G.M. Jog, and I. Brilakis, *Automated pothole distress assessment using asphalt pavement video data. Journal of Computing in Civil Engineering*, 2013. **27**(4): p. 370–378.

67 Koch, C. and I. Brilakis, *Achievements and challenges in machine vision-based inspection of large concrete structures. Advances in Structural Engineering*, 2014. **17**(3): p. 303–318.

68 Kong, X. and J. Li, *Vision-based fatigue crack detection of steel structures using video feature tracking. Computer-Aided Civil and Infrastructure Engineering*, 2018. **33**(9): p. 783–799.

69 Kong, X. and J. Li, *Non-contact fatigue crack detection in civil infrastructure through image overlapping and crack breathing sensing. Automation in Construction*, 2019. **99**: p. 125–139.

70 Koutsopoulos, H.N. and A.B. Downey, *Primitive-based classification of pavement cracking images. Journal of Transportation Engineering*, 1993. **119**(3): p. 402–418.

71 Koutsopoulos, H.N., I. El Sanhouri, and A.B. Downey, *Analysis of segmentation algorithms for pavement distress images. Journal of Transportation Engineering*, 1993. **119**(6): p. 868–888.

72 Koutsopoulos, H.N., V.I. Kapotis, and A.B. Downey, *Improved methods for classification of pavement distress images. Transportation Research Part C*, 1994. **2**(1): p. 19–33.

73 Lattanzi, D. and G. Miller, *Review of robotic infrastructure inspection systems. Journal of Infrastructure Systems*, 2017. **23**(3): p. 04017004.

74 Le, A., L. Mai, B. Liu, and H.K. Huang, *The workflow and procedures for automatic integration of a computer-aided diagnosis workstation with a clinical PACS with real world examples.* Medical Imaging 2008: PACS and Imaging Informatics (Vol. 6919, p. 69190U). 2008. International Society for Optics and Photonics.

75 Lecompte, D., J. Vantomme, and H. Sol, *Crack detection in a concrete beam using two different camera techniques. Structural Health Monitoring*, 2006. **5**(1): p. 59–68.

76 Lee, B.J. and H.D. Lee, *Position-invariant neural network for digital pavement crack analysis. Computer-Aided Civil and Infrastructure Engineering*, 2004. **19**(2): p. 105–118.

77 Lee, H. and H. Oshima, *New crack-imaging procedure using spatial autocorrelation function. Journal of Transportation Engineering*, 1994. **120**(2): p. 206–228.

78 Leu, S.S. and S.L. Chang, *Digital image processing based approach for tunnel excavation faces. Automation in Construction*, 2005. **14**(6): p. 750–765.

79 Li, G., *New weighted mean filtering algorithm for surface image based on grey entropy. Sensors and Transducers*, 2013. **161**(12): p. 21–26.

80 Li, G., Y. Xu, and J. Li, *Fuzzy contrast enhancement algorithm for road surface image based on adaptively changing index via grey entropy. Information Technology Journal*, 2013. **12**(19): p. 5309–5314.

81 Li, G., X. Zhao, K. Du, F. Ru, and Y. Zhang, *Recognition and evaluation of bridge cracks with modified active contour model and greedy search-based support vector machine. Automation in Construction*, 2017. **78**: p. 51–61.

82 Li, L., L. Sun, G. Ning, and S. Tan, *Automatic pavement crack recognition based on BP neural network. PROMET-Traffic & Transportation*, 2014. **26**(1): p. 11–22.

83 Li, Q., M. Yao, X. Yao, and X. Bugao, *A real-time 3D scanning system for pavement distortion inspection. Measurement Science and Technology*, 2009. **21**(1): p. 015702.

84 Li, Q., Q. Zou, D. Zhang, and Q. Mao, *FoSA: F∗ Seed-growing approach for crack-line detection from pavement images. Image and Vision Computing*, 2011. **29**(12): p. 861–872.

85 Li, S., Y. Cao, and H. Cai, *Automatic pavement-crack detection and segmentation based on steerable matched filtering and an active contour model. Journal of Computing in Civil Engineering*, 2017. **31**(5): p. 04017045.

86 Li, S., X. Zhao, and G. Zhou, *Automatic pixel-level multiple damage detection of concrete structure using fully convolutional network. Computer-Aided Civil and Infrastructure Engineering*, 2019. **34**(7): p. 616–634.

87 Li, W., Z. Yu, T. Chen, and Y. Xun, *Identification and detection for surface damages of cottonseed based on morphology. Nongye Jixie Xuebao/Transactions of the Chinese Society of Agricultural Machinery*, 2009. **40**(4): p. 169–172.

88 Lian, J., K. Wang, and Z.S. Yang, *Method of wavelet multi-scale vehicle edge detection combined with signal registration technology. Zhongguo Gonglu Xuebao/China Journal of Highway and Transport*, 2007. **20**(5): p. 95–100.

89 Lindquist, W., A. Ibrahim, Y. Tung, M. Motaleb, D. Tobias, and R. Hindi, *Distortion-induced fatigue cracking in a seismically retrofitted steel bridge. Journal of Performance of Constructed Facilities*, 2016. **30**(4), 04015068.

90 Liu, F.F., G.A. Xu, J. Xiao, and Y.X. Yang, *Cracking automatic extraction of pavement based on connected domain correlating and Hough transform. Beijing Youdian Daxue Xuebao/Journal of Beijing University of Posts and Telecommunications*, 2009. **32**(2): p. 24–28.

91 Liu, X. and Q. Li, *A pyramid-based cracks statistical model for massive pavement images. Geomatics and Information Science of Wuhan University*, 2008. **33**(4): p. 430–432+436.

92 Liu, Y.-F. and S. Cho, *Concrete crack assessment using digital image processing and 3D scene reconstruction. Journal of Computing in Civil Engineering*, 2014. **30**(1): p. 04014124.

93 Lokeshwor, H., L.K. Das, and S. Goel, *Robust method for automated segmentation of frames with/without distress from road surface video clips. Journal of Transportation Engineering*, 2014. **140**(1): p. 31–41.

94 Mataei, B., F.M. Nejad, H. Zakeri, and A.H. Gandomi, *Computational intelligence for modeling of pavement surface characteristics.* in *New Materials in Civil Engineering*, p. 65–77, 2020. Butterworth-Heinemann.

95 McKim, R.A. and S.K. Sinha, *Condition assessment of underground sewer pipes using a modified digital image processing paradigm. Tunnelling and Underground Space Technology*, 1999. **14**(SUPPL. 2): p. 29–37.

96 McKim, R.A. and S.K. Sinha, *Condition assessment of underground sewer pipes using a modified digital image processing paradigm. Comptes Rendus de l'Academie de Sciences – Serie IIa: Sciences de la Terre et des Planetes*, 2000. **331**(12): p. 29–37.

97 McNeil, S. and F. Humplick, *Evaluation of errors in automated pavement distress data acquisition. Journal of Transportation Engineering*, 1991. **117**(2): p. 224–241.

98 McRobbie, S., R. Woodward, and A. Wright, *Visualisation and display of automated bridge inspection results-PPR530. Visualisation and display of automated bridge inspection results*. 2011. **1**(1): p. 1–28.

99 Meenu, N., Rajaram, and C.T. Natarajan. *Visual information processing for status assessment in bridges.* Forensic Engineering 2012: Gateway to a Safer Tomorrow. 2013: p. 38–47.

100 Moghadas Nejad, F. and H. Zakeri, *A comparison of multi-resolution methods for detection and isolation of pavement distress. Expert Systems with Applications*, 2011. **38**(3): p. 2857–2872.

101 Montero, R., J.G. Victores, S. Martinez, A. Jardón, and C. Balaguer, *Past, present and future of robotic tunnel inspection. Automation in Construction*, 2015. **59**: p. 99–112.

102 Morgenthal, G. and N. Hallermann, *Quality assessment of unmanned aerial vehicle (UAV) based visual inspection of structures. Advances in Structural Engineering*, 2014. **17**(3): p. 289–302.

103 Morgenthal, G., N. Hallermann, J. Kersten, J. Taraben, P. Debus, M. Helmrich, and V. Rodehorst, *Framework for automated UAS-based structural condition assessment of bridges. Automation in Construction*, 2019. **97**: p. 77–95.

104 Murthy, S.B.S. and G. Varaprasad, *Detection of potholes in autonomous vehicle. IET Intelligent Transport Systems*, 2014. **8**(6): p. 543–549.

105 Nayyeri, F., L. Hou, J. Zhou, and H. Guan, *Foreground–background separation technique for crack detection. Computer-Aided Civil and Infrastructure Engineering*, 2019. **34**(6): p. 457–470.

106 Nejad, F.M., N. Karimi, and H. Zakeri, *Automatic image acquisition with knowledge-based approach for multi-directional determination of skid resistance of pavements. Automation in Construction*, 2016. **71, Part 2**: p. 414–429.

107 Nejad, F.M. and H. Zakeri, *The hybrid method and its application to smart pavement management. Metaheuristics in Water, Geotechnical and Transport Engineering*, 2012. **439**, 439–483.

108 Nik, A.A., F.M. Nejad, and H. Zakeri, *Hybrid PSO and GA approach for optimizing surveyed asphalt pavement inspection units in massive network. Automation in Construction*, 2016. **71**: p. 325–345.

109 Nishikawa, T., J. Yoshida, T. Sugiyama, and Y. Fujino, *Concrete crack detection by multiple sequential image filtering. Computer-Aided Civil and Infrastructure Engineering*, 2012. **27**(1): p. 29–47.

110 Oh, J.-K., G. Jang, S. Oh, J.H. Lee, B.-J. Yi, Y.S. Moon, J.S. Lee, and Y. Choi, *Bridge inspection robot system with machine vision. Automation in Construction*, 2009. **18**(7): p. 929–941.

111 Oliveira, H. and P.L. Correia, *Automatic road crack detection and characterization. IEEE Transactions on Intelligent Transportation Systems*, 2013. **14**(1): p. 155–168.

112 Paar, G., M.d.P. Caballo-Perucha, H. Kontrus, and O. Sidla, *Optical crack following on tunnel surfaces*. in *Two-and Three-Dimensional Methods for Inspection and Metrology IV*. 2006. International Society for Optics and Photonics.

113 Payab, M., R. Abbasina, and M. Khanzadi, *A brief review and a new graph-based image analysis for concrete crack quantification. Archives of Computational Methods in Engineering*, 2019. **26**(2): p. 347–365.

114 Peng, B., Y.S. Jiang, and Y. Pu, *Automated classification algorithm of pavement crack based on digital image processing. Zhongguo Gonglu Xuebao/China Journal of Highway and Transport*, 2014. **27**(9): p. 10–18 and 24.

115 Peng, B., K.C.P. Wang, C. Chen, and Y.S. Jiang, *Automatic recognition algorithm for crack seeds based on 1 mm resolution 3D pavement images. Zhongguo Gonglu Xuebao/China Journal of Highway and Transport*, 2014. **27**(12): p. 23–32.

116 Perchant, A. and I. Bloch, *Fuzzy morphisms between graphs. Fuzzy Sets and Systems*, 2002. **128**(2): p. 149–168.

117 Phillips, S. and S. Narasimhan, *Automating data collection for robotic bridge inspections. Journal of Bridge Engineering*, 2019. **24**(8): p. 04019075.

118 Protopapadakis, E., A. Voulodimos, A. Doulamis, N. Doulamis, and T. Stathaki, *Automatic crack detection for tunnel inspection using deep learning and heuristic image post-processing. Applied Intelligence*, 2019. **49**(7): p. 2793–2806.

119 Pulugurta, H., Q. Shao, and Y. Chou, *Pavement condition prediction using Markov process. Journal of Statistics and Management Systems*, 2009. **12**(5): p. 853–871.

120 Radopoulou, S.C. and I. Brilakis, *Patch detection for pavement assessment. Automation in Construction*, 2015. **53**: p. 95–104.

121 Rajab, M.I., M.H. Alawi, and M.A. Saif, *Application of image processing to measure road distresses. WSEAS Transactions on Information Science and Applications*, 2008. **5**(1): p. 1–7.

122 Ramana, P.V. and M.K. Srimali, *The health monitoring prescription by novel method*, in *Advances in Structural Engineering: Materials*, **Vol. 3**. 2015. Springer: New Delhi, India. p. 2587–2598.

123 Ritchie, S.G., *Digital imaging concepts and applications in pavement management. Journal of Transportation Engineering*, 1990. **116**(3): p. 287–298.

124 Riveiro, B., M.J. DeJong, and B. Conde, *Automated processing of large point clouds for structural health monitoring of masonry arch bridges. Automation in Construction*, 2016. **72**: p. 258–268.

125 Rodriguez-Lozano, F.J., F. León-García, J.C. Gámez-Granados, J.M. Palomares, and J. Olivares, *Benefits of ensemble models in road pavement cracking classification. Computer-Aided Civil and Infrastructure Engineering*, 2020. **35**(11): p. 1194–1208.

126 Seo, J., L. Duque, and J. Wacker, *Drone-enabled bridge inspection methodology and application. Automation in Construction*, 2018. **94**: p. 112–126.

127 Shen, Y., J.L. Goodall, and S.B. Chase, *Condition state-based civil infrastructure deterioration model on a structure system level. Journal of Infrastructure Systems*, 2019. **25**(1), 04018042.

128 Sinha, S.K. and P.W. Fieguth, *Morphological segmentation and classification of underground pipe images. Machine Vision and Applications*, 2006. **17**(1): p. 21–31.

129 Sinha, S.K. and P.W. Fieguth, *Segmentation of buried concrete pipe images. Automation in Construction*, 2006. **15**(1): p. 47–57.

130 Sinha, S.K. and P.W. Fieguth, *Automated detection of cracks in buried concrete pipe images. Automation in Construction*, 2006. **15**(1): p. 58–72.

131 Sinha, S.K. and P.W. Fieguth, *Neuro-fuzzy network for the classification of buried pipe defects. Automation in Construction*, 2006. **15**(1): p. 73–83.

132 Sinha, S.K., P.W. Fieguth, and M.A. Polak, *Computer vision techniques for automatic structural assessment of underground pipes. Computer-Aided Civil and Infrastructure Engineering*, 2003. **18**(2): p. 95–112.

133 Sinha, S.K. and F. Karray, *Classification of underground pipe scanned images using feature extraction and neuro-fuzzy algorithm. IEEE Transactions on Neural Networks*, 2002. **13**(2): p. 393–401.

134 Sollazzo, G., K.C.P. Wang, G. Bosurgi, and J.Q. Li, *Hybrid procedure for automated detection of cracking with 3D pavement data. Journal of Computing in Civil Engineering*, 2016. **30**(6): p. 04016032.

135 Song, B. and N. Wei, *Statistics properties of asphalt pavement images for cracks detection. Journal of Information and Computational Science*, 2013. **10**(9): p. 2833–2843.

136 Song, H., W. Wang, F. Wang, L. Wu, and Z. Wang, *Pavement crack detection by ridge detection on fractional calculus and dual-thresholds. International Journal of Multimedia and Ubiquitous Engineering*, 2015. **10**(4): p. 19–30.

137 Spencer Jr, B.F., V. Hoskere, and Y. Narazaki, *Advances in computer vision-based civil infrastructure inspection and monitoring. Engineering*, 2019. **5**(2): p. 199–222.

138 Tsai, Y., V. Kaul, and A. Yezzi, *Automating the crack map detection process for machine operated crack sealer. Automation in Construction*, 2013. **31**: p. 10–18.

139 Tsai, Y.C., C. Jiang, and Y. Huang, *Multiscale crack fundamental element model for real-world pavement crack classification. Journal of Computing in Civil Engineering*, 2014. **28**(4), 04014012.

140 Tsai, Y.C., V. Kaul, and C.A. Lettsome, *Enhanced adaptive filter-bank-based automated pavement crack detection and segmentation system. Journal of Electronic Imaging*, 2012. **21**(4), 043008.

141 Tsai, Y.C., V. Kaul, and R.M. Mersereau, *Critical assessment of pavement distress segmentation methods. Journal of Transportation Engineering*, 2010. **136**(1): p. 11–19.

142 Tsai, Y.J. and F. Li, *Critical assessment of detecting asphalt pavement cracks under different lighting and low intensity contrast conditions using emerging 3D laser technology. Journal of Transportation Engineering*, 2012. **138**(5): p. 649–656.

143 Tung, P.-C., Y.-R. Hwang, and M.-C. Wu, *The development of a mobile manipulator imaging system for bridge crack inspection. Automation in Construction*, 2002. **11**(6): p. 717–729.

144 Tur, J.M.M. and W. Garthwaite, *Robotic devices for water main in-pipe inspection: A survey. Journal of Field Robotics*, 2010. **27**(4): p. 491–508.

145 Umesha, P., R. Ravichandran, and K. Sivasubramanian, *Crack detection and quantification in beams using wavelets. Computer-Aided Civil and Infrastructure Engineering*, 2009. **24**(8): p. 593–607.

146 Varadharajan, S., S. Jose, K. Sharma, L. Wander, and C. Mertz, *Vision for road inspection.* in *IEEE Winter Conference on Applications of Computer Vision*. 2014. IEEE.

147 Wang, G., P.W. Tse, and M. Yuan, *Automatic internal crack detection from a sequence of infrared images with a triple-threshold Canny edge detector. Measurement Science and Technology*, 2018. **29**(2), 025403.

148 Wang, J. and R.X. Gao, *Pavement distress analysis based on dual-tree complex wavelet transform. International Journal of Pavement Research and Technology*, 2012. **5**(5): p. 283–288.

149 Wang, K.C., *Challenges and feasibility for comprehensive automated survey of pavement conditions*, in *Applications of Advanced Technologies in Transportation Engineering*. 2004. p. 531–536.

150 Wang, K.C. and W. Gong, *Real-time automated survey system of pavement cracking in parallel environment. Journal of Infrastructure Systems*, 2005. **11**(3): p. 154–164.

151 Wang, K.C.P. and X. Li, *Use of digital cameras for pavement surface distress survey. Transportation Research Record*, 1999 **1675** (1): p. 91–97.

152 Wang, N., X. Zhao, P. Zhao, Y. Zhang, Z. Zou, and O. Jinping, *Automatic damage detection of historic masonry buildings based on mobile deep learning. Automation in Construction*, 2019. **103**: p. 53–66.

153 Wang, X., C.C. Chang, and L. Fan, *Nondestructive damage detection of bridges: A status review. Advances in Structural Engineering*, 2001. **4**(2 SPEC.ISS): p. 75–91.

154 Weng, X., Y. Huang, and W. Wang, *Segment-based pavement crack quantification. Automation in Construction*, 2019. **105**: p. 102819.

155 Xu, X.J. and X.N. Zhang, *Crack detection of reinforced concrete bridge using video image. Journal of Central South University*, 2013. **20**(9): p. 2605–2613.

156 Xu, Y. and Y. Turkan, *Bridge inspection using bridge information modeling (BrIM) and unmanned aerial system (UAS)*, in *Advances in Informatics and Computing in Civil and Construction Engineering*. 2019, Springer. p. 617–624.

157 Yang, X., H. Li, Y. Yantao, X. Luo, T. Huang, and X. Yang, *Automatic pixel-level crack detection and measurement using fully convolutional network. Computer-Aided Civil and Infrastructure Engineering*, 2018. **33**(12): p. 1090–1109.

158 Yao, X., M. Yao, and B. Xu, *Automated measurements of road cracks using line-scan imaging. Journal of Testing and Evaluation*, 2011. **39**(4), 621–629.

159 Yeau, K.Y. and H. Sezen, *Load-rating procedures and performance evaluation of metal culverts. Journal of Bridge Engineering*, 2012. **17**(1): p. 71–80.

160 Yeum, C.M., J. Choi, and S.J. Dyke, *Automated region-of-interest localization and classification for vision-based visual assessment of civil infrastructure. Structural Health Monitoring*, 2019. **18**(3): p. 675–689.

161 Ying, L. and E. Salari, *Beamlet transform-based technique for pavement crack detection and classification. Computer-Aided Civil and Infrastructure Engineering*, 2010. **25**(8): p. 572–580.

162 Yun, H.B., S.H. Kim, L. Wu, and J.J. Lee, *Development of inspection robots for bridge cables. Rivista Italiana della Saldatura*, 2016. **68**(1): p. 59–76.

163 Zakeri, H., F.M. Nejad, and A. Fahimifar, *Image based techniques for crack detection, classification and quantification in asphalt pavement: A review. Archives of Computational Methods in Engineering*, 2016: p. 1–43.

164 Zakeri, H., F.M. Nejad, and A. Fahimifar, *Rahbin: A quadcopter unmanned aerial vehicle based on a systematic image processing approach toward an automated asphalt pavement inspection. Automation in Construction*, 2016. **72**: p. 211–235.

165 Zalama, E., J. Gómez-García-Bermejo, R. Medina, and J. Llamas, *Road crack detection using visual features extracted by Gabor filters. Computer-Aided Civil and Infrastructure Engineering*, 2014. **29**(5): p. 342–358.

166 Zaurin, R. and F.N. Catbas, *Integration of computer imaging and sensor data for structural health monitoring of bridges. Smart Materials and Structures*, 2010. **19**(1), 015019.

167 Zhang, G., R.S. Harichandran, and P. Ramuhalli, *An automatic impact-based delamination detection system for concrete bridge decks. NDT and E International*, 2012. **45**(1): p. 120–127.

168 Zhang, J., A. Sha, Z.Y. Sun, and H.G. Gao, *Pavement crack automatic recognition based on wiener filtering*. in *Critical Issues in Transportation System Planning, Development, and Management Proceedings of the Nineth international Conference of Chinese Transportation Professionals*. 2009.

169 Zhang, J., A.M. Sha, Z.Y. Sun, and H.G. Gao, *Pavement crack automatic recognition based on phase-grouping method. Zhongguo Gonglu Xuebao/China Journal of Highway and Transport*, 2008. **21**(2): p. 39–42.

170 Zhang, Y. and H. Zhou, *Automatic pavement cracks detection and classification using radon transform. Journal of Information and Computational Science*, 2012. **9**(17): p. 5241–5247.

171 Zhou, J., P.S. Huang, and F.-P. Chiang, *Wavelet-based pavement distress detection and evaluation. Optical Engineering*, 2006. **45**(2): p. 027007–027007-10.

172 Zhu, Z., S. German, and I. Brilakis, *Detection of large-scale concrete columns for automated bridge inspection. Automation in Construction*, 2010. **19**(8): p. 1047–1055.

173 Zhu, Z., S. German, and I. Brilakis, *Visual retrieval of concrete crack properties for automated post-earthquake structural safety evaluation. Automation in Construction*, 2011. **20**(7): p. 874–883.

174 Zou, Q., Y. Cao, Q. Li, Q. Mao, and S. Wang, *CrackTree: Automatic crack detection from pavement images. Pattern Recognition Letters*, 2012. **33**(3): p. 227–238.

6

Feature Extraction and Fragmentation Methods

6.1 Introduction

The key goal of applying image-processing techniques is to extract meaningful features in order to perform classification and evaluation operations. As mentioned in Chapter 5, the next step after separating the objects (distresses for example) in the image is to convert the information into properties with the goal of summarizing and reducing the amount of data for further processing. Objects isolated from the segmentation stage need to be turned into a feature vector in order to be classified and evaluated by classification methods. Extracting features with more data transfer capability can increase the speed and efficiency of the method. Various methods have been proposed to extract the features, which in a general framework, are divided into six main categories: statistical, transitional, edge, and peripheral characteristics, moment, appearance, and texture, as shown in Figure 6.1.

6.2 Low-Level Feature Extraction Methods

Low-level features automatically use the original image without any extracted information (spatial information) [1, 2]. Accordingly, thresholding can be considered as a kind of low-level feature extraction that is performed as a point operation. All of these approaches can also be used to extract high-level features. In other words, when we find objects in images, these methods can be applied on each object separately. For example, one type of pavement failure can be identified from its overall pattern. This is the first and most widely used low-level feature that we work with both directly and indirectly in most methods. This method is also called edge detection, and its purpose is to extract a line, such as a defective area, or to detect a bitumen particle, as shown in Figure 6.2a,d. In this chapter, we will examine some of the most common techniques, ranging from beginner to advanced.

The importance of using image-processing algorithms is to extract pertinent and meaningful features from image data, which are expected to serve as indicators for expressing image features. As shown in Figure 6.3, this part of image processing may be used multiple times in the feature processing process. For example, a vision system in infrastructure management may use this section in the infrastructure damage identification phase as well as in the classification and evaluation phase. The characteristics used in each step are not necessarily the same. For example, lower-quality features are also useful in the detection phase because speed is more important than accuracy at this stage, and vice versa. In the evaluation and classification phase, the type and severity of damage to infrastructure requires more detailed features and accuracy. The speed of analysis is more important than accuracy. More sophisticated visual systems with higher levels of analysis must be able to interpret the results

Automation and Computational Intelligence for Road Maintenance and Management: Advances and Applications, First Edition. Hamzeh Zakeri, Fereidoon Moghadas Nejad, and Amir H. Gandomi.

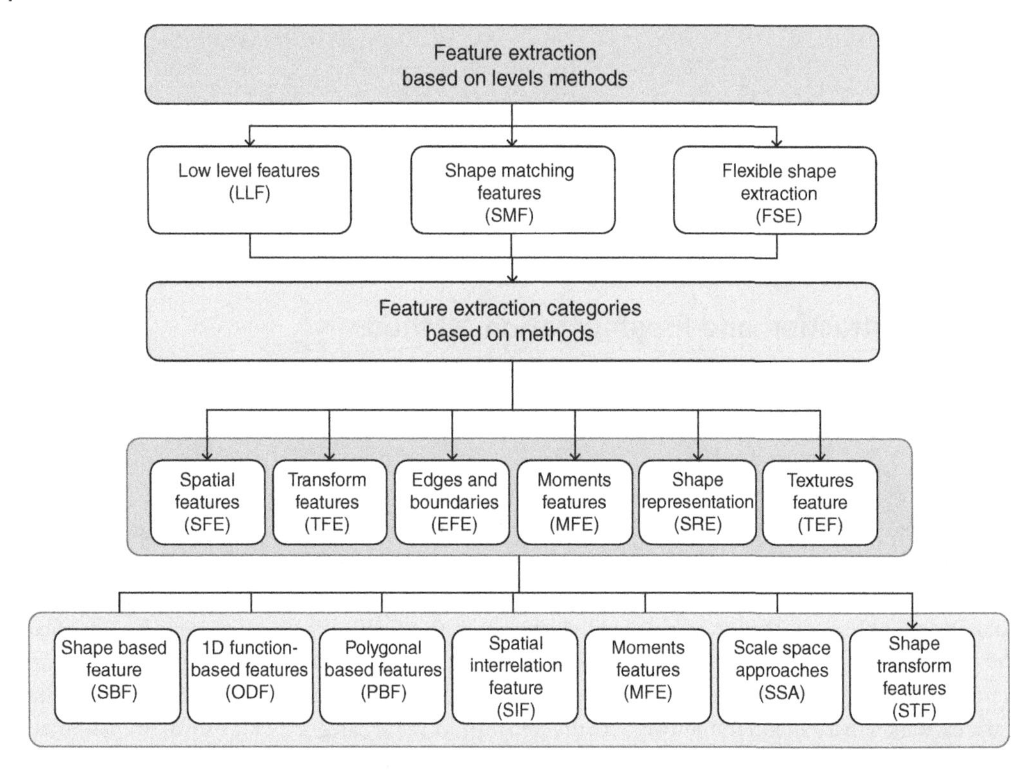

Figure 6.1 The flow of communication between features and their relationship to the target.

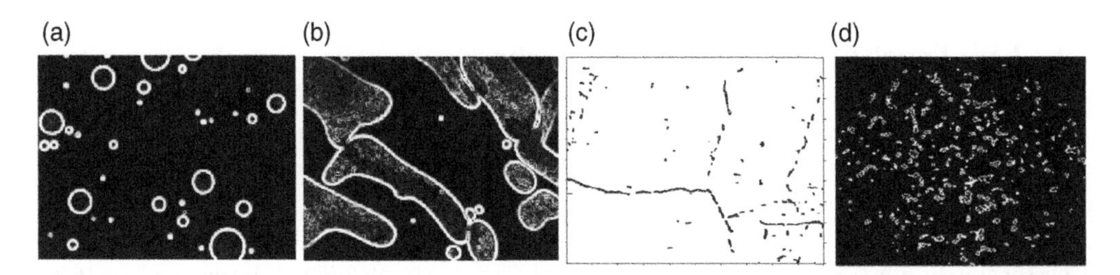

Figure 6.2 Examples of edge detection in defective area: (a) the gradient magnitude as the segmentation, opening-by-reconstruction (OBR) and closing-by-reconstruction (CBR), (b) otsu method, (c) pavement crack detection and pattern recognition, and (d) air-void phase in white asphalt core description with X-ray computed tomography and digital image processing. Source: Tapkın et al. [3]/Springer Nature.

of the analysis and description of objects in visual systems. The input image of the infrastructure is first preprocessed based on the preprocessing methods, and then its general features are extracted to divide the image into other components. For example, the crack edge is separated by extracting its boundaries or a bit of bitumen separated from the background. The fragmented image (object) is then placed in an image perception system. Image classification depicts different areas or sections in one of several objects, each marked with a label. The purpose of this section is to extract a variety of meaningful features for each object. For example, in the evaluation of cracks, we place all objects that are in a line or have a specific pattern in a category. In the analysis of the microscopic image of modified bitumen, the amount of area of each object is used as a significant parameter in evaluating the

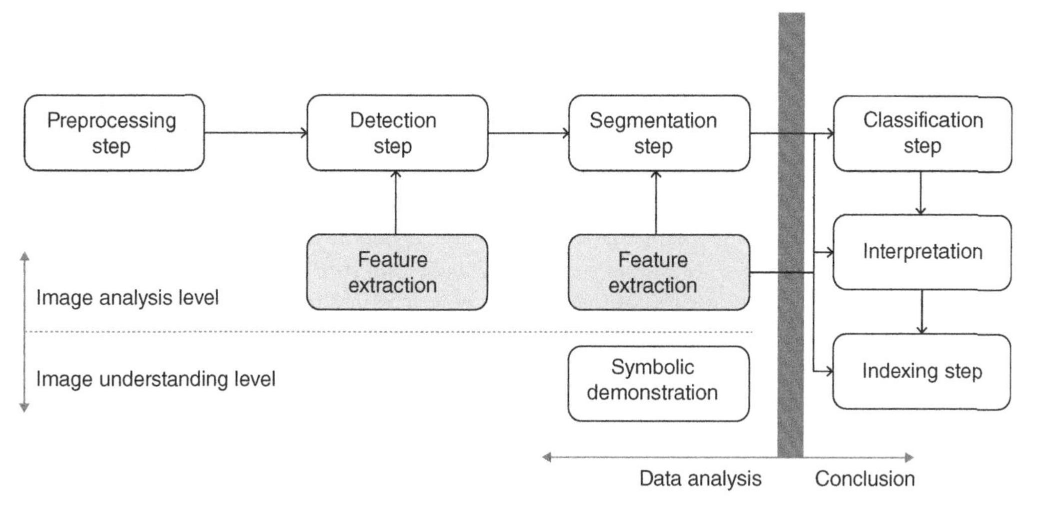

Figure 6.3 The role of feature extraction in the process of image-based systems.

performance of bitumen. Image perception systems determine the relationships between subjects to provide explanations. For instance, an image perception system must be able to interpret the report well: the field of view involves damage to the pavement of the road that is surrounded by edges. Such a system should be able to classify different textures, such as pavement texture, cracks or damage, using prior knowledge, and then use predefined rules to describe the texture.

Compared to complex methods, low-level methods automatically use the original image without any extracted information (spatial information). Accordingly, thresholding can be considered as a type of low-level feature extraction that is performed as a point operation. Of course, there are various methods of thresholding that involve different degrees of success of a feature to interpret the subject. All of these approaches can be applied to extract high-level features, particularly when there are multiple objects in images. For example, one type of pavement distress can be identified from its overall pattern.

First-order detectors are equivalent to first-order differentiation, and naturally, second-order edge detectors are equivalent to a higher level of differentiation. We will also consider corner detection of points where lines with extreme curvature bend sharply, such as the edges of a bitumen particle, as shown in Figure 6.4a,d. These examples and methods are used for low-level features that can be automatically extracted from the image.

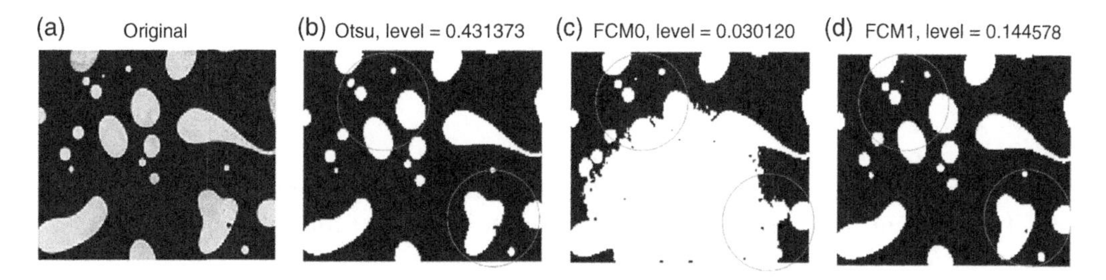

Figure 6.4 An example of binary conversion of the resulting images: (a) original image, (b) otsu method, (c) Fcm – level 0, and (d) Fcm – level 1. (White areas are assumed to be polypropylene fiber, while the black areas are bitumen base.)

In the following section, the types of feature extraction methods based on the classification in Figure 6.1 are introduced in a practical way.

6.3 Shape-Based Feature (SBF)

Generally, in the interpretation of a shape-based image, measuring the similarity of properties between shapes plays a key role. Simple geometric features are usually used to increase the efficiency of segmentation and classification algorithms as well as to describe shapes. These features do not distinguish almost identical shapes and details, in other words, simple geometric features only show sensitivity to objects with very large differences. Therefore, they are commonly used to detect shapes with large and unique geometry. These methods are not suitable for pattern-less descriptors. The feature set can be used to describe a shape with different aspects. The set of these features creates a characteristic vector that can be used in a variety of classification algorithms. These shape-based parameters include center of gravity (COG), minimum axis of inertia, digital bending energy, centrifuge, circular ratio, elliptical variance, rectangle, convexity, solidity, Euler number, prolapse, and hole area ratio [1, 4], and will be introduced in this section.

6.3.1 Center of Gravity (COG) or Center of Area (COA)

The COG is also called the focal center of an object with a binary pattern. Its position must be precisely defined in relation to the shape. If a shape with a specific function of the same object is represented, the center of that shape is calculated using the following equations and represented by the center $(Cx; Cy)$:

$$\begin{aligned} C &= (C(x), C(y)), \\ C(x) &= \left(\frac{1}{N}\right)\sum_{i=1}^{N} x_i, C(y) = \left(\frac{1}{N}\right)\sum_{i=1}^{N} y_i \end{aligned} \tag{6.1}$$

where N is the number of pixels in the shape. $C(x)$ is the COG/COA in x direction. $C(y)$ is the COG/COA in y direction. $C()$ COG/area.

If an object is represented by its contour, the position of the center of the new shape is calculated using Eq. (6.2):

$$\begin{aligned} C(x) &= \left(\frac{1}{6A}\right)\sum_{i=0}^{N-1} (x_i + x_{i+1})(x_i y_{i+1} - x_{i+1} y_i) \\ C(y) &= \left(\frac{1}{6A}\right)\sum_{i=0}^{N-1} (y_i + y_{i+1})(x_i y_{i+1} - x_{i+1} y_i) \\ A &= \left(\frac{1}{2}\right)\sum_{i=0}^{N-1} (x_i y_{i+1} - x_{i+1} y_i) \end{aligned} \tag{6.2}$$

where A is the contour area.

In order to determine the coordinates of the center of the object, the center on a contour is specified. It is pertinent to note that the position of the center in Figure 6.5 depends on how the points are distributed.

Using Eq. (6.2), the main center of a contour can be calculated separately for the relevant objects, or the sum of the objects can be considered as a general object and the main center can be calculated.

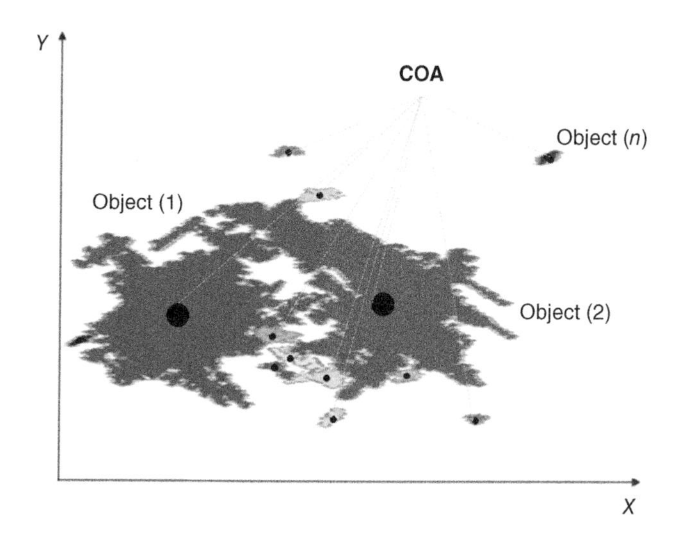

Figure 6.5 The black circle is the center of each object in the image. The internal point of each object is calculated according to the area of the same object and the distance from the origin of the coordinates. This image is an example of a cracked image obtained from Radon transmission [5, 6].

6.3.2 Axis of Least Inertia (ALI)

This section introduces a feature related to the minimum inertia axis, which acts as a unique reference line to determine the orientation of the object. The minimum ALI defines an object as a line whose square integral of distances to the boundary points of the shape is the least possible value. Since the axis of inertia passes through the main center as a line, to find ALI, we transfer the shape to the main center and, henceforth, the origin of the Cartesian coordinate system. The parametric equation $x \sin\theta - y \cos\theta = 0$ is an ALI property. Using this equation, the slope angle is calculated as follows: Let α be the angle between the ALI and the x-axis. Inertia is then calculated with (6.3):

$$\begin{aligned} \text{ALI} &= \left(\frac{1}{2}\right)(a+c) - \left(\frac{1}{2}\right)(a-c)\cos(2\alpha) - \left(\frac{1}{2}\right)(b)\sin(2\alpha) \\ a &= \sum_{i=0}^{N-1} x_i^2,\ b = 2\sum_{i=0}^{N-1} x_i y_i,\ c = \sum_{i=0}^{N-1} y_i^2 \\ \frac{dI}{d\alpha} &= (a-c)\sin(2\alpha) - b\cos(2\alpha),\ \frac{d^2I}{d\alpha^2} = 2(a-c)\cos(2\alpha) + b\sin(2\alpha) \end{aligned} \tag{6.3}$$

Let $\frac{dI}{d\alpha} = 0$, then

$$\alpha = \left(\frac{1}{2}\right)\arctan\left(\frac{b}{b-c}\right),\ -\frac{\pi}{2} < \alpha < \frac{\pi}{2}$$

Then, the slope angle (θ) can be calculated by Eq. (6.4)

$$\theta = \begin{cases} \alpha + \frac{\pi}{2} & \text{if } \frac{d^2I}{d\alpha^2} < 0 \\ \alpha & \text{otherwise} \end{cases} \tag{6.4}$$

6.3.3 Average Bending Energy

Given that the type of objects can be used to define them descriptively, the degree of circularity is used as an indicator. In this regard, it can be proved that a circle is a shape that has at least the average bending energy. On the other hand, other shapes, such as squares or shapes with infrastructure damage, and the morphology of adhesives, like bitumen or cement, can also be interpreted with this feature. The attribute is calculated using Eq. (6.5):

$$ABE = \frac{1}{N}\sum_{i=0}^{N-1} K(s)^2 \tag{6.5}$$

In this equation, $K(s)$ is a function of curvature in the corresponding object, s is the parameter of the object arc length, N is the number of points on a contour for each object.

6.3.4 Eccentricity Index (ECI)

Eccentricity is an indicator used to measure the aspect ratio, which is defined as the length of the main axis to the length of the subaxis. This property can be calculated by the method of principal axes or at least finite rectangles. In the first method, the main axes of a given shape can be uniquely defined as two sections of lines that intersect at the center of the shape and show the directions with zero correlation. Thus, a contour is considered as an example of a statistical distribution. Consider the covariance matrix C of a contour [1]:

$$C = \frac{1}{N}\sum_{i=0}^{N-1} \begin{pmatrix} x_i - g_x \\ y_i - g_y \end{pmatrix} \begin{pmatrix} x_i - g_x \\ y_i - g_y \end{pmatrix}^T = \begin{pmatrix} c_{xx} & c_{xy} \\ c_{yx} & c_{yy} \end{pmatrix} \tag{6.6}$$

where

$$\begin{aligned} c_{xx} &= \frac{1}{N}\sum_{i=0}^{N-1} (x_i - g_x)^2 \\ c_{xy} &= \frac{1}{N}\sum_{i=0}^{N-1} (x_i - g_x)(y_i - g_y) \\ c_{yx} &= \frac{1}{N}\sum_{i=0}^{N-1} (y_i - g_y)(x_i - g_x) \\ c_{yy} &= \frac{1}{N}\sum_{i=0}^{N-1} (y_i - g_y)^2 \end{aligned} \tag{6.7}$$

In this equation, $G(g_x; g_y)$ is the main center of the shape. Typically for a given shape, $c_{xy} = c_{yx}$. The lengths of the two principal axes are equal to the eigenvalues λ_1 and λ_2 of the covariance matrix C of a contour, respectively. Therefore, eigenvalues λ_1 and λ_2 can be calculated using the following equations:

$$\det(C - \lambda_{1,2} I) = \det\begin{pmatrix} c_{xx} - \lambda_{1,2} & c_{xy} \\ c_{yx} & c_{yy} - \lambda_{1,2} \end{pmatrix} = (c_{xx} - \lambda_{1,2})(c_{yy} - \lambda_{1,2}) - c_{yx}^2 = 0 \tag{6.8}$$

where:

$$\begin{aligned} \lambda_1 &= 0.5\left(c_{xx} + c_{yy} + \sqrt{(c_{xx} + c_{yy})^2 - 4(c_{xx}c_{yy} - c_{xy}^2)}\right) \\ \lambda_2 &= 0.5\left(c_{xx} + c_{yy} - \sqrt{(c_{xx} + c_{yy})^2 - 4(c_{xx}c_{yy} - c_{xy}^2)}\right) \end{aligned} \tag{6.9}$$

$$ECI = \left(\frac{\lambda_2}{\lambda_1}\right) \tag{6.10}$$

Eccentricity is calculated by the principal axis method using the above relation and the ratio of eigenvalues of small dimension to large dimension and is always a number between 0 and 1.

In the second method, the minimum constraint rectangle is also called the minimum constraint box. This is the smallest rectangle that encloses any point in the shape. For the desired shape, the centrifugal ratio of the length l and width W of the rectangle limits the minimum shape in some directions:

$$ECI = \left(\frac{W}{l}\right) \tag{6.11}$$

Other features, such as elongation and shear, can also be measured based on the indicators. Extension is a measure that has values in the range [0; 1]. A symmetrical shape on all axes, such as a circle or square, will have an elongation value of 0, while shapes with a large aspect ratio will have an elongation close to 1. This feature can be used to assess the type of failure, amount of uniform distribution, severity, the extent of the assessment, and even models for predicting the spread of damage to infrastructure (see Figure 6.6 for an example).

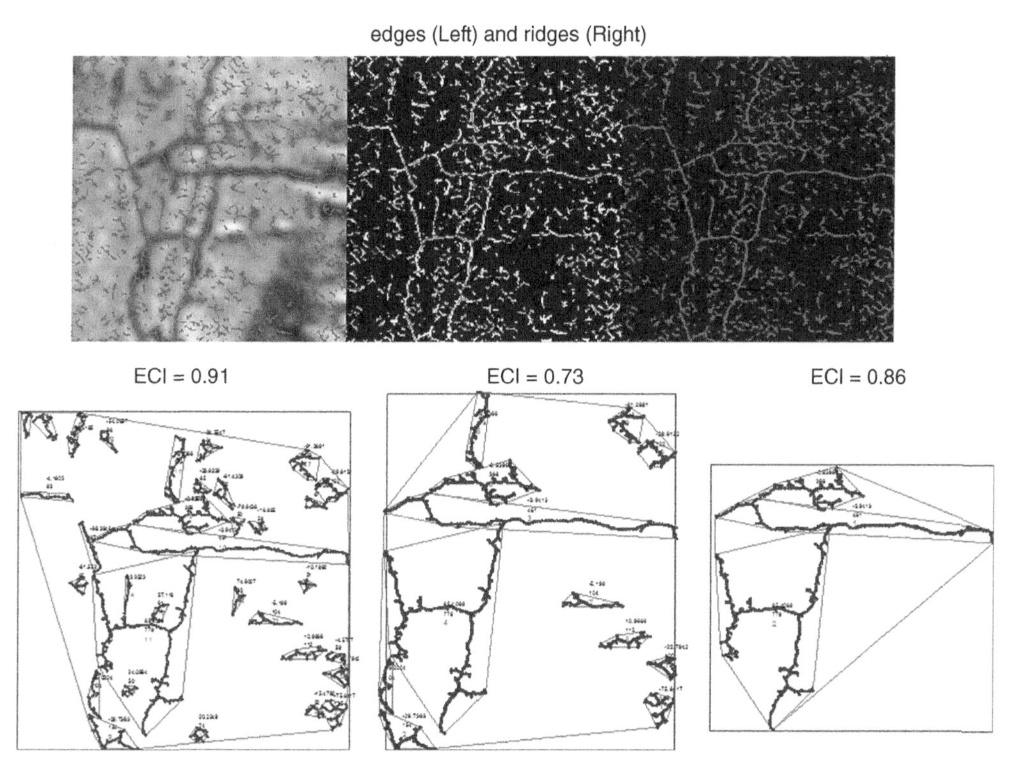

Figure 6.6 The black circle is the center of each object in the image. The internal point of each object is calculated according to the area of the same object and the distance from the origin of the coordinates. This image is an example of a cracked image obtained from Radon transmission.

6.3.5 Circularity Ratio (CIR)

The circularity index is used to determine the degree to which an object resembles a circle. This numerical index, which ranges between 0 and 1, has different definitions that depend on the amount of equivalent area, equivalent diameter, and variance. Based on this index, examples, like elongation and agglomeration of asphalt and concrete mix aggregates, determination of type one crack in asphalt pavement and bitumen development rate, and modifier distribution potential in bitumen mixture can be measured.

The index of circularity, according to definition 1, is the ratio of the area of an object (for example, the area of an asphalt aggregate) to the area of a circle with the same perimeter:

$$CIR_1 = \left(\frac{A_s}{A_C}\right) \tag{6.12}$$

where A_s is the area of object, and A_c is the area of circle having the same perimeter as the object.

Assuming that the diameter of an object is equal to d, then the index becomes $A_c = d^2/4\pi$. Knowing that the value of 4π is constant, then the second index is calculated according to this definition, which correlates the ratio of the area of the object to the square of the diameter, as follows:

$$CIR_2 = \left(\frac{A_s}{d^2}\right) \tag{6.13}$$

where A_s is the area of object, and d^2 is the perimeter of circle having the same perimeter as the object powered by 2.

Similarly, using the relationships of variance and the mean of Eq. (6.13), CIR becomes

$$\begin{aligned}
\mathrm{CIR}_3 &= \left(\frac{\mathrm{STD}_R}{\mathrm{Mean}_R}\right)\\
\mathrm{STD}_R &= \frac{1}{N}\sum\nolimits_{i=1}^{N} d_i\\
\mathrm{Mean}_R &= \sqrt{\frac{1}{N}\sum\nolimits_{i=1}^{N}(d_i - \mathrm{STD}_R)^2}\\
d_i &= \sqrt{(x_i - g_x)^2 + \left(y_i - g_y\right)^2}
\end{aligned} \tag{6.14}$$

where STD_R is the standard deviation of the radial distance from the center of the object to boundary coordinates, and Mean_R is the mean of object.

The higher the density of an object, the closer to having a circular shape, and ultimately the higher the triple indices, the closer to 1. For example, in a set of aggregates related to the asphalt sample, the denser the aggregates, the greater the total characteristics associated with this index. Thus, the nature of the asphalt mixture aggregates is more rounded, and the amount of fracture and other characteristics, such as elongation, is related to this feature (Figure 6.7).

6.3.6 Ellipse Variance Feature (EVF)

The ellipse variance property is a characteristic of a mapping error in an object located in the ellipse and has a covariance matrix with the relation: $C_{\text{ellipse}} = f$ Eq. (6.15). In practice, this method is used to determine important characteristics like elongation and thinning. For example, this feature can

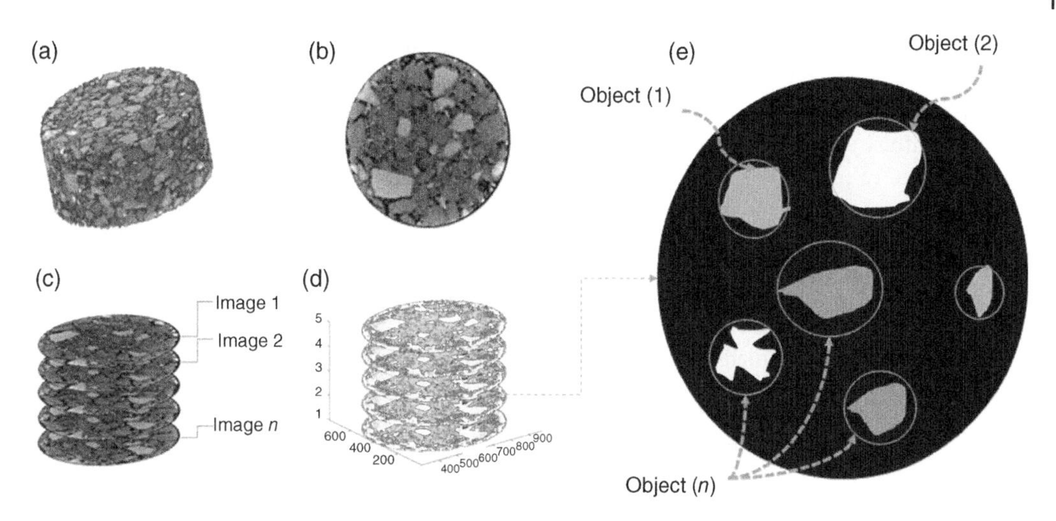

Figure 6.7 An example of the application of the extraction feature of variant circuit for modeling asphalt sample aggregates in an X-ray image: (a) three-dimensional image of asphalt sample, (b) a sample of a slice of cross-asphalt section, (c) separation of slices for analysis at different heights, (d) separation of aggregates from the sample image at h height, and (e) extracted shape features related to each object based on the CIR index.

be used to assess the failure of infrastructure to determine the type of longitudinal or patterned cracks. This characteristic is calculated using the following equations:

$$C = \frac{1}{N}\sum_{i=0}^{N-1}\begin{pmatrix} x_i - g_x \\ y_i - g_y \end{pmatrix}\begin{pmatrix} x_i - g_x \\ y_i - g_y \end{pmatrix}^T = \begin{pmatrix} c_{xx} & c_{xy} \\ c_{yx} & c_{yy} \end{pmatrix}$$

$$V_i = \begin{pmatrix} x_i - g_x \\ y_i - g_y \end{pmatrix}$$

$$d_i' = \sqrt{\begin{pmatrix} x_i - g_x \\ y_i - g_y \end{pmatrix}^T . C_{\text{ellipse}}^{-1} . \begin{pmatrix} x_i - g_x \\ y_i - g_y \end{pmatrix}}$$

$$\mu_R' = \frac{1}{N}\sum_{i=0}^{N-1}\sqrt{\begin{pmatrix} x_i - g_x \\ y_i - g_y \end{pmatrix}^T . C_{\text{ellipse}}^{-1} . \begin{pmatrix} x_i - g_x \\ y_i - g_y \end{pmatrix}} = \frac{1}{N}\sum_{i=0}^{N-1} d_i'$$

$$\sigma_R' = \sqrt{\frac{1}{N}\sum_{i=0}^{N-1}\left(\begin{array}{c}\sqrt{\begin{pmatrix} x_i - g_x \\ y_i - g_y \end{pmatrix}^T . C_{\text{ellipse}}^{-1} . \begin{pmatrix} x_i - g_x \\ y_i - g_y \end{pmatrix}} - \\ \frac{1}{N}\sum_{i=0}^{N-1}\sqrt{\begin{pmatrix} x_i - g_x \\ y_i - g_y \end{pmatrix}^T . C_{\text{ellipse}}^{-1} . \begin{pmatrix} x_i - g_x \\ y_i - g_y \end{pmatrix}}\end{array}\right)^2} = \sqrt{\frac{1}{N}\sum_{i=0}^{N-1}(d_i' - \mu_R')^2}$$

$$\text{EVF} = \left(\frac{\sigma_R'}{\mu_R'}\right) \tag{6.15}$$

where C is the covariance matrix, σ_R' is standard deviation of the radial distance from the center of the object to boundary coordinates, μ_R' is the mean, EVF is the ellipse variance feature.

In general, method B is more accurate than method A. Of course, it should be noted that this result depends a lot on the application and how this feature is used. However, regarding the two applications related to the analysis of aggregates via the X-ray method and the analysis of cracks related to infrastructure structures, including road pavement and concrete surface, index B shows more accurate results.

6.3.7 Rectangularity Feature (REF)

The rectangularity index shows how similar a shape is to a rectangular pattern, in other words, it shows at least the constrained rectangle of that object. This index is calculated using Eq. (6.16):

$$REF = \left(\frac{A_s}{W \times L}\right) = \left(\frac{A_s}{A_R}\right) \tag{6.16}$$

where A_s is the area of a shape, L is the length of a rectangle, w is the width of a rectangle, and A_R is the minimum area of a rectangle.

An example of the results of this method is shown in Figure 6.8.

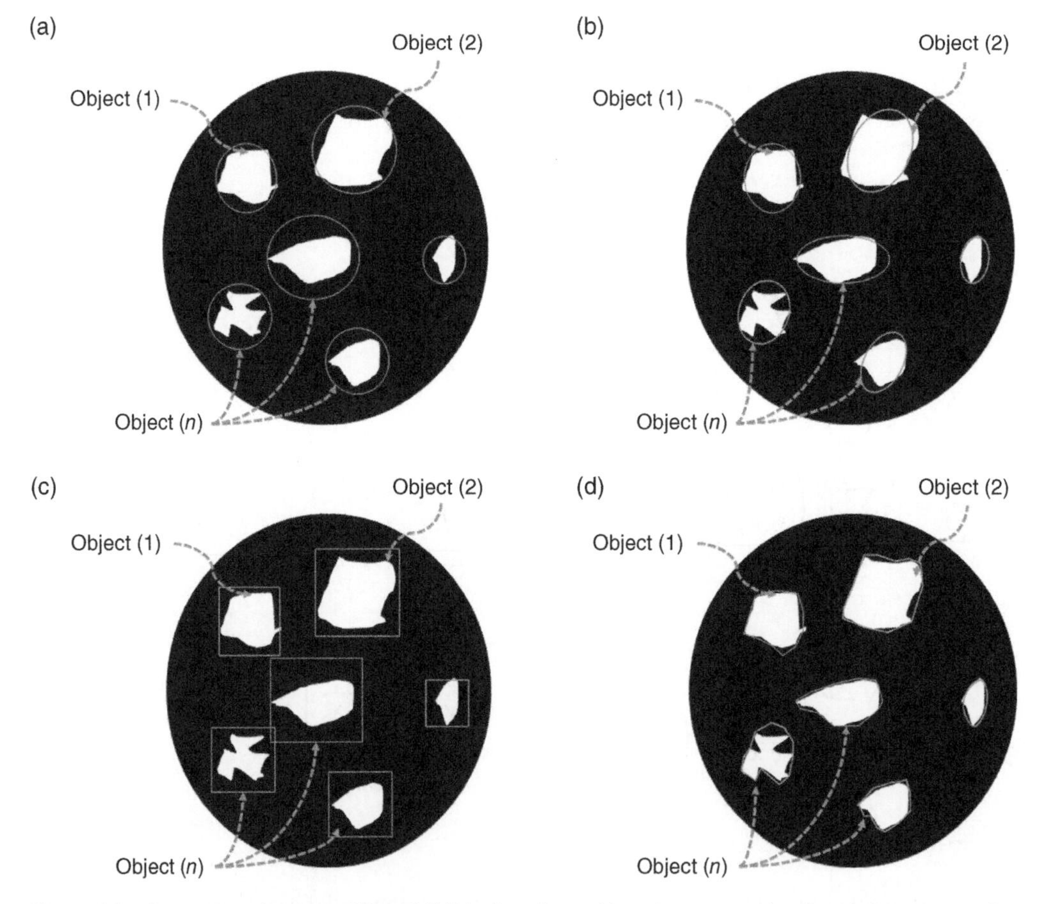

Figure 6.8 Comparison between CIR and EVF indices for an X-ray image sample of asphalt pavement in section 6.3.6: (a) CIR, (b) EVF, (c) REF, and (d) COF of each aggregate.

6.3.8 Convexity Feature (COF)

The convexity characteristic, in the definition of a convex body, is the shape of the smallest convex set it contains. A convex body is defined as the intersection of all convex sets containing a subset of Euclidean space. Based on this index, a set of all points of objects in the subset is formed. For a subset of objects, the convex enclosed shape is created by a rubber band stretched around the subset. Also, for finite point sets, the convex body is used for simple polygons, space curves, and epigraphs of functions. Convex bodies have wide applications of image processing and its application in infrastructure management based on automatic analysis.

A convex body is a collection of objects that is enclosed in the smallest convex area and encompasses all parts of the collection. If you consider a two-dimensional set of objects as nails in a wall, the convex body of that set is defined by holding an allowable strip to enclose all the nails. The COF is defined as the ratio of the perimeter of convex hull over the original contour, as follows:

$$COF_1 = \left(\frac{P_{\text{cof}}}{P}\right) \text{and } COF_2 = \left(\frac{A_s}{A_{\text{cof}}}\right)$$

where A_s is the area of a shape or object, A_{cof} is the area of a shape or object enclosed in the smallest convex hull, P is the perimeter of an object, and P_{cof} is the perimeter of an object enclosed in the smallest convex hull.

An example of the results of this method is shown in Figure 6.8d, and an application for pavement crack analysis based of the COF index is illustrated in Figure 6.9. The COF_2 is sometimes called solidity and is equal to 1 for convex objects (Figure 6.10).

6.3.9 Euler Number Feature (ENF)

The Euler number attribute is a property that subtracts the number of continuous objects in the area minus the number of holes in that object. This feature is a significant feature for 2D input tag matrices, such as images. In this index, eight connections are used to calculate the Euler number. In many infrastructure management issues, there are objects in the image that determine the number of holes to the number of objects, which is very important in interpreting features. Complementary characteristics can also be calculated using this method. For example, from the ratio of

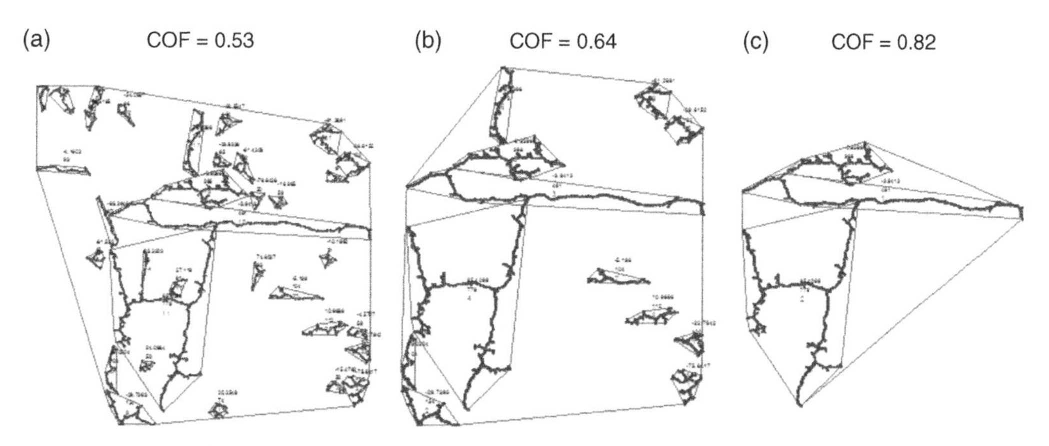

Figure 6.9 Comparison COFs of: (a) vast cracking area, (b) restricted cracking area, and (c) a main crack in cracking area.

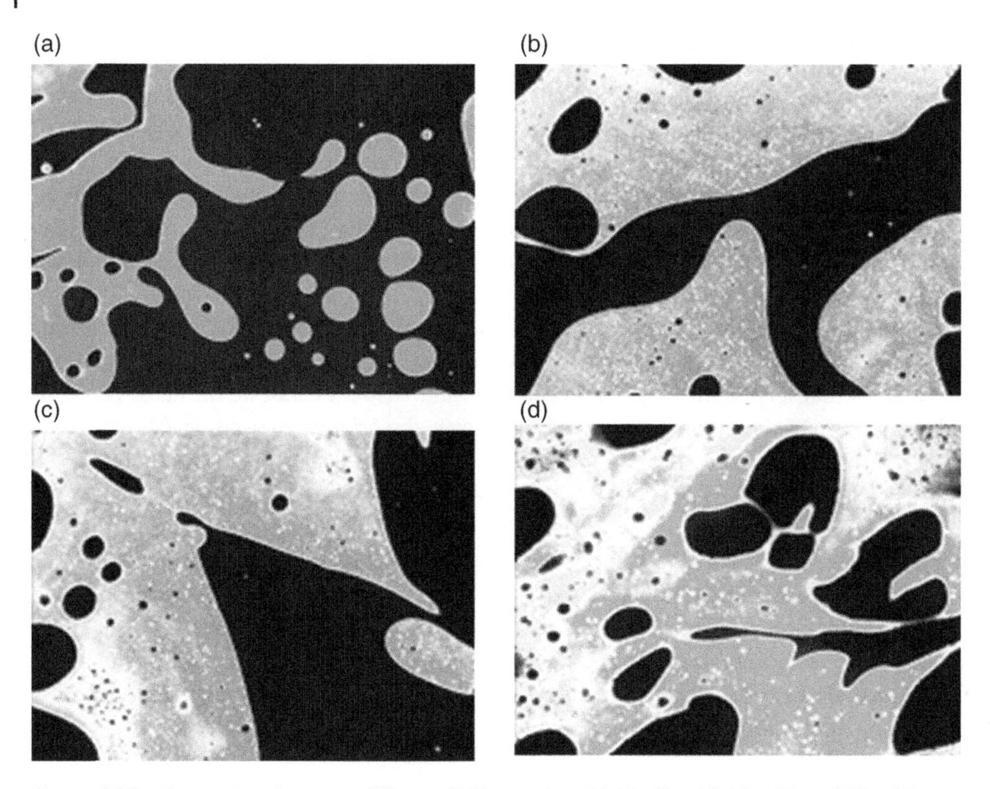

Figure 6.10 Comparison between different ENFs equal to (a) 10, (b) −14, (c) −15, and (d) −53.

the area of the internal cavities of an object to the area of the object, the stiffness index associated with the Euler number is calculated. Also, the convexity index for cavities and the main object can be calculated separately, where the ratio of which can be used to measure the Euler convexity index. Also, as another example in assessing the cracking of asphalt pavement, this number can be used to measure the amount of spall. In measuring the bitumen intensity of asphalt pavement, the closer this index is to zero, the greater the bitumen intensity is. Also, in the homogeneity assessment of the modified bitumen mixture, this number is closer to zero. The implication is that the modifier additive is better dissolved in the bitumen and has a higher homogeneity index. The Euler number is calculated using the Eq. (6.17):

$$\mathrm{ENF} = \left| \left(\sum_{i=0}^{N} \mathrm{Count}(S_i) \right) - \left(\sum_{i=0}^{N} \mathrm{Count}(H_i) \right) \right| \tag{6.17}$$

where $\mathrm{Count}(S_i)$ is the number of continuous parts of an object, $\mathrm{Count}(H_i)$ is the number of holes of an object, and ENF is the Euler number of an object.

6.3.10 Profiles Feature (PRF)

Profiles are X-shaped and Y-shaped in the Cartesian coordinate system, and based on simple equations, such as area, maximum, minimum, variance, center of mass, center of surface, different properties depend on the main directions. In this regard, we obtain two one-dimensional functions for extracting characteristics in the X- and Y-directions using the Eq. (6.18):

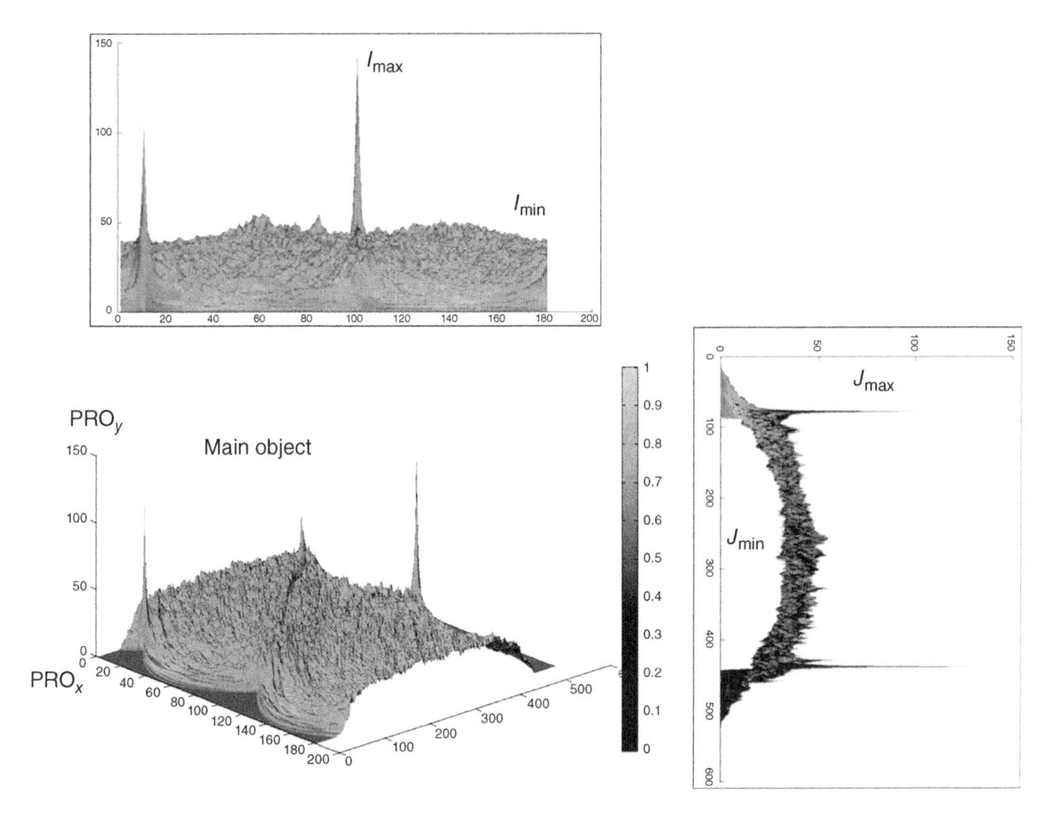

Figure 6.11 An example of a 3D transmitted image on the X- and Y-axes.

$$PRF_x(i) = \sum_{j=-j\min}^{j\max} f(i,j)$$

$$PRF_y(j) = \sum_{i=-i\min}^{i\max} f(i,j)$$

$$PRF = f\left(PRF_x(i), PRF_y(j)\right) \tag{6.18}$$

where $f(i, j)$ is the area of an object, $PRF_x(i)$ is the projection area of an object in X-direction, and $PRF_y(j)$ is the projection area of an object in Y-direction.

An example of a transmitted image on the X- and Y-axes for a 3D function is shown in Figure 6.11. By reflecting the shadow of the object in different directions, a new image is created that has a new nature, and various features can be extracted from these details.

6.4 1D Function-Based Features for Shape Representation

A one-dimensional function that is extracted from object boundary coordinates to analyze and decide on the properties of a shape is often called a shape signature and has key applications in the feature extraction section. This index can be one-dimensional, two-dimensional, and multidimensional, and its purpose is to provide a number to introduce the general features of an image.

Shape signature is related to the generality of the object and usually shows the perceptual property of the shape, which has many applications in solving problems related to automatic infrastructure management. Mixed coordinates, centroid distance function, tangent angle (rotation angle), curvature function, area function, triangle area display, and chord length function are some of the common characteristics used by this function. Shape signature in some applications can represent the shape feature alone. These features are often used for preprocessing other feature extraction algorithms or in combination with other methods. For example, some Fourier descriptors, wavelet transform, Shearlet transform, and curvelet transform are presented as shape signatures in this section.

6.4.1 Complex Coordinates Feature (CCF)

As a simple definition, a complex coordinates function is a complex number that is produced from the coordinates of the boundary points and has a real part and imaginary part using Eq. (6.19).

$$ccf(z) = [x(n) - g_x] + i\left[y(n) - g_y\right] \tag{6.19}$$

where (g_x, g_y) is the center of an object; $ccf(z)$ is the CCF of an object; $[x(n) - g_x]$ is the real part of an object; and $[y(n) - g_y]$ is the imaginary part of an object.

6.4.2 Extracting Edge Characteristics Using Complex Coordinates

Many models and algorithms have been proposed by examining the history of edge analysis, detection, and mathematical analysis. Particularly, wavelength-based multilevel methods have been widely used in this field. For example, the mixed wavelet per year concept was used for edge extraction. A similar method for mixed wavelet based on rice conversion was proposed by Felsberg and Sommer in 2007. In 2013, a new type of complex wavelet called the monotonic curvelet was introduced, replacing the old curvelet with a monotonic wave and showing better results in rotation and curved subjects. The section discusses the fixed difference method for edge detection based on the principle of phase correlation and presents some applications of the complex method using the Shearlet method to extract the edge feature.

Definition of complex wavelet-based phase measure (CW_PCM): Assuming $f \in L^2(R)$, $J \in N$ and considering the complex wavelet $\psi^c{}_{a_j,x}$ has a size using the parameter $x \in R^+$, the center around $x \in R$ is calculated from Eq. (6.20):

$$\begin{aligned} \psi^c{}_{a_j,x} &= \psi_{a_j,x} + iH\psi^c{}_{a_j,x}, \\ \psi_{a_j,x} &= a_j{}^{-1/2}\psi\left(\frac{\cdot - x}{a_j}\right), \end{aligned} \tag{6.20}$$

where $\psi^c{}_{a_j,x}$ is the complex wavelet.

Assuming $\psi \in L^2(R)$, the discrete wavelet is symmetric, and $H\psi$ is introduced for the real part of Hilbert transform. Then, using the one-dimensional mixed wavelet, where f is the estimated point $x \in R_{:}$, PC is calculated using Eq. (6.21):

$$PC_{\psi^c}(f, x) = \frac{\left|\sum_{j=1}^{J}\left\langle f, \psi^c{}_{a_j,x}\right\rangle\right|}{\sum_{j=1}^{J}\left|\left\langle f, \psi^c{}_{a_j,x}\right\rangle\right| + \epsilon}, \tag{6.21}$$

where $\epsilon > 0$ is the complex wavelet.

In the case of symmetry, where ψ has an absolutely real value, by this argument, ψ is made up of cos waves only, while $H\psi$ only contains sine waves. Thus, assuming complete phase coordination at the point x, $\langle\langle f, {\psi^c}_{a_j,x}\rangle = \left|f, {\psi^c}_{a_j,x}\right| e^{i\varphi}$ have e ^ iφ for some fixed angles $\varphi \in [0, 2\pi]$ and $j \in N$ using Eq. (6.22):

$$PC_{\psi^c}(f,x) = \frac{\left|\sum_{j=1}^{J}\left\langle f, {\psi^c}_{a_j,x}\right\rangle\right|}{\sum_{j=1}^{J}\left|\left\langle f, {\psi^c}_{a_j,x}\right\rangle\right|} = \frac{\left|e^{i\varphi}\right|\left|\sum_{j=1}^{J}\left\langle f, {\psi^c}_{a_j,x}\right\rangle\right|}{\sum_{j=1}^{J}\left|\left\langle f, {\psi^c}_{a_j,x}\right\rangle\right|} = 1. \tag{6.22}$$

Naturally, most of the phase angles are different at different scales j, and the value of the complex coefficients $\left\langle f, {\psi^c}_{a_j,x}\right\rangle$ further reduces the other coefficients, this factor causes close to the $PC_{\psi^c}(f,x)$ becomes zero.

$PC_{\psi^c}(f,x)$ is a number between 0 and 1 and used as a criterion for distinguishing contrast. In addition, by considering a two-dimensional mixed wavelet in different scales, similar to Shearlet with different orientations, it can be generalized in two dimensions.

Definition of two-dimensional complex wavelet-based phase congruency measure (2DCW_PCM): Assuming $f \in L^2(R^2)$, $J \in N$, $K \in N$with complex Wavelet Wave ${\psi^c}_{a_j,\varphi_k,x}$. Measured using the parameter $a_j \in R^+$and rotated $\varphi_k = [0, 2\pi]$ centered around$x \in R^2$ is calculated from Eq. (6.23) and Eq. (6.24):

$${\psi^c}_{a_j,\varphi_k,x} = \psi_{a_j,\varphi_k,x} + iH{\psi^c}_{a_j,\varphi_k,x}, \tag{6.23}$$

where

$$\psi_{a_j,\varphi_k,x} = {a_j}^{-1}\psi\left(R_{\varphi_k}\frac{\cdot - x}{a_j}\right) \tag{6.24}$$

Assuming $\psi \in L^2(R^2)$ is the real part of the discrete wavelet and is symmetric and $H\psi$ is introduced for the individual part of the Hilbert transform, then the rotation matrix is defined as Eq. (6.25):

$$R_{\varphi_k} = \begin{pmatrix} \cos(\varphi_k) & -\sin(\varphi_k) \\ \cos(\varphi_k) & \sin(\varphi_k) \end{pmatrix} \tag{6.25}$$

Using the two-dimensional mixed wavelet adaptation phase, f is plotted at the estimated point $x \in R^2$ and is calculated using Eq. (6.26):

$${PC^{2D}}_{\psi^c}(f,x) = \frac{\sum_{k=1}^{K}\left|\sum_{j=1}^{J}\left\langle f, \psi_{a_j,\varphi_k,x}\right\rangle\right|}{\sum_{k=1}^{K}\sum_{j=1}^{J}\left|\left\langle f, \psi_{a_j,\varphi_k,x}\right\rangle\right| + \epsilon} \tag{6.26}$$

These definitions can be generalized to Shearlet transform. It may even be thought that because of the superiority of Shearlet over wavelet for the choice of two-dimensional geometric features, Shearlet-based phase measurements create a better and clearer situation. While it is possible to generalize this method, as shown in Figure 6.12, mixed sections for the measured phase section index do not show improvement. In fact, using this method gives the opposite results due to the creation of geometric properties even though the method itself has nothing to do with geometric properties. The purpose of using a mixed wavelet is to estimate the local behavior of the Fourier atom when the frequency changes. When using Shearlet for nonisotropic dimensions, it will support the frequency of mixed chert atoms, which will lead to better adaptation. In the following section, the application

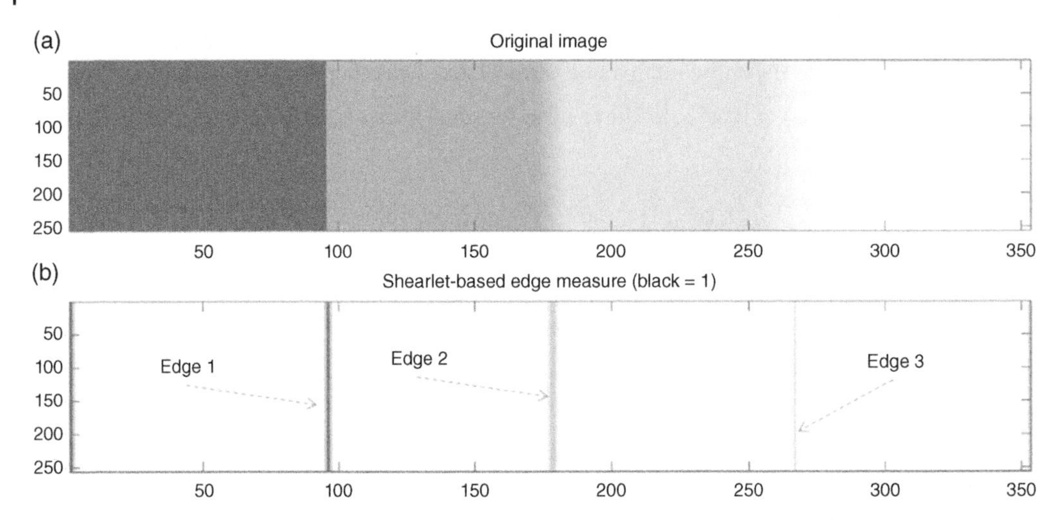

Figure 6.12 Measuring the edge of the image based on the mixed Shearlet method to detect the edge of the object separation without removing local information: (a) main image and (b) image analyzed using the Shearlet method.

of Shearlet in edge detection as well as the use of mixed Shearlet components and real and imaginary parts using Hilbert transform are discussed.

6.4.3 Edge Detection Using Even and Odd Shearlet Symmetric Generators

In this method, two generators of Even-Symmetric Shearlet Generator (ESSG) pair and symmetric Shearlet Complex Odd-Symmetric Shearlet Generator (OSSG) individual are used. In the pavement image space, which is a nonisotropic two-dimensional environment, it is necessary to select different values in different directions in order to select the appropriate generator. However, due to the volume of calculations, compression and extraction of a single index will be more efficient. While we can easily use the idea to extract some meaningful output, there are at least two logical approaches to selecting a value with a specific direction. The first and simplest method is to maximize all considered routes. This method gives a definite and accurate answer. Another method is to select a principal direction that can be obtained by the largest coefficient on the scale by OSSG. In other words, for each point in a two-dimensional plane, the absolute values of all coefficients associated with an OSSG section are calculated, and the maximum of these coefficients is determined. Then, the corresponding section direction is selected as the main direction. It should be noted that this method, due to the use of approximation and generalization in the whole image, reduces the accuracy of extracting an indirect computational law of contrast, but using a preselected process, high-contrast edges are installed with contrasting edges. In addition to computational time, this method has been successfully used to improve the extraction of edge and corner point localization. Because a discrete Shearlet method is not the same in all directions, High-Frequency Shearlet Coefficients (HFSC) areas have more elongation than Low-Frequency Shearlet Coefficients (LFSC) areas. For this purpose, changing the parameters j and k in the definition of Shearlet can be used as reference parameters for comparison. By limiting the number of eligible coefficients and directions, the direction of the number of different atoms is cut to the smallest scale. This means that for any path with predefined conditions, it may be necessary to examine more than one section at a higher scale. In this regard, the largest coefficient in the relevant scale is determined and evaluated.

A simplified example of this method is provided in the high-frequency relation in the vertical direction. The same definition can be used for other directions in the same way.

Definition of edge determination index based on two parameters OSSG and ESSG two-dimensional complex Shearlet: Assuming $\varphi^{even} \in L^2(R^2) \bigcap L^1(R^2)$ for the ESSG section with horizontal cones and $\psi^{Odd} \in L^2(R^2) \bigcap L^1(R^2)$ for OSSG section with horizontal cone, using Eq. (6.27)

$$C_{\psi^{Odd}} = \int_{-\infty}^{0} \int_{R} \psi^{Odd}(x, y)dydx \neq 0 \tag{6.27}$$

where $\|\psi^{Odd}\|L^1 = \|\psi^{even}\|L^1 = 1$ is the complex Shearlet.

Assuming the J_{min}, $j_{max} \in N$ and $J_{min} \leq j_{max}$, using Eq. (6.28)

$$\begin{aligned} \psi_{j,k,x}{}^{even} &= 2^{\frac{3j}{2}} \psi^{even}\left(S_k A_j(-x)\right), \\ \psi_{j,k,x}{}^{odd} &= 2^{\frac{3j}{2}} \psi^{odd}\left(S_k A_j(-x)\right) \end{aligned} \tag{6.28}$$

where $|k| \leq \left\lceil 2^{\frac{j}{2}} \right\rceil$, $j \in \{J_{min}, \ldots, J_{max}\}$ is the matrix and size of complex Shearlet.

For each $f \in L^2(R^2)$ and at one point, $x \in (R^2)$ using the law of direction $k^*_{\psi^{even}\psi^{odd}}(f, x) \in \left\{-\left\lceil 2^{\frac{j_{min}}{2}} \right\rceil, \ldots, \left\lceil 2^{\frac{j_{min}}{2}} \right\rceil\right\}$ is calculated using Eq. (6.29)

$$\begin{aligned} &k^*_{\psi^{even}\psi^{odd}}(f, x) = \operatorname{argmax}\left(\tilde{k} \in \left\{-2^{\frac{j_{min}}{2}}, \ldots, 2^{\frac{j_{min}}{2}}\right\}\right), \\ &\max\left(j \in \{J_{min}, \ldots, J_{max}\}, k \in \left\{-2^{\frac{j_{min}}{2}}, \ldots, 2^{\frac{j_{min}}{2}}\right\}, \left|\tilde{k} - k 2^{\frac{j_{min}}{2}} 2^{\frac{j}{2}-1}\right| \leq \frac{1}{2}\right), \\ &\left|f, \psi_{j,k,x}{}^{odd}\right| \end{aligned} \tag{6.29}$$

Also, the edge with an angle difference of less than 45° is sensitive to the vertical axis and is calculated from Eq. (6.30):

$$E_{\psi^{even}\psi^{odd}}(f, x) = \frac{\left|\sum_{j=J_{min}}^{J_{max}} \left\langle f, \psi_{j,k_j,x}{}^{odd} \right\rangle\right| - \sum_{j=J_{min}}^{J_{max}} \left\langle f, \psi_{j,k_j,x}{}^{even} \right\rangle}{(J_{max} - J_{min} + 1) \ \max_{j \in \{J_{min}, \ldots, J_{max}\}} \left|\left\langle f, \psi_{j,k_j,x}{}^{odd} \right\rangle\right| + \in}, \tag{6.30}$$

In this regard, to avoid ambiguity (zero denominator), the relationship with $\in > 0$, which is a small number, has been modified by Eq. (6.31):

$$k_j = \operatorname{argmax}\left(k \in \left\{-\left\lceil 2^{\frac{j_{min}}{2}} \right\rceil, \ldots, \left\lceil 2^{\frac{j_{min}}{2}} \right\rceil\right\}\right), \left|k^*_{\psi^{even}\psi^{odd}}(f, x) - k\left\lceil 2^{\frac{j_{min}}{2}} \right\rceil \left\lceil 2^{\frac{j}{2}} \right\rceil^{-1}\right| \leq 0.5 \tag{6.31}$$

In order to ensure that the index is in the range between 0 and 1, the relationship has been modified as Eq. (6.32):

$$E_{\psi^{even}\psi^{odd}}(f, x) = \max\left\{\tilde{E}_{\psi^{even}\psi^{odd}}(f, x), 0\right\} \tag{6.32}$$

According to this equation, if $E_{\psi^{even}\psi^{odd}}(f, x)$ for the function $f \in L^2(R^2)$ and $x \in (R^2)$ is close to 1, the value of $k^*_{\psi^{even}\psi^{odd}}(f, x)$ can be used to approximately change the angle of the edges at the point *x*. An example of the results obtained using this indicator is shown in Figure 6.12. The ESSG and OSSG shells related to the symmetric coefficients of the even size function as well as the even and odd wave function in a one-dimensional environment have already been investigated. The wavelet method has operational limitations, and the results are local. This method usually works with high-pass and symmetrical filters. As shown in the Figure 6.12, soft thresholding for smoothing and noise reduction can be done using this method using Eq. (6.33):

$$\widetilde{E}_{\psi^{even}\psi^{odd}}(f,x) = \frac{\left|\sum_{j=J_{min}}^{J_{max}} \left\langle f, \psi_{j,k_j,x}^{odd} \right\rangle\right| - \sum_{j=J_{min}}^{J_{max}} \left|f, \psi_{j,k_j,x}^{even}\right| - (J_{max} - J_{min} + 1)T}{(J_{max} - J_{min} + 1)\ \max_{j\in\{J_{min},\dots,J_{max}\}} \left|\left\langle f, \psi_{j,k_j,x}^{odd}\right\rangle\right| + \epsilon} \tag{6.33}$$

For a gray image with a range between (black) 0–255 (white), the *T* parameter is assumed to be $2\left|C_{\psi^{odd}}\right| \approx 0.466$. One of the advantages of this method is the possibility of identifying subjects and separating them without losing local information at the border between the two subjects. An example of this analysis for topics that have ambiguity at the common boundary is shown in Figure 6.17. This method is suitable for analyzing crack boundaries that are fuzzy in nature because it retains the original information about the crack itself, which is generally fuzzy, and the suffering of changing coefficients from the center of the crack to the edges is reduced. Comparatively, the methods of edge or multilevel detection always result in a high percentage of information and data deletion. Local information is preserved using the Shearlet method, and better results in interpretation can be obtained using more accurate methods.

6.4.4 Object Detection and Isolation Using the Shearlet Coefficient Feature (SCF)

The two main steps in implementing and using Shearlet include multilevel analysis and then multidirectional analysis. Different separation steps are performed using the appropriate thresholding operation. Low frequency is used to isolate noise and improve quality. High-frequency imaging can be used to check the main features. For example, in road pavement images, the cracking coefficients are different in different directions, while the Shearlet coefficient for noise is the same in all directions. Therefore, by using the variance characteristic of the Shearlet coefficient variance discriminator (SCVD), objects related to pavement damage can be separated from noise. Using the Eq. (6.34), the pavement damage object is separated from the noise and background:

$$S'_{x,y}(L(l),j) = \begin{cases} S_{x,y}(L(l),j) \text{ if } \sigma_{j,x,y} > T \\ 0 \qquad\qquad\qquad \text{else} \end{cases},$$
$$\sigma_{j,x,y} = \frac{1}{L_j}\sum_{l=1}^{L_j}\left(S_{x,y}(L(l),j) - \text{avg}(j,x,y)\right)^2, \tag{6.34}$$
$$\text{avg}(j,x,y) = \frac{1}{L_j}\sum_{l=1}^{L_j}\left(S_{x,y}(L(l),j)\right)$$

where $S'_{x,y}(L(l),j)$ is the Shearlet in the in the x, y coordinate with l and j. $\sigma_{j,x,y}$ is the SCVD in the pixel (x,y).

Various researches have been done in order to extract and isolate objects through the properties of Shearlet coefficients. Using binary coefficients, the image of the desired object can be extracted from the image, such as visual damage, to infrastructure, including pavement cracking, bitumen, peeling, etc. This separation is calculated using a threshold by Eq. (6.35):

$$Sb'_{x,y}(L(l),j) = \begin{cases} 1 & if\ abs\left(\sigma_{j,x,y}\right) > Ts \\ 0 & else \end{cases} \tag{6.35}$$

where $S'_{x,y}(L(l),j)$ is the binary value of the Shearlet coefficient in the x, y coordinates in the size j and l of the direction; $\sigma_{j,\ x,\ y}$ is the value of SCVD in pixels (x, y); and Ts is the value of the experimental threshold.

As mentioned in the previous section, Shearlet-based methods operate on a threshold basis, and the selection of the optimal threshold plays an important role in the success of this method. At the level of available information, all available research and methods have been used for the fixed threshold experimentally. The features extracted from this method are highly dependent on the light conditions and image quality, which is one of the disadvantages of this method (Figure 6.13).

6.5 Polygonal-Based Features (PBF)

This feature is designed to draw the overall shape of an object. The polygon approximation can be adjusted to eliminate minor changes along the edge and extract the overall shape of the image. This is useful because it eliminates the destructive effect of noise and reduces the contour. In general, there are two ways to understand PBF: merging and splitting. In this chapter, new and advanced topics regarding feature extraction are covered, while basic principles of the topics are available in various references.

6.6 Spatial Interrelation Feature (SIF)

This method describes the features that relate the area or line of an object to its pixels or curves. In general, the display is created using an object's geometric features, including length, curvature, orientation and position, area, distance and center of the surface, among others (see references for more information).

6.7 Moments Features (MFE)

These features are described in detail in Section 6.5 for selecting the fault detection feature.

6.8 Scale Space Approaches for Feature Extraction (SSA)

This feature is based on the principles of object simplification, whereby a shape becomes simpler in each step. See references for more information.

6.9 Shape Transform Features (STF)

6.9.1 Radon Transform Features (RTF)

The patterns obtained from an image due to polar rotation around the center are called Radon transitions. As per its definition, this conversion creates a new pattern of meaning by stacking at different angles. By thresholding on this pattern, the most important use of this method in

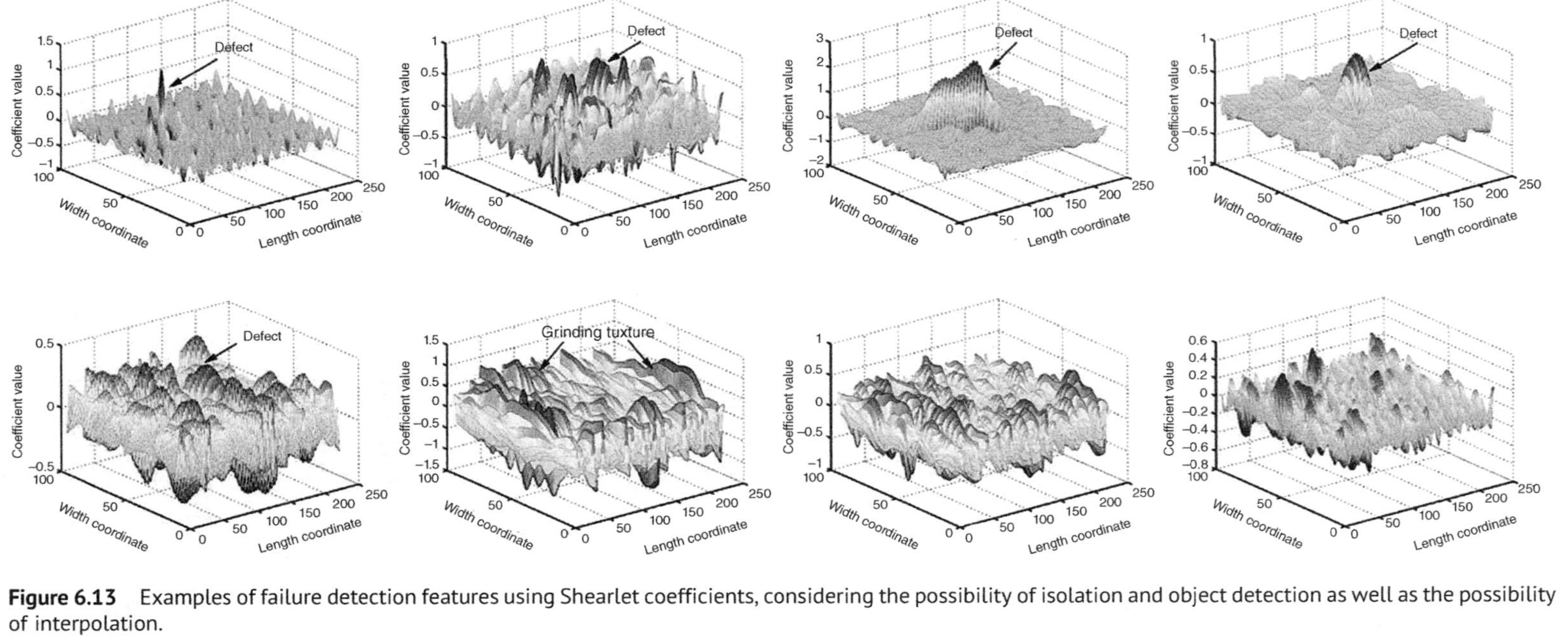

Figure 6.13 Examples of failure detection features using Shearlet coefficients, considering the possibility of isolation and object detection as well as the possibility of interpolation.

medical engineering is for three-dimensional imaging using a coat scanner. This method is used in image processing to recognize features, such as edges, determine line parameters, and detect direction. The parameter θ is the angle between the principal and the transmitted components, which is called the transfer angle. The Radon transfer $R_\theta(x')$ is a function of $f(x, y)$, which is calculated from the linear sum $f(x, y)$ along the y' axis using Eq. (6.36):

$$\begin{aligned} R_\theta(x') &= \int_{-\infty}^{+\infty} f(x,y)dy', \\ R(p,\tau) &= \int_{-\infty}^{+\infty} \mathrm{f}(\mathrm{x}, \mathrm{px} + \tau)\mathrm{dx} \end{aligned} \tag{6.36}$$

where R(p, τ) is the Radon transform for the two-dimensional function f(x, y).

The transform is calculated from the sum of the values of f along the line slope. The location of the line is determined using the slope parameter p and the output τ by using Eq. (6.37):

$$R(p,\tau) = \int_{-\infty}^{+\infty}\int_{-\infty}^{+\infty} f(x,y)\delta(y - px - \tau)\mathrm{d}x\mathrm{d}y \tag{6.37}$$

The rotational components in the polar domain is determined using Eq. (6.38)

$$\begin{bmatrix} x \\ y \end{bmatrix} = \begin{bmatrix} \cos\theta - & \sin\theta \\ \sin\theta & \cos\theta \end{bmatrix} \begin{bmatrix} x' \\ y' \end{bmatrix} \tag{6.38}$$

the function then transformed using Radon transfer (RT). The RT is obtained using the function $f(x, y)$, which is calculated by the following Eq. (6.39):

$$\begin{aligned} R_\theta(x') &= \int_{-\infty}^{+\infty} f(x'\cos\theta - y'\sin\theta, x'\sin\theta + y'\cos\theta)dy' \\ R(\rho,\theta) &= \int_{-\infty}^{+\infty}\int_{-\infty}^{+\infty} \mathrm{f}(\mathrm{x},\mathrm{y})\delta(\rho - \mathrm{x}\cos\theta - \mathrm{y}\sin\theta)\mathrm{dxdy} \end{aligned} \tag{6.39}$$

In the Eq. (6.39), δ is a function of the Dirac delta. The larger the size of an image in one direction with active pixels within the specified threshold, the higher the Radon value. In the Radon method, the angle varies from 0 to 180. Radon transfer is important because of its linear role in integrating image intensity in all directions. A line in the image is converted into a max or min point in the Radon domain (RD). As displayed in (6.14) the samples, the objects often have a specific direction (such as cracks in the pavement surface). Therefore, when placed in the dominant direction, the size (crack intensity) is added together and converted into a max point in the RD environment by the RT Radon converter. In other words, a max point exists if there is an object in the Radon domain. For a more accurate analysis, the third dimension, which is the value of the integral for each (ρ, θ), can be used. Examples of this transform for different patterns are shown in the Figure 6.14.

The properties of Radon transform can be divided into linear, translation, scaling, and point and line representation based on their relations.

6.9.2 Linear Radon Transform

Accordingly, Radon transform is the weighted sum of the functions, which is equivalent to the weighted sum of each of the functions converted using the Radon conversion separately. This feature plays an important role in transmitting information about objects (such as cracks):

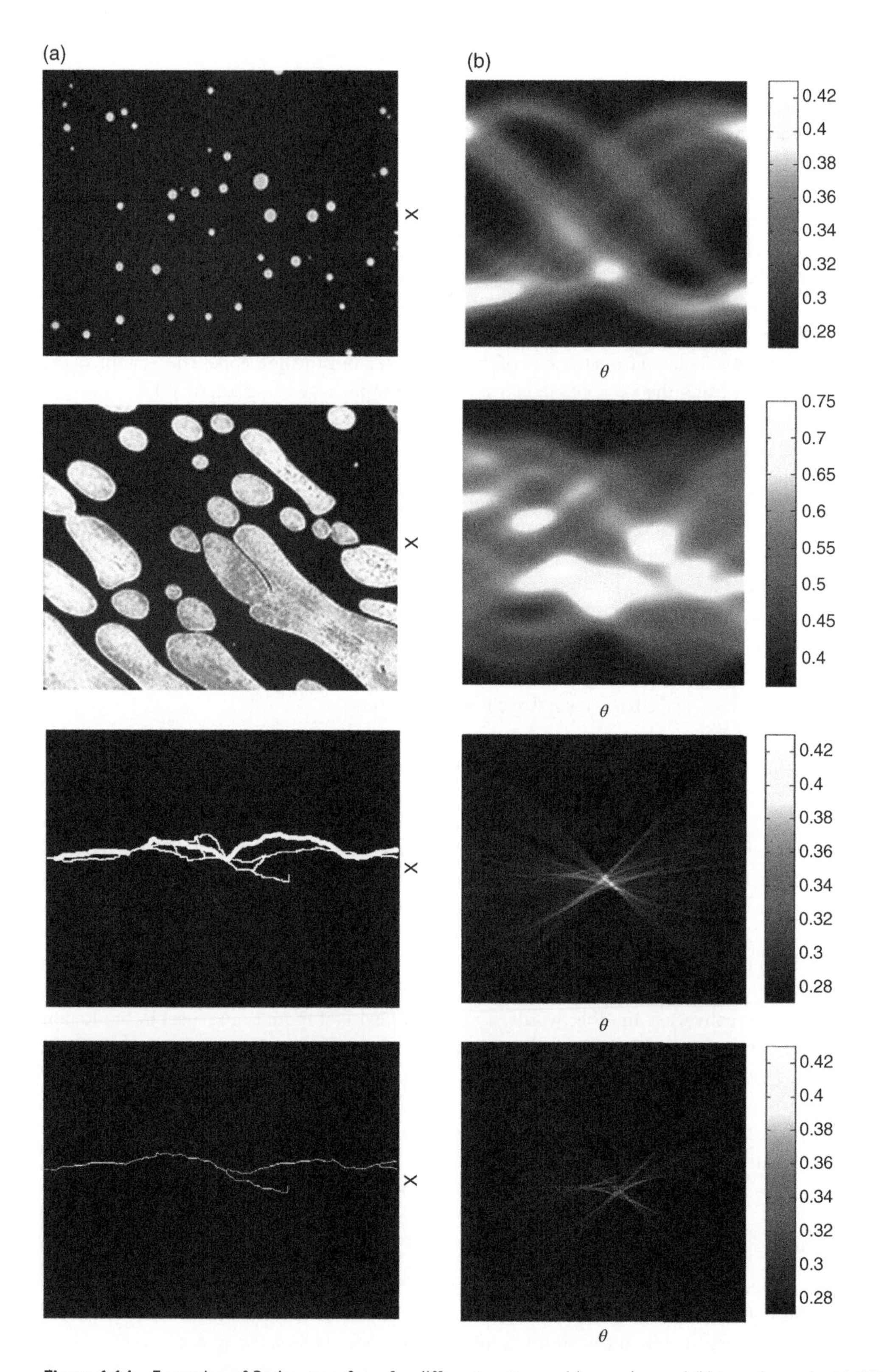

Figure 6.14 Examples of Radon transform for different patterns. (a) samples and (b) transformed with RT.

$$f(x,y) = \sum_i w_i g_i(x,y),$$
$$R(p,\tau) = \sum_i w_i \int_{-\infty}^{+\infty}\int_{-\infty}^{+\infty} g_i(x,y)\delta(y - px - \tau)dxdy = \sum_i w_i g_i(p,\tau) \tag{6.40}$$

in the Eq.(6.40), the R(p, τ) is the Radon transform for the two-dimensional function f(x, y), and w_i is the weighted sum of the functions.

6.9.3 Translation of RT

Another important feature of Radon transform is the translation property, which allows the location of a feature to be detected without changing the original properties of Radon. In terms of geometric properties, the results are fixed, and only the location changes. The slope of the line in this feature does not change with the transition, and only the position changes. Using this feature, the main characteristics of a crack for different situations can be evaluated using Eq. (6.41):

$$f(x,y) = g(x - x^*, y - y^*) \rightarrow R(p,\tau),$$
$$R(p,\tau) = \int_{-\infty}^{+\infty} g(x - x^*, px + \tau - y^*)\, dx = \int_{-\infty}^{+\infty} g(\tilde{x}, p(x + x^*) + \tau - y^*)d\tilde{x} = \tilde{g}(p, \tau - y^* + px^*) \tag{6.41}$$

6.9.4 Scaling of RT

The scaling feature can be defined as follows:

$$f(x,y) = g\left(\frac{x}{a}, \frac{y}{b}\right) \rightarrow$$
$$R(p,\tau) = \int_{-\infty}^{+\infty} g(x/x^*, px + \tau/b)\, dx = a\int_{-\infty}^{+\infty} g(\tilde{x}, pa(\tilde{x})) + \tau/b)d\tilde{x} = \left(\tilde{x} = \frac{x}{a}\right) = a\tilde{g}(pa/b, \tau/b) \tag{6.42}$$

Based on the Eq. (6.42), the properties of objects, such as lines and patterns, can be extracted. The position of the line as well as the degree of deviation from the origin can be seen in these relationships. Based on these relationships, each slope is scaled by a ratio as well as a Radon transform. This feature is used to determine the type of distress and measure its severity and extent.

6.9.5 Point and Line Transform Using RT

This property is modeled by multiplying two delta functions, as follows in Eq. (6.43):

$$f(x,y) = \delta(x)\delta(y) \rightarrow$$
$$R(p,\tau) = \int_{-\infty}^{+\infty} \delta(x)\delta(px + \tau)dx = \delta(\tau).$$
$$f(x,y) = \delta(x - x^*)\delta(y - y^*) \rightarrow \tilde{f}(p,\tau) = \delta(\tau - y^* + px)$$
$$R(p,\tau) = \int_{-\infty}^{+\infty}\int_{-\infty}^{+\infty} g(x^*, y^*)\delta(x - x^*)\delta(y - y^*)dx^*dy^* = .$$

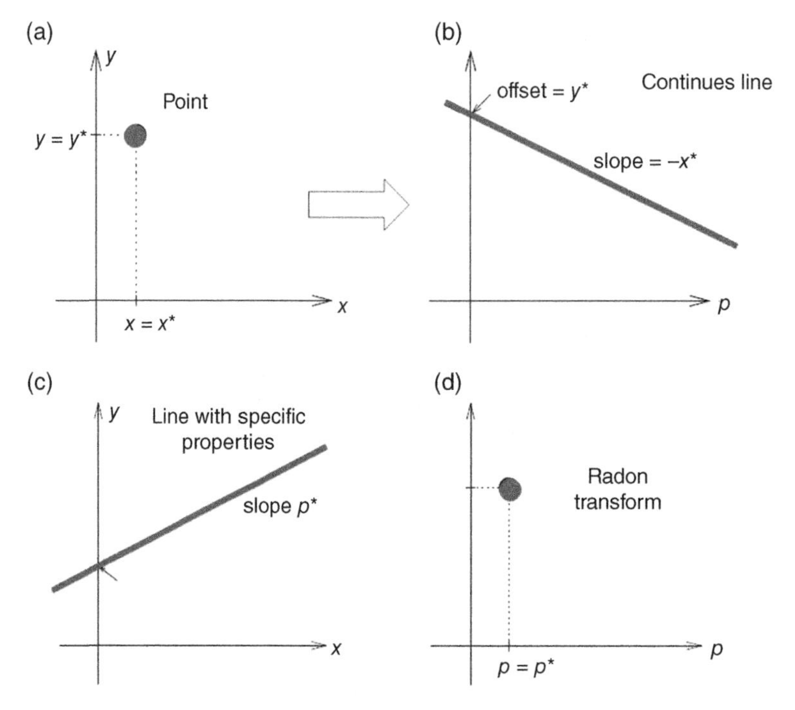

Figure 6.15 The feature extracted from different patterns using Radon transform: (a) two-dimensional function as a point $(x, y) = (x*, y*)$, (b) Radon conversion representation (result of overlapping data on a specified slope), (c) nonzero two-dimensional function as a line, and (d) conversion to a peak point using Radon conversion.

$$
\begin{aligned}
&= \int_{-\infty}^{+\infty}\int_{-\infty}^{+\infty}\int_{-\infty}^{+\infty} g(x^*, y^*)\delta(x - x^*)\delta(\tau + px - y^*)dx^*dy^*dx \\
&= \int_{-\infty}^{+\infty}\int_{-\infty}^{+\infty} g(x^*, y^*)\delta(y^* - \tau - px^*)dx^*dy^*
\end{aligned}
\tag{6.43}
$$

According to the Eq. (6.43), each point becomes a continuous line with a slope and location parameter. This relationship for a point is shown in the Figure 6.15.

Figure 6.16 displays the output results of two-dimensional Radon transform operations for a point schematically as a continuous line with specifications.

One of the most practical features of Radon transform is the transformed lines and the identification of its important properties. Each line is modeled using Eq. (6.44):

$$
\begin{aligned}
&f(x, y) = \delta(y - p^*x - \tau), \\
&R(p, \tau) = \int_{-\infty}^{+\infty}\int_{-\infty}^{+\infty} \delta(y - p^*x - \tau)\,\delta(y - px - \tau)dxdy = \int_{-\infty}^{+\infty} \delta((p - p^*)x + \tau - \tau^*)\,dx \\
&= \begin{cases} \dfrac{1}{|p - p^*|} & \text{for } p \neq p^* \\ 0 \text{ for } p = p^* \text{ and } \tau \neq \tau^* \\ \displaystyle\int_{-\infty}^{+\infty} \delta(0)\,dxd \quad \text{for } p \neq p^* \text{ and } \tau = \tau^* \end{cases}
\end{aligned}
\tag{6.44}
$$

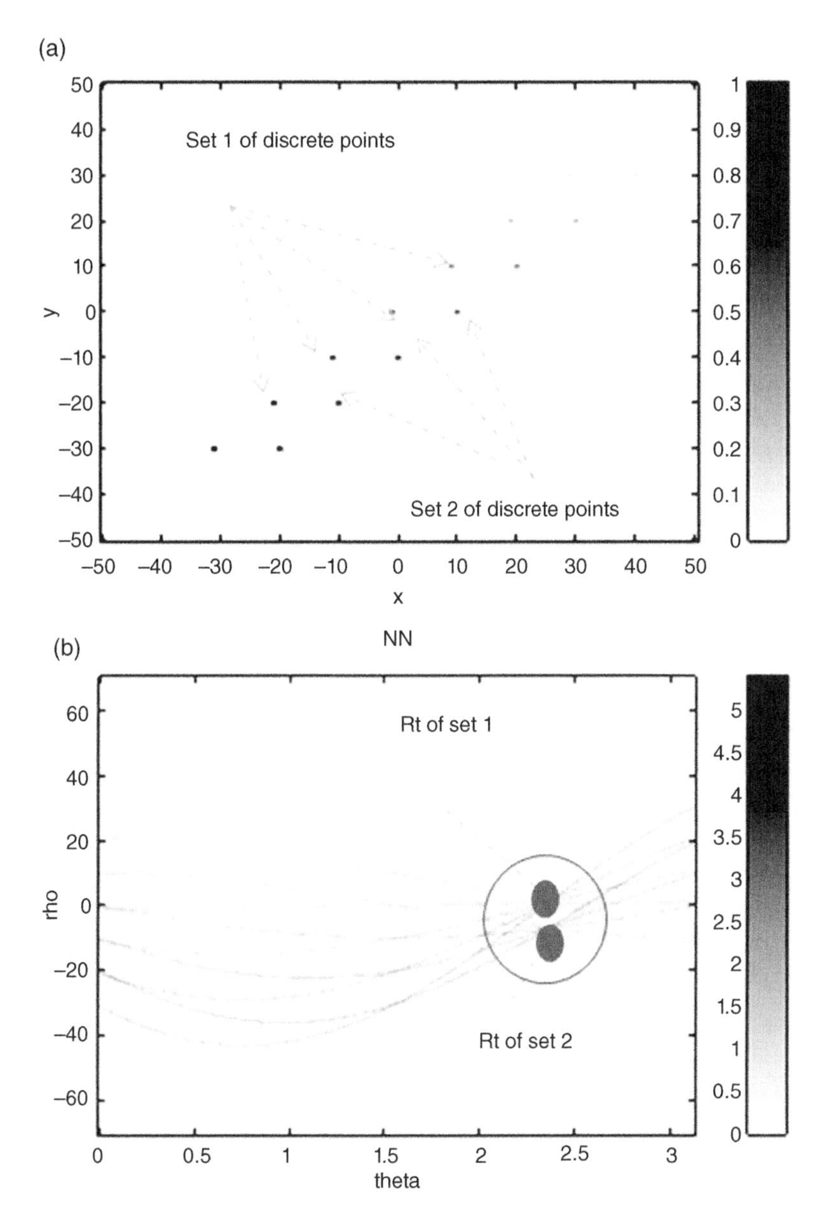

Figure 6.16 (a) Two-dimensional function of a discrete set of points in a specific direction and different locations at 14 points $(x, y) = (x^*, y^*)$; (b) Normal Radon conversion for two categories and overlapping data at two peak points.

According to the above equation, the Radon conversion for a line creates a local maximum point that has basic characteristics, such as location and slope. This feature is the basis for identifying the main parameters and its characteristics. In general, it can be concluded that a line becomes a point in a Radon environment. This feature can be generalized to a crack, where each crack in the Radon environment becomes a point with a variable boundary. Using this method, the main features of the object can be interpreted more easily based on the desired pattern.

6.9.6 RT in Sparse Objects

According to the main characteristics of Radon transform, Radon conversion for points is defined as follows using Eq. (6.45):

$$f(x,y) = \delta(x-x^*)\delta(y-y^*) \rightarrow ,$$
$$R(\rho,\theta) = \int_{-\infty}^{+\infty}\int_{-\infty}^{+\infty} \delta(x-x^*)\delta(y-y^*)\delta(\rho - x\cos\theta - y\sin\theta)dxdy$$
$$= \delta(\rho - x^*\cos\theta - y^*\sin\theta). \tag{6.45}$$

In Figure 6.17, the reference point shows the normal Radon transform, from which two types of shapes can be extracted. Wavy lines are created for points, while maximum points indicate lines. Given that the goal is to detect lines that sometimes appear discontinuous and even a set of hybrid points (continuous and discontinuous), the principle of Radon conversion and accumulation is a good tool for analyzing and compressing information in these patterns. In fact, in most infrastructure issues, objects appear discretely in images, making it difficult to create continuity. On the other hand, Radon capabilities make it possible to extract features and evaluation based on general characteristics.

6.9.7 Point and Line in RT

In order to gain information using Radon transform, the main properties of the object are obtained using the transfer of the primary element of the active image component or pixel. For example, the coordinates of an active pixel in an image are assumed to be equal to (x ^ ∗, y ^ ∗), which is derived from the following equations:

$$f(x,y) = \int_{-\infty}^{+\infty}\int_{-\infty}^{+\infty} f(x^*,y^*)\delta(y)\delta(x-x^*)\delta(y-y^*)dx^*dy^* \rightarrow ,$$
$$R(\rho,\theta) = \int_{-\infty}^{+\infty}\int_{-\infty}^{+\infty} f(x^*,y^*)\delta(\rho - x^*\cos\theta - y^*\sin\theta)dx^*dy^*,$$
$$R(\rho,\theta) = \int_{-\infty}^{+\infty}\int_{-\infty}^{+\infty} f(x^*,y^*)\delta(\rho - \rho^*\cos(\theta-\theta^*))dx^*dy^*. \tag{6.46}$$

In Eq. (6.46), $\rho^*\cos(\theta^*) = x^*$and $\rho^*\sin(\theta^*) = y^*$. These relationships show that points are converted into a sine wave using a Radon converter. Using this conversion, points can be interpreted using certain parameters. From the above equations compared with the basic Radon conversion equation, the property of accumulation in a specific direction for Radon transfer can be extracted using Eq. (6.47):

$$f(x,y) = 0 \text{ for } \sqrt{x^2+y^2} > \rho_{max} \rightarrow R(\rho,\theta) = 0 \text{ for } |\rho| > \rho_{max}. \tag{6.47}$$

With the generalization of Radon conversion for the point, it is possible to evaluate and detect line and patterns. Using the parameters (ρ^*, θ^*) and the delta function, the lines are modeled by Radon conversion as follows Eq. (6.48):

$$f(x,y) = \delta(\rho - x^*\cos\theta - y^*\sin\theta) \rightarrow ,$$
$$R(\rho,\theta) = \int_{-\infty}^{+\infty} \delta(\rho^* - (\rho\cos\theta - s\sin\theta)\cos\theta^* - (\rho\sin\theta - s\cos\theta)\sin\theta^*)\, ds,$$

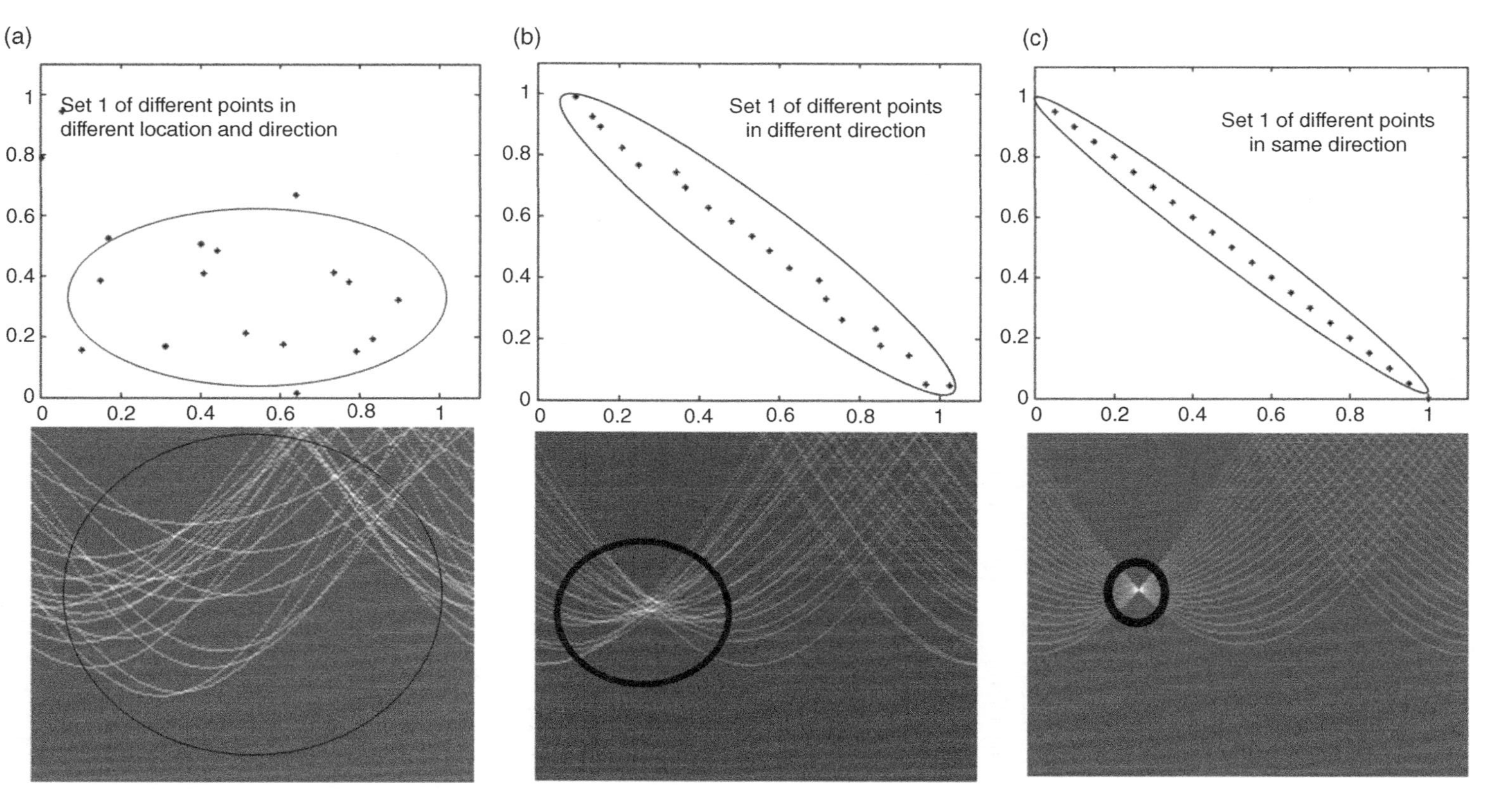

Figure 6.17 (a) Radon transform of points in a specific direction for a set of 20 points with turbulence; (b) Radon transform of points in a specific direction for a set of 20 points with deviation; and (c) Radon transform of points in a specific direction for a set of 20 points in one direction without deviation.

$$= \int_{-\infty}^{+\infty} \delta(\rho^* - \rho \cos(\theta - \theta^*) - (s \sin(\theta - \theta^*)))ds = \frac{1}{|\sin(\theta - \theta^*|},$$

$$if\ \theta - \theta^*\ then\ R(\rho, \theta) = \int_{-\infty}^{+\infty} \delta(\rho^* - \rho))\ ds =)\begin{cases} 0, & \text{if } \rho \neq \rho^* \\ \int_{-\infty}^{+\infty} \delta(0)\, ds\, \rho = \rho^* \end{cases}. \tag{6.48}$$

If $\rho = \rho^*$and $\theta = \theta^*$, then a peak point is created. Using this method, there is no limit to the Radon transform method in any direction of the line in the data overlap method.

The greater the number of points in one direction, the higher the cumulative size. If the direction of all points is in the same direction, the value of RT at that angle increases relative to the total and takes on more weight. If the active points are not in the same direction or in the same range, depending on the angle of rotation, the value of RT is distributed and becomes an obscure amplitude. By selecting the appropriate threshold, important indicators of this distribution and coefficient can be extracted, which indicates the degree of deviation from the center of the linear pattern and is considered a kind of turbulence. In order to compare this relationship, the following figure reduces the maximum Radon size by increasing the amount of turbulence in the active pixels and decreases the Radon transform factor by the same amount.

Complete linear models with low degree of turbulence are very rare in nature. In the case of infrastructure assessment (types of damage such as cracking) that are random in nature due to the nature of discontinuity, existing Radon models are associated with error detection and evaluation. This feature is called Wiggle patterns in morphological analysis models. In order to consider this feature, each image is modeled using Eq. (6.49):

$$f(m, n) = \delta(n - [\alpha^* m + \beta^* + \lambda]), \lambda \in \aleph(0, \sigma^2) \tag{6.49}$$

where $\delta(.)$ is the Kronecker delta function.

Based on the Gaussian noise distribution, the λ parameter is used along the line. This parameter indicates the amount of deviation from the specified angle. In this relation, α^* and β^* are fixed parameters that indicate the slope and starting point. In these relations, it is important to determine the coordinates of the maximum Radon conversion size and the morphological shape of the optimal value. Using these characteristics, the amount of Wiggle can be obtained relative to the regular state without disturbance Eq. (6.50):

$$P\{f(m, n)\} = P(n = [\alpha^* m + \beta^* + \lambda]) = P(0 = [\alpha^* m + \beta^* + \lambda]) = P\left(-\frac{1}{2} < n - [\alpha^* m + \beta^* + \lambda] < 1/2\right) \tag{6.50}$$

where (m, n) is the the coordinates of the point.In this equation, a point with coordinates (m, n) is converted into $[n = [\ \alpha m + \beta]$using Radon transform, which with increasing Wiggle rate, becomes

$$P\{f(m, n) = 1\} = P\left(-\frac{1}{2} < [\alpha m + \beta] - [\alpha^* m - \beta^* - \lambda] < 1/2\right)$$

$$P\{f(m, n) = 1\} = P\left(-\frac{1}{2} < \alpha m + \beta + \omega - \alpha^* m - \lambda < 1/2\right)$$

$$P\{f(m, n) = 1\} = P\left(-\frac{1}{2} < \zeta + \omega - \lambda < 1/2\right), \text{where } \zeta = (\alpha - \alpha^*)m + \beta - \beta^*$$

$$= \Phi\left(\frac{\frac{1}{2} + \zeta + \omega}{\sigma}\right) - \Phi\left(\frac{-\frac{1}{2} + \zeta + \omega}{\sigma}\right)$$

$$\Phi(x) \approx \frac{1}{2}\left(1 + \tanh\left(\frac{x}{\gamma}\right)\right) \text{ where } \gamma = \sqrt{\frac{\pi}{2}}$$

$$\int \Phi(x)dx \approx \frac{\gamma}{2}\left(\frac{1+x}{\gamma} + \log\cosh\left(\frac{x}{\gamma}\right)\right), \tag{6.51}$$

where according to Eq. (6.51) the (m, n) is the the coordinates of the point; $\Phi\,(.)$ is a function of the Gaussian distribution; ζ is the line displacement with the parameters $\alpha^*$and β^*; and P the linear pattern with the parameters α and β has a wiggle Figure 6.19.

The probability distribution function is approximated using the above equation. An example of wiggle changes and the effect on the Radon coefficient is shown in Figures 6.19–6.20. Research suggests that creating a wiggle in the pattern can reduce the amount of Radon by up to 50%.

The Radon coefficient is proportional to the creep rate of the linear pattern, where the higher the wiggle rate, the lower the maximum Radon coefficient. But in real life, the amount of noise in the image is an important factor in achieving practical results. Therefore, combining two factors of ambiguity in the image produces different results. These two factors cause the first and second types of error, the effects of which are shown in Figures 6.18 and 6.19, respectively.

In order to extract, the characteristics of linear patterns, such as cracking, increasing the amount of noise to the image, which is generally part of the nature of pavement images, cause changes in the amount of Radon Transform Maximum Value (RTMV) and cracking parameters. Choosing the right threshold is difficult in this case, and thus, it is necessary to use a method that retains useful information about cracking while reducing the impact of noise on the pavement surface.

Figure 6.22 illustrates the effects of different thresholds in separating meaningful coefficients for a line. This analysis shows that the optimal threshold can retain line information in addition to eliminating noise. At this stage, it is important to find the location of data with coefficients greater than the threshold. The morphological properties of the set of coefficients above the threshold are also used as optimal properties. For this purpose, Eq. (6.52) is used to estimate the location parameters of the peak point:

$$P_{\text{det all}} \cong \prod_{i=2}^{L} P_{\text{det }2} \cong \left(\frac{1}{2}\left(1 + \left(\operatorname{erf}\left(\frac{\lambda}{2}\right)\right)^2\right)\right)^{L-1} \cong 1 - \frac{2L}{\lambda\sqrt{\pi}} e^{-\frac{\lambda^2}{4}} \tag{6.52}$$

In the last sentence of approximation, if last sentence of approximation is close to 1, the probability of line detection is high. Using this method, membership functions related to the presence of cracks or linear patterns in the image can be extracted. In this relation, L is the deformation parameter of the estimation function. If the goal of high probability of detection is required, the value of L in the equation should be kept low. An example of these functions is shown in Figure 6.22.

6.10 Various Case-Based Examples in Infrastructures Management

6.10.1 Case 1: Feature Extraction from Polypropylene Modified Bitumen Optical Microscopy Images

This section presents a practical example using semiautomatic and automatic methods of image features extracted from modified polypropylene bitumen (PMB). The main purpose of this example is to apply feature extraction to PMB interpretation. In this application, an image-based system was developed for interpretation. Statistical criteria include 10 features based on a combination of the

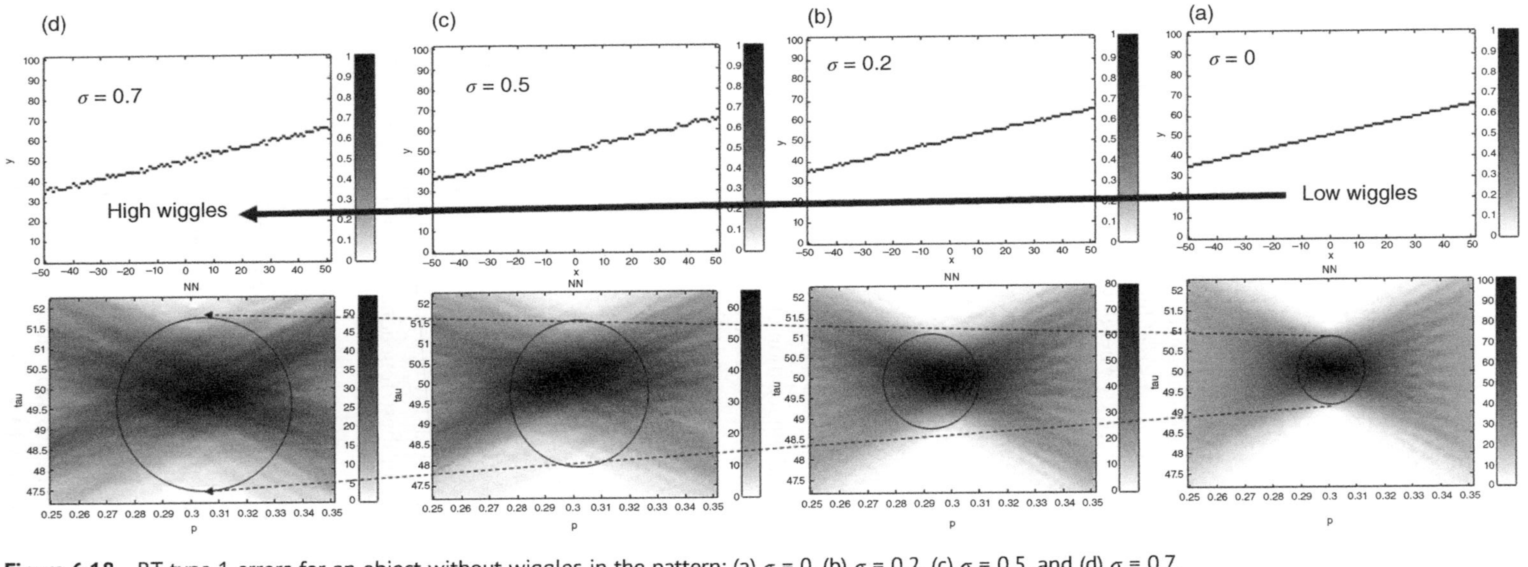

Figure 6.18 RT type 1 errors for an object without wiggles in the pattern: (a) $\sigma = 0$, (b) $\sigma = 0.2$, (c) $\sigma = 0.5$, and (d) $\sigma = 0.7$.

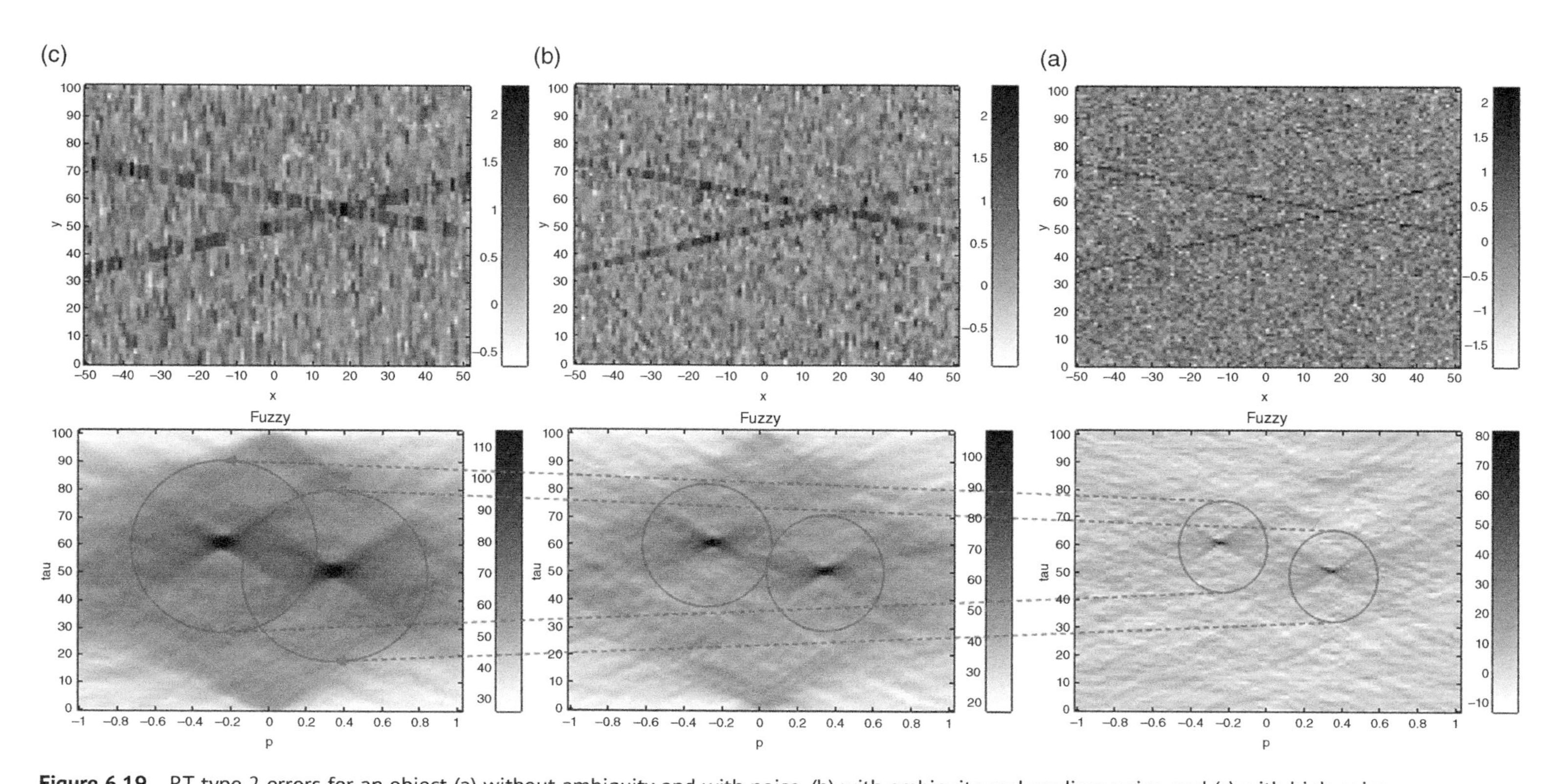

Figure 6.19 RT type 2 errors for an object (a) without ambiguity and with noise, (b) with ambiguity and medium noise, and (c) with high noise.

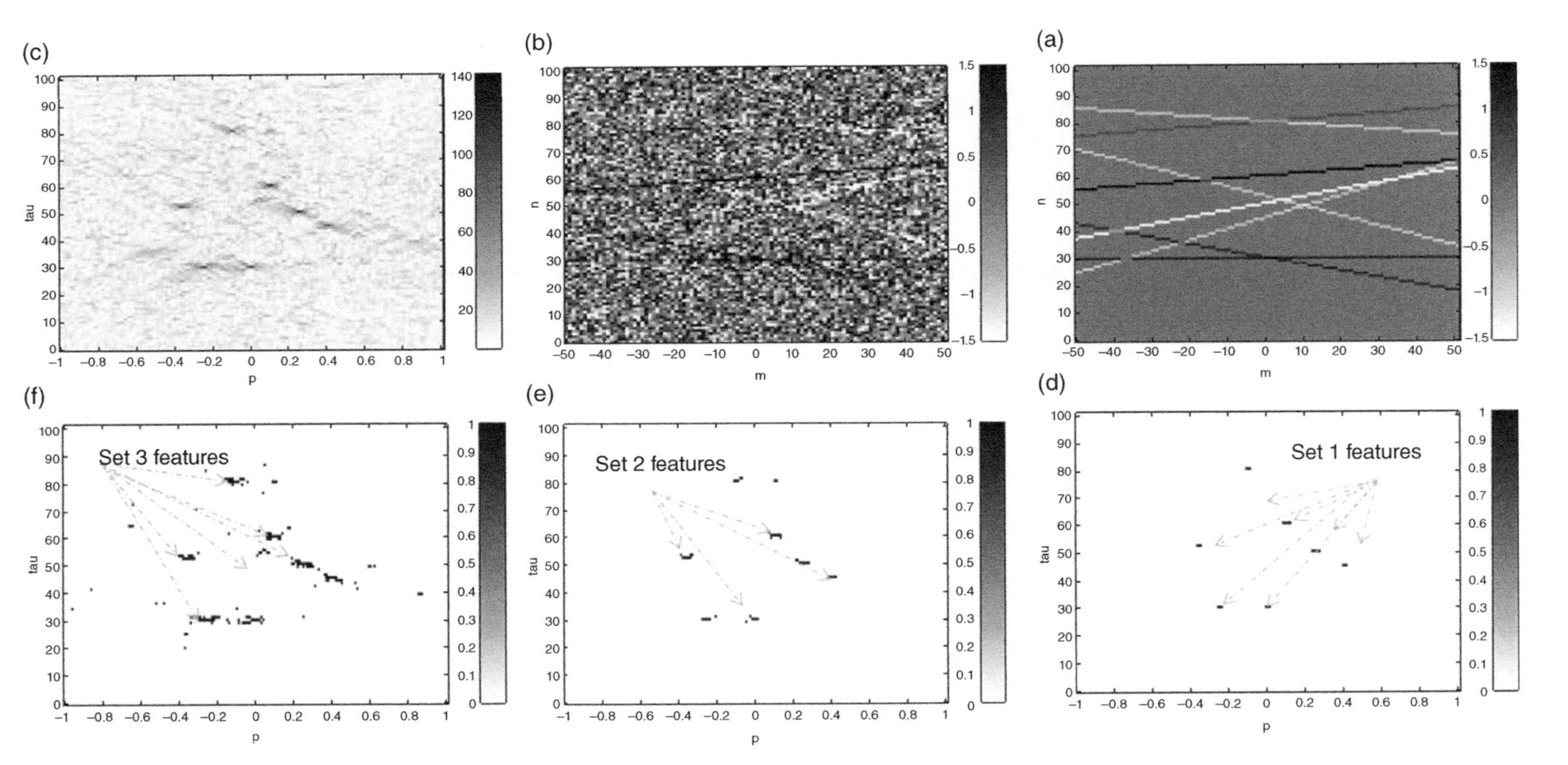

Figure 6.20 Example of static thresholding to extract the characteristics of RTMV linear patterns: (a) with different lines, (b) with increased type 2 noise, (c) with Radon conversion coefficients and noise, (d) with constant threshold equal to 0.2, (e) with fixed threshold equal to 0.5, and (f) with fixed threshold equal to 0.7.

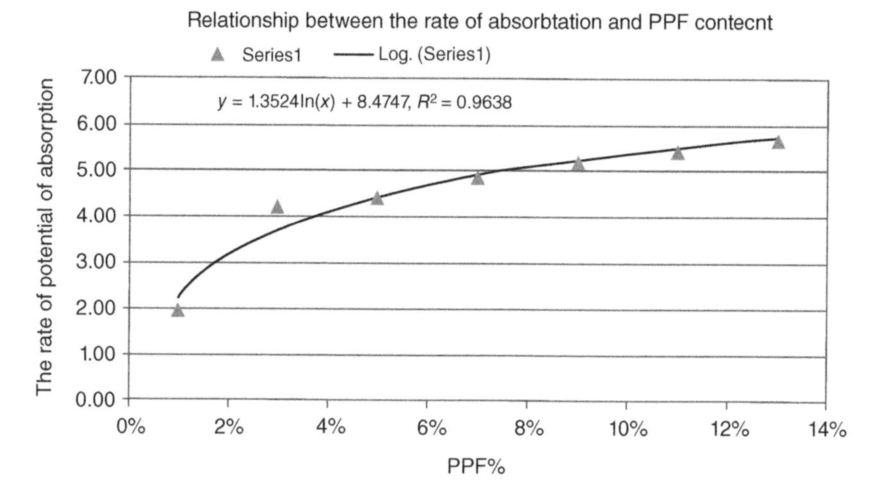

Figure 6.21 Relationship between polymer content and potential of absorption (%) of PPF-modified bitumen samples $y = 1.3524\ln(x) + 8.4747$, $R^2 = 0.9638$.

fuzzy segmentation method (FCM) and Radon transform extracted for polypropylene properties. The most important of which include high-amplitude Radon (HARP), high-energy Radon (HERP), and standard deviation (STD). Morphological features (number, position, area, value, etc.) of images taken by light microscopy were used to capture images of thin films of polypropylene fiber modified bitumen (PFMB) samples magnified on a scale of 100×. While these features can be used for other applications, it is necessary to explain that each image or application requires the selection of a specific feature to the same database. In the following, 10 features extracted from PMB images are presented as examples, 5 features from the binary image, and the other 5 from the 3D RT (Figure 6.21). These features are described in detail below (further information can be found in reference [30]).

6.10.2 Ratio of Number of Black Pixels to the Number of Total Pixels (RBT)

In the first stage, the image converted into binary form. Then, the number of pixels with a value of 1 (i.e. black pixels) was counted (indicated by IBb), and the ratio of the number of black pixels to the total number of pixels was calculated as the first feature. This morphologically based feature is simple to calculate and has many applications in similar problems using Eq. (6.53):

$$F1 = RBT = \left| \frac{count\left(\sum_{i=1}^{m}\sum_{j=1}^{n} p(i,j) > 0\right)}{count\sum_{i=1}^{m}\sum_{j=1}^{n}(p(i,j) \geq 0)} \right| * 100 \tag{6.53}$$

As can be seen in Figure 6.30, initially small polypropylene fibers (PPF) content increased slightly, such as from 1 to 2% by RBT, while initially greater PPF rapidly increased, such as from 13% to more than 70%, by RBT. With respect to RBT, it can be inferred that PPF can absorb two times its weight of the bitumen components for 1%, up to four times its weight of the bitumen components for 3%, 5%, and 7% and up to five times its weight of the bitumen components for 9%, 11%,

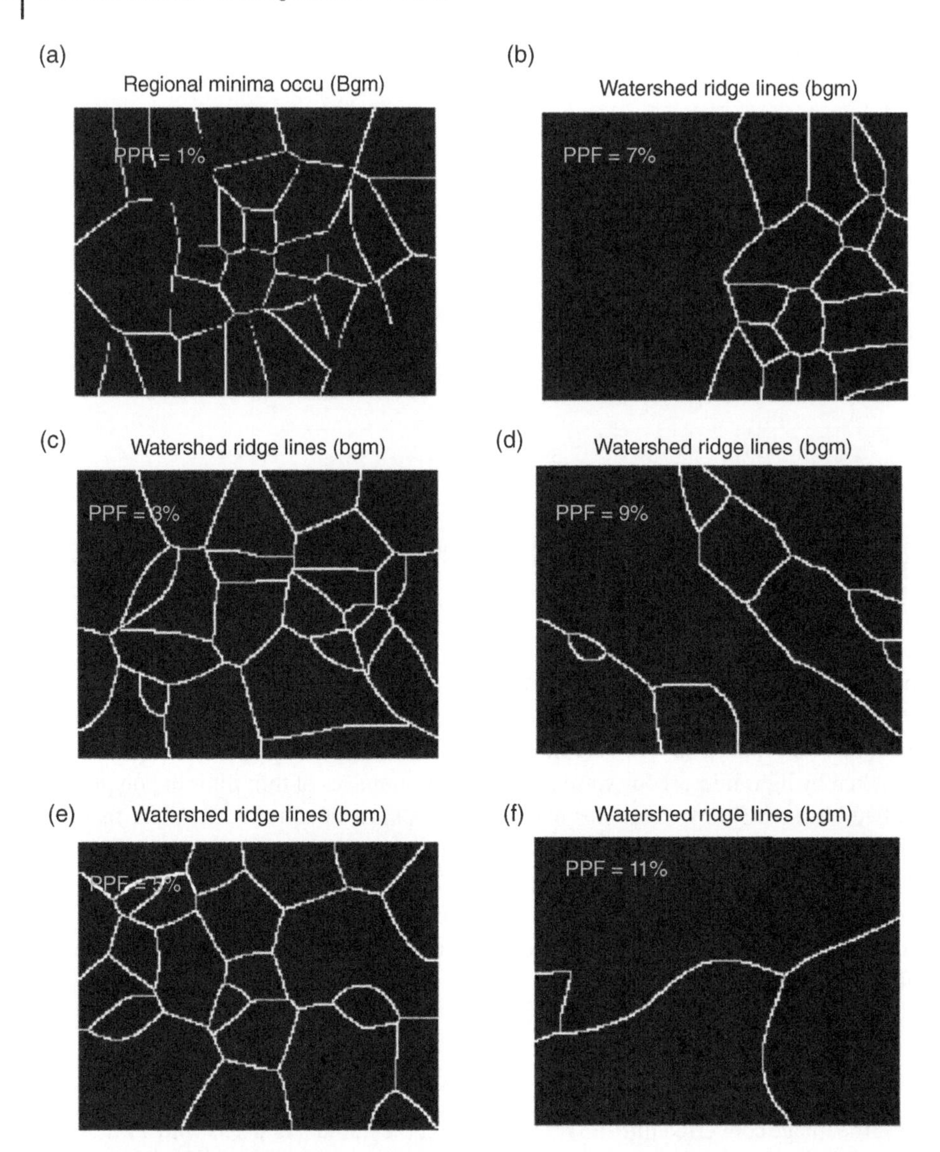

Figure 6.22 Results of watershed segmentation of PPF(s) content for static thresholding of skeleton of watershed segmented minimal path with (a) 1% PPF content, (b) 3% content, (c) 5% content, (d) 7% content, (e) 9% content, and (f) 11% content.

and 13%. Figure 6.22 indicates that the potential of absorption rapidly increased from around 2.0 in 1% PPF to 5.0 in 3% PPF(s) polymer content.

6.10.3 Ratio of Number of Black Pixels to the Number of Total Pixels in Watershed Segmentation (RWS)

After obtaining the watershed image IB, the number of pixels having a value of 1 (i.e. black pixels) were counted (denoted as IBw), and the ratio of number of black pixels to the number of total pixels was calculated as the second feature using Eq. (6.54) (see Figure 6.23 for different PPFs):

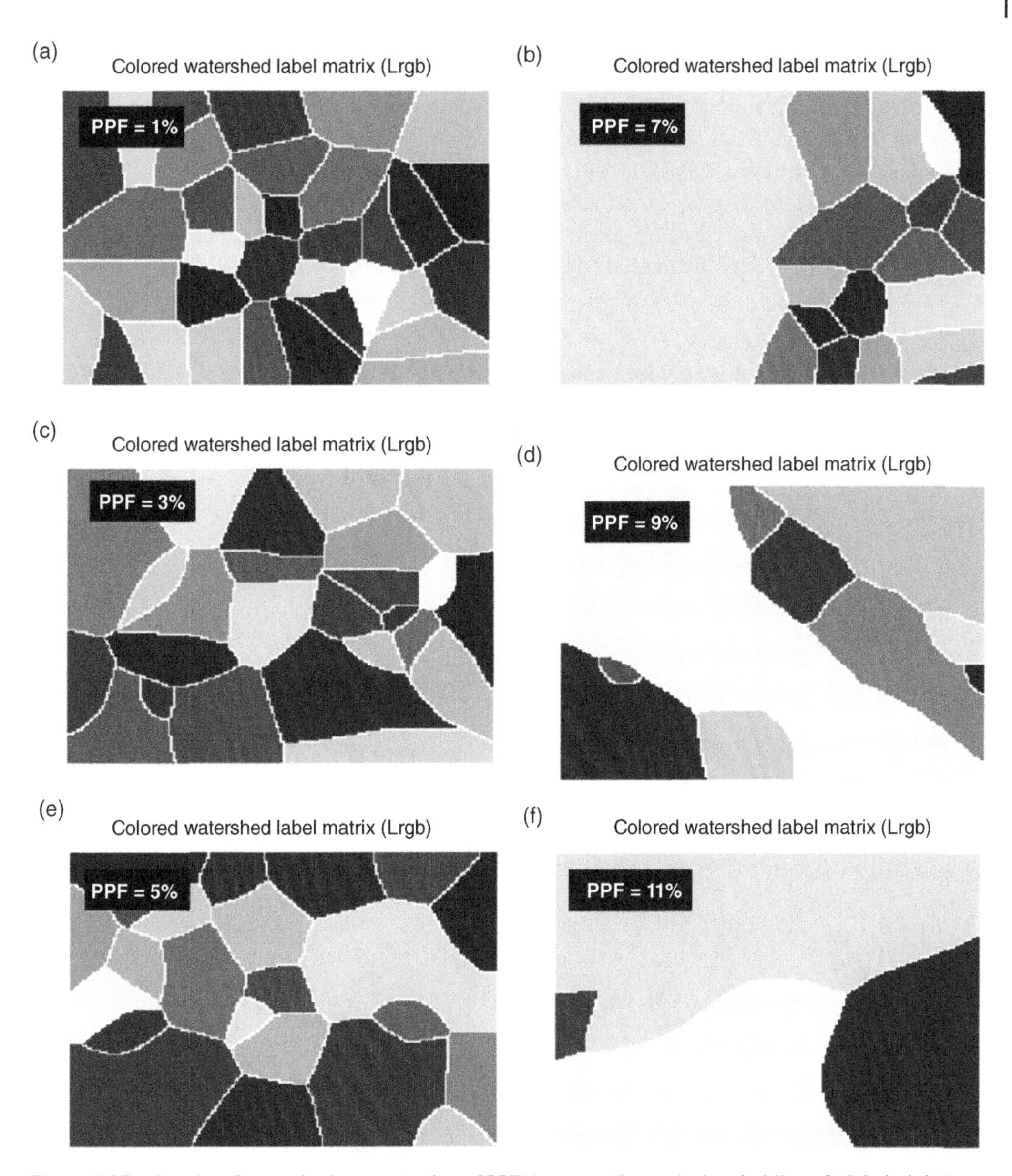

Figure 6.23 Results of watershed segmentation of PPF(s) content for static thresholding of a labeled skeleton of watershed segmented minimal path with (a) 1% PPF content, (b) 3% content, (c) 5% content, (d) 7% content, (e) 9% content, and (f) 11% content.

$$F1 = RWS = \left| \frac{count\left(\sum_{i=1}^{m}\sum_{j=1}^{n} w(i,j) > 0\right)}{count\sum_{i=1}^{m}\sum_{j=1}^{n}(w(i,j) \geq 0)} \right| * 100$$

$$y = -1.1845x^2 - 0.4224x + 0.0763$$

$$R^2 = 0.9941. \tag{6.54}$$

The binary image of watershed utilizes white pixels (i.e. the gray level is equal to 1) to represent the interfaces among the different PPF(s) objects. It is established that the number and the total length of interfaces are changing abruptly corresponding to PPF(s) content (see Figure 6.31).

In order to study the effect of PPF content on the absorption ability of a mixture, the watershed segmented minimal path (WSMP) of PPF was removed and replaced by the skeleton path in Figure 6.24. The PPF content (%) is correlated with the ratio of number of black pixels to the number of total pixels in watershed segmentation, as shown in Figure 6.25. It can be seen from Figure 6.25 that the RWS is influenced by the PPF content (%), while it also presents some statistical proximity with Y.

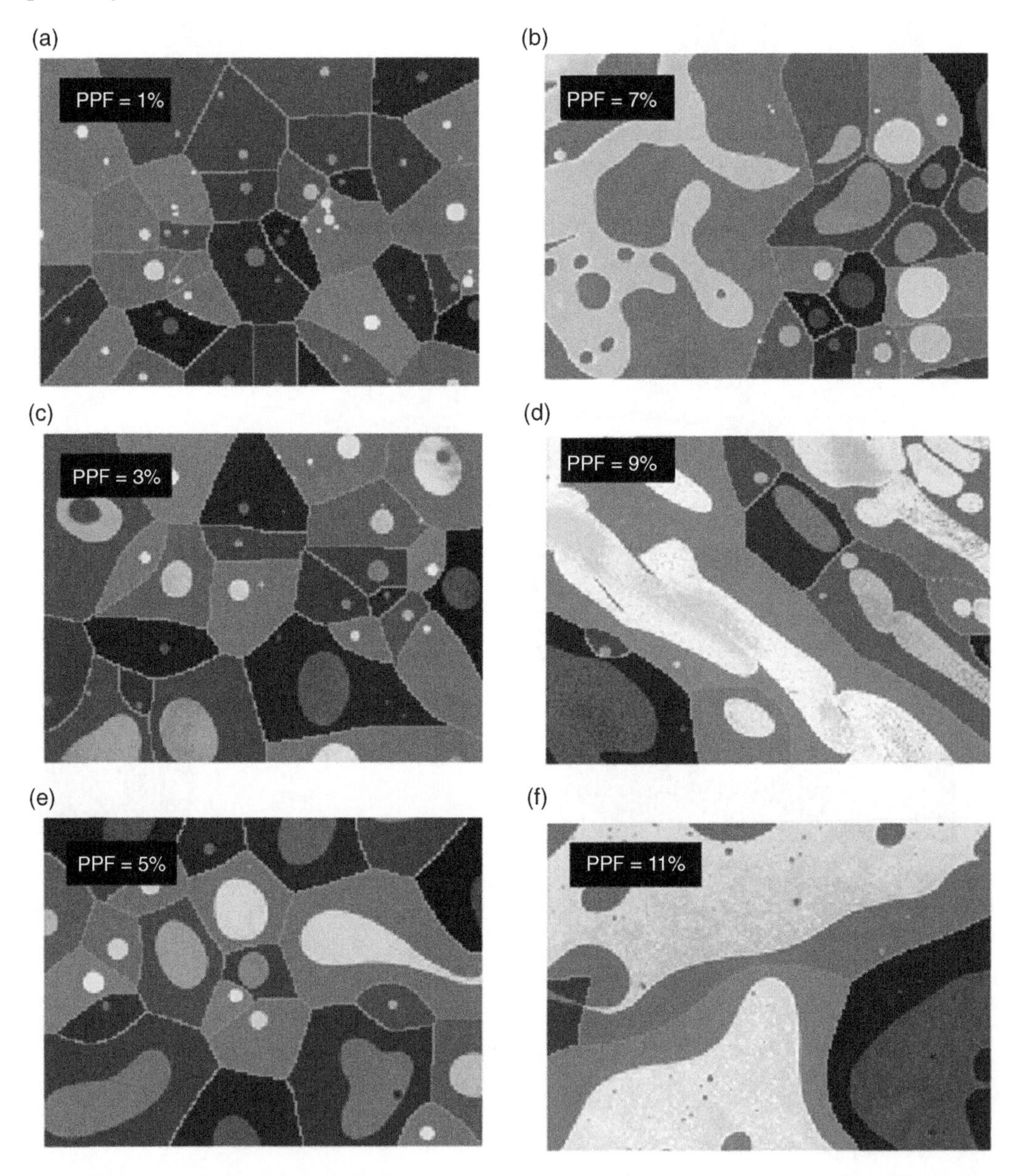

Figure 6.24 Results of clustering of PPF(s) content for static thresholding of labeled skeleton of watershed segmented minimal path with: (a) 1% PPF content, (b) 3% content, (c) 5% content, (d) 7% content, (e) 9% content, and (f) 11% content.

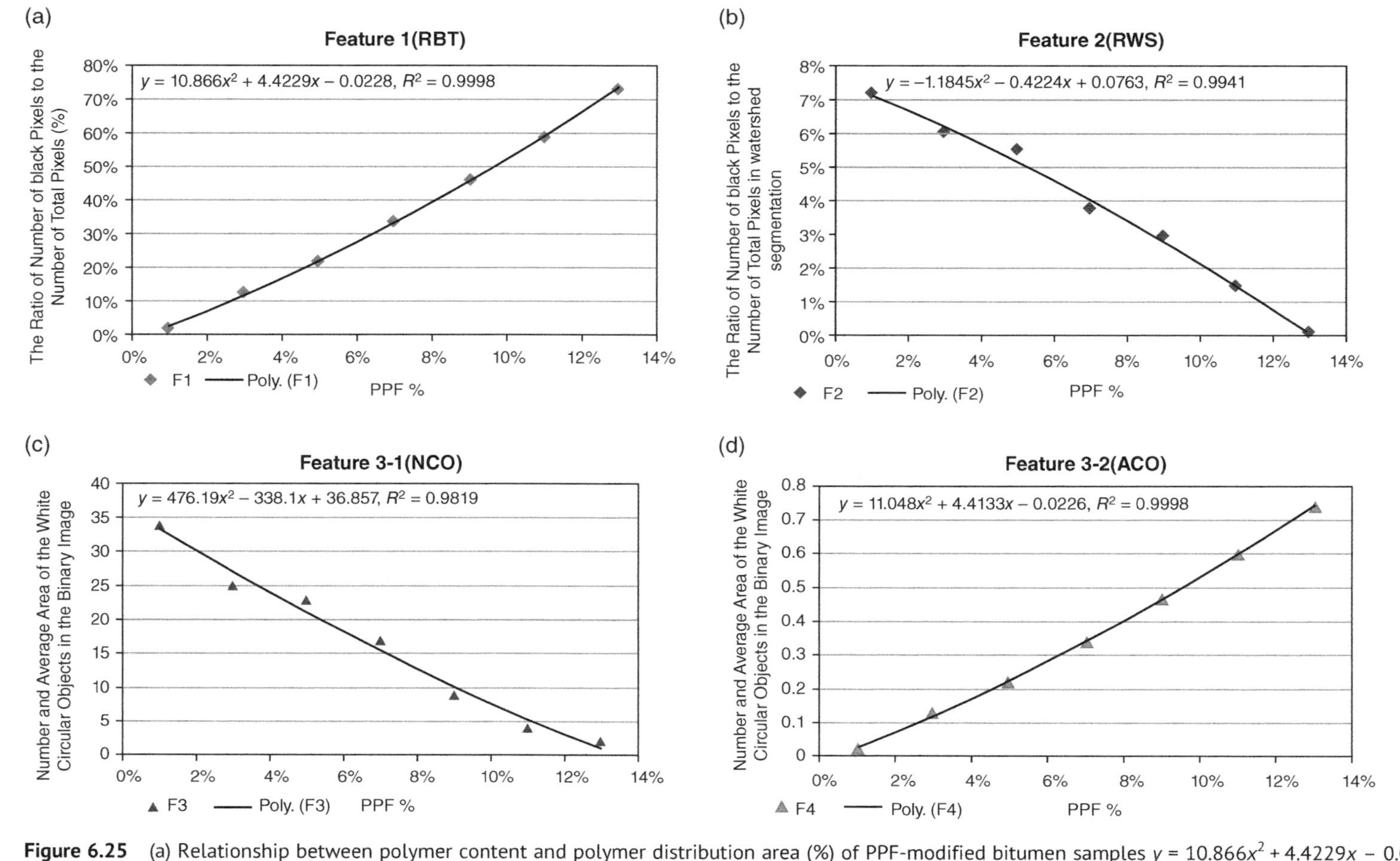

Figure 6.25 (a) Relationship between polymer content and polymer distribution area (%) of PPF-modified bitumen samples $y = 10.866x^2 + 4.4229x - 0.0228$, $R^2 = 0.9998$; (b) Feature 2: The ratio of number of black pixels to the number of total pixels in watershed segmentation $y = -1.1845x^2 - 0.4224x + 0.0763$, $R^2 = 0.9941$; (c) Number of the white circular objects in the binary image $y = 476.19x^2 - 338.1x + 36.857$, $R^2 = 0.9819$; and (d) Average area of the white circular objects in the binary image $y = 11.048x^2 + 4.4133x - 0.0226$, $R^2 = 0.999$.

6.10.4 Number and Average Area of the White Circular Objects in the Binary Image (The number of circular objects [NCO] & ACO)

The number of white circular objects in the image is labeled as Nwc. The average area of the white circular objects was calculated using Eq. (6.55):

$$F3 = AAW = \left| \frac{\left(\sum_{i=1}^{m}\sum_{j=1}^{n}(p(i,j) \approx r) > 0\right)}{count \sum_{i=1}^{m}\sum_{j=1}^{n}((p \approx r))} \right| *100, \tag{6.55}$$

where ($p \approx r$) is the area of the black match circular PPF. We estimate each PPF's area and perimeter and then use the results to form a simple metric indicating the roundness of an object using Eq. (6.56):

$$\rho_k = \frac{4\pi * A_k}{p^2} \tag{6.56}$$

This metric is equal to 1 for a circle and is less than 1 for any other PPF(s). The discrimination process can be controlled by setting an appropriate threshold. In this example, a threshold of 0.90 was used so that only the PPF rounded objects will be classified as round.

The relationships between the calculated NCO and PPF rate are shown in Figure 6.26, for 1%, 3%, 5%, 7%, 9%, 11%, and 13% PPF content. The results suggest that PPF(s) content has influence on NCO and Ant Colony Optimization (ACO), whereby the absorption rate will rise with the increase of ACO and decrease with the growth of NCO. The increase rate of absorption is related to the distribution of circle shape objects in various PPF(s). With the increase of NCO, the absorption needs more remarkable free energy to overcome the power of separation, and the effect of separation is more obvious. ACO increased by 0.3% at 1% PPF and by 3.5% at 13% PPF. Based on the results presented in Figure 6.26, the RBT increased more obviously when the PPF increased from 1 to 13%, but was less remarkably for RWS change. NCO has a significant impact on the distribution of the PPF. When NCO >20, the blend shifted to a homogeneous mixture, and the concentration was almost the same in the local part. When NCO<10 and the size of objects become bigger, the absorption speed ratio increased. For the contacting section of watersheds margins, the RWS values were the same, while NCO and ACO were different. When NCO exceeded 20, the average area of the PPF(s) exhibited little variation. The ratio between ACO and NCO varied only from 0.019 to 0.22 when NCO was greater than 20, from 1 to 7% PPF, while the ratio varies from 0.34 to 0.73% when NCO is lower than 20.

Based on Figure 6.26, it can be determined that a good relationship ($R^2 = 0.9998$) exists between the polymer content and the polymer distribution area and the average area of the white circular PPF(s) ($R^2 = 0.9998$). It should be noted that the relationships illustrated in Figure 6.26 are valid for PPF content of 1–13%. As seen in Figure 6.27, PPF content, such as 9%, was in 46% of the material and increased by the rate of 6. The results of RBT vs. Average area white (AAW) were compared with the proposed method using 1, 3, 5, 7, 9, and 11% PPF(s).

6.10.5 Entropy of the Image

Entropy is the degree of amount of detail in an image. Entropy is a statistical measure of randomness that can be used to characterize the texture of the input image and is calculated by Eq. (6.57)

$$f4 = entropy = -\sum_{i=1}^{M}\sum_{j=1}^{N} p(i,j)\, log_2^{(p(i,j))} \tag{6.57}$$

where M and N are the number of pixels in the horizontal and vertical axes, respectively.

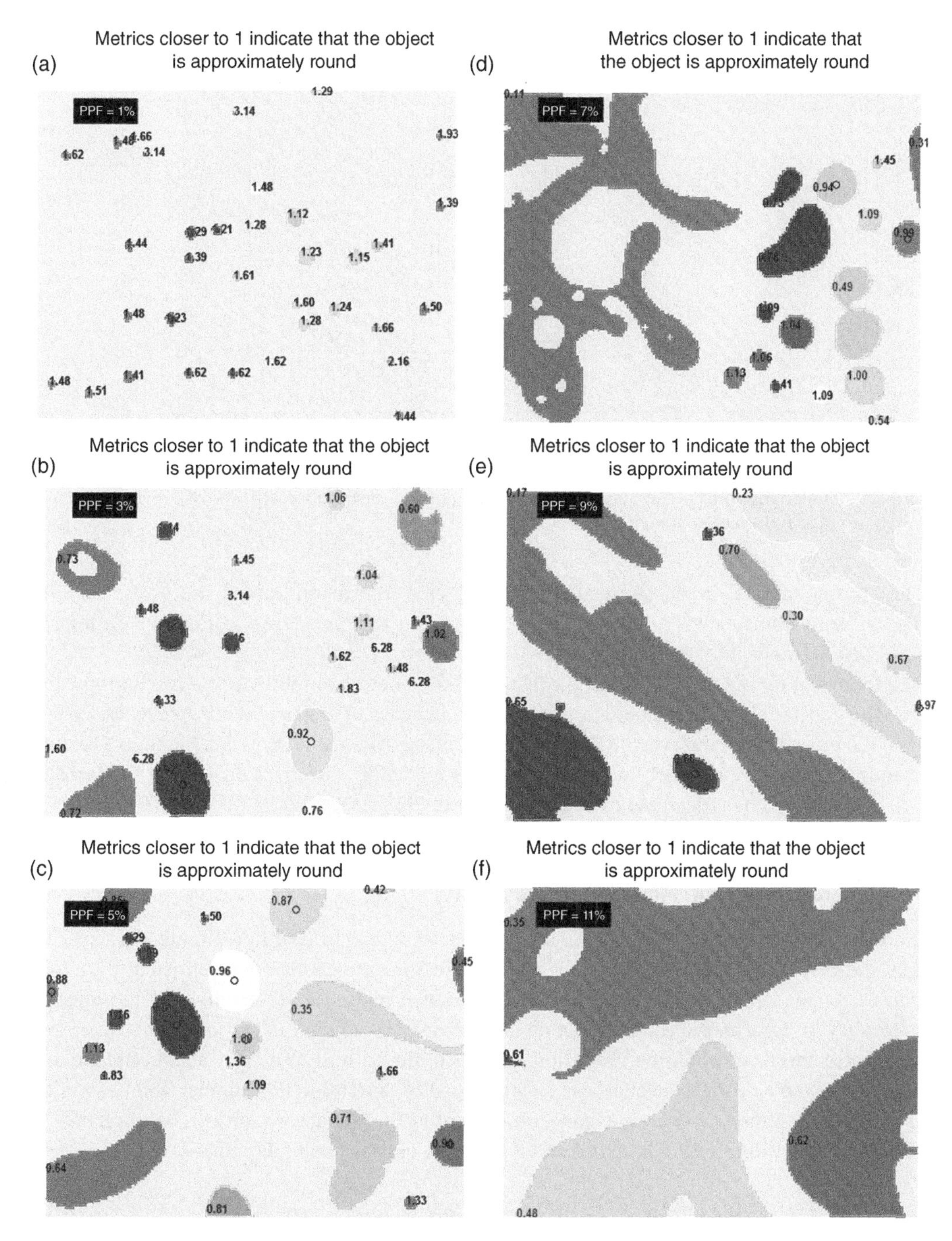

Figure 6.26 Relationship between polymer content, polymer number, and average area of the white circular objects in the binary image for PPF of: (a) 1%, (b) 3%, (c) 5%, (d) 7%, (e) 9%, and (f) 11%.

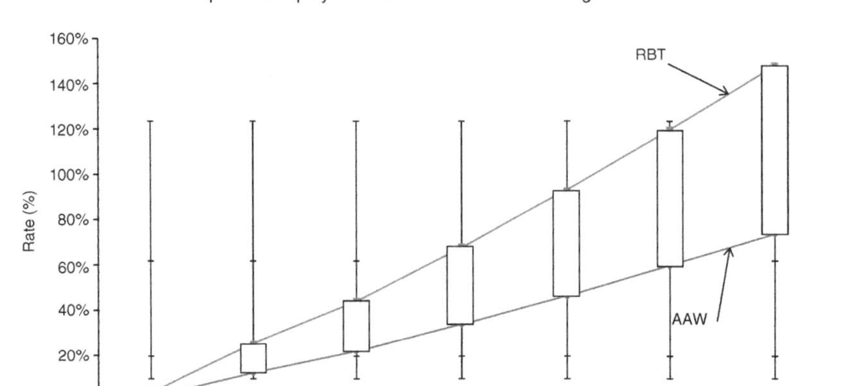

Figure 6.27 Relationship between polymer content and polymer distribution area (%) of PPF-modified bitumen samples and average area of the white circular PPF.

The entropy amount is closely related with extracted information about the texture of PPF images. The relations between the calculated entropy and PPF are shown in Figure 6.27, for 1%, 3%, 5%, 7%, 9%, and 11% PPF content.

The following results were determined. The PPF(s) content has influence on the entropy and ACO. The general absorption rate will rise with the increase of entropy number (EN) and ACO. When PPF exceeded 9%, the average EN of the PPFs images slightly decreased. EN and ACO exhibited equivalent behavior, which means that ACO is related to entropy. Considering the (γ=EN/ACO), the ratio of γ will decrease by increasing the PPF(%).

6.10.6 Radon Transform Maximum Value (RTMV)

Radon transform is the projection of the image intensity along a radial line oriented at a specific angle. The projection of a 2D function, p(i,j), is a line integral along a direction defined by the angle. When a watershed image is transformed into RT for a given angle, a watershed line will be projected into a valley in RT. The projection direction is perpendicular to the orientation of the watershed margins. For instance, when the projection angle is θ, the pattern of margin as an effective tool is θ+90. The angle of the projection can be employed to determine the number and absorption of PPFs and, therefore, estimate the concentration of PPFs. The absorption can be estimated by thresholding the values in the Radon domain and area of peaks using Eq. (6.58).

$$R_{\theta}(x') = \int_{-\infty}^{+\infty} f(x,y)\mathrm{d}y' = \int_{-\infty}^{+\infty} f(x' cos\theta - y' \sin\theta, x' \sin\theta - y' \cos\theta)\mathrm{d}y' \tag{6.58}$$

where $R_{\theta}(x')$ Radon transform of object function $f(x, y)$ is the linear integral of $f(x, y)$ along the y' axis.

The peaks in the Radon domain are related to the number of PPFs, and the watershed border is a good tool to determine the type and absorption of PPFs. According to the RT for the wider watershed, the larger the peaks extracted. The area of peaks can be used to determine the width of watershed and to further determine the absorption of PPFs (see Figure 6.28). The relationship between

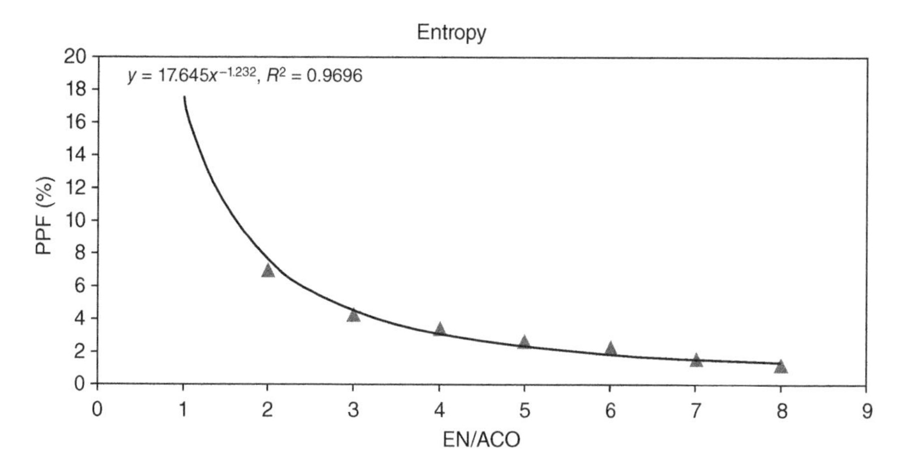

Figure 6.28 Relationship between γ = (entropy/ACO) of the image and PPF(%), $y = 17.645x - 1.232$, $R^2 = 0.9696$.

Table 6.1 Relationship between the patterns of peaks and properties of PPF(s).

	Set 1		Set 2	
Pattern of peaks	**Number**	**Area**	**ERT**	**RTMV**
Properties of PPF(s)	1%, 3%, 5% 7%, 9%, 11%, 13%	The width of watershed, border, and, number of PPF, absorbation	Randomness	Length of watershed margin Severity of absorbation

the properties of peaks and properties of PPFs is summarized in Table 6.1. A simple search algorithm is used to find the RTMVs, which is related to the peaks.

RTMV and the ratio of number of black pixels to the number of total pixels in watershed segmentation (RWS) are intercorrelated and related based on Eq. (6.59)

$$RTMV = \left(\int_{-\infty}^{+\infty} f(x' \cos\theta - y' \sin\theta, x' \sin\theta - y' \cos\theta) \mathrm{d}y'\right) * NCO \tag{6.59}$$

RTMV is also correlated with NCO in different PPF(s) content, as shown in Figure 6.29, where RTMV∗ is influenced by PPF(%). It can be concluded that a relationship exists between the polymer content, polymer distribution area, and RTMV∗. The relationship illustrated in Figure 6.29 is for PPF content of 1–13%, The relationship illustrated in Figure 6.30 is used for PPF content of 1–13%,

6.10.7 Entropy of Radon Transform (ERT)

The entropy of RT is a statistical measure of randomness of a transformed watershed pattern in RT that can be used to characterize the RT of the input image. The entropy of RT is calculated using Eq. (6.60)

$$(Entropy(R)) = (ERT) = -\sum_{x'=1}^{r} \sum_{\theta=0}^{180} RT(x', \theta) \log_2^{(RT(i,j))}$$

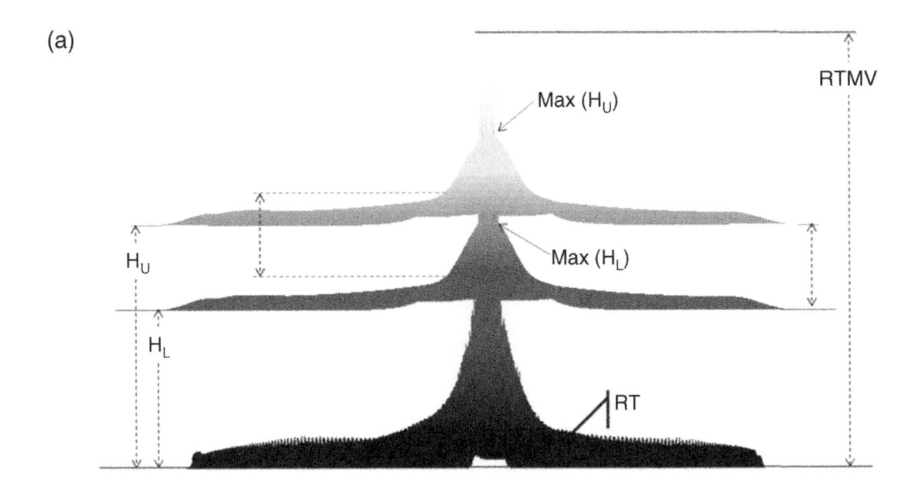

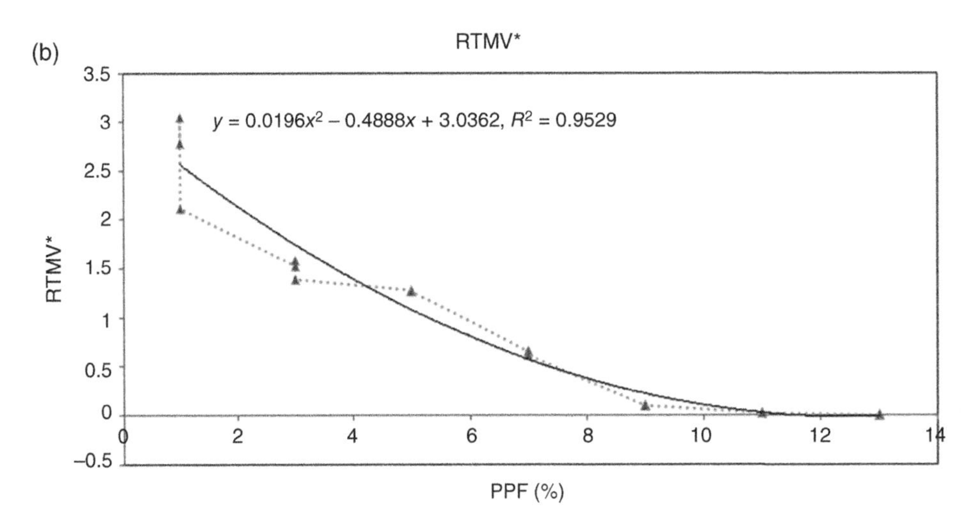

Figure 6.29 (a) Relationship between RT, RTMV, hight upper (H_U), hight lower (H_L), max(H_U), and max(H_L); (b) Relationship between RTMV* and PPF(%). $y = 0.0196x^2 - 0.4888x + 3.0362$, $R^2 = 0.9529$.

$$F6 = (Entropy(R^*)) = (Entropy(R)) \times NCO, \tag{6.60}$$

where θ is the projection angle; $R_\theta(x')$ is Radon transform; $Entropy(R^*)$ is modified entropy; NCO is the number of the white circular objects in the binary PPF images.

The entropy (R∗) decreases as PPF content increases. Specifically, entropy (R∗) in 1–5% PPF will decrease about −33%. Meanwhile, decreasing the watershed margins reflected in RT will influence the decreasing effect of RT, depending on the spreading and amount of PPF content. The relationships between the calculated entropy (R∗) and PPF (%) are shown in Figure 6.30.

The following results also extracted:

The PPF (%) has influence on the entropy (R∗), whereby the mixture entropy (R∗) increases with a decrease in PPF content and increase in NCO. Therefore, the increase in entropy (R∗) is dependent on NCO. With the increase of NCO, the image exhibits more remarkable entropy (R∗), and the

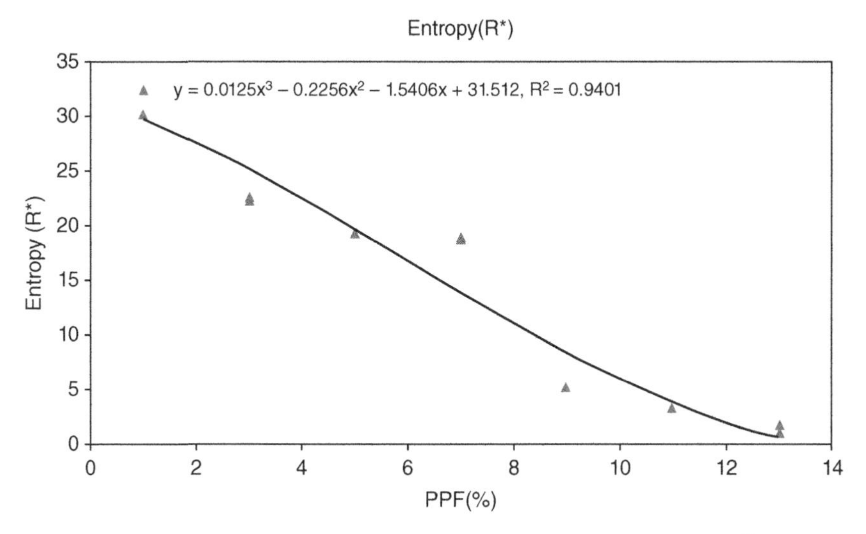

Figure 6.30 Relationship between entropy ($R*$) and PPF(%). $y = 0.0196x^2 - 0.4888x + 3.0362$, $R^2 = 0.9529$.

effect of NCO is more obvious. The mixture absorption will increase by 43% at PPF(3%), by 16% at PPF(5%), and by 1.9% at PPF(7%). Based on PPF(%)-Entropy(R∗) results, the mixture absorption increases more obviously when the PPF is increased from 1 to 3% and from 7 to 9%, but less remarkably between 3–7% and 9–13% PPF(s). The contacting PPF(s) have an equivalent Entropy(R∗) = 0, which indicates that the PPF is completely scattered on the blend.

6.10.8 High Amplitude Radon Percentage (HARP)

Irregular margins, such as 1, 3, and 5%, are transformed into high amplitude coefficients, while 13, 11, and 9% are transformed into low amplitudes. Therefore, counting the coefficients larger than a threshold is a good index for quantifying the properties of PPF(s). The RT is shown in 3D by expanding the RT value as a component. After this step, 3D RT is binarized with an adoptive threshold hyperplane surface using Eq. (6.61):

$$D_1(x', \theta) = \begin{cases} 1 \; if \; \mathrm{NRT}_{(x',\theta)} \geq C_{thu(x',\theta)} \\ 0 \; if \; \mathrm{NRT}_{(x',\theta)} < C_{thl(x',\theta)} \end{cases} \tag{6.61}$$

where $D_1(x', \theta)$ is the binarized RT; $C_{\mathrm{thu}(x',\theta)}$ is the adaptive upper threshold of Radon transform; $C_{thl(x',\theta)}$ is the adaptive lower threshold; and $\mathrm{NRT}_{(x',\theta)}$ is the normalized Radon transform.

We used cubic smoothing spline (CSS) to generate the upper and lower thresholding surface functions because of the nonuniform illumination of 3D RTs. The estimated $C_{\mathrm{thu}(x',\theta)}$ and $C_{\mathrm{thl}(x',\theta)}$ were examined by fitting a CSS to the 3D RT $f(x', \theta)$:

$$M = P \cdot \sum_{\theta = 0}^{179} \sum_{x' = 1}^{n} \left(f(x', \theta) - s(x', \theta) \right)^2 + (1 - p) \int\int \left(D^2 S(x', \theta) \right)^2 dx' d\theta \tag{6.62}$$

where $D_1(x', \theta)$ is the binarized RT by Eq. (6.62).

The smoothing factor, p, controls the balance between an interpolating spline crossing all data points (with $p = 1$) and a strictly smooth spline (with $p = 0$). The interesting range for p is often near

$1/(1 + h3/6)$, where h is the average spacing of the data sites, and it is in this range that the default value for p is chosen. The calculation of the smoothing spline requires the solution of a linear system whose coefficient matrix has the form $p* f(x', \theta) + (1-p)* s(x', \theta)$, where matrices A and B depend on the PPF image. The default value of p makes $p*$trace $(f(x', \theta))$ equal $(1-p)*$trace $(s(x', \theta))$. As p moves from 0 to 1, the smoothing spline changes from one extreme to the other (Figure 6.30).

The upper and lower thresholding are determined using $C_{thu(x',\theta)}$ and $C_{thl(x',\theta)}$ using Eq. (6.16). A reference smoothing factor ($p = 0.001$) was obtained empirically for constructing $C_{th(x',\theta)}$ in the upper bound and ($p = 0.9e{-}5$) for constructing $C_{th(x',\theta)}$ in the lower bound. For example, in the case of PPF thresholding, after testing several thresholds, the general rule can be extracted from 3D RT thresholds for upper and lower bounds based on the optimum selection of α and β using Eq. (6.63).

$$C_{thu(x',\theta)} = \left(\frac{1}{1 + \frac{(min\,(N,M))^3}{\alpha \times 600}}\right)_U \; and \; C_{thl(x',\theta)} = \left(\frac{1}{1 + \frac{(min\,(N,M))^3}{\beta \times 600}}\right)_L . \tag{6.63}$$

Normalization of 3D RT surface involves two steps: (i) calculate max 3D RT, and (ii) divide each 3D RT component by its max 3D RT. The new set must be in [0,1]. Given a 3D RT surface, its normalized components in the upper and lower bands are calculated by Eq. (6.64):

$$NRT_{(x',\theta)} = \left(\left[RT_{(x',\theta)}\right]^h \Big/ max\left[RT_{(x',\theta)}\right]^h\right)^{\frac{1}{h}} = \left(\left[RT_{(x',\theta)}\right]^h \Big/ RTMV^h\right)^{\frac{1}{h}}$$

$$C_{th(x',\theta)} = \begin{cases} \text{Upper band, } C_{thu(x',\theta)} = f\left(RT_{(x',\theta)}, \alpha, p\right) + H_U \\ \text{Lower band, } C_{thl(x',\theta)} = f\left(RT_{(x',\theta)}, \beta, p\right) + H_L \end{cases} \tag{6.64}$$

where M and N denote the size of the 3D RT platform; H is a high platform H∊[0,1]; h∊(1, ∞), and $V_{(3DRT)}{}^{h}_{(i,j)}$ are the 3D RT values in positions i and j, respectively.

A bigger h is worth a more enhanced margins, for example, in the PPF(s) smoother. HARP is then calculated using Eq. (6.65):

$$HARP = \sum_{x'=0}^{m} \sum_{\theta=0}^{179} D_1(x', \theta)/(M \times 180), \tag{6.65}$$

where M is the length of $NRT_{(x',\theta)}$.

The HARP has a value between 0 and 1, where 0 indicates concentrated PPF and is an index for high potential absorption (11, 13%, and greater), and a value of 1 represents small PPF content (1, 3, and 5%) with low potential absorption and concentration. In other words, HARP is an index to identify the potential of absorption and can be used as a measure for the separation of margins of PPFs and bitumen base (BB) and, therefore, as a criterion for absorption quantification of PPF.

For different PPF contents, the calculated HARP are presented in Figure 6.31, which reveals that HARP will increase to some degree with the reduction of PPF(%) in different images. As PPF(%) increased, the decrease in HARP exhibited less remarkable differences, and the effect of PPF content was less obvious. It can be concluded that a relationship exists between the polymer content and HARP. It should be noted that the relations illustrated in Figure 6.31 apply only to PPF content of 5–13%. Taking the outcomes into consideration, Δ(HARP) decreases less noticeably for 5–13% PPF, but less remarkably for 1–3% PPF.

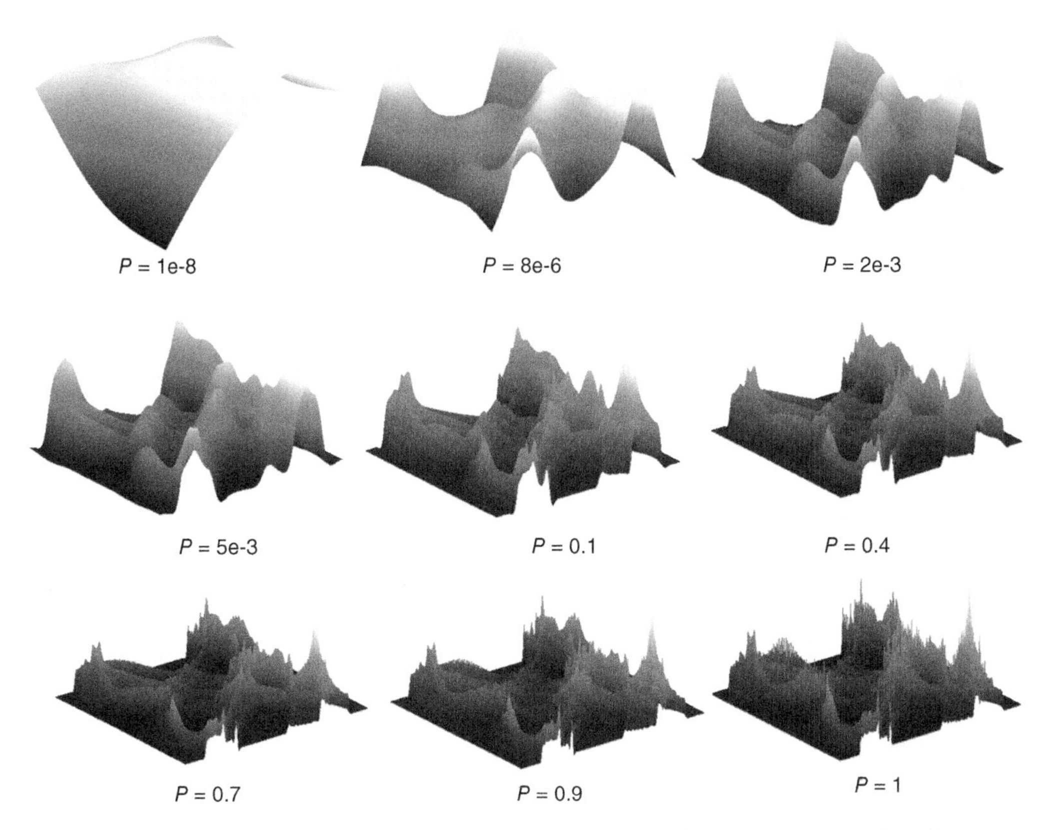

Figure 6.31 As *p* moves from 0 to 1, the smoothing spline changes from one extreme to the other.

6.10.9 High-Energy Radon Percentage (HERP)

Energy is similarly identified as consistency, uniformity of energy, and angular second moment. The smaller PPF content is, the higher the energy level will be. Watershed margins are transformed into high level of RT domain. The HERP is calculated by summing squares of all RT coefficients and then normalized by the total energy of an original image. The RT of watershed image can be calculated using Eq. (6.66):

$$HERP = \sum\nolimits_{P=0}^{x'} \sum_{\theta=0}^{180} \left[\left(\int_{-\infty}^{+\infty} f(x'cos\theta - y'sin\theta, x'sin\theta - y'cos\theta)dy'\right) \geq C_{thu(x',\theta,l)}\right]^2 \Big/ \sum\nolimits_{x=0}^{p} \sum\nolimits_{y=0}^{k} [w(x,y)]^2$$

$$HERP = \sum\nolimits_{P=0}^{x'} \sum\nolimits_{\theta=0}^{180} \left[RT_{(x',\theta)} \geq C_{thu(x',\theta,l)}\right]^2 \Big/ \sum\nolimits_{P=0}^{x'} \sum\nolimits_{\theta=0}^{180} \left[RT_{(x',\theta)}\right]^2. \tag{6.66}$$

where $w(x, y)$ is the value at a pixel (x,y) of a watershed margin; and p and k are the width and length of the image, respectively.

Figure 6.34 shows HERP for different PPFs, from which it can be concluded that the dense PPFs with high absorption have a very small value of HERP. The PPF content of 1, 3, and 5% have close values to that of 21, 20, and 8%. Comparatively, the average HERP for 11 and 13% PPF is approximately 0.14 times that of PPF. The value of HERP ranges between 0 and 100, where 0 indicates a high compactness or absorption of PPF and 100 means a perfect separation.

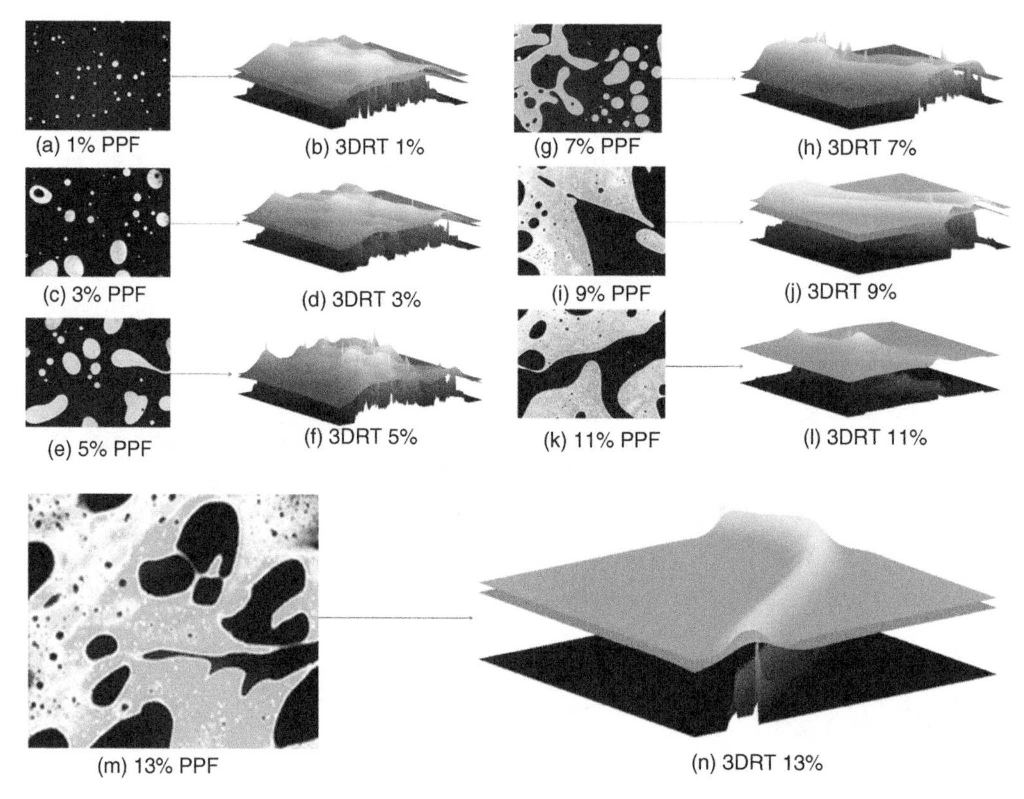

Figure 6.32 RT of the various PPF(s) based on proposed thresholding method: (a) 1% PPF, (b) 3DRT and hyperplane surface for 1% PPF, (c) 3% PPF, (d) 3DRT and hyperplane surface for 3% PPF, (e) 5% PPF, (f) 3DRT and hyperplane surface for 5% PPF, (g) 7% PPF, (h) 3DRT and hyperplane surface for 7% PPF, (i) 9% PPF, (j) 3DRT and hyperplane surface for 9% PPF, (k) 11% PPF, (l) 3DRT and hyperplane surface for 11% PPF, (m) 13% PPF, and (n) 3DRT and hyperplane surface for 13% PPF.

The leave-one-out method as a cross-validation technique was conducted to find the best parameter (l) for the HERP. In order to select the optimal threshold, we used L(0.3)/L(0.8) as a feature for HERP (see Figure 6.32).

It can be found from Figure 6.32 that the results acquired by HERP tests and lab outcomes are approximate both for ($l = 0.3$) and for ($l = 0.8$). This transferring process is shown in Figure 6.32, where the red objects represent the $RT_{(x',\theta)} \geq C_{thu(x',\theta,l)}$ and the remaining parts represent the BB.

6.10.10 Standard Deviation of Radon Transform (STDR)

The spread of the histogram can be used to characterize the scattering and absorption of PPFs. The wider the spread of histogram is, the worse the PPF(s) conditions are. Statistical parameters, including STD, mean, and max are used to describe the shape of histogram. In Figure 6.33, STD of RT, which is a measure of separation (absorption) of PPF(s), is selected for the interpretation of 3D RT. Let $RT_{(x',\theta)}$ be the Radon transform of watershed regions, then the mean μ can be determined as using Eq. (6.67):

$$\mu = \sum_{i=-x}^{+x} RT_{(x',\theta)}.C_{thu(x',\theta)},$$

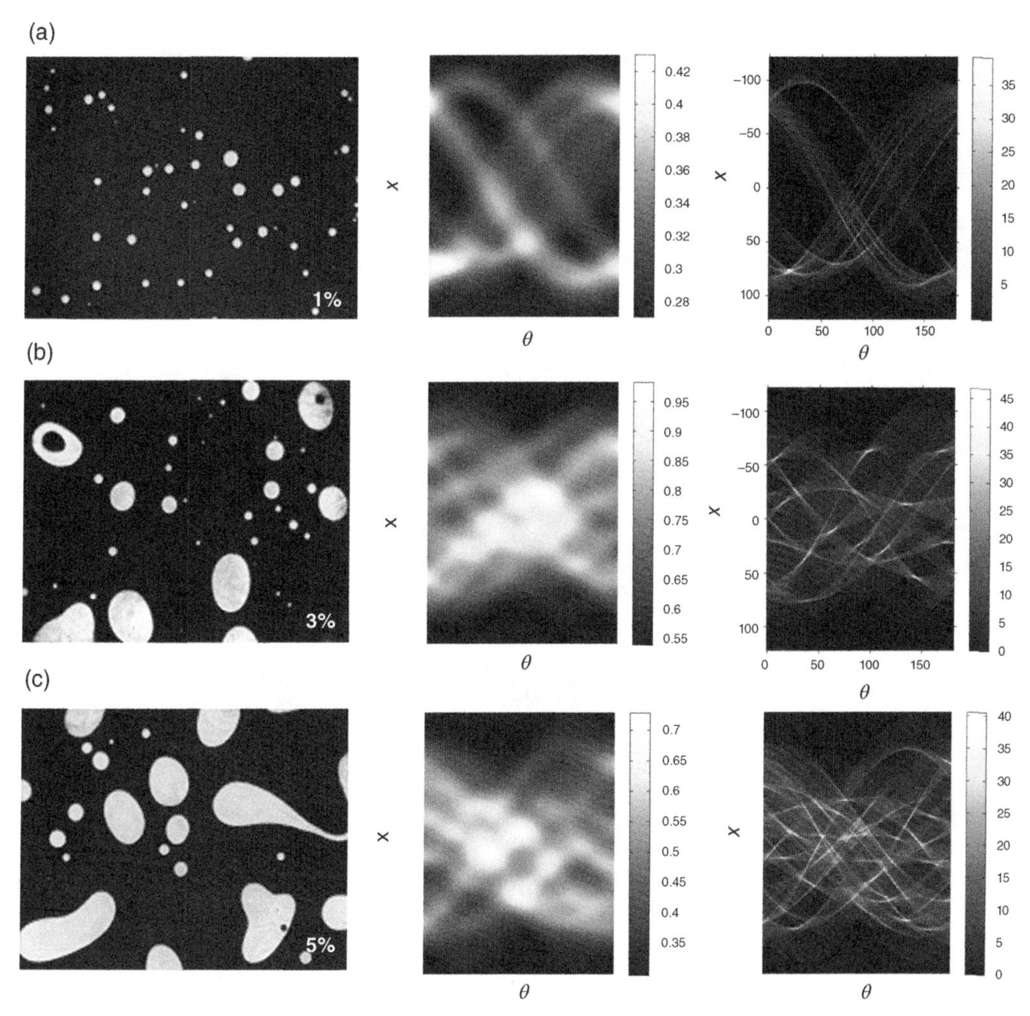

Figure 6.33 RT of the various PPF(s) %. (a) for 1% PPF, (b) for 3% PPF, (c) for 5% PPF, (d) for 7% PPF, (e) for 9% PPF, (f) 11% PPF, (g) 13% PPF.

$$\sigma = \left\{\sum_{i=-x}^{+x}\left[RT_{(x',\theta)} - \mu\right]^2 \times RT_{(x',\theta)}\right\}^2, \tag{6.67}$$

where σ is the variance of PPF(s).

The max STDR in various PPF(s) is related to PPF = 1% and the minimum SRTR is related to PPF = 13%. STDR is a fair criterion for the identification of polypropylene fiber absorption of modified bitumen based on Radon transform and watershed segmentation. Based on the assessment of, it can be concluded that a correlation exists between the polymer content (PPF) and STDR. It should be noted that the relationships illustrated in Figure 6.33 are not valid for PPF content of 1–13% (such as 7% PPF). Moreover, STDR decreases as PPF content increases, depending on the distribution and amount of PPF and the distribution of PPF and BB in the blend. For example, STDR decreased by about 42% as PPF increase from 1–13%.

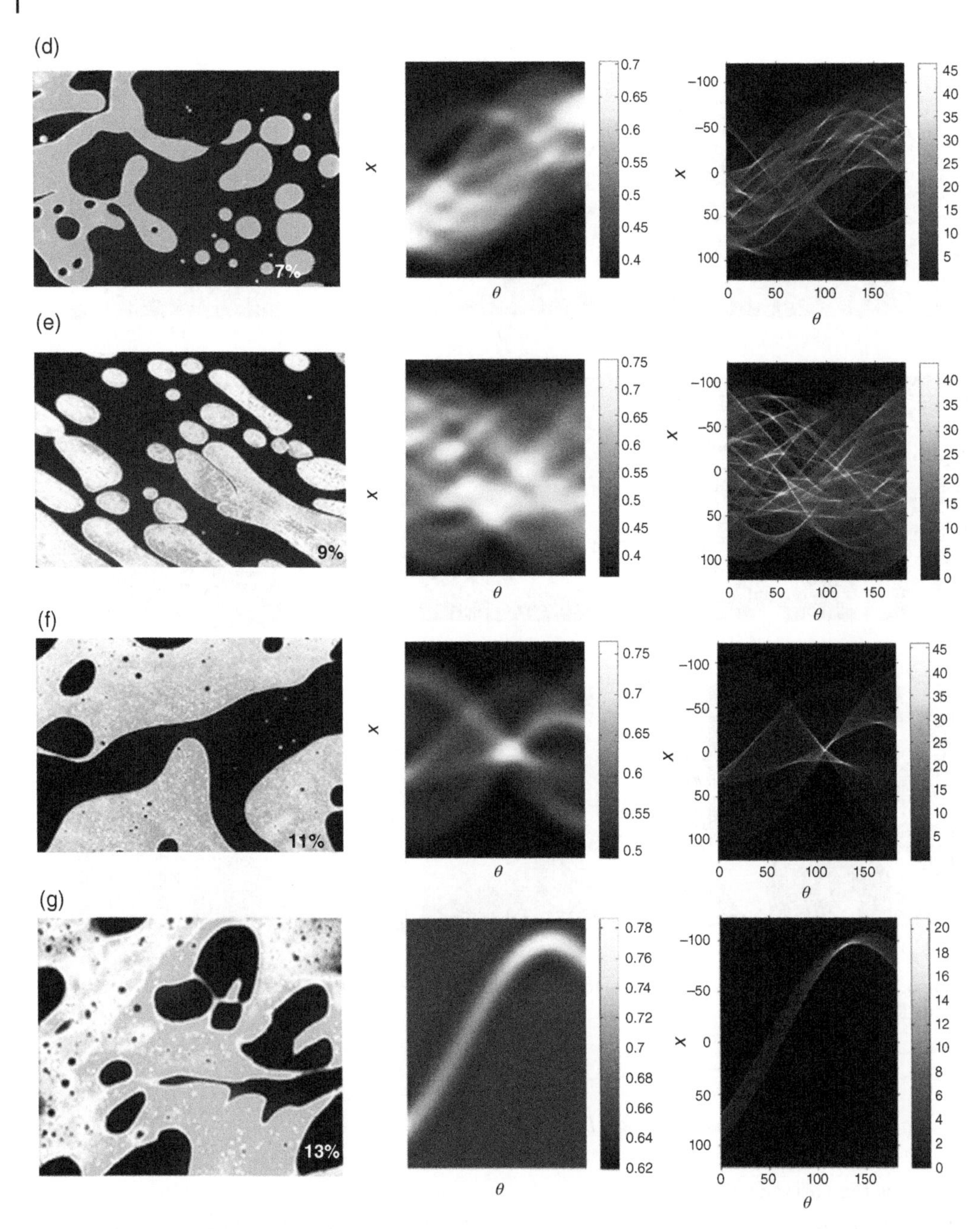

Figure 6.33 (Continued)

The relationships between the calculated STDR and PPF(%) are shown in Figure 6.34, which demonstrates the influence of PPF (%)on the mixture STDR. The decreased rate of STDR also depends on the thresholding. For example, with the increase of the smoothness parameter (p) presented, the $NRT_{(x',\theta)}$ shows more remarkable STDR. In addition, the mixture STDR decreased by 12% at 1 and 3% PPF and by 34% at 11% PPF.

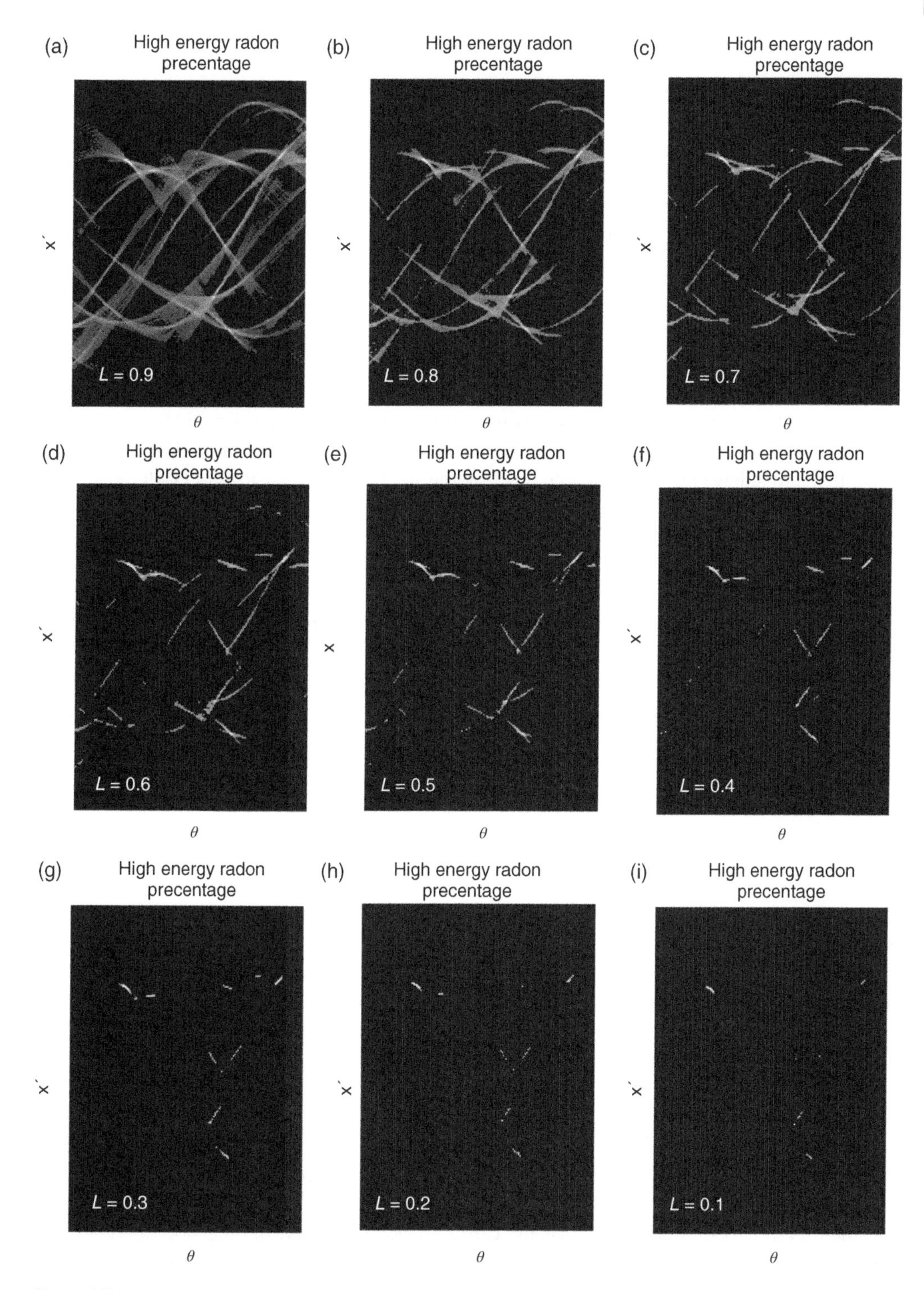

Figure 6.34 Results of HERP of PPF (5%) content for L of: (a) 0.9, (b) 0.8, (c) 0.7, (d) 0.6, (e) 0.5, (f) 0.4, (g) 0.3, (h) 0.2, and (i) 0.1.

6.10.11 Q_{th}-Moment of Radon Transform (QMRT)

The QMRT is an important characteristic to describe the spreading of PPF(s) transformed in RT. There are two types of moments that are used to characterize the histograms: (i) qth central moment and (ii) qth moment. For matrices, like images, the moment (X, order) returns to the central moment of the specified order for each column. The central first moment is zero, and the second central moment is the variance computed using a divisor of n rather than $n - 1$, where n is the length of the vector x or the number of rows in the matrix X. The central moment of order q of a distribution is defined using Eq. (6.68):

$$cm_k = \sum_{i=-\infty}^{+\infty} \left[RT_{(x',\theta)} - \mu\right]^q . RT_{(x',\theta)} \tag{6.68}$$

where $RT_{(x',\theta)}$ is Radon transform.

A meaningful criterion for PPF absorption evaluation is q = 4. Kurtosis is a measure of distribution that is outlier-prone. The kurtosis of normal distribution is 3. Distributions that are more outlier-prone than the normal distribution have kurtosis greater than 3, while distributions that are less outlier-prone have kurtosis less than 3. Using this definition, the kurtosis of QMRT evaluated for different PPF is defined using Eq. (6.69):

$$K = \frac{E(x-\mu)^4}{\sigma^4} \tag{6.69}$$

where μ is the mean of x; σ is the standard deviation of x; and E(t) represents the expected value of the quantity t.

Kurtosis computes a sample of population value. Skewness is another feature that can be used for interpretation of QMRT data, which is a measure of the irregularity of the data around the sample mean. If skewness is negative, the data are spread out more to the left of the mean than to the right. If skewness is positive, the data are spread out more to the right. The skewness of normal distribution is zero. The skewness of a distribution is defined by Eq. (6.70):

$$K = \frac{E(x-\mu)^3}{\sigma^3}, \tag{6.70}$$

where μ is the mean of x; σ is the standard deviation of x; and E(t) represents the expected value of the quantity t.

Skewness computes a sample version of this population value. The relationships between the calculated QMRT and PPF (%) are shown in Figure 6.33, from which it can be concluded that a relationship exists between the PPF content and kurtosis. It should be noted that the relationships illustrated in . are valid only for PPF content of 7–13%.

6.10.12 Case 2: Image-Based Feature Extraction for Pavement Skid Evaluation

The amount of friction on the surface of pavement reflects the degree of road safety against slipping and contributes greatly to the reduction of accidents and off-road hazards, especially when the pavement is wet. This parameter allows the car to accelerate, maneuver, and stop safely if it has the minimum requirements. The accurate and fast assessment of road roughness and friction can save hundreds of thousands of lives a year. To this end, researchers continuously aim to develop automated methods for assessing the surface friction of road surfaces. The solutions offered in this field are mostly in the range of noncontact methods, such as the use of lasers and digital images, which have grown significantly in recent years. The results of this research show that image processing as a noncontact method with acceptable accuracy and high speed can provide an effective method to determine the sliding resistance of pavement. Another new index and method that has been recently developed is the plate analysis of pavement texture for pavement slip resistance considering the effects of horizontal, vertical, and diagonal components of pavement texture. More precisely, the pavement texture in the form of a plate in both the fine and coarse texture dimensions is beneficial to obtain reproducible results in relation to the slip resistance of the pavement. In order to automate the assessment of slip resistance of roads and pavement, several methods and devices have been proposed. Image-based methods have not only many features available compared to other methods but also needs a higher processing speed and require extracted surface information to solve this problem.

In this case study, wavelet transform capabilities are considered to enhance and compress the image. Noise reduction is broken down into horizontal, vertical, and diagonal components. As an innovative method in pavement texture analysis using wavelet, three directions of pavement surface texture were identified. Given that each of these directions has different effects on road SR in different conditions, a separate indicator is defined for each. Finally, the main index is presented with an experimental combination of three main indicators, where the initial indicator is calculated as Eq. (6.71):

$$TH_i = mean\left(var\left(\psi^i(x,y)\right)\right),$$
$$SRI_i = \left(\frac{\sum(\sum(\psi^i(x,y) > TH_i))}{\sum(\sum(\psi^i(x,y) > 0))}\right) * 100, \tag{6.71}$$

where $\psi^i(x, y)$ is the image of the wavelet transform (i = H,V & D); TH_i is the mean variance of $\psi^i(x, y)$; SRI_i is the primary SR index of $\psi^i(x, y)$;$\sum(\sum(\psi^i(x, y)))$is the size of $\psi^i(x, y)$.

In this section, the choice of filter plays an important role in obtaining acceptable results. Due to the fact that medium and fine texture images are required for analysis, the large family of wavelet filters is very useful. To determine the optimal filters and "k" of each image, different filters and different values of k have been analyzed using horizontal, vertical, and diagonal components. The initial characteristics are determined by the special index provided. An example of the filters used in this method is presented in Table 6.2.

Using the diverse family of filters, different models can be used. It is even possible to combine filters and build a new complex filter. In order to select the optimal final filter and compare the results of different filters and different values of k, the mean square error (MSE) is used. In this method, the average squares of "errors" are measured, which is based on the difference between the SR index and the corresponding British pendulum number (BPN). Therefore, the results of the initial indicators for each image are determined and provide the basis for the decision. It is necessary to explain the amount of threshold, and the filter in this case study depends on the type of

Table 6.2 Wavelet filter families.

Wavelet family	Filter
Daubechies	"db1" or "haar", "db2", "db3", "db5", "db10", "db15", "db20", "db25", "db30", "db35", "db40", "db45"
Coiflet	"coif1", "coif2", "coif4", "coif5"
Biorthogonal	"bior1.3", "bior1.5", "bior2.4", "bior2.6", "bior3.1", "bior3.3", "bior3.7", "bior3.9", "bior5.5", "bior6.8"

Table 6.3 An example of wavelet filters family.

Haar	Mean square error		
k	MSEh	MSEv	MSEd
0	176.02	173.24	89.56
0.1	149.27	153.84	117.96
0.2	145.70	162.03	117.72
0.3	63.64	75.89	70.07
0.4	100.20	78.12	166.18
0.5	175.72	76.05	171.28
0.6	139.87	98.31	888.43
0.7	104.14	145.53	1741.60
0.8	52.26	122.72	1956.75
0.9	36.94	122.00	1282.44
1	146.64	193.75	1288.65

image, amount of moisture, color of the asphalt mixture, type of pavement, and even the age of the pavement. Then, for the value of k in each filter, the average square error is calculated. For example, Table 6.3 provides an example of an image database taken from a pavement section.

According to the results of the case study, in order to determine the appropriate filter and optimal *k* value for each of the presented features, the average of the square error values of each feature is given separately. The four filters and their *k* values are shown with the lowest MSE for each index according to Table 6.4. An example of a horizontal, vertical, and diagonal features extracted from

Table 6.4 An example of selected filters and their *k* values with the lowest mean square error for each feature for multidirectional evaluation in SR of pavements.

SRI	Filter	*k*	MSE
SRIh	Haar	0.9	36.94
	db45	0.4	42.71
	db45	0.1	48.88
	Haar	0.8	52.26
SRIv	bior3.7	0.5	50.36
	db30	0.4	50.53
	db40	0.7	55.75
	db10	0.6	57.59
SRId	db45	0.6	54.10
	coif5	0.7	54.27
	db40	1	57.56
	bior3.3	0.6	59.13

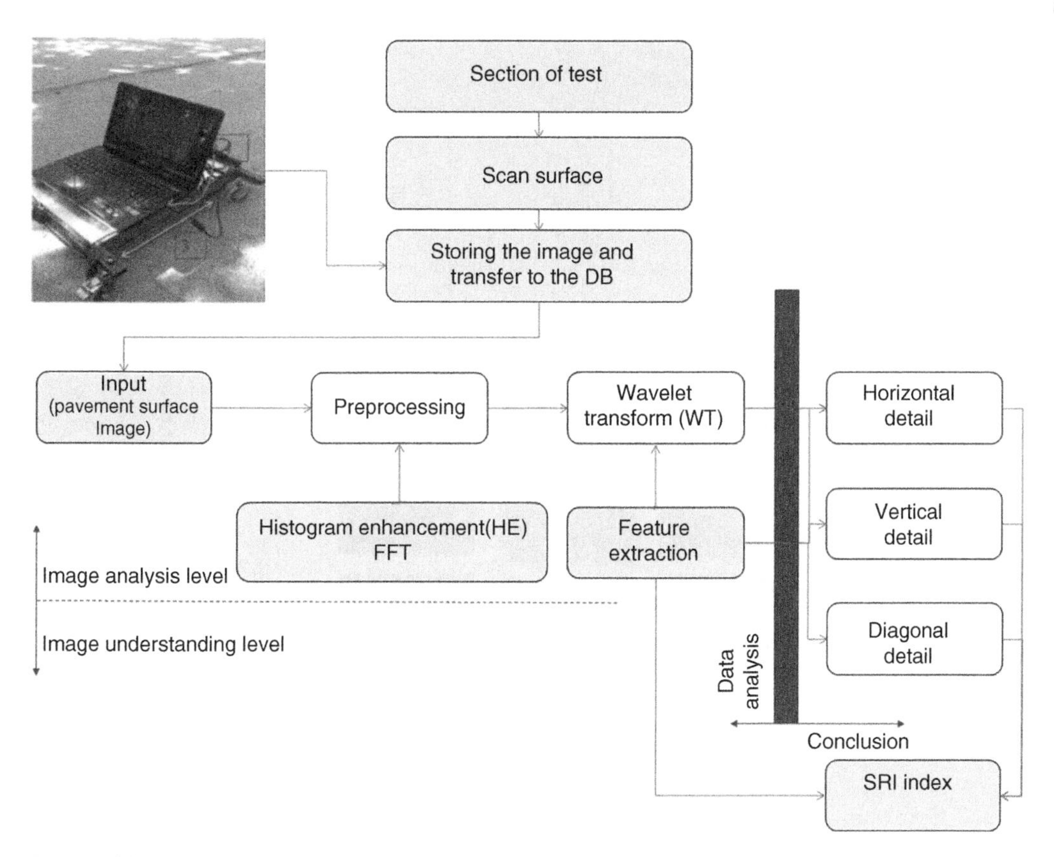

Figure 6.35 Hardware of automated image acquisition system (IAS) for pavement skid surface evaluation, and the process of knowledge-based approach (KBA) for multidirectional decision on pavement skid resistance (SRI = SR index).

the haar filter is given for a set of images taken with a pavement texture analyzer. Each graph in Figure 6.35 corresponds to one of the values of *k*. The results show that for each image, a change in the value of *k* leads to a corresponding change in the value of skid resistance index (SRI). Therefore, the optimal choice of these values plays an important role in obtaining the correct results. In addition, the results of horizontal and vertical SRIs approach the actual value by increasing the values of *k*. Also, for most images, the results of oblique SRIs are closer to the exact results (Figure 6.36).

To select the optimal filter from the four selected filters with improved features, calibration can be done more easily based on the following Eq. (6.72):

$$\begin{aligned} TH_i &= mean\left(var\left(\psi^i(x,y)\right)\right), \\ Fo &= SRIi^a/A, \\ A &= \frac{\sum_{i=1}^{22}(SRI_i - BPN_i)}{22}, \\ a &= \frac{\sum_{i=1}^{22} a_i}{22}, \end{aligned} \tag{6.72}$$

where *Fo* is the optimal of the wavelet transform filter (i = H,V & D).

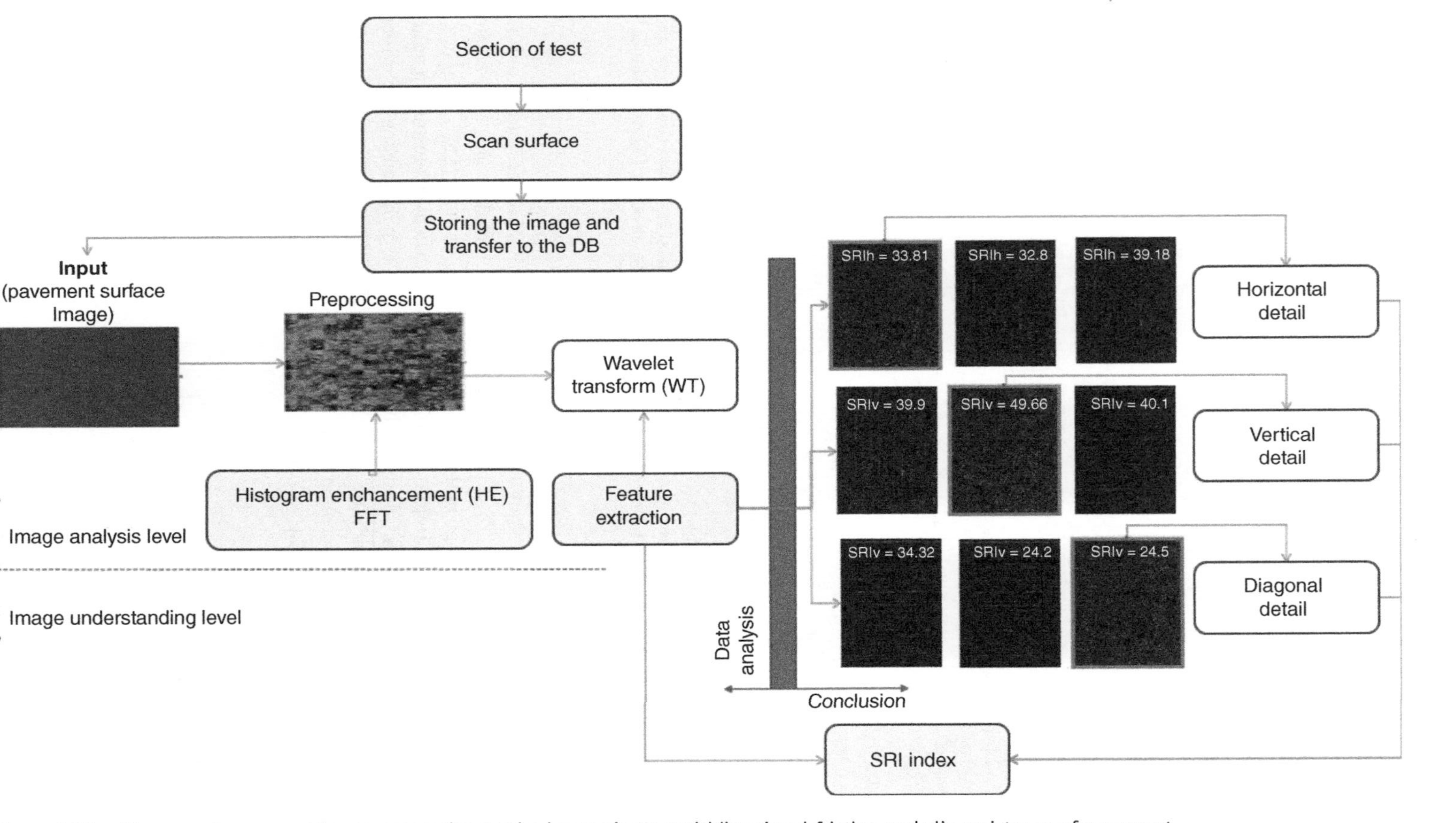

Figure 6.36 Diagram of pavement image processing method to evaluate multidirectional friction and slip resistance of pavement.

Table 6.5 Results of multidirectional judgment method in SR pavement based on selected filter using "*a*" and "*A*" parameters.

SRI	Filter	*K*	SRIia/A	MSE
SRIh	Haar	0.9	(SRIh^1.175)/1.876	32.61
	db45	0.4	(SRIh^1.43)/4.01	45.31
	db45	0.1	(SRIh^1.27)/2.28	70.07
	Haar	0.8	(SRIh^1.35)/4.12	34.43
SRIv	bior3.7	0.5	(SRIv^1.48)/4.18	76.7
	db30	0.4	(SRIv^1.5)/4.81	60.52
	db40	0.7	(SRIv^1.49)/4.76	48.39
	db10	0.6	(SRIv^0.495)/0.168	30.25
SRId	db45	0.6	(SRId^1.25)/2.25	58.04
	coif5	0.7	(SRId^1.326)/2.776	34.91
	db40	1	(SRId^1.43)/3.96	83.41
	bior3.3	0.6	(SRId^0.167)/0.05	46.4

For each image, there is a special a_i whose SRI partially improves the image using the Fo method and is used for the effective value of "a." Then, the SRI features are modified based on the following equations. For example, the results and the corresponding mean squares in Table 6.5 show that some results were improved while others were qualitatively reduced.

In order to extract the feature and its related parameters, each group of filters is compared in terms of the smallest MSE. The three filters with the closest output to real and accurate results include haar ($k = 0.9$) for the horizontal feature SR (SRIh), db10 (k = 0.6) for the vertical feature SR (SRIv), and coif5 (k = 0.7) for the diagonal feature SR (SRId).

The next step is to improve the quality of the extracted features and apply the calibration coefficients using the variance of the captured images. Based on the proposed method, the feature is calculated as Eq. (6.73):

$$Foc = SRIi^{\gamma_i a_i} / \beta_i A_i \tag{6.73}$$

where *Fo* is the Optimal details of the wavelet transform filter (i = H,V & D);*I* is the horizontal, vertical, and diagonal; γ_i is the calibration coefficient for $a_i$that changes by image variance; β_i is the calibration coefficient for $A_i$that changes by image variance.

Regardless of the round lines, the results were enhanced by applying the adjustment coefficients in Table 6.5, where MSE1 and MSE2 are the MSEs before and after the application of the calibration coefficients, respectively, which resulted in the extraction of more reliable features than the conventional method.

By designing parameters "a" and "A" and using calibration coefficients, the results of the extracted features are improved and closer to the outputs corresponding to reality. The main difference between this method and other methods in assessing the surface friction of road infrastructure, such as road surface and airports, is the presentation of significant features in different directions. To obtain a hybrid index as a feature for evaluating and determining the status of pavement, a combination of the three main features can be used, as Eq. (6.74):

$$SRI = Ln\left(e^{SRI_h} + e^{SRI_v} + e^{SRI_d}\right)$$
$$SRI = Ln\left(SRI_h + e^{SRI_v} + SRI_d\right)$$
$$SRI = 3.5*Ln(SRI_h{}^e + SRI_v{}^e + SRI_d{}^e)$$
$$SRI = Ln\left(\left(SRI_h + SRI_h + SRI_d\right)*\pi^{(SRI_h + SRI_h + SRI_d)}\right)^{0.75} \tag{6.74}$$

where SRI is the ultimate SR index; and SRI_i is the horizontal, vertical, and diagonal SR index.

A variety of methods and equations can be used to select the relation related to the features extracted from the image. Based on the MSE feature index, each combination equation is calculated, then the best equation with less error is selected. It should be noted that this method is only a case study, and to develop it for all pavements and operational works, calibration should be performed based on the extracted database, and each section may have its own relationship.

For example, considering a constant coefficient of 3.5 in Equation SRI, the calibration coefficient is calculated based on the variance of the database images using Eq. (6.74):

$$SRI = \alpha*3.5*Ln(SRI_h{}^e + SRI_v{}^e + SRI_d{}^e)$$

$$\alpha = \begin{cases} 3.33 & var_i < 0.013 \\ 3.64 & 0.013 \le var_i < 0.02 \\ 4.06 & 0.02 \le var_i < 0.022 \\ 4.21 & 0.022 \le var_i \end{cases}$$

where var_i is the variance of I = horizontal, vertical, and diagonal components of each pavement image.

Figure 6.38 summarizes the pavement image-processing method for extracting features related to pavement texture. According to this method, pavement image elements are transformed into horizontal, diagonal, and vertical components with the help of the wavelet method and the selected filter family. Then, based on the optimal k values and the first level, the SRI-related features, including horizontal, vertical, and diagonal SRI features, are extracted based on the selected filters. Extracted features horizontal, vertical, and diagonal SRIs based on the tuned coefficient, optimized filters are used to improve the quality of the extracted results.

In this case study, applied features of an automated system based on wavelet transform are provided to evaluate the slip resistance of pavement. The purpose of this practical example is to demonstrate the importance of using meaningful features to interpret problems. As shown, features are first extracted from the image and then calibrated using the correction multiplier to obtain better results. The results reveal the acceptable performance of these methods for the optimal interpretation of objects.

In addition, wavelet analysis is recognized as a powerful tool for fast tissue analysis, which reduces noise and simultaneous image compression. Therefore, this method can be used to obtain texture features. Comparison of accurate and real data based on image processing shows that these methods provide greater detail, such as the possibility of directional analysis, compared to the old methods. Moreover, these methods also offer faster analysis speed and less human error than traditional methods. In general, the process of data collection and analysis is very fast and allows quick access to a certain feature by the decision-maker. Performing analysis at the surface and in variable directions (e.g. theory of separate effects of vertical, horizontal, and diagonal components of pavement texture on SR index) to acquire reproducible results is the most important advantage of this method. Like image analysis, pavement texture extracted various features in different humidity and temperature conditions.

6.10.13 Case 3: Image-Based Feature Extraction for Pavement Texture Drainage Capability Evaluation

As another case study on the extraction of applied features, the surface drainage of road pavement at the network level, which is associated with a reduction in the number of accidents, is evaluated. One of the most important factors for off-road vehicles in rainy weather conditions is the reduction of friction due to the low quality of drainage of pavement surfaces. Since it is not possible to evaluate the infrastructure in real conditions, it is best to perform pavement drainage evaluation system in simulated conditions (rainfall) using an automated drainage evaluation system. For this purpose, the saturation state of the pavement surface is created using an innovative device, and at different times, the pavement surface is imaged and features related to drainage quality are extracted using image processing. As mentioned in the previous case study, using a suitable method to increase and prepare image quality in order to extract features has a very important role in obtaining accurate and practical results. For this purpose, multilevel analysis methods have been used to increase the quality level of the features extracted from pavement images. For this purpose, various features, such as TIME, peak signal-to-noise ratio (PSNR), SNR, MSE, mean absolute error (MAE), MSE, universal image quality index (UQI), and structural similarity image (SSIM), are analyzed in order to determine the optimal method for assessing the drainage level of road pavement. Comparison between the obtained results shows that the Shearlet conversion method outperforms other methods in providing image and processed features. In addition, it was found that regatta conversion is a more suitable method for directional and multilevel image analysis for applications that require time than other features. In the following section, each of these features is introduced, followed by a brief comparison of each method.

To conduct this case study, a special piece of hardware was created to obtain images of the pavement surface, which is registered as an international patent (more information is available in U.S. Patent No. 10,197,486.). In short, the system utilizes a digital camera that analyzes and documents the texture of the pavement surface and the drainage process. A dynamic frame was designed to simulate rainfall conditions and saturation of the pavement along with imaging and positioning components to record the location inside the device. The main focus of this work is on processing and improving images. Figure 6.35 presents the hardware and data collection method and the pavement drainage assessment framework.

To capture pavement information using a special device, images are often affected by noise or irregular objects. The purpose of noise cancellation is to preserve the important features of image data associated with surface drainage as much as possible. In addition, image compression means reducing the amount of data required to display an image with a significant impact on data quality while maintaining the important features. Various features, such as PSNR, MSE, MAE, SNR, UQI, SSIM, and TIME, are used to measure the performance of the various methods shown in Figure 6.37. All relevant features are summarized below.

Time is an important and effective feature for measuring the performance of various algorithms and selecting optimal features. In infrastructure management, designers prefer to use fast methods and features with less computational load. Since most of the computational load is related to the processor hardware, it is necessary to clearly state the type of processor in the reports, such as the Sony VAIO VGN-FW455D that is used for the surface drainage processing of the pavement.

As defined in Chapter 5, the PSNR reveals the maximum signal-to-noise ratio in decibels between two images. This ratio is often used as a measure of the quality difference between the original and compressed images. The higher the PSNR, the better the compressed or reconstructed image quality. This feature can also be used to assess the density of water-coarse pavement texture. In this

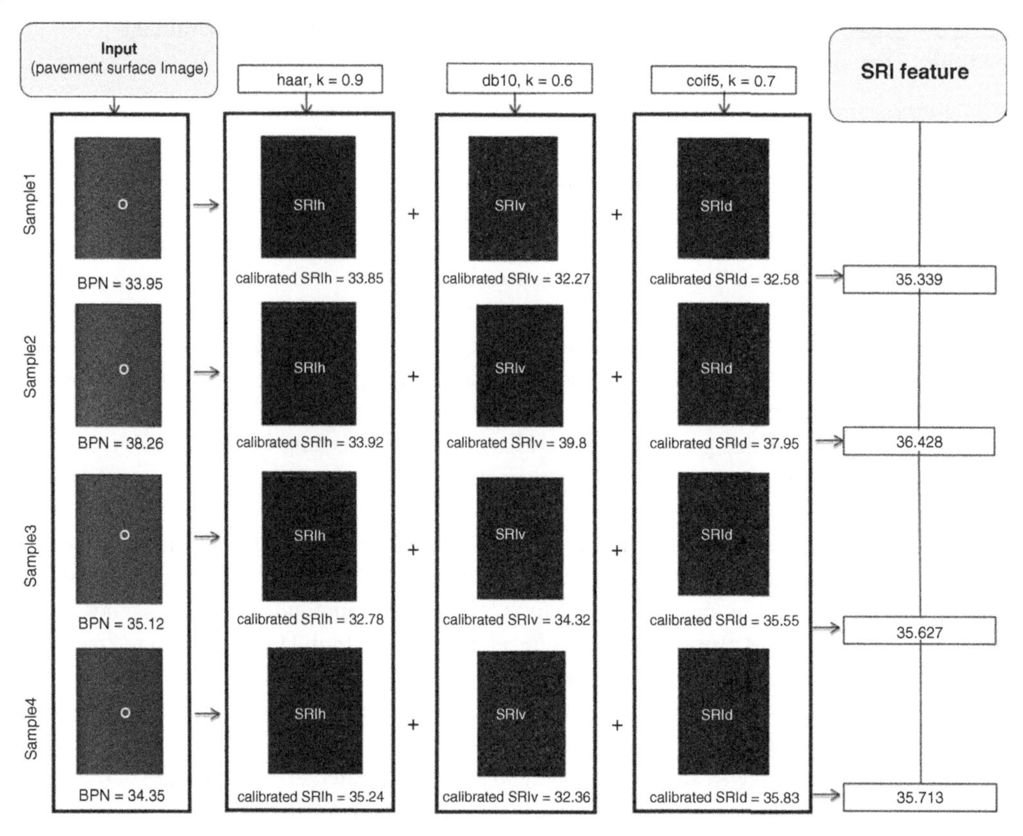

Figure 6.37 Examples of multidirectional features in pavement images to evaluate the slip resistance and friction of pavement texture using image analysis.

regard, the image without water is compared with the image in which part or all of the pavement is saturated with water caused by artificial rain. The PSNR feature is calculated by Eq. (6.75):

$$PSNR = 10\,log\,10\,\frac{\sum_{i=1}^{row}\sum_{j=1}^{col}255^2}{\sum_{i=1}^{row}\sum_{j=1}^{col}[I(i,j)-D(i,j)]^2} \tag{6.75}$$

where $I\,(i,j)$ is the original image; and $D\,(i,j)$ is the denoised image.

The signal-to-noise ratio (SNR) feature is a used to evaluate an image and the amount of signal in the original image relative to that in the image saturated with artificial rainwater created by a special device. The cross-sectional signal level in dry conditions compared to saturation conditions in different percentages can be evaluated with this feature. This feature is the ratio of signal power to noise power (effect of cross-section saturation) and is defined as follows by Eq.(6.76):

$$SNR = 20\,log\,10\left(\frac{a_{max}-a_{min}}{s_n}\right),$$

$$s_n = \sqrt{\frac{1}{\Lambda-1}\sum_{(m,n)\epsilon R}(a[m,n]-m_a)^2},$$

$$m_a = \frac{1}{\Lambda}\sum_{(m,n)\epsilon R} a[m,n], \tag{6.76}$$

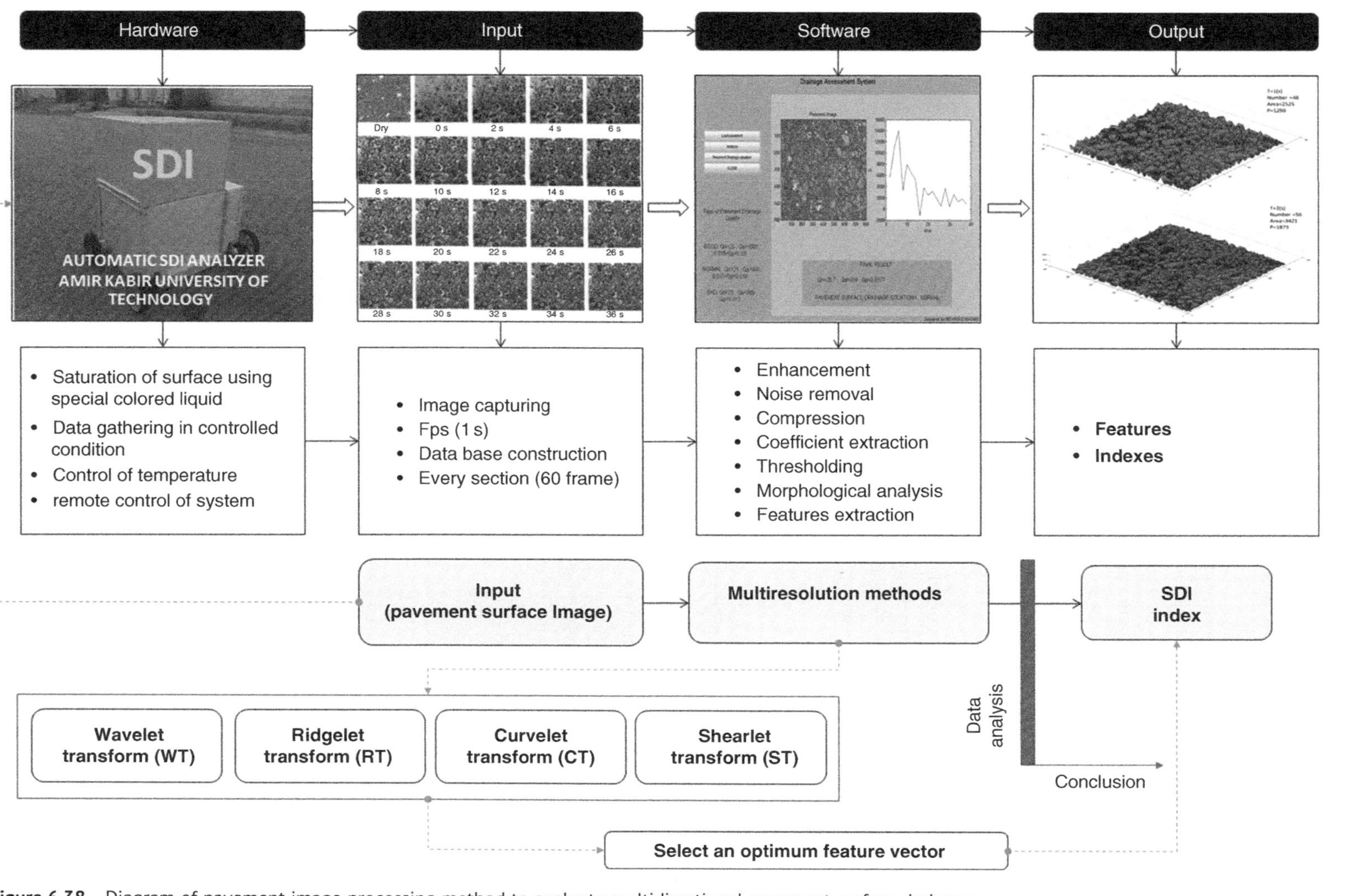

Figure 6.38 Diagram of pavement image processing method to evaluate multidirectional pavement surface drainage.

where a_{max} is the original image; $D(i,j)$ is the denoised image; a_{min} is the the pixel value with minimum intensity in the image of interest; s_n is the the standard deviation of noise; m_a is the sample mean of the pixel brightness in region R; Λ is the number of pixels in region R; and $a[m, n]$ is the pixel value.

The MSE feature is the cumulative square error between the compressed and original images, which indicates the amount of variation between the dry image and the image with different saturation percentages. The lower the MSE value, the lower the saturation rate and the lower the error rate after compression and noise cancellation. The MSE feature is related to the MSE using Eq. (6.77):

$$MSE = \sum_{i=1}^{row} \sum_{j=1}^{col} [I(i,j) - D(i,j)]^2 \tag{6.77}$$

where row and col are the numbers of rows and columns, respectively; and $I(i, j)$ is the original image.

The MAE feature gives the cumulative absolute error between the compressed and original images. Accordingly, this feature indicates the difference between the dry image and image with different saturation percentages. The lower the MAE, the less error there is, and the implication is that drainage is better and water levels remain low. The MAE feature equation is calculated as follows using Eq. (6.78):

$$MAE = \sum_{i=1}^{row} \sum_{j=1}^{col} |I(i,j) - D(i,j)| \tag{6.78}$$

where row and col are the numbers of rows and columns, respectively; and $I(i, j)$ is the original image.

The UQI is a feature that measures the distortion caused by the loss of correlation, brightness, or contrast. In this regard, the effect of saturation or insufficient drainage can affect the results, and this index is sensitive to the ability of the pavement surface to drainage. The feature is in the range between 0 and 1, whereby the better the drainage, the closer this index is to zero, and the slower the drainage, the closer the number is to 1. The UQI feature is defined using the following Eq. (6.79):

$$\begin{aligned} \text{UQI} &= \frac{4\sigma_{1,K}\overline{I} \times \overline{K}}{(\sigma_1^2 + \sigma_K^2)\left(\overline{I}^2 + \overline{K}^2\right)}, \\ \sigma_{1,K} &= \frac{1}{n-1}\sum_{k}^{n}(I_k - \overline{I})\,(K_k - \overline{K}) \end{aligned} \tag{6.79}$$

where $\overline{I}$ and $\overline{K}$ are the mean; and σ_1 and σ_K are the variance of original and denoised images.

The SSIM feature is used to measure the similarity of two images, whereby two images can be evaluated when they are dry, and can measure the amount of drainage at different times with different degrees of saturation. The closer this index is to 0, the closer the similarity to the drying time and the greater the pavement capability in surface drainage. Using the SSIM feature, any distortion is represented by a combination of three different factors: correlation loss, brightness distortion, and contrast distortion. The dynamic range of SSIM is $[-1, 1]$. The SSIM feature is calculated using Eq. (6.80):

$$\begin{aligned} SSIM &= \left(\frac{\sigma_{xy}}{\sigma_x\sigma_y}\right)^{\alpha} \times \left(\frac{2\overline{XY}}{\overline{X}^2 \times \overline{Y}^2}\right)^{\beta} \times \frac{2\sigma_x\sigma_y}{\sigma_x^2 \times \sigma_y^2}^{\gamma}, \\ &\left(\frac{\sigma_{xy}}{\sigma_x\sigma_y}\right)^{\alpha}, \left(\frac{2\overline{XY}}{\overline{X}^2 \times \overline{Y}^2}\right)^{\beta} \text{and } \frac{2\sigma_x\sigma_y}{\sigma_x^2 \times \sigma_y^2}^{\gamma} \end{aligned} \tag{6.80}$$

where $X = \{x_i|i = 1, 2, ..., N\}$; $Y = \{y_i|i = 1, 2, ..., N\}$; $\alpha > 0$, $\beta > 0$ and $\gamma > 0$ are parameters used to adjust the relative importance of the three phrases of the equation; and σ_x and σ_y are estimates of the contrast in x and y directions, respectively.

In order to evaluate the performance of the various methods mentioned, a review was performed on the images using the extracted features. In this part of the case study, experimental results and comparisons are presented to demonstrate the performance of different processing algorithms, which are based on the database collected through the device. It is necessary to explain the imaging and lighting conditions that are highly significant.

First, a comparison was made using the extracted features for each of the methods, and then a general comparison was performed to select the optimal method. In this case study, 30 images were used, which are included in Appendix 4.

Table 6.6 shows a comparison of the results of wavelet transform performance using the different filters introduced in Chapter 5. As can be deduced from the table, some features work well, while others do not make much sense using the method.

In this case study, by comparing different filters, it is clear that the demeyer filter performs better than other filters for most criteria. Haar daubechies, biorthogonal, and reverse biorthogonal filters performed similarly, and symlet filters in PSNR, SNR, UQI, and SSIM were more sensitive to changes. The coiflet filter works slightly better than the symlet filter. Images during the time should not change significantly if pavement is properly drained, considering that the greater the variation, the greater the inability to adequately drain.

It should be noted that there is no difference in MSE and MAE between each filter. Therefore, it can be concluded that changing the filters does not make any difference in the calculated errors. Moreover, symlet, inverted biorthogonal and haar filters had less time than the demeyer filter, but this difference is miniscule (0.02–0.58 seconds). The dmeyer family has the highest SSIM. It also has the highest UQI, PSNR, and SNR of 0.03, 21, and 14 over the coiflet filter, respectively (see table 6.7).

Table 6.6 A comparison of the results of wavelet transform performance using the different filters.

Filter	*K*	MSE1	MSE2
Haar	0.9	32.61	13.45
db10	0.6	30.25	22.21
coif5	0.7	34.91	18.5

Table 6.7 An example result of wavelet domain in drainage surface index (DSI) pavement.

Filter	Time (s)	Psnr (db)	SNR (db)	MSE	MAE	UQI	SSIM
Haar	2.34	19.3719	10.3241	6.2025e + 004	232.2252	0.8982	0.9985
Db	2.49	19.3719	10.3241	6.2025e + 004	232.2252	0.8982	0.9985
Coif	2.48	22.0557	13.0080	6.2025e + 004	232.2252	0.9426	0.9990
Bior	2.88	19.3719	10.3241	6.2025e + 004	232.2252	0.8982	0.9985
Rbio	1.78	19.3719	10.3241	6.2025e + 004	232.2252	0.8982	0.9985
Sym	2	21.8821	12.8343	6.2025e + 004	232.2252	0.9399	0.9990
dmey	2.36	25.5104	16.4626	6.2025e + 004	232.2252	0.9708	0.9995

Table 7.6 compares the performance of ridglet transform in reducing noise and image compression using four filters. Based on the results of the case study in the Appendix 4 image bank, all MSE, MAE, UQI, and SSIM show equal values. In addition, haar, daubechies, biorthogonal, and biorthogonal filters showed similar results and higher performance compared to the other filters in terms of PSNR and SNR parameters. The use of symlet and coiflet filters in the ridgelet method exhibited lower performance, but the results obtained with dmey filter were the worst of all other methods. The time difference across all filters was about 0.8 seconds, where the daubechies filter took a minimum of 1.7 seconds. Therefore, considering all the features obtained from the filters, the feature extracted from the daubechies filter in a set of filter families demonstrated better performance in the ridgelet transform.

Four subsets of the daubechies filters were compared based on the better performance of this filter in the Ridgelet method (Table 6.8). Here, db1 has the highest PSNR and SNR compared to the average of other filters. Also, in this comparison, the lowest time feature is related to "db1." Therefore, based on this comparison, db1 is the best filter for ridgelet transform.

Table 6.9 shows a comparison of the performance of the transform transformer in reducing noise impact and image compression capability using the filter family. This capability is also used to measure pavement drainage potential.

Like the wavelet, the haar family, daubechies, biorthogonal and reverse biorthogonal showed similar performance. In order to evaluate the differentiation for the dry section and the drainage potential of the pavement, the relevant indices can be compared. In this case, different filters can be compared. In addition, symlet and coiflet filters represent close competition with slight superiority for coiflet in PSNR, SNR, and UQI. MSE and MAE are equal in all filters. The dmeyer filter offers the best results compared to other filters and performs better on four parameters (PSNR, SNR, UQI, and

Table 6.8 An example results of ridgelet domain in DSI pavement.

Filter	Time (s)	Psnr (db)	SNR (db)	MSE	MAE	UQI	SSIM
Haar	1.8	101.2772	92.2320	6.1974e + 004	232.1057	1	1
Db	1.7	101.2772	92.2320	6.1974e + 004	232.1057	1	1
Coif	2.5	100.4977	91.4525	6.1974e + 004	232.1057	1	1
Bior	1.8	101.2772	92.2320	6.1974e + 004	232.1057	1	1
Rbio	1.8	101.2772	92.2320	6.1974e + 004	232.1057	1	1
Sym	1.8	100.2639	91.2187	6.1974e + 004	232.1057	1	1
dmey	2.2	98.0846	89.0394	6.1974e + 004	232.1057	1	1

Table 6.9 The performance of ridgelet transform in using four filters.

Filter	Time (s)	Psnr (db)	SNR (db)	MSE	MAE	UQI	SSIM
Db1	1.7	101.2772	92.2320	6.1974e + 004	232.1057	1	1
Db2	2.9	101.1839	92.1306	6.1971e + 004	232.1310	1	1
Db10	2.1	98.6610	89.6077	6.1971e + 004	232.1310	1	1
Db15	3.1	99.2742	90.2209	6.1971e + 004	232.1310	1	1

Table 6.10 The performance of curvelet transform in using different filters.

Filter	Time (s)	Psnr (db)	SNR (db)	MSE	MAE	UQI	SSIM
Haar	89.5	27.3484	18.3006	6.2025e + 004	232.2252	0.9768	0.9995
Db	108	27.3484	18.3006	6.2025e + 004	232.2252	0.9768	0.9995
Coif	136.5	31.6616	22.6138	6.2025e + 004	232.2252	0.9909	0.9998
Bior	111	27.3484	18.3006	6.2025e + 004	232.2252	0.9768	0.9995
Rbio	93	27.3484	18.3006	6.2025e + 004	232.2252	0.9768	0.9995
Sym	104	31.0382	21.9905	6.2025e + 004	232.2252	0.9895	0.9998
dmey	120	34.5235	25.4757	6.2025e + 004	232.2252	0.9952	0.9999

SSIM). The highest UQI, PSNR, and SNR with 0.005, respectively, with 9 and 12 show superiority over the coiflet. However, according to TIME, the best performance belongs to the haar with about 30 seconds less than the demeyer family. Hence, it can be concluded that in a specific task that determines time, haar is a good choice, but if higher quality is desired, the demeyer filter is the best choice.

Table 6.10 shows a comparison of the use of Shearlet converter for noise reduction and image compression with four filters for resolution including 7–9, maxflat, pyr, and pyrexc. MSE, MAE, UQI, and SSIM values are the same for all filters. Pyrex has the highest PSNR and SNR, which is 2% better than the average of other filters and the minimum computational time (2 seconds less than the average time of other filters) is better than other filters. Therefore, it is used in the shearlet method as an optimal filter to analyze pavement images and drainage potential and pavement texture.

Table 6.11 shows an evaluation of different filter families to reduce noise and compress images with contourlet transform as a filter for the Laplacian pyramid. It should be noted that MSE, MAE, UQI, and SSIM are similar in all filters. In addition, Coiflet and symlet filters performed poorly on PSNR and SNR. Although the haar, daubechies, biorthogonal, and reverse biorthogonal filters all performed similarly, the daubechies filter (2.8 seconds difference from the average time of other filters) was selected as the appropriate filter for converting the cannula because of the less time required.

As the optimal filters were determined using other methods, it is possible to evaluate all the features for each change in the image. For this purpose, a comparison has been made between the types of transform methods with the best filters, which were determined in the previous sections. The minimum time required and the optimal MSE and MAE belong to the ridgelet transform, but

Table 6.11 The performance of shearlet transform in using different filters.

Filter	Time (s)	Psnr (db)	SNR (db)	MSE	MAE	UQI	SSIM
9-7	8.7	305.6655	296.6177	6.2025e + 004	232.2252	1	1
maxflat	9.5	300.2109	291.1631	6.2025e + 004	232.2252	1	1
pyr	6.6	309.9713	300.9235	6.2025e + 004	232.2252	1	1
pyrexc	6.3	310.0177	300.9699	6.2025e + 004	232.2252	1	1

Table 6.12 The performance of Ridgelet transform in using different filters.

Filter	Time (s)	Psnr (db)	SNR (db)	MSE	MAE	UQI	SSIM
Haar	15.6	297.4252	288.3705	6.2028e + 004	232.2581	1	1
Db	13.7	297.4252	288.3705	6.2028e + 004	232.2581	1	1
Coif	15.5	239.1084	230.0538	6.2028e + 004	232.2581	1	1
Bior	17.9	297.4252	288.3705	6.2028e + 004	232.2581	1	1
Rbio	16.9	297.4252	288.3705	6.2028e + 004	232.2581	1	1
Sym	16.8	242.3197	233.2651	6.2028e + 004	232.2581	1	1

the Shearlet and Contourlet show higher SNR and PSNR. Based on these results, Ridgelet and Shearlet transform are better filters to evaluate the surface drainage potential of the pavement. The sensitivity of MSE and MAE features in Ridgelet is less than that of Shearlet and requires less time to calculate. However, on the other hand, Cheryl has a higher PSNR and SNR. Finally, it can be concluded that in projects where time is the most important parameter and is important, ridgelet transform works better, but in cases where image quality is inadequate, shearlet works better. Therefore, from the set of proposed methods and extracted features, Shearlet methods can be used as an optimal method to interpret the pavement texture and surface drainage potential of the pavement. time ($TIME_{Ridgelet} = 0.27\ TIME_{Shearlet}$). Based on the results pretested in tables (6.12) and (6.13), the other hand, however, shearlet has higher PSNR and SNR ($PSNR_{Ridgelet} = 0.33 PSNR_{Shearlet}$, $SNR_{Ridgelet} = 0.31\ SNR_{Shearlet}$).

In this section, various features have been used to calculate the surface drainage index of pavement based on images taken from pavement texture in conditions of saturation with artificial rain fall using Eq. (6.81):

$$\begin{aligned}
F1 &= -\sum_{i=1}^{M}\sum_{j=1}^{N} p(i,j)\log_2 p(i,j) \\
F2 &= \sum_{i,j} p(i,j)^2 \\
F3 &= C - H \\
F4 &= \left|\frac{count\sum_{i=1}^{m}\sum_{j=1}^{n} p(i,j) = 0}{count\sum_{i=1}^{m}\sum_{j=1}^{n} p(i,j) \geq 0}\right| \\
F5 &= \left|\frac{\sum_{i=1}^{m}\sum_{j=1}^{n} p(i,j) = 0}{count\sum_{i=1}^{m}\sum_{j=1}^{n} p(i,j) = 0}\right| \\
F6 &= F(i \mid D) = \text{Prob}\,[S_D(x,y) = i] \\
D &= (\Delta x, \Delta y)\ S_D(x,y)) = \mid S(x,y) - S(x+\Delta x, y+\Delta y) \mid
\end{aligned} \tag{6.81}$$

where $F1$ is the entropy of the pavement section; $F2$ is the energy of the pavement section; $F3$ is the Euler number of the pavement section; $F4$ is the ratio of number black pixels to the number of total pixels (RBT) of the pavement section; $F5$ is the average area of the black objects in the binary image (AABO) of the pavement section; $F6$ is the probability density function (PDF) of the pavement section.

In this case study, a method based on image processing and image features for automatic evaluation of asphalt surface drainage is presented. Specifically, morphological features have been

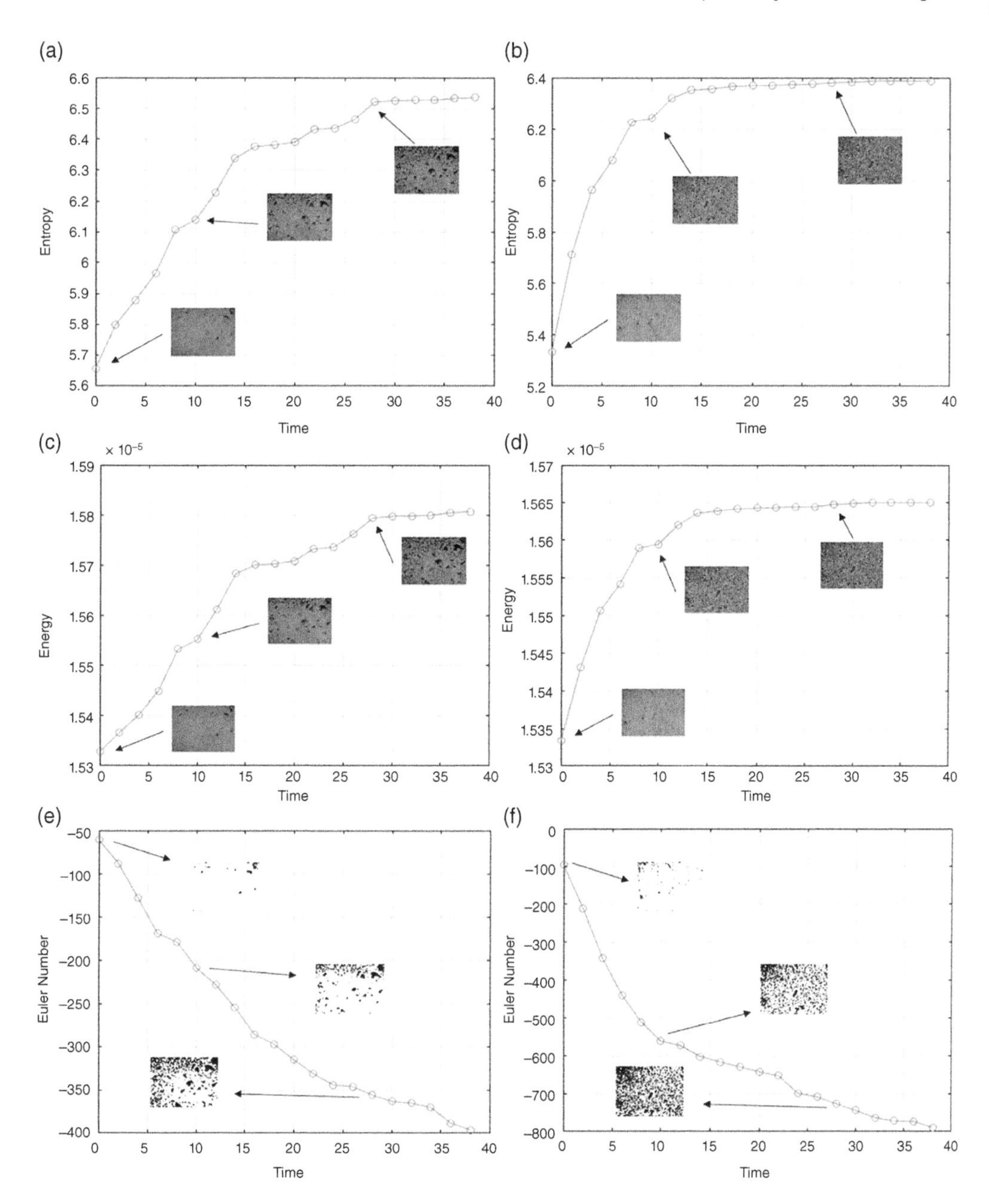

Figure 6.39 Comparison of feature changes with time based on pavement images over time: (a) entropy for sample 1, (b) entropy for sample 2, (c) energy for sample 1, (d) energy for sample 2, (e) Euler number for sample 1, and (f) Euler number for sample 2.

extracted and introduced to evaluate the drainage conditions of the pavement surface. Figures 6.39 and 6.40 shows Comparison of feature changes with time based on pavement images over time.

In this section, six features are introduced for surface drainage analysis. In general, advanced image-based methods can be used to assess and manage infrastructure, including safety management and assessment of road and airport pavement friction potential.

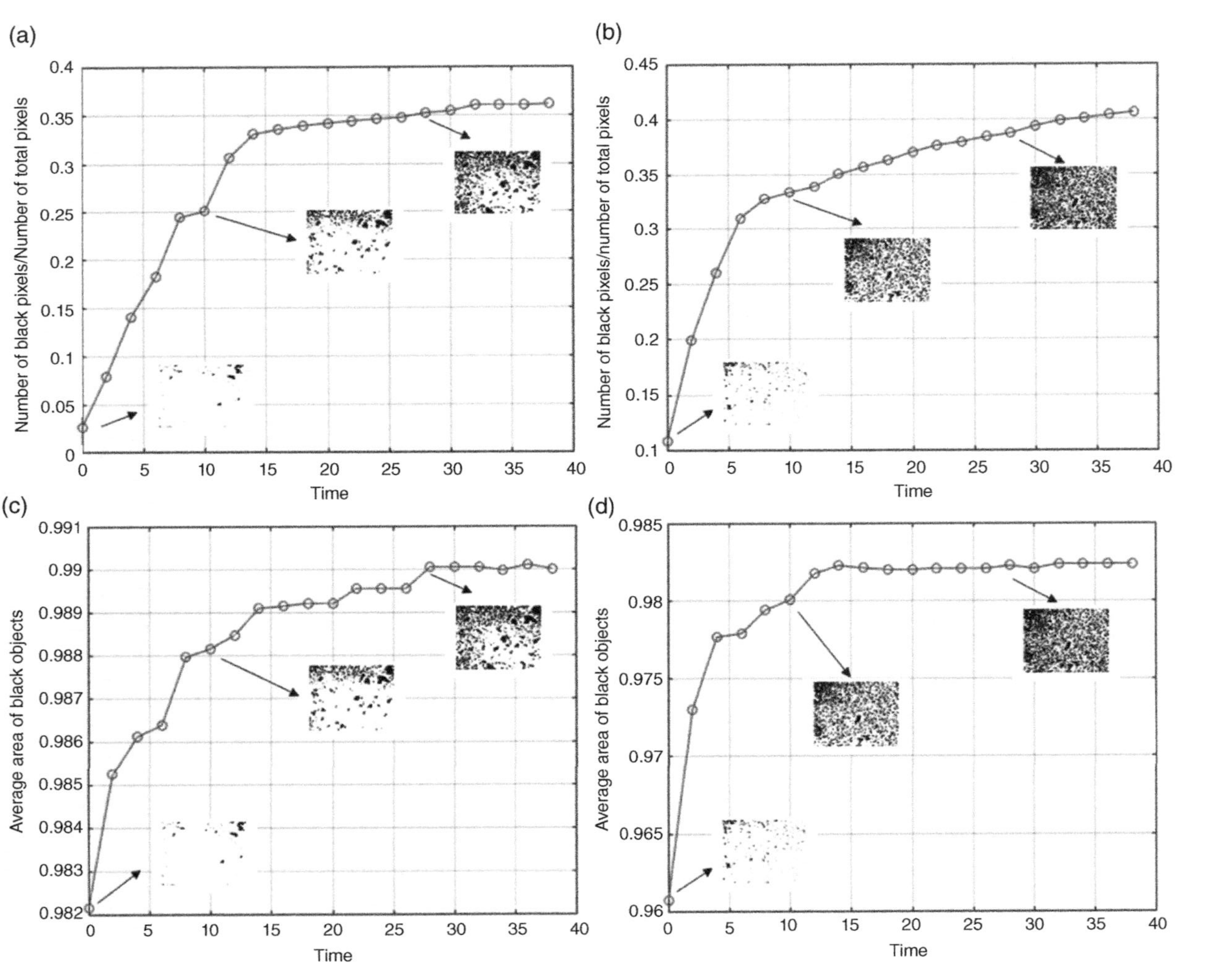

Figure 6.40 Comparison of feature changes with time based on pavement images over time: (a) ratio of number black pixels to the number of total pixels (RBT) for sample 1, (b) ratio of number black pixels to the number of total pixels (RBT) for sample 2, (c) Average area of the black objects in the binary image (AABO) for sample 1, and (d) Average area of the black objects in the binary image (AABO) for sample 2.

Table 6.13 The performance of contourlet transform in using different filters.

Filter	Time (s)	Psnr (db)	SNR (db)	MSE	MAE	UQI	SSIM
Db1	13.7	297.4252	288.3705	6.2028e + 004	232.2581	1	1
Db2	15.5	242.3366	233.2888	6.2028e + 004	232.2581	1	1
Db10	19	225.8498	216.8021	6.2028e + 004	232.2581	1	1
Db15	25.6	241.6502	232.6024	6.2028e + 004	232.2581	1	1

6.10.14 Case 4: Image-Based Features Extraction in Pavement Cracking Evaluation

The morphological characteristics of cracks are the most important feature of infrastructure assessment. Depending on the type and geometric characteristics, each crack has important characteristics that, in order to automate the inference, need to be individually analyzed and accurately extracted to control the overall crack failure assessment index. Manual measurement of such characteristics, discrete due to disconnected nature and additionally human error does not provide acceptable results. Therefore, it is necessary to use automated methods to extract these characteristics. An example of cracking, along with the main characteristics of the distress skeleton, is provided in Figure 6.40. As shown in the figure, each crack member has its own characteristics, the combination of which reflects the main nature of the crack, including type, severity, and extent.

Much research has been done to measure crack width, which is one of the main features of cracking. Most of these methods are based on the distance from the edge of crack. One of the most important drawbacks of this method is the uncertainty in determining the crack edge thresholds due to the fuzzy nature of the image. Therefore, based on these hypotheses, the distance from the edge perpendicular to the crack edge is considered as the crack width. The desired width according to standards, including ASTM6433, is the average crack width along the crack piece. However, the measurement of the average width in the manual method is done discretely. In general, due to the consideration of the direction of movement, the size of the width measured manually is different from the exact width. One of the objectives of this study is to provide an accurate method for measuring crack width in each section and to provide statistical details of the section with cracks.

Correct identification of the crack width and its location can be used to assess the location of the failure. As a general rule, with the growth of cracking, the crack width increases compared to the initial width, and the strain related to the crack width is considered as an indicator to identify the location of maximum stress and minimum resistance. This rule is different for sections with spalls, suffer small fractures at the joint due to traffic, and the width of the crack is reduced by the return, broken parts and its sinking into the pavement tissue. Therefore, the width of the crack alone is not a suitable criterion for determining the severity of failure and reduction of service. Spall is one of the most important failures in cracking assessment that most inspection regulations have given little attention, considering the lack of a clear method of quantitative assessment, and have generally dealt with it as a qualitative parameter in the category of failure severity assessment. There are two main hypotheses to identify and qualitatively evaluate of spalls: (i) Spall refers to the parts of cracks that are created by the secondary cracks adjacent to the main crack and cause cracking and reduction of crack resistance. According to this definition, the main cause of failure is caused by loading and fatigue conditions around the crack. In many cases, this breakdown leads to fatigue cracking. It should be noted that in the assessment of failure, fatigue cracks appear in various type of cracks, while the spalls are used as a main component in assessing the severity of

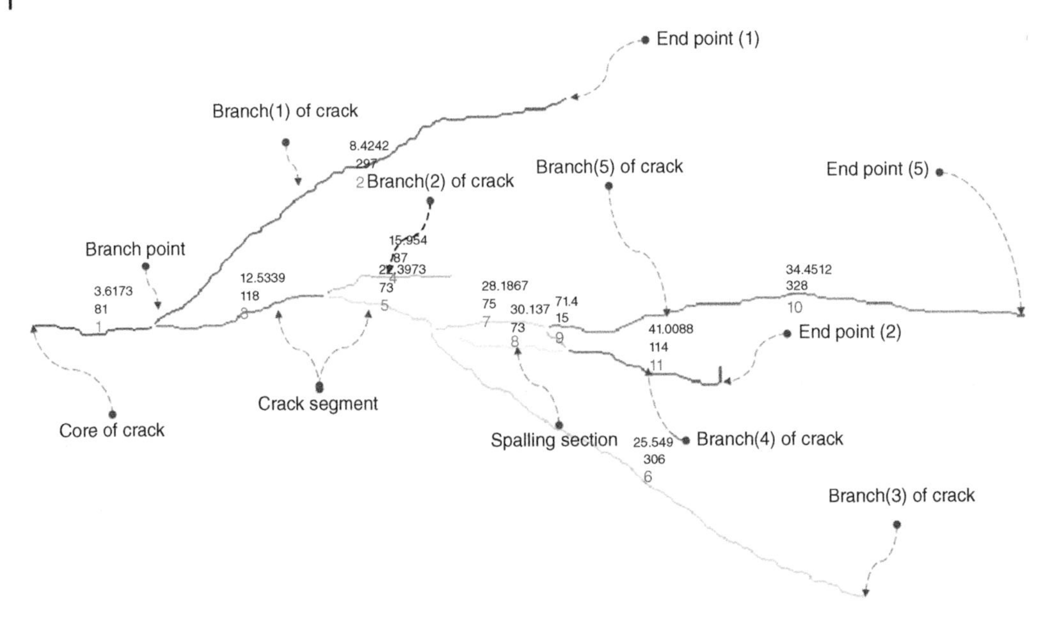

Figure 6.41 Various components of spalling and its relationship to cracking on the crack skeleton.

failure. (ii) Spall refers to corrosion around the crack edge due to weather conditions and separation, broken grains due to weathering, and reduced resistance of the adhesive. According to this definition, a spall is a progressive failure based on weather conditions that is the main cause of error in the evaluation of cracks. Figure 6.41 shows the different components of a spall and its relationship to cracking on the crack skeleton. In this study, it is hypothesized that spall cracks at the boundary of the connection with the main crack have breakage wedges that cause the piece to separate at a depth less than the thickness of the layer with the ruptured plate of the main crack. This secondary rupture plate reduces the bearing capacity of the pavement section and leads to progressive damage in the pavement section, which can be compared to pavement cancer. This hypothesis has been evaluated in the field research section, and its accuracy has been proven regarding the layers and how to connect with the main crack. Figure 6.42 shows a schematic view of the spall crack and its relationship to the main crack.

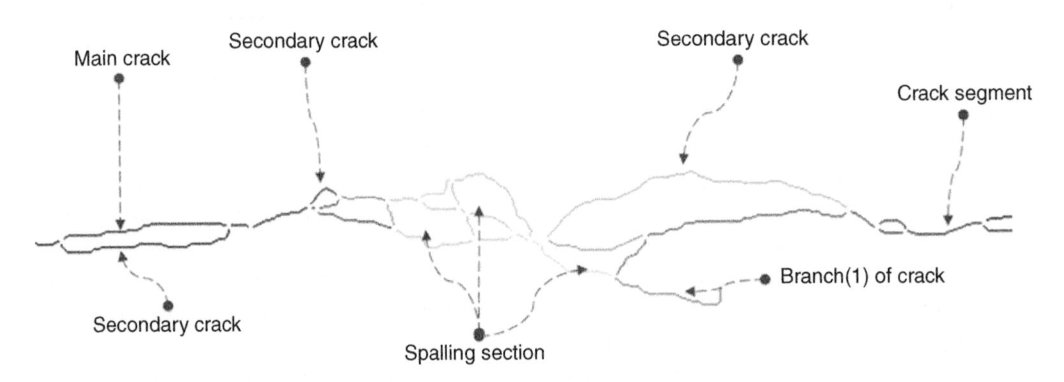

Figure 6.42 Various components of cracking on the crack skeleton.

As can be seen, the spall cracks attach to the main crack plate at points less than the cross-sectional height. But in the second phase, by reducing the bearing capacity of the pavement by reducing the cross-sectional modulus, by repeating the load, there is a significant reduction in the remaining life of the pavement. The main difference between a spall crack and a fatigue crack is the distance of this crack from the main crack as well as the height of *hs*. The larger this value, the closer the characteristics and behavior of spall cracking to fatigue cracking. In single fatigue, the *hs* point is equal to the height of the cross section, and there may be no connection point.

6.10.15 Automatic Extraction of Crack Features

In this section, using the image-processing tool, the features, and the characteristics of the crack are extracted automatically. These features include width of crack (WC) width, ratio of circular objects (RC) root, block cracking (BC) crack branches, and SC scales (see Figure 6.43). Based on these features, the main characteristic of cracks, including type, intensity, and extent, can be evaluated. After determining the key specifications, various types of features, such as uniformity index, spall person type 1 and 2 as well as general failure status index, can be extracted. Figure 6.44 presents the main steps for extracting key features in pavement cracking detection.

6.10.16 Extraction of Crack Skeleton Using Shearlet Complex Method

Preprocessing and the method selection stages are presented in Chapter 5. In order to reduce the background effect and noise, as well as to improve the quality of the cracking section using multilevel methods, the Shearlet method was selected according to the appropriate results in the processing stage and used in this section. After preprocessing, crack segmentation is performed using complex Shearlet coefficients. By extracting the Shearlet coefficients in the real and imaginary part, the Eq. (6.82) is used in order to segment the crack:

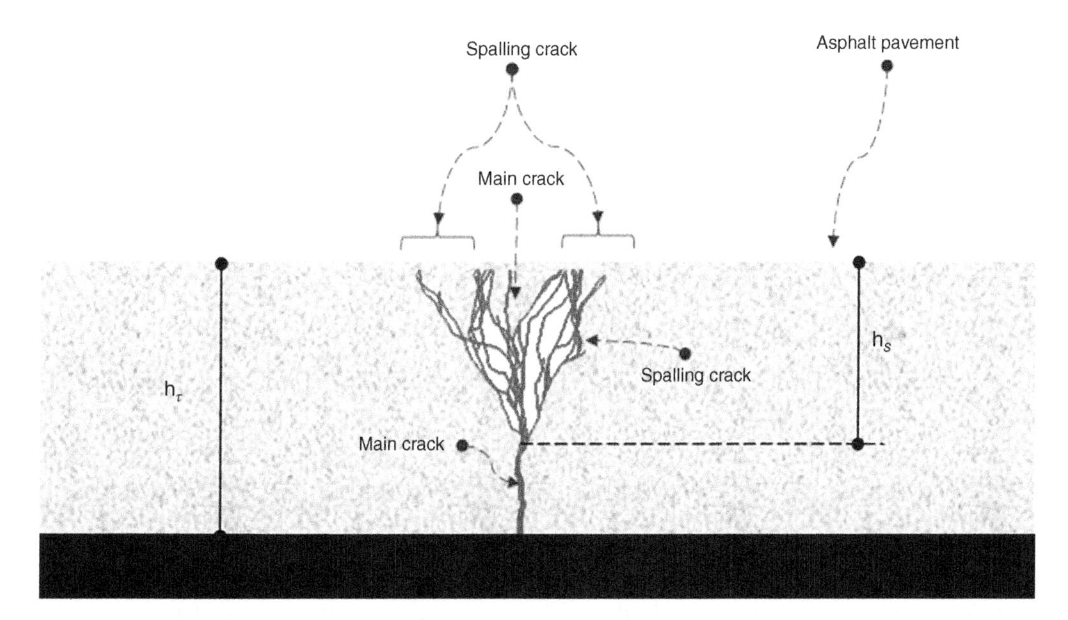

Figure 6.43 The pavement section displaying spalls and connected to the main crack shear plate at h_s height.

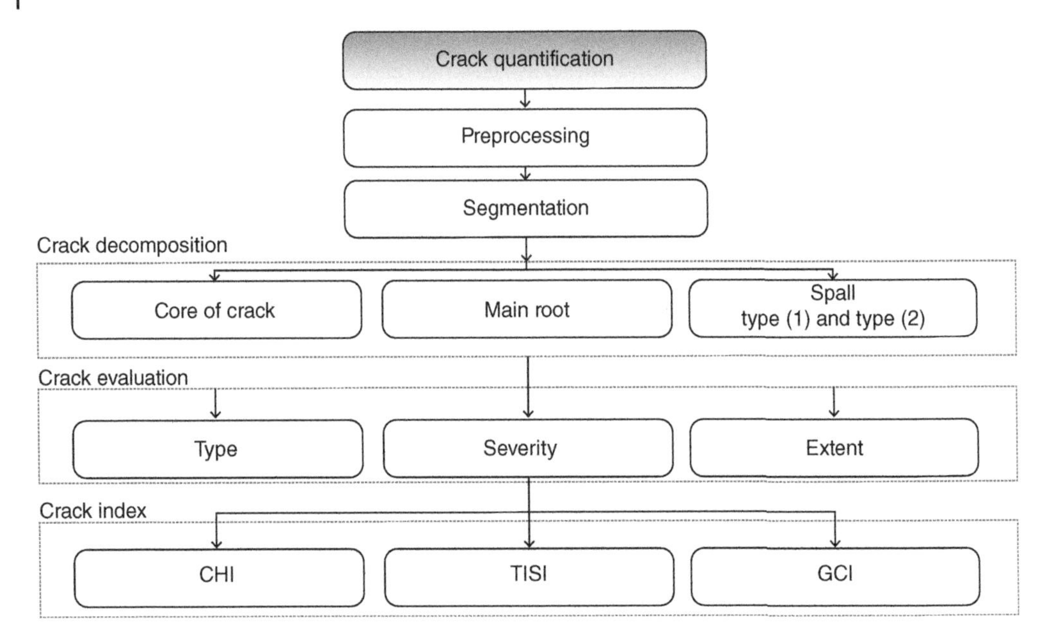

Figure 6.44 The main steps of extracting key features, pavement cracking assessment, and indexing.

$$E_{\psi^{\text{even}}\psi^{\text{odd}}}(f,x) = \frac{\left|\sum_{j=J_{\min}}^{J_{\max}} \left\langle f, \psi_{j,k_j,x}{}^{odd} \right\rangle\right| - \sum_{j=J_{\min}}^{J_{\max}} \left\langle f, \psi_{j,k_j,x}{}^{\text{even}} \right\rangle}{(J_{\max} - J_{\min} + 1) \max_{j\in\{J_{\min},\ldots,J_{\max}\}} \left|\left\langle f, \psi_{j,k_j,x}{}^{odd} \right\rangle\right| + \in}, \tag{6.82}$$

In this regard, to avoid ambiguity, the relationship with $\in > 0$, which is a small number, is considered. An example of segmented pavement images using this method is shown in Figure 6.44, in which the main cracking characteristics can be modified by changing the coefficients of the complex Shearlet. Therefore, the complex Shearlet is a flexible and efficient method for crack detection and separation. In addition, due to other functional features, such as noise reduction, crack quality improvement, and image analysis in different directions, it is known as an effective method in assessing pavement distress. After crack extraction, the cracked skeleton is extracted using the extraction morphological operators, and the main crack skeleton is used in a binary manner. Other general characteristics of the crack, such as the maximum amount of curvature and maximum placement direction, can be calculated. For example, the maximum angle is 177.5°, and the angle of curvature is 46.18 for Figure 6.44.

6.10.17 Calculate Crack Width Feature Using External Multiplication Method

Determining the crack width feature is always an important issue in assessing the severity of cracking. Although various methods have been proposed to calculate crack width, these methods are time-consuming and do not provide accurate information on crack width distribution. In this section, a mathematical method is used to calculate the crack width. By multiplying the crack skeleton by the distance vector, the vertical image is created from half the crack width, which is then doubled to calculate the crack width for each point of the skeleton (see Figure 6.45). This rule can be proved mathematically as follows:

Rule 1- Twice the external multiplication of the distance vector in the crack skeleton is equal to the crack width at each point. In order to prove this rule, it is assumed that the distance is calculated based on the Euclidean relation as follows:

$$\begin{aligned} a(i) &= \frac{a(b \times c)}{|b \times c|^2}(b \times c), a_\perp = a - a_v, ba_\perp = |b|.|a_\perp| \cos\theta \\ c(i) &= \frac{c(a \times b)}{|a \times b|^2}(a \times b), c_\perp = c - c_v, ac_\perp = |a|.|c_\perp| \cos\theta \end{aligned} \tag{6.83}$$

By multiplying the distance vector by the skeleton vector, the value of the crack width distribution perpendicular to the skeleton can be displayed on the crack. The greater the width, the greater the amount of skeleton absorbed at the same point and, consequently, the greater the width. Figure 6.46 shows the external multiplication rule used to calculate crack width.

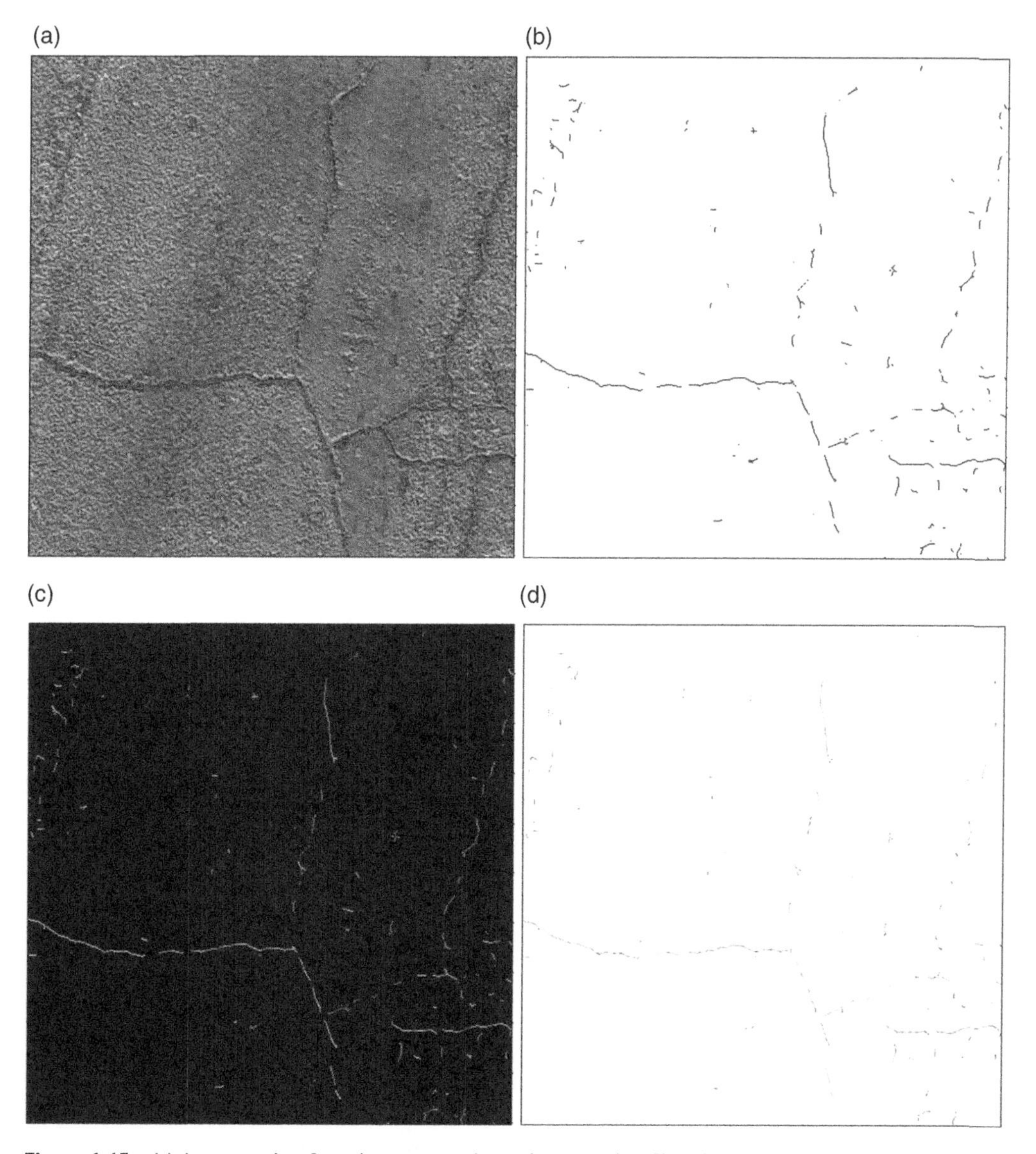

Figure 6.45 (a) An example of crack segmentation using complex Shearlet transform; (b) Extraction of the cracking skeleton using morphological operators; (c) Creating an image of crack angle changes; and (d) Image of crack curvature.

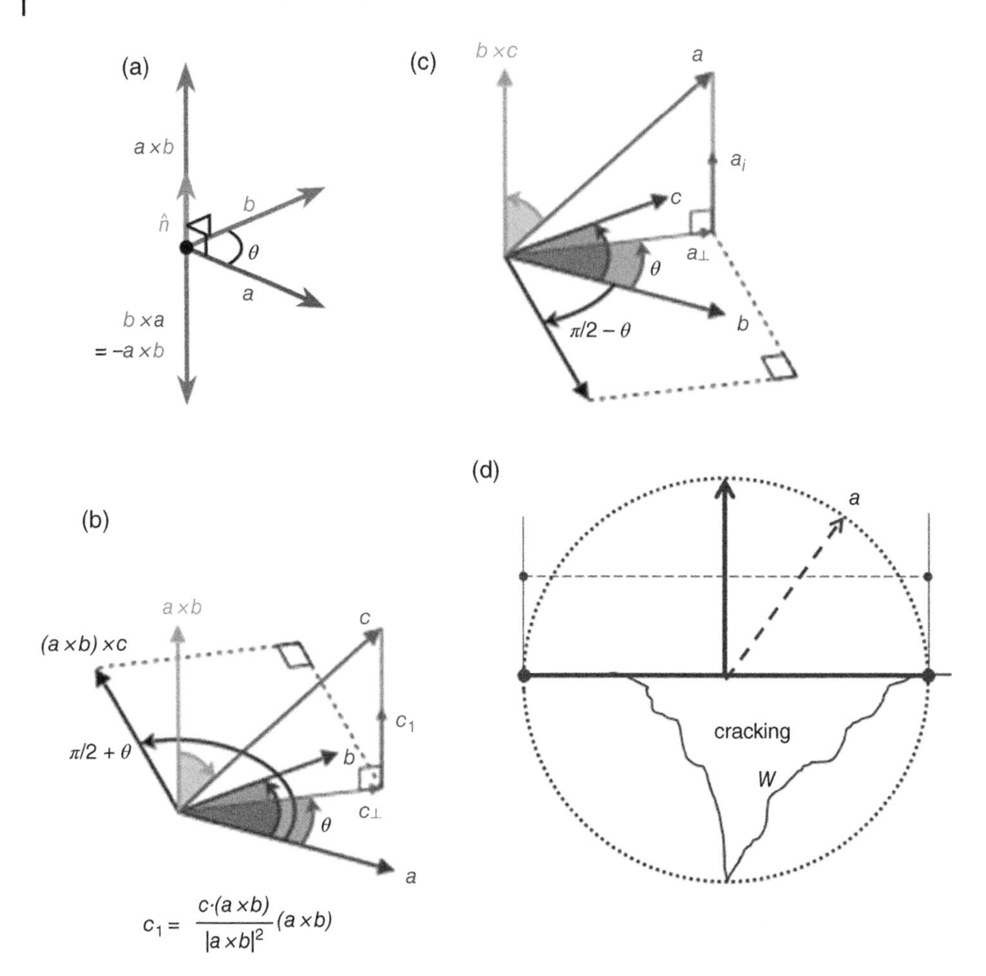

Figure 6.46 (a) Rules related to external multiplication; (b) Relationship between the angle θ and the vertical vector; (c) Relationship between the vertical vector and the angle θ; and (d) Relationship between the vector perpendicular to the crack structure with width.

Figure 6.47 displays the distance relative to the edge and external multiplication of the distance in the crack skeleton. Figure 6.48 is an example of the theoretical results of crack width measured with different methods (Euclidean, block, quasi-Euclidean, and checkerboard), revealing the different results. According to the study, among these methods, the Euclidean method calculates the most accurate measurement of crack width and has no computational error (see Figure 6.49).

6.10.18 Detection of Crack Starting Feature (Crack Core) Using EPA Emperor Penguin Metaheuristic Algorithm

Due to the complex patterns of cracking and its combination with different types of secondary cracks and spalls, identifying the root of the crack and the place of onset of cracking is a problem of high complexity. One of the most important features influencing the selection of C_(root) cracks is the distribution of crack width changes in an increasing direction in the development of cracking. According to Rule No. 1 in the previous section, the continuous crack width distribution can be measured. In order to determine the main roots of cracking and to study the effect of base cracks

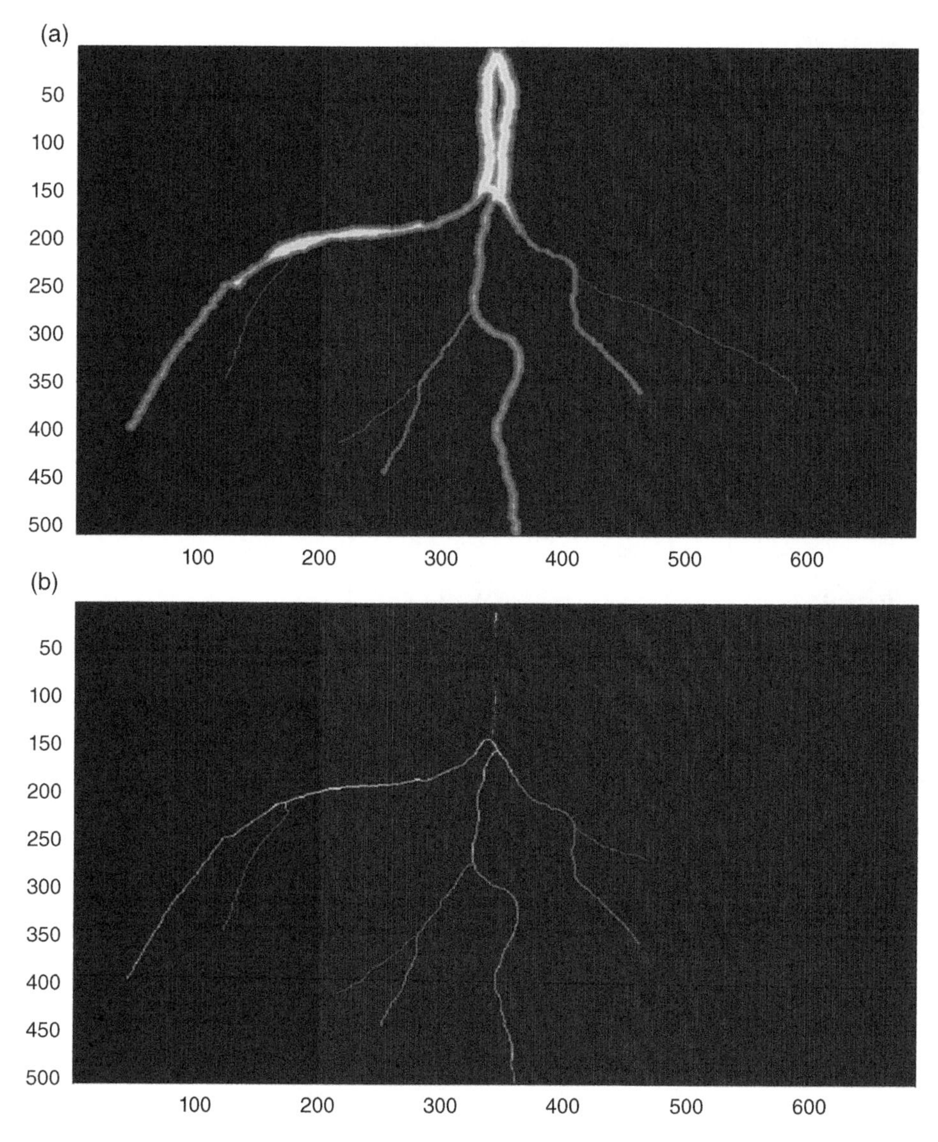

Figure 6.47 (a) The distance image relative to the edge; and (b) The external multiplication of the distance image in the crack skeleton.

and spalls, it is necessary to determine the C_{core} coordinates. This reduces the computational dimensions and enhances the accuracy of pavement failure assessment indicators. The crack detection algorithm is shown in Figure 6.50. The most important part of this algorithm is determining $C_{core} = (x_{core}, y_{core})$.

On the other hand, due to the fact that the cracks are calculated discretely in the image in many cases, this discontinuity makes it difficult to evaluate the route. Therefore, the number of start and end points is much higher than the experimental samples, and the complexity of the computational space is greatly increased. For this reason, meta-innovative methods are widely used in solving such optimization problems with high complexity. A new algorithm called EPA is presented in Chapter 10, which has a special application in determining the crack core feature.

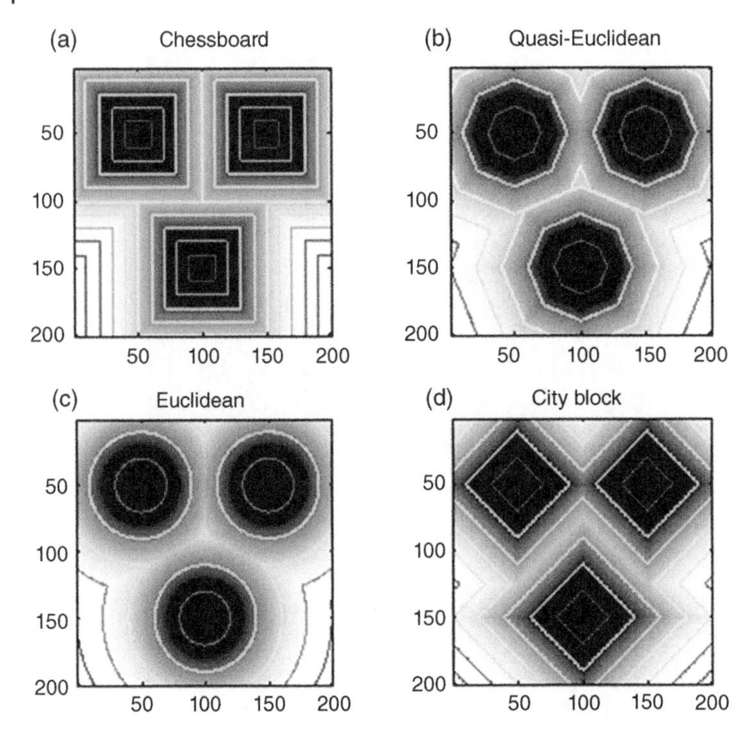

Figure 6.48 Distance to the edge by: (a) checkered methods, (b) quasi-Euclidean, (c) Euclidean, and (d) block.

6.10.19 Selection of Crack Root Feature Based on Geodetic Distance

In this section, using the meta-heuristic algorithm and the relationships of geometric distance and entropy, the failure core can be determined, which is the starting point of failure and generally the weakest point or coordinates of cracking. By identifying these coordinates through geodetic relations, the shortest path related to cracking on the crack path can be determined and the branch of each crack can be evaluated separately. The geodetic distance is used on binary images. In this method, using the point-to-point distance function based on the specified path and with the help of Euclidean relations, the distance of each section to the next point is calculated. An example of a crack image is displayed in Figure 6.50. This Euclidean distance is left equal to *lc* or branch length. Assuming point a as the starting point and point b (i) as the end points, the geodetic distance is calculated based on the number of branches and on the single corrosion path.

After determining the distance of all points, the main patterns are selected and feature entropy is used to select the core. The coordinates with the lowest entropy value are selected as the main core coordinates of the crack, and the numbers of the main pattern, extraction, and image of the crack roots are specified. Different patterns are calculated using this method, and the skeletons of the new pattern are internally multiplied by the crack thickness at the same point. Examples of calculations for the set $\{b_1, b_2, b_3, b_4, b_5, b_6\}$ are shown in Figure 6.51. As can be seen, different patterns of cracks are created for the faulty section. The optimal pattern is selected using a set of entropy indices, whereby $\arg\min(\text{Ent})$ is used as the main rule in template selection. Since active cracks are extracted using this method, the path with the least overlap has the lowest entropy value and is selected as the superior pattern. The entropy vector is calculated for each pattern, then the set

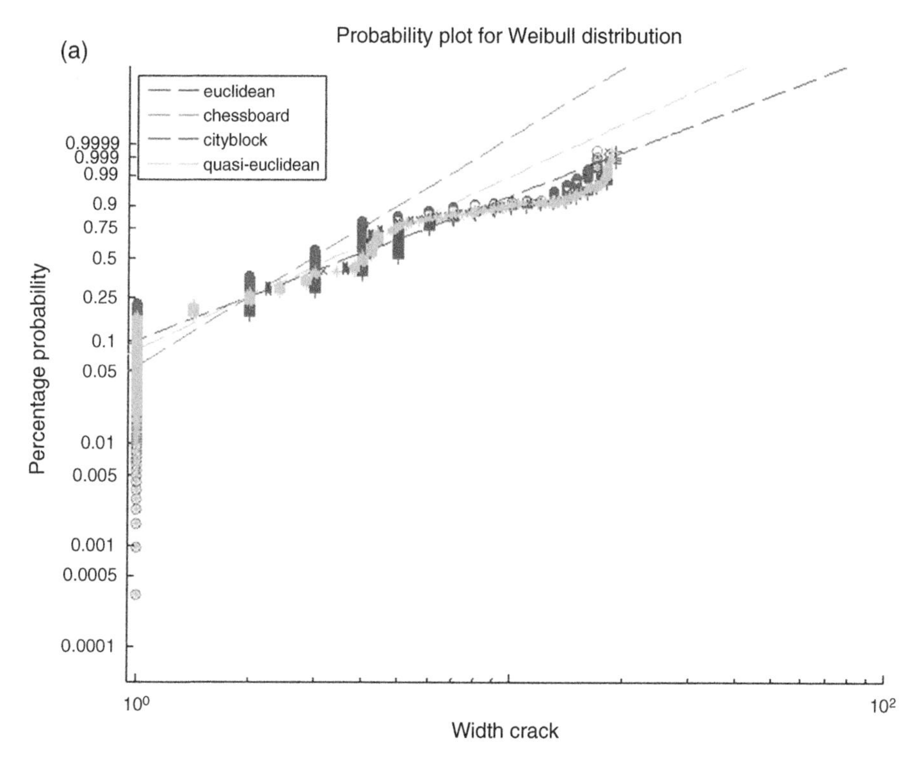

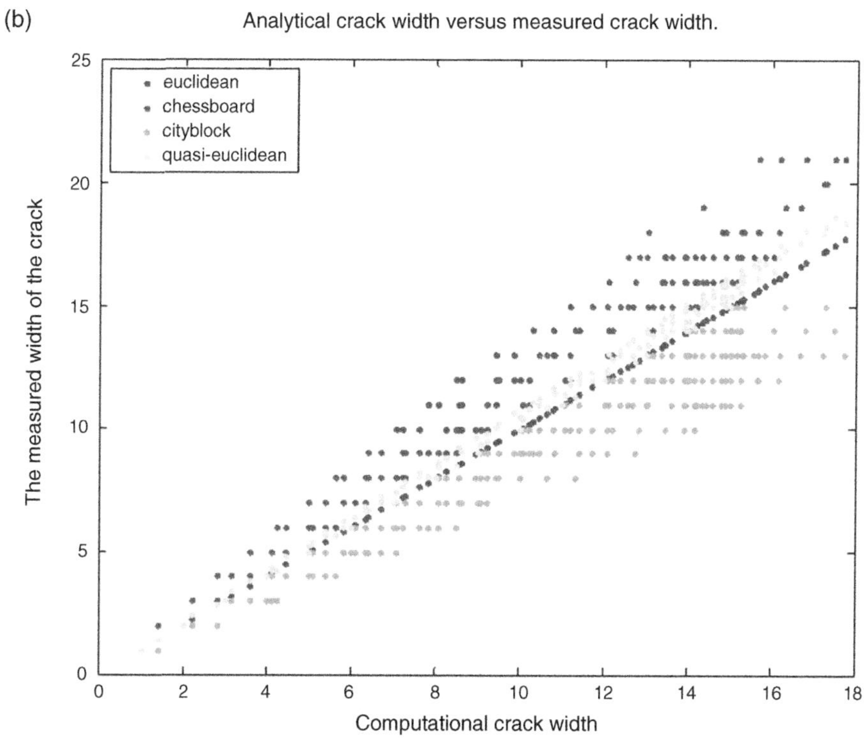

Figure 6.49 (a) The ratio of crack width and the percentage of probability of its occurrence in the image (abundance of crack width); and (b) Investigation of the effect of distance calculation method on the computational error of real width.

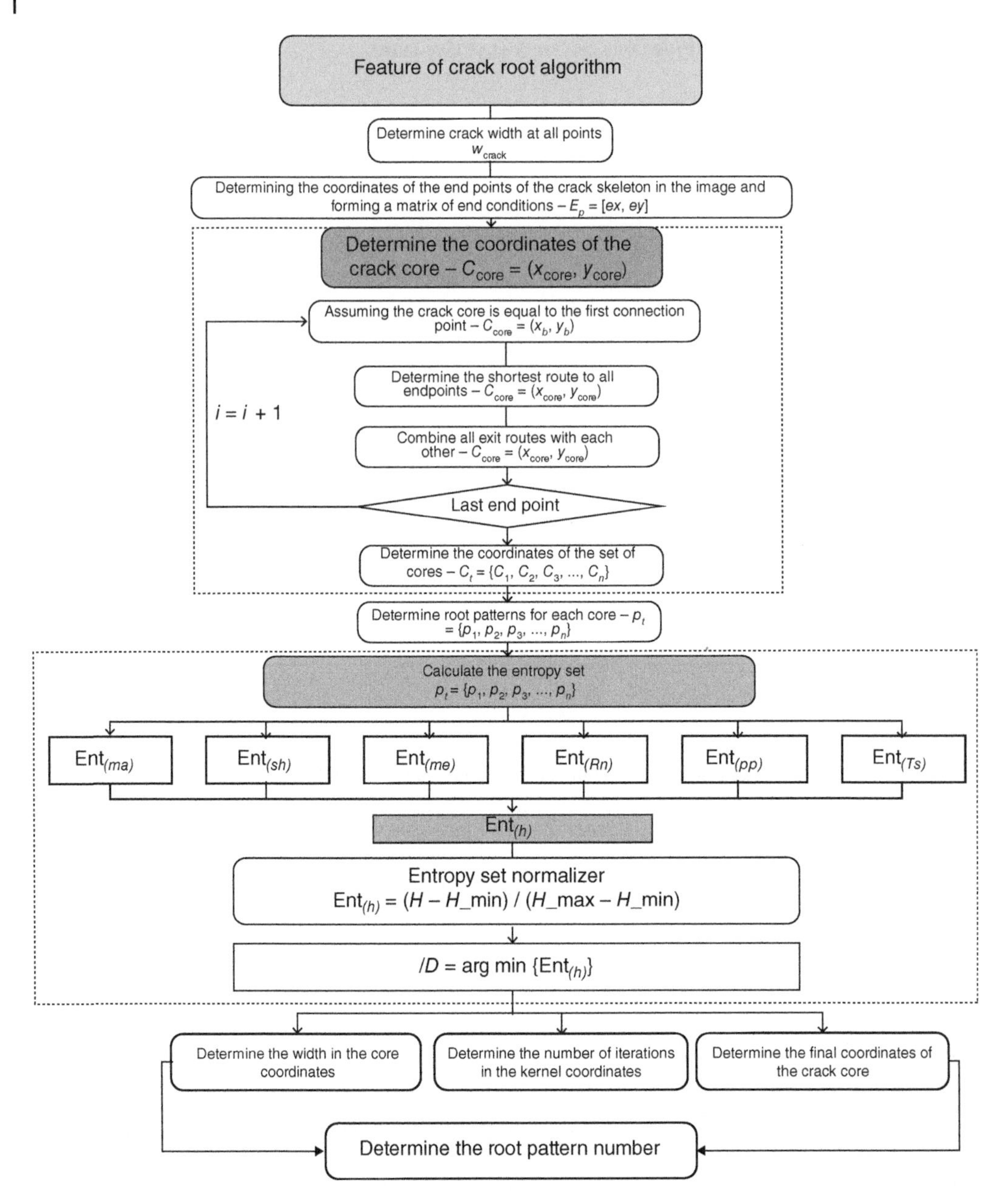

Figure 6.50 Algorithm for identifying crack core features.

of entropies of each pattern is determined. Finally, the sum of the entropies is chosen as the minimum value for the superior model as the superior model for introducing the crack root. The higher the repetition rate on the crack, the higher the entropy value, and for the sections with the lowest entropy value, the pattern is optimized (Figure 6.52).

Figure 6.53 shows an example of the EPA method and its performance in approaching greater repetition. The distribution of entropy based on entropy relations in 19 cases is normalized in the image. An example of the extraction of entropy feature vectors to determine the optimal pattern is provided in Table 6.14. In this example, pattern number 18 is selected as the optimal pattern.

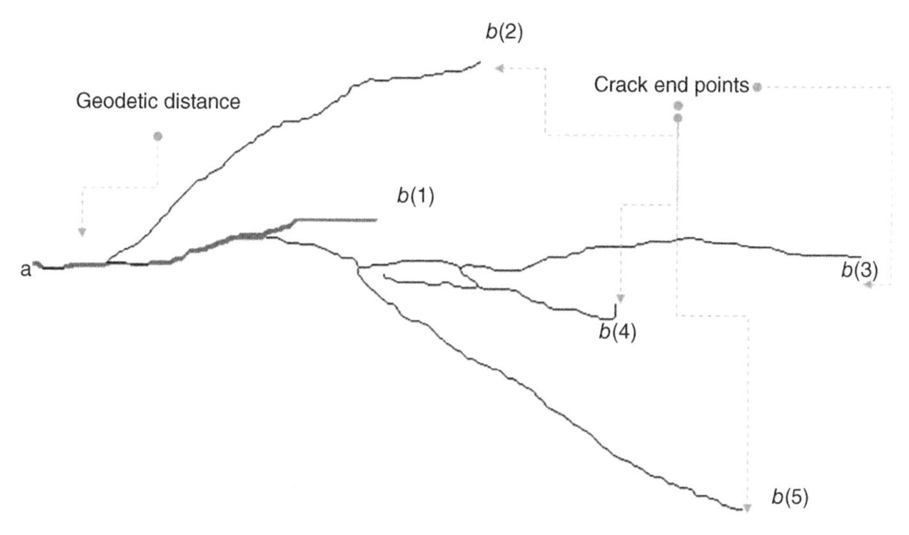

Figure 6.51 Example of calculating the geodetic distance between two cracking points from coordinates a to b (i), Is i = 1–5.

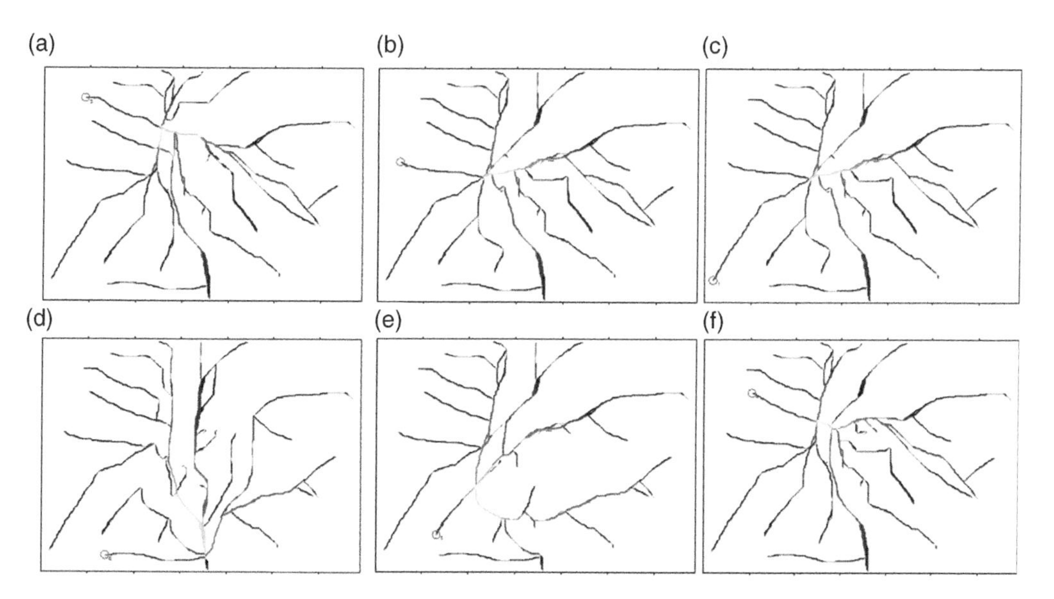

Figure 6.52 Separation of crack roots based on endpoints: (a) b1 and all points, (b) b2 and all points, (c) b3 and all points, (d) b4 and all points, (e) b5 and all points, and (f) b6 and all points.

6.10.20 Determining Coordinates of the Crack Core as the Optimal Center at the Failure Level using EPA Method

According to the EPA method, the set of coordinates of points that have the same width or width greater than the optimal width with the highest frequency is selected and identified as possible points of the crack core. For example, for the crack shown in Figure 6.53, the population of 12 EPs equal to the number of crack end points is automatically selected and, after 10 replications, is closer to the crack core point.

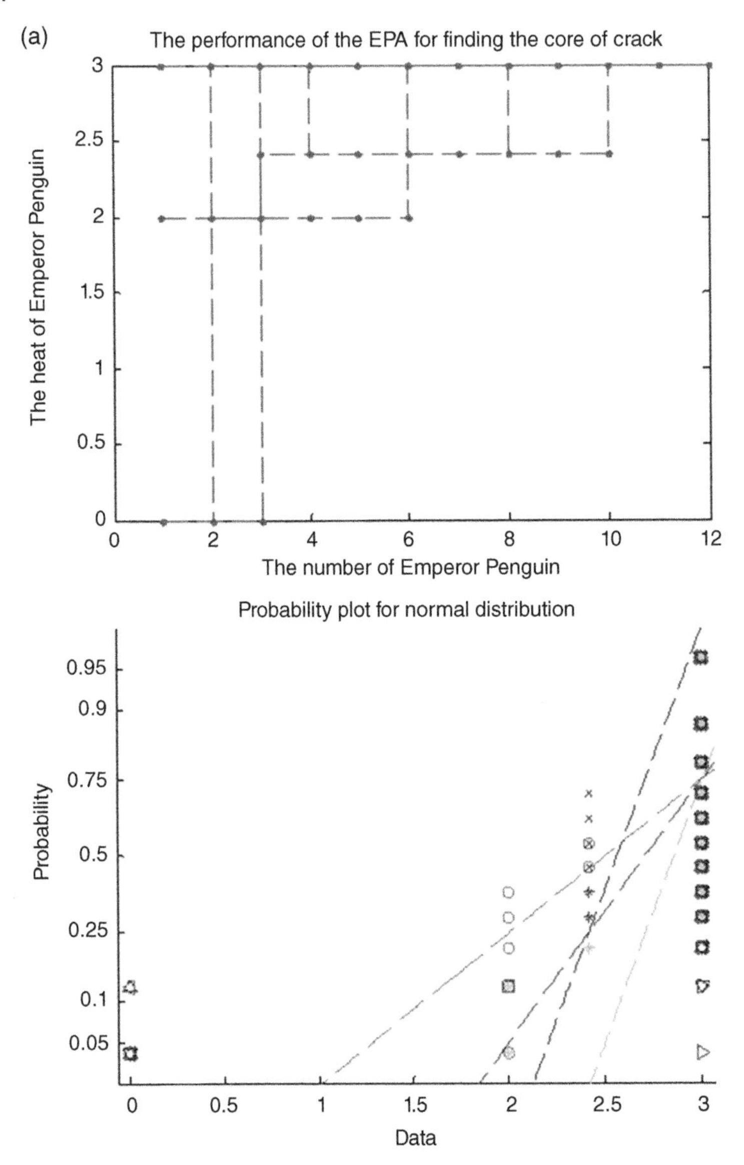

Figure 6.53 Examples of the performance of the EPA method: (a) Resize, by repeating and display the change in angle of the fitted line, (b) Show resize by repetition.

The higher the repetition rate, the lower the number of endpoints for entropy evaluation. The main structure of convergence based on geodetic distance is shown in Figure 6.56. As shown in the figure, each PE moves in a spiral path in the direction of the core, and after finding the maximum value, all PE points move toward those coordinates along the spiral path. At each stage of convergence, the average radius decreases and overall convergence and densities occur. The computational time of this method depends on the type of failure, whereby the more the crack grows toward the fatigue crack, the computational time of growth increases. Also, the number

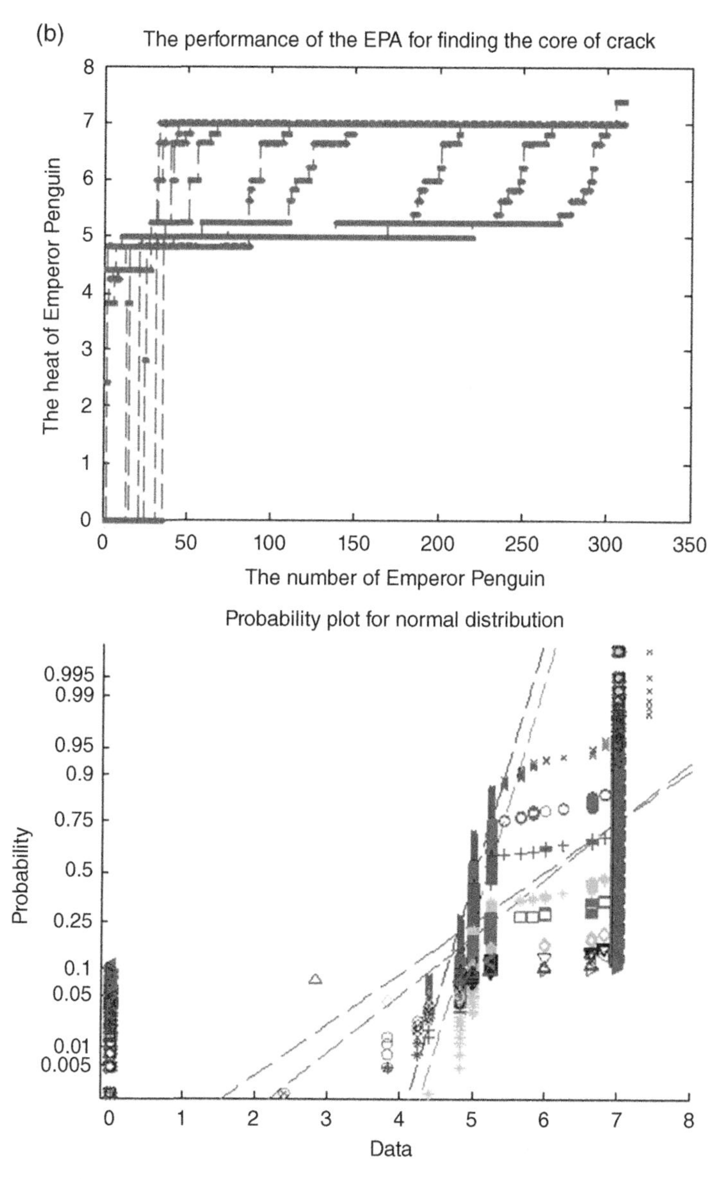

Figure 6.53 (Continued)

of operating points has a significant effect on computational time. In this section, a maximum of 100 actions were used for search. An example of the convergence of operating points on a fatigue crack to the crack core is shown in Figures 6.54 and 6.55.

The average computational time for 10 repetitions is 0.53 seconds for fatigue cracks, block, and spall and is 0.25 for linear and surface cracks. After determining the crack core in the defective section, the main crack pattern is extracted based on the geodetic distance. An example of these patterns for surface cracks is shown in Figure 6.57. Based on these patterns, spall cracks are automatically removed, and by decomposing this type of cracks, it is possible to check the root of the cracks as well as the amount and level of spall.

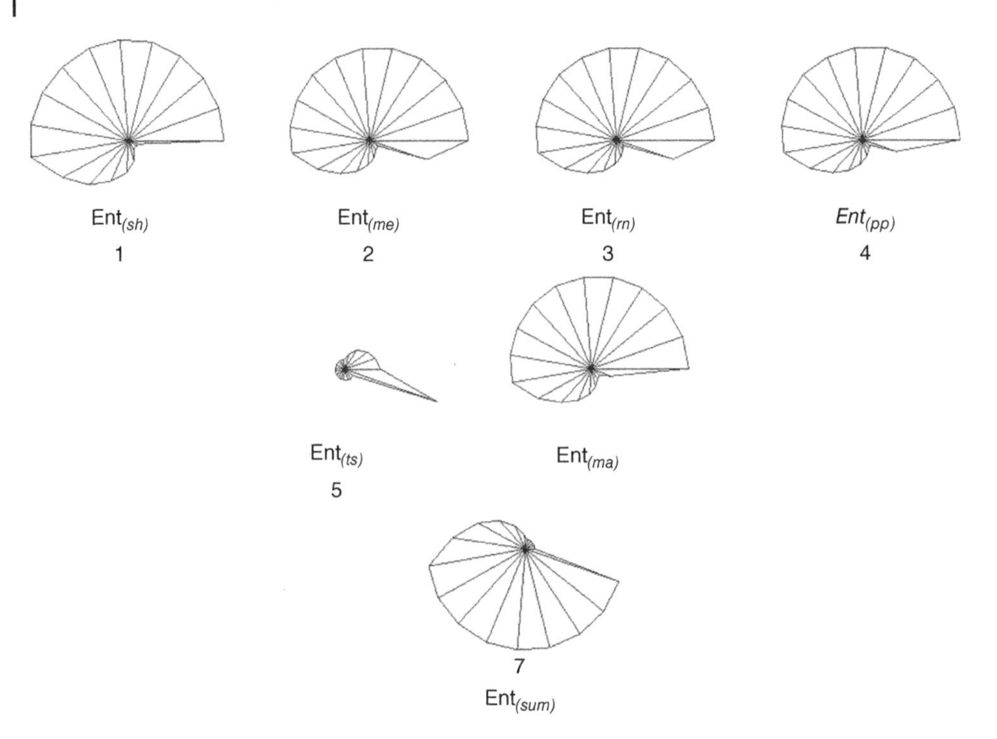

Figure 6.54 An example of the normalized entropy distribution with respect to the extracted patterns and the selection of the minimum entropy as the final pattern.

6.10.21 Development of New Features for Crack Evaluation Based on Graph Energy

Cracks can be modeled as a model of graph properties. According to this theory, each crack has branches that are connected to each other by arms. Guttmann first proposed the concept of graph energy using two parameters $G = (V, E)$. The energy of each graph, denoted by $E(G)$, is defined using the eigenvalues of the adjacency matrix using Eq. (6.84):

$$E = E(G) = \sum\nolimits_{i=1}^{n} |\lambda_i| \tag{6.84}$$

In this model, a crack with complex bifurcation points is considered as a G graph. If there are cracks between the nodes, the two vertices are considered adjacent to each other. The graph obtained using this method is called the crack-spall graph. For a crack-spall graph G, $E\,(G)$ is equal to the total crack energy calculated from Hookel's theory. Two crack-spall graphs are also called energies if they have equal energies. Basically, crack-spall graphs are nonhomogeneous graphs that can be energetic.

An adjacency matrix can be used for each crack and its branches. The proximity matrix is obtained using Eq. (6.85):

$$a_{ij} = \begin{cases} 0, & \text{if } i = j \text{ or } v_i \neq v_j \\ 1, & \text{if } i = j \text{ or } v_i \leftrightarrow v_j \end{cases} \tag{6.85}$$

In this regard, if two adjacent points are connected to each other by a crack, the value of the number is equal to 1, and if there is no connection, the corresponding number a_ij is considered equal to

Table 6.14 Comparison of different directional filters in contourlet transform.

Filter	Time (s)	Psnr (db)	SNR (db)	MSE	MAE	UQI	SSIM
5-3	14.2	297.3974	288.3497	6.2028e + 004	232.2581	1	1
9-7	21.5	276.2821	267.2343	6.2028e + 004	232.2581	1	1
dc	21	276.2821	267.2343	6.2028e + 004	232.2581	1	1
avkp	13.7	297.4252	288.3705	6.2028e + 004	232.2581	1	1

Table 6.15 An example of the extraction of entropy feature vectors to determine the optimal pattern.

		ENT1	ENT2	ENT3	ENT4	ENT5	ENT6	sum
1	Patt1	1.020	0.390	0.320	1.450	0.560	0.070	3.810
2	Patt2	1.020	0.390	0.320	1.450	0.560	0.070	3.810
3	Patt3	1.010	0.390	0.320	1.450	0.560	0.070	3.800
4	Patt4	1.010	0.390	0.320	1.450	0.560	0.070	3.800
5	Patt5	1.010	0.390	0.320	1.450	0.560	0.070	3.800
6	Patt6	1.010	0.390	0.320	1.450	0.560	0.070	3.800
7	Patt7	1.010	0.390	0.320	1.450	0.560	0.070	3.790
8	Patt8	1.000	0.390	0.320	1.450	0.560	0.070	3.790
9	Patt9	1.000	0.390	0.320	1.450	0.560	0.070	3.790
10	Patt10	1.000	0.390	0.320	1.450	0.560	0.070	3.780
11	Patt11	1.000	0.390	0.320	1.450	0.560	0.070	3.780
12	Patt12	0.990	0.390	0.320	1.450	0.560	0.070	3.780
13	Patt13	0.990	0.390	0.320	1.450	0.560	0.070	3.770
14	Patt14	0.990	0.390	0.320	1.450	0.560	0.070	3.770
15	Patt15	0.980	0.390	0.320	1.450	0.560	0.070	3.770
16	Patt16	0.980	0.390	0.320	1.450	0.560	0.070	3.760
17	Patt17	0.970	0.390	0.320	1.450	0.560	0.070	3.760
18	Patt18	0.970	0.390	0.320	1.450	0.560	0.070	3.750
19	Patt19	0.970	0.390	0.320	1.450	0.560	0.070	3.760
Argmin		0.970	0.390	0.320	1.450	0.560	0.070	3.750
Mode		18	18	18	18	18	18	18

zero. The greater the number of crack components, the greater the fracture, and the larger the dimensions of the a_ij proximity matrix. In this study, the proximity matrix for cracking is calculated using Eq. (6.86):

$$\begin{aligned} Adj_{wideij} &= a_{ij} \times w_{ij} \\ w_{ij} &= \{w_{\max}, w_{\text{mean}}, w_{\min}, w_{\text{mode}}\} \\ Adj_{lenij} &= a_{ij} \times lenght_{ij} \\ Adj_{conij} &= a_{ij} \times ID_{ij} \end{aligned} \tag{6.86}$$

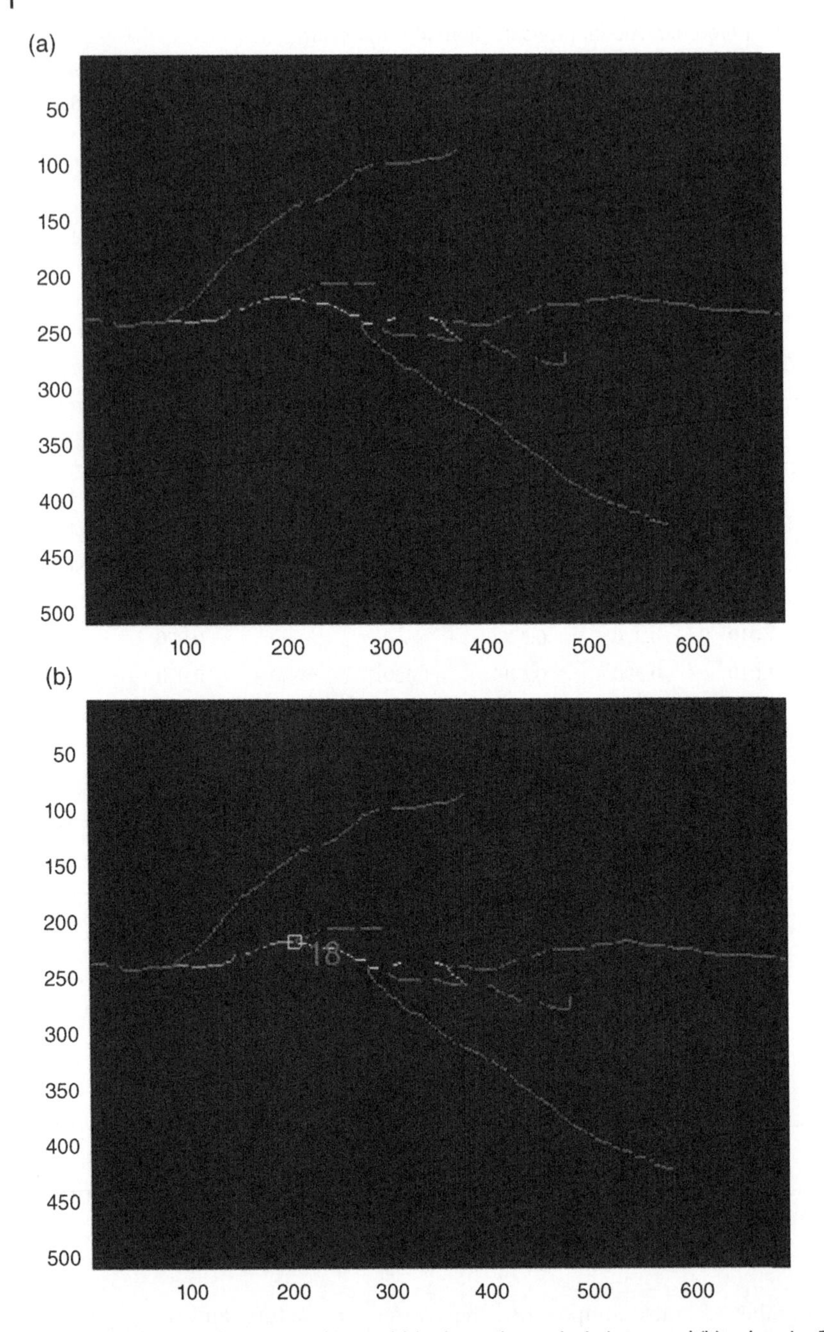

Figure 6.55 Locating the crack core: (a) in the main crack skeleton and (b) using the EPA method and entropy committee.

Using these relationships, the adjacency matrix for cracking is calculated and by applying the characteristics of each piece, the characteristic adjacency matrix is obtained for each crack. In these relationships, $\mathrm{Adj}_{\mathrm{wide}ij}$ is the crack width adjacency matrix, $\mathrm{Adj}_{\mathrm{wide}ij}$ is the crack length adjacency matrix, and $\mathrm{Adj}_{\mathrm{con}ij}$ is the crack component (see Tables 6.16 and 6.17). Also, ID_{ij} is the crack

Table 6.16 Example 1 of the extraction of adjacency matrix.

		C1	C2	C3	C4	C5	C6
1	C1	0	1	0	0	0	0
2	C2	1	0	1	1	0	0
3	C3	0	1	0	0	0	0
4	C4	0	1	0	0	1	1
5	C5	0	0	0	1	0	0
6	C6	0	0	0	1	0	0

Table 6.17 Example 2 of the extraction of adjacency matrix.

		C1	C2	C3	C4	C5	C6
1	C1	1	0	0	0	0	0
2	C2	1	2	3	0	0	0
3	C3	0	2	0	0	0	0
4	C4	0	0	3	4	5	0
5	C5	0	0	0	4	0	0
6	C6	0	0	0	0	5	0

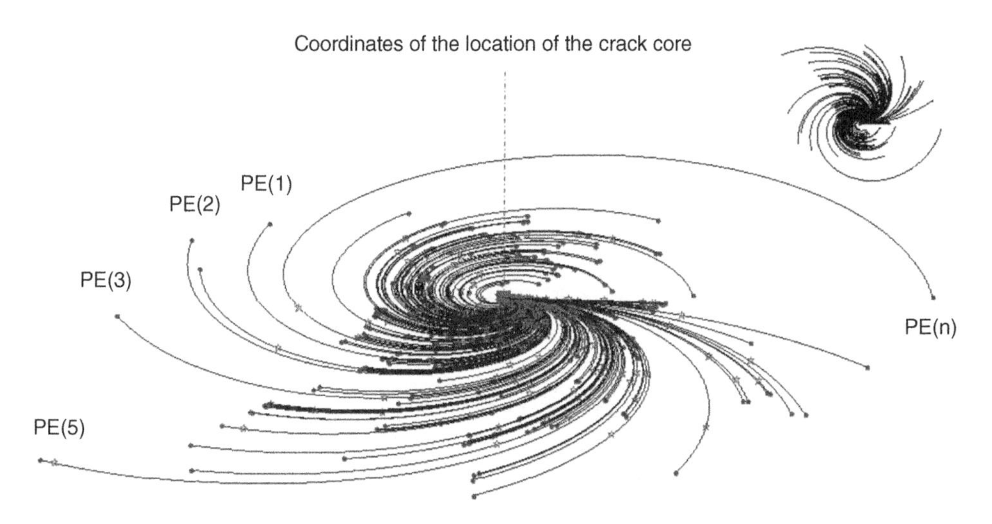

Figure 6.56 Determining the coordinates of the cracked core, locating the cracked core using the EPA method and the entropy committee.

member number, lenght$_{ij}$ is the crack length for the ij component, and w_ij is the width of the crack component. Since crack width is a continuous and variable characteristic, various parameters, such as maximum $w_{\min}$ width, minimum w_{ij} width, and maximum w_{mode} width, can be used as an effective component in crack evaluation. The example in Figure 6.59 shows a longitudinal crack with

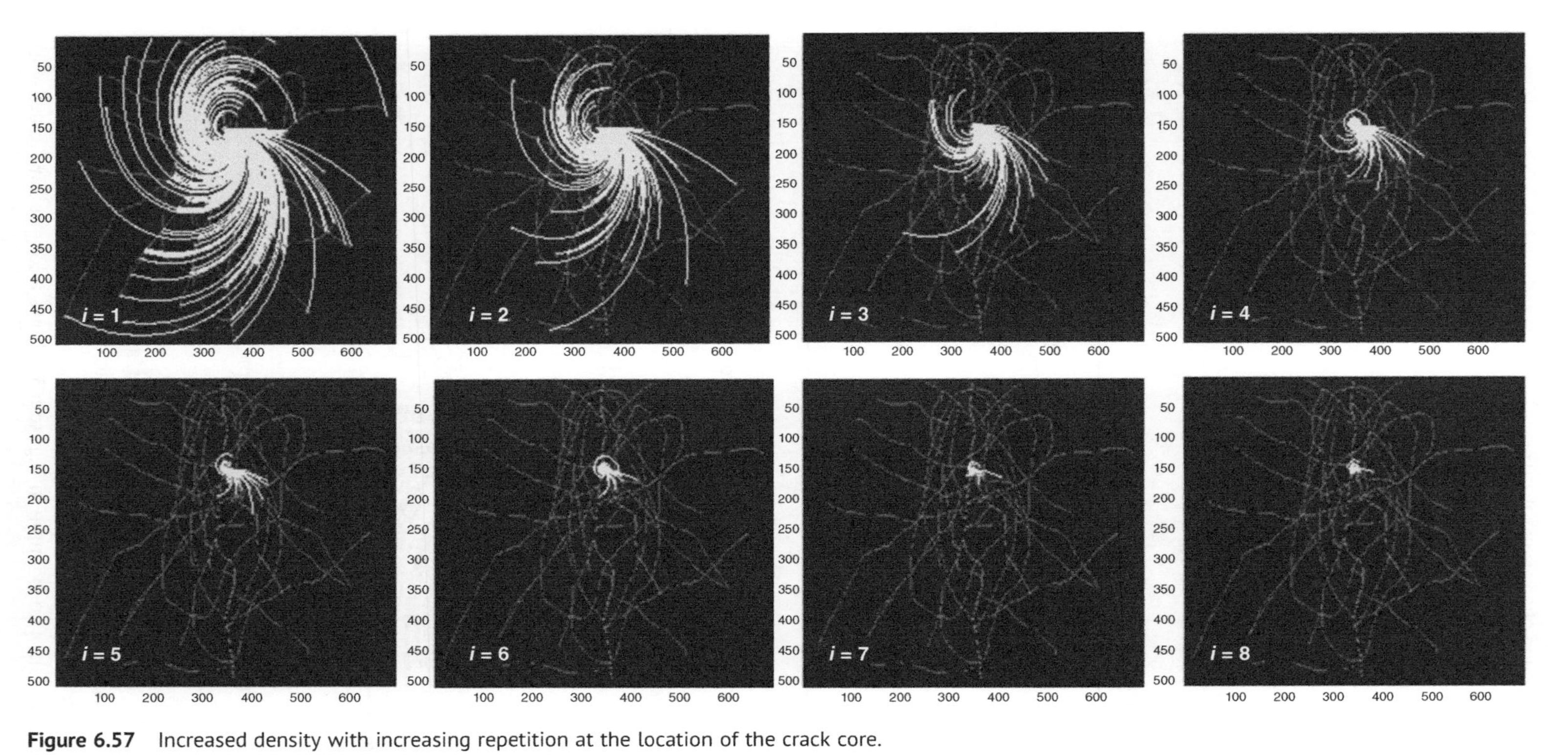

Figure 6.57 Increased density with increasing repetition at the location of the crack core.

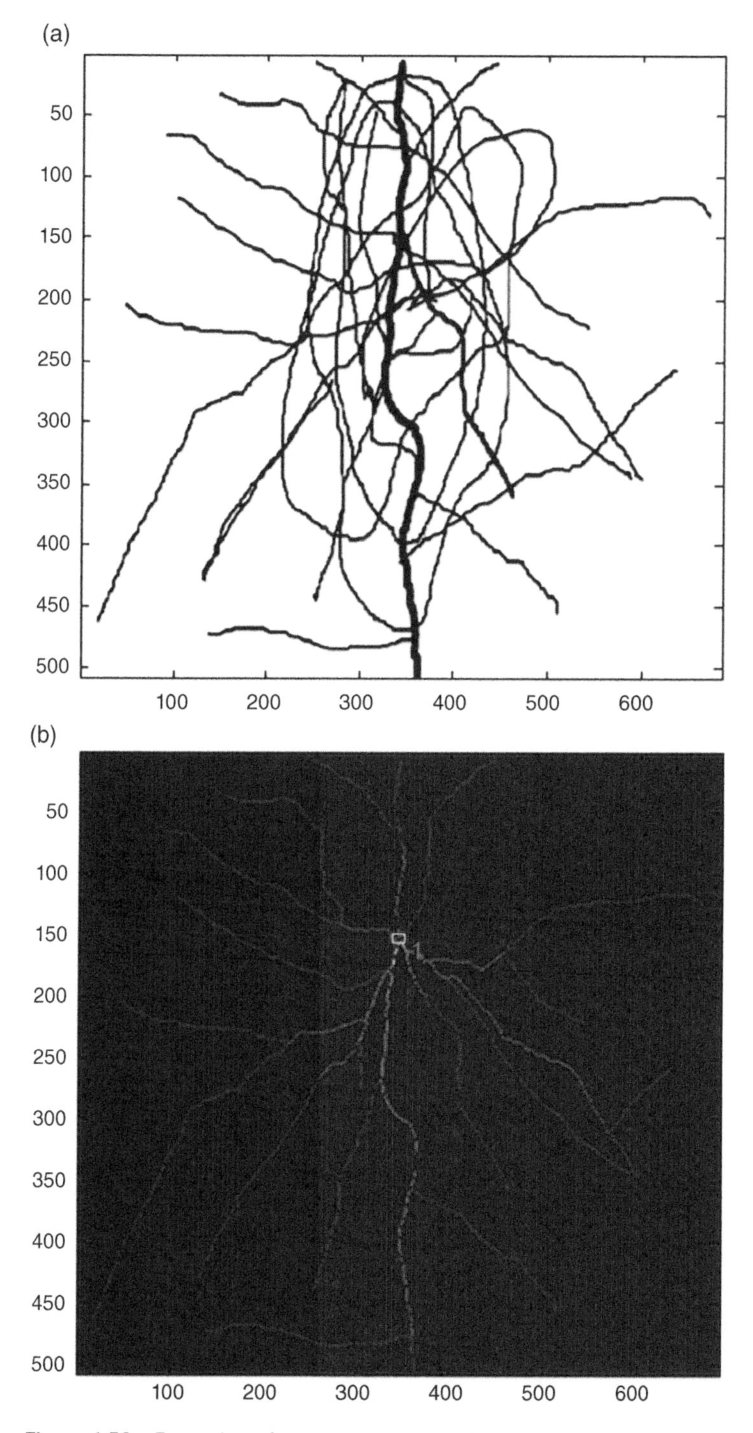

Figure 6.58 Extraction of core and crack root feature for sample image: (a) main image and (b) crack core and crack root pattern.

Table 6.18 Example 3 the extraction of adjacency matrix.

		C1	C2	C3	C4	C5	C6
1	C1	0	202	0	0	0	0
2	C2	202	0	87	151	0	0
3	C3	0	87	0	0	0	0
4	C4	0	151	0	0	130	328
5	C5	0	0	0	130	0	0
6	C6	0	0	0	328	0	0

roots and without spalls. These cracks make up an interconnected networks and have different widths in each section. The length of each component is different, and the connection number and member are specified. Using morphological operations, the components are first decomposed into separate members with a special number. The purpose of this decomposition is to determine the adjacency matrix and extract the main features for this type of failure. Using the proximity matrix, the vector of the adjacency matrix can be extracted to calculate the energy of each of the properties, including the unit matrix, width, length, and connection matrices. The adjacency matrix for this crack are given in Table 6.15.

According to Table 6.18, the general characteristics of crack components for sample failure are shown in Figure 6.58. All information about the crack image is summarized in the Adj matrix, and the main characteristics can be extracted based on the graph equation based using the following equation. Each of these adjacency matrices has an equation, eigenvalues λ_i, and energy $E(G)$. Using the Eq. (6.87), the energy value of each characteristic can be obtained:

$$\begin{aligned}
E(G)_{wide_{max}} &= \sum_{i=1}^{n} |\lambda_i|_{wide_{\max}} \Rightarrow E_{wide_{\max}} = \frac{\sum_{i=1}^{n} |\lambda_i|_{wide_{\max}}}{\sum_{i=1}^{n} |\lambda_i|}, \\
E(G)_{wide_{\min}} &= \sum_{i=1}^{n} |\lambda_i|_{wide_{\min}} \Rightarrow E_{wide_{\min}} = \frac{\sum_{i=1}^{n} |\lambda_i|_{wide_{\min}}}{\sum_{i=1}^{n} |\lambda_i|}, \\
E(G)_{wide_{median}} &= \sum_{i=1}^{n} |\lambda_i|_{wide_{median}} \Rightarrow E_{wide_{median}} = \frac{\sum_{i=1}^{n} |\lambda_i|_{wide_{median}}}{\sum_{i=1}^{n} |\lambda_i|}, \\
E(G)_{length} &= \sum_{i=1}^{n} |\lambda_i|_{length} \Rightarrow E_{wide_{\max}} = \frac{\sum_{i=1}^{n} |\lambda_i|_{length}}{\sum_{i=1}^{n} |\lambda_i|}
\end{aligned} \tag{6.87}$$

In these equations, normalized energies $E_{\text{wide}_{\max}}$, $E_{\text{wide}_{\min}}$, $E_{\text{wide}_{\text{median}}}$, $E_{\text{wide}_{\text{mode}}}$, and $E_{\text{wide}_{\max}}$ are crack characteristics. The adjacency matrix is shown in Figure 6.59.

The spall energy can be calculated by measuring the energy of the whole image using Eq. (6.88) as well as measuring the crack root energy:

$$E_{\text{spall}} = E(G)_t - E(G)_{\text{root}} = \left(\left(\sum_{i=1}^{n} |\lambda_i| \right)_t - \left(\sum_{i=1}^{n} |\lambda_i| \right)_{\text{root}} \right) \tag{6.88}$$

The equation (6.88) can be implemented for all features. In this section, three new indicators are presented using graph theory to evaluate cracks. An example of a spall energy calculation for a

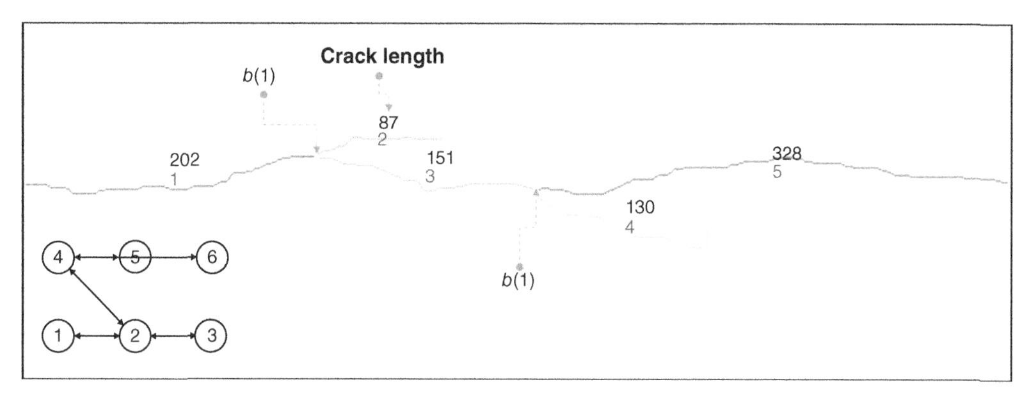

Figure 6.59 An example of a longitudinal crack with roots and without spalls.

cracked section is shown in Figures 6.61 and 6.62. As shown in Figure 6.60, cracks have severe spalls and a large number of joints and fractures. By removing these fractures and selecting the root and core of the crack, a simpler pattern of crack is created that has less energy than the root pattern of the crack. Using the constant energy of the main graph in the Figure 6.60c, the amount of spall energy is calculated.

Based on the eigenvalues, a set of general crack characteristics such as crack uniformity, spall type 1 index, and general cracking index can be evaluated. This classification and its computational algorithm are shown in Figures 6.63–6.65.

6.10.22 Crack Homogeneity Feature Based on Graph Energy Theory

If the crack changes in the conduction matrix are the same, then $|\lambda_i|_{\text{wide}_{\max}} = |\lambda_i|_{\text{wide}_{\min}}$ in the two halves of the adjacent matrix, resulting in no energy change being zero. The higher the value, the greater the change in crack width. This numerical homogeneity index of cracking is between 0 and 1. The larger the number, the greater the variation in crack width per crack. If the width is constant, this index is equal to zero.

$$\text{CHI} = \left(\left(\sum_{i=1}^{m}\sum_{i=1}^{n}\left(|\lambda_i|_{\max} - |\lambda_i|_{\min}\right)/\sum_{i=1}^{m}\sum_{i=1}^{n}|\lambda_i|_{\max}\right)_t\right) \tag{6.89}$$

where in the Eq. (6.89), *CHI* is the homogeneity index of cracking, $|\lambda_i|_{\max}$ is the maximum eigenvalues, and $|\lambda_i|_{\min}$ is the minimum eigenvalues.

6.10.23 Spall Type 1 Feature: Crack Based on Graph Energy Theory in Crack Width Mode

This feature depends on the number of main cracks or the crack width. If the total energy is greater than the energy value of the adjacent matrix graph $\text{Adj}_{\text{wide}_{\text{mode}}}$, it has a type 1 spall. If the width mode is adjacent to the matrix $\text{Adj}_{\text{wide}_{\max}}$, then there is no type 1 spall. This numerical index *T1SI* is the type 1 feature between 0 and 1. The larger the number, the greater the effect of type 1 spalls using Eq. (6.90):

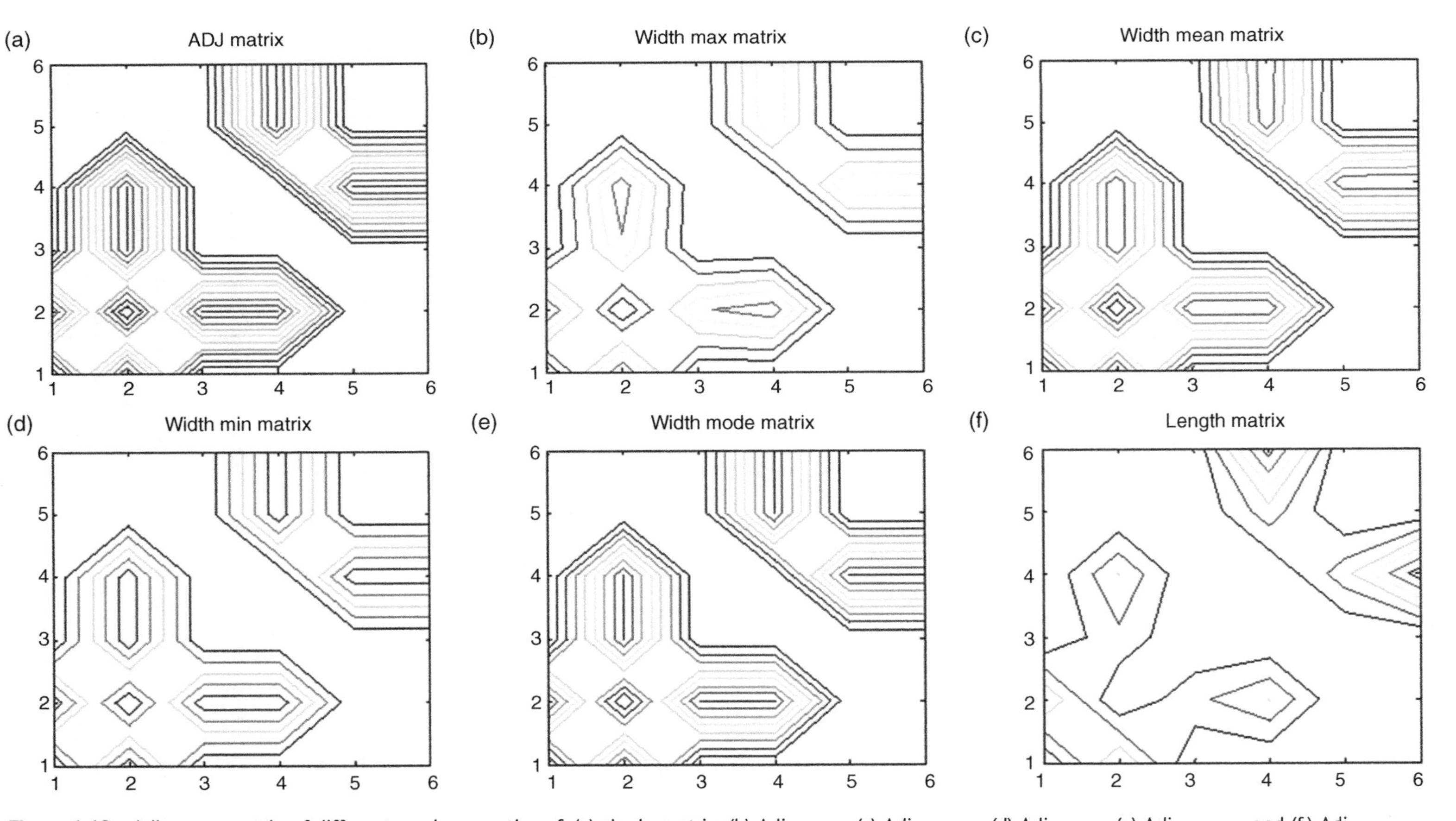

Figure 6.60 Adjacency matrix of different crack properties of: (a) single matrix, (b) $Adj_{widemax}$, (c) $Adj_{widemean}$, (d) $Adj_{widemin}$, (e) $Adj_{widemode}$, and (f) Adj_{length}.

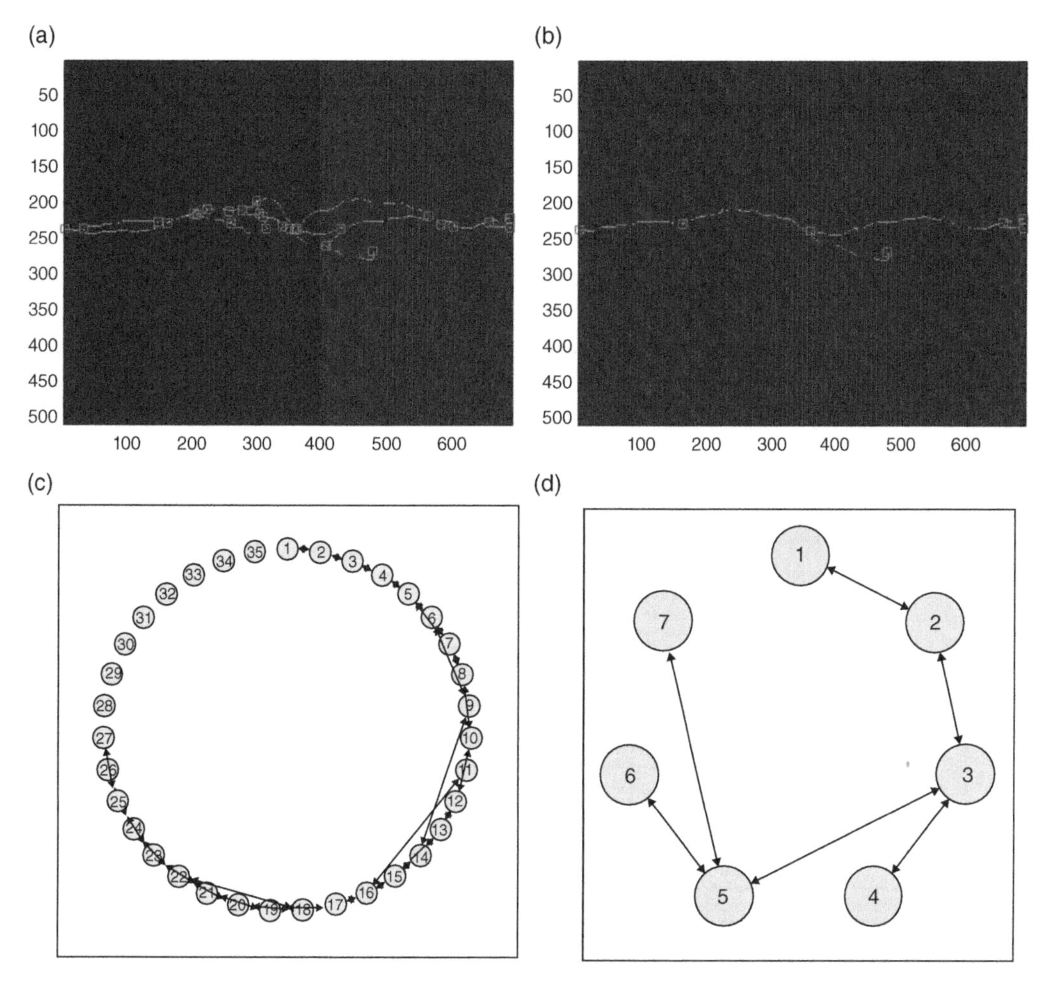

Figure 6.61 Root and spall separation, application of EPA and entropy and application of graph theory in identifying the main features, (a) a crack with spall, (b) a crack without spall, (c) the graph energy of spalled crack, and (d) the graph energy of simple crack.

$$T1SI = \left(\left(\sum_{i=1}^{m}\sum_{i=1}^{n}\left(|\lambda_i|_{\max} - |\lambda_i|_{\text{mode}}\right)\Big/\sum_{i=1}^{m}\sum_{i=1}^{n}|\lambda_i|_{\max}\right)_t\right) \tag{6.90}$$

where according to Eq. (6.90), $T1SI$ is the type 1 feature; $|\lambda_i|_{\max}$ is the maximum eigenvalues; and $|\lambda_i|_{\min}$ is the minimum eigenvalues.

6.10.24 General Crack Index Based on Graph Energy Theory

This index is used to assess the overall state of failure. If the crack is linear, the value of this index is equal to 0.5, and the greater the crack patterns, the more this index will decrease.

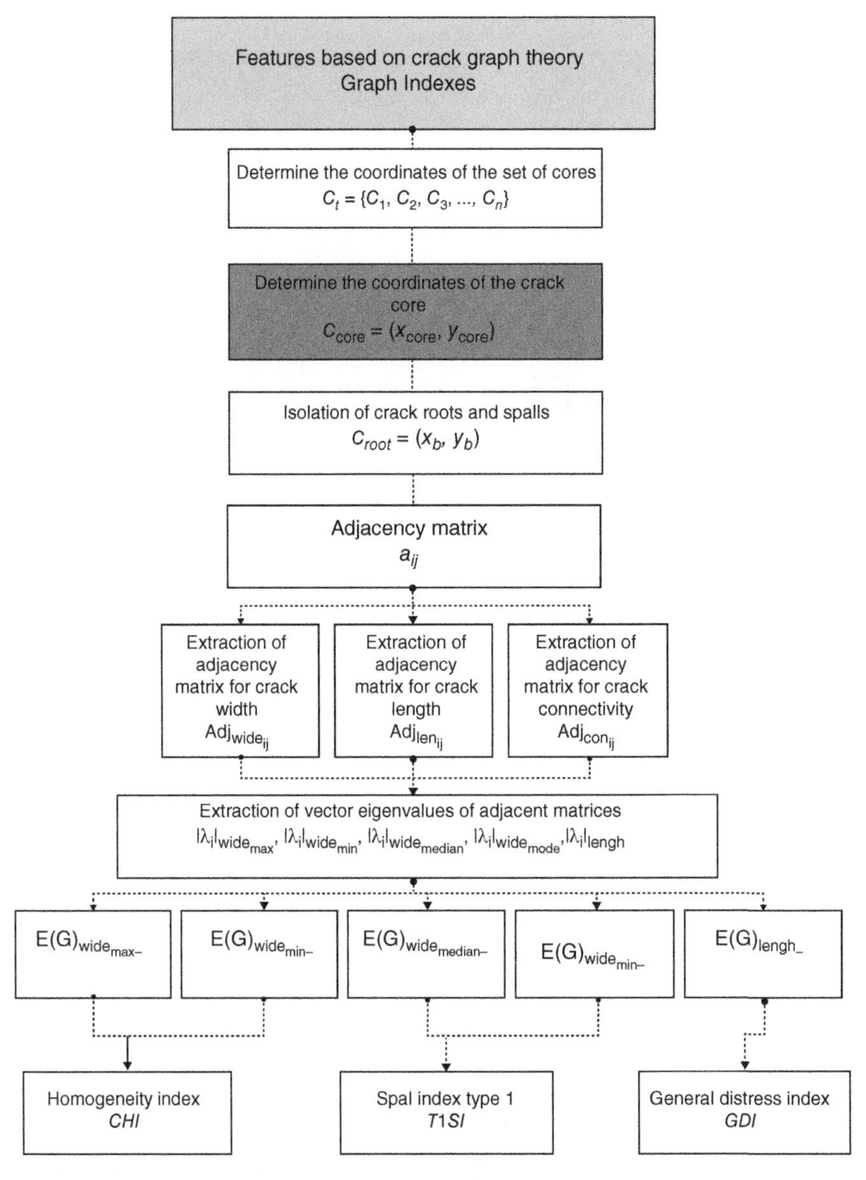

Figure 6.62 Classification and computational indices of crack evaluation using graph theory and calculation of indices obtained from eigenvalues of adjacency matrices.

$$
\begin{aligned}
\mathrm{GCI1} &= \sum_{i=1}^{n}\left(\left(\mathrm{Adj}_{\mathrm{widemean}} \Big/ \sum_{i=1}^{m}\sum_{i=1}^{n}\mathrm{Adj}_{\mathrm{widemean}}\right) \times \left(\mathrm{Adj}_{\mathrm{lengthmean}} \Big/ \sum_{i=1}^{m}\sum_{i=1}^{n}\mathrm{Adj}_{\mathrm{lengthmean}}\right)\right) \\
\mathrm{GCI2} &= \sum_{i=1}^{m}\sum_{i=1}^{n}\left(|\lambda_i|_{\mathrm{mean}}\right) \\
\mathrm{GCI3} &= \sum_{i=1}^{m}\sum_{i=1}^{n}\left(|\lambda_i|_{\mathrm{length}}\right)
\end{aligned} \tag{6.91}
$$

where in Eq. (6.91), the GCI1 is the general indicator of cracking; GCI2 is the overall crack index for average width; and GCI3 is the overall crack index for component length.

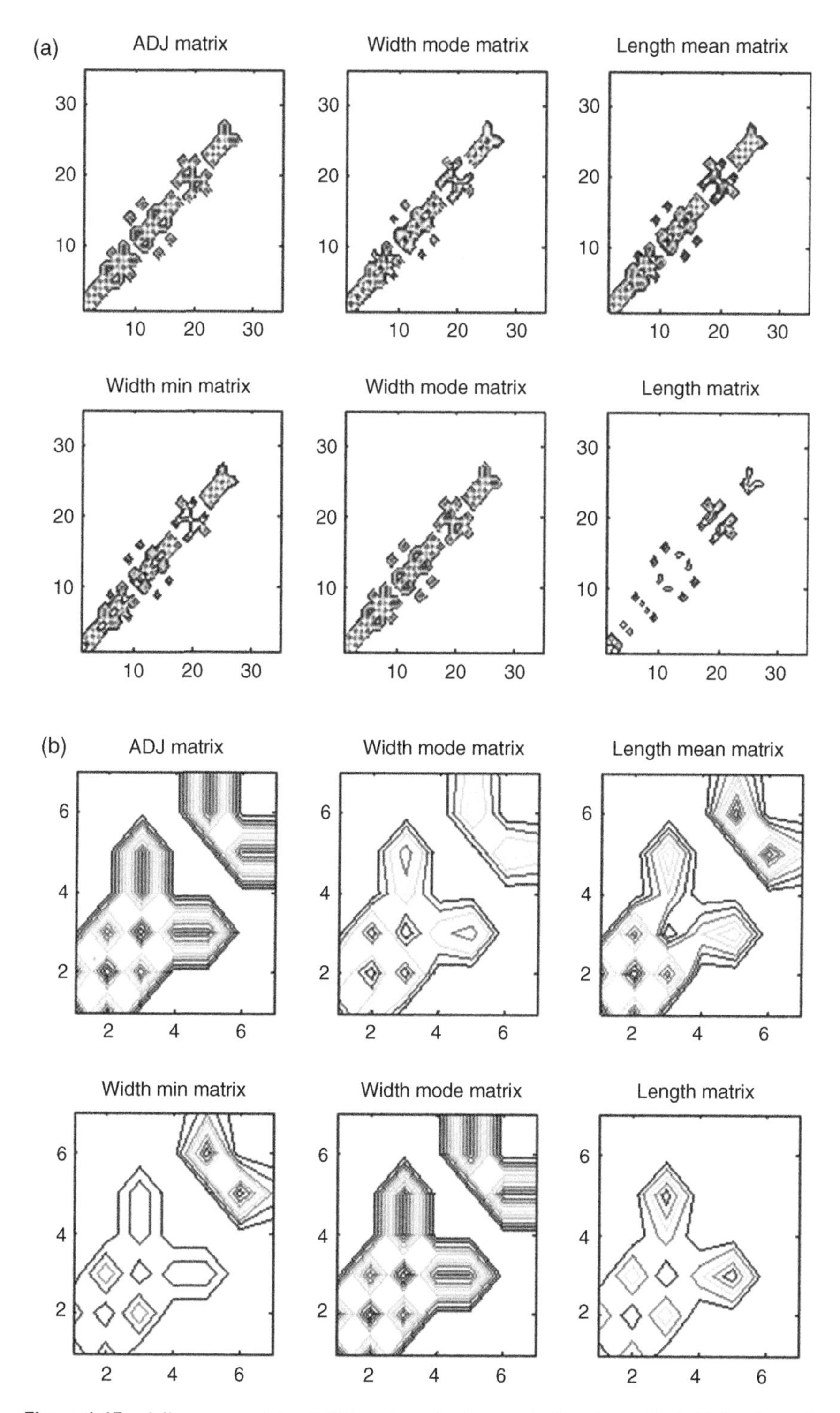

Figure 6.63 Adjacency matrix of different crack characteristics of sample 1: (a) for the main crack and (b) for the crack root.

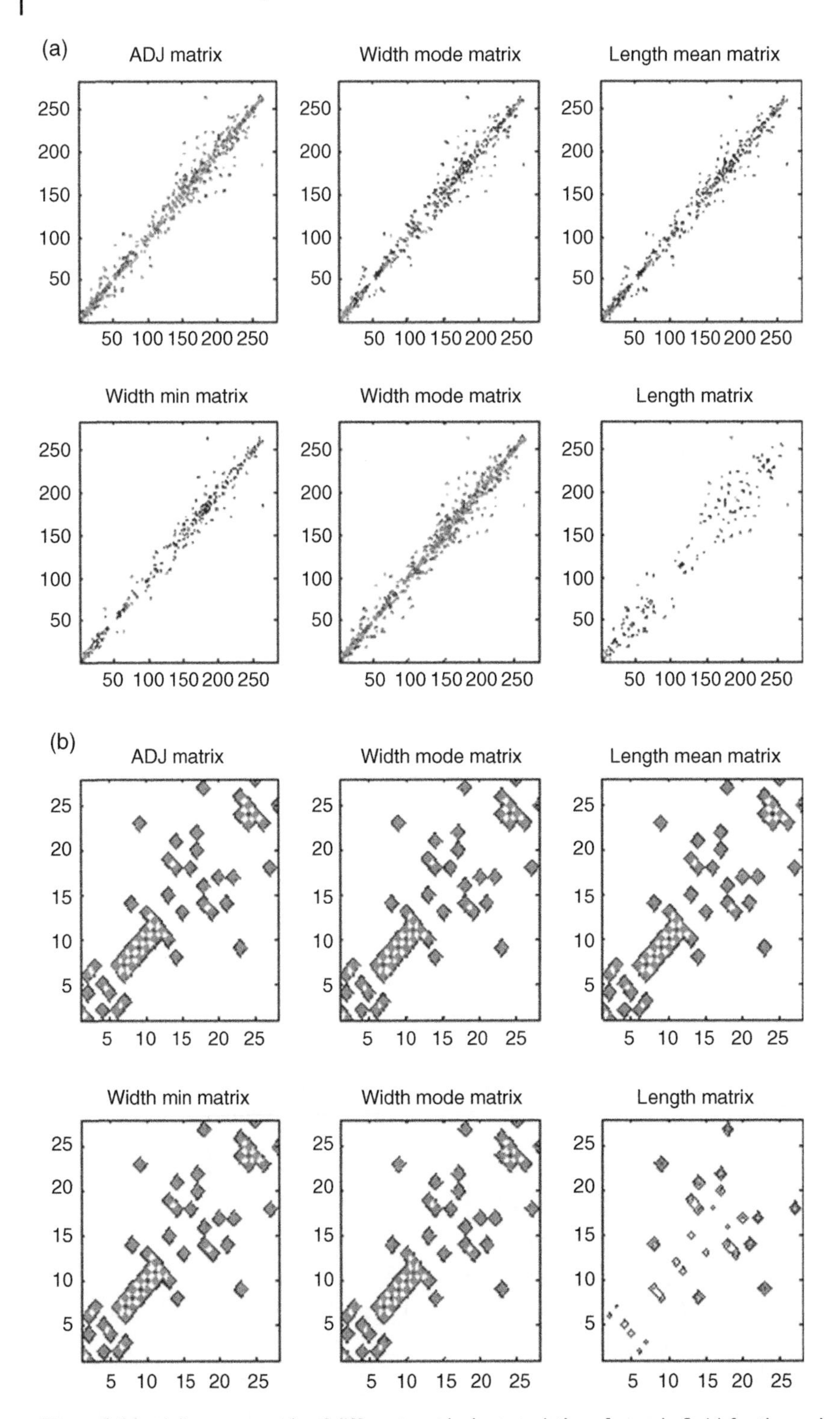

Figure 6.64 Adjacency matrix of different crack characteristics of sample 2: (a) for the main crack and (b) for the crack root.

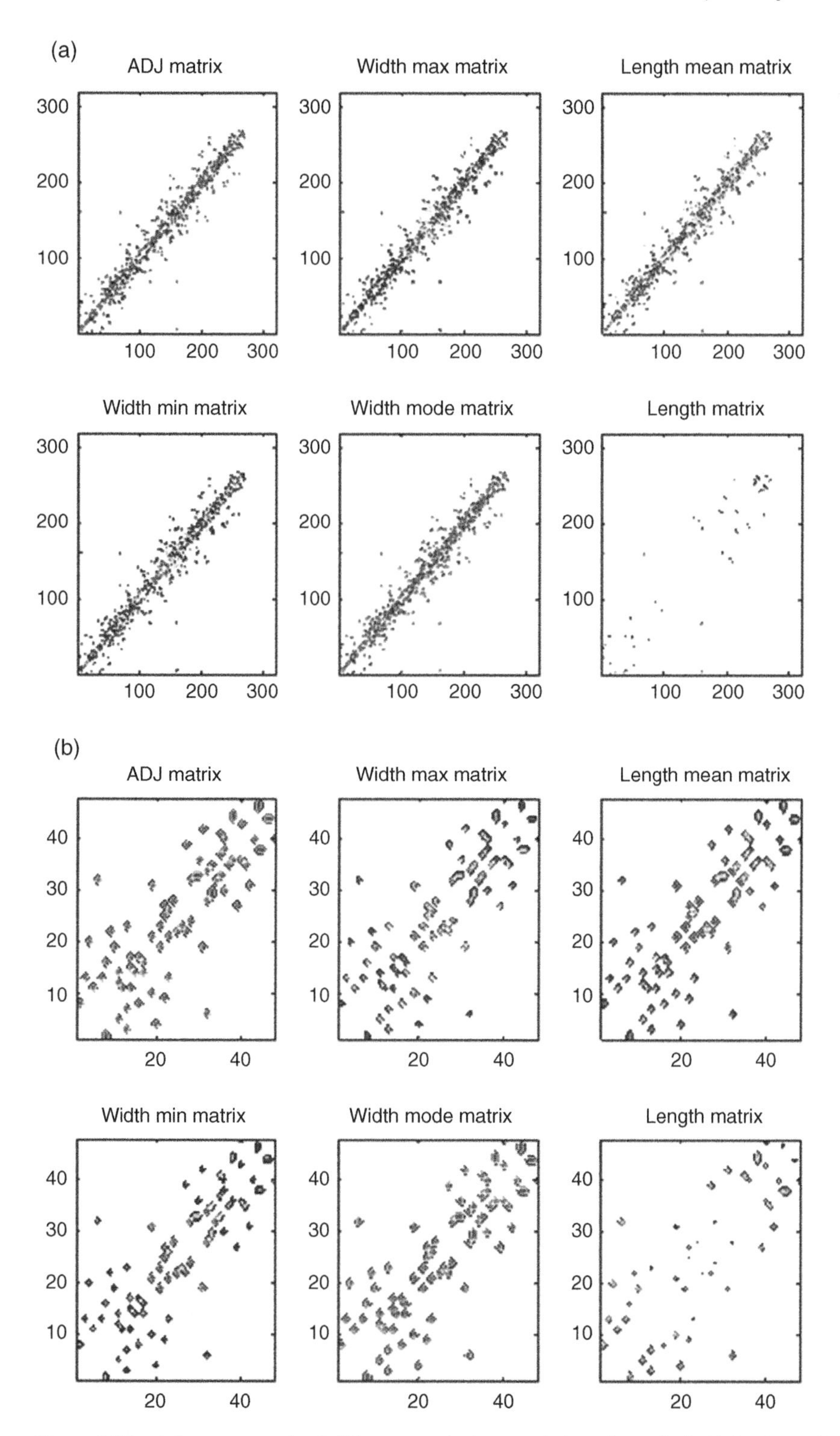

Figure 6.65 Adjacency matrix of different crack characteristics of sample 3: (a) for the main crack and (b) for the crack root.

Table 6.19 General specifications of crack components for sample failure Figure 6.58.

	C1	C2	C3	C4	C5	Minimum	Mean	Maximum
1	6	4.828	5.657	4.828	4.828	4.828	5.162	6
2	2.828	2.828	2.828	2.828	2.828	2.828	2.828	2.828
3	4.143	3.971	3.938	3.899	3.969	3.899	3.97	4.143
4	4	4	4	4	4	4	4	4
5	4	4	4	4	4	4	4	4
6	202	87	151	130	328	87	164	328

Table 6.20 Graph energy based features and adjacency matrix eigenvalues.

		CHI	T1SI	GCI1	GCI2	GCI3
1	CRACK1	0.460	0.230	0.070	23.900	1185
2	CRACK2	0.440	0.210	0.040	53.000	2385
3	CRACK3	0.320	0.090	0.010	274.400	1704
4	CRACK4	0.280	0.290	0.000	1620.000	9016
5	CRACK5	0.000	0.000	0.500	4.000	1454
6	CRACK6	0.630	0.230	0.040	77.900	1052
7	CRACK7	0.330	0.170	0.010	186.500	1296
8	CRACK8	0.470	0.410	0.030	121.900	1070
9	CRACK9	0.320	0.270	0.010	175.900	1850
10	CRACK10	0.310	0.020	0.070	145.000	1362

These features are used to evaluate the crack pattern. In order to assess the proposed methods, a set of cracks with different characteristics was examined in terms of type 1 spall and overall crack index, as shown in Table 6.19 and Graph energy based features and adjacency matrix eigenvalues shown in Table 6.20.

6.11 Summary and Conclusion

In this chapter, the types of feature extraction methods and image-related features were examined. The transformation of information into meaningful and useful features into different categories was also presented with the aim of summarizing and increasing the speed and accuracy of detection and also reducing the amount of image data for further processing. Extraction of optimal features to increase the ability to transfer more visual information or high information gain has remained a pertinent research focus. Therefore, various methods for extracting features are proposed in this chapter. On a case-by-case basis, optimal methods based on the existing database are proposed, which can be used to manage infrastructure. Moreover, effective and practical features in automatic detection of damaged images are described along with case examples of their application in infrastructure management.

In general, the eight main categories of feature extraction include the following: Low-level feature extraction methods, SBF, 1d Function-Based Features for Shape Represent, PBF, SIF, MFE, SSA, and STF. In order to evaluate the proposed methods, four case studies in various applications of infrastructure management were examined and discussed with the interpretation of the relevant features. These four case studies are the following:

Case 1: Feature extraction from polypropylene-modified bitumen optical microscopy images
Case 2: Image-based feature extraction for pavement skid evaluation
Case 3: Image-based feature extraction for pavement texture drainage capability evaluation
Case 4: Image-based feature extraction for pavement cracking evaluation

Chapter 7 is devoted to feature selection as an important step to reduce dimensions, with the approach of removing less important features, which is ultimately limited to a small subset of main features. Choosing the optimal feature can improve learning performance, enhance accuracy, and lower computational costs.

6.12 Questions and Exercises

1. Name the types of feature extraction methods. Create a category to extract content based on content. Create an image database of 100 images for use in the infrastructure manager. Tag images and create a database for feature analysis and extraction. Is it possible to make decisions about infrastructure based only on the image? Build special hardware to solve a problem in infrastructure management and prepare your data using this hardware.

2. Explain the role of feature extraction in the process of image-based systems. Sort the types of methods based on the accuracy of the database in Question 1. Which method yields the best result?

3. Name the types of shape-based methods and create examples of them with hybrid relationships. Does using combination features with simple shape-based relationships give better results?

4. Solve the following equation related to the ellipse variance feature (EVF) for a square, rectangular, and circular shape and prove that

$$\mathrm{EVF} = \left(\frac{\sigma'_R}{\mu'_R}\right)$$

5. Calculate the rectangularity feature (REF) for a square, triangle, and octagon and report the result relative to the convexity feature (COF).

6. Calculate the Euler number feature (ENF) for a data base of pavement distress (consist of 100 images). Based on the extracted features, provide a predictive model for mixture homogeneity.

7 Prove the complex Shearlet coefficients based on even and odd functions. Prove the following equation, and calculate the edges of the various examples using the following equation:

$$S'_{x,y}(L(l),j) = \begin{cases} S_{x,y}(L(l),j) \text{ if } \sigma_{j,x,y} > T \\ 0 \quad \text{else} \end{cases}$$

8 Prove the relationships of Radon transform:

$$\text{if } \theta - \theta^* \text{ then } R(\rho,\theta) = \int_{-\infty}^{+\infty} \delta(\rho^* - \rho))\ \mathrm{d}s =)\begin{cases} 0, \text{if } \rho \neq \rho^* \\ \int_{-\infty}^{+\infty} \delta(0)\ \mathrm{d}s\ \rho = \rho^* \end{cases}$$

$$\int \Phi(x)\mathrm{d}x \approx \frac{\gamma}{2}\left(\frac{1+x}{\gamma} + \log\ \cos h\left(\frac{x}{\gamma}\right)\right)$$

Further Reading

1 Mingqiang, Y., K. Kidiyo, and R. Joseph, *A survey of shape feature extraction techniques. Pattern Recognition*, 2008. **15**(7): p. 43–90.

2 Lindeberg, T., *Discrete derivative approximations with scale-space properties: A basis for low-level feature extraction. Journal of Mathematical Imaging and Vision*, 1993. **3**(4): p. 349–376.

3 Tapkın, S., H. Zakeri, A. Topal, F. M. Nejad, A. Khodaii, and B. Şengöz, *A brief review and a new automatic method for interpretation of polypropylene modified bitumen based on fuzzy radon transform and watershed segmentation. Archives of Computational Methods in Engineering*, 2020. **27**(3): p. 773–803.

4 Hu, Y., C.-x. Zhao, and H.-n. Wang, *Automatic pavement crack detection using texture and shape descriptors. IETE Technical Review*, 2010. **27**(5): p. 398–405.

5 Zakeri, H., F.M. Nejad, and A. Fahimifar, *Rahbin: A quadcopter unmanned aerial vehicle based on a systematic image processing approach toward an automated asphalt pavement inspection. Automation in Construction*, 2016. **72**: p. 211–235.

6 Ranjbar, S., F.M. Nejad, and H. Zakeri, *An image-based system for pavement crack evaluation using transfer learning and wavelet transform. International Journal of Pavement Research and Technology*, 2021. **14**(4): p. 437–449.

7 Adu-Gyamfi, Y.O., T. Tienaah, N. O. Attoh-Okine, and C. Kambhamettu, *Functional evaluation of pavement condition using a complete vision system. Journal of Transportation Engineering*, 2014. **140** (9), 04014040, 1–10, https://doi.org/10.1061/(ASCE)TE.1943-5436.0000638.

8 Ahmed, M., C.T. Haas, and R. Haas, *Toward low-cost 3D automatic pavement distress surveying: The close range photogrammetry approach. Canadian Journal of Civil Engineering*, 2011. **38**(12): p. 1301–1313.

9 Amhaz, R., S. Chambon, J. Idier, and V. Baltazart, *A new minimal path selection algorithm for automatic crack detection on pavement images.* in *2014 IEEE International Conference on Image Processing, ICIP 2014.* 2014. Institute of Electrical and Electronics Engineers Inc, (pp. 788–792). IEEE.

10 Andaló, F.A., P. A. Miranda, R. D. S. Torres, and A. X. Falcão, *Shape feature extraction and description based on tensor scale. Pattern Recognition*, 2010. **43**(1): p. 26–36.

11 Bello-Salau, H., A. M. Aibinu, E. N. Onwuka, J. J. Dukiya, and A. J. Onumanyi, *Image processing techniques for automated road defect detection: A survey.* in *11th International Conference on Electronics, Computer and Computation, ICECCO 2014.* 2014. Institute of Electrical and Electronics Engineers Inc., (pp. 1–4). IEEE.

12 Chambon, S. *Detection of road cracks with multiple images.* in International Joint Conference on Computer Vision Theory and Applications, VISAPP *(p. sp).*(S. Chambon. Detection of road cracks with multiple images. International Joint Conference on Computer Vision Theory and Applications, VISAPP, May 2010, France. hal-00878872). 2010. Angers.

13 Chambon, S., C. Gourraud, J. M. Moliard, and P. Nicolle, *Road crack extraction with adapted filtering and Markov model-based segmentation: Introduction and validation.* in *International joint conference on computer vision theory and applications, visapp* (p. sp). (S. Chambon, C. Gourraud, J. M. Moliard, P. Nicolle. Road crack extraction with adapted filtering and Markov model-based segmentation: introduction and validation. International Joint Conference on Computer Vision Theory and Applications, VISAPP, May 2010, France.hal-00612537). 2010. Angers.

14 Chambon, S. and J.M. Moliard, *Automatic road pavement assessment with image processing: Review and comparison. International Journal of Geophysics*, 2011. **2011**, 1–20, https://doi.org/10.1155/2011/989354.

15 Chavda, S. and M. Goyani, *Recent evaluation on content based image retrieval. International Journal of Computer Sciences and Engineering*, 2019. **7**(4): p. 325–329.

16 Coenen, T.B.J. and A. Golroo, *A review on automated pavement distress detection methods. Cogent Engineering*, 2017. **4**(1): p. 1374822.

17 Cord, A. and S. Chambon, *Automatic road defect detection by textural pattern recognition based on AdaBoost. Computer-Aided Civil and Infrastructure Engineering*, 2012. **27**(4): p. 244–259.

18 Ferguson, R.A., D. N. Pratt, P. R. Turtle, I. B. MacIntyre, D. P. Moore, P. D. Kearney, et al., *Road pavement deterioration inspection system.* 2003, Google Patents.

19 Gavilán, M., D. Balcones, O. Marcos, D. F. Llorca, M. A. Sotelo, I. Parra, et al., *Adaptive road crack detection system by pavement classification. Sensors*, 2011. **11**(10): p. 9628–9657.

20 Golparvar-Fard, M., V. Balali, and J.M. De La Garza, *Segmentation and recognition of highway assets using image-based 3d point clouds and semantic Texton Forests. Journal of Computing in Civil Engineering*, 2015. **29**(1), 04014023.

21 Guan, H., J. Li, Y. Yu, M. Chapman, and C. Wang, *Automated road information extraction from mobile laser scanning data. IEEE Transactions on Intelligent Transportation Systems*, 2015. **16**(1): p. 194–205.

22 Guan, H., J. Li, Y. Yu, M. Chapman, H. Wang, C. Wang, et al., *Iterative tensor voting for pavement crack extraction using mobile laser scanning data. IEEE Transactions on Geoscience and Remote Sensing*, 2014. **53**(3): p. 1527–1537.

23 Han, H., H. Deng, Q. Dong, X. Gu, T. Zhang, and Y. Wang, *An advanced Otsu method integrated with edge detection and decision tree for crack detection in highway transportation infrastructure. Advances in Materials Science and Engineering*, 2021. **2021**: p. 1–12, https://doi.org/10.1155/2021/9205509.

24 Iyer, S. and S.K. Sinha, *A robust approach for automatic detection and segmentation of cracks in underground pipeline images. Image and Vision Computing*, 2005. **23**(10): p. 921–933.

25 Iyer, S. and S.K. Sinha, *Segmentation of pipe images for crack detection in buried sewers. Computer-Aided Civil and Infrastructure Engineering*, 2006. **21**(6): p. 395–410.

26 Jahanshahi, M.R., F. J. Karimi, S. F. Masri, and B. Becerik-Gerber, *Autonomous pavement condition assessment.* 2013, Google Patents.

27 Jahanshahi, M.R., J. S. Kelly, S. F. Masri, and G. S. Sukhatme, *A survey and evaluation of promising approaches for automatic image-based defect detection of bridge structures. Structure and Infrastructure Engineering*, 2009. **5**(6): p. 455–486.

28 Jahanshahi, M.R. and S.F. Masri, *Adaptive vision-based crack detection using 3D scene reconstruction for condition assessment of structures. Automation in Construction*, 2012. **22**: p. 567–576.

29 Jahanshahi, M.R. and S.F. Masri, *A new methodology for non-contact accurate crack width measurement through photogrammetry for automated structural safety evaluation. Smart Materials and Structures*, 2013. **22**(3), 035019. (pp.1–10), https://doi.org/10.1088/0964-1726/22/3/035019.

30 Jiang, J., H. Liu, H. Ye, and F. Feng, *Crack enhancement algorithm based on improved EM. Journal of Information and Computational Science*, 2015. **12**(3): p. 1037–1043.

31 Jiao, C., M. Heitzler, and L. Hurni, *A survey of road feature extraction methods from raster maps. Transactions in GIS*, 2021. **25**(6): 2734–2763.

32 Jing, L. and Z. Aiqin. *Pavement crack distress detection based on image analysis.* in *2010 International Conference on Machine Vision and Human-Machine Interface, MVHI 2010.* 2010. (pp. 576–579). IEEE: Kaifeng.

33 Koutsopoulos, H.N., I. El Sanhouri, and A.B. Downey, *Analysis of segmentation algorithms for pavement distress images. Journal of Transportation Engineering*, 1993. **119**(6): p. 868–888.

34 Laurent, J. and M. Doucet, *Vision system and a method for scanning a traveling surface to detect surface defects thereof.* 2010, Google Patents.

35 Li, G., *Improved pavement distress detection based on contourlet transform and multi-direction morphological structuring elements*, in *2012 International Conference on Intelligent System and Applied Material, GSAM 2012.* 2012: Taiyuan, Shanxi. p. 371–375.

36 Li, L., L. J. Sun, S. G. Tan, and G. B. Ning, *An efficient way in image preprocessing for pavement crack images.* in *CICTP 2012: Multimodal Transportation Systems—Convenient, Safe, Cost-Effective, Efficient.* 2012. (pp. 3095–3103) Beijing.

37 Li, Q., Q. Zou, D. Zhang, and Q. Mao, *FoSA: F∗ seed-growing approach for crack-line detection from pavement images. Image and Vision Computing*, 2011. **29**(12): p. 861–872.

38 Lokeshwor, H., L.K. Das, and S. Goel, *Robust method for automated segmentation of frames with/without distress from road surface video clips. Journal of Transportation Engineering*, 2014. **140**(1): p. 31–41.

39 Moghadas Nejad, F. and H. Zakeri, *A comparison of multi-resolution methods for detection and isolation of pavement distress. Expert Systems with Applications*, 2011. **38**(3): p. 2857–2872.

40 Moussa, G. and K. Hussain. *A new technique for automatic detection and parameters estimation of pavement crack.* in *4th International Multi-Conference on Engineering and Technological Innovation, IMETI 2011.* 2011. Orlando, FL.

41 Nishikawa, T., J. Yoshida, T. Sugiyama, and Y. Fujino, *Concrete crack detection by multiple sequential image filtering. Computer-Aided Civil and Infrastructure Engineering*, 2012. **27**(1): p. 29–47.

42 Oliveira, H. and P.L. Correia, *Automatic road crack detection and characterization. IEEE Transactions on Intelligent Transportation Systems*, 2013. **14**(1): p. 155–168.

43 Ouyang, A. and Y. Wang, *Edge detection in pavement crack image with beamlet transform.* in *2nd International Conference on Electronic & Mechanical Engineering and Information Technology.* 2012. (pp. 2036–2039) Atlantis Press: Shenyang, Liaoning.

44 Payab, M., R. Abbasina, and M. Khanzadi, *A brief review and a new graph-based image analysis for concrete crack quantification. Archives of Computational Methods in Engineering*, 2019. **26**(2): p. 347–365.

45 Payab, M. and M. Khanzadi, *State of the art and a new methodology based on multi-agent fuzzy system for concrete crack detection and type classification. Archives of Computational Methods in Engineering*, 2021. **28**(4): p. 2509–2542.

46 ping Tian, D., *A review on image feature extraction and representation techniques. International Journal of Multimedia and Ubiquitous Engineering*, 2013. **8**(4): p. 385–396.

47 Sinha, S.K. and P.W. Fieguth, *Segmentation of buried concrete pipe images. Automation in Construction*, 2006. **15**(1): p. 47–57.

48 Sinha, S.K. and P.W. Fieguth, *Automated detection of cracks in buried concrete pipe images. Automation in Construction*, 2006. **15**(1): p. 58–72.

49 Sinha, S.K. and P.W. Fieguth, *Neuro-fuzzy network for the classification of buried pipe defects. Automation in Construction*, 2006. **15**(1): p. 73–83.

50 Tsai, Y., V. Kaul, and A. Yezzi, *Automating the crack map detection process for machine operated crack sealer. Automation in Construction*, 2013. **31**: p. 10–18.

51 Xu, B. and Y. Huang, *Automated surface distress measurement system*. 2010, Google Patents.

52 Xu, W., Z. Tang, D. Xu, and G. Wu, *Integrating multi-features fusion and gestalt principles for pavement crack detection. Jisuanji Fuzhu Sheji Yu Tuxingxue Xuebao/Journal of Computer-Aided Design and Computer Graphics*, 2015. **27**(1): p. 147–156.

53 Yadav, B.A.M. and B. Sengar, *A survey on:'Content based image retrieval systems,'. International Journal of Emerging Technology and Advanced Engineering*, 2014. **4**(6): p. 22–26.

54 Ying, L. and E. Salari, *Beamlet transform-based technique for pavement crack detection and classification. Computer-Aided Civil and Infrastructure Engineering*, 2010. **25**(8): p. 572–580.

55 Zakeri, H., F.M. Nejad, and A. Fahimifar, *Image based techniques for crack detection, classification and quantification in asphalt pavement: A review. Archives of Computational Methods in Engineering*, 2017. **24**(4): p. 935–977.

56 Zhou, J., P.S. Huang, and F.P. Chiang. *Wavelet-based pavement distress detection and evaluation.* in *Wavelets: Applications in Signal and Image Processing X*. 2003. (Vol. **5207**, pp. 728–739). International Society for Optics and Photonics: San Diego, CA.

57 Zhu, Z., S. German, and I. Brilakis, *Visual retrieval of concrete crack properties for automated post-earthquake structural safety evaluation. Automation in Construction*, 2011. **20**(7): p. 874–883.

58 Zou, Q., Y. Cao, Q. Li, Q. Mao, and S. Wang, *CrackTree: Automatic crack detection from pavement images. Pattern Recognition Letters*, 2012. **33**(3): p. 227–238.

7

Feature Prioritization and Selection Methods

7.1 Introduction

Feature selection is an important step to reduce dimensions, whereby less important features are removed and ultimately limited to a small subset of main features. Choosing the optimal feature can improve learning performance, accuracy, and lower computational costs. This section is devoted to optimal feature selection methods and provides an overview of the types of features, methods, and techniques.

7.2 A Variety of Features Selection Methods

There are various methods for selecting effective features through which a subset of input variables is selected that can describe the input data more effectively and the negative effects of additional variables and input errors through topics, such as noise or additional variables. Subsequently, better predictive results, higher speeds, and lower computational complexity can be achieved. One analytical application of feature selection is to influence the properties of a category, or the extent to which features overlap, to describe an issue. Standard data can contain hundreds of attributes, many of which may be closely related to other variables (for example, when two attributes are perfectly related, only one attribute is sufficient to describe the data, and one additional attribute needs to be removed). Dependent variables do not provide any useful information about classification and are, therefore, considered a redundant feature for classification. This means that the entire content of the information can be obtained by minimizing the number of unique independent attributes that contain the maximum separating information. Therefore, by keeping the variables independent, the number of features can be reduced, which will improve the performance of the classifier (speed and accuracy). In many applications, variables that are more relevant to the class are retained, and other attributes known as noise may reduce classification performance if used. Therefore, choosing the right feature for classifier designers and researchers is a prominent and key strategy.

When there is not enough information about the process under study and the problem, highly dependent parameters are used. Feature selection techniques can provide insight into the process and improve the need for calculation and forecasting accuracy. To keep an attribute relevant, an attribute selection or attribute selection criterion is required that can accurately measure the relevance of each attribute to the output class/tags. If completely related variables are used in the design of the classifier, it is expected that optimal results will be obtained.

It should be noted that there are other methods for dimensional reduction, such as principal component analysis (PCA), that should not be confused with these methods. Importantly, the purpose

Automation and Computational Intelligence for Road Maintenance and Management: Advances and Applications, First Edition. Hamzeh Zakeri, Fereidoon Moghadas Nejad, and Amir H. Gandomi.

of these methods is to eliminate variables with similarity and overlap if similar PCA methods are used to reduce dimensionality. It is also pertinent to mention that good and optimal features can be used independently from the rest of the data.

Input features are used in feature selection methods to reduce their number. When determining the attribute selection criteria, it is critical to use a method that separates the subset of the best useful attributes and considers them as a new set. This complexity can be interpreted in such a way that the direct evaluation of attribute subsets (2^N) for existing data becomes a complex *NP-complete problems* (NP) problem as the number of attributes increases. To solve this problem, optimization and metaheuristic methods are used, which can increase the speed but suffer from longer computational time compared to the exact methods. In this chapter, we review some of these methods that have been developed for this purpose and address simple applications of the infrastructure-related problems presented in Chapter 5 with new methods.

There are three general types of relations in the feature model extracted from the data: related (relevant), additional (redundant), and irrelevant (unrelated). The first attribute describes the main purpose; the second attribute overlaps with the first attribute; and the third feature contains irrelevant features that do not have significant information about the purpose (Figure 7.1).

Variable removal methods generally fall into four categories:

1) Filter methods
2) Wrapper methods
3) Embedded methods
4) Hybrid methods

These methods are classified. In filter methods, the operator is used as preprocessing to rank properties, in which high-ranking properties are selected and used for prediction. In this method, other characteristics play an important role to classification, speed, and accuracy increase with the optimal feature vector.

In wrapper methods, the criterion for selecting a feature depends on the performance of the predictor, i.e. the predictor is placed on a search algorithm to find a subset that provides the highest prediction performance. This optimization method aims to maximize performance and minimize the number of features.

Embedded methods involve selecting variables as part of the training process that operate without dividing the data into training and testing sets. Finally, hybrid methods use a combination of the above three methods.

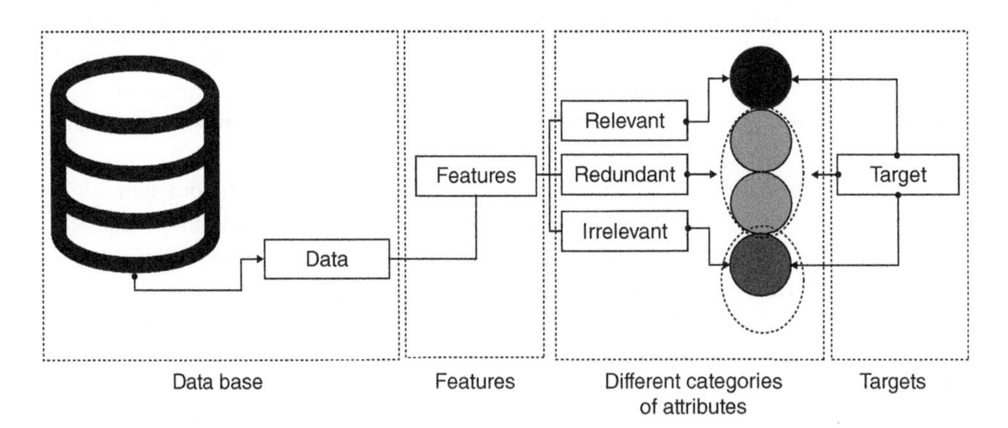

Figure 7.1 The flow of communication between features and their relationship to the target.

In this chapter, feature selection methods using supervised, unsupervised, and semi-supervised learning algorithms are reviewed, then a new semi-supervised method based on the fuzzy method for sample feature vectors is presented. It should be noted that the choice of method depends on the type of issue and quality of the characteristics. The best method to select a feature is by developing a method specific to the problem that requires proper identification of the relationships and an understanding of the concept of the features (Figure 7.2).

7.2.1 Filter Methods

In filter methods, various ranking methods, such as Correlation Criteria (CC) and Mutual Information (MI), are the main criteria for feature selection. These indicators, based on some inherent and statistical characteristics, assign ratings to each attribute, and an appropriate ranking method is selected and used to rank the attributes. After this step, the appropriate threshold is selected based on trial-and-error. Depending on the threshold value, the properties are deleted one after the other. Here, the subject threshold value is a challenge, and the success of a method highly depends on the correct and optimal choice of this parameter. The threshold value is selected using heuristic algorithms or by the trial-and-error method. Filter methods select features based on the internal properties of the attributes. These methods do not employ a classifier to select features and only use a binary separator threshold. The input data $[F_{ij}, C_j]$ consist of N samples $i = 1$ to N with D variables $j = 1$ to D; where F_i,is the ith sample, and C_k is the class label $k = 1$ to Y.

7.2.2 Correlation Criteria

The Pearson correlation coefficient method is one of the most practical feature selection methods, which is defined as follows:

$$R(i) = \left(\frac{\text{cov}(F_i, C)}{\sqrt{\text{var}(F_i * \text{var}(C)}}\right). \tag{7.1}$$

where

F_i is the i_{th} variable.
C is the class labels.
Cov() is the covariance.
Var() is the variance.

According to this definition, correlation ranking recognizes the linear dependencies between the variable and the target and, finally, based on this correlation, removes the variables that are dependent.

7.2.3 Mutual Information (MI)

Theoretical information ranking criteria use the degree of dependency between the variable and target. To describe MI, one of the most well-known methods in this regard is Shannon's definition of entropy, which is calculated using the following equation:

$$H(Y) = -\sum p(y) \log (p(y)) \tag{7.2}$$

where

$H(Y)$ is the uncertainty (information content) in output Y.

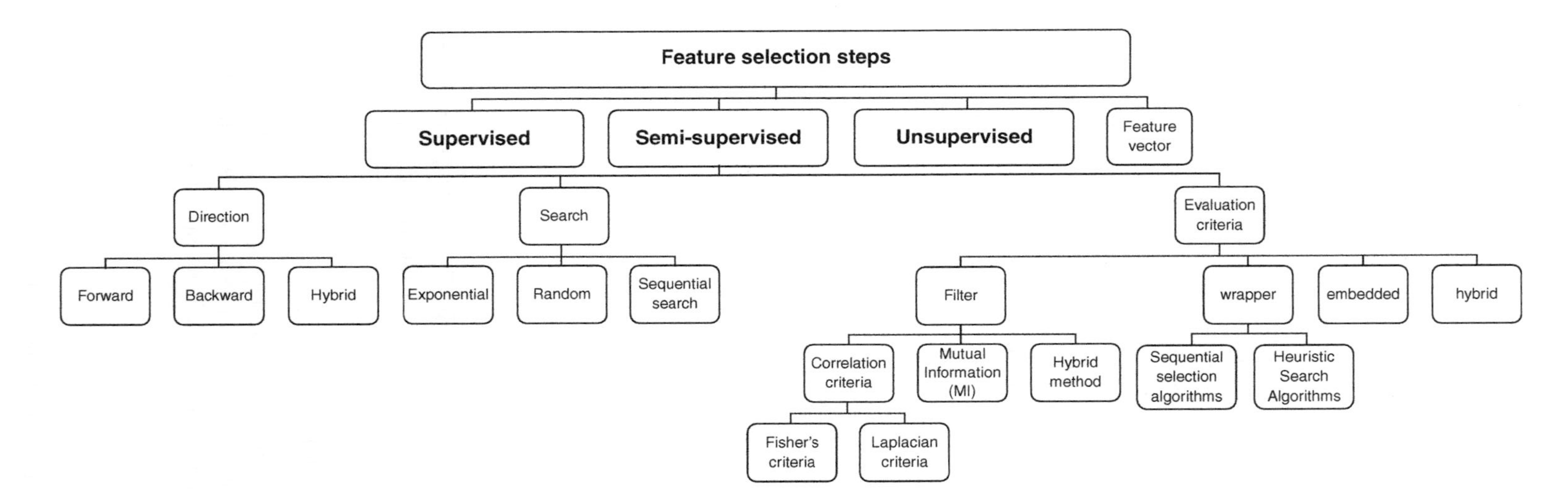

Figure 7.2 General framework of main characteristics of feature selection and methods classification.

The conditional entropy is given by:

$$H(Y \mid X) = -\sum_x \sum_y p(x,y) \log (p(y \mid x)). \tag{7.3}$$

where

$H(Y|X)$ indicates that by perceiving a variable X.
The vagueness in the output Y is reduced

The decrease in uncertainty is given as:

$$I(Y,X) = H(Y) - H(Y \mid X). \tag{7.4}$$

This method presents an MI index between the two variables Y and X. Accordingly, if X and Y are independent, MI will be zero, and if they are dependent, MI will be greater than zero. Interpretation of this index means that one variable can contain part of the information related to another variable, and as a result, dependence can be measured using this index. The definitions provided above can also be extended to continuous variables. The MI index can also be used as a distance measurement parameter as:

$$K(f,g) = \int f(y) \log (f(y)/g(y)) \tag{7.5}$$

where

$K(f, g)$ is the Kullback–Leibler divergence among two variables and is used as a measure of MI.

In the above equation, the probability density function (PDF) of the variables is used to calculate MI. It should be noted that, since the data obtained from the samples are not complete, the PDF is calculated approximately and ambiguously. This point should be considered in the calculations and selection of attribute dependencies. Other methods for calculating MI have been proposed by other researchers. When a method for calculating MI is selected, the feature selection methodology is used to identify the MI between each attribute and the output class labels and, from which the attributes are ranked based on the amount of MI. After ranking using a threshold that is usually set via trial-and-error, which has a value $d < D$, a number of features that largely overlap are removed. Although this method is simple to use, the results are not very robust, since MI is not considered between the attributes and only the corresponding attributes of both variables are measured. Although this method is not robust enough, MI is generally a highly practical and important concept with important applications in other methods. Various methods have been proposed to measure the similarity of attribute ranking criteria based on MI. These indicators update the feature set repeatedly by maximizing or minimizing the placement. The score in each iteration is calculated using the following equation:

$$\mathrm{SM}(n) = \min_{l<k} \hat{I}\left(Y; \left(X_n \mid X_{v(l)}\right)\right) \tag{7.6}$$

where

$\mathrm{SM}(n)$ is similarity index which is updated at each iteration.
X_n is the current evaluated feature.
$X_{v(l)}$ is the set of already selected features.

Attributes are selected repeatedly based on MI index maximization, while similar attributes are not selected using this method. The basic principles of this method are the opposition between dependence and complete independence of characteristics. Regarding feature ranking, it should be noted that important features that are less independent can perform well with other features that are removed. This is a primary weakness of this method. Choosing an optimal learning algorithm is also difficult and varies depending on the problem. It should be noted that the smaller the dimensions of the feature vector, the easier it is to choose the model, but the more likely it is that the optimal answer will not completely overlap.

7.2.4 Wrapper Methods

In the wrapper method (WM), a classifier is used as a target, and the function of the objective based on classification accuracy and error rate is used as a goal to prioritize the subset of variables. The following figure shows the general framework of WM, which is based on minimizing errors and increasing the accuracy of the classifier. The main disadvantages of this method are high computational time and various performances of different classifiers. In most applications, support vector machine (SVM), Naïve Bayes, or Random Forest classifiers are used for classification (Figure 7.3).

The wrapper method is classified into two algorithms: Sequential Selection (SS) and Heuristic Search algorithms (HSA). In general, SS algorithms start with an empty set (complete feature set) then add features (remove features one by one) to achieve the optimal (maximum) objective function. In order to increase the speed, a criterion is selected that gradually increases the target function to reach the target with the least number of features.

HSA search algorithms evaluate different subsets to optimize the target function. Different subcategories are created either by searching around in a search space or by creating solutions to the optimization problem. In the following, each of these methods and their respective theories are reviewed.

7.2.5 Sequential Feature Selection (SFS) Algorithm

Due to their reproducibility, Sequential Feature Selection (SFS) algorithms are commonly used. SFS starts with an empty set (size of the feature set) and adds a feature that has the best specification to the empty matrix for the first step. In this case, the attribute has the maximum value for the

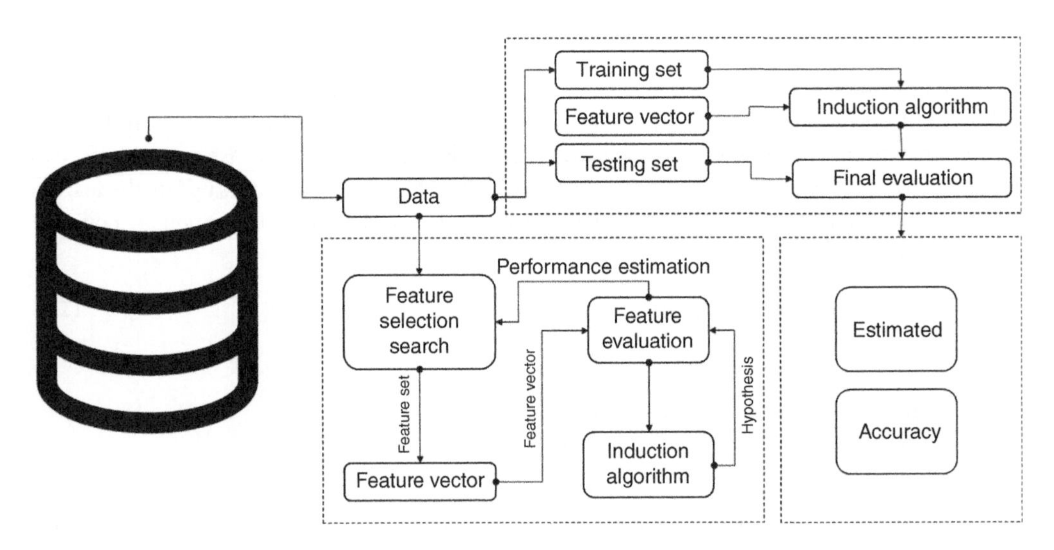

Figure 7.3 General framework of main characteristics of feature selection wrapper methods.

objective function. Then, the remaining features are separately added one after the other until the last set, and a new subset is created and evaluated for its optimality. At this point, if the new set has a higher accuracy than the previous set, it remains as the optimal feature vector, otherwise the last feature of the previous hay is stored in memory as the optimal feature vector. This process is performed until either the condition or the number of iterations reaches the defaults. It should be noted that this method is a simple SFS algorithm because the dependence between features is not considered.

> In this method the mission is finding a feature subset X_d of size $d(<D)$ from the original feature set Y of size D.

The Sequential Backward Selection (SBS) algorithm is similar to SFS, except that it starts with a complete set of variables and removes one attribute at a time, eliminating the slightest reduction in prediction performance. The Sequential Floating Forward Selection (SFFS) algorithm is more flexible than the simple SFS because it introduces an additional retraction step. The general framework of the SFS method is shown in Figure 7.4, in which k is the optimal attribute vector, and d is the required dimension (maximum number of acceptable attributes).

In this step, the subset obtained in the first step is deleted and then the new subset is evaluated. When an attribute is deleted, which increases the value of the objective function, then that attribute is deleted and removed from the new subset and returns to the first step. This process is repeated for the new set. This process is repeated until the required number of features is added or the required functionality is reached. One of the most important limitations of SFS and SFFS methods is the generation of subsets, in which the relationship between the two characteristics is not evaluated internally. To avoid this, an adaptive version of SFFS is introduced, namely the Adaptive Sequential

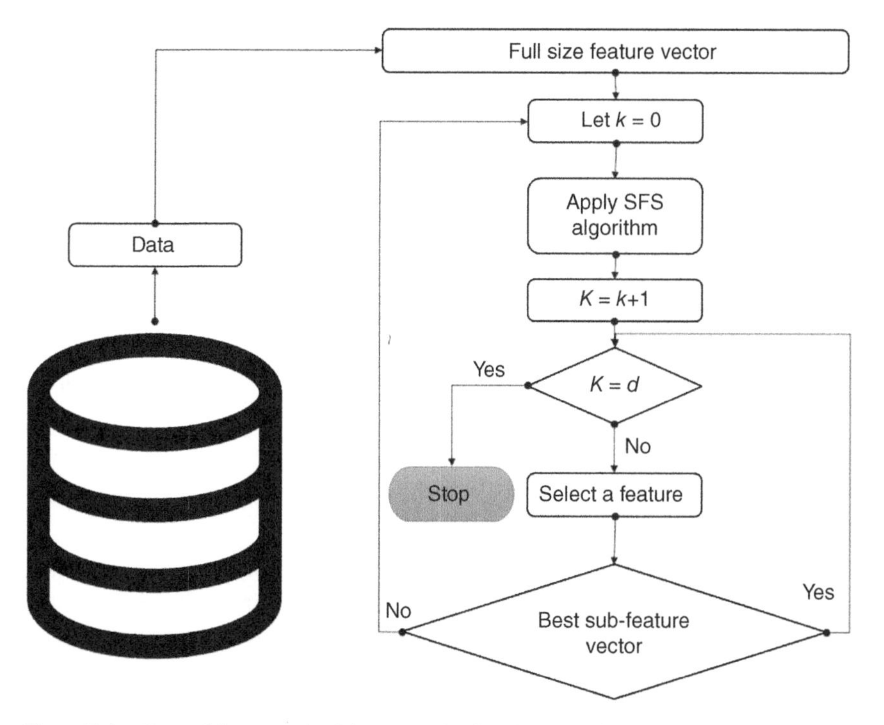

Figure 7.4 General framework of feature selection Sequential Feature Selection (SFS) algorithm.

Forward Floating Selection (ASFFS) algorithm. ASFFS uses the r parameter to specify the number of features added in the input step, which is calculated adaptively. One of the most important advantages of the ASFFS method is the selection of a subset less than the number of SFFS algorithms to provide a better subset of SFFS, depending on the objective function and distribution of data. The general framework of the ASFFS method is shown in Figure 7.5, where k is the optimal attribute vector, and d is the required dimension.

The adaptive sequential backward floating selection (ASBFS) is similar to the ASFFS algorithm, except that $k = D$ and $X_d = Y$. In other words, ASBFS is the "top-down" complement to the ASFFS method.

7.2.6 Heuristic Search Algorithm (HAS)

Heuristic and metaheuristic methods, such as the genetic algorithm (GA), particle swarm optimizations (PSO), ant colony optimization (ACO), artificial bee colony algorithm (ABCA), bat algorithm (BA), and spiral optimization (SPO) algorithm, can also be used to find a subset of features, and its can be used to extract optimal features. Feature selection via HSAs are described in this section, while a more detailed discussion of the types of metaheuristic methods is given in Chapter 10. To determine the optimal feature vector subset, a binary set of features is formed in the heuristic roots, then the algorithm is applied, from which the result shows whether this feature includes the optimized vector set or not. The maximum range for the objective function can be found to provide the best optimal subset. The parameters and operators of metaheuristic methods have sufficient flexibility, while an evolutionary algorithm tailored to the type of problem and data can achieve the best performance.

For example, the structure of the GA method for feature selection is shown in Figure 7.6. In this method, the best characteristics N are selected from the group of parents and children, that is, better children replace less suitable parents. A semi-uniform crossover operator is used for this purpose. In the reproductive stage, each parent member is randomly selected without replacement and is then mated. Not all couples cross each other, and the distance between the parents is calculated before mating – if half of this distance does not exceed the d threshold, they will not mate. The threshold for the GA method is usually assumed to be L = 4, where L is the length of the chromosome. If no children are born, the threshold is reduced by one. If a child is not produced and its threshold reaches zero, a new population will be created. The best person in the current parent population is considered as a model for creating a new population. The rest of the people are obtained by randomly rotating the bits of the template. The normal mutation is ignored after the crossover each time, and the mutations are performed if necessary. In general, heuristic methods converge faster than other methods due to the use of functions that adapt to the solution, and with the help of variable functions, a better search is performed in the answer space. Other population-based methods, such as the PSO algorithm, can also be used to perform feature selection, and these algorithms can be used in combination (Table 7.1).

7.2.7 Embedded Methods

In embedded methods, the goal is to reduce the computation time for classifying different subsets. The main approach of this method is to consider features as part of the training process. As mentioned earlier, MI is an important and key indicator, but ranking using MI does not provide acceptable results alone due to the lack of consideration for the relationship between the indicators. For this purpose, the greedy search algorithm is used to evaluate subsets. In this method, the objective function is designed to maximize the selection of an MI attribute between the attribute and output

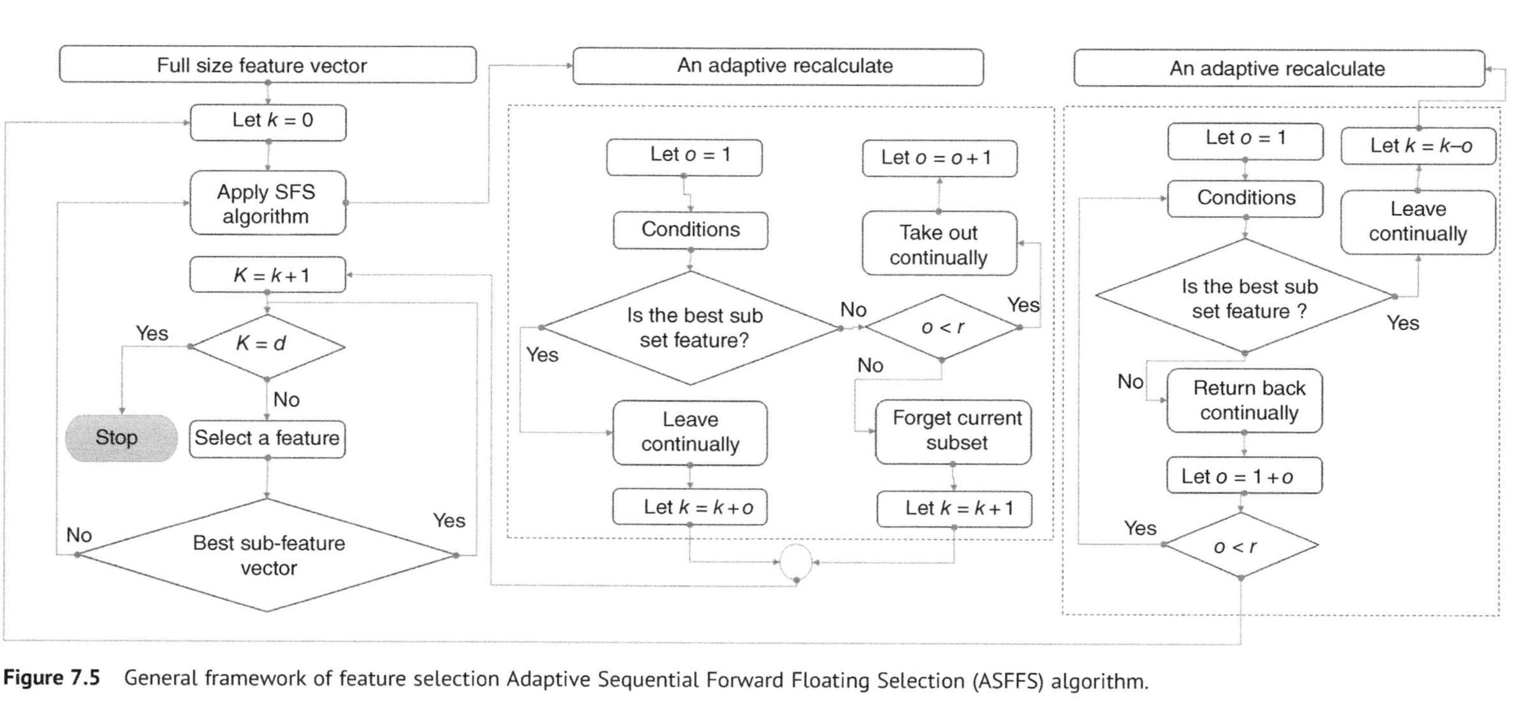

Figure 7.5 General framework of feature selection Adaptive Sequential Forward Floating Selection (ASFFS) algorithm.

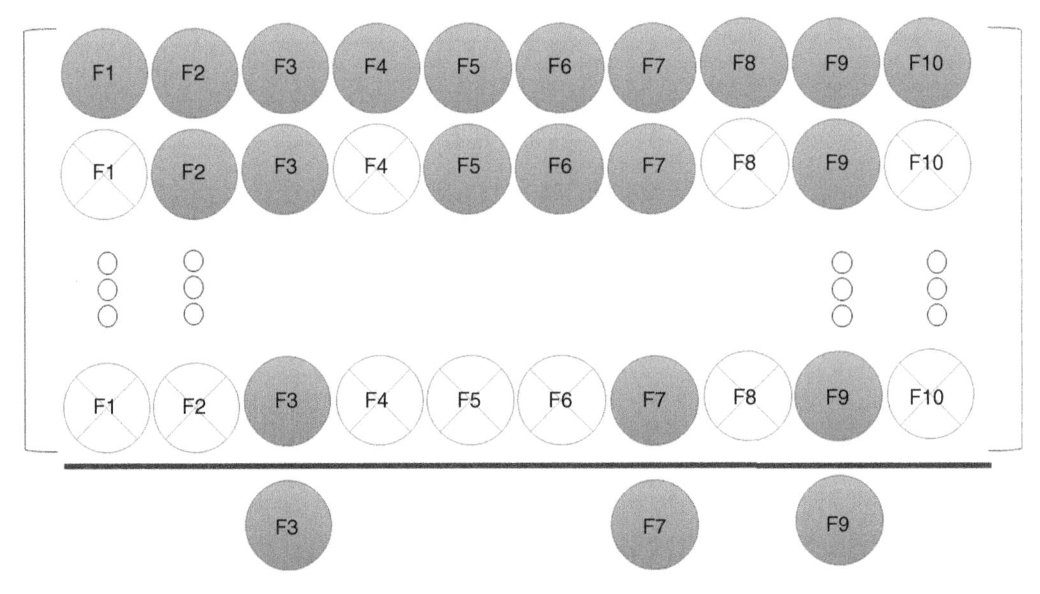

Figure 7.6 An example of feature selection process by GA.

Table 7.1 Heuristic search algorithm over a ranking for feature selection.

Input: training feature set
M: measure,
C: classifier
O: Best sub feature vector

- list l = {}
- for each $Fi \in$ F
- $S(i)$ = compute (fi, M)
- position Fi into l according to $S(i)$
- Best_C = 0
- O = ∅
- for each $Fi \in$ l
- Temp_O = BestSubset $\cup Fi$
    - Temp_C = WrapperClassification (Temp_C, C)
- if (Temp_C = Best_C)
- Best_O = TempSubset
- Best_C = Temp_C

of the class, while the MI between the selected attribute and subset of the selected attribute is minimal. This rule can be defined as follows:

$$I(Y,f) - \beta \sum_{s \in S} I(f : s) \tag{7.7}$$

where

Y is the output.
f is the current selected feature.
s is the feature in the already selected subset.

S and b control the rank of the MI between the current feature f and the features in the subset S.

The output subset is based on a classifier, such as a neural network or SVM. Equation (7.5) allows a better set to be selected for categorization, because the MI between attributes in the calculation is used to select additional attributes. This method can be improved according to the type of problem. For example, MI can be estimated using the Parzen window method, and the maximum correlation is used as the objective function. Another example of evaluation criterion is as follows:

$$I(x_j, C) - \left(\frac{1}{M-1}\right)\sum\nolimits_{x_L \epsilon s_{L-1}} I(x_j : x_L) \tag{7.8}$$

where

x_j is the mth feature in subset S.
s_{m-1} is the so far selected subset with $m-1$ features.

In this method, instead of a greedy algorithm, a two-step approach is used. First, a number k is selected, from which the desired number of features is determined. These numbers have the fewest errors based on the classifier. Embedded methods (EM) methods are used to evaluate different subsets of size k or different features to find the subset that consistently has the smallest classification error. Other EM methods are used to classify weights to rank and remove their characteristics. According to this method, the weight of wj is defined as follows:

$$\begin{aligned} x_j &= \left(m_j(+) - m_j(-)\right)/\left(v_j(+) + v_j(-)\right) \\ D(x) &= w(x-m) \end{aligned} \tag{7.9}$$

where

$m_j(+)$ is the mean of the samples in class (+).
$m_j(-)$ is the mean of the samples in class (−).
$v_j(+)$ is the variance of the class (+) and $j = 1$ to D.
$v_j(-)$ is the variance of the class (−) and $j = 1$ to D.
W is the rank of the features or weight.
$D(x)$ is the decision rule.
m is the mean of the data.

This equation is used as a ranking criterion to sort the features and make the subset. The rank vector w is employed to classify features' ranks to the relationship, and the weight wj is modified to remove a feature j.

7.2.8 Hybrid Methods

Hybrid methods use a combination of the above three feature selection methods, taking advantage of each of their benefits. In the filtering stage in hybrid methods, the properties are ranked or selected using measurements based on the intrinsic properties of the data, which is a kind of preselection. Then, using the wrapper method, specific subsets are evaluated to find the best result through a special clustering algorithm. In general, hybrid methods can be divided into two types: (i) methods based on feature ranking and (ii) other methods. In this section, some methods of both categories related to the combined approach are reviewed. Figure 7.7 shows the wrapper method and filter method, which are very similar to each other, except that the wrapper method includes a learning algorithm in the measurement phase. This is the main reason why wrapper methods are slower than filter method. On the other hand, the wrapper method is obtained due to the use of a better learning algorithm for feature selection. In both methods, the stop criteria related to error

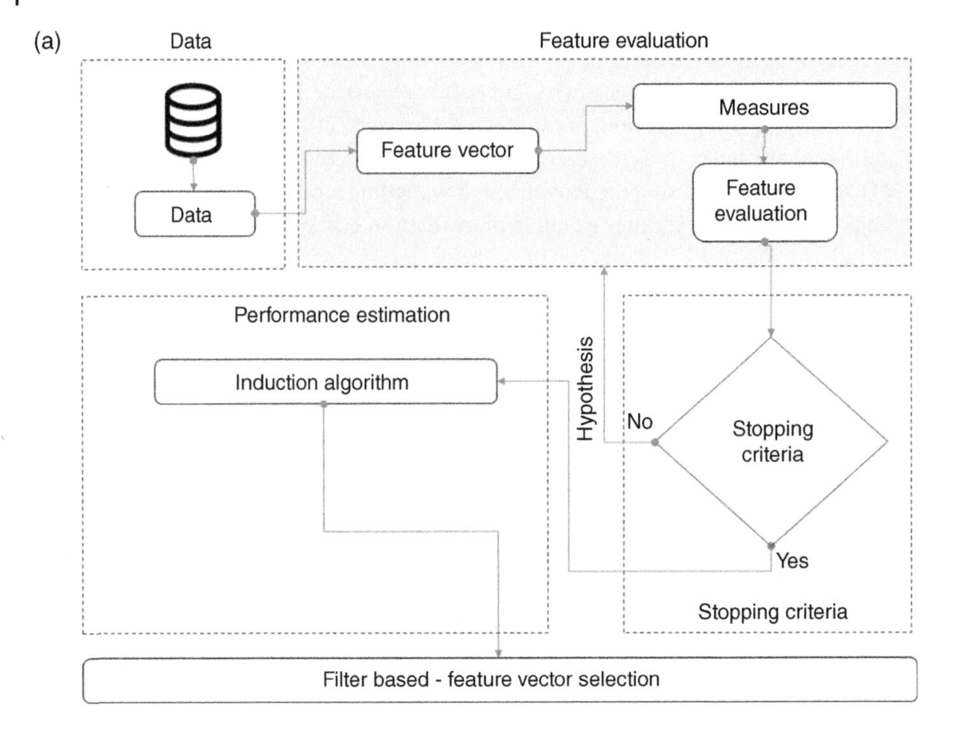

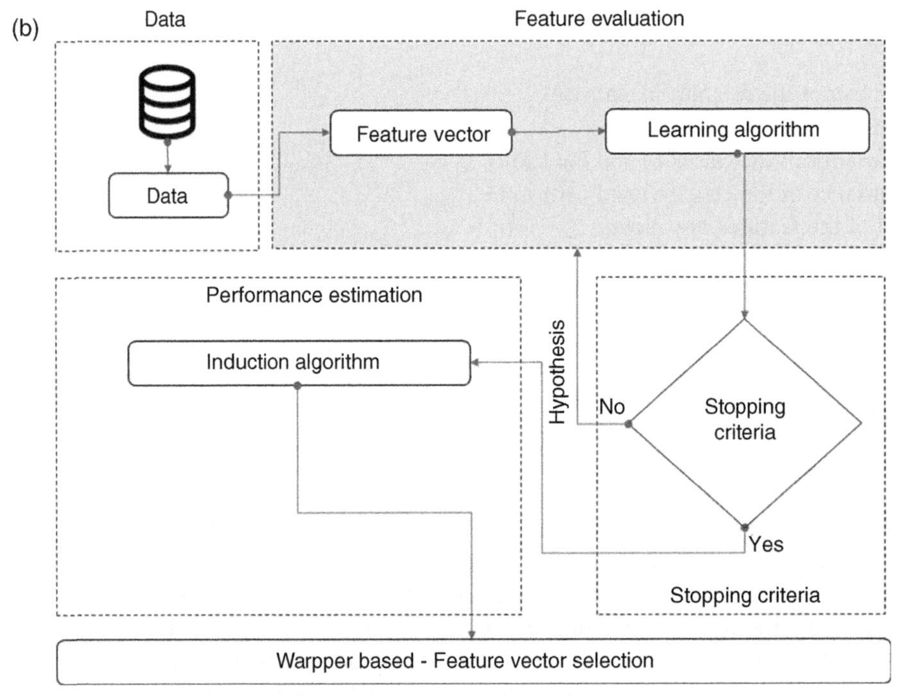

Figure 7.7 General framework for feature selection: (a) filter method and (b) wrapper method.

and reaching the maximum number of thresholds. If the parameters are reached, the search is stopped. Wrapper methods have high classification accuracy but slow processing speed, while filter methods can calculate basic information from the properties themselves. In the filter method, the results of its feature selection depend on the measured properties of the properties. Comparatively, the data mining index plays a key role in the filtering method, whereas learning algorithms are significant in the wrapper method. Different learning methods can be used, and the outcome is highly dependent on the level of learning of the learning algorithm. Specifically, the wrapper method has a slow processing speed due to the use of a learning algorithm, but has good feature selection accuracy because of the implementation of an adaptive method. In the hybrid method, the feature selection mechanism takes advantage of both the filter and wrapper methods. By combining sizes and learning algorithms, we can improve the accuracy of filter classification and reduce the processing time of the wrapper method.

Figure 7.8 shows the framework and general idea of the hybrid feature selection method. In this method, two filter models are used as initial screening to remove additional or irrelevant features of the selection in the design. Based on this method, the *F*-score index and the acquisition of core information are used as primary screening methods. These two feature sets are combined as a pre-processed feature set for fine-tuning. This process is called a hybrid model since separate methods are applied. Moreover, classification methods are used to fine-tune the selected features based on the following relations (7.10). In the first step, the pre-processing of the filter relationship described in the previous section is used. *F*-scores and information gain are good filtering criteria for removing extra and irrelevant features. The *F*-score is applied to classify filter methods by

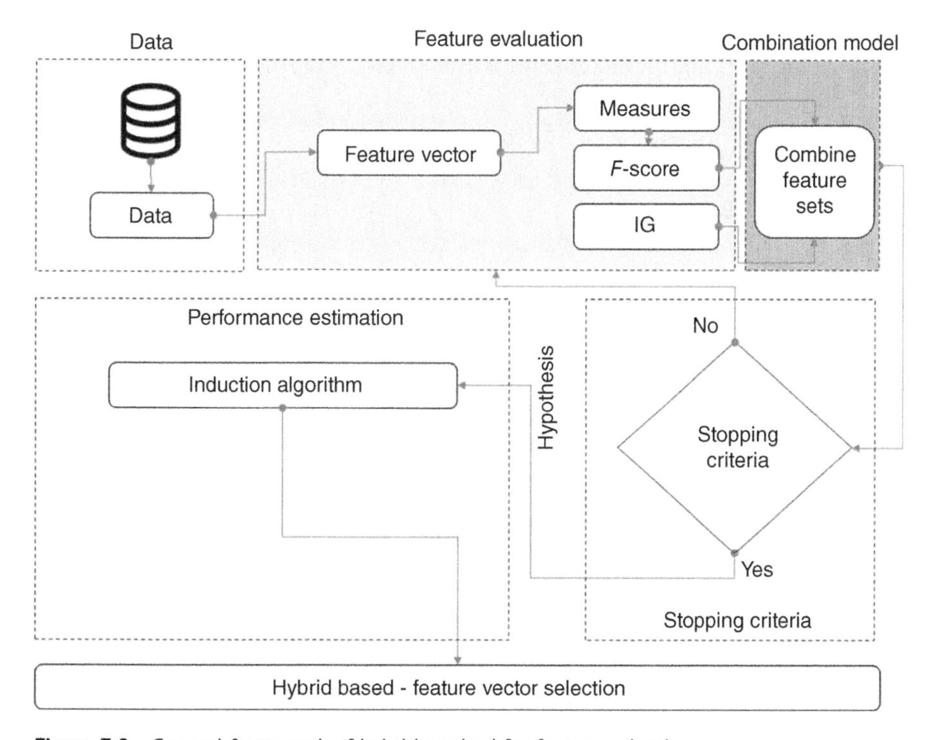

Figure 7.8 General framework of hybrid method for feature selection.

calculating the ability and potential of each feature to differentiate. The higher the *F*-score, the better the segregation ability in classification issues, and the lower the additional features. The F-score is defined by the following equation:

$$F(i) = \left(\frac{\left((m_j(+) - m_j(.))^2 + (m_j(-) - m_j(.))^2 \right)}{\left(\left(\frac{1}{n(+)-1} \sum_{k=1}^{n(+)} \left(f_{k,j}(+) - m_j(+) \right)^2 \right) + \left(\frac{1}{n(-)-1} \sum_{k=1}^{n(-)} \left(f_{k,j}(-) - m_j(-) \right)^2 \right) \right)} \right) \tag{7.10}$$

where

$m_j(+)$ is the mean of the samples in class (+).
$m_j(-)$ is the mean of the samples in class (−)
$m_j(.)$ is the mean of the samples (−) and (+).
$f_{k,\,j}(+)$ is the ith feature of the *k*th positive instances.
$f_{k,\,j}(-)$ is the ith feature of the *k*th negative instances.
$v_j(-)$ is the variance of the class (−) and $j = 1$ to D.
$F(i)(F-\text{score})$ is the stronger discriminative ability the feature i.

However, the *F*-score can only distinguish one attribute and is not effective for distinguishing several attributes. Therefore, the feature with low scores are ignored, even if they have optimal answers in combination with other features.

For this purpose, another index called Information Gain (IG) is used, which is a type of complementary filter that selects related features. This index compares and selects candidate features with more information. Specifically, IG depends on how much information each feature contains, which can increase the accuracy and speed of calculations. Therefore, the more information a feature has, the higher the IG, the lower the value, and the more likely it is to be removed from the list of deleted features.

$$\begin{aligned} E(N) &= \sum_{i=1}^{k} P_i \log_k \left(\frac{1}{P_i} \right) = -\sum_{i=1}^{k} P_i \log_k (P_i) \\ E(D_j) &= \sum_{i=1}^{|D_j|} \frac{D_{ji}}{N} \times E(D_{ji}) \\ \text{IG} &= E(N) - E(D) \end{aligned} \tag{7.11}$$

where

P_j is the probability of class (i) in N point of data domain.
D_{ji} is the *j*th feature contains I kinds of different value.
IG is the information gain of the *j*th feature.

7.2.9 Feature Selection Using the Fuzzy Entropy Method

In many issues related to infrastructure evaluation, features are uncertain, in which case it is necessary to use an index that is able to distinguish between vague and inaccurate data. The degree of ambiguity or uncertainty can be modeled using a fuzzy set. In general, each fuzzy set shows the degree of compliance and membership of each data with the exact answer to the problem. The amount of fuzzy set information is modeled based on the entropy probability theory using the following equation:

$$\text{Ent}_{\text{sh}}(w) = -\sum_{j=1}^{n} \left(\mu_w(x_j)\log\mu_w(x_j) + (1-\mu_w(x_j))\log(1-\mu_w(x_j))\right) \tag{7.12}$$

where

$\mu_w(x_j)$ is the fuzzy membership function.
$\text{Ent}_{\text{sh}}(w)$ is the entropy of feature w.

The value of fuzzy entropy is a function to evaluate the degree of ambiguity and certainty of the properties of the set. The closer this value is to the number 1 or zero, the less ambiguity it has. For a definite set, the value $\text{Ent}_{\text{sh}}(w) = 0$, and for a set with a membership function $\mu_w(x_j) = 0.5$, $\text{Ent}_{\text{sh}}(w) = \max$. There are several functions for measuring entropy. In this section, a new method for selecting a feature using a set of entropies and another method for measuring the similarity index based on fuzzy sets are presented. Fuzzy hybrid entropy values are designed using the Backward Sequential Search (BSS) method to select the appropriate property to remove from the feature set. In this method, the answer vector $V_i = \{v_i(f1), \ldots, v_i(fn)\}$ contains a set of categories that are precisely defined, and for each set of vector features, $W_i = \{w_i(f1), \ldots, w_i(fn)\}$ is numerically assigned. Each of these attribute vector sets is assigned to a C_i class. After assigning the attribute vector to a specific category, the similarity index $\text{SM}\langle W, V\rangle$ is calculated for the input data W and output vector V. In order to assign an attribute vector to each category, assuming the independence of each attribute, the value of the index $\text{SM}\langle W, V\rangle$ is used. In this research, different methods for the $\text{SM}\langle W, V\rangle$ index were developed, where the coefficient committee method is used to select then remove the main feature from the set of main features. In unambiguous sets, the similarity value is $\text{SM}\langle W, V\rangle = 1$, and if the set does not belong to the category, $\text{SM}\langle W, V\rangle = 0$. In general, using this method, $\text{SM}\langle W, V\rangle_{F(i)}$ is calculated for each property, and a set of indicators per property based on the similarity of the feature vector and the ideal output, $\text{EN}(i) = \{\text{SM}\langle W, V\rangle_{F(1)}, \text{SM}\langle W, V\rangle_{F(2)}, \ldots, \text{SM}\langle W, V\rangle_{F(i)}\}$ is calculated. According to the results, if the value of fuzzy entropy is low, then the amount of ambiguity is minimal, and the similarity to the ideal vector is high. If the similarity index is close to 0.5, the maximum entropy value is $\arg\max(\text{EN}(i))$, and the amount of ambiguity is maximized. Using this theory, two entropy parameters $\text{EN}(i)$and similarity $\text{SM}\langle W, V\rangle_{F(i)}$ are used to evaluate the properties $F(i)$ and the classification capability. After deleting each feature, this method can be repeated on the remaining features until the last feature of the steps.

Using this method, the main features are ranked and based on the approach of selecting the category, in which the accuracy and time are evaluated using the complexity matrix. The flowchart of the proposed method for feature selection is shown in Figure 7.9. The main difference between this method and other methods concerns the use of similarity indices and various fuzzy functions in the decision core. In the method presented in this section, a committee of similarity indices, called similarity vectors, was applied to evaluate the features.

7.2.10 Hybrid-Based Feature Selection Using the Hierarchical Fuzzy Entropy Method

In the Luukka method, a similarity index is used for evaluation, while Shannon entropy is also considered to estimate the degree of uncertainty. In this section, 12 similarity indicators are introduced, from which a set of indicators is used to build a search engine and a general index. The set of results of these indices constitutes the similarity vector, and based on the selection of the maximum frequency, the main property is separated from the other properties and placed in the final vector list. The similarity criteria used in this section include the Luukka (LUCA), Yu (YU), Weber, Dubois, Yager, Schweizer, Haraacher, Frank, and Dombi methods, which are defined below.

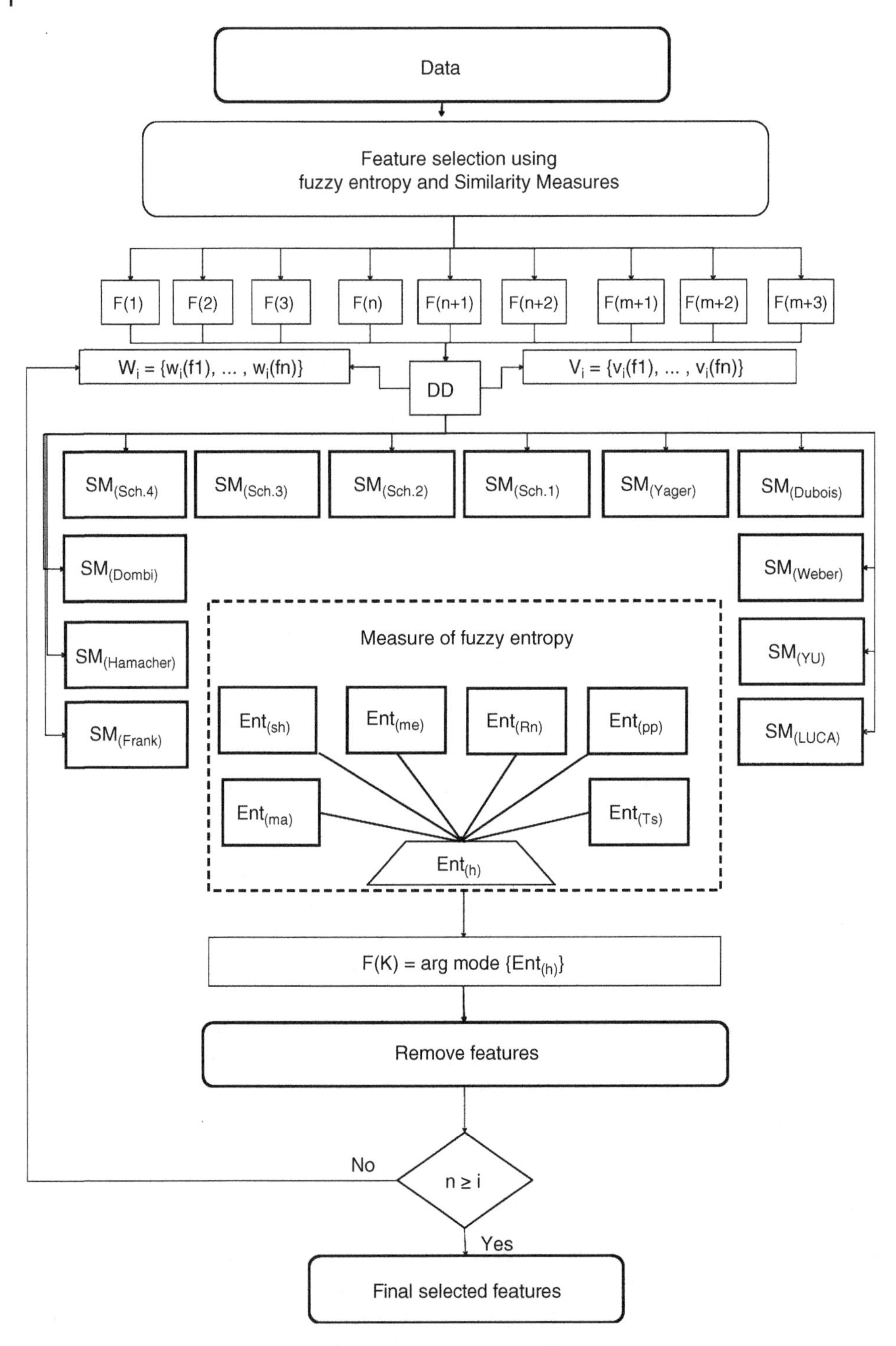

Figure 7.9 General framework of hybrid method using hierarchical fuzzy entropy and similarity measure.

$$SM_{\text{LUCA}} = \left(\frac{1}{t}\sum_{r=1}^{t} w_r(1-|x(f_r)^p - v(f_r)^p|)\right)^{\frac{m}{p}})^{(\frac{1}{m})} \quad ,v \in [0,1]^t, p = 1 \tag{7.13}$$

$$SM_{\text{YU}} = \max[0,(1+\lambda)(a+b-1)-\lambda ab] \tag{7.14}$$

$$SM_{\text{Weber}} = \max\left[0, \frac{(a+b+\lambda ab-1)}{(1+\lambda)}\right] \tag{7.15}$$

$$SM_{\text{Dubios}} = \left[\frac{(ab)}{\max(a,b,\alpha)}\right], \alpha \in [0,1] \tag{7.16}$$

$$SM_{\text{Yager}} = 1 - \min\left\{1,[(1-a)^w + (1-b)^w]^{1/w}\right\} \tag{7.17}$$

$$SM_{\text{Schweizer1}} = \{\max(0, a^p + b^p - 1\}^{1/p}, p \neq 0 \tag{7.18}$$

$$SM_{\text{Schweizer2}} = 1 - [(1-a)^w - (1-b)^w - (1-a)^w(1-b)^w]^{1/w} \tag{7.19}$$

$$SM_{\text{Schweizer3}} = \exp\left(-(|\ln a|^p + |\ln b|^p)^{1/p}\right) \tag{7.20}$$

$$SM_{\text{Schweizer4}} = \left[\frac{(ab)}{[a^p + b^p - a^p b^p]^{\frac{1}{p}}}\right] \tag{7.21}$$

$$SM_{\text{Hamacher}} = \left[\frac{(ab)}{r + (1-r)(a+b-ab)}\right] \tag{7.22}$$

$$SM_{\text{Frank}} = \log_r\left(1 + \frac{(s^a-1)(s^b-1)}{s-1}\right) \; s > 0, s \neq 1 \tag{7.23}$$

$$SM_{\text{Dombi}} = \left\{1 + \left[\left(\frac{1}{a}-1\right)^{\lambda} + \left(\frac{1}{b}-1\right)^{\lambda}\right]^{1/\lambda}\right\}^{-1}, \lambda > 0 \tag{7.24}$$

Parameters a and b are made for two sets of membership functions, and the fuzzy membership coefficients are related to these equations. These parameters, which actually min operators, are calculated using the following equations.

$$SM_{\text{YU}} = \min[1, a+b+\lambda ab] \tag{7.25}$$

$$SM_{\text{Weber}} = \min[1,(a+b-(\lambda/(1-\lambda))ab)] \tag{7.26}$$

$$SM_{\text{Dubios}} = \left[1 - \frac{(1-a)(1-b)}{\max((1-a)(1-b),\alpha)}\right], \alpha \in [0,1] \tag{7.27}$$

$$SM_{\text{Yager}} = \min\left\{1,[(a)^w + (b)^w]^{1/w}\right\} \tag{7.28}$$

$$SM_{\text{Schweizer1}} = 1 - \{\max(0,(1-a)^p + (1-b)^p - 1\}^{1/p}, p \neq 0 \tag{7.29}$$

$$SM_{\text{Schweizer2}} = 1 - [(a)^w - (b)^w - (a)^w(b)^w]^{1/w} \tag{7.30}$$

$$SM_{\text{Schweizer3}} = 1 - \exp\left(-(|ln(1-a)|^p + |ln(1-b)|^p)^{1/p}\right) \tag{7.31}$$

$$SM_{\text{Schweizer4}} = \left[\frac{((1-a)(1-b))}{[(1-a)^p + (1-b)^p - (1-a)^p(1-b)^p]^{\frac{1}{p}}}\right] \tag{7.32}$$

$$SM_{\text{Hamacher}} = \left[\frac{(a+b+(r-2)ab)}{r+(r-1)(ab)}\right] \tag{7.33}$$

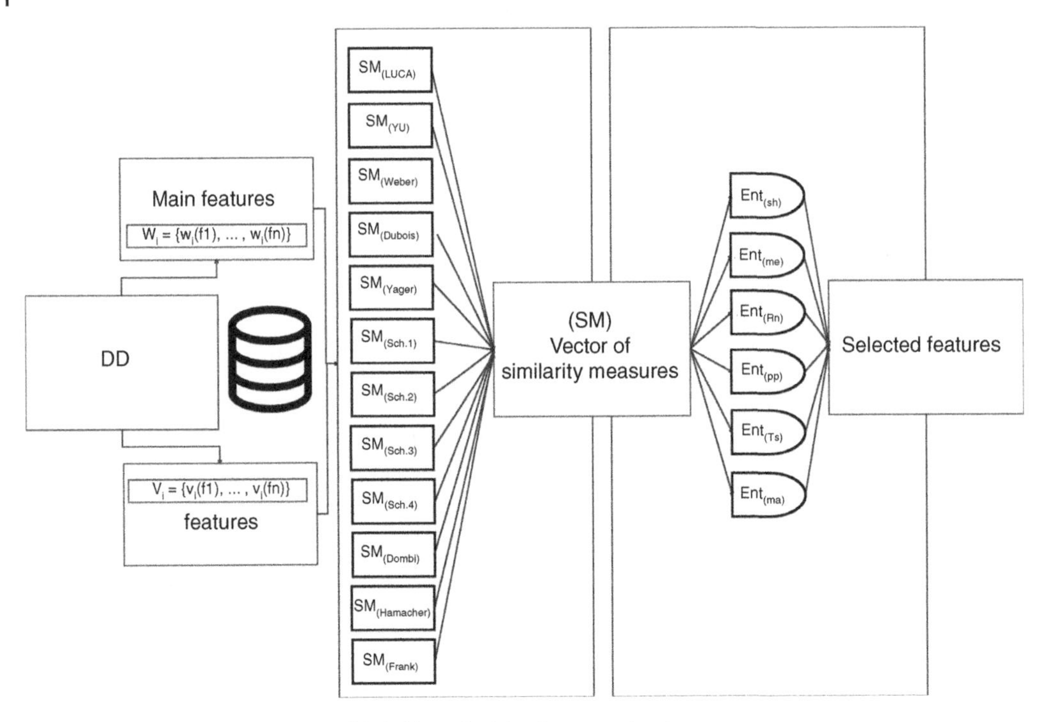

Figure 7.10 General framework of hybrid method for feature selection.

$$\mathrm{SM}_{\mathrm{Frank}} = \log_r \left[1 + \frac{(s^a - 1)(s^b - 1)}{s - 1} \right] s > 0, s \neq 1 \tag{7.34}$$

$$\mathrm{SM}_{\mathrm{Dombi}} = \left\{ 1 + \left[\left(\frac{1}{a} - 1 \right)^{\lambda} + \left(\frac{1}{b} - 1 \right)^{\lambda} \right]^{1/\lambda} \right\}^{-1}, \lambda > 0 \tag{7.35}$$

The process of analysis performed using the combined method with similarity indices and the 12 criteria was performed for a database of images with pavement failure and extracted indices, as shown in Figure 7.10.

In order to evaluate the ambiguity and uncertainty in the set of features and its relationship with the extracted main features, the entropy index is used. Much research has been done on the use of entropy and the introduction of new methods for measuring information gain. Shannon, who introduced Shannon entropy $\mathrm{Ent}_{(\mathrm{sh})}$,and his colleagues first proposed a method based on logarithmic function employing a probability power parameter to control the shape and probabilistic distribution of variables. Renyi entropy $\mathrm{Ent}_{(\mathrm{RN})}$ utilizes the α parameter to expand upon $\mathrm{Ent}_{(\mathrm{sh})}$. Tsallis entropy ($\mathrm{Ent}_{(\mathrm{TS})}$) was introduced in 2001 based on a similar relationship between $\mathrm{Ent}_{(\mathrm{RN})}$ and the α parameter. The information gain power function was introduced in 1992 to solve the problem of the logarithmic function $\mathrm{Ent}_{(\mathrm{sh})}$, using the relation Pal and Pal entropy ($\mathrm{Ent}_{(\mathrm{pp})}$). Also, a new type of entropy, called Hanman–Anirban entropy $\mathrm{Ent}_{(\mathrm{ma})}$, was introduced with the aim of creating free and controllable parameters to adjust the outputs, in which p_i is the probability of occurrence of a probability variable created for x_i. Also, the set of free parameters $\{a, b, \alpha, \beta, \gamma\}$ are the setting values for $\mathrm{Ent}_{(\mathrm{ma})}$. In order to normalize the entropy function $\mathrm{Ent}_{(\mathrm{ma})}$, the minimum and maximum values of $\mathrm{Ent}_{(\mathrm{ma})_{\max}}$ and $\mathrm{Ent}_{(\mathrm{ma})_{\min}}$ are used, whereby the higher the probability, the lower the IG. The relationships of the types of entropies used in this section are defined below.

$$\mathrm{Ent}_{\mathrm{sh}}(w) = -\sum_{j=1}^{n} \left(\mu_w(x_j)\log \mu_w(x_j)\right) \tag{7.36}$$

$$\mathrm{Ent}_{\mathrm{RN}}(w) = \left(\log \sum_{j=1}^{n} \left(\mu_w(x_j)\right)^{\alpha}\right)/(1-\alpha) \tag{7.37}$$

$$\mathrm{Ent}_{\mathrm{ME}}(w) = \sum_{j=1}^{n} \left((W_{\mathrm{j}})\left(\mathrm{Sin}\frac{\pi\mu_w(x_j)}{2} + \sin\frac{\pi(1-\mu_w(x_j))}{2} - 1\right)\right) \tag{7.38}$$

$$\mathrm{Ent}_{\mathrm{TS}}(w) = \left(1 - \left(\sum_{j=1}^{n} \left(\mu_w(x_j)\right)^{\alpha}\right)\right)/(1-\alpha) \tag{7.39}$$

$$\mathrm{Ent}_{\mathrm{pp}}(w) = -\sum_{j=1}^{n} \left(\mu_w(x_j)\mathrm{e}^{1-\mu_w(x_j)}\right) \tag{7.40}$$

$$\mathrm{Ent}_{m}(w) = \sum_{j=1}^{n} \left(\mu_w(x_j)e^{-\left(a\mu_w(x_j)^3 + b\mu_w(x_j)^2 + c\mu_w(x_j) + d\right)}\right) \tag{7.41}$$

$$\mathrm{Ent}_{\mathrm{ma}}(w) = \frac{\mathrm{Ent}_{m}(w) - e^{-(a+b)^{\beta}}}{n^{(1-\gamma)}e^{-\left(\frac{a}{n^{a}}+b\right)^{\beta}} - e^{-(a+b)^{\beta}}} = \frac{\mathrm{Ent}_{m}(w) - \mathrm{Ent}_{(\mathrm{ma})_{\min}}}{\mathrm{Ent}_{(\mathrm{ma})_{\max}} - \mathrm{Ent}_{(\mathrm{ma})_{\min}}} \tag{7.42}$$

Equations (7.36)–(7.42) can be applied to a variety of information gain (IG) measurement methods, including committees of entropies Ent = {$\mathrm{Ent}_{\mathrm{sh}}(w)$, $\mathrm{Ent}_{\mathrm{me}}(w)$, $\mathrm{Ent}_{\mathrm{rn}}(w)$, $\mathrm{Ent}_{\mathrm{pp}}(w)$, $\mathrm{Ent}_{\mathrm{ma}}(w)$}. Specifically, feature selection with the help of a hybrid algorithm to prioritize the main features is highly efficient (Figure 7.11).

Using this method, the computational dimension is reduced, and features with a higher degree of ambiguity are removed from the feature bank. Moreover, features with less entropy remain in the characteristic bank in order to continue the evaluation process.

For example, the hybrid method of selecting a feature for a database of damaged pavement images is as follows. In this algorithm, m is a sample of cracking images, and t is the main characteristic extracted from the feature extraction method (Chapter 6) for l categories. In this method, the similarity index is prioritized and the largest value of the similarity index is determined then the fuzzy entropy value is calculated for each characteristic. After this step, each property with the highest entropy value is identified and removed from the feature database. If the number of remaining attributes is greater than n, the steps are repeated with the new attribute base (Table 7.2).

To evaluate the efficiency of this method in infrastructure management in the field of management automation, three sets with different numbers of data were considered. Specifically, the characteristic set is divided into three main categories, namely db1, db2, and db3 image sets, where the set of images obtained from the Laser Crack Measurement System (LCMS) machine display characteristics of the aerial imagery bank. These three sets are used for 2 categories with 33 features and 62, 720, and 1000 data, respectively (the database is included in the book appendix).

7.2.11 Step 1: Measure Similarity Index and Evaluate Features

In order to evaluate the characteristics, different similarity indices are calculated. A fuzzy membership function is used for each characteristic and image. In each step, one of the main features is removed from the feature bank then analyzed with a smaller feature bank in the next step. It should be noted that the main features are re-prioritized and analyzed with a new arrangement as a new feature bank. Due to the reduction in the size and number of features, it is predicted that the computational time to calculate the similarity and entropy will be reduced and the algorithm will move faster toward the effective features. The value of the membership function for the crash database to remove the database indices in steps 1, 10, and 15 in Figures 7.12–7.14, respectively.

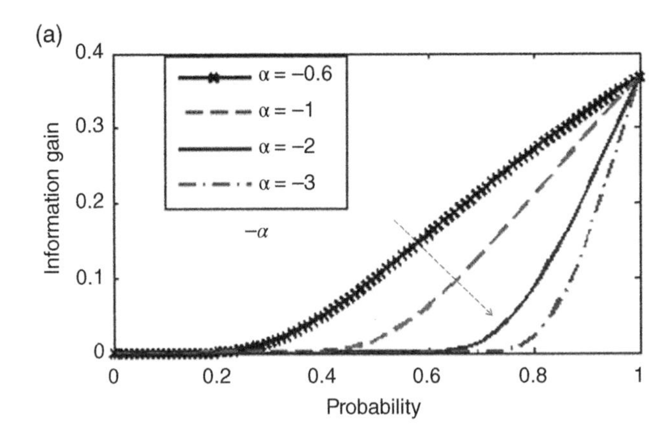

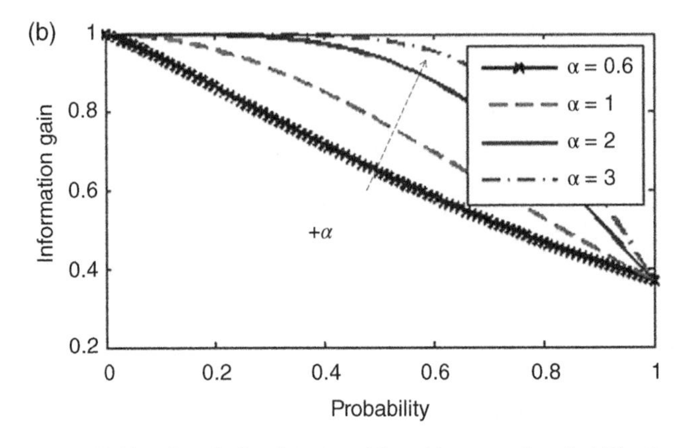

Figure 7.11 Correlation between IG and increased probability for: (a) negative parameter values and (b) positive parameter values of $Ent_{(ma)}$.

Table 7.2 Hybrid feature selection algorithm based on hierarchical fuzzy entropy.

```
Input : idealvec[1, ..., l], Datalearn[1, ..., m], n
# for j=1 to m do
        for j=1 to m do
                for i=1 to t do
                        for k=1 to l do
                        %similarity
                        For s=1:12
                                Sim[i][j][k] = {SM_model(s)}
                        end for
                end for
        end for
        Sort Sim[i][j][k] based on feature U
                for i=1 to 6
                        Ent[i] = {Ent_model(i)}
                end for
                for j=1 to 6
                        for i=1 to 6
                                J[i]=arg max{Ent_model(i)}
                        end for
                Wf= arg mode(J[j])
        end for
        if m <n then go to #
end for
Remove Wf from data set
```

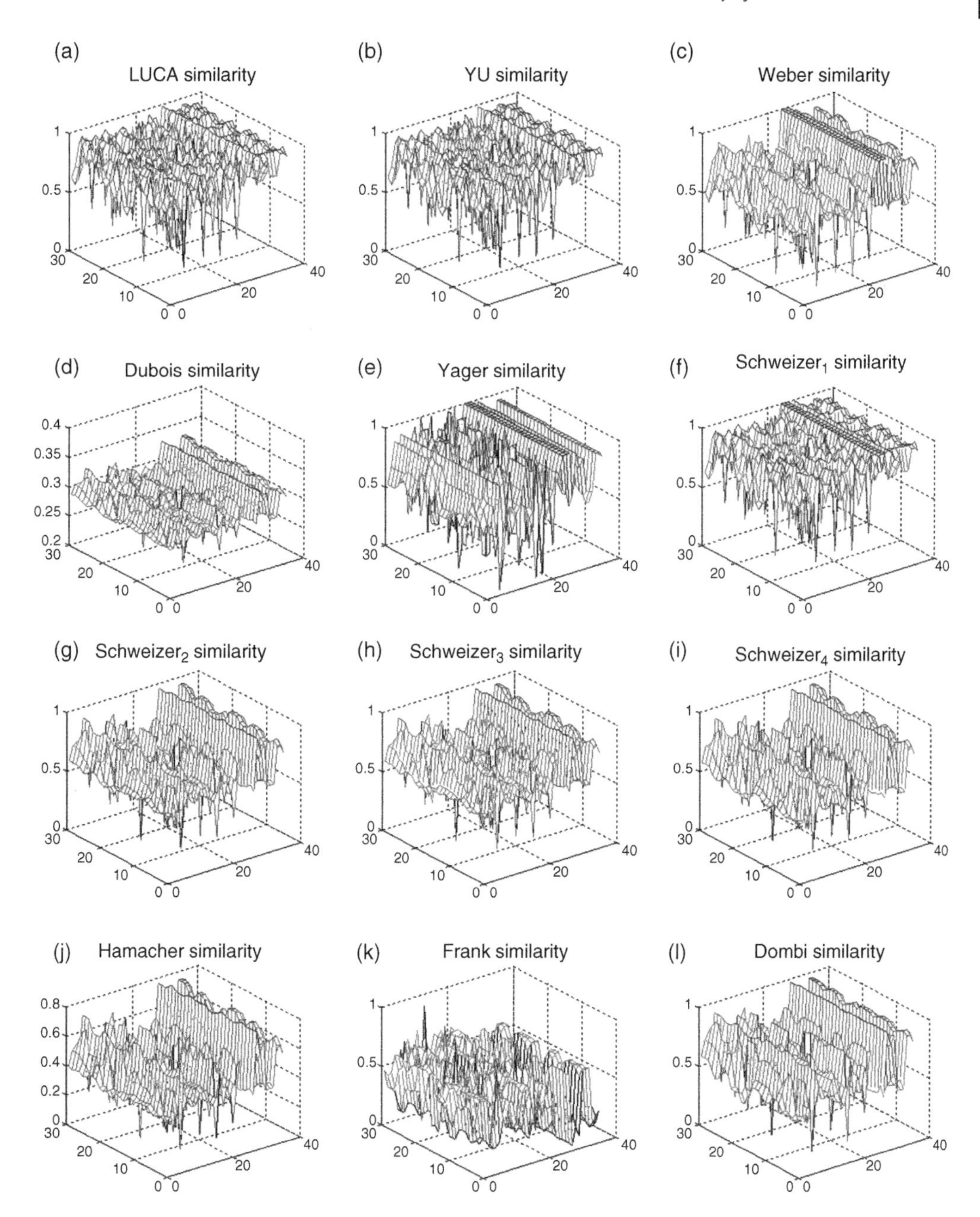

Figure 7.12 An example of fuzzy functions extracted using different similarity index methods in step (1): (a) LUCA, (b) YU, (c) Weber, (d) Dubois, (e) Yager, (f) Schweizer 1, (g) Schweizer 2, (h) Schweizer 3, (i) Schweizer 4, (j) Hamacher, (k) Frank, and (l) Dombi methods.

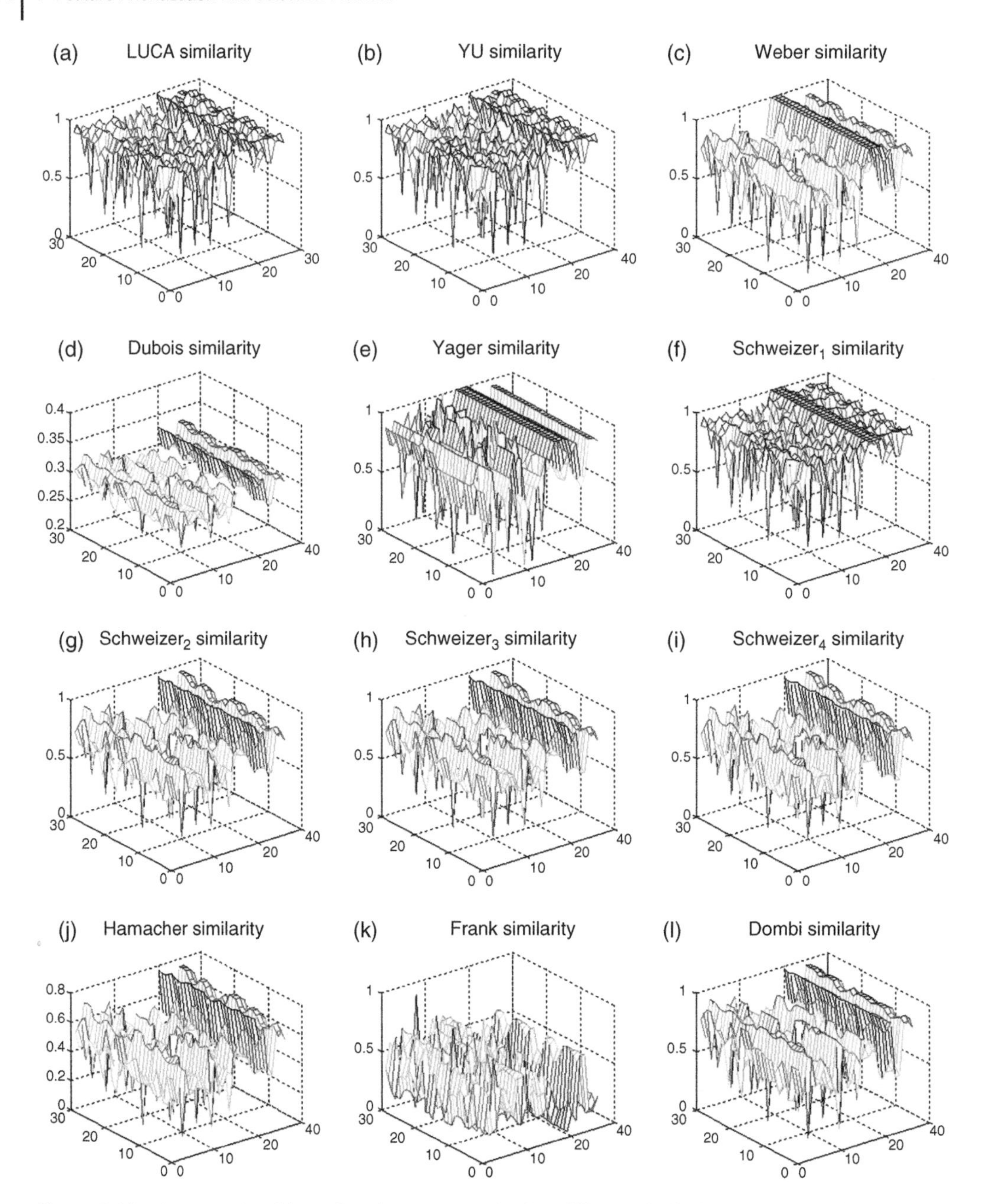

Figure 7.13 An example of fuzzy functions extracted using different similarity index methods in step (10): (a) LUCA, (b) YU, (c) Weber, (d) Dubois, (e) Yager, (f) Schweizer 1, (g) Schweizer 2, (h) Schweizer 3, (i) Schweizer 4, (j) Hamacher, (k) Frank, and (l) Dombi methods.

The vector of the remaining properties in each step is indicated by P_k. In this vector, each number $v(i)$ represents the ith property of the k-matrix. For example, [4 3 3 1] the vector of the residual properties is for step 4, where $v(1) = 4$, represents the fourth property of the first round property matrix, $v(2) = 3$ represents the third property of the second round property matrix, $v(3) = 3$ represents the third property of the third round property matrix, and $v(4) = 1$ represents the first property

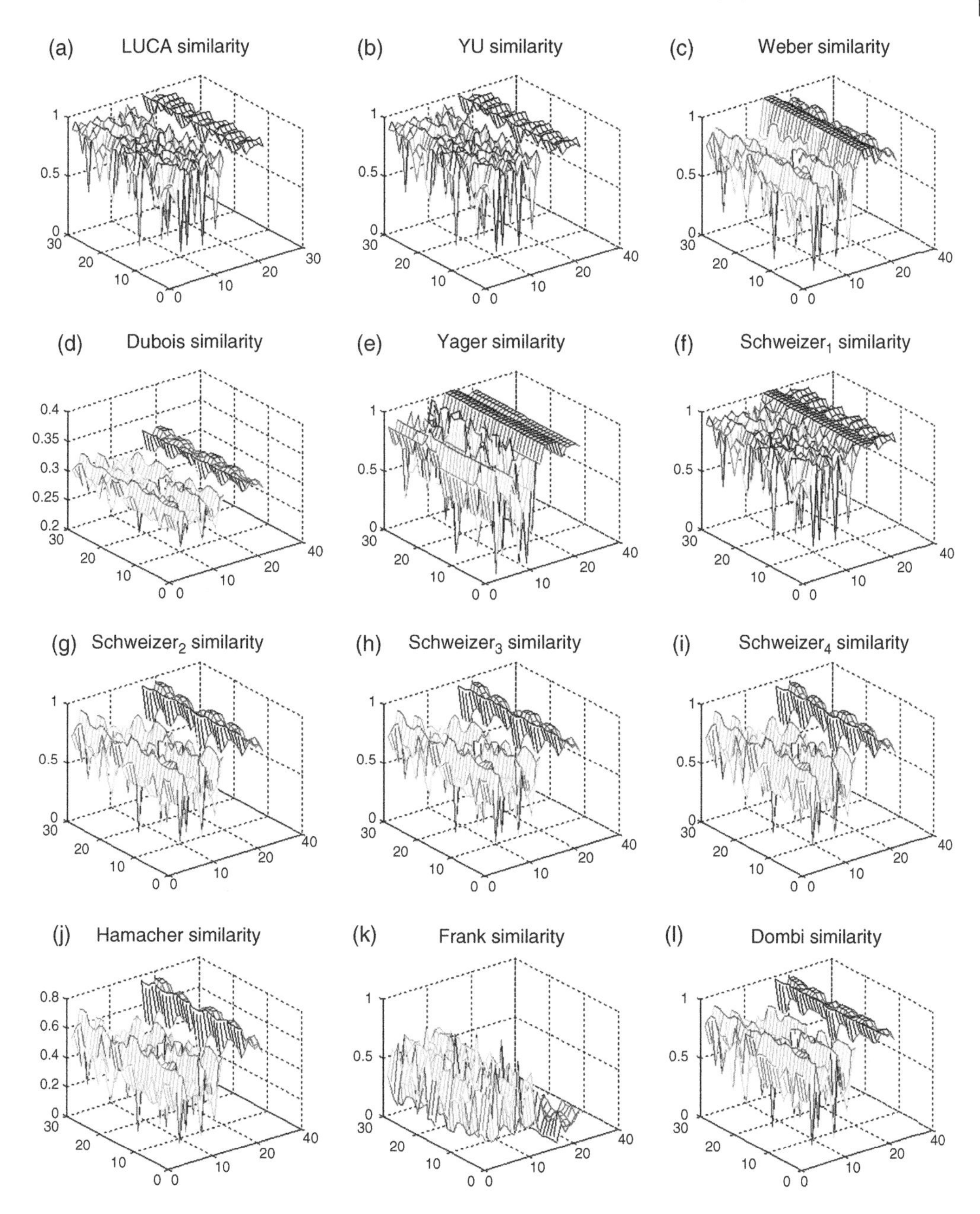

Figure 7.14 An example of fuzzy functions extracted using different similarity index methods in step (15): (a) LUCA, (b) YU, (c) Weber, (d) Dubois, (e) Yager, (f) Schweizer 1, (g) Schweizer 2, (h) Schweizer 3, (i) Schweizer 4, (j) Hamacher, (k) Frank, and (l) Dombi methods.

of the fourth round property matrix. Results for step 1, in which the feature with maximum entropy and 50% similarity are extracted, are shown in Table 7.3.

As shown in Tables 7.3 and 7.4, in the first step for db1 with 100 images, the decision matrix is formed using similarity and entropy indices, and the number of selected attributes is selected based on the maximum iteration mode Is. In the first step, feature number 24, with the highest entropy

Table 7.3 The similarity–entropy matrix for db1 bank – in step 1.

	$Ent_{(sh)}$	$Ent_{(me)}$	$Ent_{(rn)}$	$Ent_{(pp)}$	$Ent_{(ts)}$	$Ent_{(sh)}$	
Similarity/entropy	**ENT1**	**ENT2**	**ENT3**	**ENT4**	**ENT5**	**ENT6**	**Mode**
SE1	22	22	22	22	22	22	22
SE2	22	22	22	22	22	22	22
SE3	24	24	24	24	24	24	24
SE4	25	25	25	25	25	25	25
SE5	24	24	24	24	24	24	24
SE6	22	22	22	22	22	22	22
SE7	29	29	29	29	29	29	29
SE8	29	29	29	29	29	29	29
SE9	24	24	24	24	24	24	24
SE10	12	12	12	12	12	12	12
SE11	3	3	3	3	3	3	3
SE12	24	24	24	24	24	24	24
Mode	**24**	**24**	**24**	**2**	**24**	**24**	**24**

Table 7.4 The similarity-entropy matrix for db2 bank – in step 10.

	$Ent_{(sh)}$	$Ent_{(me)}$	$Ent_{(rn)}$	$Ent_{(pp)}$	$Ent_{(ts)}$	$Ent_{(sh)}$	
Similarity/entropy	**ENT1**	**ENT2**	**ENT3**	**ENT4**	**ENT5**	**ENT6**	**Mode**
SE1	2	2	2	2	2	2	2
SE2	2	2	2	2	2	2	2
SE3	10	10	10	10	10	10	10
SE4	16	16	16	16	16	16	16
SE5	10	10	10	10	10	10	10
SE6	2	2	2	2	2	2	2
SE7	10	10	10	10	10	10	10
SE8	10	10	10	10	10	10	10
SE9	10	10	10	10	10	10	10
SE10	8	8	8	8	8	8	8
SE11	11	11	11	11	11	11	11
SE12	10	10	10	10	10	10	10
Mode	**10**	**10**	**10**	**2**	**10**	**10**	**10**

value and 50% similarity, is selected as a candidate index by the indicators committee to be removed from the evaluation cycle in the next characteristic bank. With each step, a feature is selected. The Table 7.5 shows the step outputs for the proposed method. The residual feature vector of each step for the characteristic bank db1 is shown in Figure 7.15.

Table 7.5 An example of a feature vector for selecting priority features.

Features	Good	Distress	Spall	Long.	Trans.	Digo.	Block	Fatige	Complex	Pothole
	Nd	Dd	Sp	Lc	Tc	Dc	Bc	Fc	Cc	Pd
N	903	1257	576	249	303	132	177	171	120	114
Dr	—	—	0.46	0.20	0.24	0.11	0.14	0.14	0.10	0.09
%	0.42	0.58	0.27	0.12	0.14	0.06	0.08	0.08	0.06	0.05
Dr (class%)	—	—	—		0.54			0.37		0.05

N: Number, Dr: Road pavement damage ratio, %: Damage ratio to total%, Dr (class%): Road pavement damage ratio based on class of distress

7.2.12 Step 2: Final Feature Vector

According to the new method presented for feature selection, this section provides a general review of the relationship between each feature and the main features, the detection failure, and the relationship of the image feature with the main features, including the type of cracking. In this section, a database with 2160 images is used. The main characteristics of the images are shown in Table 7.6.

According to the prioritization method, based on the fuzzy similarity method and entropy committee, different prioritization features and two top features for categorization are categorized with 31 and 30 repetitions, respectively, as shown in Table 7.6. The remaining two features are the top features with the least amount of ambiguity and maximum information gain.

Based on the results obtained from the feature vector analysis, the top extracted three features are summarized in Table 7.6. Diagnosis of infrastructure-related damage, including road pavement damage, using high amplitude shearlet coefficient percentage (HASHCP) and high frequency shearlet energy percentage (HFSHEP) characteristics shows better results, where the entropy of ambiguity is minimal and the degree of similarity is maximum. Based on Figure 7.15, features 16 and 25 (correlation and Zernike moment 1, respectively) are selected for spalling failures, while features 20 and 25 (correlation and Zernike moment 5, respectively) are chosen for surface distress.

It should be noted that one of the most important parts in selecting a feature is using an efficient classification algorithm. In this regard, due to the fact that the data related to feature selection are always correlated and overlap with each other, inherent ambiguity is always present. As such, all categories have errors at this stage. Using a classifier allows control of ambiguity, which can be very useful. For this purpose, the various classification methods presented in Chapter 8 can be used.

7.3 Classification Algorithm Based on Modified Support Vectors for Feature Selection – CDFESVM

In order to select superior features, accuracy indicators are evaluated using categorical outputs. The classification of images containing distress, the two main features classifier, the SVM and the coordinate descent fuzzy ENTROPY support vector machine (CDFESVM) method, are explained in this section. In recent years, the use of support vector method has grown significantly in classification and regression. This method is based on statistical learning, considering experimental risk, and is intended to maximize the distance between the categories and the fit of the dividing line between

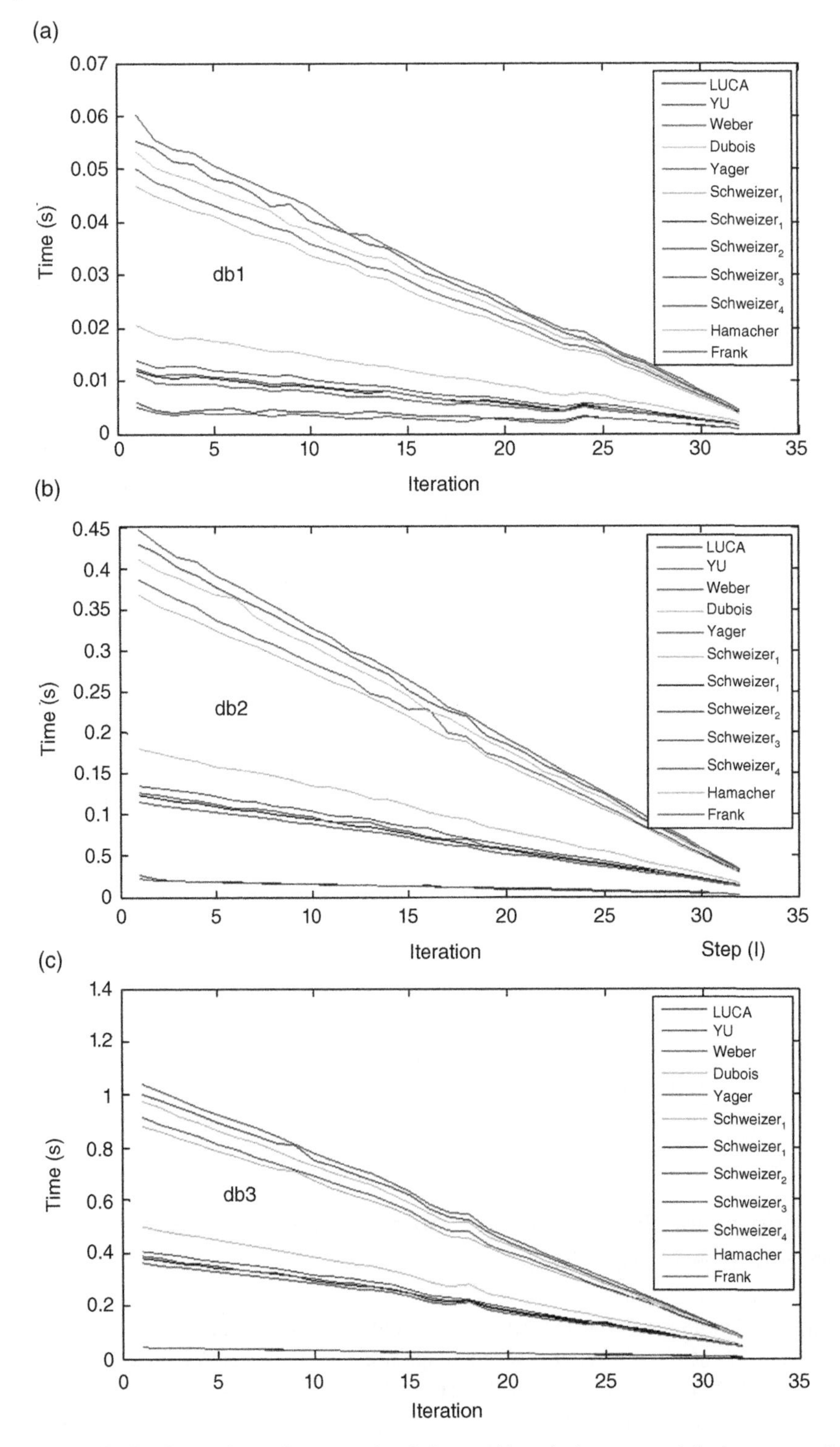

Figure 7.15 Comparison of computational time and its relation to step 1 for image sets: (a) db1, (b) db2, and (c) db3.

Table 7.6 Summary of priority characteristics according to the type of cracks and structural damage.

Features	Distress	Spall	Long.	Trans.	Digo.	Block	Fatige	Complex	Pothole
	Dd	Sp	Lc	Tc	Dc	Bc	Fc	Cc	Pd
F3	13.00	5.00	31.00	16.00	5.00	16.00	16.00	5.00	5.00
F2	2.00	16.00	16.00	20.00	16.00	20.00	20.00	16.00	16.00
F1	1.00	25.00	25.00	25.00	25.00	25.00	25.00	25.00	25.00
TR(%)	58%	27%	12%	14%	6%	8%	8%	6%	5%

F1: Feature 1, F2: Feature 2, F3: Feature 3, TR(%): Total Ratio

the main margin of the two categories. In the two-class category, the twin SVM proposed in 2006 aims to solve two Quadratic Programming Problems (QPPs) simultaneously. In the SVM method, all data are considered using a quadratic programming (QP) under restrictive conditions. In the twin support vector machine (TSVM) method, the data and patterns of one category will limit the QP conditions for the other category, and vice versa. This method has both increased computational speed and accuracy compared to the original SVM method. Also, due to the uncertainty in defining the classifier boundary and the existence of slacks, it is not possible to use a specific method to solve QPPs. For this purpose, the fuzzy based-SVM (FSVM) method with the theory of assigning a fuzzy membership function to a more important category is presented. In this section, two innovations in the classification algorithm using the backup vector method are presented. In the first part, different fuzzy membership functions are performed using fuzzy relations using four classical methods (Sugeno, Yeger, Continuous, and the entropy Committee), and the entropy committee, including six entropies $\text{Ent} = \{\text{Ent}_{sh}(w), \text{Ent}_{me}(w), \text{Ent}_{rn}(w), \text{Ent}_{pp}(w), \text{Ent}_{ma}(w)\}$, is used in the second part. In Section 7.2, an entropy-based twin classifier was introduced for fast classification with higher efficiency than conventional methods. The algorithm used for classification is shown in Figure 7.16. Based on the proposed method, two rules proposed for selecting the optimal class centres:

Rule 1: The optimal center of the classes has a minimum entropy.
Rule 2: The optimal center of each class has the maximum compression index.

In order to show the type of performance of this class, a sample of the database with 100 images was collected, as shown in Figure 7.16, using the proposed indices and the optimal feature vector. According to this set, the purpose of classifying two classes is based on the existence of distress in the section.

Vectors of unhealthy pavement sections with characteristic 1 and healthy vectors with characteristic −1 have been determined. In this example, based on the proposed algorithm, the entropy value is calculated using six methods, the minimum entropy value is determined, and the corresponding vectors are extracted. The storage vector number and basic information are extracted for each vector. If only one vector is extracted, the coordinates of this data are stored and considered as the center. If the number of feature vectors is $n > 1$, then according to Rule 2, the data density index is used. An example of membership coefficients that are modified for two features 5 and 25 using the proposed algorithm is shown in Figure 7.17. In this section, three image banks were used to evaluate the proposed method. The characteristics and indicators of these three banks are summarized in Table 7.7. The extracted optimal characteristics in the feature selection stage were also tested by this method, and the results are given Table 7.7.

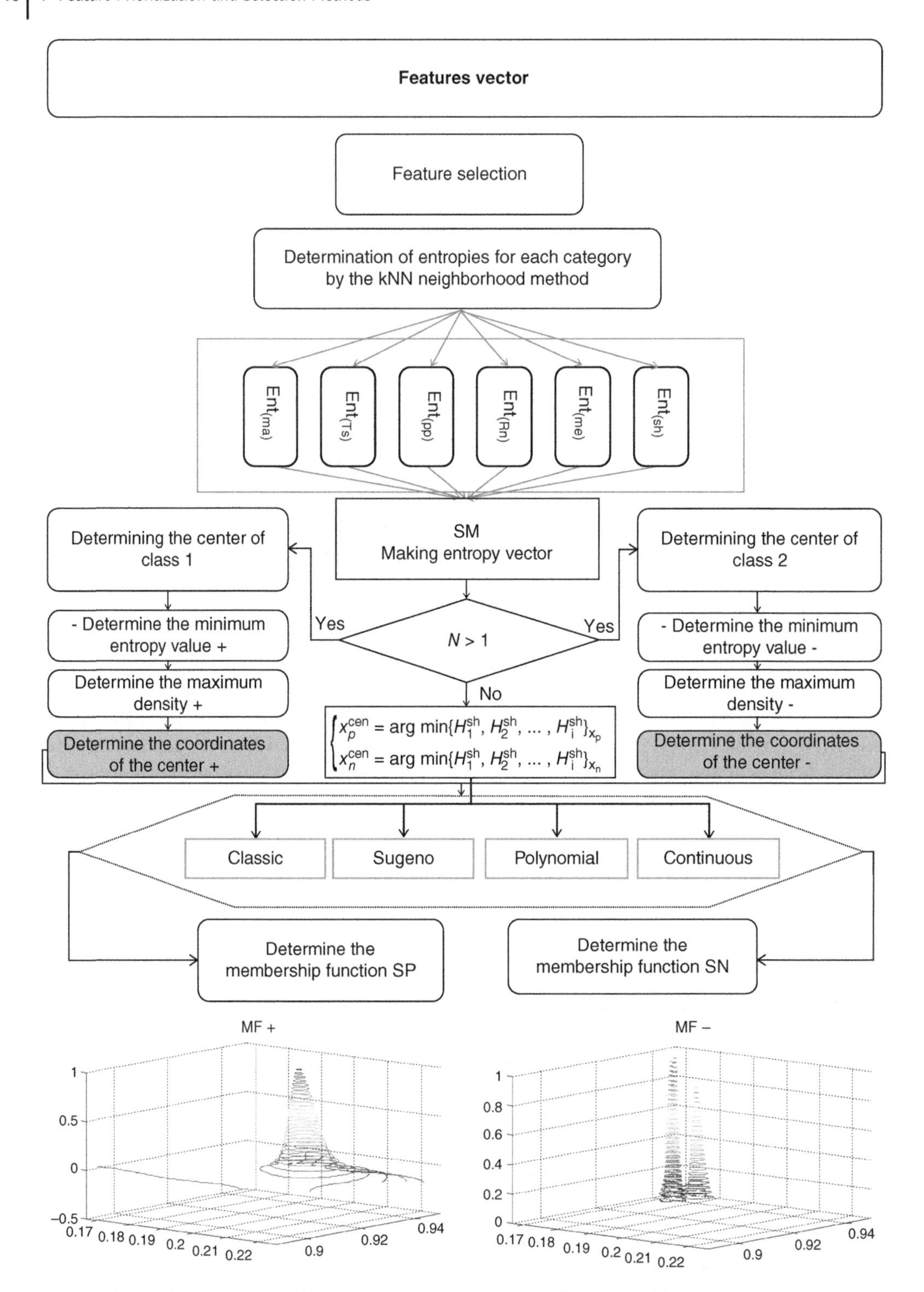

Figure 7.16 The framework of CDFESVM classification algorithm for feature selection and pavement distress classification.

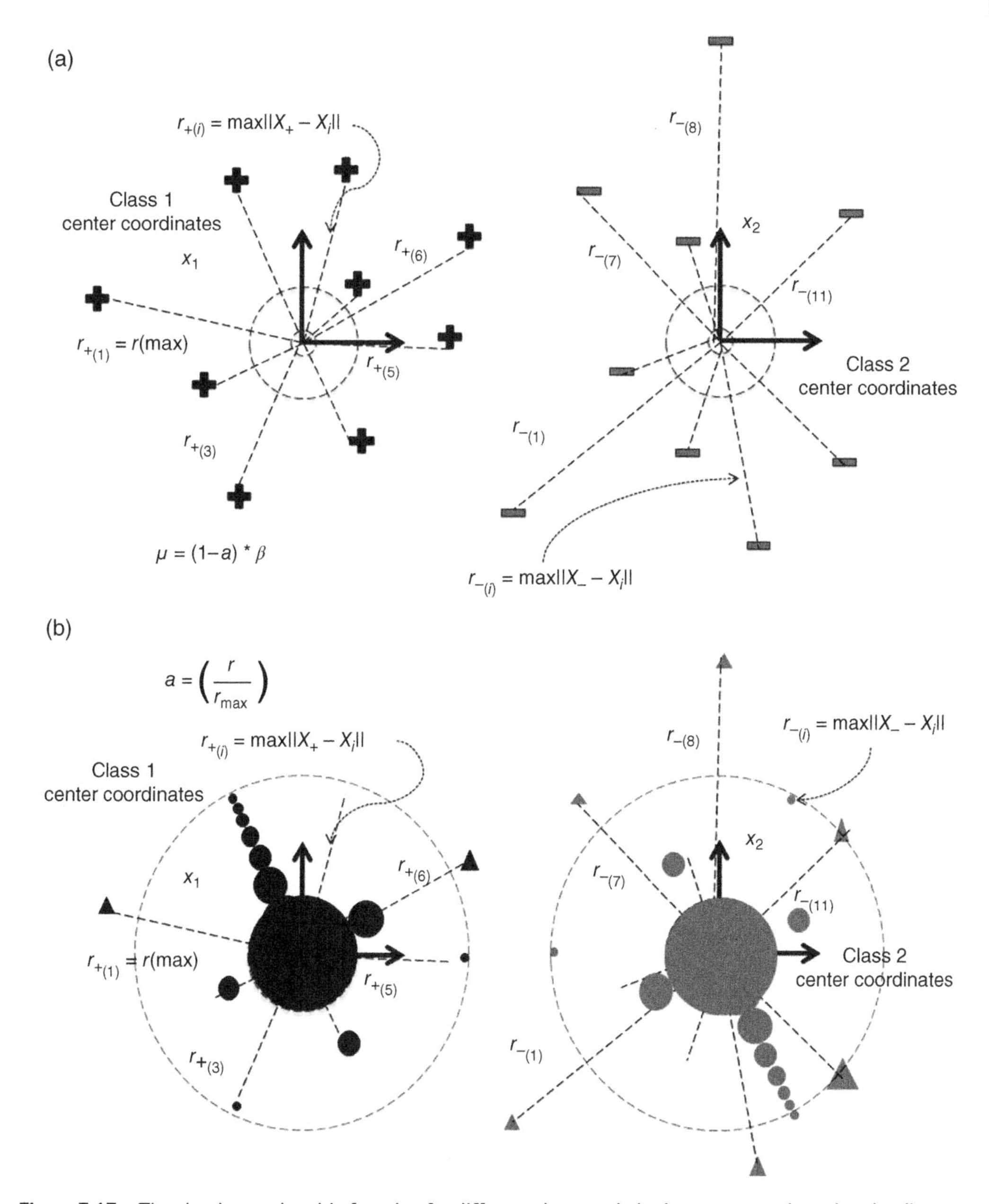

Figure 7.17 The classic membership function for different characteristics in two categories using the distance function. a) Select the optimal center based on distance and b) Weighting according to knn method.

7.3.1 Methods for Determining the Fuzzy Membership Function in Feature Selection

In this section, the subject of the membership function in feature selection is explained with a practical example in the field of pavement failure detection. To diagnose pavement damage, an image may not be definitively in the same class, meaning that the center of a particular category does not clearly exist to distinguish it or may have many errors. As mentioned in Chapter 4, in order to solve this problem, fuzzy membership functions are designed for each input feature vector. Due to the

Table 7.7 A set of features and characteristics of pavement images for training and structure determination, evaluation and validation.

Features sets	Distress	Health	Number of images
Training and structure determination			
	Dd	Hs	NI
Db1 (train)	260	140	400
Db2 (train)	476	324	800
Db3 (train)	1257	903	2100
Evaluation and validation			
Db1 (test)	54	46	100
Db2 (test)	40	110	150
Db3 (test)	211	189	400

existence of different fuzzy methods and with the aim of comparing and selecting the optimal method in this study, four fuzzy methods, including the classical, Yager, Sugeno, and continuous methods, were used. Membership functions were extracted using these methods and employed as a criterion for categorizing defective images. Figure 7.18 shows the theory of optimal center selection based on the kNN method and related weighting.

As can be seen in the figure, with decreasing distance from the optimal center, the value of the membership coefficient increases and is positive with a high-impact coefficient at the boundary with β-radius. If the spatial distance of the new category is greater than the specified value, the coefficients are reduced and the amount of data impact is greatly reduced. For example, for data located in the optimal center, the membership coefficient is 1, and the coefficient $\beta = 0.9$ is considered. In these relations, $a = \left(\frac{r}{r_{\max}}\right)$ is the ratio of distance to maximum distance. Using this ratio and adjusting parameters, λ, ω, and m, membership functions are extracted for each image feature vector.

When $\omega = 1$, the Yager method becomes the classical membership function. The changes and effect of the relationship parameters are shown in Figure 7.18. In order to compare the membership coefficients using different fuzzy functions, the probability of distribution of coefficients and probability distribution for features 16 and 28 in db1 with 400 images from the instructional samples of Table 7.8 are shown in Figure 7.18. These coefficients are positive and negative for both categories. Data are split among the first and second category, while some data have a certainty of zero in the other category (Figure 7.19).

For example, a small number of data with a high probability of 99% have 0.9 or more in the + category, and about 75% of the data in the + group have a membership factor of zero. In other words, in the negative category, the same number of data has a membership coefficient greater than zero. Based on the explanations provided, a method called CDFESVM based on fuzzy and entropy classification has good performance in the fault detection stage. This method is more accurate, faster, and thus outperforms other support vector-based classification methods. The different stages of implementation of this method are shown in Figure 7.16. This method also benefits from a higher accuracy than the classical method in terms of using the fuzzy method, a higher efficiency due to the use of entropy method and density measurement, and faster computational speed because of the coordinate descent (CD) method. Therefore, this hybrid method has been used as a practical method with high reliability and appropriate speed for fast processing of this research.

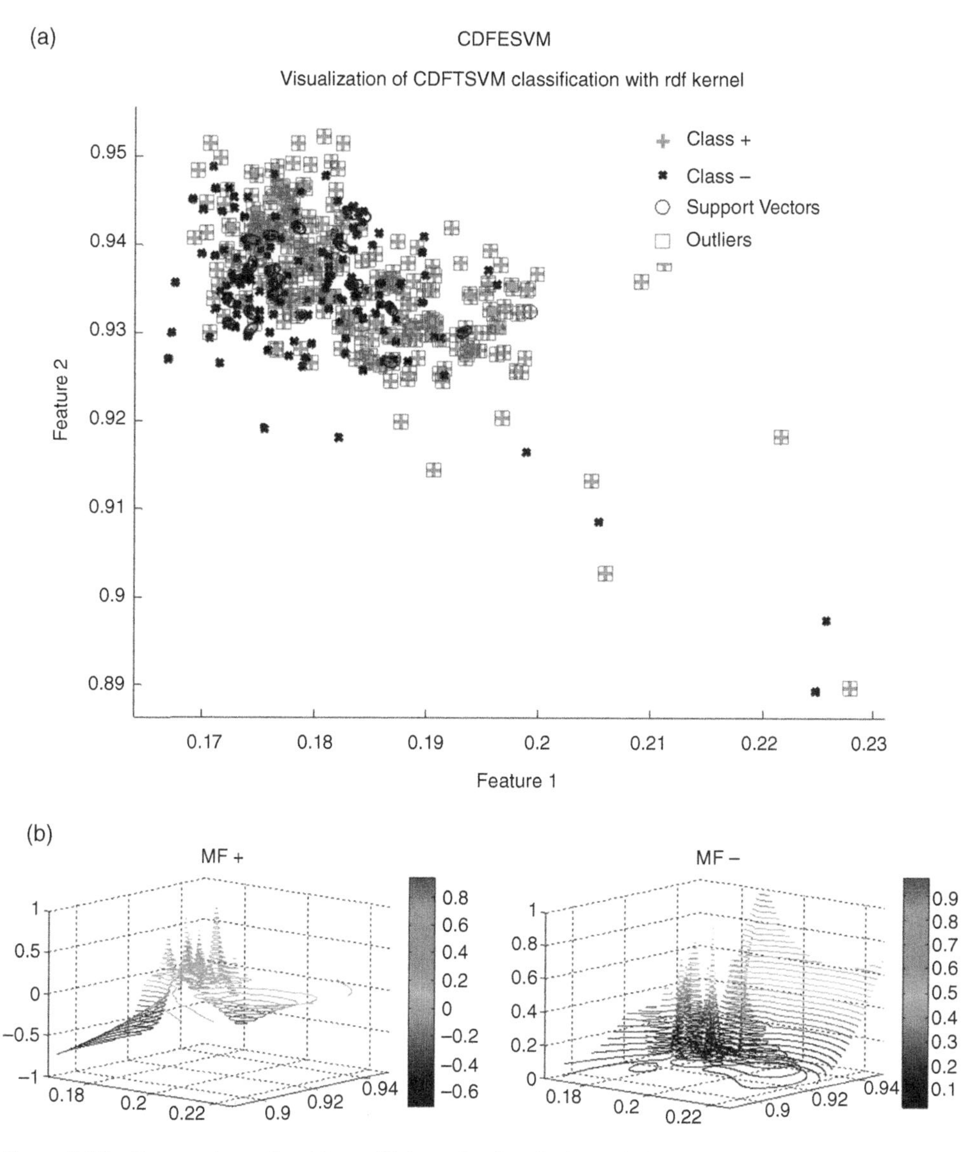

Figure 7.18 Extracted membership coefficients for data in the training feature bank containing 100 images db1: (a) class 2 characteristics and (b) class 1 characteristics.

Table 7.8 Different types of fuzzy generator methods to Select feature, based on distance and regulator parameters, for fuzzy support vector machine (FSVM).

Fuzzy generator	Formula	Tuning parameters			
Classic	$a = \left(\frac{r}{r_{\max}}\right), \mu = (1-a)*\beta$	β	0.6	0.75	0.9
Sugeno class	$\mu_\lambda(a) = \left(\left(\frac{1-a}{1+\lambda a}\right)\right), \lambda \in (-1, \infty)$	λ	−1	5	10
Yager class	$\mu_\omega(a) = (1-a^\omega)^{\frac{1}{\omega}}, \omega \in (0, \infty)$	ω	0.1	0.2	0.5
Continuous	$\mu(a) = \frac{1}{2^m}(1 + \cos(\pi a))$	m	1	0.5	2

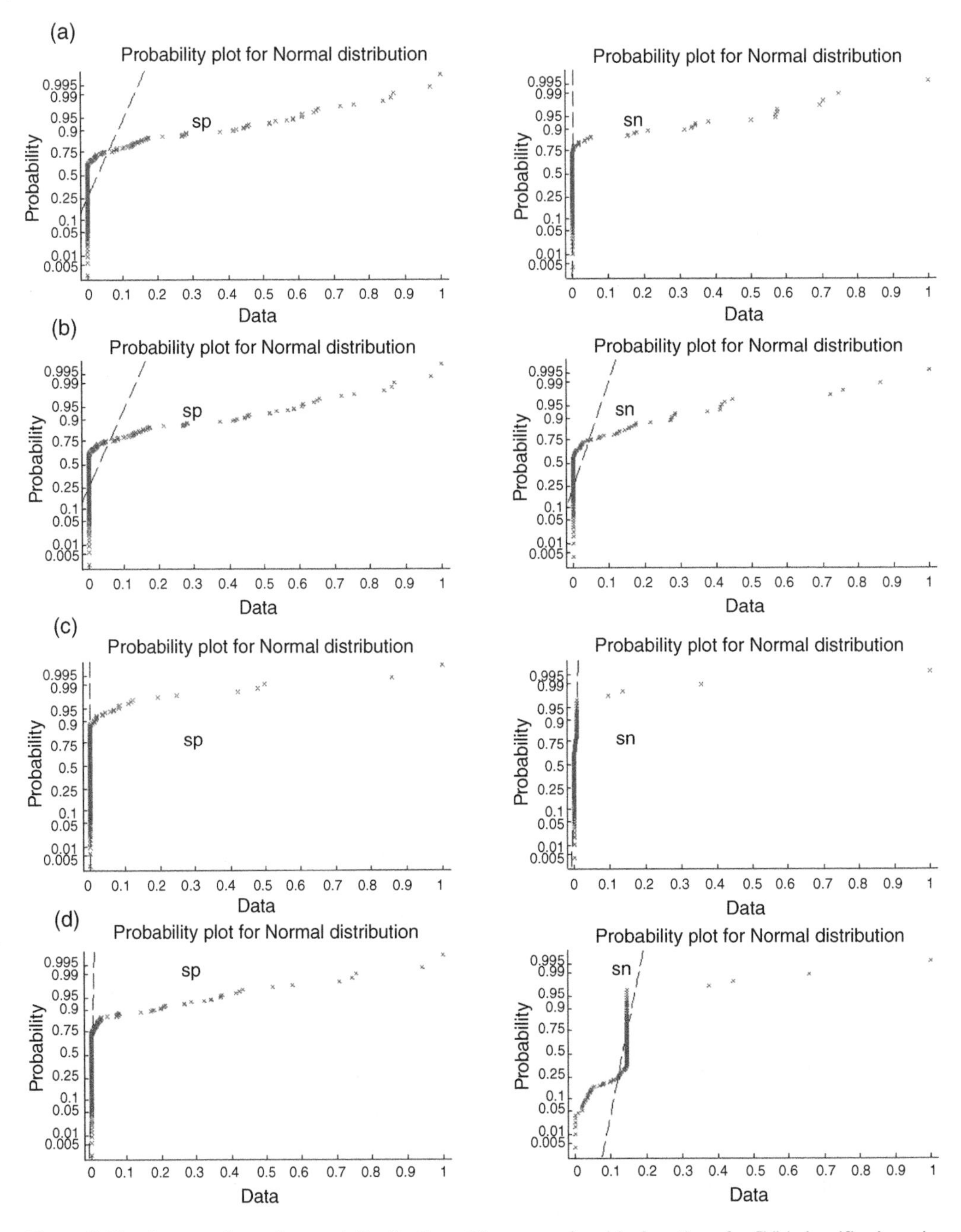

Figure 7.19 A comparison of normal distribution of fuzzy membership functions for SVM classifier based on density measurement and determination of entropy fuzzy support vector machine (EFSVM) class center for: (a) simple, (b) Sugeno, (c) other, and (d) continuous fuzzy structures.

Vectors with higher density have higher uncertainty, whereas lower density vectors have more ambiguity. Herein, data density was calculated based on the k-nearest neighbours algorithm (KNN) method, neighborhood, and the minimum value of the data collection distance, which is selected as the optimal center among the candidate vectors. Data farther from the center have a lower

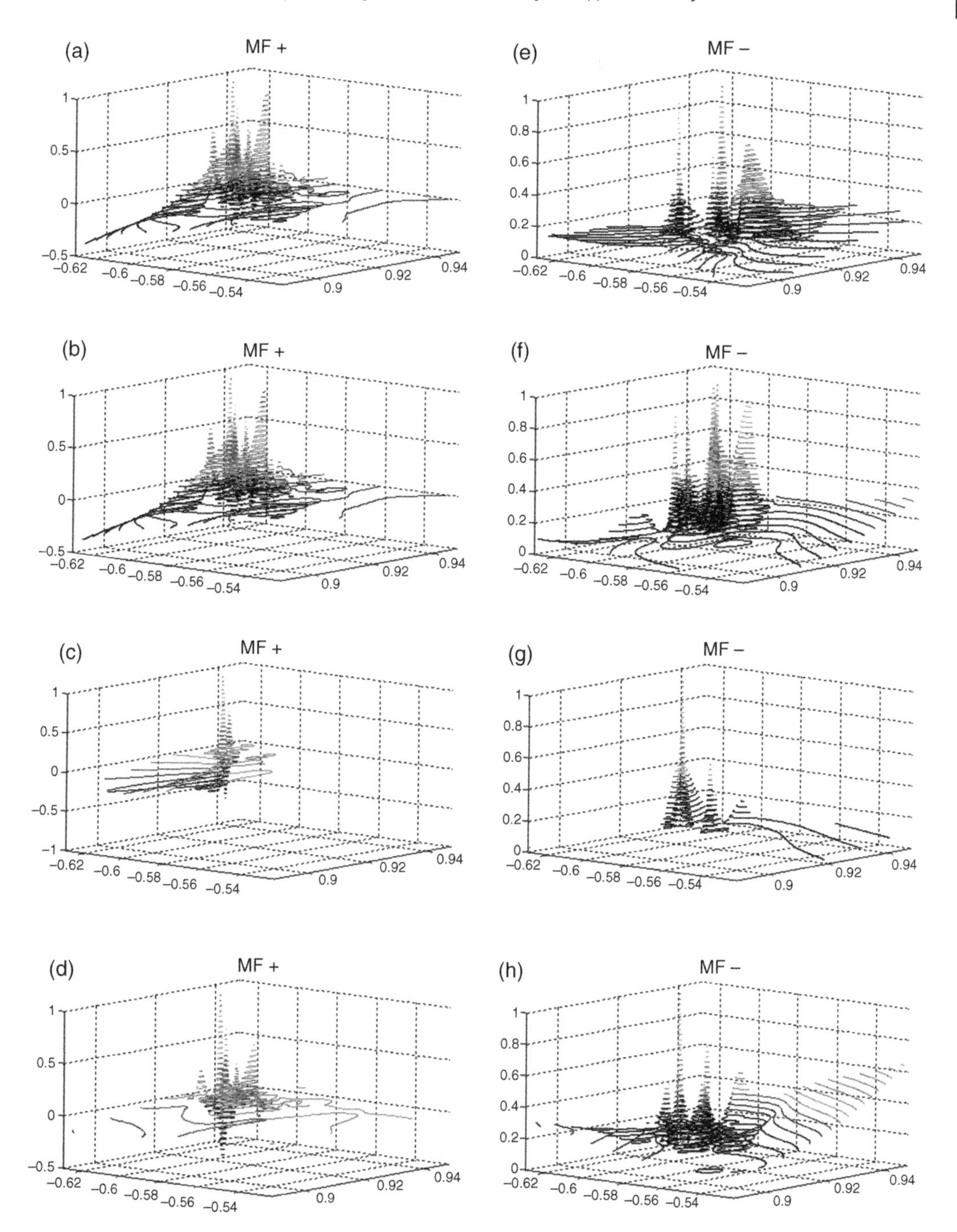

Figure 7.20 The fuzzy membership coefficients for categories 1 and 2 using: (a) classic for class 1, (b) Sugeno for class 1, (c) Yager for class 1, (d) continuous fuzzy method for class 1, (e) classic for class 2, (f) Sugeno for class 2, (g) Yager for class 2, (h) continuous fuzzy method for class 2.

membership coefficient, and data farther away have a membership coefficient close to 1. As shown in Figure 7.20, as distance increases in the classical method, the membership coefficient decreases. However, this relation has a different trend for other methods, such as Yager and Sugeno.

Table 7.9 Classification of features and characteristics of pavement images using SVM.

Classic	Kernel function	Linear	Quadratic	RBF	MLP
Db1	Quadratic programming	58.42	55.45	64.36	58.42
	Sequential minimal optimization	52.48	56.44	55.45	67.33
	Least squares	58.42	56.44	64.36	70.30
Db2	Quadratic programming	59.20	60.20	59.20	59.60
	Sequential minimal optimization	59.20	61.19	58.21	52.48
	Least squares	59.20	59.70	58.71	60.20
Db3	Quadratic programming	60.20	46.27	49.75	60.20
	Sequential minimal optimization	59.20	45.27	60.20	59.20
	Least squares	60.70	46.27	50.25	58.71

According to the basics provided, calculations for the membership coefficients, examples of these coefficients, and adjusted parameters based on the accuracy of the classifier and true positive (TP) ratio for the database of characteristics db1 to db3 are presented in the Table 7.9 related to the image bank. To compare the performance of the proposed method with the SVM method, the characteristic banks are evaluated using the SVM method for different kernels and methods, as given in the table below. In this research, the 10-fold method was used to evaluate the results. For db1 collection, 360 training and validation tests of 40 images were performed, and the average results are given in the table. For the db2 collection, 720 samples were used for training and 80 samples for validation. For the db3 collection, 1890 samples were designed for training and 210 samples for validation.

As shown in Table 7.9, for the SVM classifier, the highest performance (above 70%) of the multilayer perceptron kernel (MLP) and least squares (LS) method was obtained in db1. As the number of features and data expanded, this value decreased to 61%. This indicates that in the database, the properties do not have a linear model and the existing noise is not well separated by the Surface Kernel. On the other hand, center was separated using slack. The results obtained using a fuzzy classifier with a variety of fuzzy functions. These method combined with a support vector as a new classifier for three databases with 10-fold. Computational time for each method is also provided. It should be noted that the index under study is the ratio of tp + tn to the total data presented in Chapter 9. Therefore, considering that the purpose of presenting this classifier was to reduce the computational load and the computational complexity ratio, after comparison, it was found that the existing error ratio is related to fn samples that have practically no effect on computational accuracy, and using the existing methods on all images helped to identify and separate the faulty image. A number of nondamaged images may also be misdiagnosed as having a defective section due to the presence of shadows or large, shaded textures. However, this type of image is automatically deleted in the next step (fault assessment) by calculating other indicators, according to the status index and spall (Table 7.10).

In this section, a general comparison of the four methods, namely SVM, FSVM, coordinate descent fuzzy support vector machine (CDFSVM), and CDFESVM, is provided. These methods were compared in terms of efficiency to identify the optimal method. According to Table 7.11,

Table 7.10 Classification of features and characteristics of pavement images using FSVM.

Classic	Data base	Classic	Sugeno	Yager	Continue
Linear	Db1	63.1	65.6	64.6	60.1
	Db2	66.0	86.0	86.0	81.0
	Db3	67.0	98.0	93.0	92.0
Rbf	Db1	79.0	99.0	99.0	100.0
	Db2	86.0	100.0	100.0	100.0
	Db3	97.0	98.0	97.0	98.0

Table 7.11 Classification of features and characteristics of pavement images using CDFTSVM.

Classic	Data base	Classic	Sugeno	Yager	Continue
Linear	Db1	63.1	63.4	65.6	60.1
	Db2	66.0	85.0	86.0	60.0
	Db3	67.0	96.0	98.0	96.0
Rbf	Db1	79.0	100.0	99.0	73.0
	Db2	86.0	99.0	100.0	73.0
	Db3	97.0	97.0	98.0	96.0

the maximum accuracy index of 70% corresponds to db1. In general, as the number of data in the database increases, it is expected that in the training phase, the accuracy of the method will increase and the super-page lines will be categorized with a more accurate backup vector. But as you can see, the efficiency is reported to be around 60% at best. The QP, sequential minimal optimization (SMO), and LS methods have an average computational lead of 1.35, 1.28, and 0.89 seconds, respectively. Moreover, the average accuracy of the LS method is higher than QP and SMO. Regarding the method presented in Table 7.9, the average accuracy of these respective methods for the image banks is 91, 94, and 97%. In comparison, the computational time of FSVM is three times that of the computational method using the CD method. For this purpose, the CDFSVM method is used for classification. Although the computational speed of the CDFSVM method is lower than the FSVM method, the average accuracy of the method is 3% less than the FSVM method. Meanwhile, the CDFESVM method shows good performance in terms of both accuracy and speed, achieving the same accuracy of the FSVM method and computational speed similar to the CDFSVM method. It should be noted that the calculations were performed on an Intel® processor, core™ i7-3612QM CPU@2.1GHz with 12GB internal memory. The computational time for processing was 0.02 seconds, and the results demonstrate the high efficiency of CDFSVM method for detecting and isolating faulty images (Table 7.12).

Table 7.12 Classification of features and characteristics of pavement images using CDFETSVM.

Classic	Data base	Classic	Sugeno	Yager	Continue
Linear	Db1	64.6	60.1	65.6	63.4
	Db2	86.0	60.0	81.0	85.0
	Db3	93.0	96.0	92.0	96.0
Rbf	Db1	99.0	73.0	100.0	100.0
	Db2	100.0	73.0	100.0	99.0
	Db3	97.0	96.0	98.0	98.0

7.4 Summary and Conclusion

This chapter presents the latest developments in various feature selection methods. Filter, wrapper, and hybrid methods utilizing a combination of different methods were examined through examples and applications in the field of road infrastructure assessment (road pavement). A summary of the important results of this section includes the following:

- Effective features in identifying and isolating distressed sections are suggested and evaluated for different samples. In this chapter, from different indicators in the Shearlet transform domain, in real and imaginary space, edge detection by complex shearlet transform, momentum family characteristics, including central moment, Hu moment, and Zernik, energy, wavelength (period), contrast, correlation, homogeneity, entropy, local range, deviation from local standard, and fractal characteristics were used to isolate and detect sections contain distress.
- The feature vector consisting of 33 properties was extracted and feature vectors for each image is provided. In order to simplify the selection of effective features, three image banks were analyzed using feature extraction algorithms, and the results were stored as feature vectors for each image.
- Using the modified feature selection method, the similarity index, and entropy committee regarding feature selection, an overview of the relationship of each parameter with the main features of the cracked section, the existence of failure to detect, and the relationship of the visual feature with the main features, including the type of cracking, was presented.
- According to the prioritization method based on the fuzzy similarity method and entropy committee, different prioritization features and two superior features were presented for the class. Distress detection using HASHCP and HFSHEP characteristics showed better results with minimal entropy of ambiguity and a maximum degree of similarity. For spall failures, properties 16 and 25, including correlation and Zernic moment 1, were selected, and for surface failures, characteristics 20 and 25, including correlation and Zernic moment 5, were selected.
- The average accuracy of the methods for image banks db1, db2, and db3 was 91, 94, and 97%, respectively. However, the computational time for FSVM was three times that of the CD method. For this purpose, the CDFSVM method was used. Although the computational speed of the CDFSVM method is lower than the FSVM method, the average accuracy of the method is 3% less than the FSVM method. Comparatively, the CDFESVM method showed good performance in terms of both accuracy and speed, meeting the accuracy of the FSVM method and higher computational speed of the CDFSVM method, which further demonstrates its advantages.

7.5 Questions and Exercises

1 Define the various types of feature selection methods. Draw the general relationship and framework for selecting the feature. What are the characteristics of the features in general? Describe a feature selection by giving an example in the field of infrastructure management? Why is feature selection so important in classification? If the properties of the two attributes overlap, what criteria do the selection methods use to select them?

2 Define the three main features of feature selection, considering three databases containing characteristics in the field of infrastructure management in accordance with the conditions:

The input data [F_ij, C_j] consists of N samples i = 1 to N with D variables j = 1 to D; F_i, is the ith sample and C_kis the class label k = 1 to Y.

Prepare all three general characteristic vector banks. Is it possible to use a systematic method to determine the attribute vector for each database? In one of the programs, such as matlab, design an automatic feature vector.

3 Evaluate the pavement texture information and pavement failure using the parameters of Eq. (7.1). Compare the performance of this index using trial and error indicators. Is the use of this index scalable? What is the alternative solution for better results using this theory? Prove the following relation parametrically:

$$R(i) = \left(\frac{\text{cov}(F_i, C)}{\sqrt{\text{var}(F_i * \text{var}(C)}} \right).$$

4 Define the general framework of the wrapper method. Mention the different solutions of this method and give an example for each. What is the main structure and mission of the SFS method? Draw this method for a hypothetical property vector containing 10 properties. How can these methods be made intelligent and adaptable?

5 What is the main difference between filtering and wrapper methods? List the applications of hybrid filters with examples of applications in the field of infrastructure management. Define the main indicators in each method and the error associated with each method. Describe the benefits of each method. Design an algorithm that combines two methods to retain index information in such a way that there are at least three features used as neural network classifier input and two categories, including fault and healthy, with numbers −1 and +1.

6 Using hybrid algorithms, calculate the important features of pavement texture in three modes: fine, medium, and large. Choose a metaheuristic algorithm and select the optimal parameters. Compare the extracted indices with the filter and wrapper methods. Use a good classifier like SVM for classification. Does the result change as the classifier parameters change? How much do setting the classifier parameters affect the feature selection results? Calculate this for the five types of classifiers mentioned in Chapter 8.

7 Define the entropy relation below, and define its relation with information gain. Prove the information interest relationship. Prove this relation for different entropies. Define a new entropy in your name to introduce characteristic information and information gain then prove. For your own entropy, give an example of an application of infrastructure management and compare it with other methods. Can more modern methods be provided to determine the amount of information on each characteristic?

$$E(N) = \sum_{i=1}^{k} P_i \log_k \left(\frac{1}{P_i}\right) = -\sum_{i=1}^{k} P_i \log_k (P_i)$$

8 Compare the 12 similarity indicators introduced in this section for the parameters of a hypothetical set in infrastructure management. Use a set of these indicators to build a search engine. Determine the similarity vector formation. Separate the main feature from the other features by selecting the maximum frequency. Determine the final vector list based on an automated algorithm. Define the set of similarity criteria used in this section, including the Luukka (LUCA), Yu (YU), Weber, Dubois, Yager, Schweizer, Haraacher, Frank, and Dombi methods, and prove the relationships of each.

9 Consider a set of free parameters $\{a, b, \alpha, \beta, \gamma\}$ and determine the setting values Ent _ ((ma)) for the set of entropies. In order to normalize the entropy function Ent _ ((ma)), use the minimum and maximum values of Ent _ ((ma) _max) and Ent _ ((ma) _min). Use a new method to normalize and select the index. What is the relationship between probability and information gain? Provide relationships for known entropy types and suggest a hybrid entropy.

10 Calculate the accuracy and timing of the FSVM calculation using the CD method for the feature bank in Question 1. Use the CDFSVM method for separation. Compare the CDFSVM method with the FSVM method. Compare the efficiency of the methods as well as the computational time with the CDFESVM method. Summarize the most important advantages and disadvantages of the above methods in the table.

Further Reading

1 Abpeykar, S., M. Ghatee, and H. Zare, *Ensemble decision forest of RBF networks via hybrid feature clustering approach for high-dimensional data classification. Computational Statistics & Data Analysis*, 2019. **131**: p. 12–36.

2 Abualigah, L., A. Diabat, S. Mirjalili, M. Abd Elaziz, and A. H. Gandomi, *The arithmetic optimization algorithm. Computer Methods in Applied Mechanics and Engineering*, 2021. **376**: p. 113609.

3 Abualigah, L., D. Yousri, M. Abd Elaziz, A. A. Ewees, M. A. Al-qaness, and A. H. Gandomi, *Aquila optimizer: A novel meta-heuristic optimization algorithm. Computers & Industrial Engineering*, 2021. **157**: p. 107250.

4 Ahmad, F., N. A. M. Isa, Z. Hussain, M. K. Osman, and S. N. Sulaiman, *A GA-based feature selection and parameter optimization of an ANN in diagnosing breast cancer. Pattern Analysis and Applications*, 2015. **18**(4): p. 861–870.

5 Anowar, F., S. Sadaoui, and B. Selim, *Conceptual and empirical comparison of dimensionality reduction algorithms (PCA, KPCA, LDA, MDS, SVD, LLE, ISOMAP, LE, ICA, t-SNE). Computer Science Review*, 2021. **40**: p. 100378.

6 Bennasar, M., Y. Hicks, and R. Setchi, *Feature selection using joint mutual information maximisation. Expert Systems with Applications*, 2015. **42**(22): p. 8520–8532.

7 Bolón-Canedo, V., N. Sánchez-Maroño, and A. Alonso-Betanzos, *Recent advances and emerging challenges of feature selection in the context of big data. Knowledge-Based Systems*, 2015. **86**: p. 33–45.

8 Cai, J., J. Luo, S. Wang, and S. Yang, *Feature selection in machine learning: A new perspective. Neurocomputing*, 2018. **300**: p. 70–79.

9 Chandrashekar, G. and F. Sahin, *A survey on feature selection methods. Computers and Electrical Engineering*, 2014. **40**(1): p. 16–28.

10 Chao, G., Y. Luo, and W. Ding, *Recent advances in supervised dimension reduction: A survey. Machine Learning and Knowledge Extraction*, 2019. **1**(1): p. 341–358.

11 Chen, R.-C., C. Dewi, S.-W. Huang, and R. E. Caraka, *Selecting critical features for data classification based on machine learning methods. Journal of Big Data*, 2020. **7**: p. 1–26.

12 Dash, M. and H. Liu, *Feature selection for classification. Intelligent Data Analysis*, 1997. **1**(1–4): p. 131–156.

13 Dong, X., L. Zhu, X. Song, J. Li, and Z. Cheng, *Adaptive collaborative similarity learning for unsupervised multi-view feature selection.* arXiv preprint arXiv:1904.11228, 2019.

14 Drotár, P., M. Gazda, and L. Vokorokos, *Ensemble feature selection using election methods and ranker clustering. Information Sciences*, 2019. **480**: p. 365–380.

15 Emary, E., H.M. Zawbaa, and A.E. Hassanien, *Binary grey wolf optimization approaches for feature selection. Neurocomputing*, 2016. **172**: p. 371–381.

16 Faramarzi, A., M. Heidarinejad, S. Mirjalili, and A. H. Gandomi, *Marine predators algorithm: A nature-inspired metaheuristic. Expert Systems with Applications*, 2020. **152**: p. 113377.

17 Gan, M. and L. Zhang, *Iteratively local fisher score for feature selection. Applied Intelligence*, 2021. **51**: p. 6167–6181.

18 Gandomi, A.H., *Interior search algorithm (ISA): A novel approach for global optimization. ISA Transactions*, 2014. **53**(4): p. 1168–1183.

19 Gandomi, A.H. and A.H. Alavi, *Krill herd: A new bio-inspired optimization algorithm. Communications in Nonlinear Science and Numerical Simulation*, 2012. **17**(12): p. 4831–4845.

20 Hancer, E., B. Xue, and M. Zhang, *A survey on feature selection approaches for clustering. Artificial Intelligence Review*, 2020. **53**(6): p. 4519–4545.

21 Hoque, N., D.K. Bhattacharyya, and J.K. Kalita, *MIFS-ND: A mutual information-based feature selection method. Expert Systems with Applications*, 2014. **41**(14): p. 6371–6385.

22 Hsu, H.-H., C.-W. Hsieh, and M.-D. Lu, *Hybrid feature selection by combining filters and wrappers. Expert Systems with Applications*, 2011. **38**(7): p. 8144–8150.

23 Kotsiantis, S., *Feature selection for machine learning classification problems: A recent overview. Artificial Intelligence Review*, 2011. **42**(1): p. 157–176.

24 Li, J., K. Cheng, S. Wang, F. Morstatter, R. P. Trevino, J. Tang, et al., *Feature selection: A data perspective. ACM Computing Surveys (CSUR)*, 2017. **50**(6): p. 1–45.

25 Li, J., L. Wu, H. Dani, and H. Liu, *Unsupervised personalized feature selection.* in *Proceedings of the AAAI Conference on Artificial Intelligence*. 2018.

26 Li, Y., T. Li, and H. Liu, *Recent advances in feature selection and its applications. Knowledge and Information Systems*, 2017. **53**(3): p. 551–577.

27 Lin, Y., Y. Li, C. Wang, and J. Chen, *Attribute reduction for multi-label learning with fuzzy rough set. Knowledge-Based Systems*, 2018. **152**: p. 51–61.

28 Mafarja, M., I. Aljarah, A. A. Heidari, A. I. Hammouri, H. Faris, A.-Z. Ala'M, et al., *Evolutionary population dynamics and grasshopper optimization approaches for feature selection problems. Knowledge-Based Systems*, 2018. **145**: p. 25–45.

29 Manoj, R.J., M.A. Praveena, and K. Vijayakumar, *An ACO–ANN based feature selection algorithm for big data. Cluster Computing*, 2019. **22**(2): p. 3953–3960.

30 Mirjalili, S., A. H. Gandomi, S. Z. Mirjalili, S. Saremi, H. Faris, and S. M. Mirjaliliet al., *Salp swarm algorithm: A bio-inspired optimizer for engineering design problems. Advances in Engineering Software*, 2017. **114**: p. 163–191.

31 Mitra, P., C. Murthy, and S.K. Pal, *Unsupervised feature selection using feature similarity. IEEE Transactions on Pattern Analysis and Machine Intelligence*, 2002. **24**(3): p. 301–312.

32 Moran, M. and G. Gordon, *Curious feature selection. Information Sciences*, 2019. **485**: p. 42–54.

33 Rao, H., X. Shi, A. K. Rodrigue, J. Feng, Y. Xia, M. Elhoseny, et al., *Feature selection based on artificial bee colony and gradient boosting decision tree. Applied Soft Computing*, 2019. **74**: p. 634–642.

34 Rezaee, M.R., B. Goedhart, B. P. Lelieveldt, and J. H. Reiber, *Fuzzy feature selection. Pattern Recognition*, 1999. **32**(12): p. 2011–2019.

35 Rong, M., D. Gong, and X. Gao, *Feature selection and its use in big data: Challenges, methods, and trends. IEEE Access*, 2019. **7**: p. 19709–19725.

36 Shi, Y., J. Miao, Z. Wang, P. Zhang, and L. Niu, *Feature selection with {2, 1-2} regularization. IEEE Transactions on Neural Networks and Learning Systems*, 2018. **29**(10): p. 4967–4982.

37 Solorio-Fernández, S., J.A. Carrasco-Ochoa, and J.F. Martínez-Trinidad, *A review of unsupervised feature selection methods. Artificial Intelligence Review*, 2020. **53**(2): p. 907–948.

38 Talatahari, S., M. Azizi, and A.H. Gandomi, *Material generation algorithm: A novel metaheuristic algorithm for optimization of engineering problems. Processes*, 2021. **9**(5): p. 859.

39 Tu, J., H. Chen, M. Wang, and A. H. Gandomi, *The colony predation algorithm. Journal of Bionic Engineering*, 2021. **18**(3): p. 674–710.

40 Vieira, S.M., L. F. Mendonça, G. J. Farinha, and J. M. Sousa, *Modified binary PSO for feature selection using SVM applied to mortality prediction of septic patients. Applied Soft Computing*, 2013. **13**(8): p. 3494–3504.

41 Vieira, S.M., J.M. Sousa, and U. Kaymak, *Fuzzy criteria for feature selection. Fuzzy Sets and Systems*, 2012. **189**(1): p. 1–18.

42 Vignolo, L.D., D.H. Milone, and J. Scharcanski, *Feature selection for face recognition based on multi-objective evolutionary wrappers. Expert Systems with Applications*, 2013. **40**(13): p. 5077–5084.

43 Yang, Y., H. Chen, A. A. Heidari, and A. H. Gandomi, *Hunger games search: Visions, conception, implementation, deep analysis, perspectives, and towards performance shifts. Expert Systems with Applications*, 2021. **177**: p. 114864.

44 Zakeri, H., F.M. Nejad, and A. Fahimifar, *Image based techniques for crack detection, classification and quantification in asphalt pavement: A review. Archives of Computational Methods in Engineering*, 2017. **24**(4): p. 935–977.

45 Zamani, H., M.H. Nadimi-Shahraki, and A.H. Gandomi, *QANA: Quantum-based avian navigation optimizer algorithm. Engineering Applications of Artificial Intelligence*, 2021. **104**: p. 104314.

8

Classification Methods and Its Applications in Infrastructure Management

8.1 Introduction

Classification is a type of prediction and calculation method in which a method is designed to guess the placement of data in a category, whereby the output is categorized and a class is located. Comparatively, in a regression method, the output is numerical. Descriptive modeling, or clustering, assigns each input to a cluster so that similar inputs are placed in the same cluster. Finally, by observing the relationship between similar data, one can discover rules in the form of association rules about the relationship between inputs.

By common definition, classification is an important application in data science that predicts the attribution of a variable (goal or class) based on the construction of a model using one or more numerical variables (predictors or attributes) and so on. The basis of all variables depends on the categories to which they best belong. The amount of affiliation is calculated by different methods, which fall under four general categories, including:

1) Frequency table
2) Covariance matrix
3) Similarity functions
4) Others

Classification is the most commonly used large dataset classification technique in most research and practical applications of infrastructure management and is widely employed in a variety of applications, such as detection, feature extraction, feature selection, type classification, severity classification, and extent evaluation. In this method, data analysis is based on learning-based algorithms that are supervised and consistent with data quality. In these algorithms, the goal is to detect and deduce a relationship that qualitatively relates the desired variable to other observed variables that can be predicted. As a result, extracting this relationship is a kind of learning from data behavior that is considered as a branch of data mining.

By definition, the classifier is the algorithm responsible for classification, and all data are sample observations. It should be noted that classification is used when the desired variable is qualitative. For example, in infrastructure management, the failure is based on its severity (low, medium, and high), type (longitudinal, transverse, oblique, block, and fatigue), and extent (low, medium, high, and very high).

The classification method employs the Mecca algorithms shown in Figure 8.1 to obtain useful information. In infrastructure management, this method is used to inform behavior or conditions of roads, bridges, tunnels, and technical buildings. Using classification methods to classify is a kind of intelligence in management. Through classification, a distinction can be made between data that

Automation and Computational Intelligence for Road Maintenance and Management: Advances and Applications, First Edition. Hamzeh Zakeri, Fereidoon Moghadas Nejad, and Amir H. Gandomi.

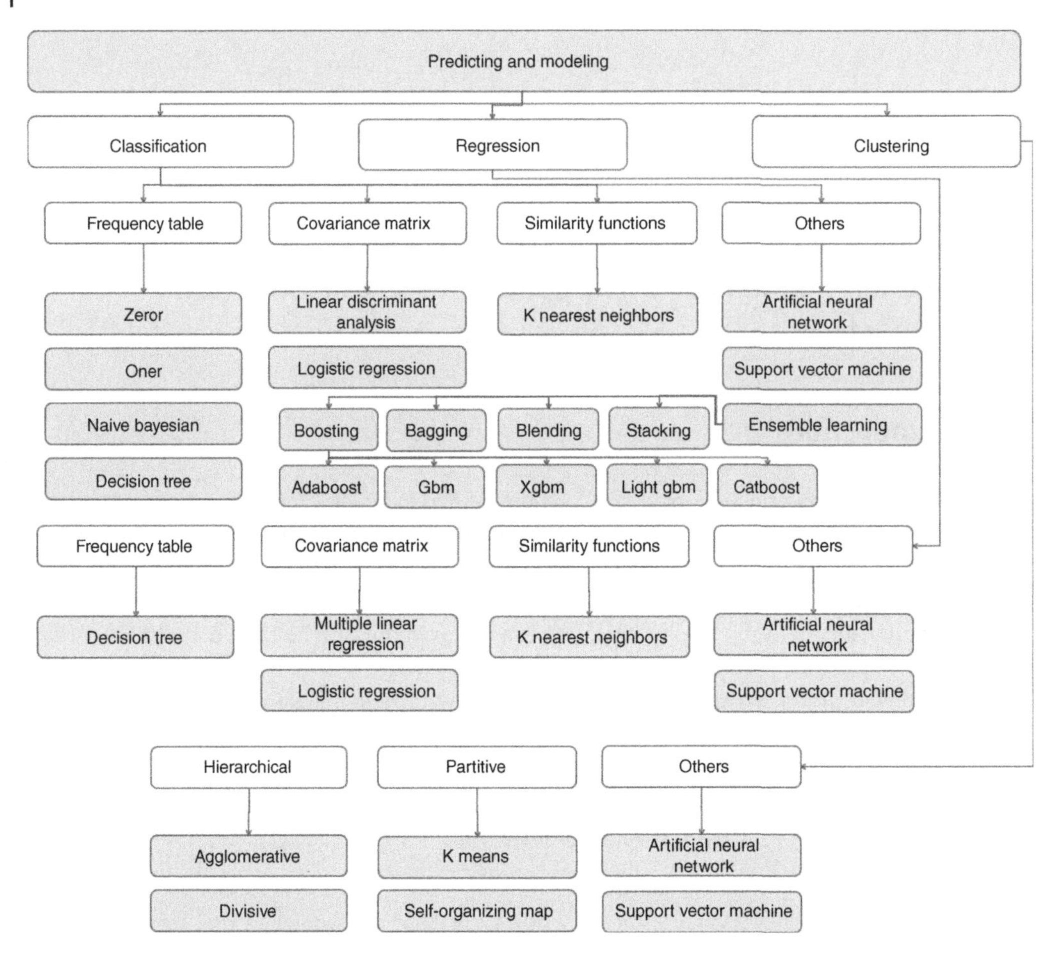

Figure 8.1 The various categories of learning modeling methods and classification.

are useful for the purpose (presence of cracks in infrastructure/detection of healthy sections) and irrelevant data (as in other types of failures). As an example, we can identify the area with distress compared to healthy sections in two categories of distress (yes/no).

This chapter briefly describes the most common classification methods used in infrastructure management and provides examples of applications. At the end of this chapter, new hybrid methods, primarily combinations of classification methods, are presented to solve various problems. In this chapter, the advanced method of classifying the support vector in polar coordinates is clearly described. Also, the basics of the fuzzy classification method used to classify ambiguous data are given as an extended method with examples.

8.2 Classification Methods

In this section, various classification methods and a summary of the basic theories of each method are provided along with simple examples to understand the method theory.

8.2.1 Naive Bayes Classification

Bayesian simple classification is based on the Bayesian theory, assuming independence between predictors. In general, designing and using a simple Bayesian model is easy. This model is created without estimating complex iterative parameters that make it useful for very large datasets. It has also shown considerable performance in classifying complex topics and solving engineering and classification problems due to its efficiency and simplicity of modeling in topics with high complexity.

Naive Bayes is a simple model for creating a functional classifier, in which class labels are assigned to problem cases and represented as vector values of attributes. The labels of each class of behavior are taken from a limited set of data and patterns.

In general, there is no specific algorithm for teaching these classifiers, and it is not possible to use a specific step-by-step process. Instead, a set of algorithms is designed based on a common principle: all base class classifiers think, and the value of a particular attribute is independent of the value of any other attribute with respect to the class variable. For example, if the crack has an irregular pattern and the size of each piece of fracture is less than 400 cm^2, the crack is considered fatigue type. A simple Bayesian classification considers each of these features to be independent of the probability of fatigue failure, regardless of any possible association between pattern, regularity, and fracture size.

For some types of possible models, simple base classifiers can be trained in a highly efficient supervised learning environment. In many practical applications, parameter estimation for simple Bayesian models is performed with the maximum probability method. Despite their naive design and seemingly simple assumptions, simple Bayesian classifiers perform very well in many complex real-world situations.

The most important advantage of this method is that it requires a small amount of training data to estimate the necessary parameters for classification, which makes this method very different from other methods.

Bayes' theorem provides a simple way to calculate the posterior probability, $P(c \mid x)$, of $P(c)$, $P(x)$, and $P(x \mid c)$. The naive Bayes classifier is designed assuming that the effect of the predictor value (x) on a given class (c) is independent of the values of other predictors. This assumption is called conditional independence, which is a major part of Bayes' classification algorithm.

$$P(c \mid x) = \frac{P(x \mid c)P(c)}{P(x)} \tag{8.1}$$

$$P(c \mid X) = P(x_1 \mid c) \times P(x_2 \mid c) \times P(x_3 \mid c) \times \cdots \times P(x_n \mid c) \times P(c)$$

where

$P(c|x)$ is the posterior probability of a class (target) given a predictor (feature).
$P(c)$ is the prior probability of a class.
$P(x \mid c)$ is the likelihood, or the probability of a predictor given a class.
$P(x)$ is the prior probability of a predictor.

The naive Bayesian classifier includes all predictors using the simple Bayesian principles, assuming the independence of each predictor.

In this method, the posterior probability must first be calculated, and then a table is created for each feature in front of the target. Then, the frequency tables are converted into probability tables, and finally, the simple Bayesian equation is used to calculate the posterior probability for each class. The class with the highest posterior probability is the result of the prediction for classification.

$$P(c \mid X) = P(x_1 \mid c) \times P(x_2 \mid c) \times P(x_3 \mid c) \times \cdots \times P(x_n \mid c) \times P(c)$$

$$P(c \mid x) = \frac{P(x \mid c)P(c)}{P(x)}$$

$$P(x \mid c) = P(L \mid F1) = \left(\frac{4}{5}\right) = 0.8$$

$$P(c) = P(F1) = \left(\frac{5}{20}\right) = 0.25$$

$$P(x) = P(L) = \left(\frac{8}{20}\right) = 0.4$$

$$P(F1 \mid L) = \frac{P(L \mid F1)P(F1)}{P(L)} = \frac{0.8 \times 0.25}{0.4} = 0.5$$

$$P(F1 \mid M) = \frac{P(M \mid F1)P(F1)}{P(M)} = \frac{0.2 \times 0.25}{0.2} = 0.2$$

$$P(F1 \mid M) = \frac{P(H \mid F1)P(F1)}{P(H)} = \frac{0 \times 0.25}{0.4} = 0.$$

$$P(F2 \mid L) = \frac{P(L \mid F1)P(F1)}{P(L)} = \frac{0.33 \times 0.3}{0.4} = 0.25$$

$$P(F2 \mid M) = \frac{P(M \mid F2)P(F2)}{P(M)} = \frac{0.33 \times 0.3}{0.2} = 0.5$$

$$P(F2 \mid H) = \frac{P(H \mid F2)P(F2)}{P(H)} = \frac{0.33 \times 0.3}{0.4} = 0.25 \tag{8.2}$$

$$P(F3 \mid L) = \frac{P(L \mid F3)P(F3)}{P(L)} = \frac{0.25 \times 0.2}{0.4} = 0.125$$

$$P(F3 \mid M) = \frac{P(M \mid F3)P(F3)}{P(M)} = \frac{0.25 \times 0.2}{0.2} = 0.15$$

$$P(F3 \mid H) = \frac{P(H \mid F3)P(F3)}{P(H)} = \frac{0.5 \times 0.2}{0.4} = 0.25$$

$$P(F4 \mid L) = \frac{P(L \mid F4)P(F4)}{P(L)} = \frac{0 \times 0.1}{0.4} = 0.0$$

$$P(F4 \mid M) = \frac{P(M \mid F4)P(F4)}{P(M)} = \frac{0 \times 0.1}{0.2} = 0.0$$

$$P(F4 \mid H) = \frac{P(H \mid F4)P(F4)}{P(H)} = \frac{1 \times 0.1}{0.4} = 0.25$$

$$P(F5 \mid L) = \frac{P(L \mid F5)P(F5)}{P(L)} = \frac{0 \times 0.1}{0.4} = 0.0$$

$$P(F5 \mid M) = \frac{P(M \mid F5)P(F5)}{P(M)} = \frac{0 \times 0.1}{0.2} = 0.0$$

$$P(F5 \mid H) = \frac{P(H \mid F5)P(F5)}{P(H)} = \frac{1 \times 0.1}{0.4} = 0.25$$

In this example, we have five inputs (predictors). The final posterior probabilities can be standardized between 0 and 1. Using the Table 8.2, the probability of each is calculated based on three levels, and whichever has the highest value is selected as the result. The likelihood tables for all five predictors presented in Table 8.3.

For example, using the relationships provided, the extent to which the following set of samples belong to each of the five categories in Table 8.1 is shown below:

Feature	1	2	3	4	5	Class
	Low	Medium	High	Low	Low	?

Note: L = Low, M = Medium, H = High

Table 8.1 An example of classification based of 5 input and 5 class.

	Feature 1	Feature 2	Feature 3	Feature 4	Feature 5	Output
1	L	M	L	L	H	1
2	L	M	L	L	M	3
3	L	M	L	L	H	1
4	M	L	L	L	L	2
5	M	L	M	H	L	2
6	M	H	M	H	L	3
7	L	L	M	H	M	1
8	H	L	H	H	M	2
9	H	M	H	L	M	5
10	H	M	H	L	L	4
11	L	M	L	L	M	2
12	L	L	L	M	H	2
13	H	L	L	M	L	3
14	M	L	H	M	M	1
15	L	H	H	H	H	3
16	H	H	H	H	L	4
17	H	H	M	H	L	5
18	H	M	M	M	M	3
19	H	M	M	M	M	2
20	L	M	M	M	M	1

L = Low, M = Medium, H = High.

Table 8.2 The Likelihood of classification based of 5 input and 5 class.

Frequency table feature 1	Class					Class					
	1	**2**	**3**	**4**	**5**	**1**	**2**	**3**	**4**	**5**	***P*(*x*)**
L	4	2	2	0		0.80	0.33	0.25	0.00	0.00	0.40
M	1	2	1	0	0	0.20	0.33	0.25	0.00	0.00	0.20
H	0	2	2	2	2	0.00	0.33	0.50	1.00	1.00	0.40
N(C)	5	6	4	2	2	0.3	0.3	0.2	0.1	0.1	1

Frequency table feature 2	Class					Class					
	1	**2**	**3**	**4**	**5**	**1**	**2**	**3**	**4**	**5**	***P*(*x*)**
L	2	4	1			0.80	0.33	0.25	0.00	0.00	0.40
M	3	2	2	1	1	0.20	0.33	0.25	0.00	0.00	0.20
H			2	1	1	0.00	0.33	0.50	1.00	1.00	0.40
N(C)	5	6	5	2	2	0.3	0.3	0.2	0.1	0.1	1

Frequency table feature 3	Class						Frequency table feature 3	Class					
	1	**2**	**3**	**4**	**5**	***n***		**1**	**2**	**3**	**4**	**5**	***P*(*x*)**
L	2	3	2			7	L	0.40	0.67	0.20	0.00	0.00	0.35
M	2	2	2		1	7	M	0.60	0.33	0.40	0.50	0.50	0.45
H	1	1	1	2	1	6	H	0.00	0.00	0.40	0.50	0.50	0.20
N(*C*)	5	6	5	2	2	20	*P*(*c*)	0.25	0.3	0.25	0.1	0.1	1

Frequency table feature 4	Class						Frequency table feature 4	Class					
	1	**2**	**3**	**4**	**5**	***n***		**1**	**2**	**3**	**4**	**5**	***P*(*x*)**
L	2	2	1	1	1	7	L	0.40	0.33	0.20	0.50	0.50	0.35
M	2	2	2			6	M	0.40	0.33	0.40	0.00	0.00	0.30
H	1	2	2	1	1	7	H	0.20	0.33	0.40	0.50	0.50	0.35
N(*C*)	5	6	5	2	2	20	*P*(*c*)	0.25	0.30	0.25	0.10	0.10	1.00

Frequency table feature 5	Class						Frequency table feature 5	Class					
	1	**2**	**3**	**4**	**5**	***n***		**1**	**2**	**3**	**4**	**5**	***P*(*x*)**
L		2	2	2	1	7	L	0.00	0.33	0.40	1.00	0.50	0.35
M	3	3	1		1	8	M	0.60	0.50	0.20	0.00	0.50	0.40
H	2	1	2			5	H	0.40	0.17	0.40	0.00	0.00	0.25
N(*C*)	5	6	5	2	2	20	*P*(*c*)	0.25	0.30	0.25	0.10	0.10	1.00

Table 8.3 The likelihood tables for all five predictors.

	Class				
Feature 1	**1**	**2**	**3**	**4**	**5**
L	0.5	0.25	0.125	0	0.00
M	0.25	0.5	0.25	0	0
H	0	0.25	0.25	0.25	0.25
	Class				
Feature 2	1	2	3	4	5
L	0.25	0.5	0.15625	0	0.00
M	0.75	0.5	0.625	0.25	0.25
H	0	0	0.3125	0.125	0.125
	Class				
Feature 3	1	2	3	4	5
L	0.286	0.571	0.143	0.000	0.000
M	0.333	0.222	0.222	0.111	0.111
H	0.000	0.000	0.500	0.250	0.250
	Class				
Feature 4	1	2	3	4	5
L	0.286	0.286	0.143	0.143	0.143
M	0.333	0.333	0.333	0.000	0.000
H	0.143	0.286	0.286	0.143	0.143
	Class				
Feature 5	1	2	3	4	5
L	0.000	0.286	0.286	0.286	0.143
M	0.375	0.375	0.125	0.000	0.125
H	0.400	0.200	0.400	0.000	0.000

$$\begin{aligned}
P(C1 \mid L) &= P(x_1 \mid c) \times P(x_2 \mid c) \times P(x_3 \mid c) \times \cdots \times P(x_n \mid c) \times P(L) \\
&= 0.5 \times 0.75 \times 0 \times 0.286 \times 0.286 \times 0 = 0 \\
P(C2 \mid M) &= P(x_1 \mid c) \times P(x_2 \mid c) \times P(x_3 \mid c) \times \cdots \times P(x_n \mid c) \times P(L) \\
&= 0.25 \times 0.5 \times 0 \times 0.286 \times 0.286 = 0.0 \\
P(C3 \mid H) &= P(x_1 \mid c) \times P(x_2 \mid c) \times P(x_3 \mid c) \times \cdots \times P(x_n \mid c) \times P(L) \\
&= 0.125 \times 0.625 \times 0.5 \times 0.1436 \times 0.286 = 0.0016 \\
P(C4 \mid L) &= P(x_1 \mid c) \times P(x_2 \mid c) \times P(x_3 \mid c) \times \cdots \times P(x_n \mid c) \times P(L) \\
&= 0 \times 0.25 \times 0.25 \times 0.143 \times 0.286 = 0 \\
P(C4 \mid L) &= P(x_1 \mid c) \times P(x_2 \mid c) \times P(x_3 \mid c) \times \cdots \times P(x_n \mid c) \times P(L) \\
&= 0.0 \times 0.25 \times 0.25 \times 0.143 \times 0.143 = 0
\end{aligned} \tag{8.3}$$

As can be seen from the results, this example belongs to the third category. In other words, considering that all four categories, 1, 2, 4, and 5, have a probability of zero, with a very high probability, we can say that this sample is in category 3.

Numerical variables must be converted into their counterparts before creating frequency tables. This method is an alternative to the numerical variable distribution of the probability density function for the normal distribution with two parameters (mean and standard deviation).

$$\mu = \frac{1}{n}\sum_{i=1}^{n} x_i$$

$$\sigma = \sqrt{\left[\frac{1}{n-1}\sum_{i=1}^{n} (x_i - \mu)^2\right]}$$

$$f(x) = \frac{1}{\sqrt{2\pi}\sigma} e^{\left(\frac{-(x-\mu)^2}{2\sigma^2}\right)} \tag{8.4}$$

where

μ is the mean.
σ is the standard deviation.
$F(x)$ is the normal distribution.

As another way to select the main attribute and the general category, obtaining Kononenko information as the sum of the information provided by each attribute can be an explanation of how predictor values affect class probability. According to this theory, each attribute is known to have the highest probability based on the probability of each class that has more information and belongs to the same class.

$$\log_2 P(c \mid x) - \log_2 P(c) \tag{8.5}$$

where

$P(c|x)$ is the posterior probability of a class (target) given a predictor (feature).
$P(c)$ is the prior probability of class.

Using this method, the effect of each predictor can also be shown by drawing graphs to better understand the results and contribution of each predictor. These graphs illustrate the odds ratio for each value of each prediction. The length of the lines is related to the range of the odds ratio and indicates the importance of the relevant predictor as well as the effects of the value of the predictor.

8.2.2 Decision Trees

One of the most widely used methods for classifiers is the decision tree theory, which is based on information gain. The decision tree generates and displays classification or regression models in the form of a tree structure, breaking down a large dataset into smaller subsets and roots. Then, a related decision tree is gradually developed to form a representation of data connections. The end result is the formation of a tree with decision nodes and leaf nodes. A decision node (e.g. TYPE) has two or more branches (e.g. longitudinal, transverse, diagonal, block, and fatigue). A leaf node (for example, in good condition) represents a classification or decision. The highest decision node in a tree that corresponds to the best predictor is called the root node. Since decision trees can also manage both batch and numeric data, it has no restrictions.

The original algorithm for constructing decision trees, called ID3, was developed by J. R. Quinlan, which uses a greedy top-down search in the space of potential branches without the possibility of reversal. In other words, it starts from the top and tries to move forward in the branches that have more information of interest. For this purpose, increased entropy and information are used to construct the decision tree. In the naive Bayesian method, all predictors are advanced using the Bayesian rule and the assumptions of independence, but the decision tree includes all predictors only assuming the dependence between the predictors, which increases speed and accuracy. As introduced in Chapter 7, the types of entropies can be used to calculate information gain, whereby the most common method is Shannon entropy. The relationships related to the types of entropies can be employed to calculate information gain (IG), which are described as follows:

$$\mathrm{Ent_{sh}}(S) = -\sum\nolimits_{j=1}^{n} P_j \log_2(P_j) \tag{8.6}$$

$$\mathrm{Ent_{RN}}(w) = \left(\log \sum\nolimits_{j=1}^{n} (\mu_w(x_j))^{\alpha}\right)/(1-\alpha) \tag{8.7}$$

$$\mathrm{Ent_{ME}}(w) = \sum\nolimits_{j=1}^{n} \left((W_j)\left(\mathrm{Sin}\frac{\pi\mu_w(x_j)}{2} + \sin\frac{\pi(1-\mu_w(x_j))}{2} - 1\right)\right) \tag{8.8}$$

$$\mathrm{Ent_{TS}}(w) = \left(1 - \left(\sum\nolimits_{j=1}^{n} (\mu_w(x_j))^{\alpha}\right)\right)/(1-\alpha) \tag{8.9}$$

$$\mathrm{Ent_{pp}}(w) = -\sum\nolimits_{j=1}^{n} \left(\mu_w(x_j) e^{1-\mu_w(x_j)}\right) \tag{8.10}$$

$$\mathrm{Ent}_m(w) = \sum\nolimits_{j=1}^{n} \left(\mu_w(x_j) e^{-\left(a\mu_w(x_j)^3 + b\mu_w(x_j)^2 + c\mu_w(x_j) + d\right)}\right) \tag{8.11}$$

$$\mathrm{Ent_{ma}}(w) = \frac{\mathrm{Ent}_m(w) - e^{-(a+b)^{\beta}}}{n^{(1-\gamma)} e^{-\left(\frac{a}{n^{a}}+b\right)^{\beta}} - e^{-(a+b)^{\beta}}} = \frac{\mathrm{Ent_m(w)} - \mathrm{Ent}_{(\mathrm{ma})_{\min}}}{\mathrm{Ent}_{(\mathrm{ma})_{\max}} - \mathrm{Ent}_{(\mathrm{ma})_{\min}}} \tag{8.12}$$

Equations (8.6) to (8.12) are employed in a variety of IG measurement methods, including committees of entropies Ent = $\{\mathrm{Ent_{sh}}(w), \mathrm{Ent_{me}}(w), \mathrm{Ent_{rn}}(w), \mathrm{Ent_{pp}}(w), \mathrm{Ent_{ma}}(w)\}$. Feature selection is made very efficient with the help of a hybrid algorithm in order to prioritize the main features; in other words, it is possible to prioritize the features based on their effectiveness and information gain. Using hybrid algorithms, the decision of arranging tree nodes is possible.

In this section, the entropy value is calculated using the frequency table for the two properties of T and X using the following equation for Shannon entropy. The same model as that for Shannon entropy, the entropy of other methods can be calculated.

Feature 1	Feature 2	Feature 3	Feature 4	Feature 5	Distress detection
L	L	L	L	H	Yes
M	M	L	L	H	Yes
L	L	L	L	H	No
M	L	L	L	L	No
M	L	L	H	L	Yes
M	H	M	H	L	No
L	L	M	H	M	No
H	L	H	L	M	No
H	M	H	H	L	Yes
H	M	H	L	L	No
L	M	L	L	M	No
L	L	L	M	H	Yes
H	L	L	M	L	No
M	L	H	M	M	No
L	H	H	H	H	No
H	H	H	H	L	Yes
H	H	M	H	L	No
H	M	M	M	M	No
H	H	M	M	L	Yes
L	M	M	M	M	No

Step 2

Row labels	Count healthy
No	13
Yes	7
Grand total	20

Step 2

Entropy (Healthy) = Ent(13,7) = Ent(0.65,0.35) = $-(0.65\log_2(0.65)) - (0.35\log_2(0.35)) = 0.93$

Figure 8.2 Entropy using the frequency table of one feature.

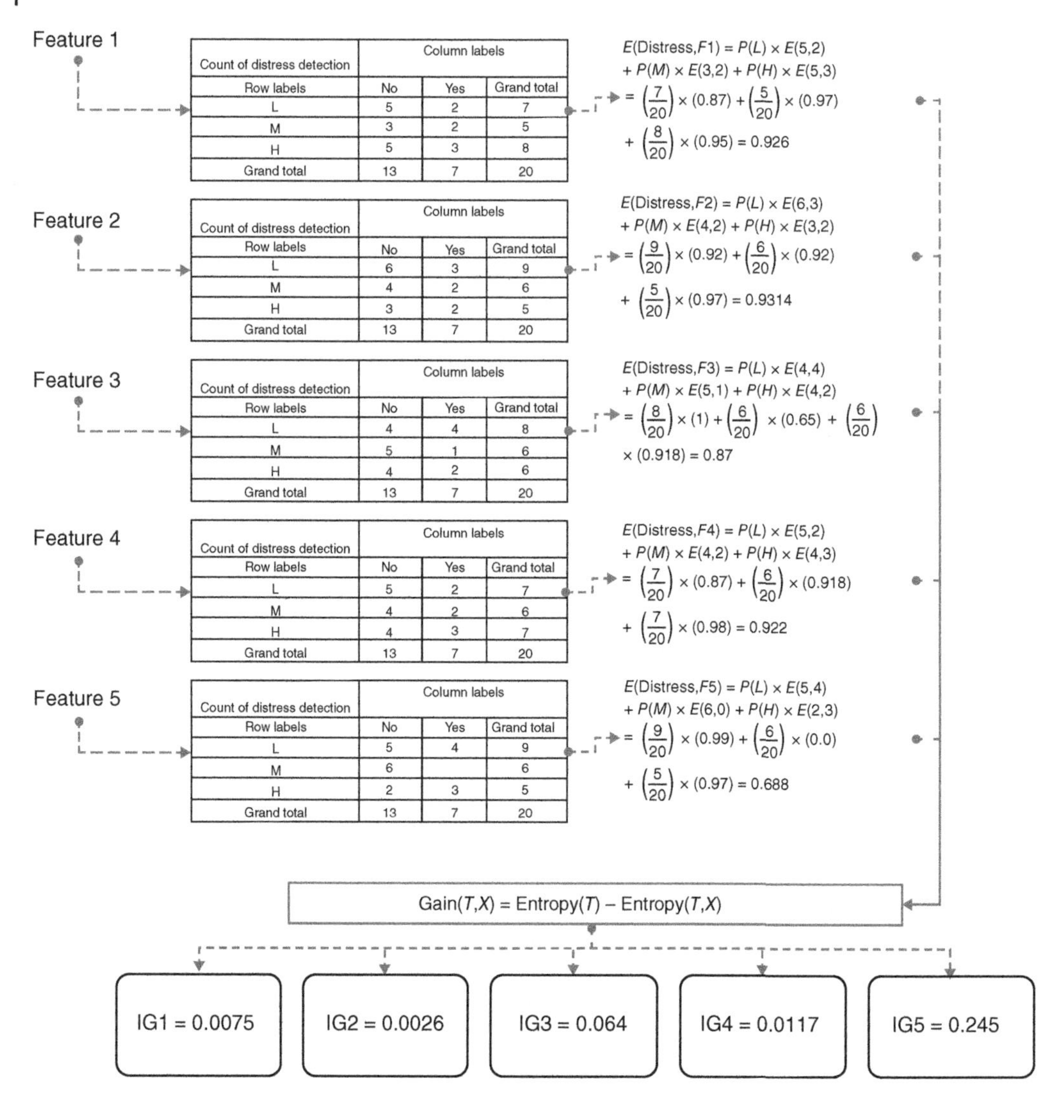

Feature 1

Count of distress detection	Column labels		
Row labels	No	Yes	Grand total
L	5	2	7
M	3	2	5
H	5	3	8
Grand total	13	7	20

Feature 2

Count of distress detection	Column labels		
Row labels	No	Yes	Grand total
L	6	3	9
M	4	2	6
H	3	2	5
Grand total	13	7	20

Feature 3

Count of distress detection	Column labels		
Row labels	No	Yes	Grand total
L	4	4	8
M	5	1	6
H	4	2	6
Grand total	13	7	20

Feature 4

Count of distress detection	Column labels		
Row labels	No	Yes	Grand total
L	5	2	7
M	4	2	6
H	4	3	7
Grand total	13	7	20

Feature 5

Count of distress detection	Column labels		
Row labels	No	Yes	Grand total
L	5	4	9
M	6		6
H	2	3	5
Grand total	13	7	20

Figure 8.3 Example of the results for information gain, or decrease in entropy.

$$\mathrm{Ent_{sh}}(T,X) = \sum\nolimits_{j=1}^{n} P(c)_j E(c) \tag{8.13}$$

where

$\mathrm{Ent_{sh}}(T, X)$ is the entropy using the frequency table of two features.
$P(c)$ is the prior probability of a class.

As shown in Figure 8.2, the dataset is divided into different attributes. The entropy is calculated for each branch and then added proportionally to obtain the total entropy for division. The entropy is subtracted from the entropy before division. The result is an increase in information or a decrease in entropy. As this ratio is calculated, an index of information gain and the effectiveness of features in achieving good results is obtained from the categorizer (Figure 8.3).

$$\mathrm{Gain}(T,X) = \mathrm{Entropy}(T) - \mathrm{Entropy}(T,X) \tag{8.14}$$

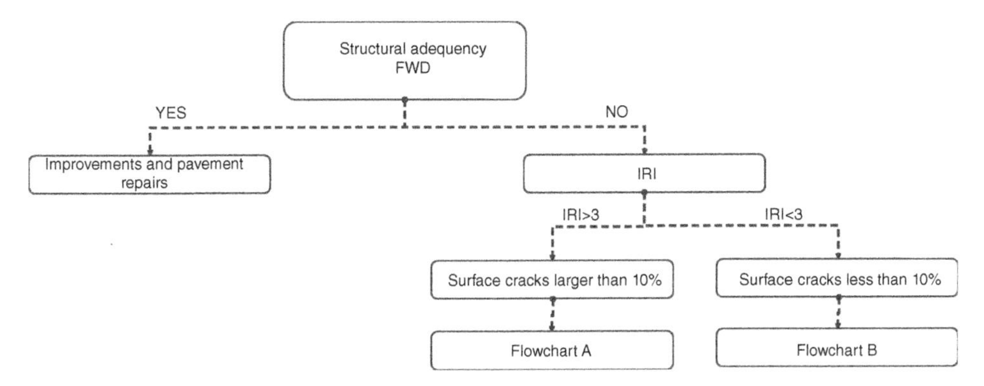

Figure 8.4 An example of the use of a decision tree in a pavement management system to select the optimal maintenance option.

where

Gain(T, X) is the information gain or decrease in entropy.

Using this index, the attribute is selected as the decision node with the highest increase of information, then the dataset is divided into its branches, and the same process is repeated in each branch to place all the attributes. As shown in Figure 8.3, the fifth attribute is selected as the decision node with the largest increase in information, then the dataset is divided into its branches, and the same process is repeated in each branch. A branch with entropy 0 is considered a leaf node, and a property with a higher entropy than other features is selected. A branch with entropy greater than 0 needs to be further decomposed. This process is repeated to form and design the entire decision tree. In the method, the ID3 algorithm is run recursively on leafless branches until all data are classified and no data remain. This method is generally designed based on information gain. A decision tree can easily be turned into a set of rules by mapping from the root node to leaf node. Figure 8.4 shows an example of a decision tree used in pavement management to segment and determine the optimal operations for maintenance.

As a practical example, the outputs of a decision tree method in a management system are presented in Figure 8.3. A pavement management system includes various components to prioritize different sections with variable indicators and multiple characteristics, so that the optimal maintenance strategy can be selected for the pavement network of related roads in a wide area. Due to the increasing rate of demolition of pavement parts and limited budget and resources, selecting the most optimal maintenance scenarios for each part of the pavement is of particular importance. In this example, according to the capabilities of the decision tree, a recommended model based on cloud decision tree theory (CDT) has been used to select the optimal maintenance repair strategy for each piece of a national road network pavement in Iran. A CDT system is provided for the road network, which includes a general decision model and different decision trees for each province. The most important valuable feature of this method is that this classification is based on the data collected in the database separately for freeways, highways, and main roads. In addition to providing a cloud network of the decision tree, three different models of the decision tree based on annual average daily traffic (AADT) road traffic are also presented. The results of this study showed that CDT has good capability and efficiency in modeling the optimal decision-making process regarding the classification and selection of appropriate pavement maintenance and repair options in infrastructure management. According to decision tree theory, fatigue cracking and the international roughness index are the most important indicators for determining the appropriate M&R option. An example of a decision tree is shown in Figure 8.5.

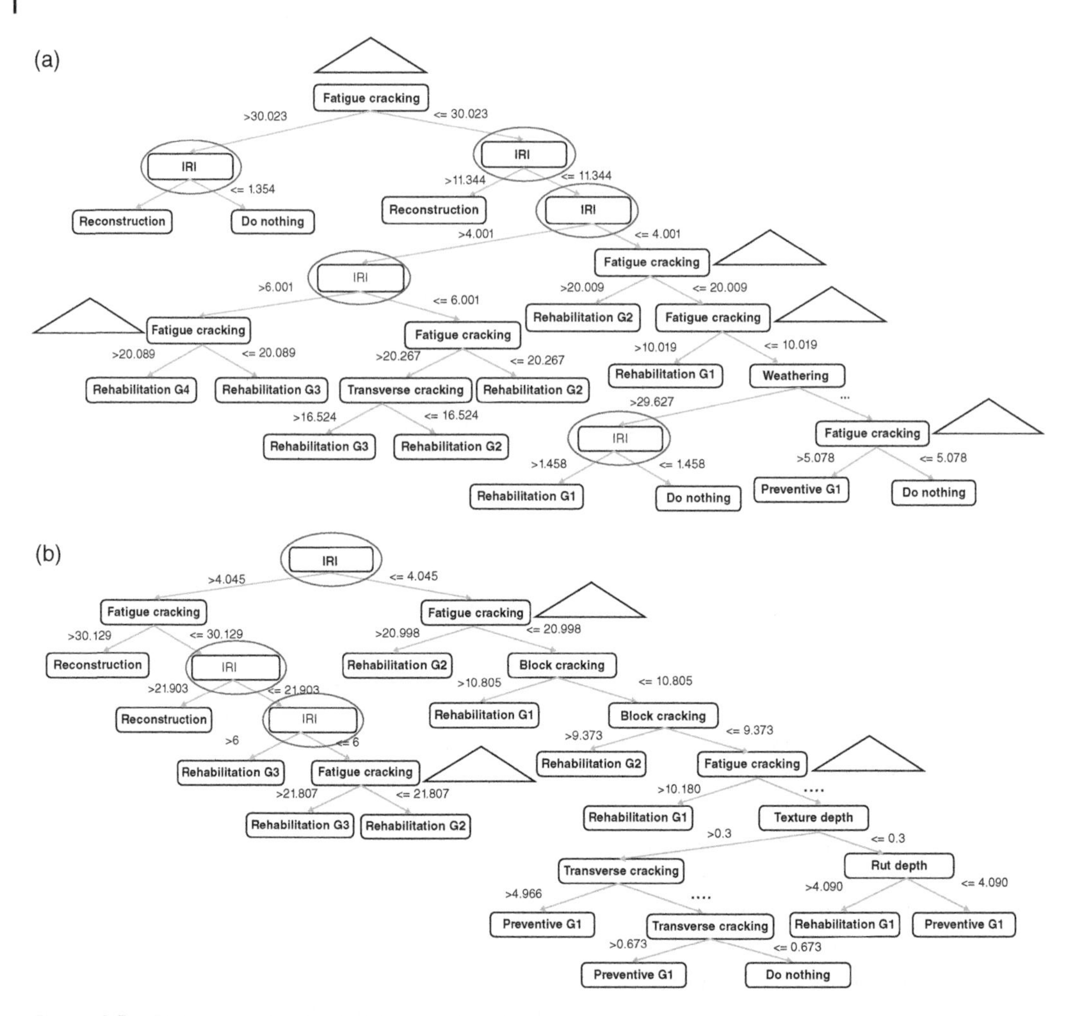

Figure 8.5 (a) An example of a decision tree in selecting the optimal repair and maintenance option for the road network in Iran in order to select the optimal class. (b) An example of a decision tree in selecting the optimal repair and maintenance option for the road network in a region or province, such as Tehran, in order to select the optimal class.

The decision tree (DT) method is selected using the data mining to determine the optimal maintenance policy of a road network in Iran. The pavement management system data include the status of road procedures, and for the data with known maintenance options, a database containing more than 10 000 data points was created to build the decision tree. This method is used to discover the relationship between pavement failure data collected by the automated pavement assessment device and dedicated options in order to discover a cloud decision tree model for selecting the optimal M&R option. In this case study, the problem has been analyzed with different dimensions. This analysis was also performed for the entire network, then based on the type of road and state area, this analysis was performed separately. In this analysis, each highway and main road has their own decision daughter, and their structure is different according to the type of data. Decision tree models also have different structures based on the type of traffic. The results of this case study showed that the cloud model of the decision tree exhibits good performance for modeling the decision-making process regarding the optimal choice of pavement network maintenance option. By examining the decision trees, it was found that the fatigue cracking index and international roughness index (IRI) are the best indicators in determining the M&R option (Figure 8.6).

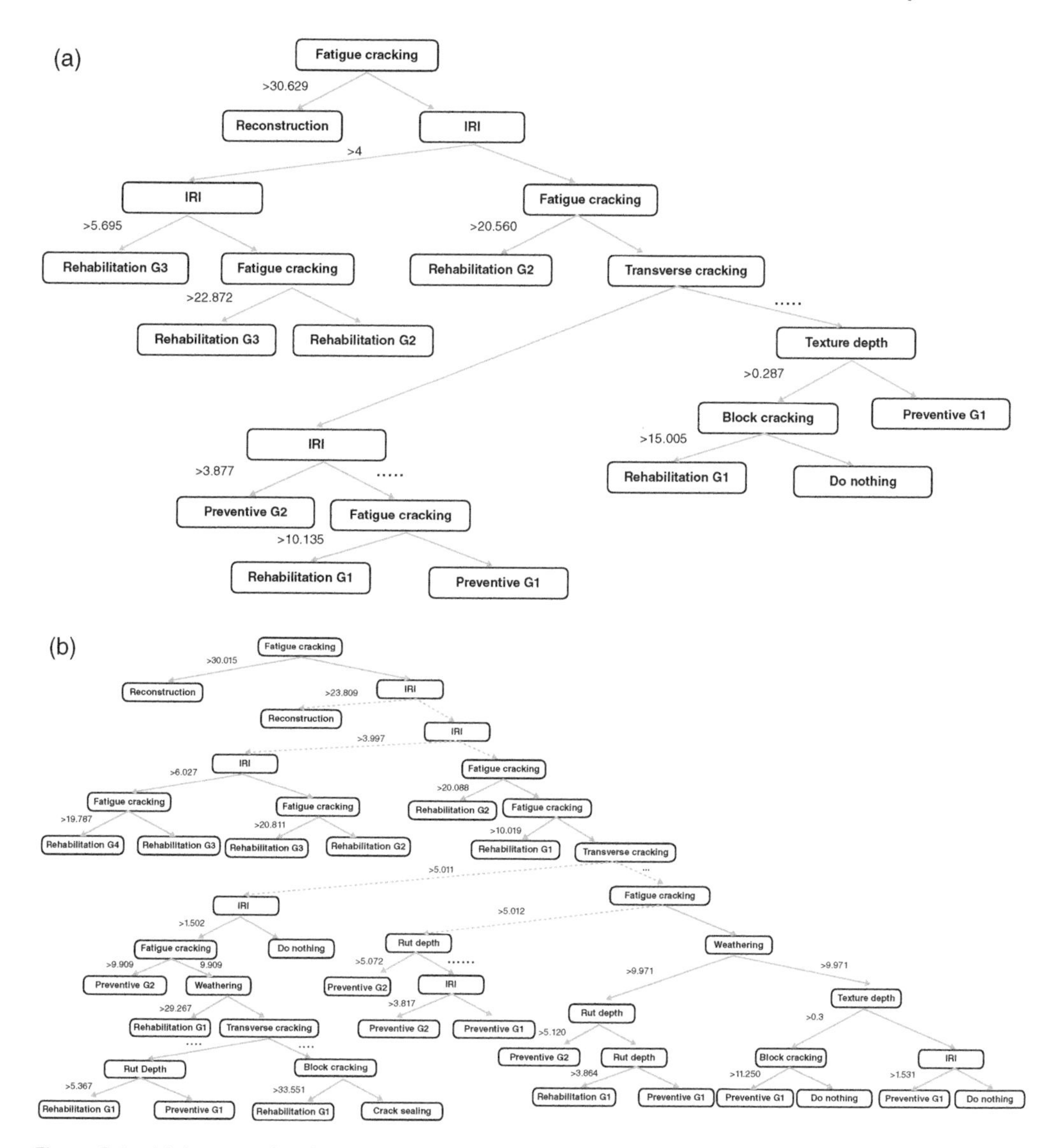

Figure 8.6 (a) An example of a decision tree in choosing the optimal repair and maintenance option for maintenance of a freeway network. (b) An example of a decision tree in selecting the optimal repair and maintenance option for maintenance of highways in the country's road network.

8.2.3 Logistic Regression

The linear discriminant analysis (LDA) method was developed in 1936 by R.A. Fisher and has a strong mathematical basis. Most models derived from this method have an accuracy that can compete with complex methods. The LDA method is based on the concept of searching for a linear combination of variables that best separate the two classes. To create segregation, Fisher defined the following score function:

$$Z = \beta_1 x_1 + \beta_2 x_3 + \dots + \beta_d x_d$$

$$S(\beta) = \frac{\beta^T \mu_1 - \beta^T \mu_2}{\beta^T C \beta} = \left(\frac{\overline{Z}_1 - \overline{Z}_2}{\text{var}(Z)}\right) \tag{8.15}$$

where

$S(\beta)$ is the score function.

To solve the problem of estimating the linear coefficients that occur due to the maximization of the score, the following equations are presented:

$$\begin{aligned} \beta &= C^{-1}(\mu_1 - \mu_2) \\ C &= \frac{1}{n_1 + n_2}(n_1 C_1 + n_2 C_2) \end{aligned} \tag{8.16}$$

where

β is the linear model coefficient.
C is the pooled covariance matrix.
C_1, C_2 are the covariance matrices.

There are several ways to evaluate the separability of categories. One way to estimate the effectiveness of the ability to separate the two groups is to calculate the Mahalanobis distance between the two groups. A distance greater than 3 means that there is a difference of more than 3 standard deviations in the two averages, and thus, the degree of overlap (probability of incorrect classification) is very small OR is negligible. The higher the probability, the higher the probability of separation.

$$\begin{aligned} &\Delta^2 = \beta^T(\mu_1 - \mu_2) \\ &\beta^T\left(x - \left(\frac{\mu_1 - \mu_2}{2}\right)\right) > -\log\left(\frac{p(c_1)}{p(c_2)}\right) \end{aligned} \tag{8.17}$$

where

β^T is the coefficient vector.
x is the data vector.
$\left(\frac{\mu_1 - \mu_2}{2}\right)$ is the mean vector.
$p(c_1)$ is the probability in class 1.
$p(c_2)$ is the probability in class 2.

The quadratic decision analysis (QDA) is a general classifier function with quadratic decision boundaries that can be used to classify a dataset with two or more classes. Normally, the QDA method has more predictive power than LDA, but it is necessary to correctly estimate the covariance matrix for each class.

$$Z_k(x) = -\frac{1}{2}(x - \mu_k)^T C_k^{-1}(x - \mu_k) - \frac{1}{2}\ln|C_k| + \ln P(c_k) \tag{8.18}$$

where

C_k is the covariance matrix for the class k (−1 means inverse matrix).
$|C_k|$ is the determinant covariance matrix.
$P(c_k)$ is the probability of class k.

Based on a similar entropy method, the classification law works simply by finding the class with the highest value of Z.

8.2.4 *k*-Nearest Neighbors (kNN)

kNN are a simple algorithm that stores all existing items and classifies new items based on size similarity (e.g. distance functions), which can be calculated according to various relationships. The simplest and most common method for calculating the distance is the Euclidean method, in which a new case class is identified by a majority vote of its neighbors. In the kNN' algorithm, each new data is assigned to a class that is measured by the distance function between its nearest neighbors k. If $k = 1$, this item is simply assigned to the nearest neighbor class. An example of k-nearest neighbors method to determine if data belong to a specific category based on distance is illustrated in Figure 8.7.

$$d_e = \sqrt{\left(\sum_{i=1}^{k}(x_i - y_i)^2\right)}$$

$$d_{\mathrm{Ma}} = \sum_{i=1}^{k}|x_i - y_i|$$

$$d_{\mathrm{Mi}} = \left(\sum_{i=1}^{k}(|x_i - y_i|)^q\right)^{1/q}$$

$$d_H = \sum_{i=1}^{k}|x_i - y_i|, X = Y \rightarrow D = 0, X \neq Y \rightarrow D = 1 \tag{8.19}$$

where

d_e is the Euclidean distance.
d_{Ma} is the Manhattan distance.
d_{Mi} is the Minkowski distance.
d_H is the Hamming distance.

It should be emphasized that all three of the above measurements can only be used to calculate distances for continuous variables. For the classification variables, Hamming distance can be used. When there is a combination of numerical and class variables in the dataset, the practical solution is to standardize the numerical variables between 0 and 1.

Selecting the appropriate value for K is done by first reviewing the data. In general, the larger the value of K, the more accurate the value obtained since noise is reduced, and thus accuracy is increased. It should be noted that there is no guarantee of accurate results. Based on past research and successful applications, the optimal value for most datasets is more than 3 and less than 10.

8.2.5 Ensemble Techniques

Ensemble learning is a machine learning algorithm with many capabilities and a variety of applications in infrastructure management. The most important idea of ensemble learning techniques is that they combine the prediction results of several machine learning models to produce a very accurate result. These group learning techniques include popular machine learning algorithms, such as XGBoost, Gradient Boosting, and more. In this section, the most important combination method, namely Boosting, is presented. You can certainly give examples. Examples for this are provided of how group learning works in one of the topics of infrastructure assessment and how useful it is.

The maximum voting method is generally used for classification issues, whereby several models are used to predict the class of each data. The predictions of each method are then considered as "votes." The final category is determined based on the frequency of predicted categories. For example, when the goal is to determine the severity of a type of distress of type (L, M, H) using seven

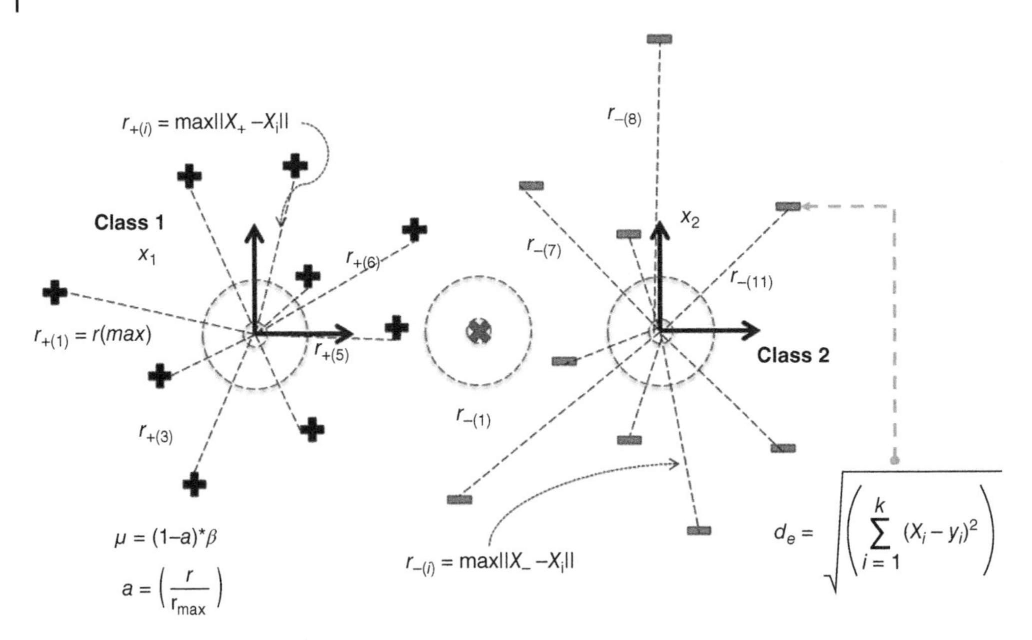

Figure 8.7 Example of *k*-nearest neighbors method to determine if data belong to a specific category based on distance.

classifiers, it detects the results of four types of M and the other two categories recognize two times for H and 1 time for L. Because they gave the majority of points to M. Based on this method, the final type of distress severity is determined to be medium.

This method is similar to the maximum voting method. Based on the averaging method, the categories of each data are determined using different classifiers, and finally, the category range is determined using the averaging method. The result is closer to each category, the same class is selected as the result. In the distress severity example, the averaging method takes the average of all values, that is $(4 \times 2 + 2 \times 3 + 1)/7 = 2.14$. So finally, the data is placed in a category with a weight factor of 2, which is equal to M (See table 8.4) .

In the averaging method, different weights are assigned to all models based on the amount of weights, which shows the importance of each model for classification. For example, if two classifiers have more power in the correct classification, while the other classifiers do not have much ability in this area, the answer of these two methods is more important than the other methods. In the above example, if classifiers 3 and 5 are more accurate than other methods, they weigh more, and calculations related to determining the final class are performed based on the Table 8.5. In this method, the weight of all methods, except 3 and 5, is 0.4, and the weight of these two methods is 0.8.

Table 8.4 The final class is determined based on the frequency of predicted categories.

Classifier	1	2	3	4	5	6	7
Severity	M	M	H	H	M	M	L
Weight	2	2	3	3	2	2	1

Table 8.5 The final class is determined based on importance weight.

Classifier	1	2	3	4	5	6	7
Severity	M	M	H	H	M	M	L
Weight	2	2	3	3	2	2	1
Importance weight	0.4	0.4	0.8	0.4	0.8	0.4	0.4

The result is calculated as $[(2\times0.4\times3)+(3\times0.8)+(3\times0.4)+(2\times0.8)+(1\times0.4)]/(5\times0.4+2\times0.8)=2.22$ for final importance weight.

More advanced methods for combining classifiers are generally classified into four general categories, including Stacking, Blending, Bagging, and Boosting.

Stacking is an assembly learning technique that uses the prediction of several classifier models (e.g. decision tree, kNN, and Support vector machine [SVM]) to build a new model. This model uses a test set to make predictions, in which the steps of the dataset is divided into 10 parts (folds).

A classifier model (assuming a decision tree, neural network [NN], or SVM) is used in nine parts and to make predictions for the 10th part. This is done for each part of the training section dataset. The base model (e.g. NN) is then applied to the entire dataset, and predictions are made on the test dataset. The above steps are repeated for another base model (e.g. DT and SVM), resulting in another set of predictions for the training dataset and the test dataset. Database predictions are used as features to build a new model, which is then applied for the final classification in the test category set.

The blending method imitates a similar approach to the stacking method, except that it uses a validator of the train set to make predictions. Unlike the stacking method, predictions are made only in the holding set in this method. The dataset is separated, and the predictions are used to create a model that runs on the test set. Based on this model, the training set is divided into training sets and validation sets, then the classification model is applied to the training set. In this model, predictions are made about the validation and test sets. The validation set and its predictions are used as features to build the new model, which utilized to make final predictions in the test set and general characteristics.

Third, the idea behind the bagging method is to combine the results of several models in order to obtain an overall result. In order to change the inputs of a database, the bootstrapping method is used to solve this feature. Bootstrapping is a sampling method in which subsets of observations are created from the original dataset. The size of the subsets is the same as the size of the original collection. The bagging method uses a simple idea based on justice, considering a complete set distribution to build a new and diverse database. The size of the subsets created for each set may be smaller than the original collection, which is normal. On the one hand, this method can solve the problem of identical output of one type of classifier. On the other hand, using this method creates diversity in obtaining the results of a database, which ultimately leads to more accurate results and evaluates the sensitivity of a method to simple classifiers.

In this method, different subsets of the original data and new banks are created by replacing the subsets. A base model, usually made up of a weak model, is created in each of these subsets. The models work in parallel and are used in each subset independently of each other. The final classification is determined by combining the predictions of all models.

In hybrid methods, if data are categorized by a first model in a class and then the next model misdiagnoses, combining the classifiers will yield poor results. This commonly occurs in most engineering issues, for which boosting methods have an amazing application. Boosting is a continuous process in which each classifier tries to reduce the error associated with the previous classifier. In other words, the performance of the next model and its type are dependent on the previous model.

In general, the following rules apply to the boosting process. First, the main set is divided into subsets, and the weight of all equal points is considered. A classifier model is applied to this subset to predict the entire dataset. Errors are calculated using principal values and values classified with the applied classifier. A to incorrectly classified data. Then, another classifier is created, and the categories are applied to the dataset. This model tries to correct the errors of the previous model by applying the basic boosting method. Several models can be created following this same process, each of which, as a boosting agent, corrects the errors of the previous model and increases the overall performance of the model. In general, the final model (which is a strong learner) is the weight average of all conventional classifier models (weak classifiers).

The structure and law of boosting creates a strong boosting learning classifier using a number of weak classifiers. In general, each of these classifiers performs well for part of the dataset but not the entire dataset. Thus, each model actually assumes part of the task of categorizing the data and ultimately increases the overall performance of the categorizing group by using a group task.

8.2.6 Adaptive Boosting (AdaBoost)

Boosting is an effective way to cleverly combine several simple weak classifiers to create a new classification committee architecture that dramatically improves performance. This section presents the most widely used model of the boosting algorithm, called AdaBoost, which stands for "boost adaptive." Boosting can lead to acceptable and surprising results in complex issues. In this method, basic classifiers are sometimes known as weak models. The main difference between the boosting methods and the classification committee described in the previous section (such as bagging) is that each basic (weak) classifier is trained in order in the former method? If so, how does this compare to the latter method. Using a weighting method, the AdaBoost also learns from the dataset, in which the corresponding weight coefficient is calculated. In this method, when the classification is misdiagnosed, more weight is given to that classifier. Once all the classifiers have been trained, their predictions are combined through a weighted majority voting scheme, the operation of which is shown schematically in Figure 8.8. Comparison of stacking and blending methods illustrated in Figure 8.9.

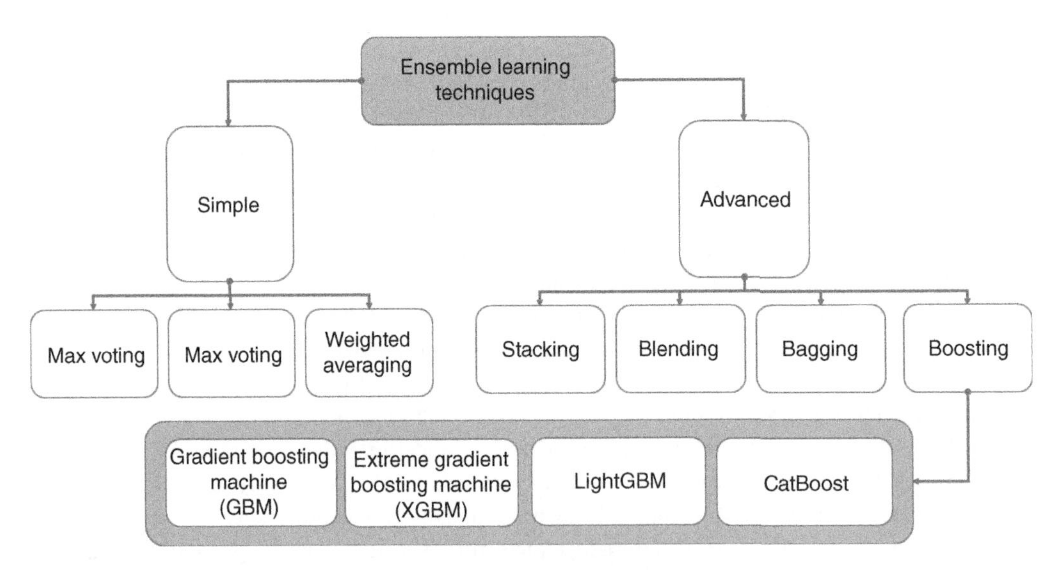

Figure 8.8 A general classification for ensemble learning techniques.

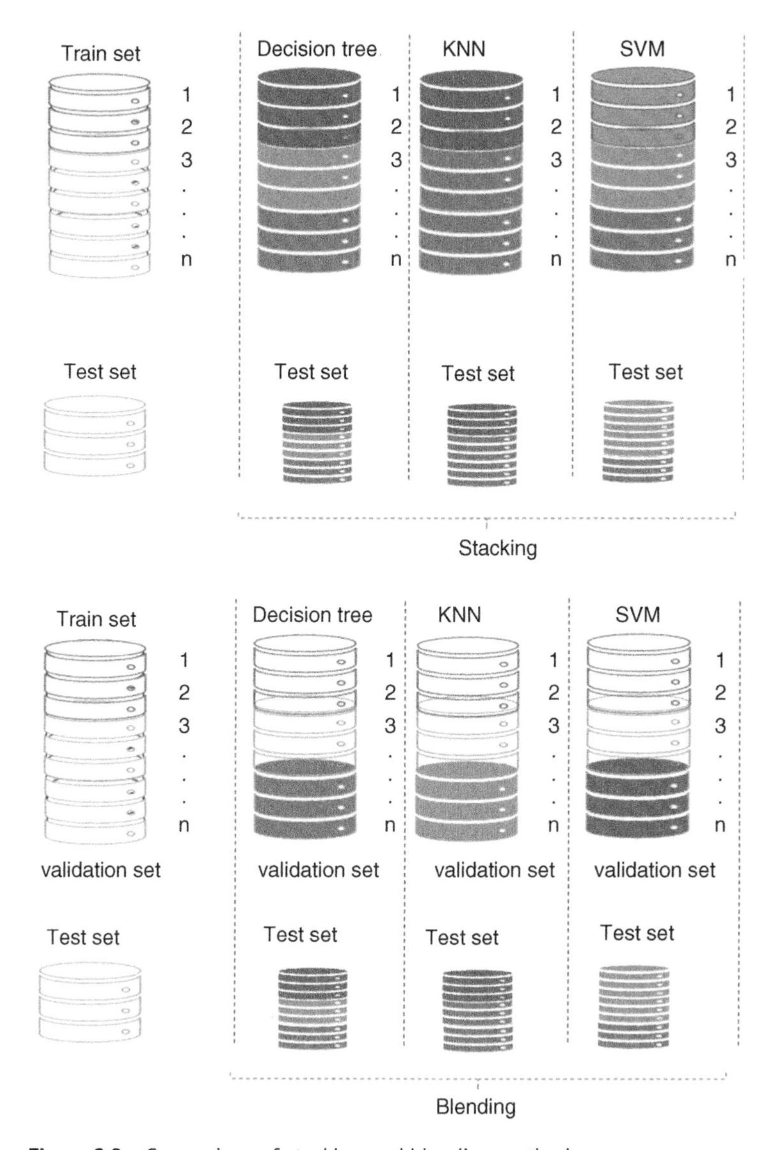

Figure 8.9 Comparison of stacking and blending methods.

Algorithm AdaBoost for Two Class

1) Initialize the observation weighting coefficients $w_n^{(1)} = 1/n, i = 1, 2, \ldots, n$.
2) For $m = 1$ to M:

   i) Fit a classifier $\mathrm{Cl}^{(m)}(x)$ to the training data using weights wi.
   ii) Compute:

$$J^{(m)} = \sum_{i=1}^{n} w_i I\left(C_i \neq \mathrm{Cl}^{(m)}(x_i)\right) / \sum_{i=1}^{n} w_i.$$

   iii) Compute:

$$\alpha^{(m)} = \log \frac{1 - J^{(m)}}{J^{(m)}}$$

iv) Update the data weighting coefficients:

$$w_i \leftarrow w_i.\exp\left(\alpha^{(m)}.I\left(C_i \neq \mathrm{Cl}^{(m)}(x_i)\right)\right)$$

For $i = 1, 2, ..., n$.

v) Renormalized w_i.

3) Make classification using the below model:

$$C(x) = \mathrm{argmax} \sum_{m=1}^{M} \alpha^{(m)}.I\left(\left(\mathrm{Cl}^{(m)}(x) = k\right)\right)$$

According to this algorithm, the first base classifier $\mathrm{Cl}^{(1)}$ is taught using the weighting coefficients $w_n^{(1)}$, which are all equal in the first step and, therefore, in accordance with the usual method for teaching a classifier. Using this algorithm, in subsequent iterations, the weighting coefficients w_i are increased for data points that are misdiagnosed and decreased for data points that are correctly classified. As per this theory, classifiers are constantly forced to place more emphasis on points that have been misclassified by previous classifiers, resulting in more weight being given to data that is still misclassified by successive classifiers. The values obtained for m indicate the error weight index of each base classifier. The schematic framework of the Ada-Boost algorithm is shown in Figure 8.11. A similar algorithm is presented for multi-class classification.

Algorithm AdaBoost for Multi-Class Based on Stagewise Additive Modeling Using a Multi-Class Exponential Loss Function

1) Initialize the observation weighting coefficients $w_n^{(1)} = 1/n, i = 1, 2, ..., n$.
2) For $m = 1$ to M:

vi) Fit a classifier $\mathrm{Cl}^{(m)}(x)$ to the training data using weights wi.
vii) Compute:

$$J^{(m)} = \sum_{i=1}^{n} w_i I\left(C_i \neq \mathrm{Cl}^{(m)}(x_i)\right) / \sum_{i=1}^{n} w_i$$

viii) Compute:

$$\alpha^{(m)} = \log \frac{1 - J^{(m)}}{J^{(m)}} + \log(K - 1)$$

ix) Update the data weighting coefficients:

$$w_i \leftarrow w_i.\exp\left(\alpha^{(m)}.I\left(C_i \neq \mathrm{Cl}^{(m)}(x_i)\right)\right)$$

For $i = 1, 2, ..., n$.

x) Renormalized w_i.

3) Make classification using the below model:

$$C(x) = \text{argmax} \sum_{m=1}^{M} \alpha^{(m)}.I\left(\left(Cl^{(m)}(x) = k\right)\right)$$

It should be noted that the multi-class algorithm is the same as the simple AdaBoost modular structure, except that the additional sentence $(K-1)$ is shared in the two-class algorithm. Obviously, when $K = 2$, the algorithm converts to AdaBoost. The term log $(K-1)$ is very important in the first algorithm for multi-classes $(K > 2)$.

8.2.7 Artificial Neural Network

Neural networks are one of the most common methods of classification and machine learning and artificial intelligence, which have grown significantly in recent years in a variety of engineering parables. Artificial Neutral Network (ANN) is a method based on biological neural networks inspired by the natural behavior of the brain. Communication between brain cells occurs through electrochemical signals and sequential transactions. The main neurons also connect through synapses, whereby each neuron receives signals continuously by communicating with other neurons. According to the law of a certain threshold, if the sum of the signals exceeds a certain limit, a reflection is sent through the axon. The ANN method simulates the behavior of the brain's response to the reflection in order to classify functional topics. To-date, neural networks with various structures, such as multilayer perceptron (MLP) networks, are used to solve many complex and different problems.

An ANN consists of a network of artificial neurons that are connected to each other through nodes and are assigned based on the strength of their connections in learning weight problems: This weight is a value between −1 to 1, which is used to determine the relationship between neurons. A higher weight indicates a stronger connection. A transfer function is designed for each node. There are three types layers in the ANN structure: input, hidden, and output layers. Neural networks are divided into simple neural networks or multilayer neural networks, such as Feedforward Networks, Learning Vector Quantization Networks, Adaptive Resonance Theory, Hamming Networks, Radial Basis Functions, Self-Organizing Maps, and Probabilistic Networks, based on the type of layers they are designed with.

Initially, the input nodes receive information from the main characteristics, which are numerical. Basic information is entered into the network as values for network activation and sensitivity, formation of connection weight, and amount of node activation. First, a number is assigned to each node at random, whereby the more input data, the higher the activation rate. This information is then quickly injected into the entire network, then the amount of excitation, transmission functions, and activation is determined based on the generated flow. In each node, the result of the activation coefficient is specified. The correction factor is then determined based on the transfer function. Hidden layers play an important role in activating the flow of information and learning to achieve output values. After performing this process, the output nodes represent the optimal results based on the type of input in a meaningful way. Based on the amount of error, which is determined by dividing difference between the predicted value and actual value by the weight of each node, the amount of error per node and the total error are calculated. Based on the total error rate, the error of each node is calculated, and finally, the weight of each node is corrected. Ultimately, the error is divided between the nodes to reduce the total error.

The transmission function is an operator that converts input signals into output signals. In thematic literature, four types of transfer functions are commonly used for conversion: unit step (threshold), sigmoid, piecewise linear, and Gaussian.

The f(x), in which the output is categorized into level 1 or level 0 depending on whether the total input is more or less than the threshold value, is calculated using the following equation:

$$f(x) = \begin{cases} 0 \text{ if } 0 > x \\ 1 \text{ if } x \geq 0 \end{cases} \tag{8.20}$$

The sigmoid function consists of logistic and tangential functions, which vary from 0 to 1 and −1 to +1, respectively. This function has more flexibility than the previous function.

$$f(x) = \frac{1}{1 + e^{-\beta x}} \tag{8.21}$$

In the next functions for piecewise and linear, the output is proportional to the total weighted output and is distributed proportionally.

$$f(x) = \begin{cases} 0 \ \text{ if } x \leq x_{\min} \\ mx + b \text{ if } x_{\max} > x > x_{\min} \\ 1 \text{ if } x \geq x_{\max} \end{cases} \tag{8.22}$$

Gaussian functions, which are common in neural networks, display continuous and bell-shaped curves. The outputs of each node are divided into 0 and 1 categories based on the class membership, depending on how close the net input is to the selected value of the mean.

$$f(x) = \frac{1}{\sqrt{2\pi\sigma}} e^{\frac{-(x-\mu)^2}{2\sigma^2}} \tag{8.23}$$

The linear activation function converts the total weighted inputs of neurons into the output using a linear function. Moreover, neural networks can generally be divided into two types: feed-forward and feed-back networks.

1) The feed-forward network is a non-duplicate network that consists of three layers, including input, output, and hidden layers. In this model, the input data can only move in one direction and are transferred to a layer of analytics elements that performs the basic calculations. Each neuron calculates based on the total weight of its inputs then the total weight of output is calculated. In this method, new values are calculated, and then new input values are created and the next layer operates using the same function. This process works for the layers until the process is complete and the output is ultimately determined. Finally, the threshold transfer function is used to determine the amount of neuron output in the output layer. The feed-forward networks include perceptron networks (linear and nonlinear) and radial functions and are often used in data mining issues.
2) The feed-back network is a repetitive neural network that uses regressive signals in both the forward and reverse directions. In this network, all possible modes of possible connections between neural cells are recognized. Until the network reaches equilibrium, there are continuous changes in the loops.

In neural network modeling, monolayer perceptron (single layer perceptron [SLP]) is a leading network based on threshold transmission performance. SLP is the simplest type of artificial neural network and can only classify linearly separable items with a binary target (1, 0). There is no prior knowledge in the SLP method, and random initial weights are used. The SLP adds all the weight inputs, and if the sum is above the threshold, the output is 1.

In this method, the input values are first entered into the perceptron, and if the predicted output is close to the desired output and there is not much difference, then acceptable efficiency and accuracy and low error rate are considered. In this case, the weights are not changed, but if the error rate is high, the weights are changed to increase accuracy.

$$\begin{aligned} & w_1x_1 + w_2x_2 + w_3x_3 + \ldots + w_1x_1 > \theta \rightarrow (\text{output} = 1) \\ & w_1x_1 + w_2x_2 + w_3x_3 + \ldots + w_1x_1 \leq \theta \rightarrow (\text{output} = 0) \end{aligned} \tag{8.24}$$

$$\Delta w = \alpha \times d \times x \tag{8.25}$$

where

d is the forecasted output.
α is the learning rate <1.
x is the input data.

MLP is made of the same structure of monolayer perceptron with one or more hidden layers. This algorithm consists of two actions: (i) the forward stage in which the movement from the input to the output layer is performed, and (ii) the return stage in which the error between the actual value observed and the predicted value in the output layer is propagated backward.

$$\begin{aligned} & s = \sum w.x \\ & f(s) = \frac{1}{1 + e^{-s}} \\ & e_0 = y \times (1 - y) \times (t - y) \\ & e_i = y_i \times (1 - y_i) \times (w_i - d_0) \\ & \Delta w_i = \alpha \times d \times x_i \end{aligned} \tag{8.26}$$

where

s is the summation.
$f(s)$ is the transformation.

An example of a neural network constructed by this method is shown in Figure 8.12. In the construction of this neural network of definitions, a set of video inputs has been used for each construction of the input vector. According to this definition, four inputs and one output are considered for the network. Any road pavement damage enters the network based on the direction in the *P* input matrix. The output is only one number and represents the type characteristic and is according to the Table 8.6. The number of middle layers will be determined based on the sample test and the minimum error. The general structure of the neural network is shown in Figure 8.10.

To teach the network, different structures [specify], learning coefficients of 0.05 and 0.001, four input layers, and one fixed output layer were used. The objective function was set to 0.001 and 0.01, and different intermediate layers were used to obtain the best result. The efficiency of the designed structure was determined and is written in the sixth column of Table 8.7.

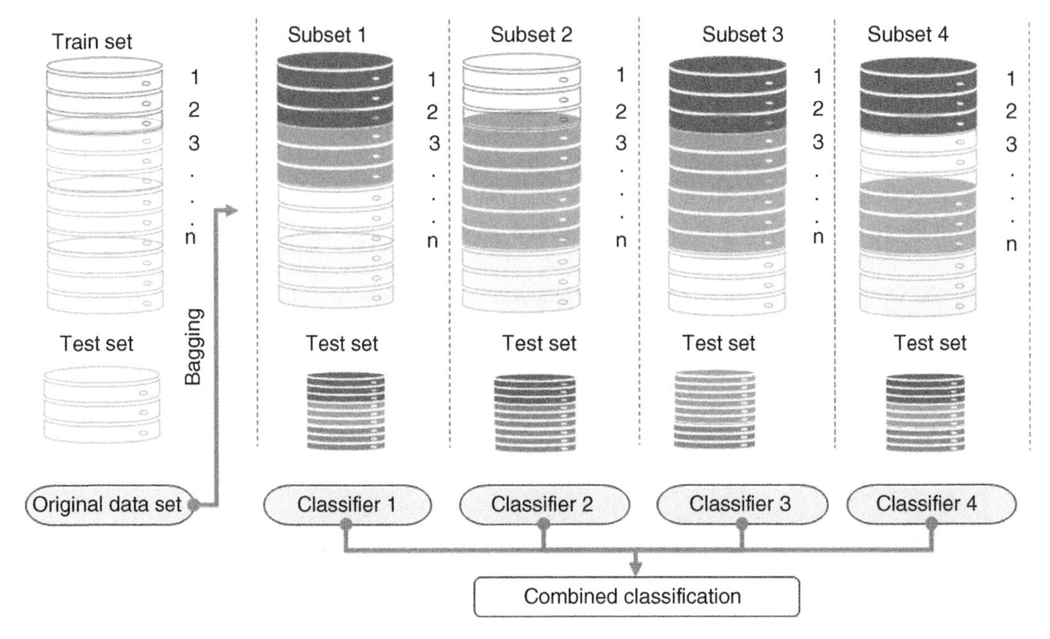

Figure 8.10 The general structure of the bagging method in the training and testing stage.

Table 8.6 The general structure of the neural network designed for use in the inference engine of the expert system based on 4 inputs and 1 output.

The number of layers	3
The number of neurons on the layers	Input: 4
	Hidden: var
	Output: 1
Activation functions	Log – sigmoid
	Log – sigmoid
	Linear
Training parameters	Back –propagation
Learning rule	Initial = 0.00001
Adaptive learning rate	Increase = 1.01
The initial weights and biases	Random
Momentum constant	0.98
Sum – squared error	0.000001

For each category of failures in the six classes, the results of each network in determining the number of distress were extracted and are written separately in Table 8.4. In this experiment, 90, 90, 90, 810, and 450 samples were used for types 1 to 6 pavement distress, respectively, according to Table 8.4. The type of class identified after using the network on the training samples for each distress is given in Table 8.8.

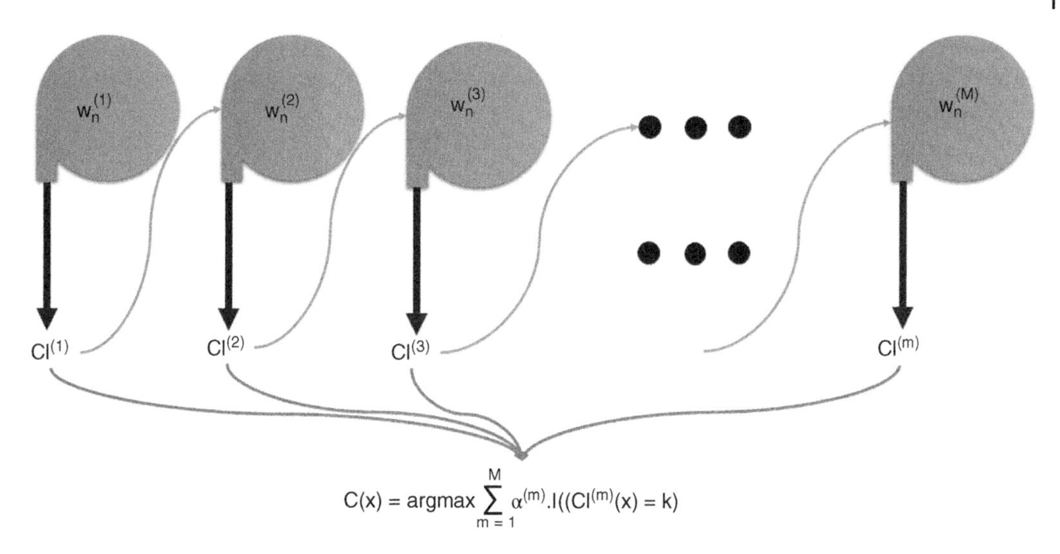

Figure 8.11 Schematic of the performance of the overall boosting framework, where each cl(*m*) classifier is taught on a weight from the training set (black continuous lines), and weight *w*(*m*) is equal to the performance of the previous classifier cl(*m* − 1)(*x*). Finally, when all the main classifiers are trained, they are combined to obtain the final classifier *c*(*x*). This method has a high efficiency in classification.

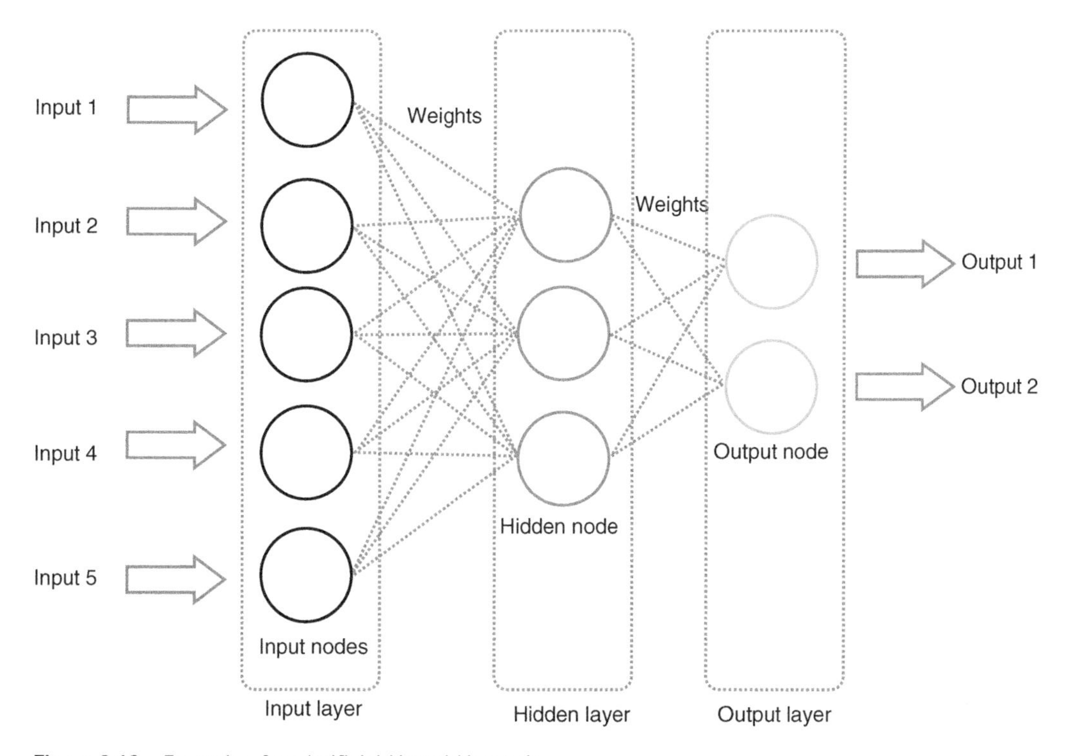

Figure 8.12 Example of an Artificial Neural Network.

The percentage of accuracy of each structure in the network test was calculated separately for each class in Table 8.9. The percentage of total accuracy of the structure was calculated and provided in the ninth column.

8.2.8 Support Vector Machine

SVM is a new and well-known method for classification, which includes a backup vector classifier (support vector classification [SVC]) and support vector regressior (SVR). This method, first proposed by Cortes and Vepnik, has proven to be one of the most accurate algorithms in the field of data mining due to the lack of sensitivity to the dimensions of the problem. SVM methods have a coherent and proven mathematical model and have many successful applications in several classification problems. The SVM method uses a secure margin to separate the data so that the problem becomes a quadratic programming (QP) optimization. In this section, the theory of the fuzzy SVM (FSVM) method is presented as a new method to classify data using the entropy method.

Assuming the instructional data $\{(X_1, y_1), ..., (X_i, y_i)\}$ for instructional data, each class is labeled $y_i \in \{-1, 1\}$. This weight (w) is considered separately. When the data is linearly separable, the SVM method can be separated with the maximum margin between the two classes without any error. This problem can be calculated by solving the following QP equation:

$$\begin{cases} \min \quad \|w\|^2 \\ y_i(w^T X_i + b) \geq i = 1, 2, 3, .., N. \end{cases} \tag{8.27}$$

where

w is the weight vector.
b is the bias value of the equation.

The nonlinear separators do not necessarily meet all the conditions of the equation. Using the slack variables, ξ_i, $i \in \{1, 2, ..., N\}$, the amount of constraint change is determined. The QP equation is converted as follows:

$$\begin{cases} \min & 0.5\|w\|^2 + C \sum_{i=1}^{N} \xi_i \\ y_i(w^T X_i + b) \geq 1 - \xi_i, & , i = 1, 2, 3, .., N. \\ \xi_i \geq 0, & i = 1, 2, 3, .., N. \end{cases} \tag{8.28}$$

where

c is the amount of the fine of the function.

In this case, the higher the value of c, the more acceptable the amount of the fine (maximum slack departure from the border of the margins). In many real problems, due to the lack of homogeneity and noise in the data, the use of linear methods is not possible. In order to solve such problems, X_i data are transferred to a higher dimension space, and the vector $\Phi(X_i)$ is used in the attribute space, according to the X_i attributes in the main space. In order to solve the QP problem, the relation $\Phi(X_i)$. $\Phi(X_j)$ is used. Because the shape of the $\Phi(X_i)$ function is generally unknown, the kernel function method is applied, which is defined using the multiplication $\Phi(X_i)$. $\Phi(X_j) = K(X_i, X_j)$. Using Lagrangian and the kernel method, the QP problem is defined as follows:

$$\begin{cases} \min & 0.5\sum_{i=1}^{N}\sum_{j=1}^{N} y_i y_j \alpha_i \alpha_j K(X_i, X_j) - \sum_{j=1}^{N} \alpha_i \\ \sum_{i=1}^{N} y_j \alpha_i = 0, & \\ 0 \le \alpha_i \le C, & i = 1, 2, 3, .., N. \end{cases} \tag{8.29}$$

Various kernel functions are provided in this regard in Table 8.10, including:
To solve multidimensional problems, $k(x, x') = \prod_{k=1}^{n} k(x^k, x^k)$.

8.2.9 Fuzzy Support Vector Machine (FSVM)

According to the basic theory of SVM, the FSVM method is based on the assumption that all data fall into all categories. In many real applications, data may not be clearly grouped, that is, at the decision-making level, there may not be a clear classification of the data. To solve this problem, fuzzy membership functions are used for each entry point. Therefore, the relationship becomes the following:

$$S = \{(X_i, y_i, s_i), i = 1, ..., N\} \tag{8.30}$$

In this equation, $X_i \in R^N$ are examples of learning; $y_i \in \{-1, 1\}$, represents the category; and $S_i\{ i = 1, ..., N\}$ is a fuzzy membership function that is a number between (0,1]. The relation $\sigma \le$ s_i≤ 1 is defined.

$$C^+ = \{(X_i \mid X_i \in S \text{ and } y_i = +1)\} \tag{8.31}$$

Also for the classification, the data X_i with $y_i = +1$ is denoted by C^+:

$$C^- = \{(X_i \mid X_i \in S \text{ and } y_i = -1)\} \tag{8.32}$$

Using the above equations, we can conclude $Q = C^+ \bigcup C^-$. The QP problem in this method then becomes:

$$\begin{cases} \min & 0.5\|w\|^2 + C\sum_{i=1}^{N} s_i \xi_i \\ y_i(w^T X_i + b) \ge 1 - \xi_i, & , i = 1, 2, 3, .., N. \\ \xi_i \ge 0 & , i = 1, 2, 3, .., N. \end{cases} \tag{8.33}$$

where

c is the fixed number.

Given that ξ_i represents the error and $s_i$represents the membership function of the SVM method, $s_i\xi_i$ is an error index with different weights. A lower value of s_i can reduce the effect of $s_i\xi_i$ on the whole function and reduce the importance of input X_i.

Selecting the appropriate fuzzy function to solve the problem plays an important role in obtaining the results and the efficiency of the categorizer. The membership function reduces the optimal center distance from the data. The radii of the categories $C^-$and $C^+$are calculated as follows:

$$\begin{aligned} r^+ &= \max \|X_+ - X_i\| \text{ where } X_i \in C^+ \\ r^- &= \max \|X_- - X_i\| \text{ where } X_i \in C^- \end{aligned} \tag{8.34}$$

In the introductory model, the relative fuzzy function is considered as a distance and is defined as follows:

$$s_i = \begin{cases} 1 - \dfrac{\|X_+ - X_i\|}{r^+ + \delta} & \text{if } X_i \in C^+ \\ 1 - \dfrac{\|X_- - X_i\|}{r^- + \delta} & \text{if } X_i \in C^- \end{cases} \tag{8.35}$$

where

δ is a very small number that is intended to avoid ambiguity.

In this regard, two parameters should be considered: (i) data center and (ii) membership function. Various methods have been proposed to obtain the fuzzy function. As mentioned, distance and center are two important and influential parameters in the results. The different fuzzy methods based on distance are summarized in Table 8.11.

In the classification, when $\omega = 1$, this method becomes the classical membership function.

8.2.10 Twin Support Vector Machine (TSVM)

If there are a small number of backup vectors, the classical SVM method has a high computational speed and a much lower time for classification than other methods. But if the data are irregular and the number of these data becomes non-uniform, the number of these vectors increases and, in practice, the computational speed decreases drastically. The fuzzy method can be used effectively to reduce the effect of noise. The TSVM method is a different form of the SVM method in that two non-parallel linear planes are used for decision-making and categorization. In this method, two lines are determined with the following equations to categorize the data:

$$\omega_+^T x + b_+ = 0 \text{ and } \omega_-^T x + b_- = 0$$

$$\begin{cases} \min & 0.5\|X_+\omega_+ + e_+b_+\|^2 + C_1e_-^T\xi_- \\ \text{s.t.} - (X_-\omega_+ + e_-b_+) + \xi_- \geq e_-, \xi_- \geq 0 \end{cases}$$

$$\begin{cases} \min & 0.5\|X_-\omega_- + e_-b_-\|^2 + C_2e_+^T\xi_+ \\ \text{s.t.} - (X_+\omega_- + e_+b_-) + \xi_+ \geq e_+, \xi_+ \geq 0 \end{cases} \tag{8.36}$$

where

ξ are the vectors of slack variables.

In these relations, C_2 and C_1 are positive numbers; and ξ are the vectors of slack variables for the positive and negative categories. Also in this relation, e is the row vector, and its size is exactly equal to the number of data. If based on these relationships the values of pairs (ω_+, b_+) and (ω_-, b_-) are extracted, then new data can be categorized via TSVM and using the following equations for linear and nonlinear states, respectively:

$$f(x) = \text{argmin}\left(\frac{|\omega_\pm^{*\ T}x + b_\pm^*|}{\|\omega_\pm^*\|}\right)_\mp$$

$$f(x) = \text{argmin}\left(\frac{|k(x,\ X^T)\,\omega_\pm^{*\ T}x + b_\pm^*|}{\sqrt{(\omega_\pm^{*\ T}k(X,X^T)\omega_\pm^*}}\right)_\mp \tag{8.37}$$

8.2.11 Fuzzy Twin Support Vector Machine (FTSVM)

One of the most important steps in determining the membership function is to select the category center to evaluate other data in the whole set. The choice of center is made using a simple assumption. For a positive category, the center of the category is determined using the following equation:

$$\varphi_{+C} = \frac{1}{l_+}\sum\nolimits_{j=1}^{l_+}\varphi(x_j), \text{for}\, x_j \in X_+ \tag{8.38}$$

For a negative category, the center is considered, and the data scatter rate is evaluated based on radius and distance as follows:

$$r_{+C} = \max \left\|\varphi(x_j) - \varphi_{+C}\right\| \text{ for } \varphi(x_j) \in X_+ \tag{8.39}$$

Using these parameters, the fuzzy membership rate for each data is extracted as follows:

$$s_{i+} = \begin{cases} \mu(1 - \sqrt{\dfrac{\|\varphi(x_i) - \varphi_{+C}\|^2}{(r_{+C} + \delta)}})\ \mathit{if}\ \|\varphi(x_i) - \varphi_{+C}\| \geq \left\|\varphi(x_j) - \varphi_{-C}\right\| \\ (1-\mu)\left(1 - \sqrt{\dfrac{\|\varphi(x_i) - \varphi_{+C}\|^2}{(r_{+C} + \delta)}}\right)\ \mathit{if}\ \|\varphi(x_i) - \varphi_{+C}\| < \left\|\varphi(x_j) - \varphi_{-C}\right\| \end{cases} \tag{8.40}$$

In this regard, δ is a small value intended to avoid creating an ambiguous state of zero in the denominator.

In order to solve *QP* problems, functions with more than one line can be developed in a similar way. For example, $Q = \left(H_+^T H_+ + C_1E_1\right)^{-1}H_-^T$ and $\overline{Q} = H - Q$, in which case the solution of the *QP* problem becomes:

$$\begin{cases} \min & f(\alpha) = 0.5\alpha^T\overline{Q}\alpha - e_-^T\alpha \\ \text{s.t.} & 0 \leq \alpha \leq C_3 s_- \end{cases} \tag{8.41}$$

In order to solve the above equation, the coordinate descent (CD) strategy method with distance reduction is used. The algorithm for using this method is given below, in which $\nabla_i^{\text{proj}} f(\alpha)$ are slope components that are defined as follows:

$$\begin{cases} \min(0,\ \nabla_i f(\alpha)) \ \text{ if } \ \alpha_i = 0 \\ \nabla_i f(\alpha), \quad \mathit{if}\ \ 0 < \alpha_i < C_3 s_{i-} \\ \max(0, \nabla_i f(\alpha)), \text{if } \alpha_i = C_3 s_{i-} \end{cases} \tag{8.42}$$

where

$\nabla_i f(\alpha)$ is the ith component of the gradient ∇f.

8.2.12 Entropy and Its Application FSVM

Considering information gain, entropy is a practical indicator for determining the degree of ambiguity. Initially, Shannon et al. used the entropy index as a negative logarithmic function to determine the probability of an event to occur. Other researchers have made extensive use of entropy for image processing and pattern recognition. Given that entropy is an effective way to assess category reliability, this method can be used as an indicator to measure ambiguity. In evaluating fuzzy membership functions and considering entropy based on the degree of fuzzy membership, entropy-based FSVM (EFSVM) has shown more capacity to solve unbalanced data problems. As a rule, in the categorizers, the positive samples are given more attention than the negative samples; therefore, the positive samples are given more fuzzy membership to ensure the importance of the positive samples.

In the initial version of SVM, all samples have the same weight. Using the membership function to weigh data allows for a better evaluation. Due to the existence of different types of fuzzy functions and parameters, a method is needed to evaluate or select the optimal parameters of membership functions. Evaluating the degree of ambiguity is related to entropy, and positive samples should have a higher membership rate than negative samples. In the first stage, each sample must have a belonging index related to each category. In each database, the number of positive and negative samples may not be equal, so an IR index is used to assess the data balance.

$$\mathrm{IR} = \frac{N_{\mathrm{neg}}}{N_{\mathrm{pos}}} \tag{8.43}$$

where

N_{neg} is the number of negative samples.
N_{pos} is the number of positive samples.

In information gain, entropy represents the average amount of information contained in each data. Thus, entropy represents the certainty of the source of information, in the sense that less entropy indicates clearer information. Using entropy, we can assess class confidence in instructional examples. Then, a fuzzy membership of training samples is assigned based on the degree of certainty of placement in each category. In practice, for more confident class samples, a low entropy index is assigned to larger fuzzy members to increase their share of the decision-making level. Based on the training examples, for each sample $\{x_i, y_i\}_{i=1}^{N}, y_i \in \{+1, -1\}$ for $y_i = 1$, the implication is that these data belong to the positive category. The probability of x_i data belonging to a positive or negative category, based on entropy, is defined as follows:

$$H_i = -p^i_+ \ln\left(p^i_+\right) - p^i_- \ln\left(p^i_-\right) \tag{8.44}$$

In this equation, ln is a natural logarithm operator. In this regard, the main key is to assess the probability of samples for each class. Since this evaluation can be defined on the basis of neighborhood, the kNN method using k neighborhood containing $\{x_{i1}, \ldots, x_{ik}\}_{i=1}^{N}$ is used. Assuming that the number of positive samples in the vicinity of the neighborhood is num_{-i} and the number of negative data is num_{+i}, the probability of p_{+i} and p_{-i} is calculated by dividing the positive and negative data by the total data. With two probabilities, the entropy can be easily calculated using Eq. (8.23).

Figure 8.13 shows the algorithm of EFSVM based on the entropy method and provides an example of the outputs obtained by the fuzzy method based on Shannon entropy. As can be seen, this method performs better than the classic SVM and FSVM methods, whereby the separator screen separates the data more accurately.

Various methods have been proposed to calculate entropy. These following relationships are used as indicators for quantifying information about each collection:

$$H^{\mathrm{SH}} = -\sum p \log p, \sum p = 1 \tag{8.45}$$

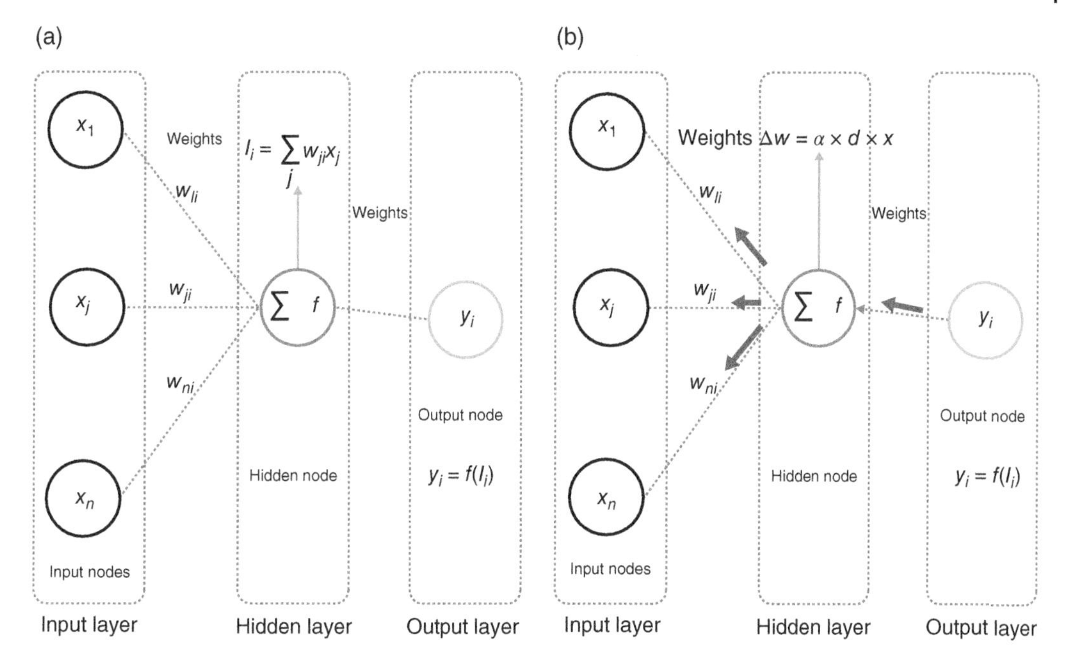

Figure 8.13 (a) Feed-forward input data ANN and (b) feed-back error propagation.

$$H^{\mathrm{RN}} = \log \sum p^{\alpha}/(1-\alpha) \tag{8.46}$$

$$H^{\mathrm{TS}} = \left(1-\sum p^{\alpha}\right)/(1-\alpha) \tag{8.47}$$

$$H^{\mathrm{PP}} = \sum pe^{1-p} \tag{8.48}$$

$$\mathrm{H}^{\mathrm{MA}} = \frac{H - e^{-(a+b)^{\beta}}}{n^{(1-\gamma)}e^{-\left(\frac{a}{n^{\alpha}}+b\right)^{\beta}} - e^{-(a+b)^{\beta}}} \tag{8.49}$$

These relations include Shannon entropy, Mamta entropy, Tsallis entropy, and Pal entropy.

8.2.13 Development of Entropy Fuzzy Coordinate Descent Support Vector Machine (EFCDSVM)

In existing methods, the x_p^{cen} and x_n^{cen} centers usually select data that is an average or averaged. However, the main point in choosing the center of the center of the category is the minimum amount of ambiguity and maximum clarity among the data. Using the entropy relation can be useful in this context. By determining the number of k neighbors and amount of entropy for the data, it is possible to determine the center of the batch with good accuracy and classify the data using the law of entropy. In this method, the separator cloud screen will be optimized and the computing time will be reduced due to the use of the Coordinate Descent Support (CDS) method. According to this method, first the number of positive and negative points in each category with radius r is counted, the entropy value (by different methods) is calculated, then the entropy vector set is constructed. The number of this set can be equal to or less than the number of data, which is introduced as $\{H_1^{\mathrm{sh}}, H_2^{\mathrm{sh}}, \ldots, H_i^{\mathrm{sh}}\}$ for Shannon entropy. Using the following relation, the optimal location to determine the center $x_p^{\mathrm{cen}}$and x_n^{cen} is determined:

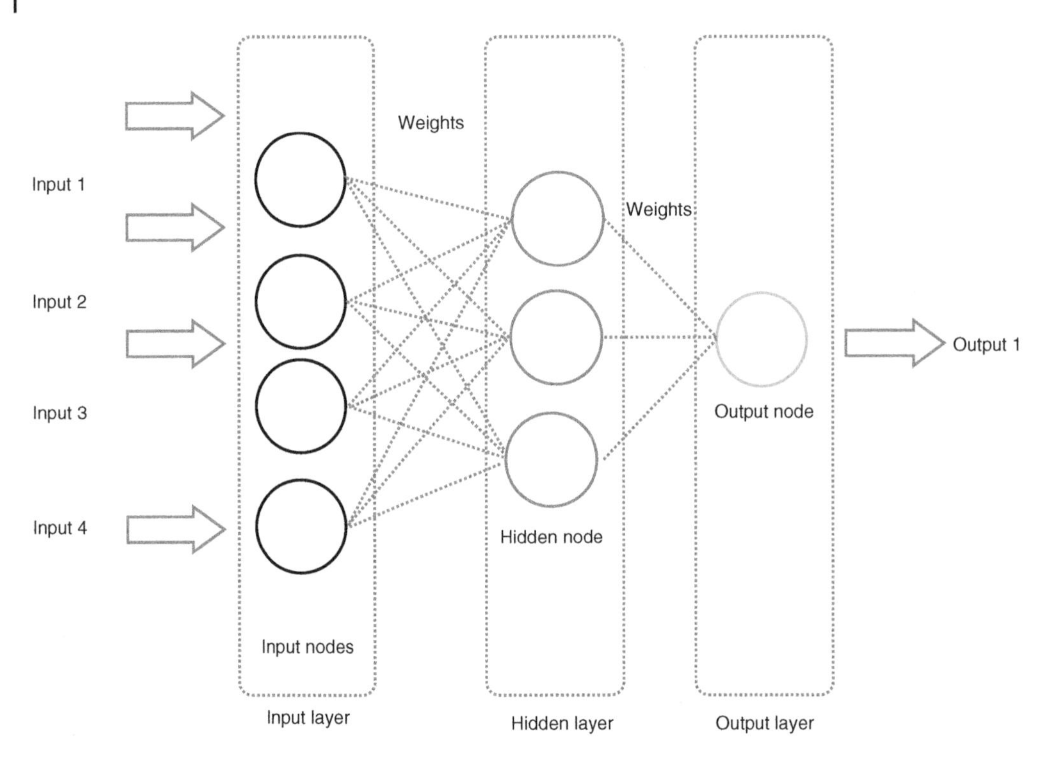

Figure 8.14 The structure of the artificial neural network used to determine the type of road surface distress.

$$\begin{cases} x_p^{\text{cen}} = \arg\min \left\{H_1^{\text{sh}}, H_2^{\text{sh}}, \ldots, H_i^{\text{sh}}\right\}_{x_p} \\ x_n^{\text{cen}} = \arg\min \left\{H_1^{\text{sh}}, H_2^{\text{sh}}, \ldots, H_i^{\text{sh}}\right\}_{x_n} \end{cases} \tag{8.50}$$

Figure 8.14 shows an example of the developed method.

8.2.14 Development of a New Support Vector Machine in Polar Frame (PSVM)

If the data are transferred in the polar domain, it is possible to reduce the computational volume and compress the extraction rules. Using different kernels, different issues can be accommodated by transferring to higher dimensions. However, by using transitions of similar dimensions in polar coordinates for a three-dimensional problem, the classification can be done more accurately than the classical relation by changing the SVM relation parameter.

In the classical method, it is assumed that the dividing line of the two categories is located at the farthest point from the margins, which is placed in the middle of the overall margin. In the proposed method, $\Delta\theta = \theta_2 - \theta_1$ is the criterion for separating the line.

The main idea of polar support vector machine (PSVM) is to construct a polar surface in three-dimensional coordinates and then solve an experimental maximization problem to determine $\Delta\theta_{\max}$. To learn from samples, the performance dependence between an input in (θ, ρ) versus an output batch is estimated, using:

$$\text{Class} = \{(\theta_i, \rho_i), \theta_i \epsilon \theta, \rho_i \epsilon M \text{ or } N \epsilon \{-1, 1\}, i = 1, 2, \ldots l\} \tag{8.51}$$

Table 8.7 An example of classification Neural network training and hidden layer selection 4 input and 1 output.

Sample	Learning coefficient	Epochs	Input-hidden-output	Goal	Performance
1	0.05	282	4,32,1	0.001	0.02388
2	0.05	77	4,33,1	0.1	0.09926
3	0.05	41	4,38,1	0.1	0.09928
4	0.001	500	4,33,1	0.001	0.0099097
5	0.001	107	4,33,1	0.001	0.02876
6	0.001	162	4,50,1	0.1	0.005731
7	0.001	634	4,60,1	0.001	0.0009901
8	0.001	338	4,90,1	0.001	0.0009986
9	0.001	184	4,120,1	0.001	0.0009827
10	0.001	265	4,120,1	0.001	0.0009105
11	0.001	296	4,30,12,1	0.0001	0.0000987
12	0.001	221	4,40,12,1	0.0001	0.00009091
13	0.001	196	4,40,16,1	0.0001	0.00009977

Table 8.8 An example of classification of training samples.

	Train					
Sample	Class 1	Class 2	Class 3	Class 4	Class 5	Class 6
1	90	90	90	792	886	434
2	102	90	78	744	954	414
3	90	90	90	722	1006	384
4	90	90	90	810	858	444
5	90	90	90	786	884	442
6	90	90	94	806	854	448
7	90	90	90	810	852	450
8	90	90	90	810	852	450
9	90	90	90	806	856	450
10	90	90	90	810	852	450
11	90	90	90	810	852	450
12	90	90	90	810	852	450
13	90	90	90	810	852	450

Table 8.9 An example of Classification of training samples.

	Test							
Sample	% Class 1	% Class 2	% Class 3	% Class 4	% Class 5	% Class 6	IRE	%Accuracy Total
1	100	100	100	98.27	95.77	95.11	72	98.19
2	86.66667	95.55	91.11	91.11	87.32	92	240	90.62
3	100	82.22	82.22	88.88	81.22	84.44	352	86.49
4	100	100	100	99.75	99.06	98.66	16	99.58
5	100	100	100	97.03	96.24	98.22	64	98.58
6	100	95.55	91.11	99.50	99.76	99.55	20	97.58
7	100	100	100	99.75	99.76	100	4	99.91
9	100	97.77	97.77	99.50	99.53	100	12	99.09

Table 8.10 Types of kernels that can be used in SVM.

Row	Kernel	Name of kernel
1	$K(x, x') = \langle x, x' \rangle$	Linear kernel
2	$K(x, x') = \exp(-\sigma\|x - x'\|)$	Radial Basis Function (RBF)
3	$K(x, x') = \text{scale}(-\sigma.\ \langle x - x' \rangle + \text{offset})^{\text{degree}}$	Polynomial kernel
4	$K(x, x') = \tanh(\text{scale}.\ \langle x, x' \rangle + \text{offset})$	Hyperbolic tangent kernel
5	$K(x, x') = \dfrac{\text{Bessel}^{n}_{(v+1)}(\sigma\|x - x'\|)}{\|x - x'\|^{-n(v+1)}}$	Bessel function of the first kind kernel
6	$K(x, x') = \exp(-\sigma\|x - x'\|)$	Laplace Radial Basis Function (RBF) kernel
7	$K(x, x') = \sum_{k=1}^{n} \exp\left(-\sigma(x^{K} - x'^{K})^{2}\right)^{d}$	ANOVA radial basis kernel
8	$K(x, x') = 1 + xx' \min(x, x') - \dfrac{x - x'}{2}\left(\min(x, x')^{2} + \dfrac{(\min(x, x')^{2}}{3}\right.$	Linear splines kernel in one dimension

Table 8.11 Types of fuzzy methods based on distance in fuzzy support vector machine (FSVM).

Num	Equation	Fuzzifier
1	$\mu = (1 - a) * \beta,\ a = \left(\dfrac{r}{r_{\max}}\right)$	Classic
2	$\mu_{\lambda}(a) = \left(\left(\dfrac{1 - a}{1 + \lambda a}\right)\right), \lambda \in (-1, \infty)$	Sugeno class
3	$\mu_{\omega}(a) = (1 - a^{\omega})^{\frac{1}{\omega}}, \omega \in (0, \infty)$	Yeager
4	$\mu(a) = \dfrac{1}{2^{m}}(1 + \cos(\pi a))$	Continuous

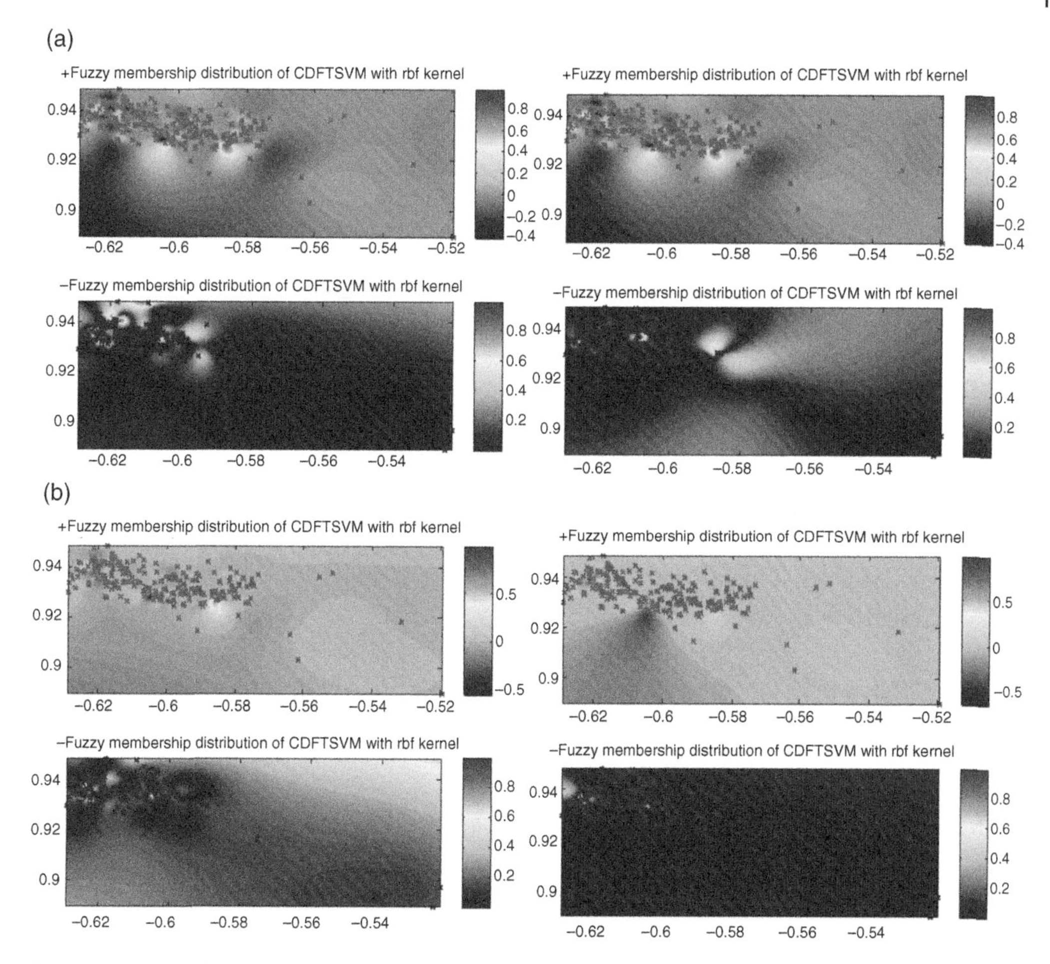

Figure 8.15 The CDS implementation algorithm for use in QP solution; classification outputs using linear method and RBF for (a) example 1 and (b) example 2.

In the above equation, (θ_i, ρ_i) has a value for each angle and radius, which is used to determine the category. On the other hand, the value of each angle/radius, which is equivalent to the value of RT, is determined by (θ_i, ρ_i), where $i = (0.360)$. The solution of the new equation is performed using the kernel function, whereby arg $\max(\Delta\theta)$decreases as the error decreases.

To solve this problem, a preliminary rule of "arc length" is used. Using this definition, the cross-sectional width of the polar coordinates $l_0 = \Delta\theta.\ \rho_0 l_1 = \Delta\theta.\ \rho_1$, which must be widened to the maximum value. The Figure 8.15 shows an example of the proposed classification based on the length of the arc.

After selecting $\max : \Delta\theta_{\text{opt}}$, which represents the acceptable separation, the middle of the segment is selected using the simple relation $\Delta\theta_{\text{opt}}/2, \rho$ and added to the initial angle. The main page cloud is extracted to separate the data and measure the polar margin. One of the main objectives of this method is to reduce the degree of the separating cloud from the nonlinear state in Cartesian coordinates to the linear state in polar coordinates.

In the PSVM method, the goal is to minimize the error function and maximize the opening value of the following function:

$$\min : \left(\frac{1}{2}\|l_1\|^2 + C\sum_{i=1}^{l} \Delta\theta\varepsilon_i\right)$$
$$\left(y_i\left[f\left((\Delta\theta)_{\text{opt}}/2, \rho\right) \geq 1 - (\Delta\theta\varepsilon)_i\right), \forall\_i = 1, 2, 3, \dots, l \tag{8.52}$$
$$\varepsilon_i \geq 0, \forall_i = 1, 2, 3, \dots, l$$

In this equation, the training set $\{(\Delta\theta_1, \rho_1), \dots(\Delta\theta_i, \rho_i)\}$, where each input $(\Delta\theta_i, \rho_i) \in R^{N-1}$, the inputs are labeled with a value of $y_i \in \{-1, 1\}$and $\Delta\theta_i$, ρ_i are separated using a linear classifier in polar coordinates. In this relation, C is a value to be adjusted, and $\Delta\theta$ is the value of the slack variable used to control the error. The modified formulation of this equation in polar coordinates is as follows:

$$\max : \left(\sum_{i=1}^{N} \alpha_i - 0.5\sum_{i=1}^{N}\sum_{j=1}^{N} \alpha_i\alpha_j y_i y_j [\text{polar } K(\Delta\theta_i, \rho_i)]\right)$$
$$\text{subject to} : \sum_{i=1}^{N} \alpha_i y_i = 0 \text{ and } 0 \leq \alpha_i \leq C$$
$$i = 1, 2, 3, \dots, l \tag{8.53}$$

Due to the linear nature of the separation plane, in polar coordinates, the linear kernel function is used as follows:

$$K(x_i, x_j) = \left(\varphi(\Delta\theta_i, \rho_i).\varphi\left(\Delta\theta_j, \rho_j\right)\right) \tag{8.54}$$

As a result, the final function $f(x)$ is defined as:

$$f(x) = \text{sgn}\left((w.\Delta\theta) + r_0\right) = \text{sgn}\left(\left[\sum_{i=1}^{N} \alpha_i^* y_i K(\Delta\theta_i, \rho_i) + r^*\right]\right) \tag{8.55}$$

Figure 8.16 shows an example of a classifier in polar coordinates.

8.2.15 Case Study: Pavement Crack Classification Based on PSVM

In order to classify the failure, the adaptive membership and threshold function have been used to extract and classify the patterns using the support vector theory in polar space. As mentioned in the previous section, if the data is transferred to a polar environment and this kernel is categorized then reduced later, the polarized patterns operate with considerable accuracy and speed. Instead of regression and lines, separators in higher space are used for this purpose. In this section, new method combining PSVM and DT, using the SVM classifier, for the types of faults transmitted in the polar domain is presented.

As shown in Figure 8.17, the main aim of using this method is to find a vertical separator plate to separate the nature of different data in the polar domain. The difference between this method and the SVM method is that the former uses arg max($\Delta\theta$) to maximize and reduce classification error.

In Figure 8.18, an initial classification is made based on the type of distress. As it turns out, each distress is in a regular range. In order to simplify the search phase, the polar domain connects the two ends of the membership functions, which allows the longitudinal distortion to be categorized between 0° and 180°. Also, for 45° and 135° diagonal cracks, using one plate puts both of them. The Table 8.12 shows examples of data extracted by the polar method for classification with PSVM. These data were extracted from the center of the threshold data surface and labeled to determine the separator screen cloud. In classification, two steps are performed: first, the longitudinal and

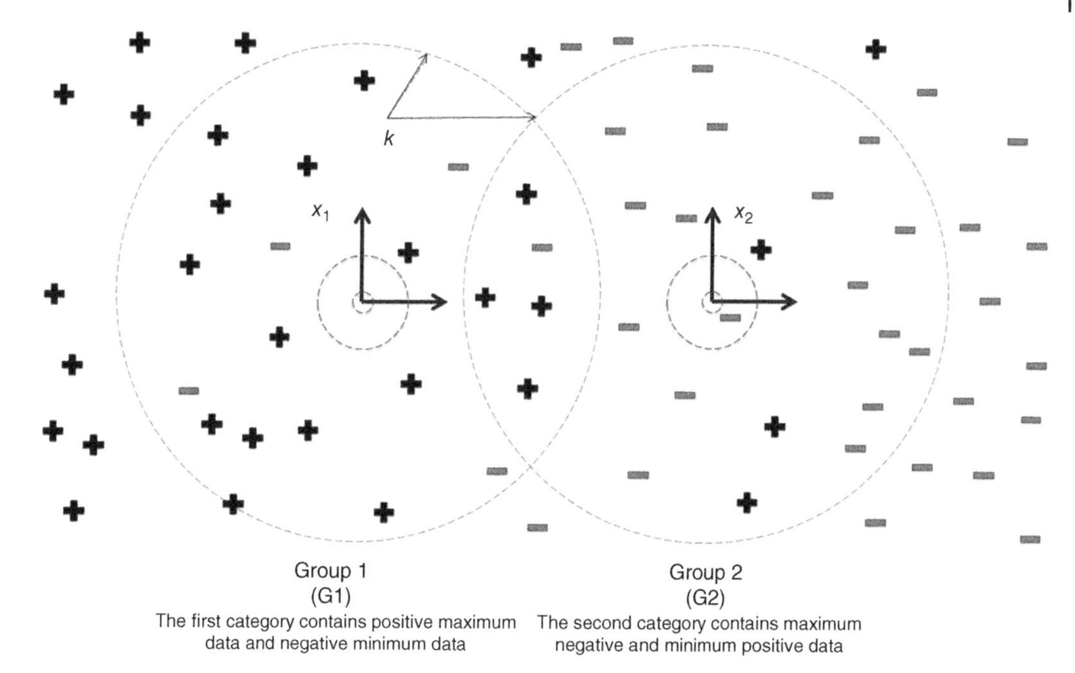

Figure 8.16 Method for determining the probability of each category based on the neighborhood k of the circle range for both positive and negative categories.

transverse cracks are distinguished using the data obtained from the database, their aggregation, and main range; and second the complex distress like alligator crack and block crack distinguished. It should be noted that some of the data related to distress, in the form of slack, is in another range. According to the desired range for each category, five angles are separated, and a range is specified for each. The purpose of this range is to de-bit existing bows. This operation is performed based on the slope of the fitted line and the separating boundary, whereby the middle of the arc Δ θ is the segment. Basic principles of PSVM classifier shows in Figure 8.19.

$$\Delta\theta_i = \begin{cases} \Delta\theta_1 = m_{12} - m_{22} \\ \Delta\theta_2 = m_{22} - m_{21} \\ \Delta\theta_3 = m_{21} - m_{11} \\ \Delta\theta_4 = m_{12} \\ \Delta\theta_5 = \pi - m_{11} \end{cases} \tag{8.56}$$

Using this method, the decision tree based on the support vector is created in the polar environment, which can be used as a knowledge base.

Using the proposed PSVM, seven main rules were extracted to classify the types of pavement cracks. Knowledge base rules, based on new image bank data and classifications, led to the production of rules in the modified polar environment. These rules are similar to the rules extracted by Zhu to classify crack failure. The difference between this method and the Zhou method is the use of the polar classification method, instead of the Cartesian space, and the reduction of separating cloud plates for separation. In the proposed method, due to the creation of a polar environment, there is no need to modify the page and change the radon phase (Figures 8.20 and 8.21).

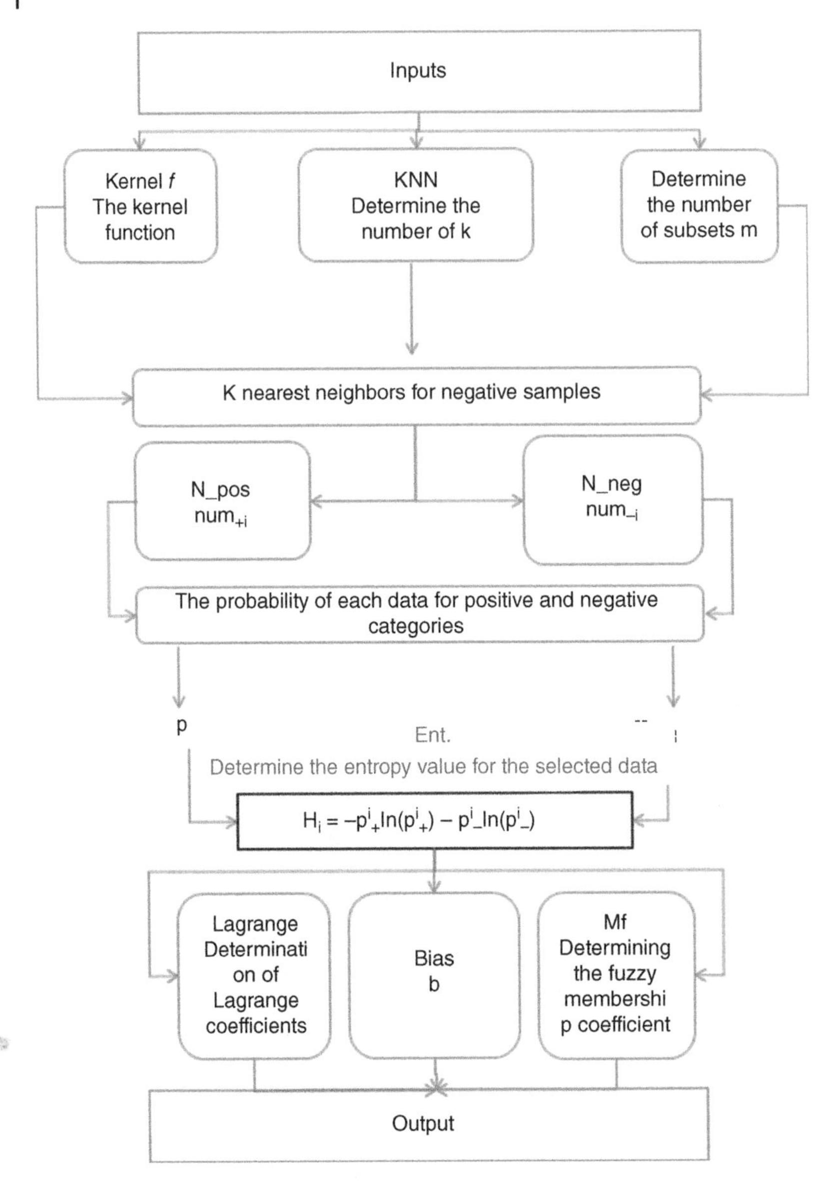

Figure 8.17 Entropy-based EFSVM classification algorithm.

Using this classification, the slope of the separating lines is extracted using the linear method, and the slope difference of the lines is used to calculate the arc length. Figure 8.22 shows the selected ranges using the proposed method for the db1 database containing 400 images and 260 cracks. It is not possible to define a boundary and separate these boundaries using a precise linear plane, and for faults that have different intensities and magnitudes, these boundaries are merged into each other in a binary manner. Based on the principles of automation, by using a method that can create a flexible boundary for accurate evaluation and high-performance classification, the accuracy and efficiency of this method in the classifier can be increased. Sample results of a combination of linear crack classifier and SVM presented in Table 8.13 and Classifier function in separating single linear cracks presented in Table 8.14. Figure 8.23 determines the separating plate and an example of a complex classifier of decision tree and PSVM is presented in Figure 8.24. An example of knowledge base and rules of a classifier complex of decision tree and PSVM is illustrated in Figure 8.25.

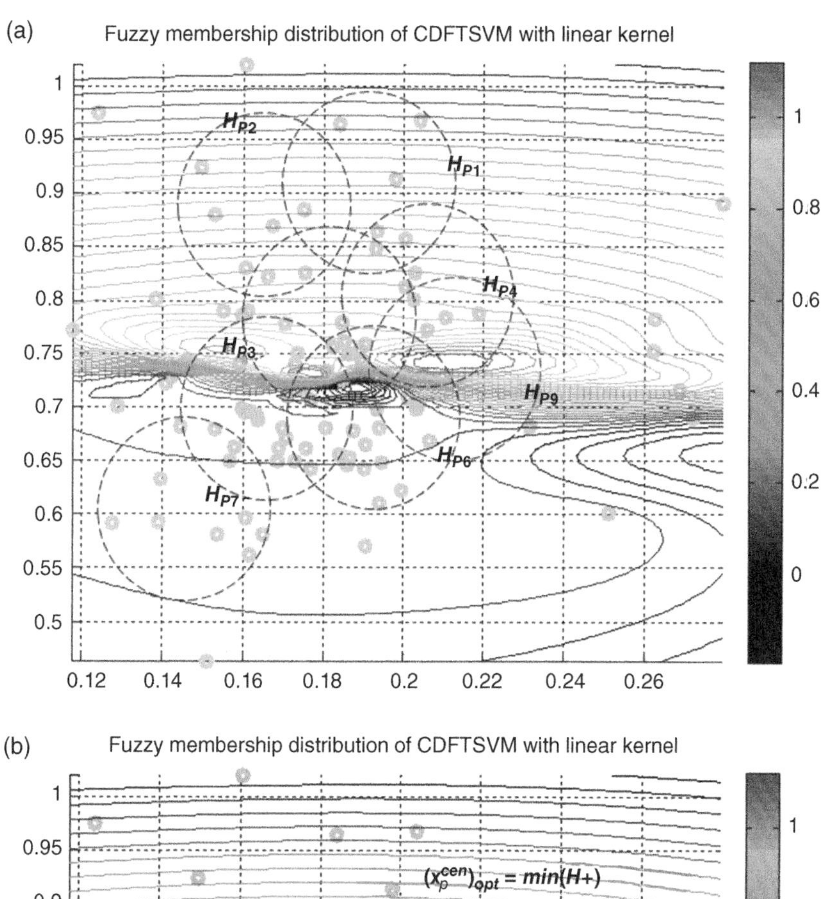

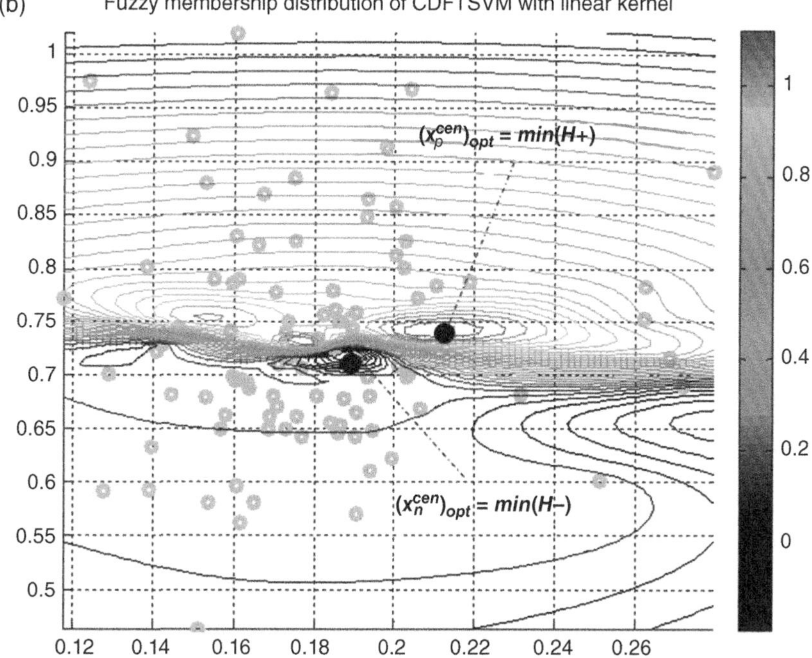

Figure 8.18 An example of the outputs of EFCDSVM based on entropy theory and coordinate descent fuzzy twin support vector machine (CDTSVM) for determining: (a) the entropy of different positions and (b) the exact location of x_p^{cen} and x_n^{cen}.

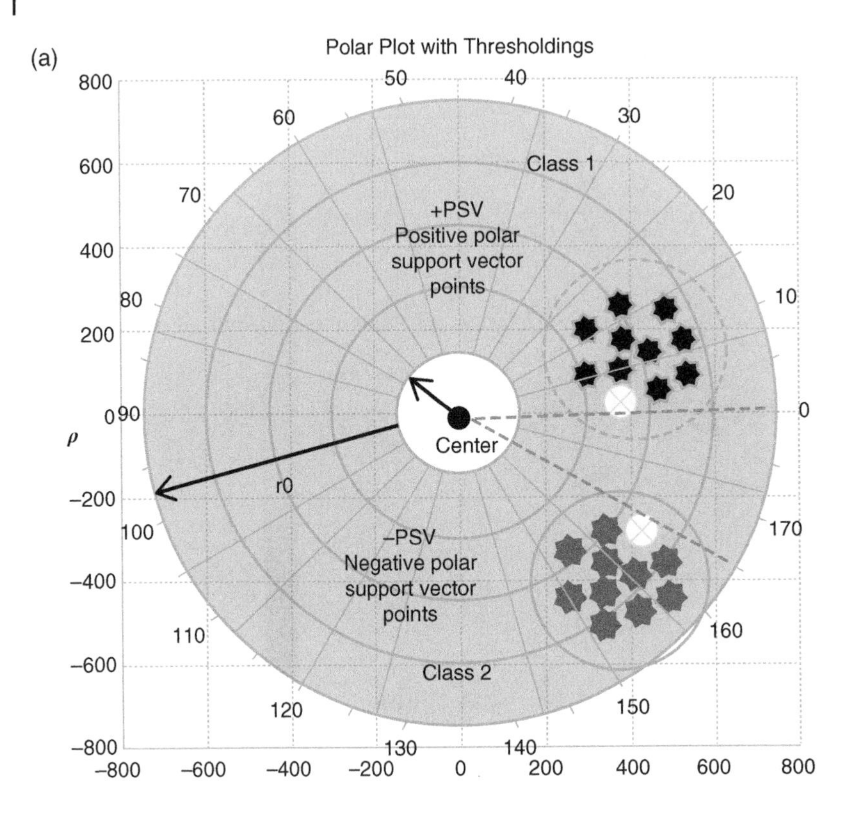

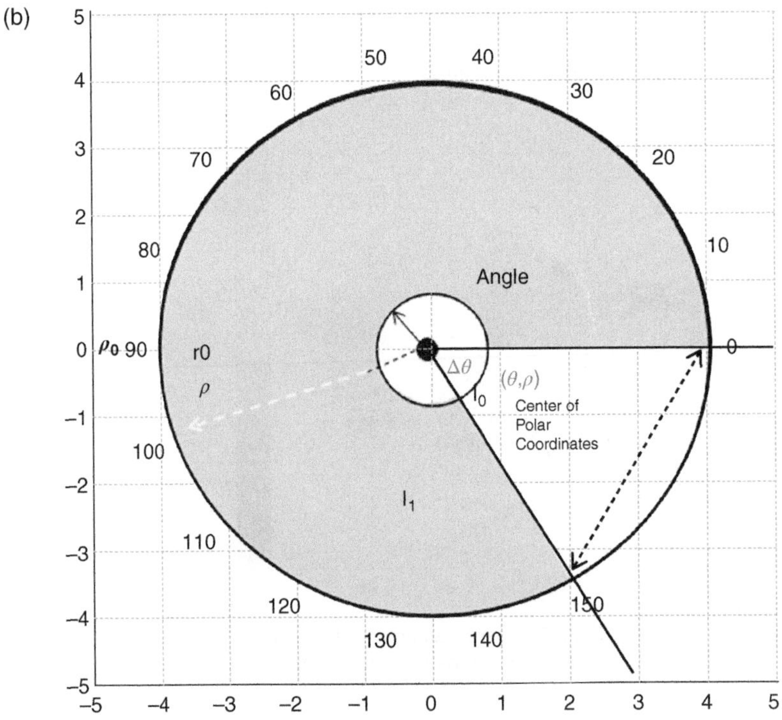

Figure 8.19 Basic principles of PSVM classifier: (a) Classification using maximizing arc length and location of slings, and (b) main parameters of PSVM.

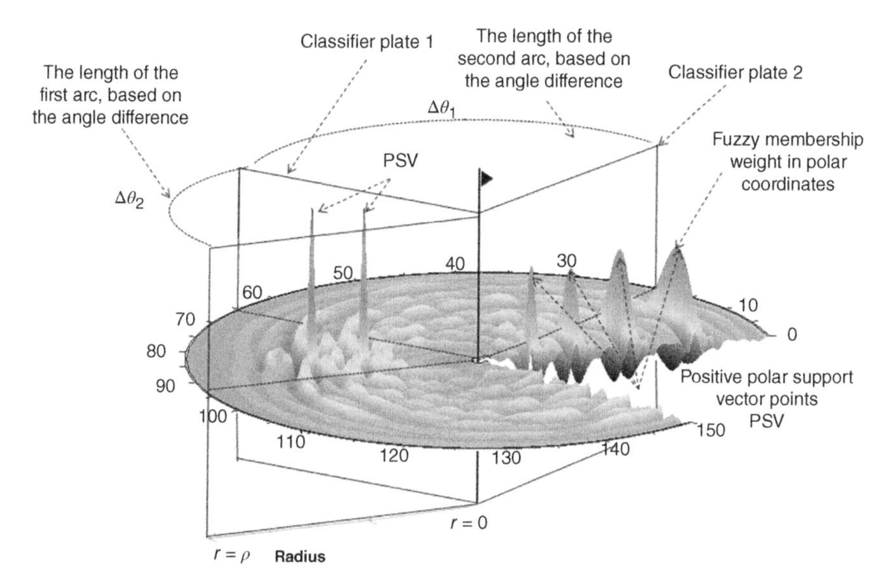

Figure 8.20 An example of classifier margins and polar support vectors, using arc length maximization in determining the separation plate.

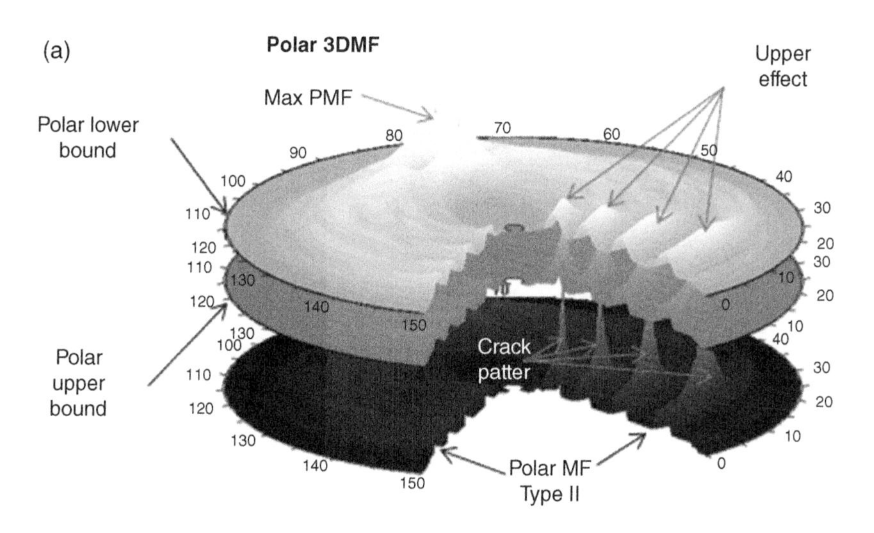

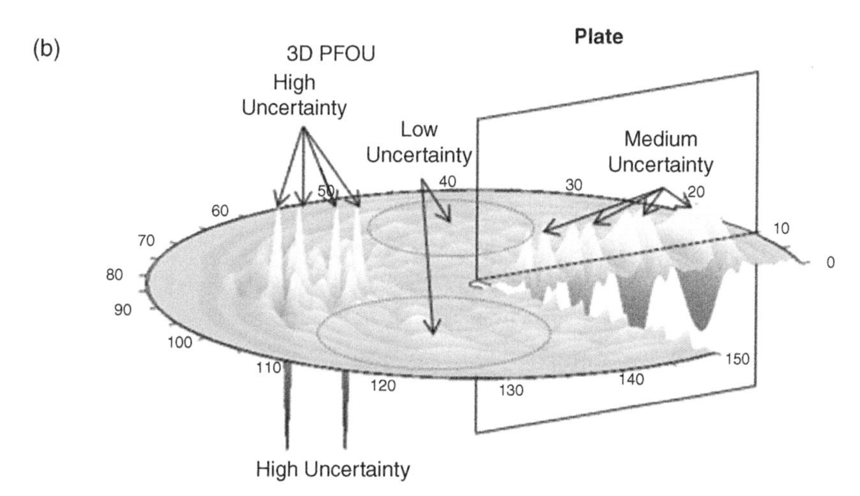

Figure 8.21 Functional structures of the proposed PSVM separator using: (a) the G3DT2FLs and (b) the 3DFOU rule and fuzzy boundary functions for classification.

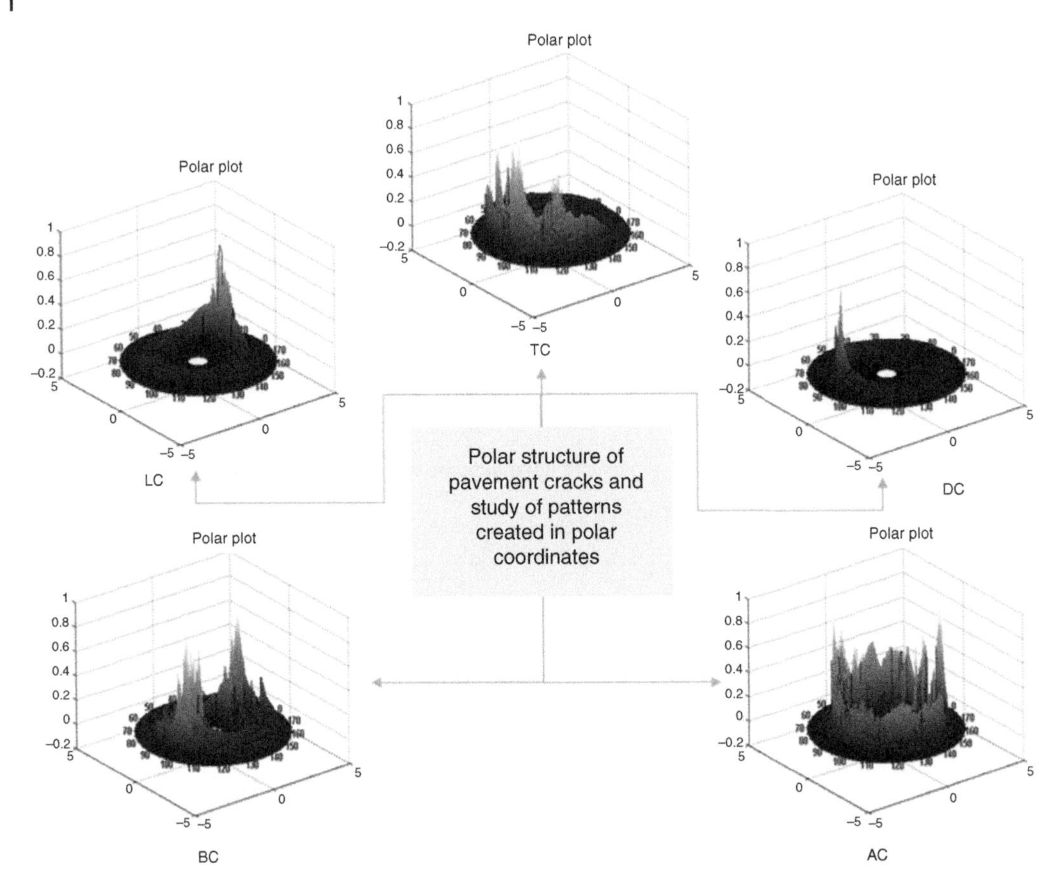

Figure 8.22 An example of cumulative patterns of crack types in asphalt pavement based on database by type of distress using SVM in the polar domain.

Table 8.12 Sample data extracted from categories with PSVM.

		N	*A*	*A*(*d*)	Mean	*θ*	*x*	*O*	*ρ*
Main parameters									
1	Max	1.00	2.00	1.60	0.80	11.50	297.50	−45.00	308.00
2	Mean	1.00	2.00	1.60	0.70	11.50	297.50	−45.00	308.00
3	Min	1.00	2.00	1.60	0.70	11.50	297.50	−45.00	308.00
4	Max	2.00	329.00	20.50	1.00	176.50	449.30	−76.10	459.00
5	Mean	2.00	250.00	17.60	0.30	91.60	303.50	−81.60	314.00
6	Min	2.00	171.00	14.80	0.00	6.70	157.70	−87.10	168.00
7	Max	1.00	13.00	4.10	0.60	15.60	152.20	−85.50	162.00
8	Mean	1.00	13.00	4.10	0.40	15.60	152.20	−85.50	162.00
9	Min	1.00	13.00	4.10	0.10	15.60	152.20	−85.50	162.00
10	Max	1.00	4.00	2.30	1.00	16.00	212.80	−50.30	223.00
11	Mean	1.00	4.00	2.30	0.90	16.00	212.80	−50.30	223.00
12	Min	1.00	4.00	2.30	0.80	16.00	212.80	−50.30	223.00

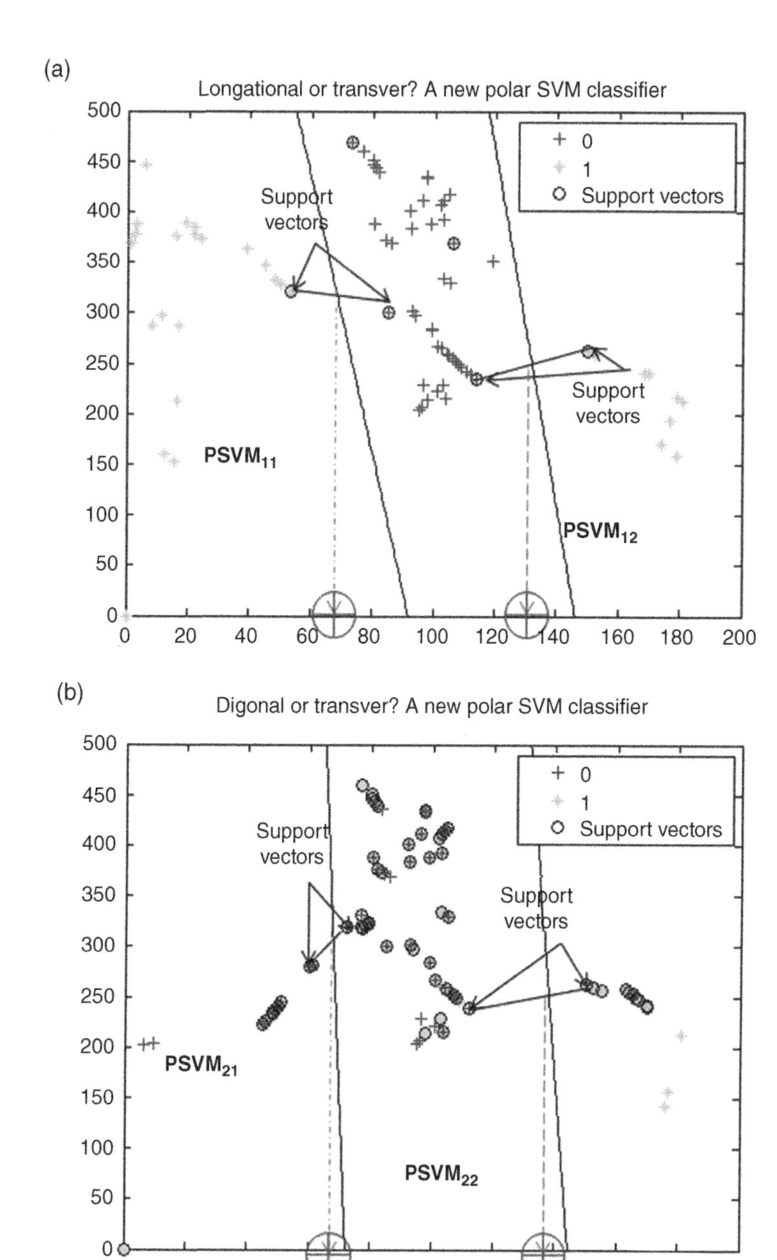

Figure 8.23 Determining the separating plate (a) Longitudinal and transverse cracks, (b) Diagonal and transverse cracks.

Table 8.13 Sample results of a combination of linear crack classifier and SVM.

		Psvm11	Psvm21	Psvm21	Psvm22
1	Number of observations	128.00	128.00	128.00	128.00
2	Correct rate	0.98	1.00	1.00	1.00
3	Error rat	0.02	0.00	0.00	0.00
4	Last correct rate	0.98	1.00	1.00	1.00
5	Last error rate	0.02	0.00	0.00	0.00
6	Specificity	0.97	1.00	1.00	1.00
7	Positive predictive value	0.97	1.00	1.00	1.00
8	Prevalence	0.49	0.49	0.49	128.00

Table 8.14 Classifier 1 function in separating single linear cracks.

		PSVM1		PSVM2	
1	Bias	−0.560	−0.030	−0.020	0.050
2	Performances	1.000	1.000	0.830	0.950

8.3 Summary and Conclusion

This chapter was devoted to a variety of classifiers, including those classified as: (i) frequency table, (ii) covariance matrix, (iii) similarity functions, and (iv) others. In addition, each classifier is explained in detail, and the relationships of the following working methods are presented: Naive Bayes Classification, Decision Trees, Logistic Regression, kNN, Artificial Neural Network, SVM, FSVM, TSVM, FTSVM, and Entropy. In particularly, the development of the entropy decision support vector model (EFCDSVM) is one of the most important algorithms presented in this chapter. Common methods used in infrastructure classification are briefly introduced, including SVM, fuzzy, entropy, and polar domain methods and combinations with decision tree methods.

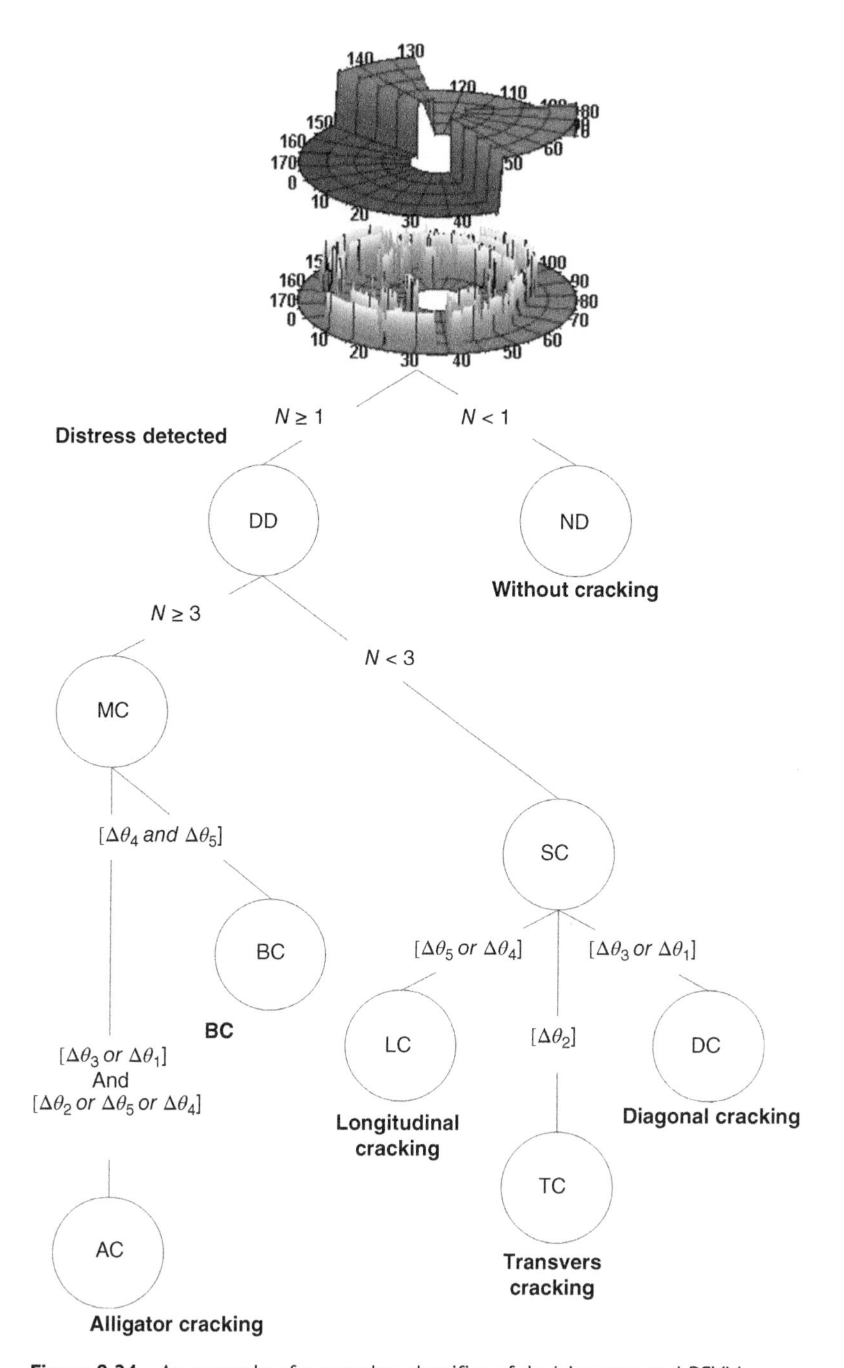

Figure 8.24 An example of a complex classifier of decision tree and PSVM.

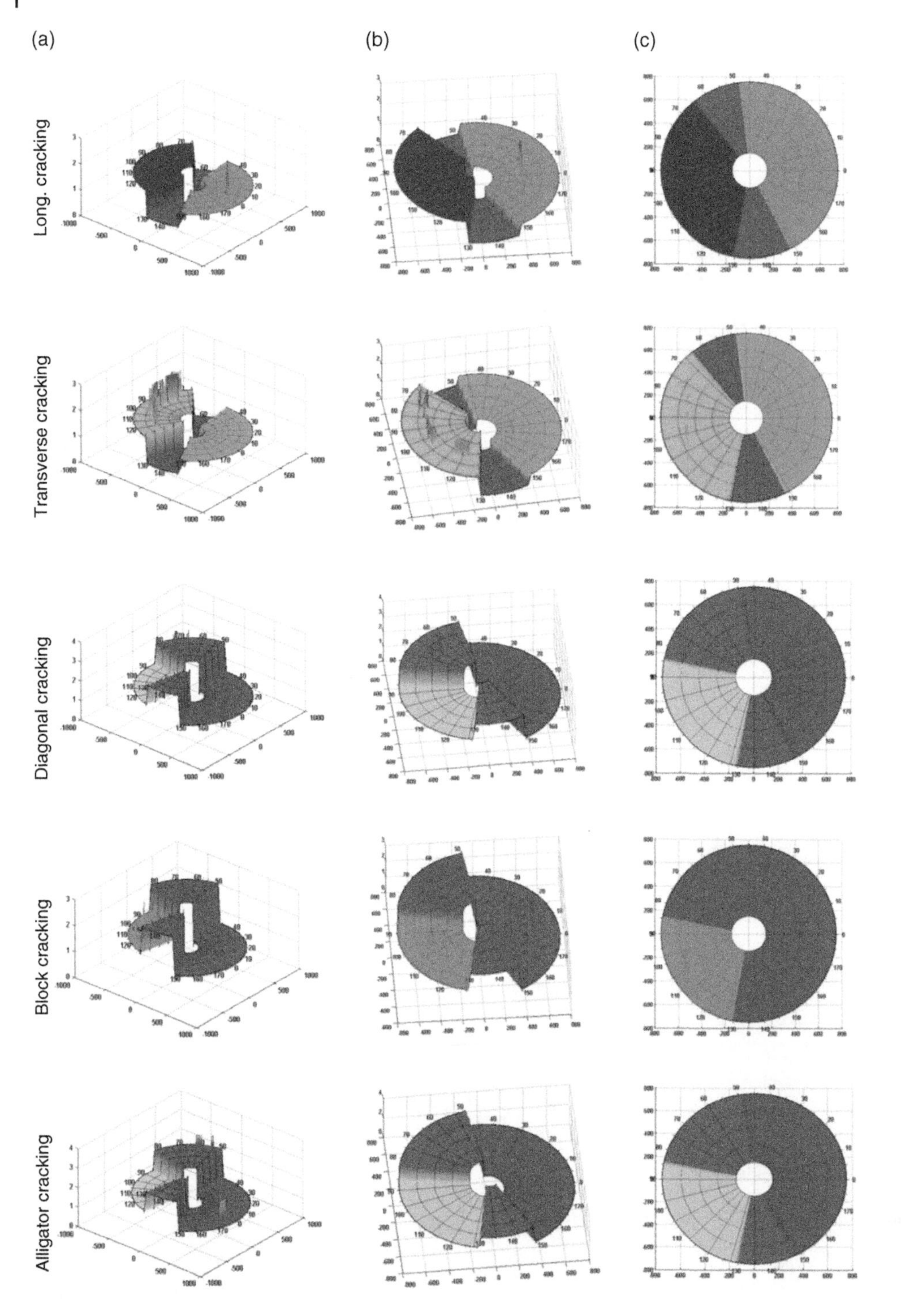

Figure 8.25 An example of knowledge base and rules of a classifier complex of decision tree and PSVM.

8.4 Questions and Exercises

1 Categorize classifiers in terms of speed and computational complexity, and compare and report the efficiency of methods for a database of 1000. Calculate the types of performance indicators presented in Chapter 9 for each classifier.

2 Analyze and categorize the data presented in the table below using the Naive Bayes classification method. Calculate the regression matrix based on the Khay index presented in Chapter 9. Compare the results with the outputs of the SVM method.

An example of classification based of 7 input and 3 class

	Feature 1	Feature 2	Feature 3	Feature 4	Feature 5	Feature 6	Feature 7	OUTPUT
1	H	M	H	M	H	L	H	L
2	M	M	M	L	M	L	M	M
3	L	M	M	L	H	L	H	H
4	M	L	L	L	L	L	L	L
5	M	L	M	H	L	H	HL	H
6	M	H	M	H	L	H	L	L
7	L	L	M	H	M	H	M	M
8	H	L	H	H	M	H	M	M
9	H	M	H	L	M	L	M	H
10	H	M	H	L	L	L	L	L

L = Low, M = Medium, H = High.

3 Define the decision trees, and compare entropy indices for feature selection and layout of nodes based on information gain in different methods for an infrastructure management database.

4 Determine the optimal neural network structure for a database with 2000 data, and use a variety of functions to design the structure. Using confusion matrix indices, examine the types of indices and compare them with SVM outputs.

5 Use the Twin Support Vector Machine (TSVM) for a 4-class database with 1000 data. Compare the results with those of the simple SVM method. Report the speed of classification and execution of each algorithm.

6 Design a new entropy index and its application in FSVM composed of other entropies, and use this index in the SVM category for a database with more than 1000 data.

Further Reading

1 Abualigah, L., A. Diabat, P. Sumari, and A. H. Gandomi, *Applications, deployments, and integration of internet of drones (IoD): A review. IEEE Sensors Journal*, 2021. **21**(22), 15.

2 Adarkwa, O.A. and N. Attoh-Okine, *Pavement crack classification based on tensor factorization. Construction and Building Materials*, 2013. **48**: p. 853–857.

3 Akhani, M., A. R. Kashani, M. Mousavi, and A. H. Gandomi, *A hybrid computational intelligence approach to predict spectral acceleration. Measurement*, 2019. **138**: p. 578–589.

4 Alavi, A., A. Gandomi, M. Gandomi, and S. Sadat Hosseini, *Prediction of maximum dry density and optimum moisture content of stabilised soil using RBF neural networks. The IES Journal Part A: Civil & Structural Engineering*, 2009. **2**(2): p. 98–106.
5 Alavi, A.H., M. Ameri, A. H. Gandomi, and M. R. Mirzahosseini, *Formulation of flow number of asphalt mixes using a hybrid computational method. Construction and Building Materials*, 2011. **25**(3): p. 1338–1355.
6 Alavi, A.H. and A.H. Gandomi, *Prediction of principal ground-motion parameters using a hybrid method coupling artificial neural networks and simulated annealing. Computers & Structures*, 2011. **89**(23–24): p. 2176–2194.
7 Alavi, A.H. and A.H. Gandomi, *A robust data mining approach for formulation of geotechnical engineering systems. Engineering Computations*, 2011. **28**(3): pp. 242–274, https://doi.org/10.1108/02644401111118132.
8 Alavi, A.H., A. H. Gandomi, A. Mollahasani, and A. Rashed, *Nonlinear modeling of soil cohesion intercept using generalized regression neural network*, "Computational Civil Engineering 2010", International Symposium Iasi, Romania, 2010.
9 Aminian, P., M. R. Javid, A. Asghari, A. H. Gandomi, and M. A. Esmaeili, *A robust predictive model for base shear of steel frame structures using a hybrid genetic programming and simulated annealing method. Neural Computing and Applications*, 2011. **20**(8): p. 1321–1332.
10 Arianasab, E., M. Maadani, and A. Gandomi. *A neural-network based gender detection algorithm on full-face photograph.* in *2015 2nd International Conference on Knowledge-Based Engineering and Innovation (KBEI)*. 2015. (pp. 892–896). IEEE.
11 Asteris, P.G., S. Nozhati, M. Nikoo, L. Cavaleri, and M. Nikoo, *Krill herd algorithm-based neural network in structural seismic reliability evaluation. Mechanics of Advanced Materials and Structures*, 2019. **26**(13): p. 1146–1153.
12 Ayyadevara, V.K., *Gradient boosting machine.* in *Pro Machine Learning Algorithms*. 2018. Springer. p. 117–134.
13 Bagheri, M., T. Borhani, A. Gandomi, and Z. Manan, *A simple modelling approach for prediction of standard state real gas entropy of pure materials. SAR and QSAR in Environmental Research*, 2014. **25**(9): p. 695–710.
14 Baliarsingh, S.K., S. Vipsita, A. H. Gandomi, A. Panda, S. Bakshi, and S. Ramasubbareddy, *Analysis of high-dimensional genomic data using MapReduce based probabilistic neural network. Computer Methods and Programs in Biomedicine*, 2020. **195**: p. 105625.
15 Bardhan, A., P. Samui, K. Ghosh, A. H. Gandomi, and S. Bhattacharyya, *ELM-based adaptive neuro swarm intelligence techniques for predicting the California bearing ratio of soils in soaked conditions. Applied Soft Computing*, 2021. **110**: p. 107595.
16 Bentéjac, C., A. Csörgő, and G. Martínez-Muñoz, *A comparative analysis of gradient boosting algorithms. Artificial Intelligence Review*, 2021. **54**(3): p. 1937–1967.
17 Chen, T., T. He, M. Benesty, V. Khotilovich, Y. Tang, and H. Cho, *Xgboost: Extreme gradient boosting.* R package version 0.4-2, 2015. **1**(4): p. 1–4.
18 Doborjeh, M., N. Kasabov, Z. Doborjeh, R. Enayatollahi, E. Tu, and A. H. Gandomi, *Personalised modelling with spiking neural networks integrating temporal and static information. Neural Networks*, 2019. **119**: p. 162–177.
19 Fakhri, S.A., M. Saadatseresht, M. Varshosaz, and H. Zakeri, *Evaluation of UAV photogrammetric capability in road pavement cracks detection. Amirkabir Journal of Civil Engineering*, 2021. 1–3.
20 Ferreira, F.G.D.C., A.H. Gandomi, and R.T.N. Cardoso. *Financial time-series analysis of Brazilian stock market using machine learning.* in *2020 IEEE Symposium Series on Computational Intelligence (SSCI)*. 2020. (pp. 2853–2860) IEEE.

21 Friedman, J.H., *Greedy function approximation: A gradient boosting machine. Annals of Statistics*, 2001. **29**(5): p. 1189–1232.

22 Gandomi, A.H. and A.H. Alavi. *Applications of computational intelligence in behavior simulation of concrete materials.* in *Computational Optimization and Applications in Engineering and Industry.* 2011. Springer. p. 221–243.

23 Gandomi, A.H. and A.H. Alavi, *A new multi-gene genetic programming approach to non-linear system modeling. Part II: Geotechnical and earthquake engineering problems. Neural Computing and Applications*, 2012. **21**(1): p. 189–201.

24 Gandomi, A.H., A. H. Alavi, M. R. Mirzahosseini, and F. M. Nejad, *Nonlinear genetic-based models for prediction of flow number of asphalt mixtures. Journal of Materials in Civil Engineering*, 2011. **23**(3): p. 248–263.

25 Gandomi, A.H. and D.A. Roke, *Assessment of artificial neural network and genetic programming as predictive tools. Advances in Engineering Software*, 2015. **88**: p. 63–72.

26 Gandomi, A.H., X.-S. Yang, S. Talatahari, and A. H. Alavi, *Metaheuristic Applications in Structures and Infrastructures*. 2013. Newnes. Elsevier.

27 Gandomi, A.H., G.J. Yun, and A.H. Alavi, *An evolutionary approach for modeling of shear strength of RC deep beams. Materials and Structures*, 2013. **46**(12): p. 2109–2119.

28 Gandomi, M., M. D. Pirooz, I. Varjavand, and M. R. Nikoo, *Application of multilayer perceptron neural network and support vector machine for modeling the hydrodynamic behavior of permeable breakwaters with porous core. Journal of Marine Engineering*, 2019. **15**(29): p. 167–179.

29 Gandomi, M., M. Soltanpour, M. R. Zolfaghari, and A. H. Gandomi, *Prediction of peak ground acceleration of Iran's tectonic regions using a hybrid soft computing technique. Geoscience Frontiers*, 2016. **7**(1): p. 75–82.

30 Govindarajan, P., R. K. Soundarapandian, A. H. Gandomi, R. Patan, P. Jayaraman, and R. Manikandan, *Classification of stroke disease using machine learning algorithms. Neural Computing and Applications*, 2020. **32**(3): p. 817–828.

31 Harandizadeh, H., D. J. Armaghani, P. G. Asteris, and A. H. Gandomi, *TBM performance prediction developing a hybrid ANFIS-PNN predictive model optimized by imperialism competitive algorithm. Neural Computing and Applications*, 2021. **33**(23): p. 16149–16179.

32 Hastie, T., S. Rosset, J. Zhu, and H. Zou, *Multi-class adaboost. Statistics and its Interface*, 2009. **2**(3): p. 349–360.

33 Heshmati, R.A.A., A. H. Alavi, M. Keramati, and A. H. Gandomi, *A radial basis function neural network approach for compressive strength prediction of stabilized soil.* in Road Pavement Material Characterization and Rehabilitation: Selected Papers from the 2009 GeoHunan International Conference. 2009.

34 Hossein Alavi, A., A. Hossein Gandomi, A. Mollahassani, A. Akbar Heshmati, and A. Rashed, *Modeling of maximum dry density and optimum moisture content of stabilized soil using artificial neural networks. Journal of Plant Nutrition and Soil Science*, 2010. **173**(3): p. 368–379.

35 Iyer, S., T. Velmurugan, A. Gandomi, V. N. Mohammed, K. Saravanan, and S. Nandakumar, *Structural health monitoring of railway tracks using IoT-based multi-robot system. Neural Computing and Applications*, 2021. **33**(11): p. 5897–5915.

36 Jayabarathi, T., T. Raghunathan, and A. Gandomi, *The bat algorithm, variants and some practical engineering applications: A review. Nature-Inspired Algorithms and Applied Optimization.* Springer International Publishing AG. 2018: p. 313–330.

37 Jothiramalingam, R., A. Jude, R. Patan, M. Ramachandran, J. H. Duraisamy, and A. H. Gandomi, *Machine learning-based left ventricular hypertrophy detection using multi-lead ECG signal. Neural Computing and Applications*, 2021. **33**(9): p. 4445–4455.

38 Kadiyala, A. and A. Kumar, *Applications of python to evaluate the performance of decision tree-based boosting algorithms. Environmental Progress & Sustainable Energy*, 2018. **37**(2): p. 618–623.

39 Kashani, A.R., M. Akhani, C. V. Camp, and A. H. Gandomi, *A neural network to predict spectral acceleration.* in *Basics of Computational Geophysics*. 2021. Elsevier. p. 335–349.

40 Kasinathan, G., S. Jayakumar, A. H. Gandomi, M. Ramachandran, S. J. Fong, and R. Patan, *Automated 3-D lung tumor detection and classification by an active contour model and CNN classifier. Expert Systems with Applications*, 2019. **134**: p. 112–119.

41 Kennedy, M.J., A.H. Gandomi, and C.M. Miller, *Coagulation modeling using artificial neural networks to predict both turbidity and DOM-PARAFAC component removal. Journal of Environmental Chemical Engineering*, 2015. **3**(4): p. 2829–2838.

42 Khari, M., A. K. Garg, A. H. Gandomi, R. Gupta, R. Patan, and B. Balusamy, *Securing data in internet of things (IoT) using cryptography and steganography techniques. IEEE Transactions on Systems, Man, and Cybernetics: Systems*, 2019. **50**(1): p. 73–80.

43 Kousik, N., Y. Natarajan, R. A. Raja, S. Kallam, R. Patan, and A. H. Gandomi, *Improved salient object detection using hybrid convolution recurrent neural network. Expert Systems with Applications*, 2021. **166**: p. 114064.

44 Kumar, A., M. Ramachandran, A. H. Gandomi, R. Patan, S. Lukasik, and R. K. Soundarapandian, *A deep neural network based classifier for brain tumor diagnosis. Applied Soft Computing*, 2019. **82**: p. 105528.

45 Kumar, A.S., L. T. Jule, K. Ramaswamy, S. Sountharrajan, N. Yuuvaraj, and A. H. Gandomi, *Analysis of false data detection rate in generative adversarial networks using recurrent neural network*, in *Generative Adversarial Networks for Image-to-Image Translation*. 2021. Elsevier. p. 289–312.

46 Kurugodu, H., S. Bordoloi, Y. Hong, A. Garg, A. Garg, S. Sreedeep, et al., *Genetic programming for soil-fiber composite assessment. Advances in Engineering Software*, 2018. **122**: p. 50–61.

47 Lary, D.J., A. H. Alavi, A. H. Gandomi, and A. L. Walker, *Machine learning in geosciences and remote sensing. Geoscience Frontiers*, 2016. **7**(1): p. 3–10.

48 Mirzahosseini, M., Y. Najjar, A. H. Alavi, and A. H. Gandomi, *ANN-based prediction model for rutting propensity of asphalt mixtures.* in *92nd Annual Meeting of Transportation Research Board, Washington DC, USA*. 2013.

49 Mirzahosseini, M., Y. M. Najjar, A. H. Alavi, and A. H. Gandomi, *Next-generation models for evaluation of the flow number of asphalt mixtures. International Journal of Geomechanics*, 2015. **15**(6): p. 04015009.

50 Mohebali, B., A. Tahmassebi, A. Meyer-Baese, and A. H. Gandomi, *Probabilistic neural networks: A brief overview of theory, implementation, and application*. in *Handbook of Probabilistic Models*. 2020: p. 347–367. Elsevier, https://doi.org/10.1016/B978-0-12-816514-0.00014-X.

51 Mollahasani, A., A. H. Alavi, A. H. Gandomi, and A. Rashed, *Nonlinear neural-based modeling of soil cohesion intercept. KSCE Journal of Civil Engineering*, 2011. **15**(5): p. 831–840.

52 Mousavi, M. and A.H. Gandomi, *Prediction error of Johansen cointegration residuals for structural health monitoring. Mechanical Systems and Signal Processing*, 2021. **160**: p. 107847.

53 Mousavi, M. and A.H. Gandomi, *Structural health monitoring under environmental and operational variations using MCD prediction error. Journal of Sound and Vibration*, 2021. **512**: p. 116370.

54 Mousavi, S.M., P. Aminian, A. H. Gandomi, A. H. Alavi, and H. Bolandi, *A new predictive model for compressive strength of HPC using gene expression programming. Advances in Engineering Software*, 2012. **45**(1): p. 105–114.

55 Nagasubramanian, G., R. K. Sakthivel, R. Patan, M. Sankayya, M. Daneshmand, and A. H. Gandomi, *Ensemble classification and IoT based pattern recognition for crop disease monitoring system. IEEE Internet of Things Journal*, 2021.

56 Nejad, F.M., A. Mehrabi, and H. Zakeri, *Prediction of asphalt mixture resistance using neural network via laboratorial X-ray images. Journal of Industrial and Intelligent Information*, 2015. **3**(1): 48–53.

57 Nejad, F.M. and H. Zakeri, *An optimum feature extraction method based on wavelet–radon transform and dynamic neural network for pavement distress classification. Expert Systems with Applications*, 2011. **38**(8): p. 9442–9460.

58 Nejad, F.M. and H. Zakeri, *An expert system based on wavelet transform and radon neural network for pavement distress classification. Expert Systems with Applications*, 2011. **38**(6): p. 7088–7101.

59 Nejad, F.M. and H. Zakeri, *A comparison of multi-resolution methods for detection and isolation of pavement distress. Expert Systems with Applications*, 2011. **38**(3): p. 2857–2872.

60 Nejad, F.M. and H. Zakeri, *The hybrid method and its application to smart pavement management. Metaheuristics in Water, Geotechnical and Transport Engineering*, 2012. **439**: 440–483.

61 Nik, A.A., F.M. Nejad, and H. Zakeri, *A survey on pavement sectioning in network level and an intelligent homogeneous method by hybrid PSO and GA. Archives of Computational Methods in Engineering*, 2020. **27**(3): p. 977–997.

62 Nosratabadi, S., A. Mosavi, P. Duan, P. Ghamisi, F. Filip, S. S. Band, et al., *Data science in economics: Comprehensive review of advanced machine learning and deep learning methods. Mathematics*, 2020. **8**(10): p. 1799.

63 Pled, F., C. Desceliers, A. H. Gandomi, and C. Soize, *Neural network prediction of cortical bone damage using a stochastic computational mechanical model.* in *3rd International Conference on Uncertainty Quantification in Computational Sciences and Engineering (UNCECOMP 2019)*. 2019.

64 Punitha, S., T. Stephan, and A.H. Gandomi, *A novel breast cancer diagnosis scheme with intelligent feature and parameter selections. Computer Methods and Programs in Biomedicine*, 2021. **214**: p. 106432.

65 Rahimi, I., F. Chen, and A.H. Gandomi, *A review on COVID-19 forecasting models. Neural Computing and Applications*, 2021: p. 1–11.

66 Ranjbar, S., F. Moghaddasnezhad, and H. Zakeri, *Pavement cracks detection and classification using deep convolutional networks. Amirkabir Journal of Civil Engineering*, 2020. **52**(9): p. 2255–2278.

67 Ranjbar, S., F.M. Nejad, and H. Zakeri, *An image-based system for asphalt pavement bleeding inspection. International Journal of Pavement Engineering*, 2021: p. 1–17, https://doi.org/10.1080/10298436.2021.1932881.

68 Shahrara, N., T. Çelik, and A.H. Gandomi, *Risk analysis of BOT contracts using soft computing. Journal of Civil Engineering and Management*, 2017. **23**(2): p. 232–240.

69 Tahmassebi, A., A. H. Gandomi, S. Fong, A. Meyer-Baese, and S. Y. Foo, *Multi-stage optimization of a deep model: A case study on ground motion modeling. PLoS One*, 2018. **13**(9): p. e0203829.

70 Tahmassebi, A., A. H. Gandomi, I. McCann, M. H. Schulte, A. E. Goudriaan, and A. Meyer-Baese, *Deep learning in medical imaging: FMRI big data analysis via convolutional neural networks*, in *Proceedings of the Practice and Experience on Advanced Research Computing*. 2018: p. 1–4. ICPS Proceedings, https://doi.org/10.1145/3219104.3229250.

71 Tapkın, S., H. Zakeri, A. Topal, F. M. Nejad, A. Khodaii, and B. Şengöz, *A brief review and a new automatic method for interpretation of polypropylene modified bitumen based on fuzzy radon transform and watershed segmentation. Archives of Computational Methods in Engineering*, 2020. **27**(3): p. 773–803.

72 Telikani, A. and A.H. Gandomi, *Cost-sensitive stacked auto-encoders for intrusion detection in the Internet of Things. Internet of Things*, 2019, **14**: p. 100122.

73 Telikani, A., A. Tahmassebi, W. Banzhaf, and A. H. Gandomi, *Evolutionary machine learning: A survey. ACM Computing Surveys (CSUR)*, 2021. **54**(8): p. 1–35.

74 Yang, X.-S., A. H. Gandomi, S. Talatahari, and A. H. Alavi, *Metaheuristics in Water, Geotechnical and Transport Engineering*. 2012. Newnes.

75 Zakeri, H., F. M. Nejad, A. Fahimifar, A. D. Torshizi, and M. F. Zarandiet al., *A multi-stage expert system for classification of pavement cracking*. in *2013 Joint IFSA World Congress and NAFIPS Annual Meeting (IFSA/NAFIPS)*. 2013. (pp. 1125–1130). IEEE.

76 Zakeri, H., F. M. Nejad, A. D. Torshizi, and M. Fazel, *A new type of fuzzy membership functions and uncertainly grade in the frame of polar systems*. Proceedings of 8th International Symposium on Intelligent and Manufacturing Systems (IMS 2012), Sakarya University Department of Industrial Engineering, Adrasan, Antalya, Turkey, 27–28 September, 2012: 117–126.

77 Zarandi, M.F., F.M. Nejad, and H. Zakeri, *A type-2 fuzzy model based on three dimensional membership functions for smart thresholding in control systems*. in *Fuzzy Controllers – Recent Advances in Theory and Applications*, 2012. Intech.

9

Models of Performance Measures and Quantification in Automation

9.1 Introduction

In order to evaluate the overall performance of a method or a classifier, a wide range of criteria with varying degrees of sensitivity have been identified to classify methods and decide on selection and application. This chapter is of particular importance due to the importance of evaluating the methods in infrastructure management. In general, these indicators are generally complementary, and the adequacy of one indicator alone does not mean the overall guarantee of the method. The results of evaluation methods of models and algorithms are usually highly dependent on the method of analysis, design, and conditions. The results explain how to extract instructions for extracting information from infrastructure, technologies used, weather conditions when extracting information, processor and speed of analysis, and many other related factors. For these reasons, different results from evaluating automated methods can be expected. In this chapter, the performance evaluation methods and indicators are first presented, followed by a summary of the types of general indicators in infrastructure evaluation. Figure 9.1 shows a general classification of the performance appraisal methods and common indicators.

Assessing the true accuracy and performance of automated infrastructure assessment algorithms is significant in choosing the path of analysis and obtaining practical results. Many indices, such as mean square error (MSE), entropy, fuzzy index, signal-to-noise ratio (SNR), peak signal-to-noise ratio (PSNR), mean absolute error (MAE), among others, have been proposed and used for evaluation. Due to the existence of hidden characteristics, like the number of positive and negative samples and categories with different numbers of samples, using the accuracy index alone is not a good indicator to evaluate the efficiency of a method.

In this chapter, evaluation methods are divided into five general categories, including (i) General statistics, (ii) Basic rations, (iii) Rations of ratios, (iv) Additional statistics, and (v) Operating characteristics. Each category is introduced and examined for each method of diagnosis and classification. Based on the information extracted from the complexity matrix, indices, such as accuracy, error, probability of detection, productive index, selectivity, reproduction, negative predicted value (NPV), false positive rate (FPR), false negative rate (FNR), false discovery rate (FDR), false omission rate (FOR), likelihood ratio for positive tests (LRPT), likelihood ratio for negative tests (LRNT), likelihood ratio for positive subjects (LRPS), and likelihood ratio for negative subjects (LRNS), are calculated. Also, using these indicators, new indicators like F-measure, balanced accuracy (BalAcc), Matthews correlation coefficient (MCC), Chisq: χ^2, difference between automatic and manual methods, and differentiation index, are used for comparison and evaluation. The overall performance of a method can be evaluated based on the index matrix extracted from the complexity matrix. This section first provides a definition of the classification matrix then presents the database

Automation and Computational Intelligence for Road Maintenance and Management: Advances and Applications, First Edition. Hamzeh Zakeri, Fereidoon Moghadas Nejad, and Amir H. Gandomi.

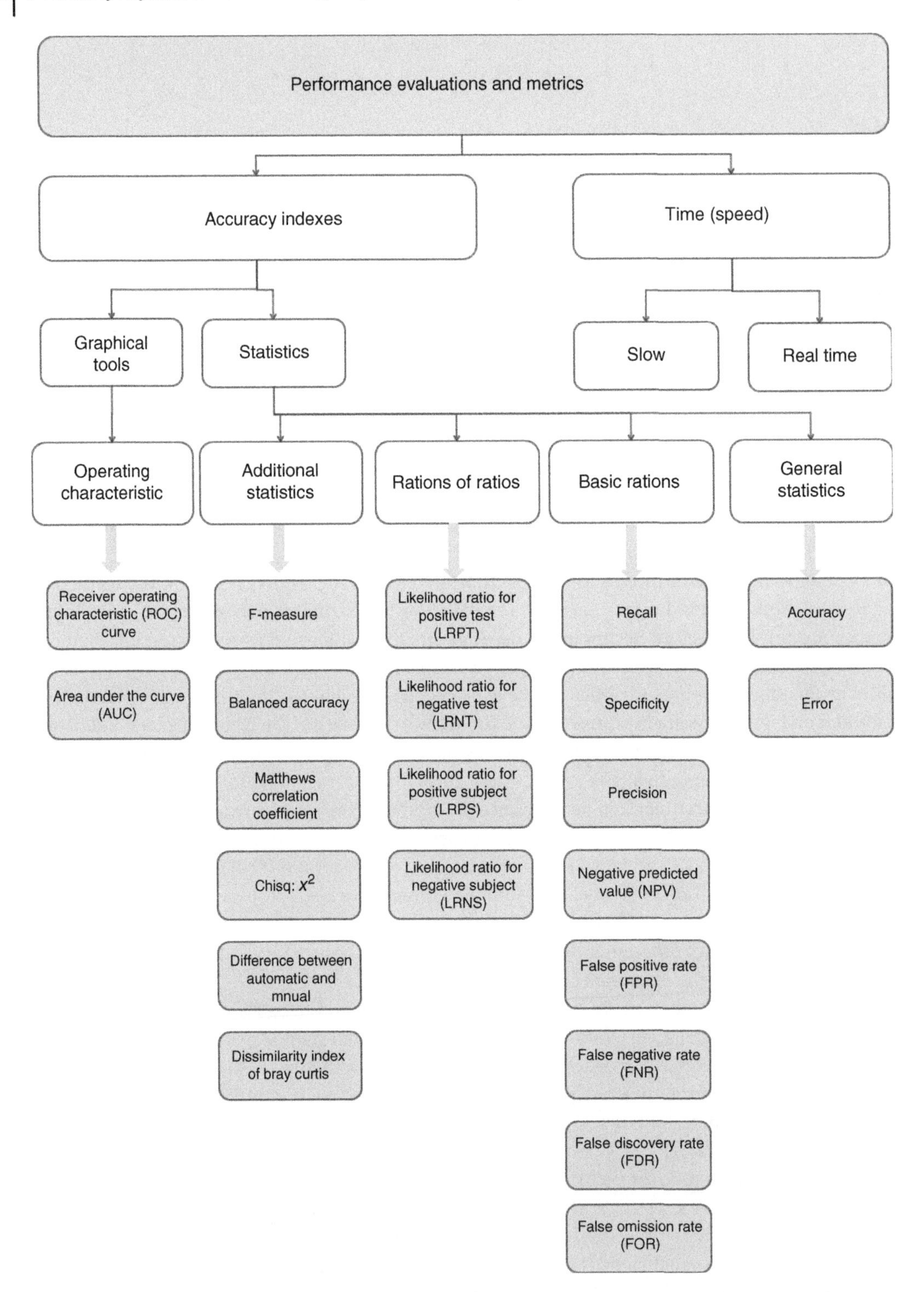

Figure 9.1 Classification of performance measurement indicators to evaluate methods of diagnosis, isolation, and classification, including statistical and graphical methods and time-based calculation.

modeling method. Based on the classification in Figure 9.1, the types of evaluation indicators are introduced, followed by an introduction to various calculation examples of these indicators. At the end, some general specialized indicators to evaluate the infrastructure are described.

9.2 Basic Definitions

In the field of binary or multiple classification, it is necessary to introduce the main criteria that are important for evaluating the performance of a model.

9.2.1 Confusion Matrix

In the world of machine learning and automated algorithms in statistical classification, the confusion matrix, also known as the error matrix, is always used for evaluation. In other words, all evaluation algorithms need to fill the cells of this matrix and are evaluated using this matrix based on the simple relationships presented below. The confusion matrix is a special table that makes it possible to estimate the performance of an algorithm. In this matrix, each row represents samples in a real class, and each column represents samples in a prediction class; it should be noted that this arrangement can be reversed. The nomenclature of the matrix is confused because, based on its outputs, it can be inferred whether the system confuses the two classes via a single point. In other words, an instance in a class may be incorrectly labeled, or an instance belonging to a class may correctly labeled. This matrix can also be extended to multi-class instances. In this case, the number of rows and columns increases based on the number of classes.

The confusion matrix in Figure 9.2a has only two conditions: positive and negative. In most applications, confusion matrices can have more categories. Figure 9.2b provides an example of the classification of road pavement failures for the five categories of distress, where zero indicates no category belonging. For a good classifier, each member should only be in its own class.

9.2.2 Main Metrics

Metrics that are commonly used to evaluate the performance of classification models generally fall into one of five categories: General statistics, Basic rations, Rations of ratios, Additional statistics, and Operating characteristics. These metrics have different functions depending on the type of problem and data related to different classes, some of which are more common than others. For

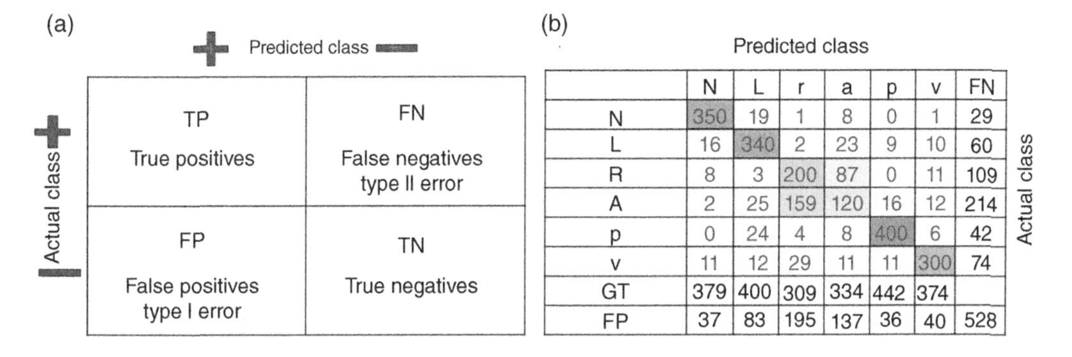

Figure 9.2 Examples of confusion matrices for a: (a) 2-class problem and (b) 5-class problem.

example, the accuracy of positive predictions and coverage of actual negative and positive samples are used for overall performance, while hybrid metrics like accuracy, precision (PRE), sensitivity, specificity, and F1-score are used for unbalanced classes. Normally, these indicators are calculated and, for simplicity, all are placed in a table for simultaneous comparison. Interpreting these metrics is important to evaluate the performance of a classifier.

9.2.3 Accuracy Indexes

One of the most important parameters is the evaluation of time and other indicators of accuracy. Based on the classification of accuracy indicators, two main categories of statistics and graphic tools are included. In the first category, statistical metrics are used for evaluation, and in the second category, graphic metrics are used based on drawing a chart and calculating the area under the chart. In both cases, it is necessary to prepare an identity card for each of these metrics and to decide which single format to use. An example of these metrics and a summary table of metrics are illustrated in Figure 9.3. In Section 9.4, each metric will be discussed in detail.

The receiver performance curve, also called receiver operating characteristic (ROC), is generated from a true positive rate (TPR) versus FPR graph with a threshold change. This diagram is a two-dimensional curve that shows the changes of two metrics relative to each other. The area below the graph, which is a number between 0 and 1, is also used for evaluation. The input for calculating these criteria and its identity card are shown in Figure 9.3.

9.2.4 Time (Speed)

Considering that some methods for evaluating algorithm are more efficient than others, the optimality of algorithms and their efficiency are also evaluated based on metrics of temporal and computational complexity. Complexity is a metric that measures the performance of a method based on its behavior in dealing with big data, which is a good descriptor for evaluating the performance and computation time of algorithms. Temporal complexity and spatial complexity metrics are defined to evaluate performance as follows:

Time complexity is a measure of the amount of time an override algorithm needs. "Time" can include various processes designed in the algorithm. It is necessary to mention several factors, such as the type of programming language, algorithm designer and their skill level, computing hardware, compilers, input data type, design methods, among others.

(a)

TP	350	340	200	120	400	300
TN	1360	1370	1510	1590	1310	1410
FP	37	83	195	137	36	40
FN	29	60	109	214	42	74
Acc	96.28%	92.28%	84.91%	82.97%	95.64%	93.75%
Er = 1-Acc	3.72%	7.72%	15.09%	17.03%	4.36%	6.25%

(b)

TP True positives rate	Recall sensitivity
FP False positives rate type I error	1-Specificity

Figure 9.3 (a) An example of these metrics and a summary table of metrics. (b) The input for calculating these criteria and its identity card for ROC and AUC.

Space complexity is a metric that depends on the amount of memory required by an algorithm, whereby the algorithm's ability to perform analysis is associated with large data volumes. Sometimes, space complexity is not considered for evaluating algorithms because it is indirectly implicit with time complexity. However, it becomes an important issue considering the complexity and inability of an algorithm to process.

9.3 Database Modeling and Model Selection

9.3.1 Different Parts of the Data

In this section, the necessary standards for building a database for training and testing algorithms are briefly presented. Each model consists of three parts in the database for validation and performance evaluation. These three main sections include training data, validation data, and test data. The sharing of each of these three sections must be properly determined in order to divide the database. Therefore, three sets must be extracted from each database to build the model, which are as follows:

Training Set: This set typically holds 80% of the data. It should be noted that the larger the training set, the more positive impact it has on the training algorithm and the better the algorithm experiences the problem environment. Therefore, the accuracy of the algorithm is expected to increase with increasing number of data.

Validation Set: This set is actually part of the same training set that is used to determine and tune the parameters. This includes a collection of data that have been already seen.

Test Set: This set is the rest of the database data, which usually contains 20% of the data and is used to test the final performance of the model. This set of data is unseen in the sense that it has not previously been used by the model in either the training or validation phase. The purpose of designing this section in the evaluation model is the predictability of the algorithm and the power of encountering and analyzing in untested conditions (Figure 9.4).

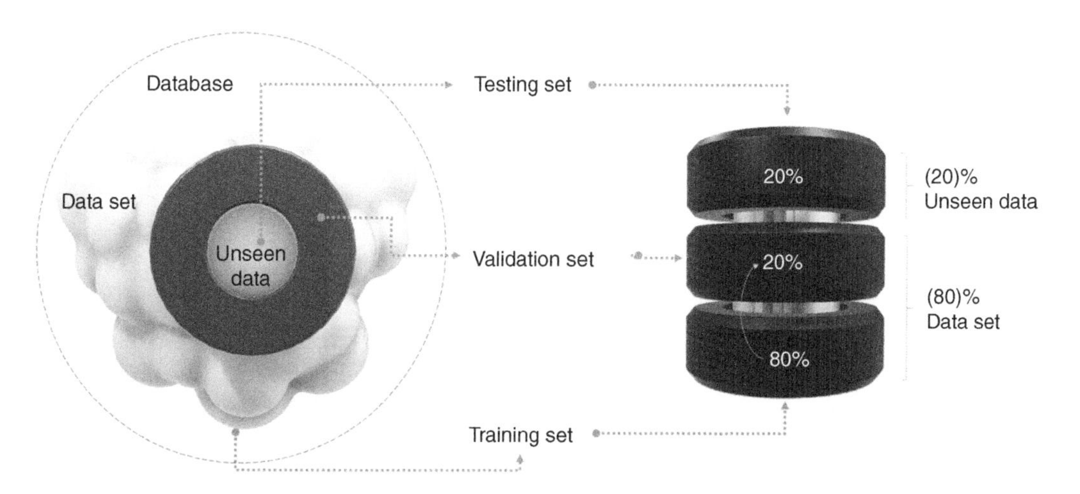

Figure 9.4 Schematic model of the database and its division into training, validation, and testing sets.

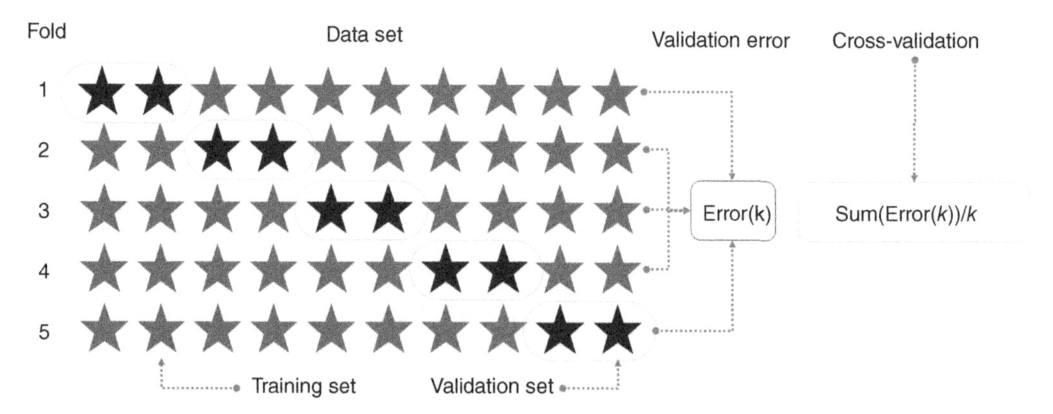

Figure 9.5 Example of the common k-fold method for $k = 5$.

9.3.2 Cross Validation

Cross validation is a method used to select a model independent of the nature of the database. Based on this method, also known as cross validation (CV), a model that is not too dependent on the initial training set is selected, which has the ability to predict and obtain appropriate results for data that has not been seen before. Some widely used cross validation methods are as follows:

k-Fold: Several folds are selected from the training set and validated with one folder, then the average of the errors is considered as the total error of the algorithm. The number of folds is usually 5–10. Figure 9.5 shows an example of the k-fold method.

Leave-p-out: The training is done with n−p sets, and the validation is performed with p sets. Similarly, the mean of the errors is considered as the algorithm error.

9.3.3 Regularization Techniques and Overfitting

Learning the training data too exactly usually leads to poor classification results on new data (see Figure 9.6). The regularization method is a step that was designed to avoid data fitting and generally

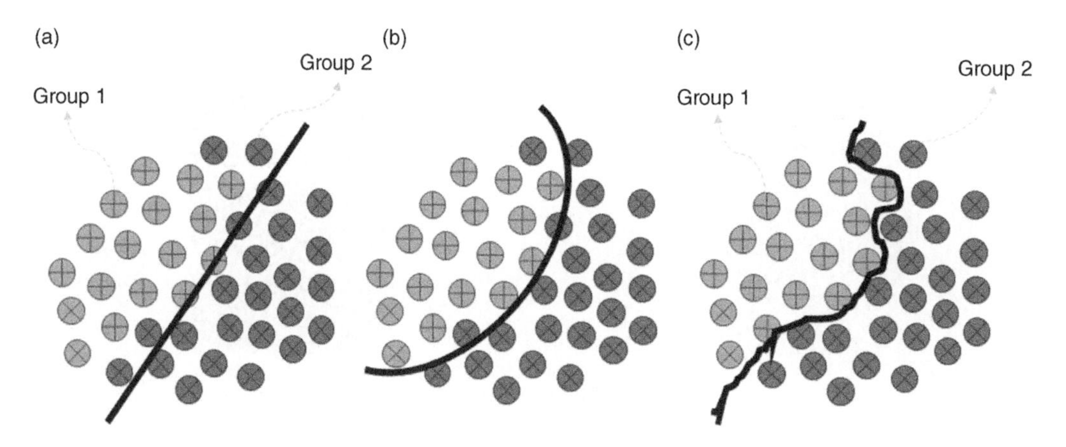

Figure 9.6 An example of dataset (a) underfitting, (b) fitting, and (c) overfitting.

deals with issues of high variance. Common methods include least absolute shrinkage and selection operator (LASSO), Ridge, and Elastic Net. The LASSO method reduces the coefficients to 0 and is selected for a good variable. The Ridge method makes coefficients smaller. The last Elastic Net selects variables with small coefficients using trial and error to prevent over-fitting.

9.4 Performance Evaluations and Main Metrics

In this section, a brief explanation of the main parameters is provided, then the five main categories of evaluation criteria are presented. The parameter true positive (TP) is the number of particles of the group of interest that are correctly classified. True negative (TN) is the number of particles of all the other groups that are classified as other groups. False positive (FP) is the number of particles of other groups classified in the group of interest. False negative (FN) is the number of particles of the group of interest classified in the other groups.

9.4.1 General Statistics

In this category, the two main statistical parameters, namely predictive error (ERR) and accuracy (ACC), show the overall performance of the algorithm in terms of the number of incorrectly classified samples. This error can be defined as the sum of all misdiagnosed predictions divided by the total number of predictions, as well as the correctness of the sum of correctly detected predictions divided by the total number of predictions.

$$\text{Accuracy} = \frac{(\text{TP} + \text{TN})}{(\text{TP} + \text{TN} + \text{FP} + \text{FN})} \tag{9.1}$$

$$\text{Error} = 1 - \text{Accuracy} = 1 - \frac{(\text{TP} + \text{TN})}{(\text{TP} + \text{TN} + \text{FP} + \text{FN})} \tag{9.2}$$

9.4.2 Basic Rations

TPR and FPR are performance metrics that are commonly used for unbalanced class problems. For example, in classifying infrastructure damage through images, we are primarily interested in identifying and removing harmless and healthy images. However, it is important to reduce the number of images that are incorrectly classified as damaged (FP) images: a situation in which an automated damage detection system loses a cross section that is faulty; "worse" refers to a situation in which the system mistakenly considers several healthy sections in the category of damaged sections. Unlike FPR, TPR provides useful information about a fraction of positive (or related) samples that have been correctly identified from the sum of positive sets.

$$\text{TPR} = \text{Recall} = 1 - \text{FNR} = \frac{\text{TP}}{\text{TP} + \text{FN}} \tag{9.3}$$

$$\text{FPR} = 1 - \text{Specificity} = \frac{\text{FP}}{\text{FP} + \text{TN}} = \text{Type I error} \tag{9.4}$$

Precision (PRE) and Recall (REC) are common criteria used in information technology to evaluate the performance of a system or algorithm and are associated with false and TPRs. In fact, this measurement is synonymous with real positive rates and is sometimes known as sensitivity.

F1-score can be considered as a combination of precision and recall, and its relationship is presented in the section on additional statistical indicators.

$$\text{PPV} = \text{Precision} = 1 - \text{FDR} = \frac{\text{TP}}{\text{TP} + \text{FP}} \tag{9.5}$$

The TNR metric is calculated based on the ratio of the number of TN samples to the total number of TN and FP samples and is often referred to as Specificity. This index is a complement to the FPR metric and is calculated using the following equation:

$$\text{TNR} = \text{Specificity} = 1 - \text{FPR} = \frac{\text{TN}}{\text{TN} + \text{FP}} \tag{9.6}$$

The NPV index is based on the ratio of TN to total TN and FN. This metric is a measure of the ratio of the number of correctly detected negative samples to the total number of negative samples. For example, the number of samples of road pavement sections that are healthy and are not the purpose of automatic identification is calculated as the total number of samples of healthy sections that are correctly and incorrectly detected using this method.

$$\text{NPV} = 1 - \text{FOR} = \frac{\text{TN}}{\text{TN} + \text{FN}} \tag{9.7}$$

The FNR index is a supplement to the REC index and is known in most references as a type II error. This metric is based on the ratio of negatively detected false samples to total samples and is calculated based on the ratio of FN to the sum of TN and FN using the following equation:

$$\text{FNR} = 1 - \text{Recall} = \frac{\text{FN}}{\text{TP} + \text{FN}} = \text{Type II error} \tag{9.8}$$

The FDR metric is calculated on the basis of evaluating the ratio of the number of positively positive samples detected to the sum of true and FP samples detected. This index complements the precision metric and is also known as the q-value. Based on this metric, the ratio of FP to the sum of TP and FP is calculated using the following equation:

$$\text{FDR} = 1 - \text{Precision} = \frac{\text{FP}}{\text{TP} + \text{FP}} = q - \text{Value} \tag{9.9}$$

The FOR metric is an index that complements the NPV index and is calculated based on the ratio of FN to the sum of FN and TN. The more negative the detected samples are the higher the index, and the higher the number of TN samples, the lower the number.

$$\text{FOR} = 1 - \text{NPV} = \frac{\text{FN}}{\text{FN} + \text{TN}} \tag{9.10}$$

9.4.3 Rations of Ratios

In this section, four special metrics that are obtained from the indicators including the LRPT, LRNT, LRPS, and LRNS are introduced. These indices are related to each other based on the definitions provided in the Basic Rations and are used to interpret the properties of an algorithm. Sensitivity analysis and changing the parameters of each algorithm are performed to optimize the parameters related to the algorithms with these indicators.

The LRPT metric is an index that is calculated as a combination of basic indicators via a combination of the ratio of the number of correctly identified samples to the sum of samples and the ratio of incorrectly detected samples to the sum of samples. Based on these parameters, the ratio of REC to specificity supplement, or the FPR index, gives LRPT as follows:

$$\text{LRPT} = \frac{\frac{\text{TP}}{\text{TP} + \text{FN}}}{\frac{\text{FP}}{\text{FP} + \text{TN}}} = \frac{\text{Recall}}{1 - \text{Specificity}} = \frac{\text{Recall}}{\text{FPR}} \tag{9.11}$$

The LRNT metric is an index that is calculated based on the relationship of FN to the sum of samples and the ratio of TN. In general, the REC supplement, or FNR, introduces this index to the Specificity index. This metric is also a practical indicator for evaluating the performance of an algorithm and optimizing the parameters, which is calculated using the following equation:

$$\text{LRNT} = \frac{\frac{\text{FN}}{\text{FN} + \text{TP}}}{\frac{\text{TN}}{\text{TN} + \text{FP}}} = \frac{(1 - \text{Recall})}{\text{Specificity}} = \frac{\text{FNR}}{\text{Specificity}} \tag{9.12}$$

The LRPS metric is another useful indicator for evaluating the sensitivity analysis of algorithms and measuring the optimality of the parameters of classification algorithms. This metric is dependent on the ratio of TP and FN, where the higher TP and FN are, the higher the index is. Also, the ratio of precision to NPV complement, or the FOR index, constitutes this metric, where the higher the precision for a set, the higher the index:

$$\text{LRPS} = \frac{\frac{\text{TP}}{\text{TP} + \text{FP}}}{\frac{\text{FN}}{\text{FN} + \text{TN}}} = \frac{\text{Precision}}{(1 - \text{NPV})} = \frac{\text{Precision}}{\text{FOR}} \tag{9.13}$$

The LRNS metric is one of the composite indicators that is calculated based on the ratio of FP to the sum of incorrect samples and also the ratio of TN to the number of true and false samples of the negative category. This index is a ratio of the precision supplement to FOR supplement, which is more easily calculated as the ratio of FDR to NPV and is obtained based on the following equation:

$$\text{LRNS} = \frac{\frac{\text{FP}}{\text{FP} + \text{TP}}}{\frac{\text{TN}}{\text{TN} + \text{FN}}} = \frac{(1 - \text{Precision})}{(1 - \text{FOR})} = \frac{\text{FDR}}{\text{NPV}} \tag{9.14}$$

9.4.4 Additional Statistics

This section is also dedicated to statistical criteria in addition to composite metrics. Six metrics calculated from the baseline indicators in Section 9.4 are used to perform more complex evaluations. These six criteria include (i) F_{mes} (which is also F1-score); (ii) harmonic mean of precision and recall; (iii) BalAcc; (iv) MCC, whereby if any sum of the denominator is 0, the total denominator can be set to 1; (v) χ^2 (Chisq, which is also called Significance); and (vi) difference between Automatic and Manual classification (Auto_Manu), where the Dissimilarity Index of Bray Curtis is part of this family.

The F_{mes} metric is an index that is obtained from the REC and precision metric parameters and both of these parameters are larger than F_{mes}. The F_{mes} index shows an index that calculated using the following equation:

$$\text{F}_{\text{mes}} = F - \text{measure} = F1 = 2 \times \frac{(\text{Precision} \times \text{Recall})}{(\text{Precision} + \text{Recall})} \tag{9.15}$$

The BalAcc metric is also calculated based on REC and FPR complement and is the average of two indicators of REC and specificity and is calculated using the following equation:

$$\text{BalAcc} = \frac{(\text{Recall} + (1 - \text{FPR}))}{2} = \frac{(\text{Recall} + \text{Specificity})}{2} \tag{9.16}$$

Based on the relationship of TP, TN, FP, and FN parameters, the MCC metric can be used as a statistical indicator to evaluate the performance of an algorithm or classifier and is calculated using the following equation:

$$\text{MCC} = \frac{(\text{TP} \times \text{TN}) - (\text{FP} \times \text{FN})}{((\text{TP} + \text{FP}) \times (\text{TP} + \text{FN}) \times (\text{TN} + \text{FP}) \times (\text{TN} + \text{FN}))^{0.5}} \tag{9.17}$$

Similar to the MCC index, the significance metric is a statistical index for evaluation. It is calculated based on the relationship between the parameters TP, TN, FP, and FN using the following equation:

$$\text{Significance} = \text{Chisq} : \chi 2 = \frac{[(\text{TP} \times \text{TN}) - (\text{FP} \times \text{FN})]^2 \times (\text{TP} + \text{TN} + \text{FP} + \text{FN})}{(\text{TP} + \text{FP}) \times (\text{TP} + \text{FN}) \times (\text{TN} + \text{FP}) \times (\text{TN} + \text{FN})} \tag{9.18}$$

The difference between automatic and manual classification is calculated using the Auto_Manu metric. This index is simply calculated using the difference of positive samples and the sum of positive samples correctly detected and not correctly detected, using the following equation:

$$\text{Auto}_{\text{Manu}} = (\text{TP} + \text{FP}) - (\text{TP} + \text{FN}) \tag{9.19}$$

The Dissimilarity Index of Bray Curtis is a statistical index that is calculated based on the Auto_Manu index and the sum of the samples of the basic indicators based on the following equation:

$$\text{Dissimilarity Index of Bray Curtis} = \frac{|\,(\text{Auto}_{\text{Manu}})\,|}{\sum(\text{TP} + \text{FP}) + \sum(\text{TP} + \text{FN})} \tag{9.20}$$

Unlike the FPR = (1 − specificity) metric, the TPR = (1 − FNR) index provides significant knowledge about the number of correctly detected samples. The sensitivity of a method is evaluated using the recall or sensitivity index. The high-sensitivity method has small type 1 and 2 errors. The F_{mes} parameter is an index composed of two characteristics: Positive predictive value (PPV) and TPR. The sensitivity index indicates the ability to detect positive options, while the detection index indicates the ability to detect negative samples. The MCC index is a powerful index in: range −1 to 1 for random sample distribution to complete correlation, respectively. For heterogeneous data in heterogeneous categories in terms of the number of data, a more valid index based on Eq. (9.16) called BalAcc is used. Using reproducibility or reproducibility capabilities makes it possible to evaluate performance using the PPV index.

Additional statistics and analyses are generated using rations of ratios (ROR) and statistical addition functions, which can be used to evaluate the overall efficiency of the methods. In recent years, these indicators have been increasing applied to actually evaluate the methods and algorithms, and more attention has been paid to the selection of methods based on scientific principles and statistical indicators.

9.4.5 Operating Characteristic

ROC provide a useful graphical tool for optimally selecting models and classification algorithms based on their performance on basic metrics using FP and positive rate indices. The ROC diagram in Figure 9.7 can be interpreted as a random estimate of an algorithm, where the classification

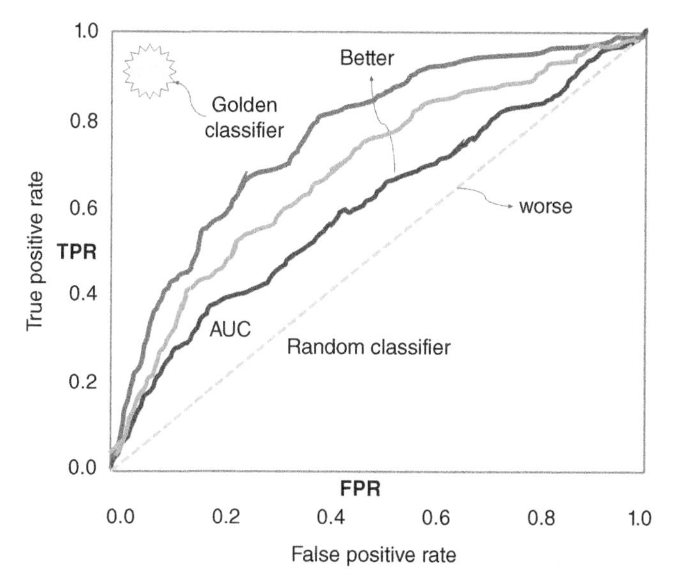

AUC	TEST quality
1–0.9	Excellent
0.9–0.8	Very good
0.8–0.7	Good
0.7–0.6	Satisfactory
0.6–0.5	Unsatisfactory

Figure 9.7 Example of a receiver operating characteristic.

models at the bottom of the diagram are worse than random guesses. A good, complete classifier can be placed in the upper left corner of the chart with a TP value of 1 and a FPR of 0. The ROC curve can be calculated by changing the decision threshold of a classifier. Based on the ROC curve, the so-called area under the curve (AUC) can also be used to determine the performance of a classification model.

ROC is also an effective tool for measuring and evaluating the performance of methods compared with numerical indicators, such as basic rations (BR), ROR, and additional statistics (AS).

The ScM measurement index is proposed on the basis of the Hausdorff distance to estimate the capability of cracking detection algorithms. This index is between 0 and 100, which is calculated from the following equation:

$$\mathrm{ScM} = \left(100 - \left(\frac{\mathrm{BH}(x_i, y_i)}{w}\right)\right) \times 100 \tag{9.21}$$

In this equation, $\mathrm{BH}(x_i, y_i)$ is the distance between two sections, and w is a number equal to 0.2 of the image width for each image.

9.5 Case Studies

Based on the statistical analysis regarding the automatic evaluation of infrastructure and road pavement, Figure 9.8a,b provides comparisons of the evaluation methods in terms of computation time and speed as well as frequency. As can be seen, the BR family has been used more than other methods.

According to the research related to pavement distress detection, classification, and quantification, the BR method has the largest share in research, which is 74% of the methods. From this study, it can be concluded that despite the high efficiency of ROR methods in identifying and evaluating infrastructure failures and automated inspection and evaluation methods, no extensive research

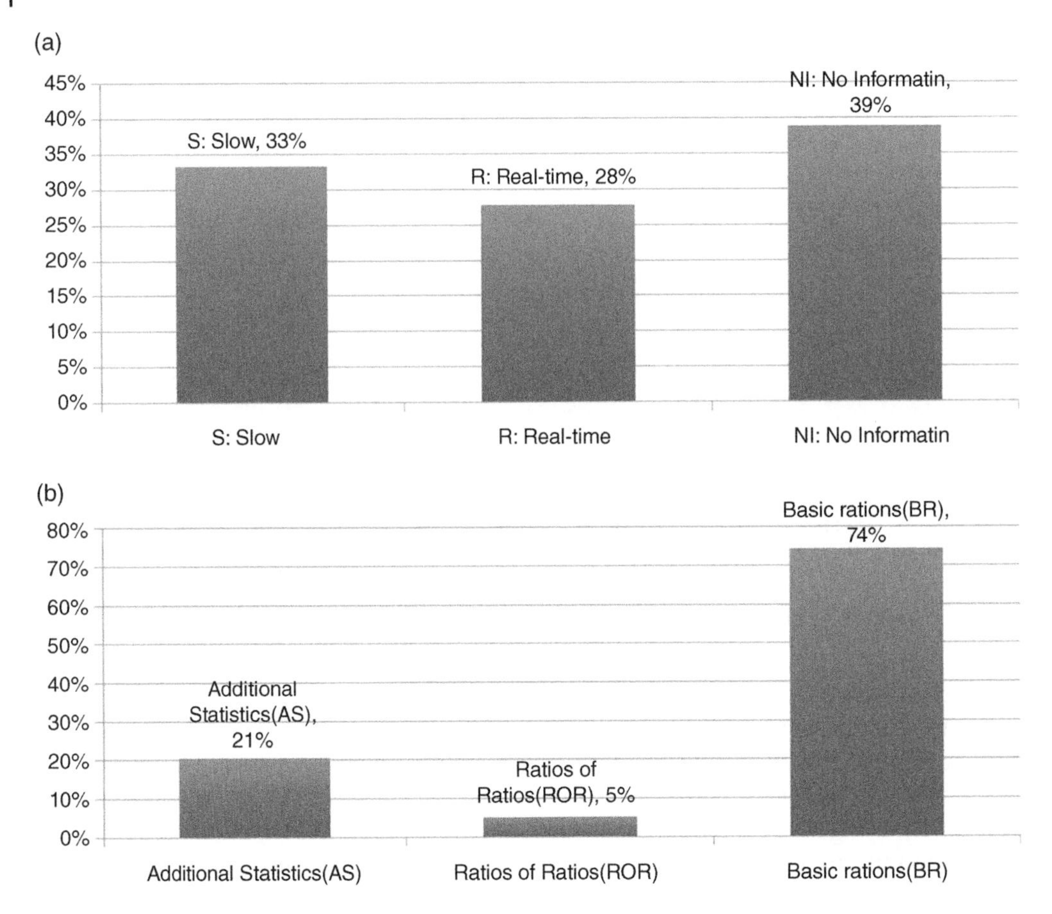

Figure 9.8 Comparison of different evaluation methods to evaluate performance: (a) comparison of methods in terms of speed and computational time, and (b) comparison of the frequency of evaluation methods in the case study to detection and classification of pavement distress.

has been done (only 5% of research). In addition, there is limited research using ROR evaluation methods in the field of computational intelligence and automated infrastructure damage detection algorithm. Due to the high capabilities of various evaluation indicators, it is necessary to use all indicators for comprehensive evaluation. Another study has also been conducted on the frequency of methods, which revealed that in 39% of the researches, despite the importance of using automated methods to reduce the computational time, information about the time and complexity of the methods was not provided. On the other hand, the frequency of high-speed, simultaneous separation and classification methods is about 28%, which is less than low-speed methods. In light of the existing conditions, the use of complementary indicators with high-speed and low-complexity methods is a priority, and for this purpose, some applications of these indicators are presented as a case study in this section.

9.5.1 Case 1: The Confusion Matrix for Evaluating Drainage of Pavement Surface

Table 9.1 presents the confusion matrix for evaluating drainage of pavement surface based on image-based analysis. For this evaluation, the drainage quality of 30 sections of pavement was

Table 9.1 The confusion matrix for evaluating drainage of pavement surface of three groups G, A, and F.

		Predicted class		
		Good (G)	**Acceptable (A)**	**Failed (F)**
True class	Good (G)	9	1	0
	Acceptable (A)	0	9	1
	Failed (F)	0	0	10

Table 9.2 Performance evaluation based on metrics of confusion matrix for evaluating drainage of pavement surface.

Different class of pavement surface drainage quality	Accuracy	Precision	Sensitivity	Specificity
Good	97	100	90	100
Acceptable	93	90	90	95
Failed	97	91	100	95
Overall system performance	95.7	93.7	93.3	96.7

classified into three categories according to experts, and the outputs were compared with the results of the classification model.

Based on the table outputs, it can be seen that 28 sections were correctly classified, and only two sections were incorrectly classified. This indicates the high performance of the model in classification. The model performed best in detecting unhealthy sections, all of which were correctly classified. This model had only one incorrect classification in faulty classes and offers acceptable performance, in which a suitable real section was classified as an acceptable section in the damaged class.

Table 9.2 provides the evaluation rate of classification metrics. Accordingly, the proposed system offers 95.7% accuracy, 93.7% accuracy, 93.3% sensitivity, and 96.7% specificity. The performance evaluation of the proposed classification model reveals the efficiency of the extracted metrics and the classification model used for the drainage evaluation of pavement surface in this case study.

9.5.2 Case 2: Metrics for Pavement Creak Detection Based on Deep Learning Using Transfer Learning

To evaluate and compare the performance of the asphalt pavement cracking using a deep learning algorithm, 16 trained models were tested for 750 images, which are shown in Table 9.3. After testing each of the models, a confusion matrix was extracted to achieve the performance criteria. Given that there are several different categories in this example, the confusion matrix and related metrics were used to evaluate the performance. The first four metrics for each class were calculated in the confusion matrix calculations, including TP, TN, FP, and FN. Due to the components of the confusion matrix, various criteria were considered to evaluate the performance of the models. In this case study, five criteria were designed and used to evaluate and compare the performance of crack detection models, including accuracy, sensitivity, specificity, accuracy, and F-score. The value of

Table 9.3 Performance evaluation based on metrics of confusion matrix for evaluating pavement cracking.

Models	Dataset	Class	Accuracy	Sensitivity	Specificity	Precision	F-score
AlexNet	A	Linear cracking	0.968	0.951	0.977	0.953	0.952
		Non-cracking	0.996	0.999	0.995	0.991	0.995
		Surface cracking	0.969	0.951	0.978	0.957	0.953
		General	0.978	0.967	0.983	0.967	0.967
	B	Linear cracking	0.958	0.939	0.968	0.937	0.937
		Non-cracking	0.995	0.999	0.993	0.987	0.993
		Surface cracking	0.961	0.935	0.975	0.949	0.941
		General	0.972	0.957	0.979	0.958	0.957
SqueezeNet	A	Linear cracking	0.986	0.989	0.985	0.970	0.980
		Non-cracking	0.999	0.999	0.999	0.997	0.998
		Surface cracking	0.987	0.969	0.995	0.991	0.980
		General	0.991	0.986	0.993	0.986	0.986
	B	Linear cracking	0.965	0.959	0.968	0.938	0.948
		Non-cracking	0.998	1.000	0.997	0.993	0.997
		Surface cracking	0.965	0.933	0.981	0.962	0.947
		General	0.976	0.964	0.982	0.964	0.964
GoogleNet	A	Linear cracking	0.984	0.980	0.985	0.971	0.975
		Non-cracking	0.999	0.997	1.000	1.000	0.999
		Surface cracking	0.984	0.973	0.990	0.980	0.977
		General	0.989	0.984	0.992	0.984	0.984
	B	Linear cracking	0.985	0.992	0.982	0.965	0.978
		Non-cracking	0.997	0.997	0.997	0.995	0.996
		Surface cracking	0.986	0.964	0.997	0.994	0.979
		General	0.990	0.984	0.992	0.985	0.984
ResNet-18	A	Linear cracking	0.968	0.964	0.971	0.943	0.953
		Non-cracking	0.997	0.999	0.996	0.992	0.995
		Surface cracking	0.972	0.943	0.986	0.971	0.957
		General	0.979	0.968	0.984	0.969	0.968
	B	Linear cracking	0.975	0.959	0.983	0.965	0.962
		Non-cracking	0.993	1.000	0.989	0.979	0.989
		Surface cracking	0.977	0.959	0.987	0.973	0.966
		General	0.982	0.972	0.986	0.973	0.972
ResNet-50	A	Linear cracking	0.961	0.932	0.976	0.951	0.941
		Non-cracking	0.996	1.000	0.994	0.988	0.994
		Surface cracking	0.965	0.952	0.972	0.944	0.948
		General	0.974	0.961	0.981	0.961	0.961
	B	Linear cracking	0.967	0.936	0.982	0.963	0.949
		Non-cracking	0.997	1.000	0.996	0.992	0.996

Table 9.3 (Continued)

Models	Dataset	Class	Accuracy	Sensitivity	Specificity	Precision	F-score
		Surface cracking	0.969	0.964	0.972	0.945	0.954
		General	0.978	0.967	0.983	0.967	0.967
ResNet-101	A	Linear cracking	0.957	0.948	0.962	0.926	0.937
		Non-cracking	0.996	0.988	1.000	1.000	0.994
		Surface cracking	0.961	0.936	0.974	0.947	0.942
		General	0.972	0.957	0.979	0.958	0.957
	B	Linear cracking	0.980	0.968	0.986	0.972	0.970
		Non-cracking	0.995	1.000	0.992	0.984	0.992
		Surface cracking	0.983	0.968	0.990	0.980	0.974
		General	0.986	0.979	0.989	0.979	0.979
DenseNet-202	A	Linear cracking	0.976	0.964	0.982	0.964	0.964
		Non-cracking	1.000	1.000	1.000	1.000	1.000
		Surface cracking	0.976	0.964	0.982	0.964	0.964
		General	0.984	0.976	0.988	0.976	0.976
	B	Linear cracking	0.976	0.964	0.982	0.964	0.964
		Non-cracking	0.993	1.000	0.990	0.980	0.990
		Surface cracking	0.983	0.964	0.992	0.984	0.974
		General	0.984	0.976	0.988	0.976	0.976
Inception-V3	A	Linear cracking	0.977	0.972	0.980	0.960	0.966
		Non-cracking	1.000	1.000	1.000	1.000	1.000
		Surface cracking	0.977	0.960	0.986	0.972	0.966
		General	0.985	0.977	0.989	0.977	0.977
	B	Linear cracking	0.980	0.968	0.986	0.972	0.970
		Non-cracking	0.996	0.996	0.996	0.992	0.994
		Surface cracking	0.984	0.976	0.988	0.976	0.976
		General	0.987	0.980	0.990	0.980	0.980

these criteria was calculated for each class in the models and is presented in Table 9.3. Also, the total value of these criteria for each model is shown in Table 9.3 (see Figure 9.8).

In this table, dataset A was used for training and dataset B for testing.

As can be seen in the results, the accuracy of the classes range from 0.957 to 1 in the model A series and from 0.958 to 0.998 in the B series. Similarly, the non-cracking class had the highest value in all models, while the linear cracking class showed the lowest accuracy. The overall accuracy of the models is in the range of 0.972–0.991 for the A series model and 0.972–0.990 for the B series.

Comparing the overall accuracy of the trained model based on dataset A, it was found that SqueezeNet and GoogleNet models with performances of 0.991 and 0.989 are better than other models, respectively, whereas ResNet-101 had a lower accuracy of 0.972 than other models. Comparing the overall accuracy of the trained model based on dataset B, it can be inferred that the GoogleNet model had the best performance accuracy of 0.99 and AlexNet had the lowest accuracy of 0.972.

Assessing the accuracy values in two series of each model (A and B) shows that the process of preparing dataset B leads to more accurately trained models than the trained model based on dataset A in GoogleNet, ResNet-18, ResNet- has been used. The data preparation process of dataset B reduced the accuracy values in AlexNet and SqueezeNet models, whereby this effect in SqueezeNet was reduced from 0.991 in model A to 0.976 in model B. The overall accuracy of DenseNet-201 was affected by the preparation process. The overall accuracy value of both models A and B is 0.984. Based on the metrics shown, the sensitivity of classes in trained models is in the range of 0.932–1, and the class without cracking had the highest sensitivity in all models with a range of 0.988–1. The evaluation of the other two classes (linear and superficial cracking) produced different results, whereby some models in the linear cracking class, such as AlexNet (B), SqueezeNet (A and B), are more sensitive: GoogleNet (A and B), ResNet-18 (A), ResNet-101 (A), Inception-V3 (A), ResNet-50 (A and B), and Inception-V3 (B) have higher sensitivity in the surface cracking class. In comparison, AlexNet (A), DenseNet (A and B), and ResNet-101 (B) have greater sensitivity in linear and surface cracking classes. The general sensitivity of the models is in the range of 0.957–0.986. SqueezeNet (A) and GoogleNet (A and B) had the highest sensitivity of 0.986 and 0.984, respectively, whereas AlexNet (B) and ResNet-101 (A) had the lowest sensitivity of 0.957.

The best performance for the detection of linear and surface cracking images was observed in SqueezeNet (A) and GoogleNet (B) with an F-score of 0.979. Comparatively, the worst performance was found to AlexNet (B) with 0.937 and 0.941 for linear and surface cracking classes, respectively. According to the general F-score, SqueezeNet (A) exhibited the best performance with 0.986, and GoogleNet (A and B) showed better performance than other models with 0.984. Moreover, AlexNet (B) had the worst performance with 0.957 (Figure 9.9).

9.5.3 Case 3: The Confusion Matrix for Evaluating Pavement Crack Classification

This case study examined the automatic detection of pavement cracks based on the type of fault classification methods using the support vector classification method in the polar support vector machine (PSVM) polar domain presented in Chapter 8. The classification results based on the type of crack are shown in Tables 9.4 and 9.5. In this study, the PSVM method was evaluated in terms of general indicators, basic indicators, productive indicators, and complementary indicators based on the relationship presented in Chapter 6. The complexity matrix based on type of failure is shown in Tables 9.6 and 9.7. These samples were visually categorized based on the expert's opinion, and the size of the original images was 2560 × 1920 pixels. In the collection of images and smaller sections of 500 × 500 fragmented images, 4253 images were tested. Of these, 1724 were healthy, and 2529 were damaged; 358 were longitudinal failures, 1366 were other failures, and 2529 were without failure. For the category of transverse cracks, out of 4253 images, 160 images showed transverse cracks, and 1564 presented other failures. For single diagonal surface cracks, this number is 264 and for other groups of pavement surface disturbances, this number is 1462. A total of 1724 images were used for the failure bank, including 780 with linear cracks and 944 with surface and composite cracks. Table 9.7 shows the comparison results of the proposed method with a neural network (NN) and support vector machine (SVM). The PSVM method exhibited 2–6% less error than the neural network and 1–3% less than the SVM. Also, the accuracy of the proposed method is in the range of 93–96%, which is a good in terms of efficiency. In comparison, NN and SVM methods showed 87–93% and 90–95% accuracy, respectively, indicating that the SVM method is generally better than

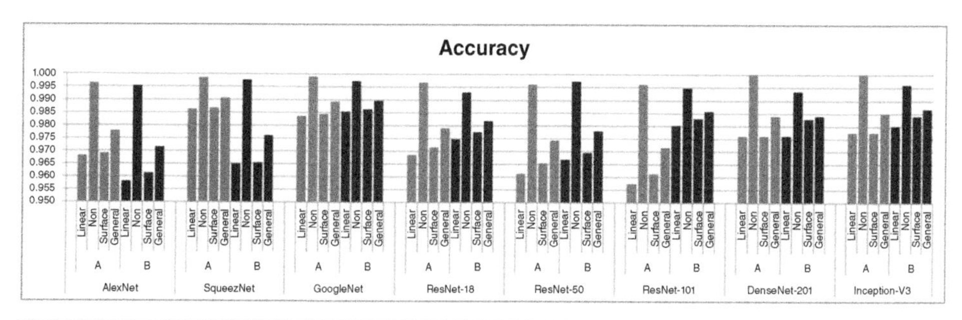

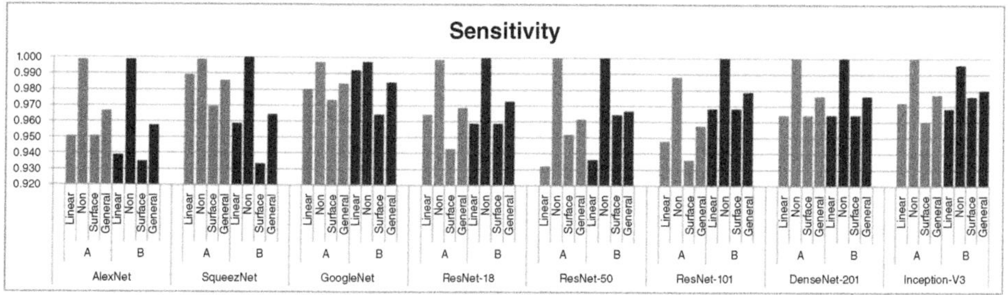

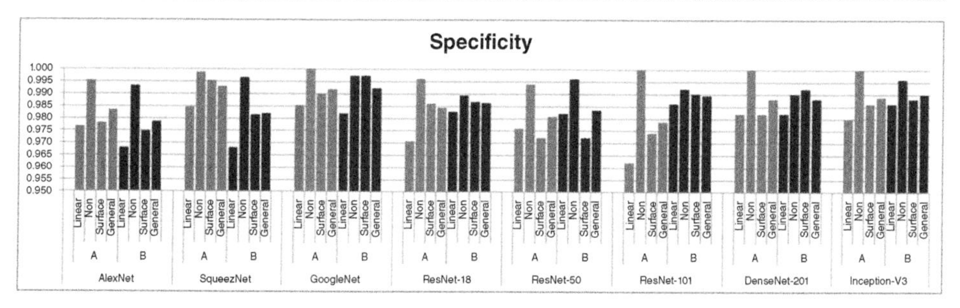

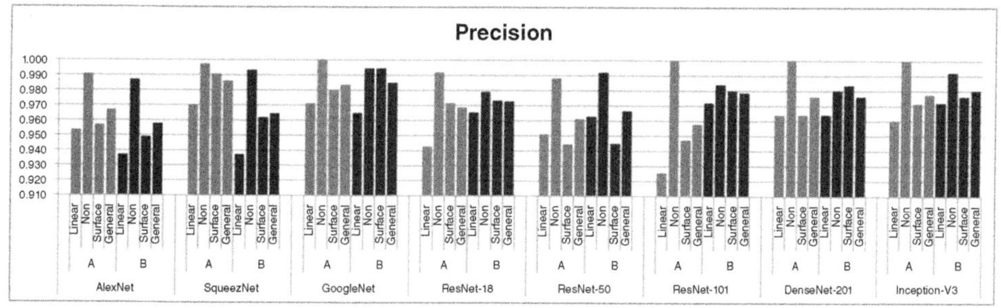

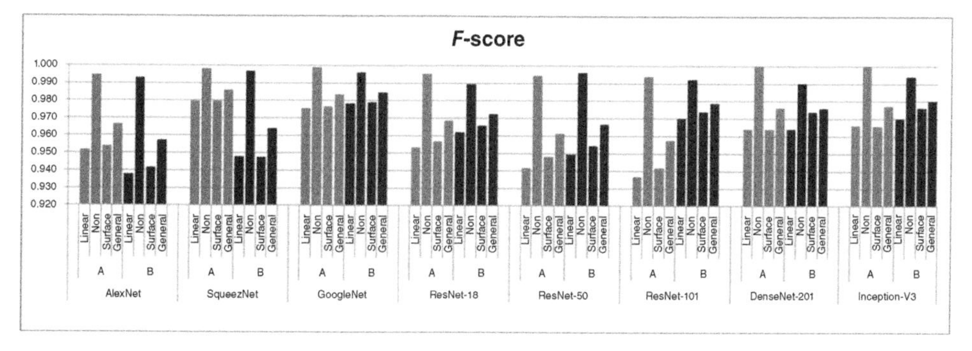

Figure 9.9 Model performances in case 2 according to various metrics.

Table 9.4 Performance evaluation based on metrics of confusion matrix for evaluating pavement cracking based on PSVM.

Metrics	Method	Longitudinal cracking (LC)	Transverse cracking (TC)	Diagonal cracking (DC)	BC	AC	Single cracking (SC)	Multi cracking (MC)
ACC	NN	0.946	0.925	0.916	0.875	0.936	0.929	0.906
	SVM	0.953	0.942	0.946	0.903	0.948	0.947	0.925
	PSVM	0.967	0.963	0.967	0.932	0.95	0.966	0.941
Error = 1 − Acc	NN	0.054	0.075	0.084	0.125	0.064	0.071	0.094
	SVM	0.047	0.058	0.054	0.097	0.052	0.053	0.075
	PSVM	0.033	0.037	0.033	0.068	0.05	0.034	0.059
TPR	NN	0.86	0.794	0.79	0.697	0.909	0.815	0.803
	SVM	0.872	0.856	0.85	0.75	0.934	0.859	0.842
	PSVM	0.908	0.888	0.878	0.829	0.953	0.891	0.891
TNR	NN	0.973	0.94	0.941	0.937	0.951	0.951	0.944
	SVM	0.977	0.952	0.965	0.955	0.955	0.965	0.955
	PSVM	0.984	0.972	0.984	0.967	0.949	0.98	0.958
PPV	NN	0.909	0.608	0.734	0.792	0.909	0.75	0.851
	SVM	0.92	0.668	0.831	0.85	0.915	0.807	0.882
	PSVM	0.942	0.776	0.916	0.894	0.9	0.878	0.897
NPV	NN	0.957	0.975	0.956	0.899	0.951	0.963	0.925
	SVM	0.962	0.983	0.97	0.918	0.966	0.972	0.942
	PSVM	0.974	0.987	0.976	0.944	0.977	0.979	0.96
FPR = Type I error	NN	0.027	0.06	0.059	0.063	0.049	0.049	0.056
	SVM	0.023	0.048	0.035	0.045	0.045	0.035	0.045
	PSVM	0.016	0.028	0.016	0.033	0.051	0.02	0.042
FNR = Type II error	NN	0.14	0.206	0.21	0.303	0.091	0.185	0.197
	SVM	0.129	0.144	0.15	0.25	0.066	0.141	0.158
	PSVM	0.058	0.224	0.084	0.106	0.1	0.122	0.103
FDR	NN	0.091	0.392	0.266	0.208	0.091	0.25	0.149
	SVM	0.08	0.332	0.169	0.15	0.085	0.194	0.118
	PSVM	0.058	0.224	0.084	0.106	0.1	0.122	0.103
FOR	NN	0.043	0.025	0.044	0.101	0.049	0.037	0.075
	SVM	0.038	0.017	0.03	0.082	0.034	0.028	0.058
	PSVM	0.026	0.013	0.024	0.056	0.023	0.021	0.04

NN in classifying this type of data. The TPR index, allocation, and sensitivity for PSVM method with 82–95%, in NN in the range of 69–90% and in the SVM method in the range of 93% to −75%. One main problems of the NN method is the need for more training data and input features. In contrast, the SVM method, which only has one classification and does not need much training data, showed higher efficiency in the classification of this type of data than NN. While PSVM and NN methods

Table 9.5 Performance evaluation based on metrics of Ratios for evaluating pavement cracking based on PSVM.

Metric	Method	LC	TC	DC	BC	AC	SC	MC
LRPT	NN	31.5	13.23	13.48	10.99	18.56	19.41	14.77
	SVM	38.41	17.7	24.54	16.61	20.87	26.88	18.74
	PSVM	56.28	31.26	55.85	25.1	18.68	47.8	21.89
LRNT	NN	0.14	0.22	0.22	0.32	0.1	0.2	0.21
	SVM	0.13	0.15	0.16	0.26	0.07	0.15	0.17
	PSVM	0.09	0.12	0.12	0.18	0.05	0.11	0.11
LRPS	NN	20.97	24.27	16.82	7.85	18.56	20.69	13.2
	SVM	24.19	39.55	27.55	10.36	26.58	30.43	18.47
	PSVM	35.77	61.26	38.57	16	38.4	45.2	27.2
LRNS	NN	0.1	0.4	0.28	0.23	0.1	0.26	0.16
	SVM	0.08	0.34	0.17	0.16	0.09	0.2	0.13
	PSVM	0.06	0.23	0.09	0.11	0.1	0.12	0.11

exhibited similar performance in the TNR index, in general, the PSVM method performed slightly better by about 0.02%.

The efficiency and repeatability of these methods were evaluated using the PPV index. The results show that the index for PSVM method is in the range of 77–94%, which is better than that of the NN and SVM methods. The reason for this is the utilization of the modified polar method and changing the angle and length of the arc, which increase efficiency of PSVM. In all types of cracks, the PSVM method achieved a better PPV index than the other two methods. Moreover, PSVM also had a better NPV index, which can be used to effectively isolate images that were entered into the damaged image bank during the detection stage. By analyzing type 1 and 2 errors, based on alpha and beta testing, PSVM showed the best performance with the lowest error rate of 1.5–5.1%. Also, the set of FNR and FPR indicators for PSVM was in the same range, which indicates the higher efficiency of this method than the NN and SVM methods. In order to evaluate the efficiency of these methods, a q test was performed. The results show that the FDR index for NN and SVM is a number greater than 1, while for the PSVM method, this index is equal to 0.41. In general, in terms of the classification of types of cracks, except for fatigue cracking, the PSVM method provided the best results. This can be attributed to the presence of spall cracks in fatigue cracking compared to other cracks, which has a negative effect on the analysis using the Radon method and causes ambiguity in the interpretation of this type of failure. By performing more experiments and also selecting the range with fuzzy nature at the boundary of the separator plates, the relationship between this type of crack and other cracks can be evaluated. The FOR index for the neural network was found to be 2.5–10%, which is greater than 1.69–8.21% for SVM and 1.27–1.59% for PSVM. In general, the PSVM method offers higher efficiency in failure separation by type and pattern compared to the other two methods.

Additional statistics were performed to evaluate the performance of the PSVM method, as shown in Table 9.6. The PSVM method has an index of 18–56 for LRPT, which is 2–4 times that of the NN and SVM methods.

In terms of the LRNT index, PSVM exhibits five times the efficiency for the category of distress. Also, according to the LRPS index, the PSVM method is in the range of 16–61, while the NN and

Table 9.6 Performance evaluation based on metrics of Additional statistics for evaluating pavement cracking based on PSVM.

Metrics	Methods	LC	TC	DC	BC	AC	SC	MC
ACC	NN	0.88	0.69	0.76	0.74	0.91	0.78	0.83
	SVM	0.9	0.75	0.84	0.8	0.92	0.83	0.86
	PSVM	0.92	0.83	0.9	0.86	0.93	0.88	0.89
Error = 1 − Acc	NN	0.92	0.87	0.87	0.82	0.93	0.88	0.87
	SVM	0.92	0.9	0.91	0.85	0.94	0.91	0.9
	PSVM	0.95	0.93	0.93	0.9	0.95	0.94	0.92
TPR	NN	0.85	0.65	0.71	0.66	0.86	0.74	0.76
	SVM	0.87	0.73	0.81	0.74	0.88	0.8	0.81
	PSVM	0.9	0.81	0.88	0.82	0.89	0.86	0.85
TNR	NN	19	49	20	50	0	16.67	25
	SVM	19	45	6	49	11	10.67	19
	PSVM	13	23	11	30	31	0.33	0.5
PPV	NN	0.006	0.011	0.003	0.006	0	0.003	0.003
	SVM	0.006	0.01	0.001	0.006	0.001	0.002	0.002
	PSVM	0.004	0.005	0.002	0.003	0.004	0	0
NPV	NN	1.67	1.28	1.27	1.06	1.72	1.41	1.39
	SVM	1.76	1.51	1.58	1.26	1.89	1.61	1.57
	PSVM	1.99	1.76	1.87	1.53	2	1.87	1.76
FPR = Type I error	NN	0.83	0.73	0.73	0.63	0.86	0.77	0.75
	SVM	0.85	0.81	0.82	0.7	0.89	0.82	0.8
	PSVM	0.89	0.86	0.86	0.8	0.9	0.87	0.85
FNR = Type II error	NN	0.88	0.69	0.76	0.74	0.91	0.78	0.83
	SVM	0.9	0.75	0.84	0.8	0.92	0.83	0.86
	PSVM	0.92	0.83	0.9	0.86	0.93	0.88	0.89

Table 9.7 The performance measures to evaluate the model for the classification of good density and low density.

	Problem 1		Problem 2	
Performance measure	**Test data**	**All data**	**Test data**	**All data**
Accuracy	0.91	0.90	0.93	0.99
Precision	0.92	0.92	0.88	0.98
MCC	0.81	0.8	0.86	0.98

SVM methods have a smaller range. The LRNS index for PSVM is lower than the other two methods. In Table 9.7, the BalAcc index is an important parameter for evaluating the categories due to the unequal number of data. This index was 90–95% for the PSVM, and 85–94% and 82–93% for NN and SVM, respectively.

The PSVM method has a lower Chisq (χ^2) ratio of with 1.4–2.7% than the other two methods. Also, the Auto_Manu index is greater than two other methods. The DIBC index is almost the same for NN and SVM and shows higher efficiency for PSVM. The Youden's index (YOI) index also shows a higher value for the PSVM method.

By examining complementary indicators, such as NPV and FDR for linear and surface cracks, it can be concluded that in the proposed method, the reproducibility of surface failures is higher than linear failures. According to the NPV index, linear failures of 97.91% and surface distress 96% correctly classify negative values. Overall the PSVM method is highly efficient in separating linear and surface faults.

Based on all evaluation and measurement parameters, including four categories of evaluation indicators, the PSVM method has high efficiency for the diagnosis and isolation pavement cracks with better performance compared to the NN method (SVM also exhibited better performance).

9.5.4 Case 4: Quality Evaluation for Determining Bulk Density of Aggregates

The purpose of this case study was to evaluate the quality of a new method for determining the bulk density of aggregates. An acoustic system was developed to measure and quantify the total density of the sample mixture. For this purpose, a hardware (an acoustic sensor) for data collection and a software (based on a NN method) is used for data analysis. Depending on how the test was performed, the steel plate was placed on the aggregate, then the sinker was dropped on the steel plate. The sound effect was analyzed using one-dimensional signal processing technology, and a neural network was used to classify the audio data into three general densities (good, medium, and low density). A classification accuracy of 77.1% was reported in the test set and 96.5% in all datasets. Another neural network was developed to classify audio data into two general densities (good density and low density). The classification accuracy was 90.6% in the experimental set and 89.6% in the entire dataset. The evaluation results are presented based on the metrics introduced in this chapter. In order to categorize metrics in this case study, a 6-30-2 neural network with traincgp function was used to identify pattern 1, a 6-35-35-3 structure was applied in the architecture of the neural network. Histogram performance indicates remote points that are known as noise and are not relevant to other data. Figure 9.10 shows the error histogram for the neural network selected to solve Problem 1. Blue, green, and red data represent training, validation, and test data, respectively. As indicated by these histograms, the number of outbound data is negligible (Figure 9.11).

The MSE curve is used to verify the network's performance. Figure 9.12 of Section B shows the average square error for training, testing, and validation data. With the increase of the validation error, which occurred in repetition 25, the training was stopped using the stop criterion. As shown in Figure 9.12, the result is reasonable due to the following considerations: (a) the average error of the final square is small; (b) test set error and validation set error have similar features; and (c) no significant additional correlation occurred with 19 iterations (where the best validation performance occurs).

The classification results are shown in a gauge matrix in Figure 9.12. A confusion matrix contains information about actual and predicted classifications performed by a classification system. Each

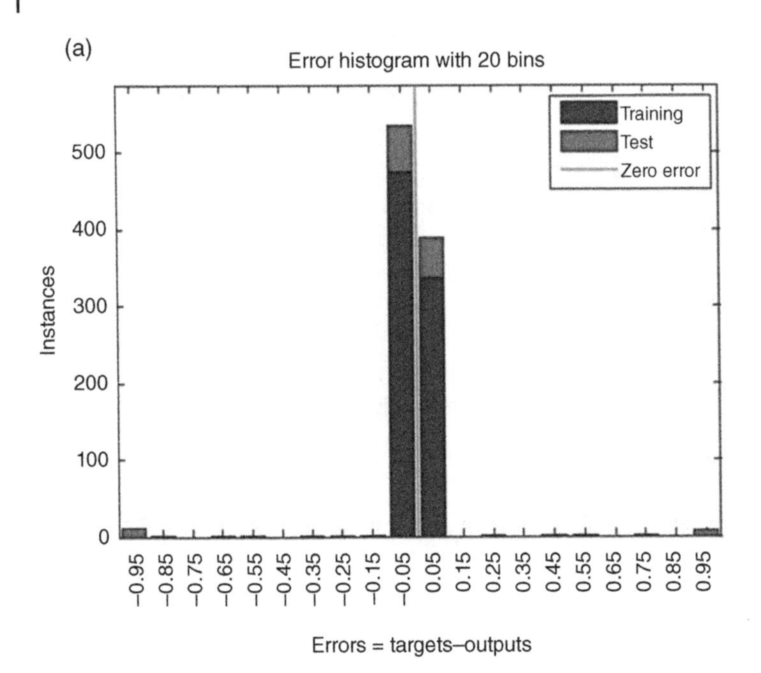

Figure 9.10 Error histogram for (a) case 1 and (b) case 2.

column of the matrix represents the samples in a predicted class, while each row represents the samples in a real class. Accurate performance criteria were used to evaluate the model for classification problems 1 and 2. The accuracy of classifying the number of correct predictions is divided by the total number of predictions made. Accuracy is the number of real positives divided by the number of real positives and FPs.

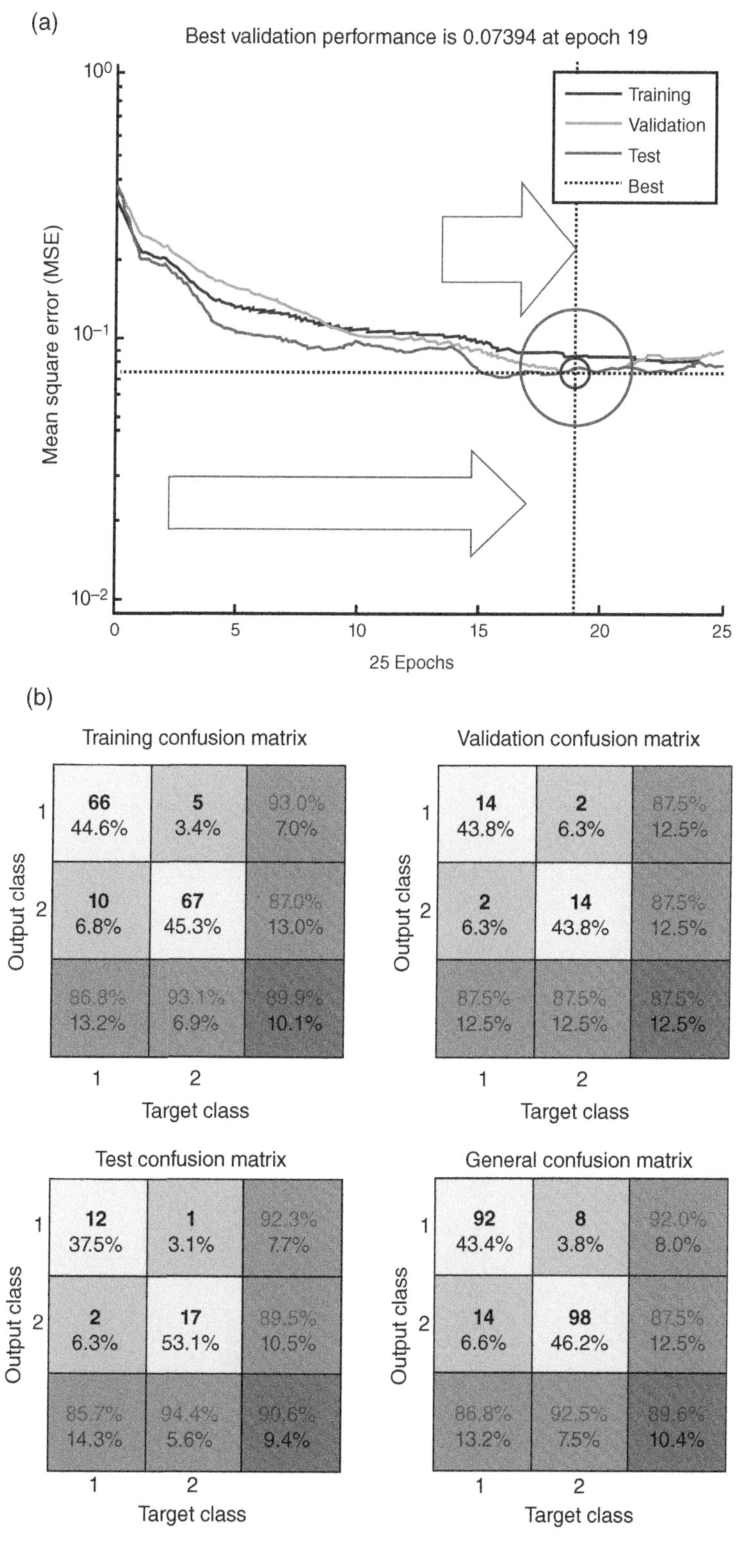

Figure 9.11 Error histogram for (a) training, validation, and test performance (MSE) and (b) training, validation, and test confusion matrixes.

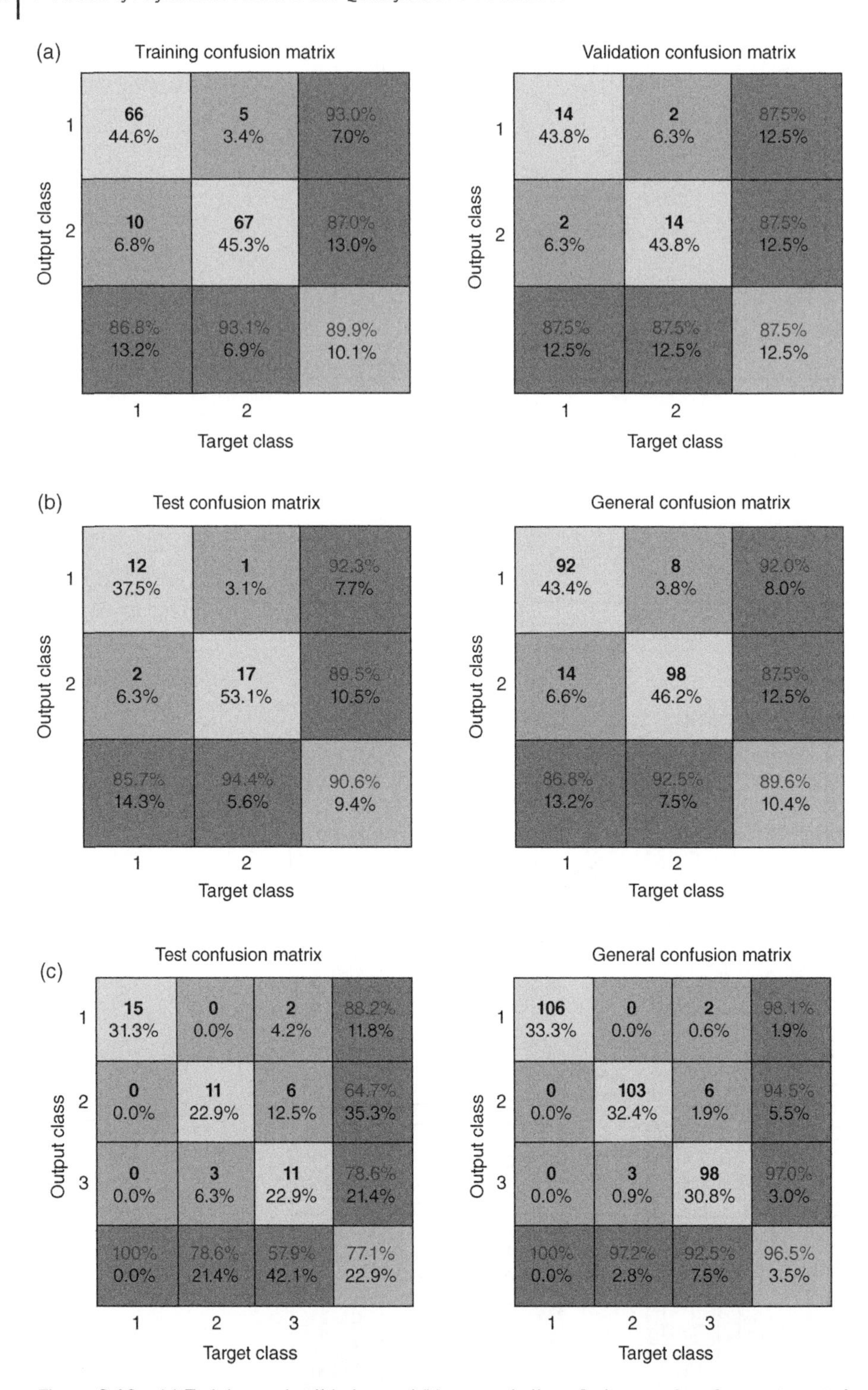

Figure 9.12 (a) Training and validation and (b) test and all confusion matrices for two target classes; (c) test confusion and all confusion matrices for three target classes. In this matrix, the brighter cells are perfectly matched, darker cells have one level of difference, gray cells have shows the matrix, and cell(4,4) shows the accuracy/error type 1.

MCC is another performance criterion for evaluating the model of classification problems. This coefficient is used as a measure of the quality of classifications. The MCC metric considers both positive and FPs and negatives.

Figure 9.12 shows the confusion matrices for training, testing, and validation and three types of data for the classification problem. According to the matrices, the prediction accuracy is 89.9%, 87.5% in the validation set, 90.6% for the test set, and 89.6% for the entire dataset example Table 9.7. Figure 9.12 also displays the test turbulence matrix and all the confusing matrices for the classification problem 2. According to these matrices, the neural network prediction accuracy is 77.1% in the experimental set and 96.5% for the entire dataset. Other useful performance metrics (MCC and precision) are listed in the table. These performance measures indicate that the density has good classification and is low. The third row in Table 9.6 presents the MCC of the two problems, which is essentially a correlation coefficient between observed and predicted classifications that returns a value between −1 and +1. A coefficient of +1 indicates a complete prediction, 0 means it is not better than a random prediction, and −1 indicates a complete mismatch between prediction and observation. As shown in the Table 9.6, MCC values for classification problems 1 and 2 are in the range of 0.8–0.98, which is acceptable.

It can further be noted that adding medium density samples does not increase or decrease the accuracy of good density classification and low density.

9.6 Summary and Conclusion

This chapter presents methods for evaluating the performance of classifiers and classification algorithms based on criteria with varying degrees of sensitivity. It can be said that the performance evaluation section as a controller of algorithm performance is one of the most important parts of deciding whether to use a method or not. In infrastructure management, this part is very important due to the importance of evaluating the methods used for smart identification and classification. As mentioned, these indicators are generally complementary, and the adequacy of one indicator alone does not mean the overall guarantee of the method. Therefore, it is necessary to use a set of these indicators to evaluate and select the method. Also, the results obtained from the evaluation methods depend on the analyst, harvest instructions, technology used, harvest conditions, analysis speed, and many other related factors. For these reasons, it is possible to obtain different results from evaluating automated methods. In this chapter, the metrics and indicators for evaluating the performance of the methods were briefly discussed and presented using various case studies in the field of infrastructure management. An attempt was made to present the application of these indicators in applied research. A general classification of the performance evaluation methods and common indicators was also provided.

In this chapter, in a general category of evaluation metrics is presented that includes (i) General statistics, (ii) Basic rations, (iii) Rations of ratios, (iv) Additional statistics, and (v) Operating characteristic. First, the method of organizing and dealing with the database to build a database, validation and testing is presented and the complexity matrix as an important structural variable, the task of summarizing and organizing the inputs to produce a variety of indicators such as accuracy, error, probability of detection, productive index, selectivity, reproduction, NPV, FPR, FNR, FDR, FOR, LRPT, LRNT, LRPS, LRNS. Also, using these indicators, new indicators such as F measurement, BalAcc, MCC, Chisq: χ^2, difference between automatic and manual methods, differentiation index are used for comparison and evaluation. Finally, to evaluate the overall performance of a method, the complexity matrix can be evaluated based on the index matrix.

9.7 Questions and Exercises

1 Define the types of evaluation metrics of classification algorithms. Mention the simplified relationships to select the algorithm. How are method selection thresholds set for each metric? Explain how to choose an algorithm by giving an example in the field of infrastructure management. If the properties of the two algorithms overlap, and there are superior differences in the metrics (i.e. if an algorithm is selected according to one metric while another algorithm is selected according to another metric), how is the final selection made?

2 Is there a priority in choosing metrics? Can each metric be used alone, or is it better to use metrics as a committee? Determine the relationship between the metrics, and synthesize hybrid metrics based on a confusion matrix.

3 Define the confusion matrix and provide a real example related to infrastructure management in at least 5 categories. Perform classification operations using three categorizers. Calculate the complexity of time and space for each class.

4 For a set of databases with 1000, 24 000, and 100 000 data, determine the three sections of training, validation, and testing. Based on the cross-validation model, create a model with 5, 10, and 20 folds, using the results of the matrix of confusion and error. Calculate the evaluation indicators in all five categories and analyze the sensitivity tables. Also, check the regulation and overfitting for each indicator and report the results based on unbalanced data.

5 Calculate the receiver operator characteristics (ROC) and area under the curve (AUC) for a database with 2000 data, and for the three preferred classifier methods, make a performance evaluation of the three classifiers based on defined thresholds. Also, calculate the results based on metrics. Compare Method 1 and Method 2 (use 10-fold method and report the results once again).

6 The sub-matrix table shows the confusion for evaluating the paving of a surface using image-based analysis. For this evaluation, 1000 pavement sections were analyzed via automated algorithms and in-depth image processing and the outputs are compared with the results of the classification model. Using the confusion matrix, calculate the types of metrics and interpret the efficiency of the algorithm. Compare the results of Algorithm A with Algorithm B and select the superior algorithm.

Table A The confusion matrix for evaluating pavement Bleeding evaluation based on classifier A.

		Predicted class		
True class		Low (L)	Medium (M)	High (H)
	Low (L)	923	37	40
	Medium (M)	23	550	27
	High (H)	37	23	640

Table B The confusion matrix for evaluating pavement Bleeding evaluation based on classifier B.

		Predicted class		
True class		Low (L)	Medium (M)	High (H)
	Low (L)	943	30	27
	Medium (M)	25	532	40
	High (H)	42	15	643

7 Prove the following equation:

$$\text{Type I Error} + \text{Type II Error} = 2 - (\text{Sensitivity} + \text{Specificity})$$

Further Reading

1 Ab Ghani, N.L., S.Z.Z. Abidin, and N.E. Abd Khalid. *Accuracy assessment of urban growth pattern classification methods using confusion matrix and ROC analysis.* in *International Conference on Soft Computing in Data Science.* 2015. (pp. 255–264), Springer: Singapore.

2 Abdel-Qader, I., O. Abudayyeh, and M.E. Kelly, Analysis of edge-detection techniques for crack identification in bridges. *Journal of Computing in Civil Engineering*, 2003. **17**(4): p. 255–263.

3 Adhikari, R., O. Moselhi, and A. Bagchi, Image-based retrieval of concrete crack properties for bridge inspection. *Automation in Construction*, 2014. **39**: p. 180–194.

4 Adu-Gyamfi, Y.O., T. Tienaah, N. O. Attoh-Okine, and C. Kambhamettu, Functional evaluation of pavement condition using a complete vision system. *Journal of Transportation Engineering*, 2014. **140**(9), 04014040. (pp. 1–10), https://doi.org/10.1061/(ASCE)TE.1943-5436.0000638.

5 Alsharqawi, M., T. Zayed, and S. Abu Dabous, Integrated condition rating and forecasting method for bridge decks using visual inspection and ground penetrating Radar. *Automation in Construction*, 2018. **89**: p. 135–145.

6 Mati, B., M. Zahedi, M. Nejad, Pavement friction and skid resistance measurement methods: A literature review. *Open Journal of Civil Engineering*, 2016. **6**: p. 537–565.

7 Barrile, V. G. Candela, A. Fotia, and E. Bernardo, *UAV survey of bridges and viaduct: Workflow and application.* in *International Conference on Computational Science and Its Applications.* 2019. (pp. 269–284), Springer: Cham.

8 Chen, L.-C., Y.-C. Shao, H.-H. Jan, C.-W. Huang, and Y.-M. Tien, Measuring system for cracks in concrete using multitemporal images. *Journal of Surveying Engineering*, 2006. **132**(2): p. 77–82.

9 Cheng, Y.M. and S.S. Leu, Constraint-based clustering model for determining contract packages of bridge maintenance inspection. *Automation in Construction*, 2008. **17**(6): p. 682–690.

10 Chun, P.J., S. Izumi, and T. Yamane, Automatic detection method of cracks from concrete surface imagery using two-step light gradient boosting machine. *Computer-Aided Civil and Infrastructure Engineering*, 2020. **36**(1), 61–72.

11 Deng, J., Y. Lu, and V.C.S. Lee, Concrete crack detection with handwriting script interferences using faster region-based convolutional neural network. *Computer-Aided Civil and Infrastructure Engineering*, 2020. **35**(4): p. 373–388.

12 Deng, W., H. Zhao, L. Zou, G. Li, X. Yang, and D. Wu, A novel collaborative optimization algorithm in solving complex optimization problems. *Soft Computing*, 2017. **21**(15): p. 4387–4398.

13 Deng, X., Q. Liu, Y. Deng, and S. Mahadevan, An improved method to construct basic probability assignment based on the confusion matrix for classification problem. *Information Sciences*, 2016. **340**: p. 250–261.

14 Dorafshan, S., R.J. Thomas, and M. Maguire, Comparison of deep convolutional neural networks and edge detectors for image-based crack detection in concrete. *Construction and Building Materials*, 2018. **186**: p. 1031–1045.

15 Dung, C.V. and L.D. Anh, Autonomous concrete crack detection using deep fully convolutional neural network. *Automation in Construction*, 2019. **99**: p. 52–58.

16 Ebrahimkhanlou, A., A. Athanasiou, T. D. Hrynyk, O. Bayrak, and S. Salamone, Fractal and multifractal analysis of crack patterns in prestressed concrete girders. *Journal of Bridge Engineering*, 2019. **24**(7): p. 04019059.

17 Ellenberg, A., A. Kontsos, F. Moon, and I. Bartoli, Bridge Deck delamination identification from unmanned aerial vehicle infrared imagery. *Automation in Construction*, 2016. **72**: p. 155–165.

18 Eschmann, C. and T. Wundsam, Web-based georeferenced 3D inspection and monitoring of bridges with unmanned aircraft systems. *Journal of Surveying Engineering*, 2017. **143**(3): p. 04017003.

19 Garcia-Balboa, J.L., M. V. Alba-Fernandez, F. J. Ariza-López, and J. Rodriguez-Avi, *Homogeneity test for confusion matrices: A method and an example.* in *IGARSS 2018–2018 IEEE International Geoscience and Remote Sensing Symposium.* 2018. (pp. 1203–1205). IEEE.

20 Gibb, S., H. M. La, T. Le, L. Nguyen, R. Schmid, and H. Pham, Nondestructive evaluation sensor fusion with autonomous robotic system for civil infrastructure inspection. *Journal of Field Robotics*, 2018. **35**(6): p. 988–1004.

21 Golparvar-Fard, M., V. Balali, and J.M. De La Garza, Segmentation and recognition of highway assets using image-based 3D point clouds and semantic Texton Forests. *Journal of Computing in Civil Engineering*, 2015. **29**(1), 04014023. (pp. 1–53).

22 Guan, H., J. Li, Y. Yu, M. Chapman, and C. Wang, Automated road information extraction from mobile laser scanning data. *IEEE Transactions on Intelligent Transportation Systems*, 2015. **16**(1): p. 194–205.

23 Guan, H., J. Li, Y. Yu, M. Chapman, H. Wang, C. Wang, Iterative tensor voting for pavement crack extraction using mobile laser scanning data. *IEEE Transactions on Geoscience and Remote Sensing*, 2014. **53**(3): p. 1527–1537.

24 Huston, D.R., J. Miller, and B. Esser. *Adaptive, robotic, and mobile sensor systems for structural assessment.* in *Smart Structures and Materials 2004: Sensors and Smart Structures Technologies for Civil, Mechanical, and Aerospace Systems.* 2004. (Vol. **5391**, pp. 189–196). International Society for Optics and Photonics.

25 Islam, M. and J.-M. Kim, Vision-based autonomous crack detection of concrete structures using a fully convolutional encoder–decoder network. *Sensors*, 2019. **19**(19): p. 4251.

26 Jahanshahi, M.R., J. S. Kelly, S. F. Masri, and G. S. Sukhatme, A survey and evaluation of promising approaches for automatic image-based defect detection of bridge structures. *Structure and Infrastructure Engineering*, 2009. **5**(6): p. 455–486.

27 Jahanshahi, M.R. and S.F. Masri, Adaptive vision-based crack detection using 3D scene reconstruction for condition assessment of structures. *Automation in Construction*, 2012. **22**: p. 567–576.

28 Jahanshahi, M.R., S. F. Masri, C. W. Padgett, and G. S. Sukhatme, An innovative methodology for detection and quantification of cracks through incorporation of depth perception. *Machine Vision and Applications*, 2013. **24**(2): p. 227–241.

29 Jiang, J., H. Liu, H. Ye, and F. Feng, Crack enhancement algorithm based on improved EM. *Journal of Information and Computational Science*, 2015. **12**(3): p. 1037–1043.

30 Jiang, Y., X. Zhang, and T. Taniguchi, Quantitative condition inspection and assessment of tunnel lining. *Automation in Construction*, 2019. **102**: p. 258–269.

31 Kim, B. and S. Cho, Image-based concrete crack assessment using mask and region-based convolutional neural network. *Structural Control and Health Monitoring*, 2019, **26**(8): p. e2381.

32 Kim, H., E. Ahn, M. Shin, and S.-H. Sim, Crack and noncrack classification from concrete surface images using machine learning. *Structural Health Monitoring*, 2019. **18**(3): p. 725–738.

33 Kim, I.-H., H. Jeon, S.-C. Baek, W.-H. Hong, and H.-J. Jung, Application of crack identification techniques for an aging concrete bridge inspection using an unmanned aerial vehicle. *Sensors*, 2018. **18**(6): p. 1881.

34 Koch, C., K. Georgieva, V. Kasireddy, B. Akinci, and P. Fieguth, A review on computer vision based defect detection and condition assessment of concrete and asphalt civil infrastructure. *Advanced Engineering Informatics*, 2015. **29**(2): p. 196–210.

35 Le, D.B., S. D. Tran, J. L. Torero, and V. T. Dao, Application of digital image correlation system for reliable deformation measurement of concrete structures at high temperatures. *Engineering Structures*, 2019. **192**: p. 181–189.

36 Lee, B.Y., Y. Y. Kim, S.-T. Yi, and J.-K. Kim, Automated image processing technique for detecting and analysing concrete surface cracks. *Structure and Infrastructure Engineering*, 2013. **9**(6): p. 567–577.

37 Li, S., Y. Cao, and H. Cai, Automatic pavement-crack detection and segmentation based on steerable matched filtering and an active contour model. *Journal of Computing in Civil Engineering*, 2017. **31**(5): p. 04017045.

38 Li, S., X. Zhao, and G. Zhou, Automatic pixel-level multiple damage detection of concrete structure using fully convolutional network. *Computer-Aided Civil and Infrastructure Engineering*, 2019. **34**(7): p. 616–634.

39 Lokeshwor, H., L.K. Das, and S. Goel, Robust method for automated segmentation of frames with/without distress from road surface video clips. *Journal of Transportation Engineering*, 2014. **140**(1): p. 31–41.

40 Luo, Q., B. Ge, and Q. Tian, A fast adaptive crack detection algorithm based on a double-edge extraction operator of FSM. *Construction and Building Materials*, 2019. **204**: p. 244–254.

41 Mataei, B., F. Nejad, H. Zakeri, and A. Gandomi, Computational intelligence for modeling of pavement surface characteristics. *New Materials in Civil Engineering*, (pp. 65–77). Butterworth-Heinemann 2020, https://doi.org/10.1016/B978-0-12-818961-0.00002-8.

42 Mathur, A. and G.M. Foody, Multiclass and binary SVM classification: Implications for training and classification users. *IEEE Geoscience and Remote Sensing Letters*, 2008. **5**(2): p. 241–245.

43 McGwire, K.C. and P. Fisher, *Spatially variable thematic accuracy: Beyond the confusion matrix*. in *Spatial Uncertainty in Ecology*. 2001. Springer. p. 308–329.

44 Montero, R., J. Victores, S. Martinez, A. Jardón, and C. Balaguer, Past, present and future of robotic tunnel inspection. *Automation in Construction*, 2015. **59**: p. 99–112.

45 Morgenthal, G. and N. Hallermann, Quality assessment of unmanned aerial vehicle (UAV) based visual inspection of structures. *Advances in Structural Engineering*, 2014. **17**(3): p. 289–302.

46 Nayyeri, F., L. Hou, J. Zhou, and H. Guan, Foreground–background separation technique for crack detection. *Computer-Aided Civil and Infrastructure Engineering*, 2019. **34**(6): p. 457–470.

47 Nejad, F.M., N. Karimi, and H. Zakeri, Automatic image acquisition with knowledge-based approach for multi-directional determination of skid resistance of pavements. *Automation in Construction*, 2016. **71(Part 2)**: p. 414–429.

48 Nejad, F.M. and H. Zakeri, The hybrid method and its application to smart pavement management. *Metaheuristics in Water, Geotechnical and Transport Engineering*, 2012. **439**.

49 Oh, J.-K., G. Jang, S. Oh, J. H. Lee, B.-J. Yi, Y. S. Moon, et al., Bridge inspection robot system with machine vision. *Automation in Construction*, 2009. **18**(7): p. 929–941.

50 Phillips, S. and S. Narasimhan, Automating data collection for robotic bridge inspections. *Journal of Bridge Engineering*, 2019. **24**(8): p. 04019075.

51 Prasanna, P., K. J. Dana, N. Gucunski, B. B. Basily, H. M. La, R. S. Lim, et al., Automated crack detection on concrete bridges. *IEEE Transactions on Automation Science and Engineering*, 2014. **13**(2): p. 591–599.

52 Protopapadakis, E., A. Voulodimos, A. Doulamis, N. Doulamis, and T. Stathaki, Automatic crack detection for tunnel inspection using deep learning and heuristic image post-processing. *Applied Intelligence*, 2019. **49**(7): p. 2793–2806.

53 Raschka, S., An overview of general performance metrics of binary classifier systems. arXiv preprint arXiv:1410.5330, 2014.

54 Russell, S.J. and P. Norvig, *Artificial Intelligence: A Modern Approach*. 2016. Malaysia: Pearson Education Limited.

55 Sokolova, M., N. Japkowicz, and S. Szpakowicz. *Beyond accuracy, F-score and ROC: A family of discriminant measures for performance evaluation*. in *Australasian Joint Conference on Artificial Intelligence*. 2006. (pp. 1015–1021). Springer: Berlin, Heidelberg.

56 Sollazzo, G., K. Wang, G. Bosurgi, and J. Liet al., Hybrid procedure for automated detection of cracking with 3D pavement data. *Journal of Computing in Civil Engineering*, 2016, **30**(6): p. 04016032.

57 Song, H., W. Wang, F. Wang, L. Wu, and Z. Wang, Pavement crack detection by ridge detection on fractional calculus and dual-thresholds. *International Journal of Multimedia and Ubiquitous Engineering*, 2015. **10**(4): p. 19–30.

58 Spencer Jr, B.F., V. Hoskere, and Y. Narazaki, Advances in computer vision-based civil infrastructure inspection and monitoring. *Engineering*, 2019. **5**(2), 199–222.

59 Tomiczek, A.P., T. J. Whitley, J. A. Bridge, and P. G. Ifju, Bridge inspections with small unmanned aircraft systems: Case studies. *Journal of Bridge Engineering*, 2019. **24**(4): p. 05019003.

60 Tsai, Y.C., V. Kaul, and R.M. Mersereau, Critical assessment of pavement distress segmentation methods. *Journal of Transportation Engineering*, 2010. **136**(1): p. 11–19.

61 Visa, S., B. Ramsay, A. L. Ralescu, and E. Van Der Knaap, Confusion matrix-based feature selection. *MAICS*, 2011. **710**: p. 120–127.

62 Wakchaure, S.S. and K.N. Jha, Review of inspection practices, health indices, and condition states for concrete bridges. *Indian Concrete Journal*, 2012. **86**(3): p. 13–26.

63 Wang, N., X. Zhao, P. Zhao, Y. Zhang, Z. Zou, and J. Ou, Automatic damage detection of historic masonry buildings based on mobile deep learning. *Automation in Construction*, 2019. **103**: p. 53–66.

64 Wang, R. and Y. Kawamura, Development of climbing robot for steel bridge inspection. *Industrial Robot: An International Journal*, 2016. **43**(4): p. 429–447.

65 Washer, G.A., Developments for the non-destructive evaluation of highway bridges in the USA. *NDT & E International*, 1998. **31**(4): p. 245–249.

66 Weng, X., Y. Huang, and W. Wang, Segment-based pavement crack quantification. *Automation in Construction*, 2019. **105**: p. 102819.

67 Wu, L., S. Mokhtari, A. Nazef, B. Nam, and H.-B. Yun, Improvement of crack-detection accuracy using a novel crack defragmentation technique in image-based road assessment. *Journal of Computing in Civil Engineering*, 2014. **30**(1): p. 04014118.

68 Yamaguchi, T. and S. Hashimoto, Fast crack detection method for large-size concrete surface images using percolation-based image processing. *Machine Vision and Applications*, 2010. **21**(5): p. 797–809.

69 Yeum, C.M., J. Choi, and S.J. Dyke, Automated region-of-interest localization and classification for vision-based visual assessment of civil infrastructure. *Structural Health Monitoring*, 2019. **18**(3): p. 675–689.

70 Yeum, C.M. and S.J. Dyke, Vision-based automated crack detection for bridge inspection. *Computer-Aided Civil and Infrastructure Engineering*, 2015. **30**(10): p. 759–770.

71 Yu, S.-N., J.-H. Jang, and C.-S. Han, Auto inspection system using a mobile robot for detecting concrete cracks in a tunnel. *Automation in Construction*, 2007. **16**(3): p. 255–261.

72 Zakeri, H., F.M. Nejad, and A. Fahimifar, Image based techniques for crack detection, classification and quantification in asphalt pavement: A review. *Archives of Computational Methods in Engineering*, 2016, **24**(4), 935–977.

73 Zakeri, H., F.M. Nejad, and A. Fahimifar, Rahbin: A quadcopter unmanned aerial vehicle based on a systematic image processing approach toward an automated asphalt pavement inspection. *Automation in Construction*, 2016. **72**: p. 211–235.

74 Zakeri, H., F.M. Nejad, and A. Fahimifar, Rahbin: A quadcopter unmanned aerial vehicle based on a systematic image processing approach toward an automated asphalt pavement inspection. *Automation in Construction*, 2016. **72**: p. 211–235.

75 Zeng, G., On the confusion matrix in credit scoring and its analytical properties. *Communications in Statistics-Theory and Methods*, 2020. **49**(9): p. 2080–2093.

76 Zhang, X., D. Rajan, and B. Story, Concrete crack detection using context-aware deep semantic segmentation network. *Computer-Aided Civil and Infrastructure Engineering*, 2019, **34**(11), 951–971.

77 Zhong, B., H. Wu, L. Ding, P. E. Love, H. Li, H. Luo, et al., Mapping computer vision research in construction: Developments, knowledge gaps and implications for research. *Automation in Construction*, 2019. **107**: p. 102919.

78 Zhou, J., P.S. Huang, and F.-P. Chiang, Wavelet-based pavement distress detection and evaluation. *Optical Engineering*, 2006. **45**(2): p. 027007–027007-10.

79 Zhu, Z., S. German, and I. Brilakis, Visual retrieval of concrete crack properties for automated post-earthquake structural safety evaluation. *Automation in Construction*, 2011. **20**(7): p. 874–883.

80 Znidaric, A., V. Pakrashi, and E.J. O'Brien, A review of road structure data in six European countries. *Proceedings of the Institution of Civil Engineers Journal of Urban Design and Planning*, 2011. **164**(4): p. 225–232.

81 Zou, Q., Y. Cao, Q. Li, Q. Mao, and S. Wang, CrackTree: Automatic crack detection from pavement images. *Pattern Recognition Letters*, 2012. **33**(3): p. 227–238.

10

Nature-Inspired Optimization Algorithms (NIOAs)

10.1 Introduction

Nature-inspired optimization algorithms (NIOAs) are a set of algorithms that are inspired by the behavior of natural phenomena, such as simulating congestion intelligence, biological systems, and physical and chemical systems. In general, biology-inspired algorithms include a class of algorithms based on crowding, swarm, and evolutionary intelligence. NIOAs are an active and important branch of artificial intelligence (AI) that continuously evolve; therefore, it is not unreasonable to expect that as you read this chapter, several new algorithms will be born.

Thus far, a large number of interesting NIOAs inspired by animal behaviors and biological, physical, and chemical systems have been proposed and applied in numerous engineering fields, especially for optimization. Some NIOAs include the genetic algorithm (GA), Particle Swarm Optimization (PSO) algorithm, Differential Evolution (DE) algorithm, Artificial Bee Colony (ABC) algorithm, Ant Colony Optimization Algorithm (ACO), Cuckoo Search (CS) algorithm, Bat Algorithm (BA), Firefly Algorithm (FA), Immune Algorithm (IA), Gray Wolf Optimizer (GWO), Gravitational Search algorithm (GSA), and Harmony Search (HS) algorithm. Much newer algorithms have also been proposed in this area recently, such as the horse herd optimization algorithm, Mayfly Optimization Algorithm, Chimp Optimization Algorithm, Coronavirus Optimization Algorithm, Water strider algorithm, Newton metaheuristic algorithm, Black Widow Optimization Algorithm, Harris hawks optimization, Sailfish Optimizer, Spider Monkey Optimization, Grasshopper Optimization Algorithm, Fractal Based Algorithm, Bacterial Foraging Inspired Algorithm, Rain-fall Optimization Algorithm, Dragonfly algorithm, Sperm Whale Algorithm, Water Wave Optimization, Ant Lion Optimizer, Symbiotic Organisms Search, Egyptian Vulture Optimization Algorithm, Dolphin echolocation, Great Salmon Run, Big Bang-Big Crunch, Flower Pollination Algorithm, Spiral Optimization Algorithm, Galaxy-based Search Algorithm, Japanese Tree Frogs, Termite Colony Optimization, CS, Glowworm Swarm Optimization, GSA, Fast Bacterial Swarming Algorithm, River Formation Dynamics, Imperialistic Competitive Algorithm, Roach Infestation Optimization, Cat Swarm Optimization, and krill herd optimization. (In Appendix, the glossary of the names of the relevant algorithms until 2021 is presented. These algorithms include 252 methods.)

In this chapter, the original theory and general idea of some NIOAs are summarized. Specifically, a discussion is provided to examine the use one or more NIOAs and describe a real-world example of a case study of how algorithms respond. At the end, a new algorithm is presented that simulates the behavior of emperor penguins along with theory and case study. This will help the reader to better understand how to simulate and model nature-inspired behaviors. By reading this chapter, you are expected to be able to design and build your own NIOA to solve optimization problems.

Automation and Computational Intelligence for Road Maintenance and Management: Advances and Applications, First Edition. Hamzeh Zakeri, Fereidoon Moghadas Nejad, and Amir H. Gandomi.

10.2 General Framework and Levels of Designing Nature-Inspired Optimization Algorithms (NIOAs)

The general framework and common design methods of most NIOAs are similar, but have different definitions for different models. Generally, the main idea of these methods is based on random search and movement in the response space. The purpose of this section is to establish a standardization method for identifying NIOAs and provide an integrated process and analysis of an idea and how the idea is implemented by NIOAs. First, a general framework for designing NIOAs is presented, then the common process in these methods is described in order to develop such models. Based on this framework, some common examples of NIOAs are given (other methods are available through various sources and references). In this regard, we suggest the very valuable book by Professor Young called "Nature-Inspired Optimization Algorithms," which is full of practical methods and examples, to the reader for further study.

In this section, the structure, general framework, and common steps of implementing different NIOA methods are shown in Figure 10.1. According to the classification proposed by previous researchers, the construction of each NIOA model has several main levels. In the first level L1, the number of populations and the initial parameters are determined based on the type of problem and expectation, the speed of reaching the answer, and the designer's experience in the NIOA model architecture. At this level, population abundance is usually created based on random methods. The population should be such that analysis time is minimal and the whole space of the problem can be covered. In all NIOA methods, time constraints, accuracy, and iteration of the process (iteration) are considered as conditions for termination of iteration and analysis.

The fitness function set is a measure of the performance of each method based on the objective function. These functions typically seek to find, at most or at least, one relation, and the algorithm error value can be calculated reliably using NIOA methods because the exact value is computationally known. In all problems, there may be two optimal search areas: (i) local optimal solution and (ii) global optimal solution. In general, optimization algorithms should be able to cover a large area of the response range and search in all response areas to avoid getting trapped in the local optimal solution.

In the second level L2, the values of the population fitness function are calculated at the end of each round. If the best global solution is implemented, the algorithm terminates and goes to the end level, L4; otherwise, level L3 aims to improve the information level and test new options and situations by performing data exchange operations among the entire population. At the end of L3, a new population as large as the L1 population, but with newer information and better response space experience are used in the L2 workflow process and repeated in the framework and structure. It should be noted that the main difference between different nature-inspired algorithms lies in creating the next population by changing the original population in level 3 – the other levels are almost the same for all algorithms. Therefore, level 3 is the most important of NIOAs and the main point of difference and simulation of nature-inspired behavior.

In order to create an integrated display and make a possible comparison, the symbols defined in Wang's et al. research (2021) are used. Specifically, D represents the dimensions of the objective functions; M is the individual number of each NIOA; N is the number of iterations; Pi(t) is the local optimal answer; Pg(i) is the universal optimal answer; $F(\cdot)$ is the objective function; and delta is the final termination condition of the algorithm.

In this section, the main idea of the most common NIOAs is briefly presented based on the levels shown in the framework in Figure 10.1. The general principles of these methods are described with simplicity so that you will become acquainted with the general concept and simplified structure of these algorithms. In general, the main steps of each NIOA pattern are:

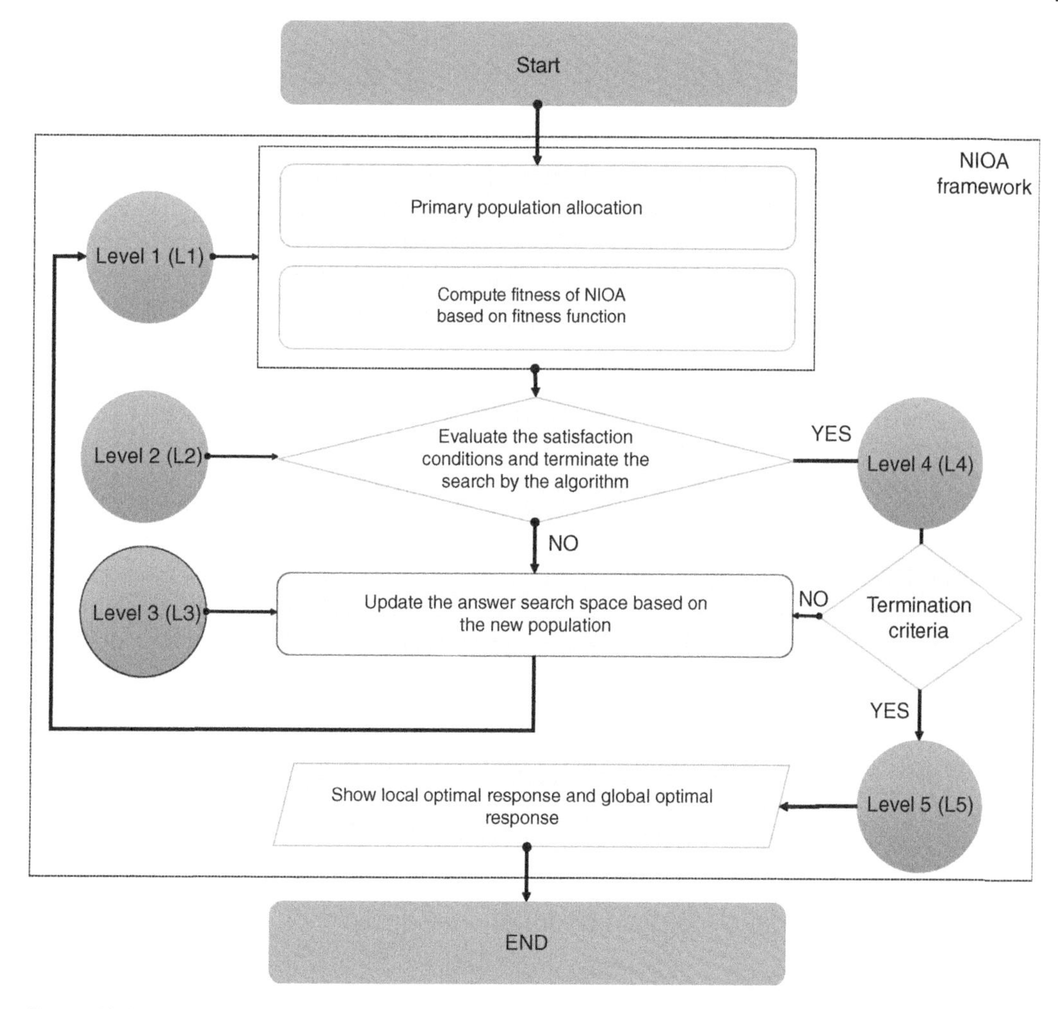

Figure 10.1 General framework and routine process of NIOAs.

Level-1: Primary population production;
Level-2: Evaluation of individuals by calculating their appropriate values;
Level-3: Creation of the next population by changing the original population;
Level-4: Termination criteria; and
Level-5: Selecting the answer with the highest value of fitness.

As mentioned, the general principles of each of these algorithms are similar, but may have different models for creating the next population or changing position due to different ideas in level 3.

10.3 Basic Principles of Important Nature-Inspired Algorithms (NIOAs)

In this section, common types of nature-inspired algorithms are summarized with emphasis on the third level for the prior-mentioned reasons.

10.3.1 Genetic Algorithm (GA)

GAs, which are based on the concepts of natural selection and genetics of living organisms, have become a widely popular method to perform the search as the main difference from other search methods, GAs work from a coded display to identify the answer to the problem, instead of doing a direct search, and assign a population of search points around a search space to find possible answers to the problem. This method approaches the problem without knowing the "gradient" information associated with the objective function. The probabilistic method is commonly applied by GAs for transitioning from one state in the problem space to another. By doing so, GAs are also considered as "multipurpose" search algorithms. Also, GAs have a special talent for searching irregular and complex search spaces.

The basic principles of GA are subsequently described. (i) Formulate the elementary population consisting of the answers to the problem: the initial and random quantification of the elementary population consists of the answers to the problem. (ii) Repeat loop: evaluate the objective function of the problem, find the right fitness function, and perform operations on populations consisting of problem-solving using genetic operators (reproduction, crossover, and mutation operators). (iii) Repeat until: the stop condition is met. In general, a "fitness function," or $F(i)$, is first formulated using the target function then used in subsequent generations after a series of genetic operations. Optimization is an indicator that measures the reproductive efficiency of chromosomes. In a GA, the fit function is used to assess the reproductive probability of chromosomes in a population. In other words, this index is a criterion for determining the goodness of chromosomes, which should usually be maximized. According to the idea of GA, the chromosomes that have the most fits will have a better chance of combining and mutating (genetic operators) than other chromosomes in the population. In maximization problems, the fit function can be defined as equivalent to the objective function as follows:

$$F(i) = O(i) \rightarrow \text{Objective function}(i) = \text{Fitness function}(i) \tag{10.1}$$

where

$O(i)$ is the objective function (i).
$F(i)$ is the fitness function (i).

In "minimization" problems, it is crucial that the main objective function of the problem be mapped to the fit function. There are several ways to do this.

$$\begin{aligned} F(x) &= \left(\frac{1}{1+f(x)}\right) \\ F(i) &= V - \frac{O(i)P}{\sum_{i=1}^{P} O(i)} \\ V &= \frac{O(i)P}{\sum_{i=1}^{P} O(i)} \end{aligned} \tag{10.2}$$

where

V is a large value that ensures that the fit values are positive.

The value of V in this operator is usually equal to the maximum value of the third expression of Eq. (10.3).

The purpose of the main operators in this method, including the reproduction, crossover, and mutation operators, is to manipulate the chromosome sample, which is the best answer to the optimization problem in the current population. In this way, a better new population of chromosomes will be produced. It is expected that the newly produced population will be continuously evaluated until level 4. In other words, the population of chromosomes or response vectors is evaluated by simultaneous manipulation operators until the condition for stopping the GA is satisfied. This process continues until the stop criterion is met. As mentioned, GA which works on the basis of three main operators:

- Selection operator (so) mimics natural selection,
- Crossover operator (co) represents genetic evolution (or mutation?), and
- Mutation operator (mo).

The roulette wheel selection, stochastic universal sampling, local selection, and tournament selection for so, one-point crossover, two-point crossover, multi-point crossover, and uniform crossover for co, and the basic mutation operator (chooses one or more genes to randomly change), the inversion operator (randomly chooses two gene points to inverse the genes between two points), for mo.

Suppose that $o1$, $o2$, and $o3$ are the probabilities of selection, crossover, and mutation, respectively, the different levels of the GA are described in Table 10.1 based on the steps shown in Figures 10.2 and 10.3.

10.3.2 Particle Swarm Optimization (PSO) Algorithm

The PSO algorithm is designed to simulate the behavior of birds in crowds and has had many applications in optimizing engineering issues, including infrastructure management. In this algorithm, based on the movement of birds, the method of movement shows the position and velocity of the ith particle in the dth dimension for the $(t+1)$th repetition, which is calculated based on the following equation:

$$\begin{aligned} v_{id}^{t+1} &= v_{id}^{t} + c_1 r_1^t \left(\text{pbest}_{id}^{t} - x_{id}^{t}\right) + c_2 r_2^t \left(\text{gbest}_{d}^{t} - x_{id}^{t+1}\right) = x_{i,d}(t) + v_{i,d}(t+1) \\ x_{id}^{t+1} &= x_{id}^{t} + v_{id}^{t+1} \end{aligned} \tag{10.3}$$

where

$c(i)$ is the learning parameter (i).
$r(i)$ is the random number (i), which is in the range [0,1].
$V_i(t)$ is the velocity $v_i(t) = (v_i,\ 1(t), \ldots, v_i,\ d(t), \ldots, v_i,\ D(t))$.

In this method, speeds are limited to the maximum speed ($V_{\max}$) in each dimension. This limitation was effectively intended to control the optimal PSO global exploration capability and improve the coverage of search areas per generation. In this method, a large $V_{\max}$ can be globally explored, but using a logical $V_{\max}$ can also detect local extreme points. Therefore, we need to consider a parameter to create a balance between exploration and exploitation that is independent of $V_{\max}$. In this regard, the weight of inertia is presented as a correction parameter for PSO. Higher values of this parameter lead to more exploration and then achieve the best global answers, while lower values are exploitable for local search. The above equation is modified using the correction parameter as follows:

Table 10.1 Pseudo code of the Emperor Penguins Algorithm (EPA).

Emperor Penguins Algorithm	
Objective function	$EP = (s_{\Delta\theta}, \ldots, s_{n.\ \Delta\theta})^T$
Generate initial population of Emperor Penguins	$EP_i\ (i = 1, 2, \ldots, n)$
Heat H_i at arc-length $s_\theta$is determined by $f(\theta_i, S_i)$	$f(\theta_i, S_i)$
Define spiral function, absorption function	Archimedes' Spiral, Circle Involute, Conical Spiral, Cornu Spiral, Cotes' Spiral, Daisy, Epispiral, Fermat's Spiral, Helix, Hyperbolic Spiral, Logarithmic Spiral, Nielsen's Spiral, Phyllotaxis, Poinsot's Spirals, Polygonal Spiral, Rational Spiral, Spherical Spiral
Initialize arbitrary constants rates A and the seeped rate τ	A, τ

```
while (t <Max number of iteration )
  while (S>ε the minum arc-length)
    for i = 1 : n all n Emperor Penguins
      for j = 1 : i all n Emperor Penguins
        if (H_j > H_i),
          Move Emperor Penguins i towards j in d-dimension spirally;
          o   diversification
          o   intensification
        end if
        Attractiveness varies with distance (arc-length) via s^(i+1)_n = s^0_n(1 - e^(-ρn))
        Evaluate new solutions and update heat
        if (f(θ_i, S_i)> f(0,ε)*)
            Accept the new solutions
              update position
             update the position and heat
        end if
      end for j
    end for i,
    Rank the Emperor Penguins and find, the current best
  end while,
end while,
```

Post process results and visualization

$$
\begin{aligned}
v_{id}^{t+1} &= w.v_{id}^{t} + c_1 r_1^t\left(\text{pbest}_{id}^t - x_{id}^t\right) + c_2 r_2^t\left(\text{gbest}_d^t - x_{id}^{t+1}\right) = x_{i,d}(t) + v_{i,d}(t+1) \\
x_{id}^{t+1} &= x_{id}^t + v_{id}^{t+1}
\end{aligned} \tag{10.4}
$$

where

w is the weight of inertia.

In most AI methods, having basic knowledge about the probability of an answer can dramatically increase the efficiency and speed of the method. In the PSO method, according to the previous relation, since the selection of $V_{\max}$ value is done randomly without using the initial information, this value is initially equal to $X_{\max}$, and the weight of inertia is used to create variety and achieve better performance of PSO. Due to the two properties of $V_{\max}$ and weight of inertia to ensure PSO

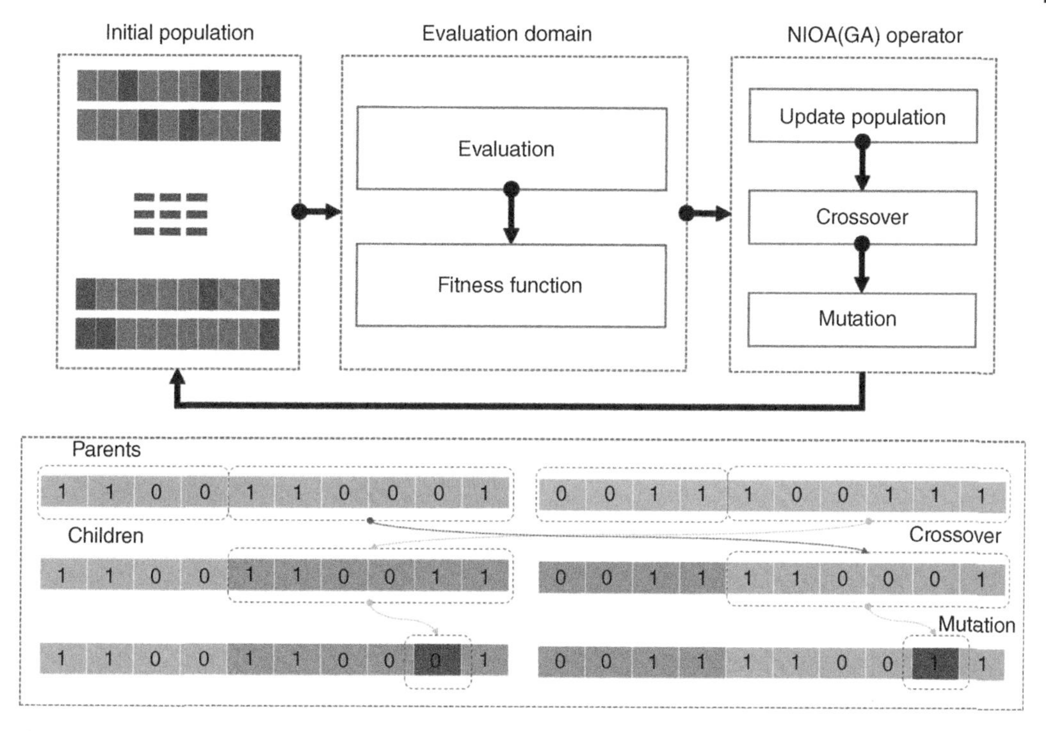

Figure 10.2 Crossover mutation chromosome in GA.

convergence, it is possible to use another parameter called constriction factor, which is calculated by the following equation:

$$
\begin{aligned}
v_{id}^{t+1} &= \chi\left[v_{id}^{t} + c_1 r_1^t\left(\text{pbest}_{id}^t - x_{id}^t\right) + c_2 r_2^t\left(\text{gbest}_d^t - x_{id}^t\right)\right] \\
x_{id}^{t+1} &= x_{id}^t + v_{id}^{t+1} \\
\chi &= \frac{2}{\left|2 - \phi - \sqrt{\phi^2 - 4\phi}\right|}
\end{aligned}
\tag{10.5}
$$

where

χ is the constriction factor, which is considered for the weight of inertia to limit the velocity.
ϕ_1, ϕ_2 are two positive constants that are experimentally equal to 2.05.

Based on the convergence factors, the optimal answer and control of the answer space need to be limited. For this purpose, three different boundary conditions are used, namely Absorbing Walls, reflective Walls, and Invisible Walls. In these cases, reflective walls are used for the proposed PSOs. At this boundary, when a particle hits the walls, the velocity sign changes, and the particle returns to the controlled response space. Binary PSO is a special case of PSO in which the position of the particles is determined by a binary vector.

The velocity vector is considered to be between 0 and 1 with the probability that each dimension will take a value based on the sigmoid function. Then, a uniform random number (r) is generated in [0, 1] and compared with the value of the sigmoid function and placed in the following equation (Figures 10.4 and 10.5):

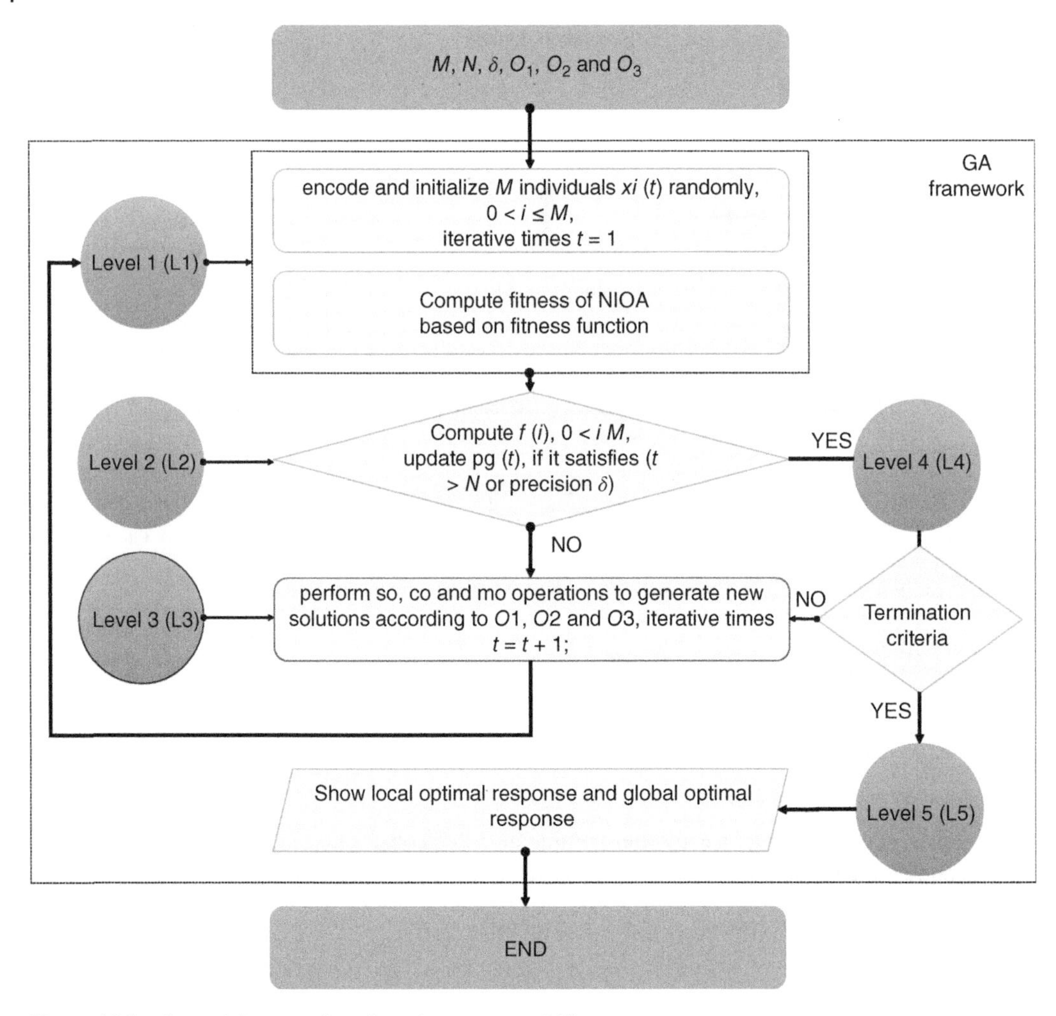

Figure 10.3 General framework and routine process of GA.

$$S(v_{id}) = \frac{1}{1 + \exp(-v_{id})}, x_{id} = \begin{cases} 1, & r < S(v_{id}) \\ 0, & \text{otherwise} \end{cases} \tag{10.6}$$

where

$V_i(t)$ is the velocity $v_i(t) = (v_i, 1(t), ..., v_i, d(t), ..., v_i, D(t))$.

10.3.3 Artificial Bee Colony (ABC) Algorithm

In order to simulate the behavior of bees, their group behavior and how they participate in food supply are considered. For this purpose, the different roles of bees simulated. In its basic model, three types of bees are considered: (i) employed forager bees (EFBs) are associated with a food source and are obliged to share it with other bees; (ii) scout bees (SBs) are responsible for searching for and finding new food sources; and (iii) onlooker bees (OBs) participate by sharing information. x_{id}^t is the position of bee i on tth iteration is x_i^t determined by the equation:

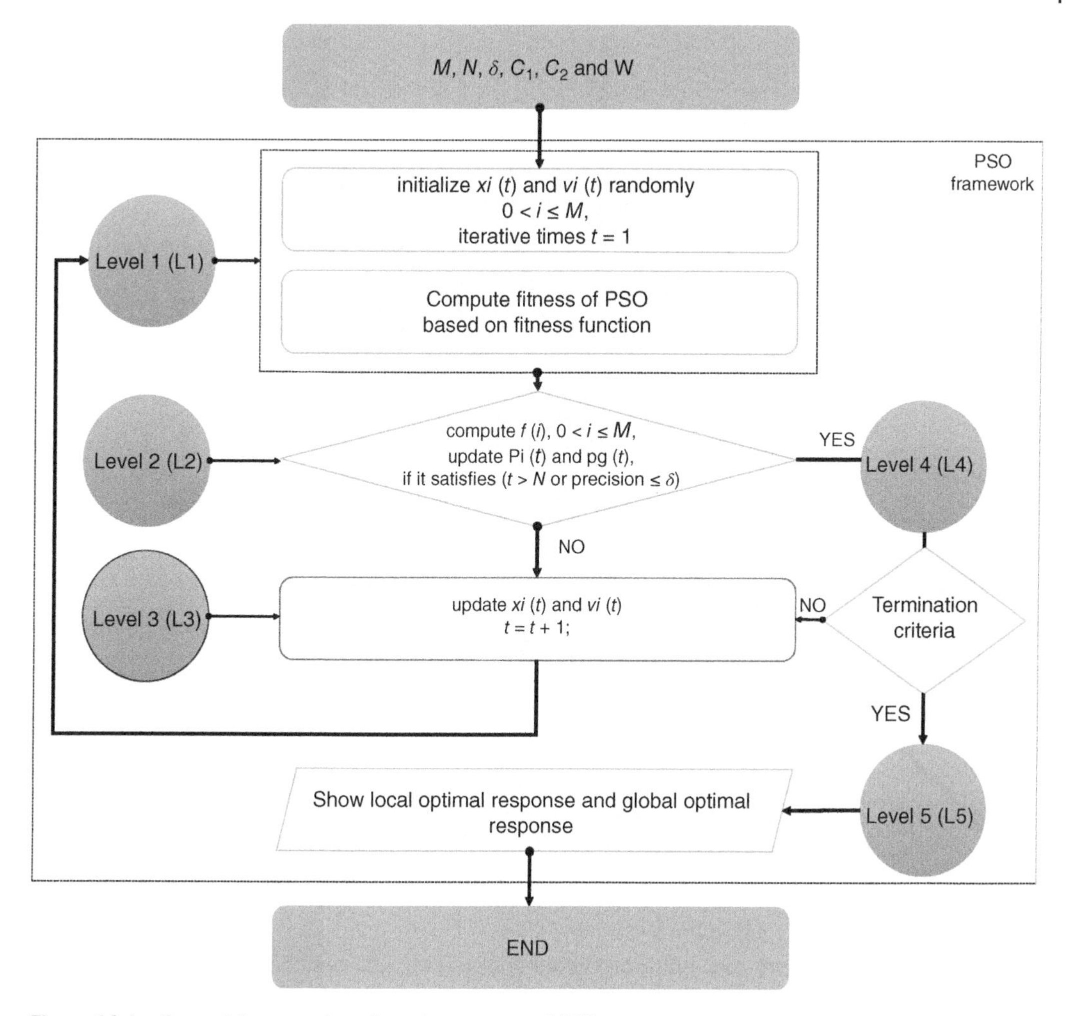

Figure 10.4 General framework and routine process of PSO.

$$x_{id}^{t} = L_d + r(0,1) \times (U_d - L_d)$$
$$v_{id}^{t+1} = x_{id}^{t} + \varphi\left(x_{id}^{t} - x_{jd}^{t}\right)\chi \tag{10.7}$$
$$p_i = \frac{f(i)t_i}{\sum_{i=1}^{N} f(i)t_i}, f(i)t_i = \begin{cases} 1/(1+f(i)), & f(i) > 0 \\ 1 + \text{abs}(f(i)), & \text{otherwise} \end{cases}$$

where

L_d is the lower bound in dth dimensional space.
U_d is the upper bound in dth dimensional space.
r is the random number uniformly distributed between (0,1).
$j, i > 0$ and $i, j \leq M$ and $j \neq i$
φ is the random number uniformly distributed between (0,1).

Employed foragers search for new food sources based on v_{id}^{t+1}, and onlookers choose food sources based on $f(i)t_i$ formula. Based on the bee's behavior, if a better food source is not found after

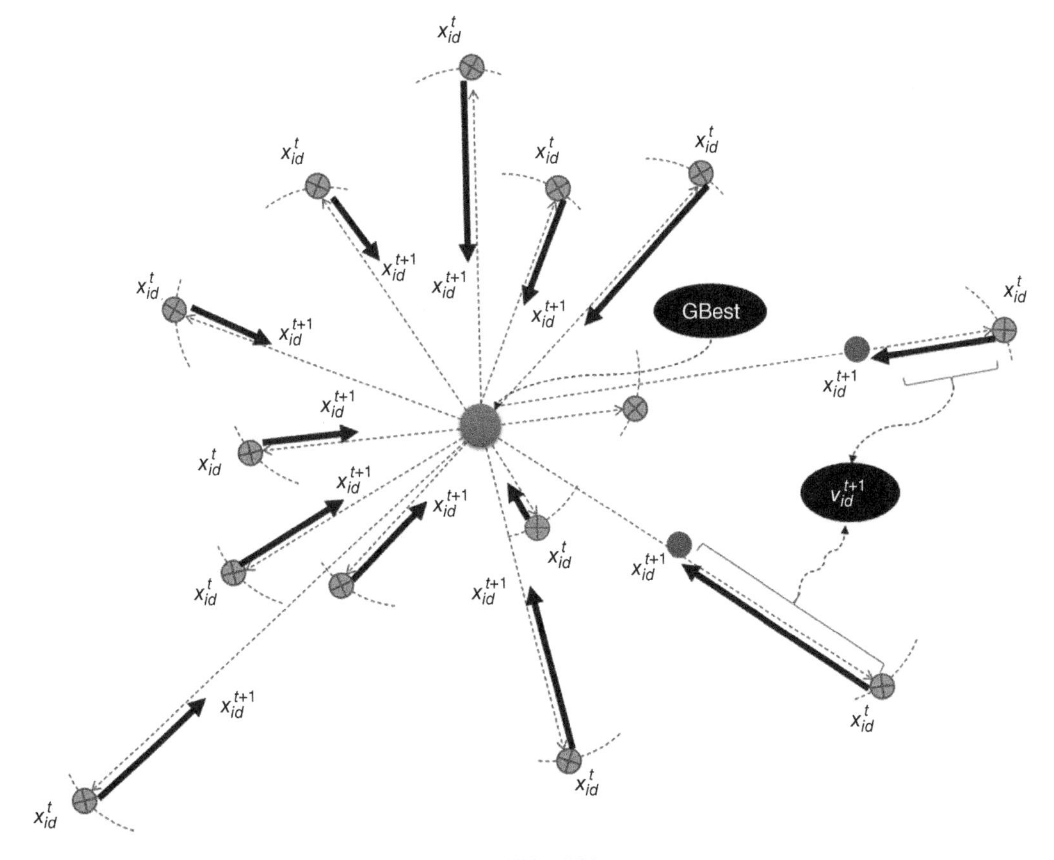

Figure 10.5 An example of particle movement in NIOA-PSO.

searching up to the maximum set limits or a better option is not selected, the ABC algorithm ignores that position then removes and changes the scout. The most important operators of this algorithm are selection, replacement of bees, and reproduction of bees. The idea of level 3 is:

1) Employed foragers search new food sources by v_{id}^{t+1} and compute $f(i)$.
2) The food source is updated if the new one is better than the old one.
3) Onlookers choose food sources of employed foragers according to p_i and $f(i)t_i$.
4) New food sources v_{id}^{t+1} are generated.
5) The food source is updated if the new one is better than the old one.
6) If there are some food sources that need to be given up, the corresponding bees become the scouts and generate new sources by x_{id}^{t}.

The framework and process of ABC are shown in Figure 10.6.

10.3.4 Bat Algorithm (BA)

Bats are equipped with a signal-emitting system, called a wave-reflecting system, to hunt, avoid obstacles, and find nests in caves in complete darkness. Bats generally live in groups and hunt as a heard. Using their special sonar ability, they emit and absorb pulses of high variability to perform different behaviors. Specifically, short frequencies are used to locate and trap prey.

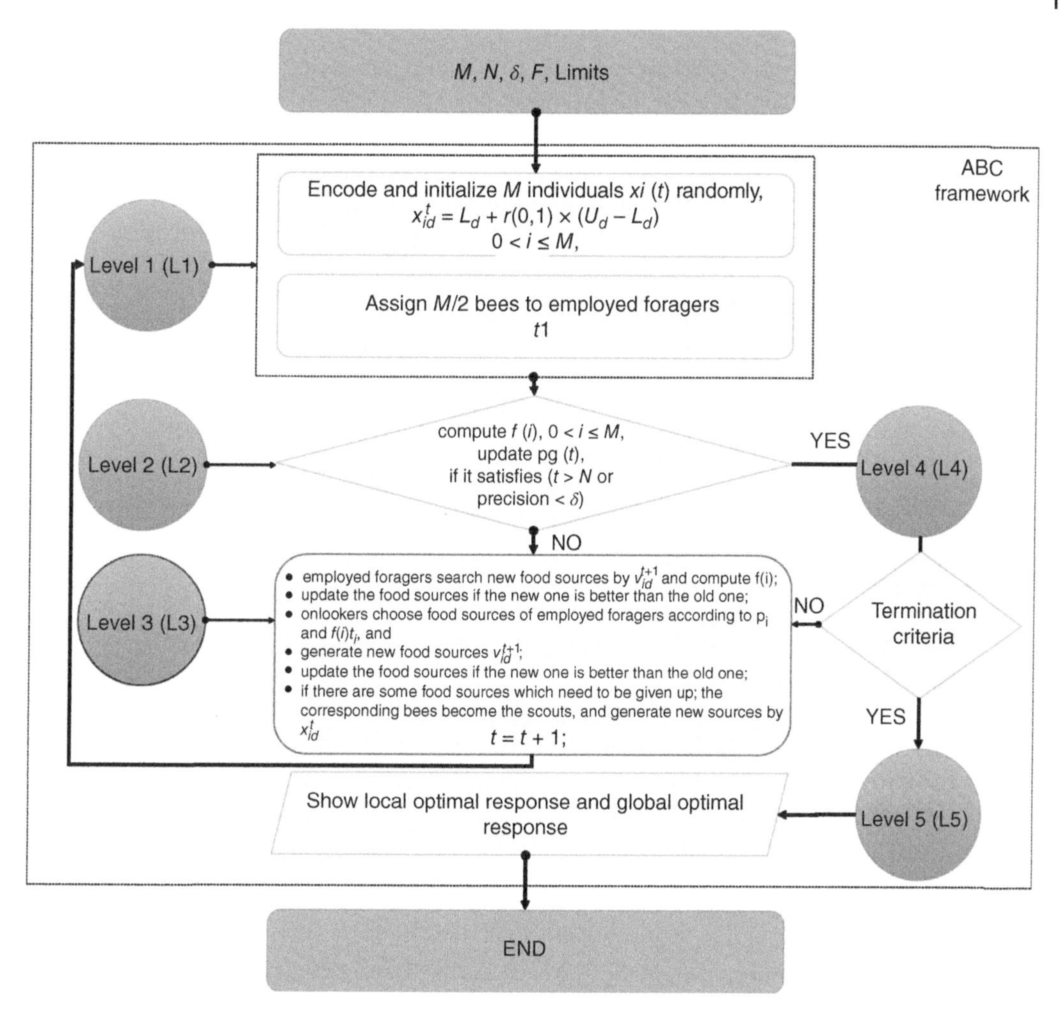

Figure 10.6 General framework and routine process of ABC.

Different algorithms can be considered to represent the ideal sound reflection properties of bats. To simplify, the following approximations are made for the idealizing rules: all bats use their sound-reflecting properties to detect distance, and bats know the difference between a target and non-target. Bats fly completely randomly at a speed of v_i in the direction of x_i with a constant frequency of $f_{\min}$. Flights to search for prey have different wavelengths λ with an intensity of A_0. Bats can automatically emit pulses with a wavelength that has a pulse ratio of between zero and one, depending on the probability of having a target.

We assume that sound is variable and has the size A_0 and a constant value $A_{\min}$. Other simplifications include ignoring the latency and three-dimensional topography of the environment. It is (what is it?) used in computing because it creates a multidimensional environment for us. To simplify the assumptions, we considered these approximations for simplification. In general, the frequency has a range between $[f_{\min}, f_{\max}]$, which is proportional to the wavelength range $[\lambda_{\min}, \lambda_{\max}]$. In real applications, different ranges can be considered for wavelengths and frequencies that must be selected according to the user. Similar to comparison methods that depend on the size and dimensions of the target, we first search for the prey then the target to reduce the size of the search space. Since we cannot necessarily use wavelengths, different frequencies with fixed wavelengths, or different wavelengths with fixed frequency, can be considered instead. This is because λ and f are interdependent and $f\lambda$ is constant. For the purpose of simplification, we can assume $f\epsilon$ [0, f_{max}]. For bats, higher frequencies for shorter wavelengths and shorter time intervals with a range of less than a few meters are generally considered. The pulse ratio can be simplified to a range between 0 and 1, where 0 indicates no pulse and 1 means the maximum pulse.

Assume that the frequency of a sound wave is in the range [F_{min}, F_{max}] and the intensity of the sound and pulse are defined as A [A_{min}, A_{max}] and r, respectively. Then, the new position is calculated using the following equation:

$$\begin{aligned} f_i &= f_{\min} + (f_{\max} - f_{\min}) \times \beta \\ v_i^t &= v_i^{t-1} + (x_i^t - x_*) f_i \\ x_i^t &= x_i^{t-1} + v_i^t \\ x_n &= x_o + \varepsilon \times \overline{A} \\ A_i^{t+1} &= \alpha A_i^t, \quad r_i^{t+1} = r_i^0[1 - \exp(-\gamma t)] \end{aligned} \tag{10.8}$$

where

f_{min} is the lower bound frequency of a sound wave.
f_{max} is the upper bound frequency of a sound wave.
β is the random number uniformly distributed between [0, 1].
ε is the random number in the range [−1, 1].
$\overline{A}$ is the average sound intensity all of bats.
x_n is the new solution generate by bats.
α is the constant parameter.
γ is the constant parameter.
r_i^0 is the initial value of r.

To simplify the equation, $\alpha = \gamma\alpha = \gamma = 0.9$ has been used in the proposed simulation. The choice of parameters depends on the experiments performed and the type of problem. Initially, each parameter must have a different form value and the pulse propagation ratio must be different, which can be chosen randomly. For example, we can consider the ratio A_i^0 to be between 1 and 2 if the ratio r_i^0 can be 0. For this value, we can consider in the range of zero and one. The volume and emission ratio can be updated when a new solution is improved, which means that bats are moving toward the optimal solution. Figure 10.7 shows the general framework and routine process of BA. Figure 10.8 provides a schematic representation of the three steps of the BA algorithm in determining the optimal global point: (i) iterate 1, (ii) iterate 2, and (iii) iterate 3.

10.3.5 Immune Algorithm (IA)

The clonal selection theory based on the artificial immune system theory was proposed to solve optimization problems. In general, the artificial immune system (IA) follows learning process similar to that of the GA; the IA method is designed as a way to guide the search process to the global optimal point. This method is based on the entropy relationship of information. The mean entropy of information H (i, j) of i and j antibodies is described based on the following equation:

$$H(i,j) = \frac{1}{D}\sum_{i=1}^{D} H_l(i,j) \tag{10.9}$$

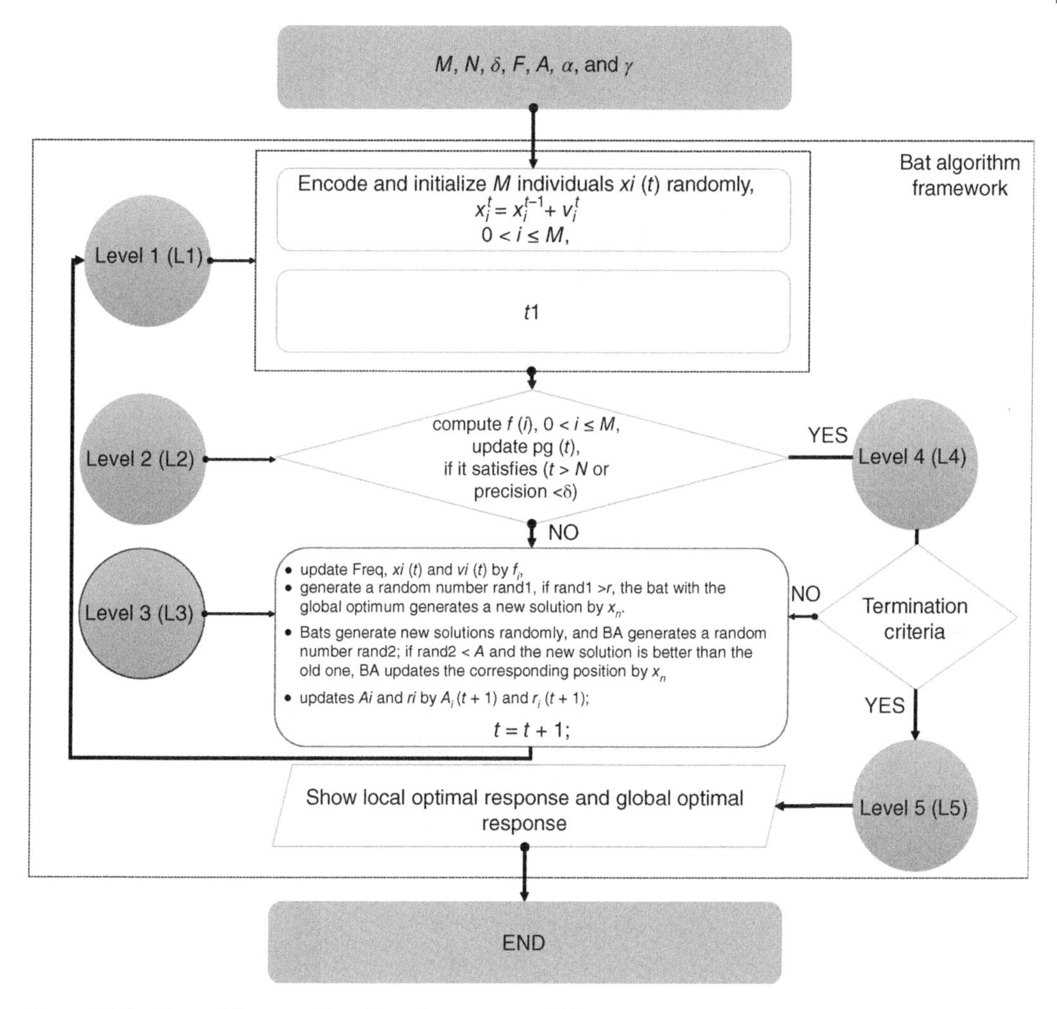

Figure 10.7 General framework and routine process of BA.

where

$H_l(i,j) = \sum_{z=1}^{S} -p_{\text{im}} \log p_{\text{im}}$ is the information entropy of the mth bit of genes for antibodies i and j.

p_{im} is the probability of regrading the mth bit of genes for antibodies i and j.

i and j are gene letters.

s is the number of gene letters.

$$A(i,j) = \frac{1}{1 + H(i,j)}$$

$$\text{Con}_i = \frac{1}{M} \sum_{i=1}^{M} c(i,j)$$

$$C(i,j) = \begin{cases} 1 & A(i,j) \geq h_1 \\ 0 & A(i,j) < h_1 \end{cases}, j = 1, 2, \ldots, M$$

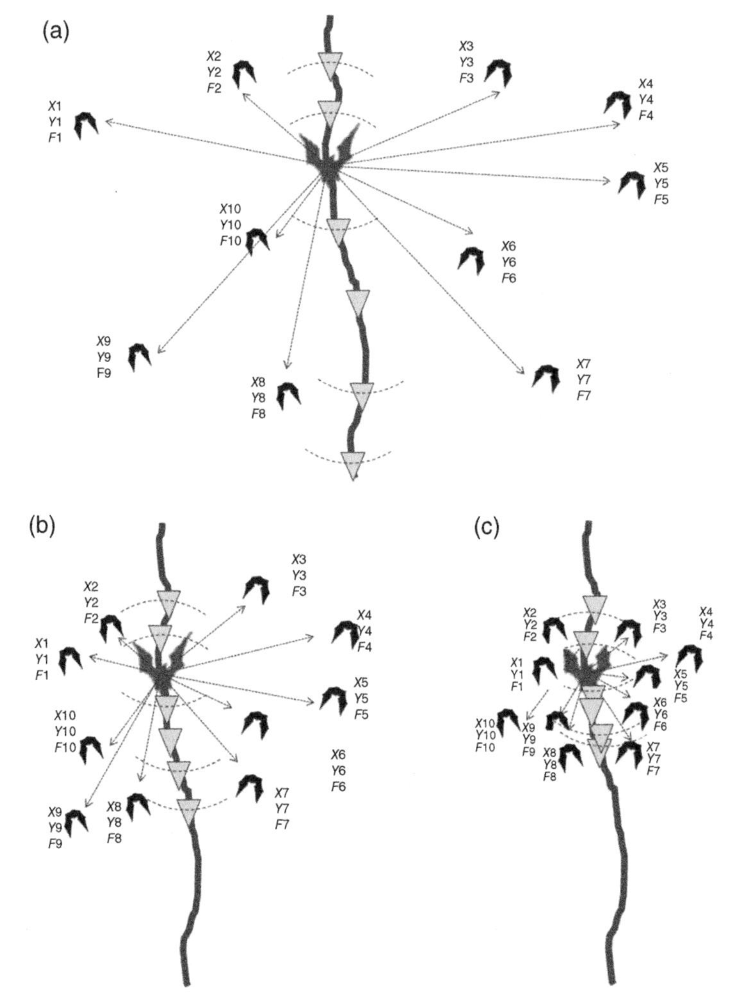

Figure 10.8 Schematic representation of the three steps of the BA algorithm in determining the optimal global point: (a) iterate 1, (b) iterate 2, and (c) iterate 3.

$$\text{Act}(i) = \begin{cases} \dfrac{\text{fit}_i(1 - \text{Con}_i)}{\sum_{i=1}^{M}\text{fit}_i*(\text{Con}_i)} & \text{Den}_i \geq h_2 \\ \dfrac{\text{fit}_i}{\sum_{i=1}^{M}\text{fit}_i*(\text{Con}_i)} & \text{Den}_i < h_2 \end{cases} \tag{10.10}$$

where

h_1 is the threshold of the affinity.
h_2 is the threshold of the antibody density.
fit_i is the fitness value of the ith antibody.
Con_i is the concentration of the *fi*th antibody.

In this method, the concentration of antibodies indicates the diversity of the whole population. The density of ith antibodies is calculated using *cij*. The degree of activity refers to the

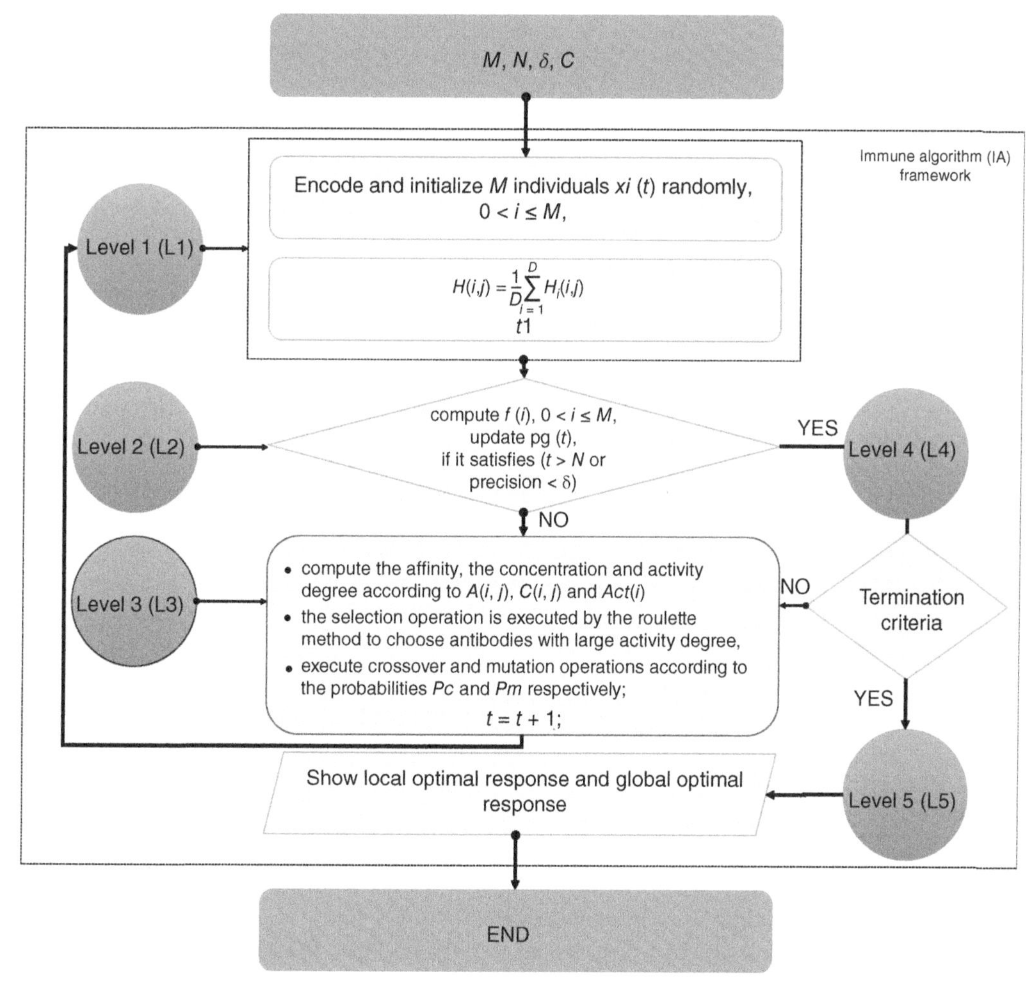

Figure 10.9 General framework and routine process of the Immune Algorithm (IA).

comprehensive ability of an antibody to respond to an antigen and be activated by other antibodies. The simulation was performed in such a way that antibodies with high affinity and low concentration would have a high degree of activity. Figure 10.9 shows the general framework and routine process of the IA.

10.3.6 Firefly Algorithm (FA)

The flashing light produced by a firefly is a natural phenomenon created by the process of bioluminescence, which is used to attract mates and prey as a hunting trap. Rhythmic flashing, blinking rate, and flash intensity are characteristics of the signal system that are used by the insect to attract the opposite, ultimately serving as a means to an optimal distance. In this phenomenon, females react to the unique pattern of males blinking, the intensity of the light at a distance r from the light source following the inverse square law. This means that the intensity of light I decreases with increasing distance r in terms of $I/1/r2$. The blinking light can be used in a convenient way to determine the objective function for optimization, which, according to this theory, creates new optimization algorithms based on the insect's absorption potential. The basic formulation of the FA depends on the amount of absorption and light intensity as well as the distance, as indicated in the following equations:

$$I = I_0 \times e^{\left(-\gamma * r_{i,j}\right)}$$
$$r_{i,j} = \left|\left|x_i - x_j\right|\right| = \left(\sum_{k=1}^{D} \left(x_{i,k} - x_{j,k}\right)^2\right)^{1/2} \quad (10.11)$$
$$\beta\left(r_{i,j}\right) = \beta_0 \times e^{\left(-\gamma * r_{i,j}^2\right)}$$

where

$x_{i,\,k}$ is the kth component of the spatial coordinate x_i of the ith firefly.
I is the luminance of the ith firefly.
γ is the attraction factor of a firefly.
I_0 is the maximum luminance of a firefly.
β is the attraction degree of a firefly.

The distance between any two fireflies i and j at x_i and x_j, respectively, is the Cartesian distance and is calculated based on $r_{i,\,j}$. The movement of a firefly i is attracted to another more nice-looking (brighter) firefly j, which is determined by:

$$x_i = x_i + \beta_0 e^{-\gamma * r_{i,j}^2}\left(x_i - x_j\right) + \alpha\left(\text{rand} - \frac{1}{2}\right) \quad (10.12)$$

where

x_i is the position of the ith firefly.

In this equation, the second expression considers the creation of gravity, and the third expression is made to create variety in the search of the answer space with random parameters. In this relation, rand is a generator of random numbers between [0, 1]. To generate random numbers, a normal distribution $N(0, 1)$, or other distributions, can be used. The design of these parameters in the equation creates both attraction and repulsion, and the value of these two behaviors is very effective in determining the convergence speed and performance of the FA algorithm (Figure 10.10).

10.3.7 Cuckoo Search (CS) Algorithm

The NIOA-CS algorithm is based on the behavior of a specific species of cuckoo and its flight characteristics. The algorithm is inspired by the following three rules for modeling. (i) Each cuckoo lays an egg, which is placed in a randomly-selected nest. (ii) The best nest with the highest egg quality is inherited to future generations. (iii) The number of host nests is fixed, and the egg is likely to be detected by the host bird pa [0, 1]. In other words, the new nest is updated by the following equation:

$$x_i = x_i + \alpha \circledast \text{Levy}(\lambda) \quad (10.13)$$

where

x_i is the host nest of the ith individual update.
α is the scaling factor of step size with a frequently equal to 1.
$\circledast$ is the multiplications, and Lévy (λ) indicates that the Lévy flight draws from the Lévy distribution.

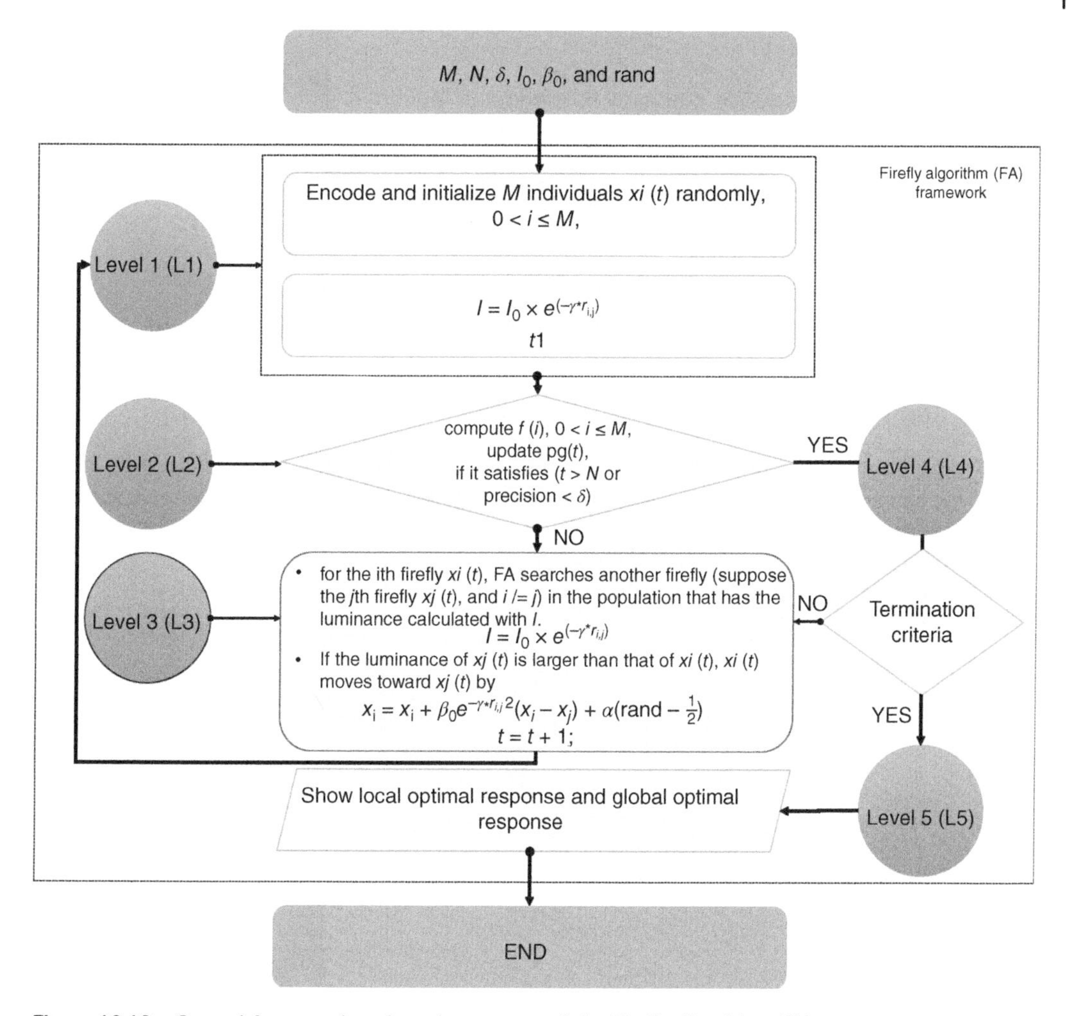

Figure 10.10 General framework and routine process of the Firefly Algorithm (FA).

$$s = \frac{u}{|v|^{1/\beta}}$$
$$u\tilde{N}(0, \sigma^2) \text{ and } v\tilde{N}(0, 1) \tag{10.14}$$
$$\sigma = \frac{ב(1+\beta)\sin\left(\frac{\pi\beta}{2}\right)}{\beta ב\left(\frac{1+\beta}{2}\right)2^{\frac{\beta-1}{2}}}$$

u and v are the Gaussian distribution parameters.
β is 1.5.

Sometimes, the cuckoo egg is dropped out of the nest by the host, in which case the cuckoo is likely to leave the nest. When the cuckoo egg is released, the cuckoo must find another location and nest to place the egg, which is calculated from the following equation:

$$x_i = x_i + \alpha s \circledast H(p_a - \varepsilon) \circledast (x_j(t) - x_k(t)) \tag{10.15}$$

where

$x_j(t)$ is the first solution selected randomly by random premutation.
$x_k(t)$ is second solution selected randomly by random premutation.
$H(p_a - \varepsilon)$ is the Heaviside function.
p_a is a random number.
ε is the step size (0, 1).
α is the scaling factor of the step size.

Figure 10.11 shows the general framework and routine process of CS Algorithm.

10.3.8 Gray Wolf Optimizer (GWO)

The gray wolf optimization method is based on the hunting behavior of wolves. According to this theory, the alpha wolf α dictates orders that must be completely obeyed by the group. Alternatively, wolf δ should respect the higher-level wolf but can command other lower-level wolves. The gray

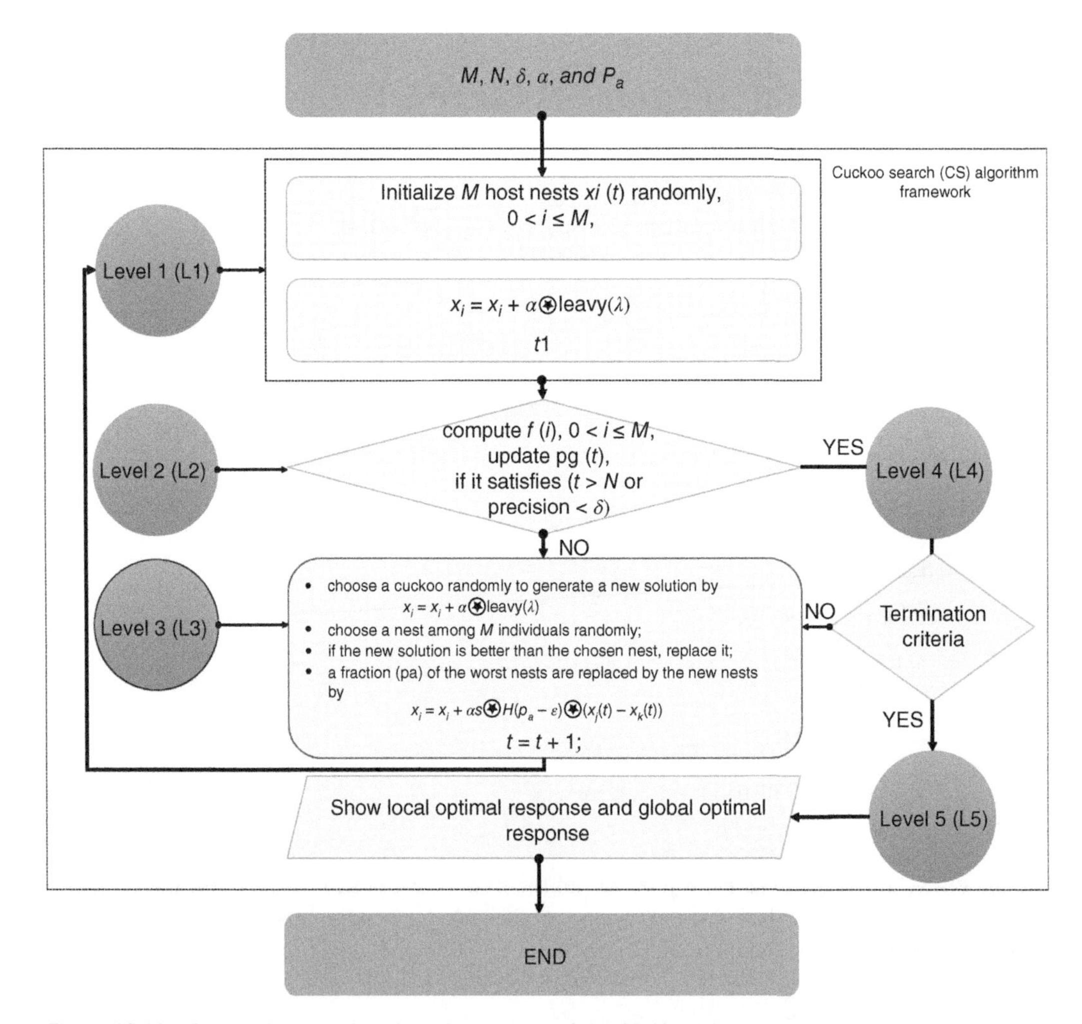

Figure 10.11 General framework and routine process of the CS Algorithm.

wolf ω has the lowest rank and position in the herd. These behavioral characteristics are modeled with the following equations:

$$\begin{aligned} D &= |C*x_p(t) - x(t)| \\ x(t+1) &= x_p(t) - A*D \\ A &= 2*a*r_1 - a \\ C &= 2*r_2 \end{aligned} \tag{10.16}$$

where

A and C are the coefficient vectors.
$x_p(t)$ is the position vector of the solution.
$x(t)$ is the position vector of the gray wolf.
a is the linearly form 2 to 0 over the iteration.
r_1 and r_2 are random parameters between [0, 1].

Assuming that the parameters α, β, and γ can provide better information about the possible location of prey, based on the first three solutions, other wolves are asked to first declare their position according to the position of the best search wolves. If they find a better case to report, the following equations are used:

$$\begin{aligned} D_\alpha &= |C*x_\alpha(t) - x(t)| \\ D_\beta &= |C*x_\beta(t) - x(t)| \\ D_\gamma &= |C*x_\gamma(t) - x(t)| \\ x_1 &= x_\alpha(t) - A_1*D_\alpha \\ x_2 &= x_\beta(t) - A_2*D_\beta \\ x_3 &= x_\gamma(t) - A_3*D_\gamma \\ x(t+1) &= \frac{1}{3}\sum_{i=1}^{3}(x_i) \end{aligned} \tag{10.17}$$

where

A_i is a coefficient generated by random values.
C_i is the coefficient generated by random values.
$x(t+1)$ is the new position vector of the solution.

In this method, the hunting technique and social hierarchy of gray wolves are modeled in order to design GWO and perform optimization. In order to mathematically model the wolf social hierarchy when designing GWO, the most appropriate solution is considered as "alpha (α)." Also, the second and third best solutions are called beta (β) and delta (γ), respectively. Other responses are called omega (x) candidates. In the GWO algorithm, hunting (optimization) is driven by α, β, and γ. The x wolves follow these three wolves. Figure 10.12 shows the general framework and routine process of the NIOA-GWO algorithm.

10.3.9 Krill Herd Algorithm (KHA)

The KHA, was developed by considering the dispersion behavior of krill herds when being attacked by predators, such as seals, penguins, and seabirds. In response to an attack, the density of the krill herd is reduced, then the herd is reformed after the predators leave. Krill flock

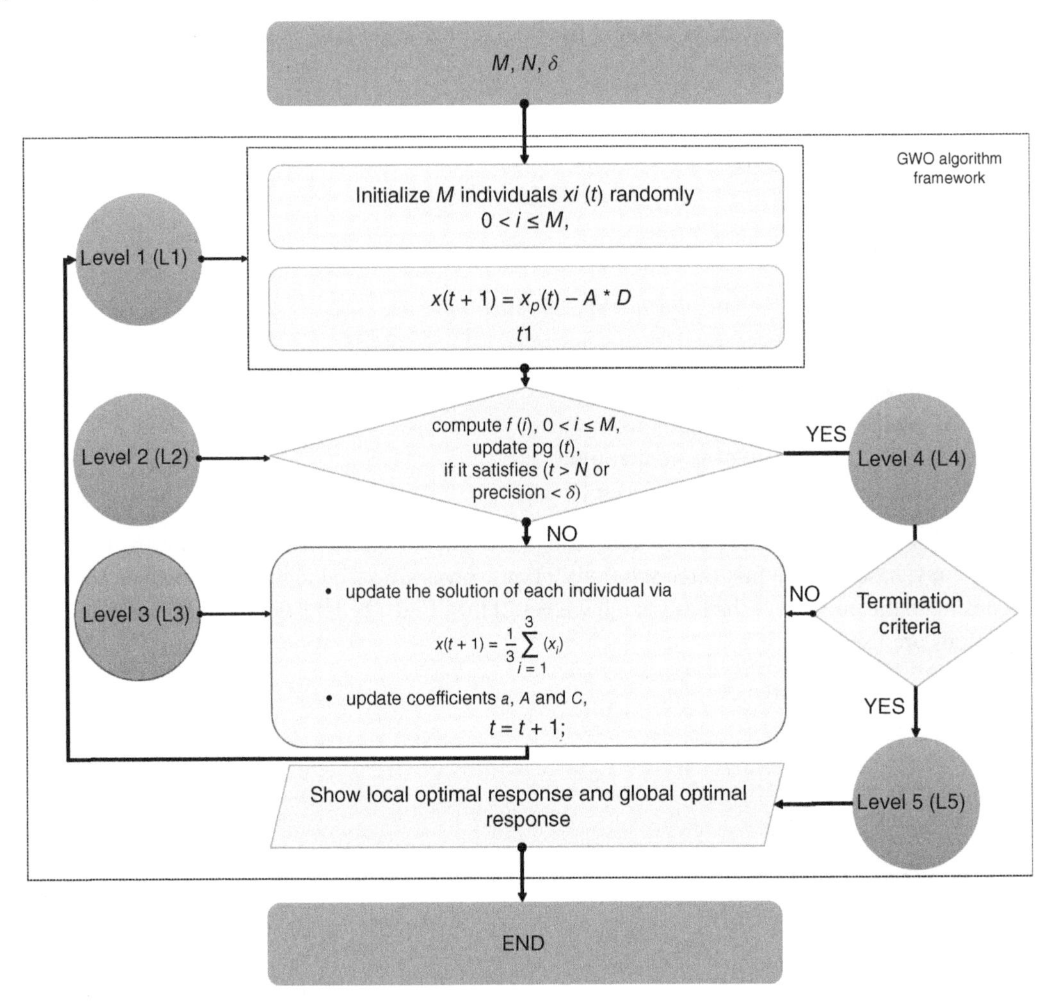

Figure 10.12 General framework and routine process of NIOA-GWO.

formation is a multi-purpose process that includes two main goals: (i) increasing krill density and (ii) reaching food. The KH algorithm is based on the simulation of the herding behavior of krill individuals. The minimum distances of each individual krill from food and from highest density of the herd are considered as the objective function for the krill movement. In this process, a krill moves to the best solution only when it is looking for the highest density and food. That is, the closer the distance to the higher density and food, the lower the objective performance. In general, some coefficients must be set to use multifunctional herd behavior for a single purpose. Each problem may have its own coefficients based on the type of optimization and the set of answers. In the natural system, fitness is a combination of distance from food and the highest density of krill. Therefore, the value distance is a function of the target. The time-dependent position of a krill is only controlled at the two-dimensional level by the following three actions: (i) movement induced by the presence of other individuals, (ii) foraging activity, and (iii) random diffusion. For more precise modeling of the krill behavior, two adaptive genetic operators are added to the algorithm.

Typically, an optimization algorithm should be able to perform the optimal global search in spaces of arbitrary dimensions. The Lagrangian model is generalized to an n-dimensional decision space to simulate KH behavior:

$$\frac{dX_i}{dt} = N_i + F_i + D_i \tag{10.18}$$

where

N_i is the motion induced by other krill individuals.
F_i is the foraging motion.
D_i is the physical diffusion of the ith krill individual.

According to the existing theory for krill, each krill tries to maintain and move at a high density due to the interaction of the herd. The direction of motion caused by ai is estimated from the local congestion density (local effect), the target congestion density (target effect) and the repulsive congestion density (repulsion effect). For a krill, this movement can be defined as follows:

$$N_i^{\text{new}} = N_i^{\max}\alpha_i + \omega_n N_i^{\text{old}}$$
$$\alpha_i = \alpha_i^{\text{local}} + \alpha_i^{\text{target}} \tag{10.19}$$

where

$N_i^{\max}$ is the maximum induced speed.
ω_n is the inertia weight of the motion [0, 1].
N_i^{old} is the last movement.
α_i^{local} is the local effect of neighboring krill.

$$\alpha_i^{\text{local}} = \sum_{j=1}^{\text{NN}} \left(\frac{X_j - X_i}{\|X_j - X_i\| + \varepsilon} \right) \left(\frac{K_i - K_j}{K^{\text{worst}} - K^{\text{best}}} \right)$$
$$d_{s,i} = \frac{1}{5N} \sum\nolimits_{J=1}^{N} \|X_I - X_J\| \tag{10.20}$$
$$C^{\text{best}} = 2\left(\text{rand} + \frac{I}{I_{\max}} \right)$$

where

K^{worst} and K^{best} are the best and worst fitness values of krill, respectively.
K_i is the fitness value of the ith krill.
X is the related position of the ith krill.
NN is the number of neighbors.
ε is the small positive number for avoiding the singularities.
rand is a random value between 0 and 1 for enhancing exploration.
I is the actual iteration number.
$I_{\max}$ is the maximum number of iterations.

The foraging movement is related the food location and the previous experience of kills. This movement can be expressed for the ith krill individual as follows:

$$F_i = V_f \beta_i + \omega_f F_i^{\text{old}}, \beta_i = \beta_i^{\text{food}} + \beta_i^{\text{best}}$$

$$\beta_i^{\text{food}} = 2\left(1 - \frac{I}{I_{\max}}\right)\hat{K}_{i,\text{food}}\hat{X}_{i,\text{food}}$$
$$\beta_i^{\text{best}} = \hat{K}_{i,Ii\text{best}}\hat{X}_{i,i\text{best}} \tag{10.21}$$
$$D_i = D^{\max}\delta$$

V_f is the foraging speed.
ω_f is the inertia weight of the foraging motion in the range [0, 1].
β_i^{food} is the food source.
β_i^{best} is the effect of the best fitness of the ith krill.
$\hat{K}_{i,Ii\text{best}}$ is the best previously visited position of the ith krill individual.
$D^{\max}$ is the maximum diffusion speed.
δ is the random directional vector with random values between −1 and 1.

In general, the movements defined in most cases change the position of a krill in a better direction. The feeding movement and the movement caused by other krill include two global strategies and two local strategies, which work in parallel, making KH a powerful algorithm.

$$X_i(t + \Delta t) = X_i(t) + \Delta t\frac{dX_i}{dt},$$
$$\Delta t = C_t\sum_{j=1}^{NV}\left(\text{UB}_j - \text{LB}_j\right) \tag{10.22}$$

NV is the total number of variables.
UB_j and LB_j are the lower and upper bounds of the jth variables (j = 1,2,...,NV).
C_t is a constant number between [0, 2].

The general framework and routine process of NIOA-KH are illustrated in Figure 10.13.

10.3.10 Emperor Penguin Algorithms (EPA)

Emperor penguins are one of the most adaptive animals on the planet in terms of their breeding habits and their ability to adjust to cold climate conditions. While most animals begin their breeding cycle in spring or summer, Emperor penguins have developed a cycle that requires them to begin "nesting" when temperatures reach as low as −45 °C. To withstand extended amounts of time in extreme cold, the penguins have various physiological and behavioral adaptations, such swarm intelligence as well as possess a thick and relatively stiff 30 mm layer of feathers, layer of blubber or fat, an efficient body shape, highly-developed counter current heat exchange vascular system, and effective metabolism. In addition, the penguins huddle together in a dense group of 10 to even hundreds of birds to create a warm center to counteract heat loss in the cold. Air temperature and wind speed affect the behavior and velocity of the colony. This social thermo-regulation reduces the total exposed surface area of the group and the total heat loss by up to 40%. In this huddle, each penguin leans forward on a neighbor, and the outside birds tend to face into the huddle and move slowly forward and induce a slow churning action. This type of spiral rotation inward allows all members of the group to enjoy the heat and access to the best condition. The behavior also enables Emperor penguins to thermoregulate their core body temperature without altering their metabolism over a wide range of temperatures. To achieve a maximum spread (diversification) over the individuals in the colony, the penguins perform a spiral-like movement toward the center, then a spiral-like movement toward the outside (logarithmic spiral) after reaching the center (intensification). This

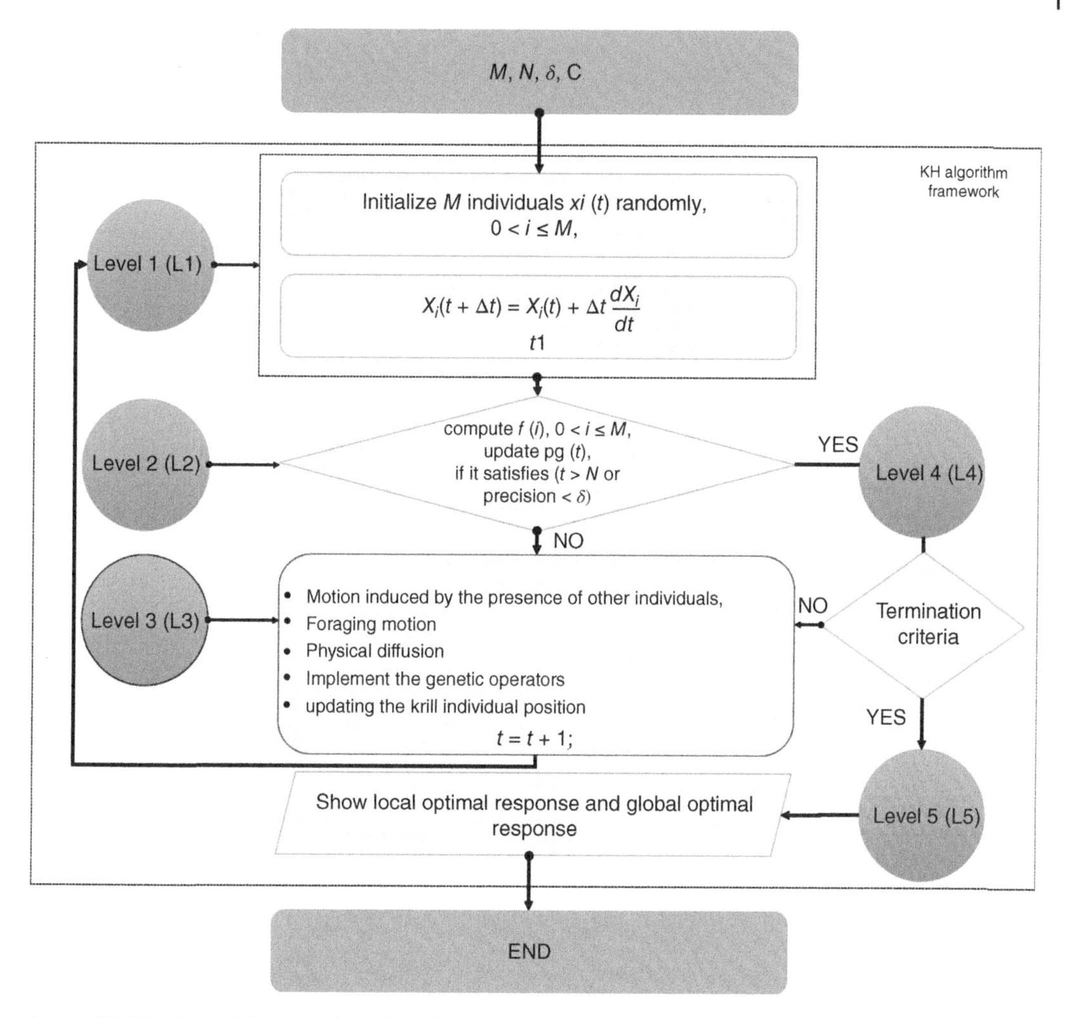

Figure 10.13 General framework and routine process of NIOA-KH.

behavior can be modeled by overlaying a spiral-like pattern on the track of the individuals, as is depicted in Figure 10.14, which can be conducted several times to create multiple spirals wave fronts.

The heat intensity at a particular distance l from the heat source obeys the inverse law, and thus, $H \propto 1/l^m$. Furthermore, as the distance increases, heat intensity decreases. The heating in the center and relation with traveling distance from the warmest place can be formulated in such a way that it is associated with the objective function to be optimized, which makes it possible to formulate new optimization algorithms. Considering the characteristics of the colony of penguins, the Emperor Penguins Algorithm is developed based the following approximate or idealized rules:

1) All penguins use spiral-like movement to sense distance, and they have movement to control the metabolism (movement by swimming, walking, and shivering).
2) Penguins converge to the center randomly with a logarithmic velocity v_i at position x_i, varying r, and arc length l_i to reach the heated center. They can automatically track the spiral route with various $r\epsilon$[function] depending on the proximity heat and swarm behavior.

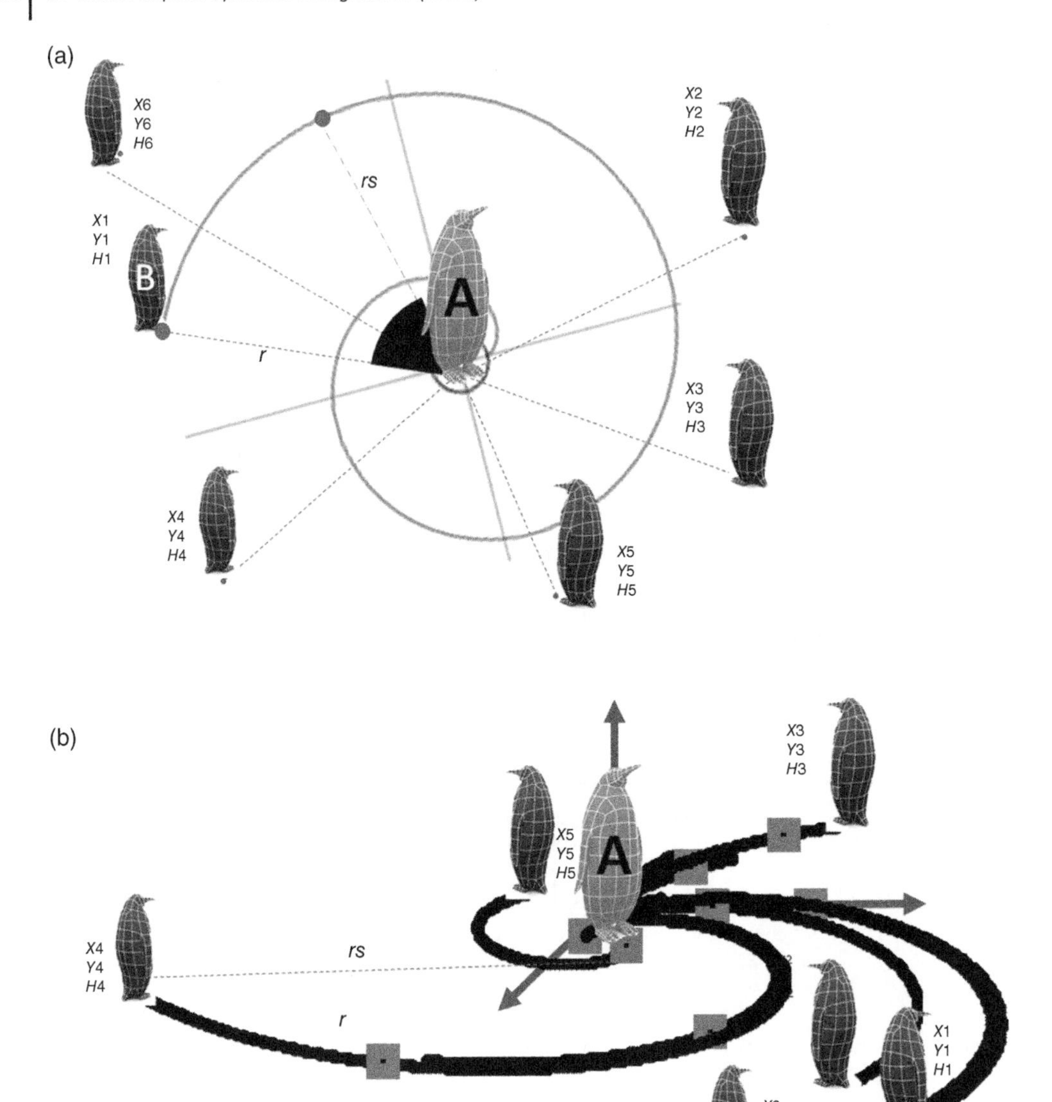

Figure 10.14 Schematic representation of (a) how a penguin moves to reach the center of the herd and (b) the behavior of the penguins around the optimal point, which has more heat (the goal of all penguins is to be in the center of the herd).

3) Although the spiral function can vary in many ways, we assume that the movement pattern varies from a large (positive) l_0 to a minimum $l_{\min} \cong 0$.
4) Attractiveness is proportional to their warmer neighbor, thus for any two Emperor penguins, the warmer one plays the role of the colony's center and the colder one will move towards the warmer one.
5) The attractiveness is proportional to the heat of the penguin's body, which decreases as the distance (arc length) increases. If no penguin is the warmest, other penguins will move randomly.

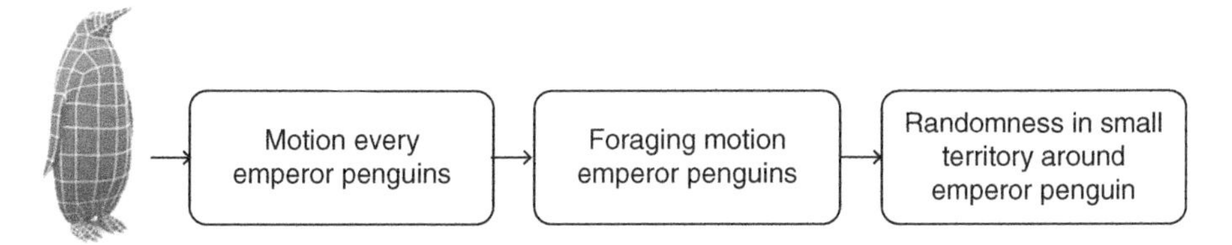

Figure 10.15 Position of any individual emperor penguin or group.

6) The warmest penguin is a benchmark of the objective function. For a maximization problem, the warm point and the intensity of heat can simply be proportional to the value of the objective function.

Based on these six rules, the Emperor Penguin Algorithm (EPA) can be summarized as the pseudo code shown in Table 10.1.

As the Emperor penguin's body heat decreases, the average colony density is reduced and increases the distance between the penguin's swarm from the warm location. This process is assumed to be the initialization phase in EPA, whereby the fitness (minimum arc length) of each individual is supposed to be a combination of the distance and warmest location from the center of colony. Therefore, the fitness (spiral arc length) is the value of the objective function. The position of any individual penguins from the warmest Emperor Penguin in 2D surface is determined by the following sequences (see Figure 10.15).

To mathematically depict the penguin movement in N dimensions for searching, the following model is presented:

$$\left\{\frac{dp}{dn}\right\}^i = f\{p_n, s_{n+1}, \text{rand}_i\} \tag{10.23}$$

p_n is the position of an Emperor penguin.
s_{n+1} is the spiral motion.
rand_i is the function that generates diversification near or inside the updated position.

Although we know $S \in [0, S_{\max}]$, to control the sequence and continuous movement, we assume $S \in [\varepsilon, S_{\max}]$. We also know that higher S have more or less heat and must travel a longer distance to reach warm. The rate of effectiveness A can simply be in the range of $A \in [0, S_{\max}]$, where 0 is the center, and $S_{\max}$ is the maximum distance in huddling Emperor Penguins colony (Figure 10.16).

In the simplest form, the heat intensity $H(r)$ varies according to the inverse square law, similar to the FA attractive function. Several functions can be employed to decrease distance.

$$\{\text{heat}(s)\}^i = \left[\frac{h_C}{r^m}\right]^n \tag{10.24}$$

For simplification, we consider $n = 1$,where h_C is heat at the source, r^m is distance r, and s is arc length (as measured from the origin, $i = \infty$). In order to avoid the singularity at $r = 0$, the convergence or intensification can be approximated using the following equation:

$$\{H(s)\}^i = \{h_C \times e^{-\tau r^m}\} \tag{10.25}$$

If $m = 2$, absorption is low, similar to the Gaussian form of the square law and absorption in FA. Figure 10.17 presents the change in function parameters, namely attractiveness and diversification, where the speed rate to converge can be controlled by parameter m and scaled with h_C.

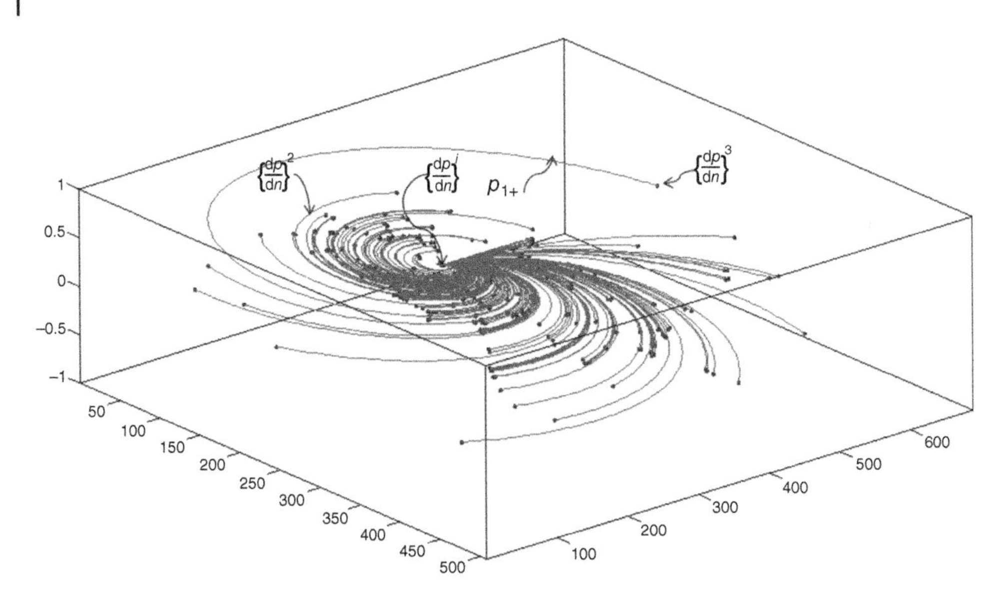

Figure 10.16 Spiral-like movement and updated position of any individual penguins or group of penguins.

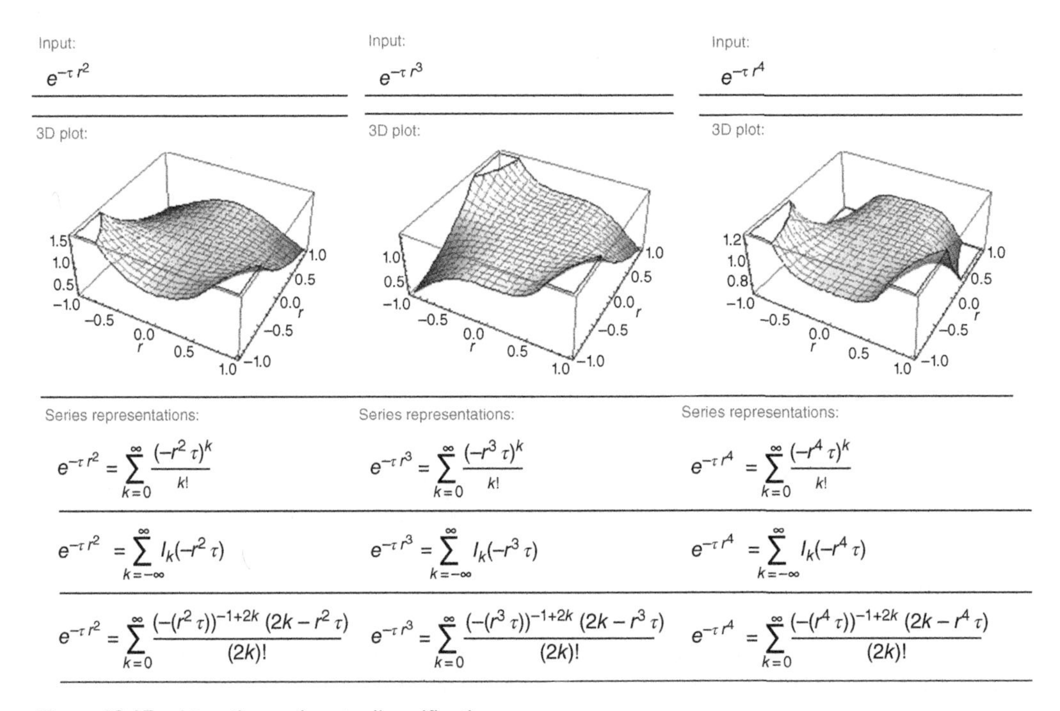

Figure 10.17 Attractive and route diversification.

The distance can be controlled by estimating the nth sentence of the series. Since the charm of the emperor penguins is proportional to the heat generated by the adjacent emperor penguins, we can model the H charm of an emperor penguin with the overall series.

$$\{H(s)\}^i = \{h_C \times e^{-\tau r}\}$$

$$e^{-\tau r^m} = \sum_{k=0}^{\infty} \frac{(-r^m\tau)}{k!} = \sum_{k=0}^{\infty} I_k(-r^m\tau) = \sum_{k=0}^{\infty} \left[\frac{(-(r^m\tau))^{-1+2k}(2k - r^m\tau)}{2k!}\right] \tag{10.26}$$

where $n!$ Is a factorial function; and $I_k(z)$ is the modified Bessel function of the first kind. The warm intensity $\{H(s)\}^i$ and h_C of heat have to be updated accordingly as the iterations proceed. As the distance usually decreases once a penguin becomes warmer, the rate of diversification increases, and the speed of intensification can be chosen as any value by scale factors.

$$h_i^{n+1} = h_i^n\gamma \quad , \qquad s_i^{n+1} = s_i^n\gamma$$

$$s_i^{n+1} = \left(s_i^0[(1 - e^{-\tau r})]\right)^m$$

$$s_i^{n+1} = (\tau r)^m\left(1 - \frac{1}{2}(mr)r + \frac{1}{24}m(3m+1)r^2\tau^2 - \frac{1}{48}\left(m^2(m+1)\tau^3\right)r^3 + \left(\frac{m(15m^3 + 30m^2 + 5m - 2)r^4\tau^4}{5760}\right) - \left(\frac{(m^2(3m^3 + 10m^2 + 5m - 2)\tau^5)r^5}{11,520}\right) + O(r^6)\right) \tag{10.27}$$

where τ and γ are constants. Parameter γ is similar to the cooling factor of a cooling schedule in the simulated annealing and is similar to the BA attraction factor. For any $0 < \gamma < 1$ and $\tau > 0$, we have

$$h_i^{n+1} \to h_c^1, \quad s_i^n \to s_i^0 \text{ as } t \to \infty \tag{10.28}$$

In EPA, $0 < \gamma < 1$ and $\tau = s$. The choice of parameters depends on the case. Initially, each Emperor penguin should have random values of heat and position with random s_i^n arc, and this can be achieved by the randomization function. The distance between any two Emperor penguins, i and j, at xi respectively, is the polar distance with respect to selected spiral function:

$$r_{ij} = s_i^n - s_j^n \tag{10.29}$$

where n is iterate number and i and j is number of Emperor Penguins. The updated position of Emperor Penguin can be calculated by:

$$x_i = r_{ij}.\cos(\theta_i), y_i = r_{ij}.\cos(\theta_i) \tag{10.30}$$

where r_{ij} is spiral distance (arc length). Arc length is defined as the length along a curve:

$$s_i^n = \int_{\gamma_n} |d\rho|_i, \tag{10.31}$$

where $|d\rho|_i$ is a differential displacement vector along a curve gamma. For example, for a route with radius r_{ij} in Figure 10.18, the arc length between two Emperor penguins with angles θ_1 and θ_2 (measured in radians) is simply:

$$s_i^n = (r_{ij})_n[(\theta_1)_n - (\theta_2)_n]$$
$$s_i^n = \int_{(\theta_1)_n}^{(\theta_2)_n} \sqrt[2]{(r_{ij})_n^{\,2} + (dr/d\theta)^2}\,d\theta \tag{10.32}$$

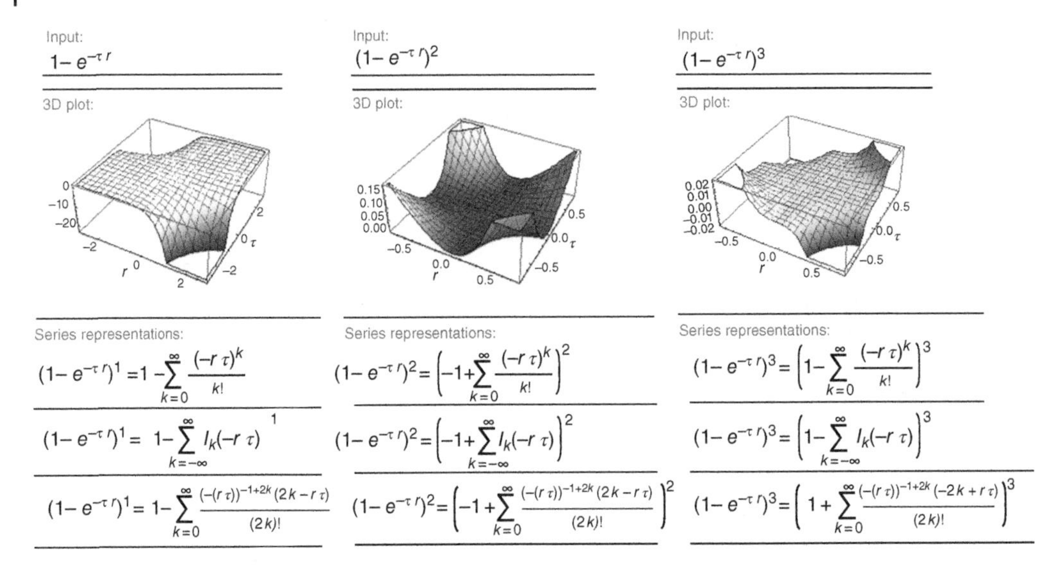

Figure 10.18 Examples of diversifier and intensifier.

The arc length of the polar curve $(r_{ij})_n[(\Delta\theta)_n]$ is given by s_i^n. The intensification and diversification of an Emperor penguin i attracted to another more attractive (warmer) Emperor penguin j are determined by:

$$s_{i+1}^n = s_i^0[(1 - e^{-\tau r})] + \omega(\text{rand} - 0.5) \tag{10.33}$$

where the first term represents intensification, and the third term is randomization with ω being the randomization parameter uniformly distributed in [0, 1]. As can be seen, the parameter r is crucial for determining the total performance of the algorithm. In addition, the use of spiral equations can create different modes of motion and make it possible to use the algorithm in many problems. Several examples of spiral equations are shown in Figures 10.18 and 10.19.

Archimedes' Spiral, Circle Involute, Conical Spiral, Cornu Spiral, Cotes' Spiral, Daisy, Epispiral, Fermat's Spiral, Helix, Hyperbolic Spiral, Logarithmic Spiral, Mice Problem, Nielsen's Spiral, Phyllotaxis, Poinsot's Spirals, Polygonal Spiral, Rational Spiral, and Spherical Spiral are spiral terms.

Moreover, EPA can find the global optima and all local optima simultaneously by a novel diversification and intensification method. In Figure 10.20, a function $r = A\theta^2$ is employed in $\theta\epsilon[0, \pi]$. Instead of using Cartesian distance, we consider the polar distance and attractiveness fired on control road ($r = \theta^2$). To update a new position, we must calculate the r' which is modified r, from arc length as (Table 10.2):

$$s_i^n = \int_{(\theta_1)_n}^{(\theta_2)_n} \sqrt[2]{(r_{ij})_n^{\;2} + (^{dr}/_{d\theta})^2} d\theta \text{ and } s_i^n = \int_{(\theta_1)_n}^{(\theta_2)_n} \sqrt[2]{(A\theta^2)_n^{\;2} + (2A\theta)^2} d\theta \tag{10.34}$$

Then

$$s_i^n(r = A\theta^2) = \frac{1}{3}A\left((\theta^2 + 4)^{\frac{3}{2}}\right)e^{\left(i\pi\left\lfloor \frac{-\arg(A)}{\pi} - \frac{\arg(\theta)}{\pi} - \frac{\arg(\theta^2+4)}{2\pi} + 0.5\right\rfloor\right)} + \text{constant}$$

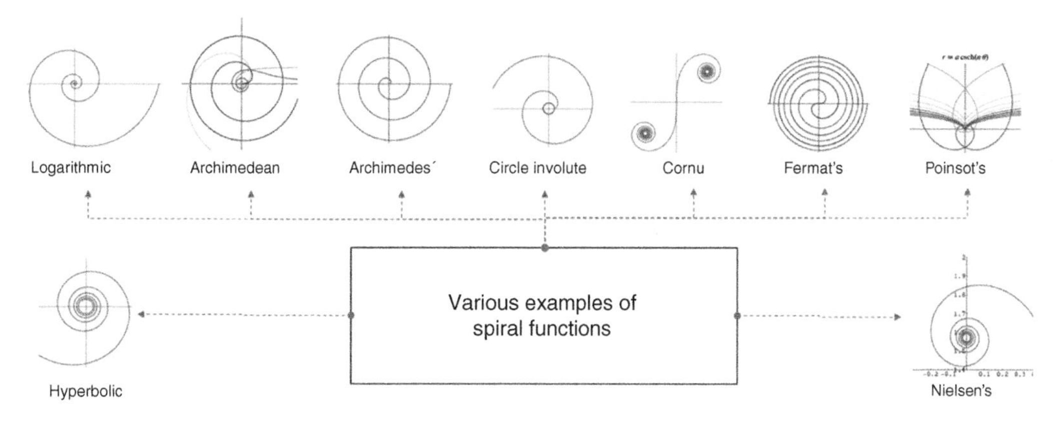

Figure 10.19 Various examples of spiral functions to create attraction and repulsion simultaneously and simulate the behavior of a penguin herd.

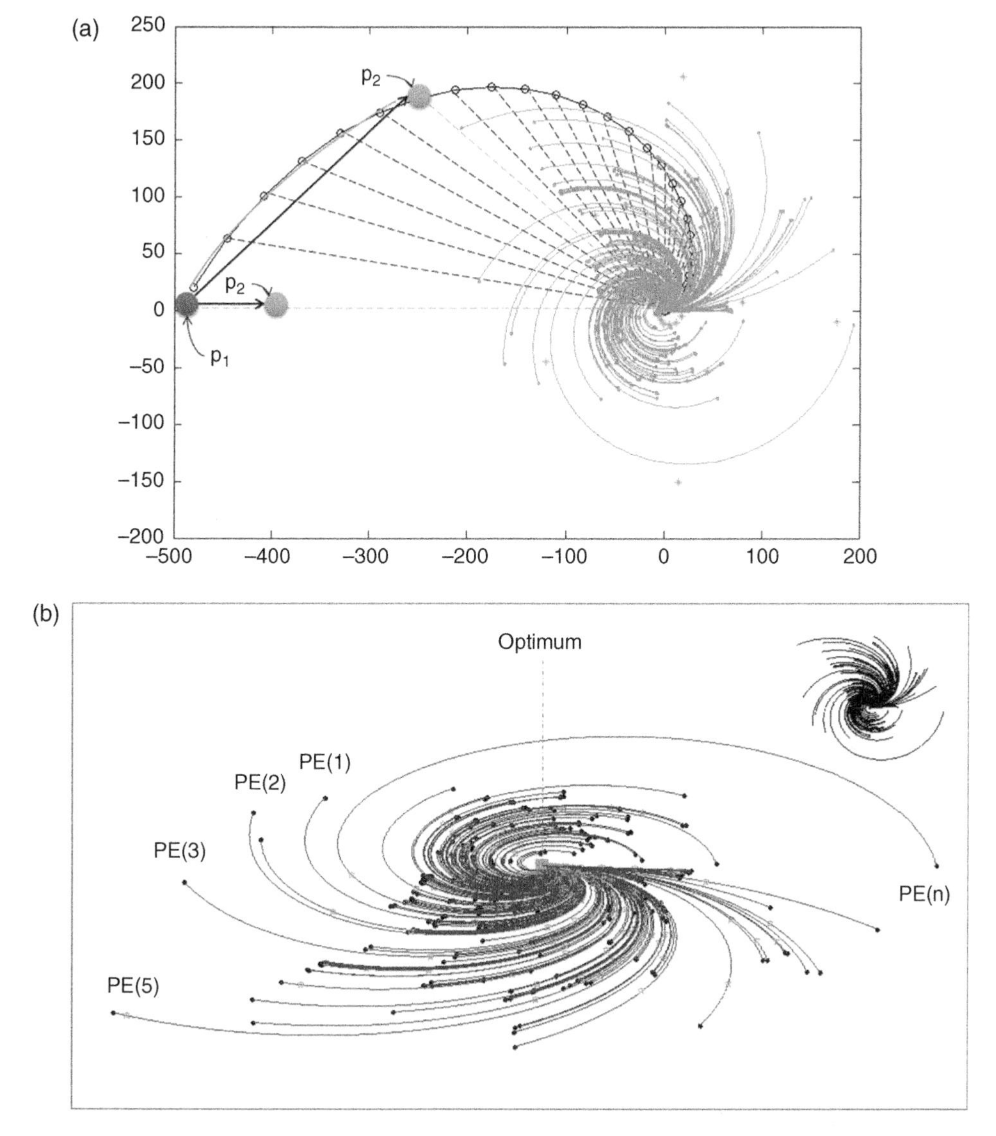

Figure 10.20 An example of the spiral motion of agents (penguins) towards the optimal point based on the spiral motion of the penguin herd and increasing the information gain of each penguin in each iteration.

Table 10.2 Spiral function for diversification and intensification of EPA.

r_{ij}	Equation given by	Arc length
Logarithmic spiral	$r = e^{b\theta}$	$s = \frac{a\sqrt{1+b^2}}{b}\, e^{b\theta}$
Archimedean spiral	$r = a\theta^{\left(\frac{1}{n}\right)}$	$s = a\theta^{\left(\frac{1}{n}\right)} {}_2F_1\left(-\frac{1}{2}, \frac{1}{2n}; 1 + \frac{1}{2n}; -n^2\,\theta^2\right)$
Archimedes' spiral	$r = a\theta$	$s = \frac{a}{2}\left\{\theta\sqrt{1+\theta^2} + \ln\left(\theta + \sqrt{1+\theta^2}\right)\right\}$
Circle involute	$r = a$	$s = \frac{at^2}{2}$
Cornu spiral	$B(t) = S(t) + i\, C(t)$	$s(t) = at$
Fermat's spiral	$r^2 = a^2\theta$	$s = \frac{1}{8}(1-i)aB\left(4\theta^2; \frac{1}{4}, \frac{3}{2}\right)$

$$
\begin{aligned}
s_i^n &= \int_{(\theta_1)_n}^{(\theta_2)_n} \sqrt[2]{\left(\frac{A}{\theta^2}\right)_n^2 + \left(\frac{-2A}{\theta^3}\right)^2}\, d\theta \\
s_i^n\left(\frac{A}{\theta^2}\right) &= \frac{\theta\sqrt{\frac{A^2(\theta^2+4)}{\theta^6}}\left(2\sqrt{\theta^2+4} + \theta^2\log(\theta) + \theta^2\log\left(\sqrt{\theta^2+4}+2\right)\right)}{4\sqrt{\theta^2+4}}.
\end{aligned}
\tag{10.35}
$$

Another function $r = \frac{A}{\theta^2}$ is employed in $\theta \epsilon [0, 2\pi]$. Because of the inherent adaptive mechanism of intensification and diversification in the spiral-like movement, performance of the algorithm is improved. To modify the position function, a new position can be calculated from:

$$
s_{i+1}^n = s_i^0[(1 - e^{-\tau r})] + R_I \tag{10.36}
$$

where R_I is the randomness of any function, the new position is r_{i+1}^n can be determined by root considering in the real domains, for function 1, $r = A\theta^2$

$$
\theta_{i+1}^n = \sqrt{-4 + \frac{3^{2/3}\left(s_{i+1}^n\right)^{2/3}}{A^{2/3}}} \tag{10.37}
$$

The new position is then estimated based on new θ_{i+1}^n:

$$
\begin{aligned}
x_{i+1} &= r_{i+1j}.\cos\left(\theta_{i+1}^n\right), \\
y_{i+1} &= r_{i+1j}.\cos\left(\theta_{i+1}^n\right)
\end{aligned}
\tag{10.38}
$$

A basic representation of EPA is presented in Figure 10.21.

Please refer to Chapter 6 for a discussion on how to determine the coordinates of the crack core as the optimal center at the crack level using the EPA method.

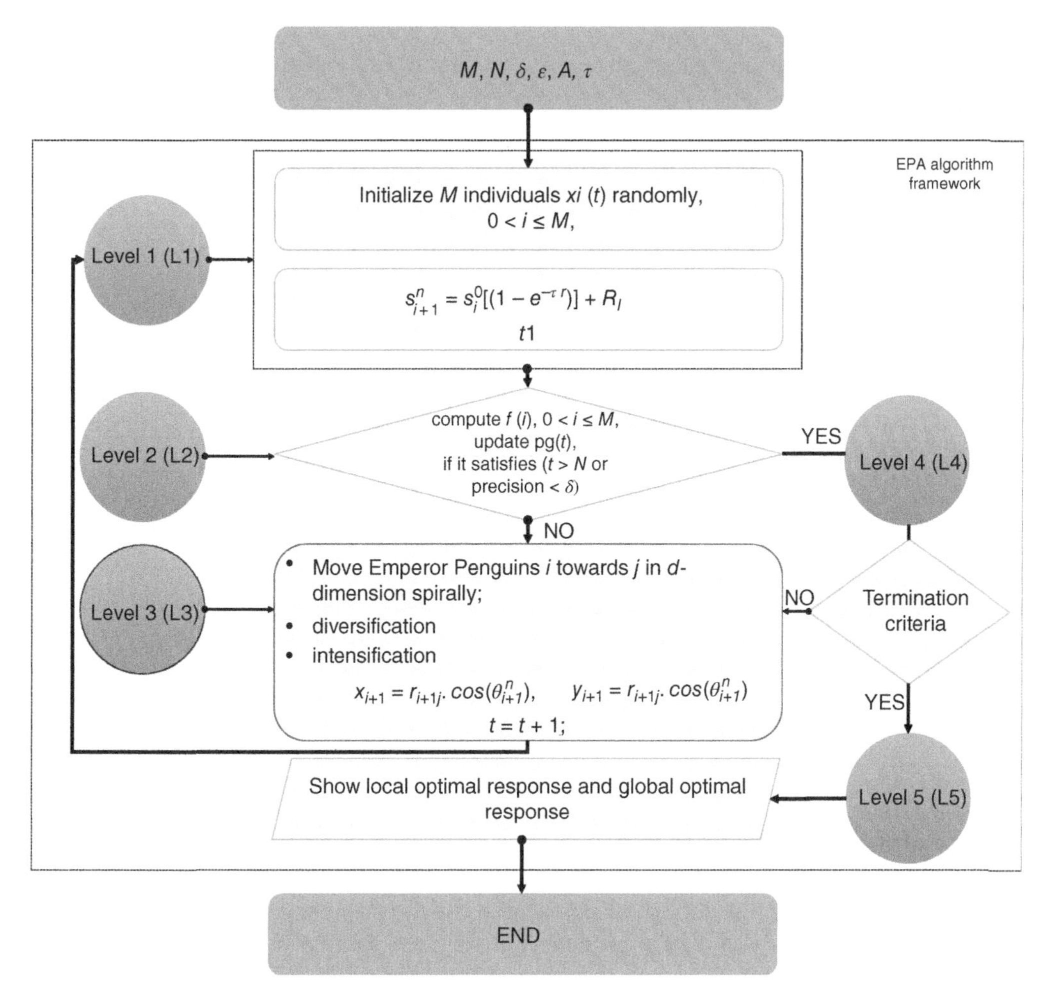

Figure 10.21 General framework and routine process of NIOA-EPA.

10.3.11 Hybrid Optimization Methods

The hybrid approach to optimization problems combines the advantages of individual methods together, to achieve better convergence, exploration, and, thus, performance to create a new method with powerful features. In this section, several simple combined PSO and GA methods are presented to classify the case study in infrastructure management (pavement segmentation for maintenance, for example). It should be mentioned that other combined methods are available in addition to those mentioned, and each of the nature-inspired methods can be used in pairs or more.

Type 1 hybridization is based on a parallel combination of two methods, such as PSO-GA, GA-PSO, PSO-PSO, and GA-GA. Type 2 hybridization refers to the use of one method within another method to optimize the parameters of the parent method, such as GA in PSO, PSO in GA, PSO in PSO, and GA in GA.

As an example of Type 1 hybridization of PSO-GA, the analysis begins with PSO then applies GA to determine the homogeneous components in pavement and to select the optimal repair and maintenance options.

The PSO-GA method begins with the production of particles and the determination of positions and velocities in a given batch. The next step is to evaluate each objective function and then determine pbest and gbest. The criterion then controls the stop of the solution: if the criterion is correct, the solutions are suitable; otherwise, the velocity and position of the particle change. If the stop criteria are met, the appropriate response is determined; otherwise, it is necessary to implement the GA selection, crossover, and mutation operators. The goal functions are then evaluated. The next steps include ranking, selecting, and replacing the population to consider new individuals. If the stop criterion is met in the last step, the appropriate solution is acceptable.

As an example of Type 2 hybridization of GA in PSO method, the parent method of PSO is selected to start, pbest and gbest are specified, and then GA is executed by the relevant operators. This method continues to update pbest and gbest to achieve the appropriate solution. The first step in this process is to produce the primary particles in a batch with their known positions and velocities, then the objective functions are evaluated to determine pbest and gbest. If the stop criterion is met, the solution is appropriate. Otherwise, GA operators and GA steps are performed to evaluate, combine, rank, select, and replace populations followed by the updating of the position and velocity of particles based on their appropriate values. After that, pbest and gbest are updated to find better solutions. Similarly, the desired solution is obtained after the stop criterion is met after applying the mentioned steps (Table 10.3).

10.4 Summary and Conclusion

In this chapter, various NIOAs that have recently been widely used in engineering were presented in a unified form for comparison and implementation. As mentioned, the structural design of these algorithms were inspired by the behavior of animals and natural phenomena. In general, these behaviors include simulating herd intelligence, simulating the behavior of biological systems, simulating the behavior of physical and chemical systems, and other types of behaviors. Accordingly, NIOAs continue to evolve on a daily basis.

Moreover, various methods, such as GA, PSO Algorithm, ABC Algorithm, BA, IA, FA, CS Algorithm, Gray Wolf Optimizer (GWO), KHA, EPA, Hybrid Optimization Methods presented. The main aim of this chapter is to simplify the description of such algorithms and prepare the reader of the book to produce his or her own innovative algorithm called NIOA-X.

In this chapter, the main theory and general idea of some NIOAs are summarized, and the final section on how to use one or more NIOAs is presented. Finally, a new algorithm based on the behavior of Emperor penguins (EPA), which was introduced in Chapter 6, is described in mathematical detail. By providing a thorough explanation of the EPA method, we hope that the readers of this book will be able to understand and develop new nature-inspired algorithms to solve specialized problems in the field of optimization, especially infrastructure management.

Table 10.3 Comparison of hybrid PSO and GA methods in network level for pavement sectioning.

	PSO-GA		GA-PSO		PSO-PSO		GA-GA		PSO in GA	
Method	**Num.**	**Time**	**Num.**	**Time**	**Num.**	**Time**	**Num.**	**Time**	**Num.**	**Time**
Average	20 061	366.77	20 110	367.20	20 139	253.62	20 576	1 379.26	20 092	401.32
Std.	93.88	42.36	122.03	43.87	57.26	25.84	146.78	259.18	54.80	71.57
Min	19 930	324.86	19 950	287.06	20 080	225.48	20 253	1 154.55	19 957	323.18
Max	20 197	448.13	20 270	414.90	20 215	300.48	20 671	1 944.13	20 145	522.83

	GA in PSO		PSO in PSO		GA in GA		GA		PSO	
Method	**Num.**	**Time**	**Num.**	**Time**	**Num.**	**Time**	**Num.**	**Time**	**Num.**	**Time**
Average	20 132	376.32	20 163	564.35	20 464	1 876.35	20 542	15 438.25	20 210	330.31
Std.	103.78	71.57	54.06	67.78	48.33	27.21	34.90	7 151.72	102.30	26.58
Min	19 951	298.18	20 115	494.98	20 401	1 822.67	20 493	1 814.74	20 102.00	298.36
Max	20 250	497.83	20 236	672.82	20 531	1 906.54	20 596	19 548.64	20 391.00	368.03

10.5 Questions and Exercises

1. Define the benchmark performance functions for the global optimization of the Ackley functions 1–4, Adjiman Function, and Alpine Function 1,2. Based on the three methods, i.e. GA, PSO, and BA, present the error rate and computational time separately.

2. Define the benchmark performance functions for the global optimization of the Bartels Conn Function, Biggs EXP2 Function, Bird Function, and Bohachevsky Function1–3. Based on the three methods, i.e. FA, KH, and BA, present the error rate and computational time separately.

3. Define the benchmark performance functions for the global optimization of the Box-Betts Quadratic Sum Function, Branin RCOS Function, and Brown Function. Based on the three methods, i.e. CS, FA, and GWO, present the error rate and computational time separately and compare one.

4. Define the benchmark performance functions for the global optimization of the Powell Singular 1,2, Powell Sum Function, and Mishra Function 1–3. Based on the three methods, i.e. ABC, CS, and EPA, present the error rate and computational time separately and compare one.

5. Based on the benchmark performance of the Helical Valley function for global and function optimization, use a combination of the PSO and GA methods for Type 1 and 2 hybridization, and compare all modes when each of these methods is used separately.

6. Based on the benchmark performance of the Egg Holder Function for global and function optimization, use a combination of the GWO and BA methods for Type 1 and 2 hybridization, and compare all modes when each of these methods is used separately.

7. Based on the benchmark performance of the Colville Function for global and function optimization, use a combination of the CS and FA methods for Type 1 and 2 hybridization, and compare all modes when each of these methods is used separately.

8. Based on the benchmark performance of the Powell Singular Function for global and function optimization, use a combination of the IA and KH methods for Type 1 and 2 hybridization, and compare all modes when each of these methods is used separately.

9. Based on the benchmark performance of the Quintic Function for global and function optimization, use a combination of the KH and CS methods for Type 1 and 2 hybridization, and compare all modes when each of these methods is used separately.

10. Based on the natural phenomena and general structure defined in the second part of this chapter, design a natural phenomenon for optimization in 5 steps, and then validate it using reference functions and compare it with other NIOA methods.

    Hint: The functions mentioned in this section can be extracted from Article "A Literature Survey of Benchmark Functions for Global Optimization Problems."

Further Reading

1 Abdollahzadeh, B., F. Soleimanian Gharehchopogh, and S. Mirjalili, *Artificial gorilla troops optimizer: A new nature-inspired metaheuristic algorithm for global optimization problems. International Journal of Intelligent Systems*, 2021. **36**(10): p. 5887–5958.
2 Abualigah, L., et al., *The arithmetic optimization algorithm. Computer Methods in Applied Mechanics and Engineering*, 2021. **376**: p. 113609.
3 Abualigah, L., et al., *Aquila optimizer: A novel meta-heuristic optimization algorithm. Computers & Industrial Engineering*, 2021. **157**: p. 107250.
4 Abualigah, L.M., et al., *A novel hybridization strategy for krill herd algorithm applied to clustering techniques. Applied Soft Computing*, 2017. **60**: p. 423–435.
5 Ahmadianfar, I., et al., *RUN beyond the metaphor: An efficient optimization algorithm based on Runge Kutta method. Expert Systems with Applications*, 2021. **181**: p. 115079.
6 Al-Betar, M.A., et al., *Coronavirus herd immunity optimizer (CHIO). Neural Computing and Applications*, 2021. **33**(10): p. 5011–5042.
7 Al-kubaisy, W.J., et al., *The red colobuses monkey: A new nature–inspired metaheuristic optimization algorithm. International Journal of Computational Intelligence Systems*, 2021. **14**(1): p. 1108–1118.
8 Azizi, M., *Atomic orbital search: A novel metaheuristic algorithm. Applied Mathematical Modelling*, 2021. **93**: p. 657–683.
9 Bairwa, A.K., S. Joshi, and D. Singh, *Dingo optimizer: A nature-inspired metaheuristic approach for engineering problems. Mathematical Problems in Engineering*, 2021. **2021**: p. 1–12.
10 Chakraborty, S., et al., *SHADE–WOA: A metaheuristic algorithm for global optimization. Applied Soft Computing*, 2021. **113**: p. 107866.
11 Chen, H., et al., *Multi-population differential evolution-assisted Harris hawks optimization: Framework and case studies. Future Generation Computer Systems*, 2020. **111**: p. 175–198.
12 Cheng, M.-Y. and D. Prayogo, *Symbiotic organisms search: A new metaheuristic optimization algorithm. Computers & Structures*, 2014. **139**: p. 98–112.
13 Das, S., et al., *Bacterial foraging optimization algorithm: theoretical foundations, analysis, and applications*, in *Foundations of Computational Intelligence Volume 3*. 2009, Springer. p. 23–55.
14 Derrac, J., et al., *Enhancing evolutionary instance selection algorithms by means of fuzzy rough set based feature selection. Information Sciences*, 2012. **186**(1): p. 73–92.
15 Dhiman, G., *ESA: A hybrid bio-inspired metaheuristic optimization approach for engineering problems. Engineering with Computers*, 2021. **37**(1): p. 323–353.
16 Dhiman, G., et al., *EMoSOA: A new evolutionary multi-objective seagull optimization algorithm for global optimization. International Journal of Machine Learning and Cybernetics*, 2021. **12**(2): p. 571–596.
17 Dokeroglu, T., et al., *A survey on new generation metaheuristic algorithms. Computers & Industrial Engineering*, 2019. **137**: p. 106040.
18 Faramarzi, A., et al., *Marine predators algorithm: A nature-inspired metaheuristic. Expert Systems with Applications*, 2020. **152**: p. 113377.
19 Fister, I., et al., *A comprehensive review of firefly algorithms. Swarm and Evolutionary Computation*, 2013. **13**: p. 34–46.
20 Gandomi, A. and M. Haider, *Beyond the hype: Big data concepts, methods, and analytics. International Journal of Information Management*, 2015. **35**(2): p. 137–144.
21 Gandomi, A.H., *Interior search algorithm (ISA): A novel approach for global optimization. ISA Transactions*, 2014. **53**(4): p. 1168–1183.

22 Gandomi, A.H. and A.H. Alavi, *Multi-stage genetic programming: A new strategy to nonlinear system modeling. Information Sciences*, 2011. **181**(23): p. 5227–5239.

23 Gandomi, A.H. and A.H. Alavi, *Krill herd: A new bio-inspired optimization algorithm. Communications in Nonlinear Science and Numerical Simulation*, 2012. **17**(12): p. 4831–4845.

24 Gandomi, A.H., et al., *Nonlinear genetic-based models for prediction of flow number of asphalt mixtures. Journal of Materials in Civil Engineering*, 2011. **23**(3): p. 248–263.

25 Gandomi, A.H. and X.-S. Yang, *Chaotic bat algorithm. Journal of Computational Science*, 2014. **5**(2): p. 224–232.

26 Gandomi, A.H., X.-S. Yang, and A.H. Alavi, *Mixed variable structural optimization using firefly algorithm. Computers & Structures*, 2011. **89**(23–24): p. 2325–2336.

27 Gandomi, A.H., X.-S. Yang, and A.H. Alavi, *Cuckoo search algorithm: A metaheuristic approach to solve structural optimization problems. Engineering with Computers*, 2013. **29**(1): p. 17–35.

28 Gandomi, A.H., et al., *Bat algorithm for constrained optimization tasks. Neural Computing and Applications*, 2013. **22**(6): p. 1239–1255.

29 Gandomi, A.H., et al., *Metaheuristic Applications in Structures and Infrastructures*. 2013. Newnes.

30 Gandomi, A.H., et al., *Firefly algorithm with chaos. Communications in Nonlinear Science and Numerical Simulation*, 2013. **18**(1): p. 89–98.

31 Gandomi, A.H., et al., *Chaos-enhanced accelerated particle swarm optimization. Communications in Nonlinear Science and Numerical Simulation*, 2013. **18**(2): p. 327–340.

32 Gharehchopogh, F.S., I. Maleki, and Z.A. Dizaji, *Chaotic vortex search algorithm: Metaheuristic algorithm for feature selection. Evolutionary Intelligence*, 2021: p. 1–32. https://doi.org/10.1007/s12065-021-00590-1.

33 Gholizadeh, S., M. Danesh, and C. Gheyratmand, *A new Newton metaheuristic algorithm for discrete performance-based design optimization of steel moment frames. Computers & Structures*, 2020. **234**: p. 106250.

34 Glover, F., *Heuristics for integer programming using surrogate constraints. Decision Sciences*, 1977. **8**(1): p. 156–166.

35 Hashim, F.A., et al., *Archimedes optimization algorithm: A new metaheuristic algorithm for solving optimization problems. Applied Intelligence*, 2021. **51**(3): p. 1531–1551.

36 Hayyolalam, V. and A.A.P. Kazem, *Black widow optimization algorithm: A novel meta-heuristic approach for solving engineering optimization problems. Engineering Applications of Artificial Intelligence*, 2020. **87**: p. 103249.

37 Ahmadianfar, I., A.A. Heidari, A.H. Gandomi, X. Chu, and H. Chen (2021). *RUN beyond the metaphor: an efficient optimization algorithm based on Runge Kutta method. Expert Systems with Applications.* **181**: p. 115079.

38 Heidari, A.A., et al., *Harris hawks optimization: Algorithm and applications. Future Generation Computer Systems*, 2019. **97**: p. 849–872.

39 Kaboli, S.H.A., J. Selvaraj, and N. Rahim, *Rain-fall optimization algorithm: A population based algorithm for solving constrained optimization problems. Journal of Computational Science*, 2017. **19**: p. 31–42.

40 Kaedi, M., *Fractal-based algorithm: A new metaheuristic method for continuous optimization. International Journal of Artificial Intelligence*, 2017. **15**(1): p. 76–92.

41 Kar, A.K., *Bio inspired computing – A review of algorithms and scope of applications. Expert Systems with Applications*, 2016. **59**: p. 20–32.

42 Kaveh, A. and A.D. Eslamlou, *Water strider algorithm: A new metaheuristic and applications.* in *Structures*. 2020. Elsevier.

43 Khanduja, N. and B. Bhushan, *Recent advances and application of metaheuristic algorithms: A survey (2014–2020).* in *Metaheuristic and Evolutionary Computation: Algorithms and Applications.* 2021: p. 207–228.

44 Khishe, M. and M.R. Mosavi, *Chimp optimization algorithm. Expert Systems with Applications*, 2020. **149**: p. 113338.

45 Kumar, N., N. Singh, and D.P. Vidyarthi, *Artificial lizard search optimization (ALSO): A novel nature-inspired meta-heuristic algorithm. Soft Computing*, 2021. **25**(8): p. 6179–6201.

46 Lones, M.A., *Mitigating metaphors: A comprehensible guide to recent nature-inspired algorithms. SN Computer Science*, 2020. **1**(1): p. 1–12.

47 Majani, H. and M. Nasri, *Water Streams Optimization (WSTO): A new metaheuristic optimization method in high-dimensional problems. Journal of Soft Computing and Information Technology*, 2021. **10**(1): p. 36–51.

48 Martínez-Álvarez, F., et al., *Coronavirus optimization algorithm: A bioinspired metaheuristic based on the COVID-19 propagation model. Big Data*, 2020. **8**(4): p. 308–322.

49 Mirjalili, S., et al., *Salp Swarm algorithm: A bio-inspired optimizer for engineering design problems. Advances in Engineering Software*, 2017. **114**: p. 163–191.

50 Mirjalili, S. and A. Lewis, *The whale optimization algorithm. Advances in Engineering Software*, 2016. **95**: p. 51–67.

51 Mirjalili, S., S.M. Mirjalili, and A. Lewis, *Grey wolf optimizer. Advances in Engineering Software*, 2014. **69**: p. 46–61.

52 Mirjalili, S.Z., et al., *Grasshopper optimization algorithm for multi-objective optimization problems. Applied Intelligence*, 2018. **48**(4): p. 805–820.

53 Mohammadi-Balani, A., et al., *Golden eagle optimizer: A nature-inspired metaheuristic algorithm. Computers & Industrial Engineering*, 2021. **152**: p. 107050.

54 Naruei, I. and F. Keynia, *Wild horse optimizer: A new meta-heuristic algorithm for solving engineering optimization problems. Engineering with Computers*, 2021: p. 1–32. https://doi.org/10.1007/s00366-021-01438-z.

55 Nik, A.A., F.M. Nejad, and H. Zakeri, *Hybrid PSO and GA approach for optimizing surveyed asphalt pavement inspection units in massive network. Automation in Construction*, 2016. **71**: p. 325–345.

56 Oyelade, O.N. and A.E. Ezugwu, *Ebola Optimization Search Algorithm (EOSA): A new metaheuristic algorithm based on the propagation model of Ebola virus disease.* arXiv preprint arXiv:2106.01416, 2021.

57 Pereira, J.L.J., et al., *Lichtenberg algorithm: A novel hybrid physics-based meta-heuristic for global optimization. Expert Systems with Applications*, 2021. **170**: p. 114522.

58 Salgotra, R., M. Gandomi, and A.H. Gandomi, *Time series analysis and forecast of the COVID-19 pandemic in India using genetic programming. Chaos, Solitons & Fractals*, 2020. **138**: p. 109945.

59 Salimi, H., *Stochastic fractal search: A powerful metaheuristic algorithm. Knowledge-Based Systems*, 2015. **75**: p. 1–18.

60 Shadravan, S., H. Naji, and V.K. Bardsiri, *The Sailfish optimizer: A novel nature-inspired metaheuristic algorithm for solving constrained engineering optimization problems. Engineering Applications of Artificial Intelligence*, 2019. **80**: p. 20–34.

61 Sharma, H., G. Hazrati, and J.C. Bansal, *Spider monkey optimization algorithm*, in *Evolutionary and Swarm Intelligence Algorithms.* 2019. Springer. p. 43–59.

62 Shunmugapriya, P. and S. Kanmani, *A hybrid algorithm using ant and bee colony optimization for feature selection and classification (AC-ABC Hybrid). Swarm and Evolutionary Computation*, 2017. **36**: p. 27–36.

63 Talatahari, S. and M. Azizi, *Chaos Game optimization: A novel metaheuristic algorithm. Artificial Intelligence Review*, 2021. **54**(2): p. 917–1004.

64 Talatahari, S., M. Azizi, and A.H. Gandomi, *Material generation algorithm: A novel metaheuristic algorithm for optimization of engineering problems. Processes*, 2021. **9**(5): p. 859.

65 Tarkhaneh, O., et al., *Golden tortoise beetle optimizer: A novel nature-inspired meta-heuristic algorithm for engineering problems.* arXiv preprint arXiv:2104.01521, 2021.

66 Tu, J., et al., *The colony predation algorithm. Journal of Bionic Engineering*, 2021. **18**(3): p. 674–710.

67 Wang, G. and L. Guo, *A novel hybrid bat algorithm with harmony search for global numerical optimization. Journal of Applied Mathematics*, 2013. **2013**.

68 Wang, G., et al., *Incorporating mutation scheme into krill herd algorithm for global numerical optimization. Neural Computing and Applications*, 2014. **24**(3): p. 853–871.

69 Wang, G.-G., A.H. Gandomi, and A.H. Alavi, *Stud krill herd algorithm. Neurocomputing*, 2014. **128**: p. 363–370.

70 Wang, G.-G., et al., *Hybridizing harmony search algorithm with cuckoo search for global numerical optimization. Soft Computing*, 2016. **20**(1): p. 273–285.

71 Wang, G.-G., et al., *Chaotic krill herd algorithm. Information Sciences*, 2014. **274**: p. 17–34.

72 Wang, Z., et al., *A comparative study of common nature-inspired algorithms for continuous function optimization. Entropy*, 2021. **23**(7): p. 874.

73 Yang, X.S. and A.H. Gandomi, *Bat algorithm: A novel approach for global engineering optimization. Engineering Computations*, 2012. **29**(5): p. 464–483.

74 Yang, X.-S., et al., *Swarm Intelligence and Bio-inspired Computation: Theory and Applications.* 2013. Newnes.

75 Yang, X.-S., et al., *Metaheuristics in Water, Geotechnical and Transport Engineering.* 2012. Newnes.

76 Yang, X.-S. and X. He, *Bat algorithm: Literature review and applications. International Journal of Bio-Inspired Computation*, 2013. **5**(3): p. 141–149.

77 Yang, X.-S., S.S.S. Hosseini, and A.H. Gandomi, *Firefly algorithm for solving non-convex economic dispatch problems with valve loading effect. Applied Soft Computing*, 2012. **12**(3): p. 1180–1186.

78 Yang, Y., et al., *Hunger games search: Visions, conception, implementation, deep analysis, perspectives, and towards performance shifts. Expert Systems with Applications*, 2021. **177**: p. 114864.

79 Zakeri, H., F.M. Nejad, and A. Fahimifar, *Rahbin: A quadcopter unmanned aerial vehicle based on a systematic image processing approach toward an automated asphalt pavement inspection. Automation in Construction*, 2016. **72**: p. 211–235.

80 Zakeri, H., F.M. Nejad, and A. Fahimifar, *Image based techniques for crack detection, classification and quantification in asphalt pavement: A review. Archives of Computational Methods in Engineering*, 2017. **24**(4): p. 935–977.

81 Zakeri, H., et al., *A multi-stage expert system for classification of pavement cracking.* in 2013 Joint IFSA World Congress and NAFIPS Annual Meeting (IFSA/NAFIPS). 2013. IEEE.

82 Zamani, H., M.H. Nadimi-Shahraki, and A.H. Gandomi, *QANA: Quantum-based avian navigation optimizer algorithm. Engineering Applications of Artificial Intelligence*, 2021. **104**: p. 104314.

83 Zervoudakis, K. and S. Tsafarakis, *A mayfly optimization algorithm. Computers & Industrial Engineering*, 2020. **145**: p. 106559.

Appendix A

Data Sets and Codes

Table A.1 Abbreviation and definition of fuzzy sets.

Abbreviations	Definitions
G3DT2PFLS	General 3D type-2 polar fuzzy logic systems
GT2PFM	General type-2 polar fuzzy membership
CSS	Cubic smoothing spline
G3DPT2	General 3D polar type-2
T1FLS	Type-1 fuzzy logic systems
IT2FLS	Interval type-2 fuzzy logic systems
GT2FLS	General type-2 fuzzy logic systems
T1FS	Type-1 fuzzy sets
T2FS	Type-2 fuzzy sets
GT2 FS	General type-2 fuzzy set
FOU	Footprint of uncertainty
3DFOU	3D footprint of uncertainty
G3DT2FLS	General 3D type-2 fuzzy logic systems
MF	Membership function
SMF	Secondary membership function
GT2 FSS	General type-2 fuzzy sets system
FLS	Fuzzy logic sets
MEP	Maximum entropy principle
UMF	Upper bound membership function
LMF	Lower bound membership function
PFOU	Polar footprint of uncertainty
PGT2FLS	Polar general type-2 fuzzy logic systems
FMF	Fuzzy membership function
T2FMF	Type-2 fuzzy membership function
FCM	Fuzzy c-means
RT	Radon transform
GF	General fuzzy
Var.	Variable

Automation and Computational Intelligence for Road Maintenance and Management: Advances and Applications,
First Edition. Hamzeh Zakeri, Fereidoon Moghadas Nejad, and Amir H. Gandomi.

Appendix B

The Glossary of Nature-Inspired Optimization Algorithms (NIOAS)

B.1 Introduction

In this Appendix, the glossary of the names of the relevant algorithms until 2021 is presented. These algorithms include 252 methods.

A
Across Neighborhood Search
Adaptive Dimensional Search
African Buffalo Optimization
African vulture's optimization algorithm
African Wild Dog Algorithm
Alienated Ant Algorithm
Anarchic Society Optimization
Animal Migration Optimization
Ant Colony Optimization
Ant Lion Optimizer
Archimedes optimization algorithm
Artificial Algae Algorithm
Artificial Bee Colony
Artificial Chemical Reaction Optimization Algorithm
Artificial Cooperative Search
Artificial Ecosystem Algorithm
Artificial Fish Swarm Algorithm
Artificial gorilla troops optimizer
Artificial lizard search optimization
Artificial Plant Optimization Algorithm
Artificial Root Foraging Algorithm
Artificial Searching Swarm Algorithm
Atmosphere Clouds Model
Atomic orbital search
B
Backtracking Search Algorithm
Bacteria Chemotaxis Algorithm
Bacterial Colony Optimization
Bacterial Evolutionary Algorithm
Bacterial Foraging Algorithm
Bacterial Foraging Inspired Algorithm
Bacterial Swarming
Bat Algorithm

(*Continued*)

Big bang-big Crunch
Biogeography Based Optimization
Bird Mating Optimizer
Bird Swarm
Black Hole Algorithm
Black Widow Optimization Algorithm
Blind, Naked Mole-rats Algorithm
Bottlenose Dolphin Optimization
Brain Storm Optimization
Bull Optimization Algorithm
Bumble Bees Mating Optimization

C

Camel Algorithm
Cat Swarm Optimization
Central Force Optimization
Chaos Game Optimization
Chaotic vortex search algorithm
Charged System Search
Chemical Reaction Optimization
Chicken Swarm Optimization
Chimp optimization algorithm
Clonal Selection Algorithm
Clonal Selection Algorithm
Cockroach Swarm Optimization
Colliding Bodies Optimization
Color Harmony Algorithm
Community of Scientist Optimization
Consultant-guided Search
Coral Reefs Optimization Algorithm
Coronavirus herd immunity optimizer
Coronavirus Optimization Algorithm
Covariance Matrix Adaptation- Evolution Strategy
Coyote Optimization Algorithm
Cricket Algorithm
Crow Search Algorithm
Crystal Energy Optimization Algorithm
Cuckoo Optimization Algorithm
Cuckoo Search
Cultural Algorithm
Cuttlefish Algorithm

D

Dialectic Search
Differential Evaluation
Differential Search Algorithm
Dingo Optimizer
Dolphin Echolocation
Dolphin Partner Optimization
Dragonfly Algorithm

E

Eagle Strategy
Ebola Optimization Search Algorithm
Eco-inspired Evolutionary Algorithm
Egyptian Vulture Optimization
Electromagnetic Field Optimization
Electro-magnetism Optimization
Elephant Herding Optimization
Elephant Search Algorithm
EMoSOA

Emperor Penguins Colony
ESA
Evolution Strategy
Evolutionary Algorithm
Evolutionary Programming
Exchange Market Algorithm
F
FIFA World Cup
Firefly Algorithm
Fireworks Algorithm
Fish-school Search
Flower Pollination Algorithm
Flying Elephants Algorithm
Forest Optimization Algorithm
Fractal Based Algorithm
Fruit Fly Optimization
Future Search Algorithm
G
Galaxy-based Search Algorithm
Gases Brownian Motion Optimization
Gene Expression Programming
General Relativity Search Algorithm
Genetic Algorithm
Giza Pyramids Construction
Glowworm Swarm Optimization
Glowworm Swarm Optimization
Golden Ball
Golden eagle optimizer
Golden Tortoise Beetle Optimizer
Good Lattice Swarm Algorithm
Grasshopper Optimisation Algorithm
Grasshopper Optimization Algorithm
Gravitational Search Algorithm
Great Deluge Algorithm
Great Salmon Run
Greedy Politics Optimization
Grenade Explosion Method
Grey Wolf Optimizer
Grey Wolf Optimizer
Group Counseling Optimization
Group Search Optimizer
Guided Local Search
H
Harmony Search
Harris Hawks Optimization
Harris hawks optimization
Heap-Based Optimizer
Heart
Hierarchical Swarm Model
Honey-bees Mating Optimization Algorithm
Hoopoe Heuristic Optimization
Human-inspired Algorithm
Hunting Search
I
Imperialistic Competitive Algorithm
Intelligent Water Drops

(*Continued*)

Interior Design and Decoration
Invasive Tumor Growth Optimization Algorithm
Invasive Weed Optimization
Ions Motion Algorithm
Iterative Local Search

J

Jaguar Algorithm
Japanese Tree Frogs Calling
Joint Operations Algorithm

K

Kaizen Programming
Keshtel Algorithm
Key Cutting Algorithm
Krill Herd Algorithm

L

League Championship Algorithm
Lichtenberg algorithm
Lightning Search Algorithm
Lion Optimization Algorithm
Local Search
Locust Swarm Algorithm

M

Magnetotactic Bacteria Optimization Algorithm
Material Generation Algorithm
Mayfly Optimization Algorithm
Memetic Algorithm
Migrating Birds Optimization
Mine Blast Algorithm
Monarch Butterfly Optimization
Monkey Algorithm
Moth-Flame Optimization
Multi-verse Optimizer
Newton metaheuristic algorithm

O

Optics Inspired Optimization
Owl Search Algorithm

P

Paddy Field Algorithm
Parliamentary Optimization Algorithm
Particle Swarm Optimization
Pattern Search
Penguins Search Optimization Algorithm
Photosynthetic Learning Algorithm
Phototropic Optimization Algorithm
Plant Growth Optimisation
Plant Propagation Algorithm
Political Optimizer
POPMUSIC: Partial Optimization Metaheuristic Under Special Intensification Conditions

Q

Queen-bee Evolution

R

Radial Movement Optimization
Rain-fall Optimization Algorithm
Raven Roosting Optimization Algorithm
Ray Optimization
Red Deer Algorithm
Reincarnation

River Formation Dynamics
Roach Infestation Optimization
Root Growth Optimizer
Rooted Tree Optimization Algorithm
Runner-root Algorithm
S
Sailfish Optimizer
Saplings Growing Up Algorithm
Scatter Search
Scientific Algorithms for the Car Renter Salesman Problem
SDA Optimization Algorithm
Seeker Optimization Algorithm
Seven-spot Ladybird Optimization
SHADE–WOA
Shark Smell Optimization
Sheep Flocks Heredity Model
Shuffled Frog Leaping Algorithm
Simplex Heuristic
Simulated Annealing
Sine Cosine Algorithm
Small-world Optimization Algorithm
Soccer Game Optimization
Social Cognitive Optimization
Social Emotional Optimization
Social Engineering Optimizer
Social Spider Algorithm
Society and Civilization
Sperm Motility Algorithm
Sperm Whale Algorithm
Spider Monkey Optimization
Spider Monkey Optimization
Spider Monkey Optimization
Spiral Dynamics Inspired Optimization
Squirrel Search Algorithm
Stochastic Diffusion Search
Stochastic Fractal Search
Stochastic Paint Optimizer
Strawberry Algorithm
Swallow Swarm Optimization Algorithm
Symbiotic Organisms Search
T
Tabu Search
Teaching-Learning-Based Optimization
Termite Colony Optimization
The Red Colobuses Monkey
V
Variable Neighborhood Search
Viral Systems
Virus Colony Search
Virus Optimization Algorithm
Vortex Search Algorithm
W
Wasp Swarm Optimization
Water Cycle Algorithm
Water Evaporation Optimization
Water Streams Optimization

(Continued)

Water strider algorithm
Water Wave Optimization
Water-flow Algorithm
Whale Optimization Algorithm
Wild horse optimizer
Wind Driven Optimization
Wolf Search Algorithm
Worm Optimization

Z

Zombie Survival Optimization

Appendix C

Sample Code for Feature Selection

```
function [data_mod, index_rem,SIM,SIM1,H]=feat_sel_sim_Schweizer_1(data, p)
%    p in (0, \infty) as default p=1. if nargin
p=1;
end

l=max(data(:,end)); % #-classes m=size(data,1); % #-samples t=size(data,2)-1;
% #-features dataold=data;
tmp=[];
% forming idealvec using arithmetic mean idealvec=zeros(1,t);
for k=1:l idealvec_s(k,:)=mean(data(find(data(:,end)==k),1:t));
end

%scaling data between [0,1] data_v=data(:,1:t); data_c=data(:,t+1); %labels
mins_v = min(data_v);
Ones = ones(size(data_v));
data_v = data_v+Ones*diag(abs(mins_v)); for k=1:l
tmp=[tmp;abs(mins_v)]; end
tmp;
idealvec_s = idealvec_s+tmp; maxs_v = max(data_v);
data_v = data_v*diag(maxs_v.^(-1)); idealvec_s=idealvec_s./repmat(maxs_v,
1,1); data = [data_v, data_c];
% sample data datalearn_s=data(:,1:t); land=0.1;
% similarities sim=zeros(t,m,l); for j=1:m

for i=1:t
for k=1:l
%%similarity Luca    with p=1
sn_1(i,j,k)=1-(max(0,((1-(1-datalearn_s(j,i)))^p+((1-idealvec_s(k,i))^p)-1)))
^(1/p);
sn_2(i,j,k)=1-(max(0,((1-(datalearn_s(j,i)))^p+((1-(1-idealvec_s(k,i)))^p)-
1)))^(1/p);
sim(i,j,k)=(max(0,((sn_1(i,j,k))^p+(sn_2(i,j,k))^p))-1)^(1/p); end
end end
```

Automation and Computational Intelligence for Road Maintenance and Management: Advances and Applications,
First Edition. Hamzeh Zakeri, Fereidoon Moghadas Nejad, and Amir H. Gandomi.

```
%class 1,2,3,4,. n
% SIM(:,:,1) = similarity for class 1
% SIM(:,:,2) = similarity for class 2 SIM=sim;
% reduce number of dimensions in sim sim=reshape(sim,t,m*l)';
SIM1=sim;

% possibility for two different entropy measures
% moodifying zero and one values of the similarity values to work with
% De Luca's entropy measure
%%%%%Hsh(1)
delta=1E-10; sim(find(sim==0))=delta; sim(find(sim==1))=1-delta;
Hsh=sum(-sim.*log(sim)-(1-sim).*log(1-sim));
Hsh=(Hsh-min(Hsh))/(max(Hsh)-min(Hsh));

% %%%%%Hme(2)
Hme=sum(sin(pi/2*sim)+sin(pi/2*(1-sim))-1); Hme=(Hme-min(Hme))/(max(Hme)-min
(Hme));

%%%%%%%Hrn(3)
alfa_rn=0.8;
Hrn=log(sum((sim.^alfa_rn)+((1-sim).^alfa_rn)))/(1-alfa_rn); Hrn=(Hrn-
min(Hrn))/(max(Hrn)-min(Hrn));

% %%%%%%Hpp(4)
Hpp=sum((sim.*exp(1.-sim))+(((1-sim).*exp(1.-(1-sim))))); Hpp=(Hpp-min(Hpp))/
(max(Hpp)-min(Hpp));

% %%%%%%%Hts(5)
alfa_ts=0.5;
Hts=1-(sum((sim.^alfa_ts)+((1-sim).^alfa_ts)))/(alfa_ts-1);
Hts=(Hts-min(Hts))/(max(Hts)-min(Hts));

%%%%%%%%Hma(6)
gama=2; a_ma=0.5; b_ma=3; alfa_ma=0.1; beta_ma=2;
Hma=(sum(((sim.^alfa_ma).*exp(a_ma.*sim.^alfa_ma+b_ma).^beta_ma)+((1-sim).
^alfa_ma).
*exp(a_ma.*(1-sim).^alfa_ma+b_ma).^beta_ma)); Hma=(Hma-min(Hma))/(max(Hma)-
min(Hma));

% find maximum feature
[i_sh, index_rem_sh]=max(Hsh); [i_me, index_rem_me]=max(Hme); [i_rn,
index_rem_rn]=max(Hrn); [i_pp, index_rem_pp]=max(Hpp); [i_ts, index_rem_ts]
=max(Hts); [i_ma, index_rem_ma]=max(Hma);

% removing feature from the data
data_mod_sh=[dataold(:,1:index_rem_sh-1) dataold(:,index_rem_sh+1:end)];
data_mod_me=[dataold(:,1:index_rem_me-1) dataold(:,index_rem_me+1:end)];
data_mod_rn=[dataold(:,1:index_rem_rn-1) dataold(:,index_rem_rn+1:end)];
data_mod_pp=[dataold(:,1:index_rem_pp-1) dataold(:,index_rem_pp+1:end)];
data_mod_ts=[dataold(:,1:index_rem_ts-1) dataold(:,index_rem_ts+1:end)];
data_mod_ma=[dataold(:,1:index_rem_ma-1) dataold(:,index_rem_ma+1:end)];
```

```
data_mod=[data_mod_sh,data_mod_me,data_mod_rn,data_mod_pp,data_mod_ts,
data_mod_ma]; index_rem=[index_rem_sh,index_rem_me,index_rem_rn,index_rem_pp,
index_rem_ts,index_rem_ma]; H=[Hsh;Hme;Hrn;Hpp;Hts;Hma];

j=1;

return

%%%%%%%%%%%%%%%%%%%%%%%


%%example
% d =[0.4600    0.3400    0.1400    0.0300    0.9218    1.0000
%      0.5000    0.3400    0.1500    0.0200    0.7382    1.0000
%      0.4400    0.2900    0.1400    0.0200    0.1763    1.0000
%      0.7600    0.3000    0.6600    0.2100    0.4057    2.0000
%      0.4900    0.2500    0.4500    0.1700    0.9355    2.0000
%      0.7300    0.2900    0.6300    0.1800    0.9169    2.0000]
% [data_mod, index_rem,SIM,SIM1,H]=feat_sel_sim1(d, 'luca', 1)

function [data_mod, index_rem,SIM,SIM1,H]=feat_sel_sim_Schweizer_2(data, p)
%

%    p in (0, \infty) as default p=1. if nargin
p=1;
end

l=max(data(:,end)); % #-classes m=size(data,1); % #-samples t=size(data,2)-1;
% #-features dataold=data;
tmp=[];
% forming idealvec using arithmetic mean idealvec=zeros(1,t);
for k=1:l idealvec_s(k,:)=mean(data(find(data(:,end)==k),1:t));
end

%scaling data between [0,1] data_v=data(:,1:t); data_c=data(:,t+1); %labels
mins_v = min(data_v);
Ones = ones(size(data_v));
data_v = data_v+Ones*diag(abs(mins_v)); for k=1:l
tmp=[tmp;abs(mins_v)]; end
tmp;
idealvec_s = idealvec_s+tmp; maxs_v = max(data_v);
data_v = data_v*diag(maxs_v.^(-1)); idealvec_s=idealvec_s./repmat(maxs_v,
l,1); data = [data_v, data_c];
% sample data datalearn_s=data(:,1:t); land=0.1;
% similarities sim=zeros(t,m,l);

for j=1:m for i=1:t
for k=1:l
%%similarity Luca    with p=1
sn_1(i,j,k)=((((((1-datalearn_s(j,i)))^p+((idealvec_s(k,i))^p))))-((1-
datalearn_s(j,i))^p*(idealvec_s (k,i))^p))^(1/p);
```

```
sn_2(i,j,k)=(((((datalearn_s(j,i)))^p+((1-idealvec_s(k,i))^p))))-
((datalearn_s(j,i))^p*(1-idealvec_s (k,i))^p))^(1/p);
sim(i,j,k)=1-(((((1-sn_1(i,j,k))^p+(1-sn_2(i,j,k))^p))-(((1-sn_1(i,j,k))^p)*
(1-sn_2(i,j,k))^p))^(1/p)); end
end end
%class 1,2,3,4,. n
% SIM(:,:,1) = similarity for class 1
% SIM(:,:,2) = similarity for class 2 SIM=sim;
% reduce number of dimensions in sim sim=reshape(sim,t,m*l)';
SIM1=sim;

% possibility for two different entropy measures
% moodifying zero and one values of the similarity values to work with
% De Luca's entropy measure
%%%%%%Hsh(1)
delta=1E-10; sim(find(sim==0))=delta; sim(find(sim==1))=1-delta;
Hsh=sum(-sim.*log(sim)-(1-sim).*log(1-sim));
Hsh=(Hsh-min(Hsh))/(max(Hsh)-min(Hsh));

% %%%%%%Hme(2)
Hme=sum(sin(pi/2*sim)+sin(pi/2*(1-sim))-1); Hme=(Hme-min(Hme))/(max(Hme)-min
(Hme));

%%%%%%%%Hrn(3)
alfa_rn=0.8;
Hrn=log(sum((sim.^alfa_rn)+((1-sim).^alfa_rn)))/(1-alfa_rn); Hrn=(Hrn-
min(Hrn))/(max(Hrn)-min(Hrn));

% %%%%%%%Hpp(4)

Hpp=sum((sim.*exp(1.-sim))+(((1-sim).*exp(1.-(1-sim))))); Hpp=(Hpp-min(Hpp))/
(max(Hpp)-min(Hpp));

% %%%%%%%%Hts(5)
alfa_ts=0.5;
Hts=1-(sum((sim.^alfa_ts)+((1-sim).^alfa_ts)))/(alfa_ts-1);
Hts=(Hts-min(Hts))/(max(Hts)-min(Hts));

%%%%%%%%%Hma(6)
gama=2; a_ma=0.5; b_ma=3; alfa_ma=0.1; beta_ma=2;
Hma=(sum(((sim.^alfa_ma).*exp(a_ma.*sim.^alfa_ma+b_ma).^beta_ma)+((1-sim).
^alfa_ma).
*exp(a_ma.*(1-sim).^alfa_ma+b_ma).^beta_ma)); Hma=(Hma-min(Hma))/(max(Hma)-
min(Hma));

% find maximum feature
[i_sh, index_rem_sh]=max(Hsh); [i_me, index_rem_me]=max(Hme);
[i_rn, index_rem_rn]=max(Hrn); [i_pp, index_rem_pp]=max(Hpp);
[i_ts, index_rem_ts]=max(Hts); [i_ma, index_rem_ma]=max(Hma);
```

```
% removing feature from the data
data_mod_sh=[dataold(:,1:index_rem_sh-1) dataold(:,index_rem_sh+1:end)];
data_mod_me=[dataold(:,1:index_rem_me-1) dataold(:,index_rem_me+1:end)];
data_mod_rn=[dataold(:,1:index_rem_rn-1) dataold(:,index_rem_rn+1:end)];
data_mod_pp=[dataold(:,1:index_rem_pp-1) dataold(:,index_rem_pp+1:end)];
data_mod_ts=[dataold(:,1:index_rem_ts-1) dataold(:,index_rem_ts+1:end)];
data_mod_ma=[dataold(:,1:index_rem_ma-1) dataold(:,index_rem_ma+1:end)];

data_mod=[data_mod_sh,data_mod_me,data_mod_rn,data_mod_pp,data_mod_ts,
data_mod_ma]; index_rem=[index_rem_sh,index_rem_me,index_rem_rn,index_rem_pp,
index_rem_ts,index_rem_ma]; H=[Hsh;Hme;Hrn;Hpp;Hts;Hma];

j=1;

return

%%%%%%%%%%%%%%%%%%%%%%%%

%%example
% d =[0.4600     0.3400    0.1400    0.0300    0.9218    1.0000
%     0.5000     0.3400    0.1500    0.0200    0.7382    1.0000
%     0.4400     0.2900    0.1400    0.0200    0.1763    1.0000
%     0.7600     0.3000    0.6600    0.2100    0.4057    2.0000
%     0.4900     0.2500    0.4500    0.1700    0.9355    2.0000
%     0.7300     0.2900    0.6300    0.1800    0.9169    2.0000]
% [data_mod, index_rem,SIM,SIM1,H]=feat_sel_sim1(d, 'luca', 1)

function [data_mod, index_rem,SIM,SIM1,H]=feat_sel_sim_Schweizer_3(data, p)
%

%    p in (0, \infty) as default p=1. if nargin
p=1;
end

l=max(data(:,end)); % #-classes m=size(data,1); % #-samples t=size(data,2)-1;
% #-features dataold=data;
tmp=[];
% forming idealvec using arithmetic mean idealvec=zeros(l,t);
for k=1:l idealvec_s(k,:)=mean(data(find(data(:,end)==k),1:t));
end

%scaling data between [0,1] data_v=data(:,1:t); data_c=data(:,t+1); %labels
mins_v = min(data_v);
Ones = ones(size(data_v));
data_v = data_v+Ones*diag(abs(mins_v)); for k=1:l
tmp=[tmp;abs(mins_v)]; end
tmp;
idealvec_s = idealvec_s+tmp; maxs_v = max(data_v);
data_v = data_v*diag(maxs_v.^(-1)); idealvec_s=idealvec_s./repmat(maxs_v,
l,1); data = [data_v, data_c];
% sample data datalearn_s=data(:,1:t); land=0.1;
% similarities sim=zeros(t,m,l);
```

```
for j=1:m for i=1:t
for k=1:l
%%similarity Luca    with p=1
a=(1-datalearn_s(j,i)); b=idealvec_s(k,i);
sn_1(i,j,k)=1-(exp(-(((abs(log(1-a))^p)+(abs(log(1-b))^p))^(1/p))));
sn_2(i,j,k)=1-(exp(-(((abs(log(1-(1-a)))^p)+(abs(log(1-(1-b)))^p))^(1/p))));
%sn_1(i,j,k)=1-exp(-1*(((abs(log(1-(1-datalearn_s(j,i))))^p)+((abs(log((1-
idealvec_s(k,i)))))
^p))))^(1/p);
%sn_2(i,j,k)=1-exp(-1*((abs(log(1-(datalearn_s(j,i))))^p)+((abs(log((1-(1-
idealvec_s(k,i)))))
^p))))^(1/p);
%sim(i,j,k)=exp(((abs(log(sn_2(i,j,k)))^p)-(abs(log(sn_1(i,j,k)))^p))^(1/p));
sim(i,j,k)=exp((-1*((((abs(log(sn_1(i,j,k))))^p))+(((abs(log(sn_2(i,j,k))))
^p))))^(1/p)); end
end end
%class 1,2,3,4,.    n
% SIM(:,:,1) = similarity for class 1
% SIM(:,:,2) = similarity for class 2 SIM=sim;
% reduce number of dimensions in sim sim=reshape(sim,t,m*l)';
SIM1=sim;

% possibility for two different entropy measures
% moodifying zero and one values of the similarity values to work with
% De Luca's entropy measure
%%%%%%Hsh(1)
delta=1E-10; sim(find(sim==0))=delta; sim(find(sim==1))=1-delta;
Hsh=sum(-sim.*log(sim)-(1-sim).*log(1-sim));
Hsh=(Hsh-min(Hsh))/(max(Hsh)-min(Hsh));

% %%%%%%Hme(2)
Hme=sum(sin(pi/2*sim)+sin(pi/2*(1-sim))-1); Hme=(Hme-min(Hme))/(max(Hme)-min
(Hme));

%%%%%%%%Hrn(3)

alfa_rn=0.8;
Hrn=log(sum((sim.^alfa_rn)+((1-sim).^alfa_rn)))/(1-alfa_rn); Hrn=(Hrn-
min(Hrn))/(max(Hrn)-min(Hrn));

% %%%%%%%Hpp(4)
Hpp=sum((sim.*exp(1.-sim))+(((1-sim).*exp(1.-(1-sim))))); Hpp=(Hpp-min(Hpp))/
(max(Hpp)-min(Hpp));

% %%%%%%%%Hts(5)
alfa_ts=0.5;
Hts=1-(sum((sim.^alfa_ts)+((1-sim).^alfa_ts)))/(alfa_ts-1);
Hts=(Hts-min(Hts))/(max(Hts)-min(Hts));
```

```
%%%%%%%%%Hma(6)
gama=2; a_ma=0.5; b_ma=3; alfa_ma=0.1; beta_ma=2;
Hma=(sum(((sim.^alfa_ma).*exp(a_ma.*sim.^alfa_ma+b_ma).^beta_ma)+((1-sim).
^alfa_ma).
*exp(a_ma.*(1-sim).^alfa_ma+b_ma).^beta_ma)); Hma=(Hma-min(Hma))/(max(Hma)-
min(Hma));

% find maximum feature
[i_sh, index_rem_sh]=max(Hsh); [i_me, index_rem_me]=max(Hme);
[i_rn, index_rem_rn]=max(Hrn); [i_pp, index_rem_pp]=max(Hpp);
[i_ts, index_rem_ts]=max(Hts); [i_ma, index_rem_ma]=max(Hma);

% removing feature from the data
data_mod_sh=[dataold(:,1:index_rem_sh-1) dataold(:,index_rem_sh+1:end)];
data_mod_me=[dataold(:,1:index_rem_me-1) dataold(:,index_rem_me+1:end)];
data_mod_rn=[dataold(:,1:index_rem_rn-1) dataold(:,index_rem_rn+1:end)];
data_mod_pp=[dataold(:,1:index_rem_pp-1) dataold(:,index_rem_pp+1:end)];
data_mod_ts=[dataold(:,1:index_rem_ts-1) dataold(:,index_rem_ts+1:end)];
data_mod_ma=[dataold(:,1:index_rem_ma-1) dataold(:,index_rem_ma+1:end)];

data_mod=[data_mod_sh,data_mod_me,data_mod_rn,data_mod_pp,data_mod_ts,
data_mod_ma]; index_rem=[index_rem_sh,index_rem_me,index_rem_rn,index_rem_pp,
index_rem_ts,index_rem_ma]; H=[Hsh;Hme;Hrn;Hpp;Hts;Hma];

j=1;

return

%%%%%%%%%%%%%%%%%%%%%%%%

%%example
% d =[0.4600    0.3400    0.1400    0.0300    0.9218    1.0000
%     0.5000    0.3400    0.1500    0.0200    0.7382    1.0000
%     0.4400    0.2900    0.1400    0.0200    0.1763    1.0000
%     0.7600    0.3000    0.6600    0.2100    0.4057    2.0000
%     0.4900    0.2500    0.4500    0.1700    0.9355    2.0000
%     0.7300    0.2900    0.6300    0.1800    0.9169    2.0000]
% [data_mod, index_rem,SIM,SIM1,H]=feat_sel_sim1(d, 'luca', 1)

function [data_mod, index_rem,SIM,SIM1,H]=feat_sel_sim_Schweizer_4(data, p)
%
%   p in (0, \infty) as default p=1. if nargin
p=1;
end

l=max(data(:,end)); % #-classes m=size(data,1); % #-samples t=size(data,2)-1;
% #-features dataold=data;
tmp=[];
```

```
% forming idealvec using arithmetic mean idealvec=zeros(1,t);
for k=1:l idealvec_s(k,:)=mean(data(find(data(:,end)==k),1:t));
end

%scaling data between [0,1] data_v=data(:,1:t); data_c=data(:,t+1); %labels
mins_v = min(data_v);
Ones = ones(size(data_v));
data_v = data_v+Ones*diag(abs(mins_v)); for k=1:l
tmp=[tmp;abs(mins_v)]; end
tmp;
idealvec_s = idealvec_s+tmp; maxs_v = max(data_v);
data_v = data_v*diag(maxs_v.^(-1)); idealvec_s=idealvec_s./repmat(maxs_v,
1,1); data = [data_v, data_c];
% sample data datalearn_s=data(:,1:t); land=0.1;
% similarities sim=zeros(t,m,l); for j=1:m

for i=1:t
for k=1:l
%%similarity Luca    with p=1
% a=(datalearn_s(j,i))
% b=(1-idealvec_s(k,i))
% 1-a=(1-(datalearn_s(j,i)))
% 1-b=(1-(1-idealvec_s(k,i)))
% (1-a)^p=((1-(datalearn_s(j,i)))^p)
% (1-b)^p=((1-(1-idealvec_s(k,i)))^p)
% (1-a)(1-b)=((1-(datalearn_s(j,i)))*(1-(1-idealvec_s(k,i))))
% (1-a)^p+(1-b)^p-(1-a)^p*(1-b)^p=(((1-(datalearn_s(j,i)))^p)
+((1-(1-idealvec_s(k,i)))^p)-(((1- (datalearn_s(j,i)))^p)*((1-(1-idealvec_
s(k,i)))^p)))^(1/p)
sn_1(i,j,k)=1-(((1-(1-datalearn_s(j,i)))*(1-(idealvec_s(k,i))))/(((1-(1-
datalearn_s(j,i)))^p)+((1-
(idealvec_s(k,i)))^p)-(((1-(1-datalearn_s(j,i)))^p)*((1-(idealvec_s(k,i)))
^p)))^(1/p));
sn_2(i,j,k)=1-(((1-(datalearn_s(j,i)))*(1-(1-idealvec_s(k,i))))/((((1-
(datalearn_s(j,i)))^p)+((1-(1-
idealvec_s(k,i)))^p)-(((1-(datalearn_s(j,i)))^p)*((1-(1-idealvec_
s(k,i)))^p)))^(1/p)));
sim(i,j,k)=(sn_1(i,j,k)*sn_2(i,j,k))/(((sn_1(i,j,k)^p)+(sn_2(i,j,k)^p)-((sn_1
(i,j,k)^p)*(sn_2(i,j,k)^p)))
^(1/p));
end end
end
%class 1,2,3,4,. n
% SIM(:,:,1) = similarity for class 1
% SIM(:,:,2) = similarity for class 2 SIM=sim;
% reduce number of dimensions in sim sim=reshape(sim,t,m*l)';
SIM1=sim;

% possibility for two different entropy measures
% moodifying zero and one values of the similarity values to work with
% De Luca's entropy measure
```

```
%%%%%%Hsh(1)
delta=1E-10; sim(find(sim==0))=delta; sim(find(sim==1))=1-delta;
Hsh=sum(-sim.*log(sim)-(1-sim).*log(1-sim));
Hsh=(Hsh-min(Hsh))/(max(Hsh)-min(Hsh));

% %%%%%%Hme(2)

Hme=sum(sin(pi/2*sim)+sin(pi/2*(1-sim))-1); Hme=(Hme-min(Hme))/(max(Hme)-min
(Hme));

%%%%%%%%Hrn(3)
alfa_rn=0.8;
Hrn=log(sum((sim.^alfa_rn)+((1-sim).^alfa_rn)))/(1-alfa_rn); Hrn=(Hrn-
min(Hrn))/(max(Hrn)-min(Hrn));

% %%%%%%%Hpp(4)
Hpp=sum((sim.*exp(1.-sim))+(((1-sim).*exp(1.-(1-sim))))); Hpp=(Hpp-min(Hpp))/
(max(Hpp)-min(Hpp));

% %%%%%%%%Hts(5)
alfa_ts=0.5;
Hts=1-(sum((sim.^alfa_ts)+((1-sim).^alfa_ts)))/(alfa_ts-1);
Hts=(Hts-min(Hts))/(max(Hts)-min(Hts));

%%%%%%%%%Hma(6)
gama=2; a_ma=0.5; b_ma=3; alfa_ma=0.1; beta_ma=2;
Hma=(sum(((sim.^alfa_ma).*exp(a_ma.*sim.^alfa_ma+b_ma).^beta_ma)+((1-sim).
^alfa_ma).
*exp(a_ma.*(1-sim).^alfa_ma+b_ma).^beta_ma)); Hma=(Hma-min(Hma))/(max(Hma)-
min(Hma));

% find maximum feature
[i_sh, index_rem_sh]=max(Hsh); [i_me, index_rem_me]=max(Hme);
[i_rn, index_rem_rn]=max(Hrn); [i_pp, index_rem_pp]=max(Hpp);
[i_ts, index_rem_ts]=max(Hts); [i_ma, index_rem_ma]=max(Hma);

% removing feature from the data
data_mod_sh=[dataold(:,1:index_rem_sh-1) dataold(:,index_rem_sh+1:end)];
data_mod_me=[dataold(:,1:index_rem_me-1) dataold(:,index_rem_me+1:end)];
data_mod_rn=[dataold(:,1:index_rem_rn-1) dataold(:,index_rem_rn+1:end)];
data_mod_pp=[dataold(:,1:index_rem_pp-1) dataold(:,index_rem_pp+1:end)];
data_mod_ts=[dataold(:,1:index_rem_ts-1) dataold(:,index_rem_ts+1:end)];
data_mod_ma=[dataold(:,1:index_rem_ma-1) dataold(:,index_rem_ma+1:end)];

data_mod=[data_mod_sh,data_mod_me,data_mod_rn,data_mod_pp,data_mod_ts,
data_mod_ma]; index_rem=[index_rem_sh,index_rem_me,index_rem_rn,index_rem_pp,
index_rem_ts,index_rem_ma]; H=[Hsh;Hme;Hrn;Hpp;Hts;Hma];

j=1;
```

```
return

%%%%%%%%%%%%%%%%%%%%%%%%

%%example
% d =[0.4600    0.3400    0.1400    0.0300    0.9218    1.0000
%      0.5000    0.3400    0.1500    0.0200    0.7382    1.0000
%      0.4400    0.2900    0.1400    0.0200    0.1763    1.0000
%      0.7600    0.3000    0.6600    0.2100    0.4057    2.0000
%      0.4900    0.2500    0.4500    0.1700    0.9355    2.0000
%      0.7300    0.2900    0.6300    0.1800    0.9169    2.0000]
% [data_mod, index_rem,SIM,SIM1,H]=feat_sel_sim1(d, 'luca', 1)

function [data_mod, index_rem,SIM,SIM1,H]=feat_sel_sim_Schweizer(data, p)
%

%    p in (0, \infty) as default p=1. if nargin
p=1;
end

l=max(data(:,end)); % #-classes m=size(data,1); % #-samples t=size(data,2)-1;
% #-features dataold=data;
tmp=[];
% forming idealvec using arithmetic mean idealvec=zeros(l,t);
for k=1:l idealvec_s(k,:)=mean(data(find(data(:,end)==k),1:t));
end

%scaling data between [0,1] data_v=data(:,1:t); data_c=data(:,t+1); %labels
mins_v = min(data_v);
Ones = ones(size(data_v));
data_v = data_v+Ones*diag(abs(mins_v)); for k=1:l
tmp=[tmp;abs(mins_v)]; end
tmp;
idealvec_s = idealvec_s+tmp; maxs_v = max(data_v);
data_v = data_v*diag(maxs_v.^(-1)); idealvec_s=idealvec_s./repmat(maxs_v,
l,1); data = [data_v, data_c];
% sample data datalearn_s=data(:,1:t); land=0.1;
% similarities sim=zeros(t,m,l);

for j=1:m for i=1:t
for k=1:l
%%similarity Luca    with p=1
sn_1(i,j,k)=1-(max(0,((1-(1-datalearn_s(j,i)))^p+((1-idealvec_s(k,i))^p)-1)))
^(1/p);
sn_2(i,j,k)=1-(max(0,((1-(datalearn_s(j,i)))^p+((1-(1-idealvec_s(k,i)))^p)-
1)))^(1/p);
sim(i,j,k)=(max(0,((sn_1(i,j,k))^p+(sn_1(i,j,k))^p))-1)^(1/p); end
end end
%class 1,2,3,4,. n
% SIM(:,:,1) = similarity for class 1
% SIM(:,:,2) = similarity for class 2 SIM=sim;
```

```
% reduce number of dimensions in sim sim=reshape(sim,t,m*l)';
SIM1=sim;

% possibility for two different entropy measures
% moodifying zero and one values of the similarity values to work with
% De Luca's entropy measure
%%%%%%Hsh(1)
delta=1E-10; sim(find(sim==0))=delta; sim(find(sim==1))=1-delta;
Hsh=sum(-sim.*log(sim)-(1-sim).*log(1-sim));
Hsh=(Hsh-min(Hsh))/(max(Hsh)-min(Hsh));

% %%%%%%Hme(2)
Hme=sum(sin(pi/2*sim)+sin(pi/2*(1-sim))-1); Hme=(Hme-min(Hme))/(max(Hme)-min
(Hme));

%%%%%%%%Hrn(3)
alfa_rn=0.8;
Hrn=log(sum((sim.^alfa_rn)+((1-sim).^alfa_rn)))/(1-alfa_rn); Hrn=(Hrn-
min(Hrn))/(max(Hrn)-min(Hrn));

% %%%%%%%Hpp(4)
Hpp=sum((sim.*exp(1.-sim))+(((1-sim).*exp(1.-(1-sim))))); Hpp=(Hpp-min(Hpp))/
(max(Hpp)-min(Hpp));


% %%%%%%%%Hts(5)
alfa_ts=0.5;
Hts=1-(sum((sim.^alfa_ts)+((1-sim).^alfa_ts)))/(alfa_ts-1);
Hts=(Hts-min(Hts))/(max(Hts)-min(Hts));

%%%%%%%%%Hma(6)
gama=2; a_ma=0.5; b_ma=3; alfa_ma=0.1; beta_ma=2;
Hma=(sum(((sim.^alfa_ma).*exp(a_ma.*sim.^alfa_ma+b_ma).^beta_ma)+((1-sim).
^alfa_ma).
*exp(a_ma.*(1-sim).^alfa_ma+b_ma).^beta_ma)); Hma=(Hma-min(Hma))/(max(Hma)-
min(Hma));

% find maximum feature
[i_sh, index_rem_sh]=max(Hsh); [i_me, index_rem_me]=max(Hme);
[i_rn, index_rem_rn]=max(Hrn); [i_pp, index_rem_pp]=max(Hpp);
[i_ts, index_rem_ts]=max(Hts); [i_ma, index_rem_ma]=max(Hma);

% removing feature from the data
data_mod_sh=[dataold(:,1:index_rem_sh-1) dataold(:,index_rem_sh+1:end)];
data_mod_me=[dataold(:,1:index_rem_me-1) dataold(:,index_rem_me+1:end)];
data_mod_rn=[dataold(:,1:index_rem_rn-1) dataold(:,index_rem_rn+1:end)];
data_mod_pp=[dataold(:,1:index_rem_pp-1) dataold(:,index_rem_pp+1:end)];
data_mod_ts=[dataold(:,1:index_rem_ts-1) dataold(:,index_rem_ts+1:end)];
data_mod_ma=[dataold(:,1:index_rem_ma-1) dataold(:,index_rem_ma+1:end)];
```

```
data_mod=[data_mod_sh,data_mod_me,data_mod_rn,data_mod_pp,data_mod_ts,
data_mod_ma]; index_rem=[index_rem_sh,index_rem_me,index_rem_rn,index_rem_pp,
index_rem_ts,index_rem_ma]; H=[Hsh;Hme;Hrn;Hpp;Hts;Hma];

j=1;

return

%%%%%%%%%%%%%%%%%%%%%%

%%example
% d =[0.4600    0.3400    0.1400    0.0300    0.9218    1.0000
%     0.5000    0.3400    0.1500    0.0200    0.7382    1.0000
%     0.4400    0.2900    0.1400    0.0200    0.1763    1.0000
%     0.7600    0.3000    0.6600    0.2100    0.4057    2.0000
%     0.4900    0.2500    0.4500    0.1700    0.9355    2.0000
%     0.7300    0.2900    0.6300    0.1800    0.9169    2.0000]
% [data_mod, index_rem,SIM,SIM1,H]=feat_sel_sim1(d, 'luca', 1)

function [data_mod, index_rem,SIM,SIM1,H]=feat_sel_sim_Hamacher(data, r)
%

%   p in (0, \infty) as default p=1. if nargin
p=1;
end

l=max(data(:,end)); % #-classes m=size(data,1); % #-samples t=size(data,2)-1;
% #-features dataold=data;
tmp=[];
% forming idealvec using arithmetic mean idealvec=zeros(l,t);
for k=1:l idealvec_s(k,:)=mean(data(find(data(:,end)==k),1:t));
end

%scaling data between [0,1] data_v=data(:,1:t); data_c=data(:,t+1); %labels
mins_v = min(data_v);
Ones = ones(size(data_v));
data_v = data_v+Ones*diag(abs(mins_v)); for k=1:l
tmp=[tmp;abs(mins_v)]; end
tmp;
idealvec_s = idealvec_s+tmp; maxs_v = max(data_v);
data_v = data_v*diag(maxs_v.^(-1)); idealvec_s=idealvec_s./repmat(maxs_v,
l,1); data = [data_v, data_c];
% sample data datalearn_s=data(:,1:t); land=0.1;
% similarities sim=zeros(t,m,l);

for j=1:m for i=1:t
for k=1:l
%%similarity Luca   with p=1
% a=(1-datalearn_s(j,i))
% b=(idealvec_s(k,i))
```

```
% 1-a=(1-(1-datalearn_s(j,i)))
% 1-b=(1-(idealvec_s(k,i)))
% (1-a)^p=((1-(1-datalearn_s(j,i)))^p)
% (1-b)^p=((1-(idealvec_s(k,i)))^p)
% (1-a)(1-b)=((1-(1-datalearn_s(j,i)))*(1-(idealvec_s(k,i))))
% (1-a)^p+(1-b)^p-(1-a)^p*(1-b)^p=(((1-(1-datalearn_s(j,i)))^p)
+((1-(idealvec_s(k,i)))^p)-(((1-(1- datalearn_s(j,i)))^p)*((1-(idealvec_s(k,
i)))^p)))^(1/p)
sn_1(i,j,k)=((1-datalearn_s(j,i))+(idealvec_s(k,i))+(r-2)*(1-datalearn_s(j,
i))*(idealvec_s(k,i)))/(r+ (r-1)*(1-datalearn_s(j,i))*(idealvec_s(k,i)));
sn_2(i,j,k)=((datalearn_s(j,i))+(1-idealvec_s(k,i))+(r-2)*(datalearn_s(j,i))*
(1-idealvec_s(k,i)))/(r+ (r-1)*(datalearn_s(j,i))*(1-idealvec_s(k,i)));
sim(i,j,k)=(sn_1(i,j,k)*sn_2(i,j,k))/(r+(1-r)*(sn_1(i,j,k)+sn_2(i,j,k)+sn_1
(i,j,k)*sn_2(i,j,k))); end
end end
%class 1,2,3,4,.n
% SIM(:,:,1) = similarity for class 1
% SIM(:,:,2) = similarity for class 2 SIM=sim;
% reduce number of dimensions in sim sim=reshape(sim,t,m*l)';
SIM1=sim;

% possibility for two different entropy measures
% moodifying zero and one values of the similarity values to work with
% De Luca's entropy measure
%%%%%%Hsh(1)
delta=1E-10; sim(find(sim==0))=delta; sim(find(sim==1))=1-delta;
Hsh=sum(-sim.*log(sim)-(1-sim).*log(1-sim));
Hsh=(Hsh-min(Hsh))/(max(Hsh)-min(Hsh));

% %%%%%%Hme(2)
Hme=sum(sin(pi/2*sim)+sin(pi/2*(1-sim))-1); Hme=(Hme-min(Hme))/(max(Hme)-min
(Hme));

%%%%%%%%Hrn(3)
alfa_rn=0.8;
Hrn=log(sum((sim.^alfa_rn)+((1-sim).^alfa_rn)))/(1-alfa_rn); Hrn=(Hrn-
min(Hrn))/(max(Hrn)-min(Hrn));

% %%%%%%%Hpp(4)
Hpp=sum((sim.*exp(1.-sim))+(((1-sim).*exp(1.-(1-sim))))); Hpp=(Hpp-min(Hpp))/
(max(Hpp)-min(Hpp));

% %%%%%%%%Hts(5)
alfa_ts=0.5;
Hts=1-(sum((sim.^alfa_ts)+((1-sim).^alfa_ts)))/(alfa_ts-1);
Hts=(Hts-min(Hts))/(max(Hts)-min(Hts));

%%%%%%%%%Hma(6)
gama=2; a_ma=0.5; b_ma=3; alfa_ma=0.1; beta_ma=2;
Hma=(sum(((sim.^alfa_ma).*exp(a_ma.*sim.^alfa_ma+b_ma).^beta_ma)+((1-sim).
^alfa_ma).
```

```
*exp(a_ma.*(1-sim).^alfa_ma+b_ma).^beta_ma)); Hma=(Hma-min(Hma))/(max(Hma)-
min(Hma));

% find maximum feature
[i_sh, index_rem_sh]=max(Hsh); [i_me, index_rem_me]=max(Hme);
[i_rn, index_rem_rn]=max(Hrn); [i_pp, index_rem_pp]=max(Hpp);
[i_ts, index_rem_ts]=max(Hts); [i_ma, index_rem_ma]=max(Hma);

% removing feature from the data
data_mod_sh=[dataold(:,1:index_rem_sh-1) dataold(:,index_rem_sh+1:end)];
data_mod_me=[dataold(:,1:index_rem_me-1) dataold(:,index_rem_me+1:end)];
data_mod_rn=[dataold(:,1:index_rem_rn-1) dataold(:,index_rem_rn+1:end)];
data_mod_pp=[dataold(:,1:index_rem_pp-1) dataold(:,index_rem_pp+1:end)];
data_mod_ts=[dataold(:,1:index_rem_ts-1) dataold(:,index_rem_ts+1:end)];
data_mod_ma=[dataold(:,1:index_rem_ma-1) dataold(:,index_rem_ma+1:end)];

data_mod=[data_mod_sh,data_mod_me,data_mod_rn,data_mod_pp,data_mod_ts,
data_mod_ma]; index_rem=[index_rem_sh,index_rem_me,index_rem_rn,index_rem_pp,
index_rem_ts,index_rem_ma]; H=[Hsh;Hme;Hrn;Hpp;Hts;Hma];

j=1;

return

%%%%%%%%%%%%%%%%%%%%%%%

%%example
% d =[0.4600    0.3400    0.1400    0.0300    0.9218    1.0000
%     0.5000    0.3400    0.1500    0.0200    0.7382    1.0000
%     0.4400    0.2900    0.1400    0.0200    0.1763    1.0000
%     0.7600    0.3000    0.6600    0.2100    0.4057    2.0000
%     0.4900    0.2500    0.4500    0.1700    0.9355    2.0000
%     0.7300    0.2900    0.6300    0.1800    0.9169    2.0000]
% [data_mod, index_rem,SIM,SIM1,H]=feat_sel_sim1(d, 'luca', 1)

function [data_mod, index_rem,SIM,SIM1,H]=feat_sel_sim_Dubois1(data,alfa)
%

%   p in (0, \infty) as default p=1. if nargin
p=1;
end

l=max(data(:,end)); % #-classes m=size(data,1); % #-samples t=size(data,2)-1;
% #-features dataold=data;
tmp=[];
% forming idealvec using arithmetic mean idealvec=zeros(1,t);
for k=1:l idealvec_s(k,:)=mean(data(find(data(:,end)==k),1:t));
end
```

```
%scaling data between [0,1] data_v=data(:,1:t); data_c=data(:,t+1); %labels
mins_v = min(data_v);
Ones = ones(size(data_v));
data_v = data_v+Ones*diag(abs(mins_v)); for k=1:l
tmp=[tmp;abs(mins_v)]; end
tmp;
idealvec_s = idealvec_s+tmp; maxs_v = max(data_v);
data_v = data_v*diag(maxs_v.^(-1)); idealvec_s=idealvec_s./repmat(maxs_v,
l,1); data = [data_v, data_c];
% sample data datalearn_s=data(:,1:t); land=0.1;
% similarities sim=zeros(t,m,l);

for j=1:m for i=1:t
for k=1:l
%%similarity Luca    with p=1
% a=(datalearn_s(j,i))
% b=(1-idealvec_s(k,i))
% 1-a=(1-(datalearn_s(j,i)))
% 1-b=(1-(1-idealvec_s(k,i)))
% (1-a)^p=((1-(datalearn_s(j,i)))^p)
% (1-b)^p=((1-(1-idealvec_s(k,i)))^p)
% (1-a)(1-b)=((1-(datalearn_s(j,i)))*(1-(1-idealvec_s(k,i))))
% (1-a)^p+(1-b)^p-(1-a)^p*(1-b)^p=(((1-(datalearn_s(j,i)))^p)
+((1-(1-idealvec_s(k,i)))^p)-(((1- (datalearn_s(j,i)))^p)*((1-(1-idealvec_s
(k,i)))^p)))^(1/p)
sn_1(i,j,k)=1-(((1-(1-datalearn_s(j,i)))*(1-(idealvec_s(k,i)))))/(max(alfa,
((((1-datalearn_s(j,i)))* ((idealvec_s(k,i)))))));
sn_2(i,j,k)=1-(((1-(datalearn_s(j,i)))*(1-(1-idealvec_s(k,i)))))/(max(alfa,
((((datalearn_s(j,i)))*((1- idealvec_s(k,i)))))));
sim(i,j,k)=(sn_1(i,j,k)*sn_2(i,j,k))./max(alfa,(sn_1(i,j,k)*sn_2(i,j,k)));
end
end end
%class 1,2,3,4,. n
% SIM(:,:,1) = similarity for class 1
% SIM(:,:,2) = similarity for class 2 SIM=sim;
% reduce number of dimensions in sim sim=reshape(sim,t,m*l)';
SIM1=sim;

% possibility for two different entropy measures
% moodifying zero and one values of the similarity values to work with
% De Luca's entropy measure
%%%%%%Hsh(1)
delta=1E-10; sim(find(sim==0))=delta; sim(find(sim==1))=1-delta;
Hsh=sum(-sim.*log(sim)-(1-sim).*log(1-sim));
Hsh=(Hsh-min(Hsh))/(max(Hsh)-min(Hsh));

% %%%%%%Hme(2)
Hme=sum(sin(pi/2*sim)+sin(pi/2*(1-sim))-1); Hme=(Hme-min(Hme))/(max(Hme)-min
(Hme));
```

```
%%%%%%%%Hrn(3)
alfa_rn=0.8;
Hrn=log(sum((sim.^alfa_rn)+((1-sim).^alfa_rn)))/(1-alfa_rn); Hrn=(Hrn-
min(Hrn))/(max(Hrn)-min(Hrn));

% %%%%%%%Hpp(4)
Hpp=sum((sim.*exp(1.-sim))+(((1-sim).*exp(1.-(1-sim))))); Hpp=(Hpp-min(Hpp))/
(max(Hpp)-min(Hpp));

% %%%%%%%%Hts(5)
alfa_ts=0.5;
Hts=1-(sum((sim.^alfa_ts)+((1-sim).^alfa_ts)))/(alfa_ts-1);
Hts=(Hts-min(Hts))/(max(Hts)-min(Hts));

%%%%%%%%%Hma(6)
gama=2; a_ma=0.5; b_ma=3; alfa_ma=0.1; beta_ma=2;
Hma=(sum(((sim.^alfa_ma).*exp(a_ma.*sim.^alfa_ma+b_ma).^beta_ma)+((1-sim).
^alfa_ma).
*exp(a_ma.*(1-sim).^alfa_ma+b_ma).^beta_ma)); Hma=(Hma-min(Hma))/(max(Hma)-
min(Hma));

% find maximum feature
[i_sh, index_rem_sh]=max(Hsh); [i_me, index_rem_me]=max(Hme);
[i_rn, index_rem_rn]=max(Hrn); [i_pp, index_rem_pp]=max(Hpp);
[i_ts, index_rem_ts]=max(Hts); [i_ma, index_rem_ma]=max(Hma);

% removing feature from the data
data_mod_sh=[dataold(:,1:index_rem_sh-1) dataold(:,index_rem_sh+1:end)];
data_mod_me=[dataold(:,1:index_rem_me-1) dataold(:,index_rem_me+1:end)];
data_mod_rn=[dataold(:,1:index_rem_rn-1) dataold(:,index_rem_rn+1:end)];
data_mod_pp=[dataold(:,1:index_rem_pp-1) dataold(:,index_rem_pp+1:end)];
data_mod_ts=[dataold(:,1:index_rem_ts-1) dataold(:,index_rem_ts+1:end)];
data_mod_ma=[dataold(:,1:index_rem_ma-1) dataold(:,index_rem_ma+1:end)];

data_mod=[data_mod_sh,data_mod_me,data_mod_rn,data_mod_pp,data_mod_ts,
data_mod_ma]; index_rem=[index_rem_sh,index_rem_me,index_rem_rn,index_rem_pp,
index_rem_ts,index_rem_ma]; H=[Hsh;Hme;Hrn;Hpp;Hts;Hma];

j=1;

return

%%%%%%%%%%%%%%%%%%%%%%%

%%example
% d =[0.4600    0.3400    0.1400    0.0300    0.9218    1.0000
%     0.5000    0.3400    0.1500    0.0200    0.7382    1.0000
%     0.4400    0.2900    0.1400    0.0200    0.1763    1.0000
%     0.7600    0.3000    0.6600    0.2100    0.4057    2.0000
%     0.4900    0.2500    0.4500    0.1700    0.9355    2.0000
```

```
%      0.7300     0.2900     0.6300     0.1800     0.9169     2.0000]
% [data_mod, index_rem,SIM,SIM1,H]=feat_sel_sim1(d, 'luca', 1)

clc clear
```

%	d = [0.4600	0.3400	1.1400	0.0300	0.9218	1.0000
%	0.5000	0.3400	0.1500	0.0200	0.7382	1.0000
%	0.4400	0.2900	0.1400	0.0200	0.1763	1.0000
%	0.7600	0.3000	0.6600	0.2100	0.4057	2.0000
%	0.5000	0.3400	0.1500	0.0200	0.7382	1.0000
%	0.4400	0.2900	0.1400	0.0200	0.1763	1.0000
%	0.7600	0.3000	1.6600	0.2100	0.4057	2.0000
%	0.4900	0.2500	0.4500	0.1700	0.9355	2.0000
%	0.5000	0.3400	0.1500	0.0200	0.7382	1.0000
%	0.4400	0.2900	0.1400	0.0200	0.1763	1.0000
%	0.7600	0.3000	0.6600	0.2100	0.4057	2.0000
%	0.4900	0.2500	0.4500	0.1700	0.9355	2.0000
%	0.4900	0.2500	0.4500	0.1700	0.9355	2.0000
%	0.5000	0.3400	0.1500	0.0200	0.7382	1.0000
%	0.4400	0.2900	0.1400	0.0200	0.1763	1.0000
%	0.7600	0.3000	0.6600	0.2100	0.4057	2.0000
%	0.4900	0.2500	0.4500	0.1700	0.9355	2.0000
%	0.7300	0.2900	0.6300	0.1800	0.9169	1.0000];

```
load d2160c d=dd(1:500,:);
mother_data=d;

for i=1:36 nd(:,i,:)=abs(mother_data(:,i,:)./max(abs(mother_data(:,i,:))));
end
d1=nd;
ddd=[d1(:, 1:23) d1(:,27:36) d(:,37)];

d=ddd;

B=zeros(1,size(d,2)-1);
BI={};
for i=1:size(d,2)-1 B(i)=i;
end BI{1}=B;
% number of level that featuer pruned for i=1:32

tictic
%% similarity LUCA
%P=1
[data_mod_luca, index_rem_luca,SIM_luca,SIM1_luca,H_luca]=feat_sel_sim_luca
(d,1); tluca(i)=toc;
%% similarity YU
%Landa(>-1)=0.1
tic
[data_mod_YU, index_rem_YU,SIM_YU,SIM1_YU,H_YU]=feat_sel_sim_luca(d,0.1);
tYU(i)=toc;
%% similarity Weber
```

```
%Landa(>-1)=3 tic
[data_mod_Weber, index_rem_Weber,SIM_Weber,SIM1_Weber,H_Weber]
=feat_sel_sim_Weber(d, 3); tWeber(i)=toc;

%% similarity Dubois
%alfa([0,1])=0.9
tic
[data_mod_Dubois, index_rem_Dubois,SIM_Dubois,SIM1_Dubois,H_Dubois]
=feat_sel_sim_Dubois1 (d,3);
tDubois(i)=toc;

%% similarity Yager
%omega(>0)=1 tic
[data_mod_Yager, index_rem_Yager,SIM_Yager,SIM1_Yager,H_Yager]
=feat_sel_sim_Yager(d, 5); tYager(i)=toc;

%% similarity Schweizer 1
%p(#0)=1
tic
[data_mod_Schweizer_1, index_rem_Schweizer_1,SIM_Schweizer_1,
SIM1_Schweizer_1, H_Schweizer_1]=feat_sel_sim_Schweizer_1(d, 3);
tSchweizer_1(i)=toc;

%% similarity Schweizer 2
%p(>0)=1
tic

[data_mod_Schweizer_2, index_rem_Schweizer_2,SIM_Schweizer_2,
SIM1_Schweizer_2, H_Schweizer_2]=feat_sel_sim_Schweizer_2(d, 2);
tSchweizer_2(i)=toc;
%% similarity Schweizer 3
%p(>0)=1
tic
[data_mod_Schweizer_3, index_rem_Schweizer_3,SIM_Schweizer_3,
SIM1_Schweizer_3, H_Schweizer_3]=feat_sel_sim_Schweizer_3(d, 2);
tSchweizer_3(i)=toc;
%% similarity Schweizer 4
%p(>0)=1
tic
[data_mod_Schweizer_4, index_rem_Schweizer_4,SIM_Schweizer_4,
SIM1_Schweizer_4, H_Schweizer_4]=feat_sel_sim_Schweizer_4(d, 2);
tSchweizer_4(i)=toc;
%% similarity Hamacher
%r(>0)=1
tic
[data_mod_Hamacher, index_rem_Hamacher,SIM_Hamacher,SIM1_Hamacher,H_Hamacher]
=feat_sel_sim_Hamacher(d, 1.4); tHamacher(i)=toc;
%% similarity Frank
%s(>0) and (#1)=2
tic
[data_mod_Frank, index_rem_Frank,SIM_Frank,SIM1_Frank,H_Frank]
```

```
=feat_sel_sim_Frank(d,6); tFrank(i)=toc;
%% similarity Dombi
%landa(>0) = 1 tic
[data_mod_Dombi, index_rem_Dombi,SIM_Dombi,SIM1_Dombi,H_Dombi]
=feat_sel_sim_Dombi(d, 4);
tDombi(i)=toc; timme(i)=toc

index_rem_luca; index_rem_YU; index_rem_Weber; index_rem_Dubois;

index_rem_Yager; index_rem_Schweizer_1; index_rem_Schweizer_2;
index_rem_Schweizer_3; index_rem_Schweizer_4; index_rem_Hamacher;
index_rem_Frank; index_rem_Dombi; A=[index_rem_luca; index_rem_YU;
index_rem_Weber; index_rem_Dubois; index_rem_Yager; index_rem_Schweizer_1;
index_rem_Schweizer_2; index_rem_Schweizer_3; index_rem_Schweizer_4;
index_rem_Hamacher; index_rem_Frank; index_rem_Dombi];
%matrix of selected feature for removing based on diffrent similarity
%measure and entropy measures(informations) AA(:,:,i)=A;

Removed_feature=mode(mode(A(1,:),1));
% ID for removed feature forn first to end RFN(:,i)=Removed_feature;
B(RFN(:,i))=[];
%Features remained in data set in level(i) BI{i+1}=B;
dataold=d;
data_new=[dataold(:,1:Removed_feature-1) dataold(:,Removed_feature+1:end)];
d=data_new;
end
%
d AA
dataold; data_new; RFN

BI
figure,
subplot(2,3,1), mesh(SIM1_luca);title('LUCA Similarity');
subplot(2,3,2), mesh(SIM1_YU);title('YU Similarity');
subplot(2,3,3), mesh(SIM1_Weber);title('Weber Similarity');
subplot(2,3,4), mesh(SIM1_Dubois);title('Dubois Similarity');
subplot(2,3,5), mesh(SIM1_Yager);title('Yager Similarity');
subplot(2,3,6), mesh(SIM1_Schweizer_1);title('Schweizer_1 Similarity'); figure,
subplot(2,3,1), mesh(SIM1_Schweizer_2);title('Schweizer_2 Similarity');
subplot(2,3,2), mesh(SIM1_Schweizer_3);title('Schweizer_3 Similarity');
subplot(2,3,3), mesh(SIM1_Schweizer_4);title('Schweizer_4 Similarity');
subplot(2,3,4), mesh(SIM1_Hamacher);title('Hamacher Similarity');
subplot(2,3,5), mesh(SIM1_Frank);title('Frank Similarity');
subplot(2,3,6), mesh(SIM1_Dombi);title('Dombi Similarity');

T=[tluca tYU tWeber tDubois tYager
tSchweizer_1 tSchweizer_2 tSchweizer_3 tSchweizer_4 tHamacher tFrank tDombi];
figure, boxplot(T);legend('LUCA','Yu','Weber','Dubois
','Yager','Schweizer_1', 'Schweizer_1','Schweizer_2','Schweizer_3',
'Schweizer_4', 'Hamacher','Frank','Dombi' )
plot(T','DisplayName','T','YDataSource','T');figure(gcf ), legend
```

```
('LUCA','Yu','Weber','Dubois ','Yager','Schweizer_1','Schweizer_1',
'Schweizer_2','Schweizer_3','Schweizer_4', 'Hamacher','Frank',
'Dombi' ), title('Time db_1');plot(T','DisplayName','T',
'YDataSource','T');figure(gcf ), legend('LUCA','Yu','Weber','Dubois
','Yager','Schweizer_1', 'Schweizer_1','Schweizer_2','Schweizer_3',
'Schweizer_4', 'Hamacher','Frank','Dombi' ), title('Time db_1')

function [data_mod, index_rem,SIM,SIM1]=feat_sel_sim(data, measure, p)
%
%    p in (0, \infty) as default p=1. if nargin
p=1;
end
if nargin
end

l=max(data(:,end)); % #-classes m=size(data,1); % #-samples t=size(data,2)-1;
% #-features dataold=data;
tmp=[];
% forming idealvec using arithmetic mean idealvec=zeros(l,t);
for k=1:l idealvec_s(k,:)=mean(data(find(data(:,end)==k),1:t));
end

%scaling data between [0,1] data_v=data(:,1:t); data_c=data(:,t+1); %labels
mins_v = min(data_v);
Ones = ones(size(data_v));
data_v = data_v+Ones*diag(abs(mins_v)); for k=1:l
tmp=[tmp;abs(mins_v)]; end
tmp;
idealvec_s = idealvec_s+tmp; maxs_v = max(data_v);
data_v = data_v*diag(maxs_v.^(-1)); idealvec_s=idealvec_s./repmat(maxs_v,
l,1); data = [data_v, data_c];
% sample data datalearn_s=data(:,1:t);

% similarities sim=zeros(t,m,l); for j=1:m
for i=1:t
for k=1:l
sim(i,j,k)=(1-abs(idealvec_s(k,i)^p-datalearn_s(j,i))^p)^(1/p); end
end end
%class 1,2,3,4,. n
% SIM(:,:,1) = similarity for class 1
% SIM(:,:,2) = similarity for class 2 SIM=sim;
% reduce number of dimensions in sim sim=reshape(sim,t,m*l)';
SIM1=sim;

% possibility for two different entropy measures if measure=='luca'
% moodifying zero and one values of the similarity values to work with
% De Luca's entropy measure delta=1E-10; sim(find(sim==0))=delta; sim(find
(sim==1))=1-delta;
H=sum(-sim.*log(sim)-(1-sim).*log(1-sim));
```

```
if measure=='hamzeh'
% moodifying zero and one values of the similarity values to work with
% De Luca's entropy measure delta=1E-10; sim(find(sim==0))=delta; sim(find
(sim==1))=1-delta;
H=sum(-sim.*log(sim)-(1-sim).*log(1-sim));

if measure=='park1' H=sum(cos(pi/2*sim)+cos(pi/2*(1-sim))-1);
end

elseif measure=='park'

H=sum(sin(pi/2*sim)+sin(pi/2*(1-sim))-1); end

% find maximum feature [i, index_rem]=max(H);

% removing feature from the data data_mod=[dataold(:,1:index_rem-1) dataold
(:,index_rem+1:end)]; j=1;

return

function [data_mod, index_rem,SIM,SIM1,H]=feat_sel_sim_Dombi(data, landa)
%
%    p in (0, \infty) as default p=1. if nargin
p=1;
end

l=max(data(:,end)); % #-classes m=size(data,1); % #-samples t=size(data,2)-1;
% #-features dataold=data;
tmp=[];
% forming idealvec using arithmetic mean idealvec=zeros(1,t);
for k=1:l idealvec_s(k,:)=mean(data(find(data(:,end)==k),1:t));
end

%scaling data between [0,1] data_v=data(:,1:t); data_c=data(:,t+1); %labels
mins_v = min(data_v);
Ones = ones(size(data_v));
data_v = data_v+Ones*diag(abs(mins_v)); for k=1:l
tmp=[tmp;abs(mins_v)]; end
tmp;
idealvec_s = idealvec_s+tmp; maxs_v = max(data_v);
data_v = data_v*diag(maxs_v.^(-1)); idealvec_s=idealvec_s./repmat(maxs_v,
l,1); data = [data_v, data_c];
% sample data datalearn_s=data(:,1:t); land=0.1;
% similarities sim=zeros(t,m,l); for j=1:m

for i=1:t
for k=1:l
%%similarity Luca    with p=1
% a=(1-datalearn_s(j,i))
% b=(idealvec_s(k,i))
```

```
%((1/a)-1)^landa=((((1/(1-datalearn_s(j,i)))-1)))^landa
%((1/b)-1)^landa=((1/(idealvec_s(k,i)))-1)^landa
sn_1(i,j,k)=(1+(((((((1/(1-datalearn_s(j,i)))-1)))^landa)+(((1/(idealvec_s(k,
i)))-1)^landa))^(-(1/landa))))^-1;
sn_2(i,j,k)=(1+(((((((1/(datalearn_s(j,i)))-1)))^landa)+(((1/(1-idealvec_s(k,
i)))-1)^landa))^(-(1/landa))))^-1;
sim(i,j,k)=(1+((((((1/sn_1(i,j,k))-1)^landa)+(((1/sn_2(i,j,k))-1)^landa))))^
(1/landa)))^(-1); end
end end
%class 1,2,3,4,. n
% SIM(:,:,1) = similarity for class 1
% SIM(:,:,2) = similarity for class 2 SIM=sim;
% reduce number of dimensions in sim sim=reshape(sim,t,m*l)';
SIM1=sim;

% possibility for two different entropy measures
% moodifying zero and one values of the similarity values to work with
% De Luca's entropy measure
%%%%%Hsh(1)
delta=1E-10; sim(find(sim==0))=delta; sim(find(sim==1))=1-delta;
Hsh=sum(-sim.*log(sim)-(1-sim).*log(1-sim));
Hsh=(Hsh-min(Hsh))/(max(Hsh)-min(Hsh));

% %%%%%Hme(2)
Hme=sum(sin(pi/2*sim)+sin(pi/2*(1-sim))-1); Hme=(Hme-min(Hme))/(max(Hme)-min
(Hme));

%%%%%%%Hrn(3)
alfa_rn=0.8;
Hrn=log(sum((sim.^alfa_rn)+((1-sim).^alfa_rn)))/(1-alfa_rn);

Hrn=(Hrn-min(Hrn))/(max(Hrn)-min(Hrn));

% %%%%%%Hpp(4)
Hpp=sum((sim.*exp(1.-sim))+(((1-sim).*exp(1.-(1-sim))))); Hpp=(Hpp-min(Hpp))/
(max(Hpp)-min(Hpp));

% %%%%%%%Hts(5)
alfa_ts=0.5;
Hts=1-(sum((sim.^alfa_ts)+((1-sim).^alfa_ts)))/(alfa_ts-1);
Hts=(Hts-min(Hts))/(max(Hts)-min(Hts));

%%%%%%%%Hma(6)
gama=2; a_ma=0.5; b_ma=3; alfa_ma=0.1; beta_ma=2;
Hma=(sum(((sim.^alfa_ma).*exp(a_ma.*sim.^alfa_ma+b_ma).^beta_ma)+((1-sim).
^alfa_ma).
*exp(a_ma.*(1-sim).^alfa_ma+b_ma).^beta_ma)); Hma=(Hma-min(Hma))/(max(Hma)-
min(Hma));

% find maximum feature
[i_sh, index_rem_sh]=max(Hsh); [i_me, index_rem_me]=max(Hme);
```

```
[i_rn, index_rem_rn]=max(Hrn); [i_pp, index_rem_pp]=max(Hpp);
[i_ts, index_rem_ts]=max(Hts); [i_ma, index_rem_ma]=max(Hma);

% removing feature from the data
data_mod_sh=[dataold(:,1:index_rem_sh-1) dataold(:,index_rem_sh+1:end)];
data_mod_me=[dataold(:,1:index_rem_me-1) dataold(:,index_rem_me+1:end)];
data_mod_rn=[dataold(:,1:index_rem_rn-1) dataold(:,index_rem_rn+1:end)];
data_mod_pp=[dataold(:,1:index_rem_pp-1) dataold(:,index_rem_pp+1:end)];
data_mod_ts=[dataold(:,1:index_rem_ts-1) dataold(:,index_rem_ts+1:end)];
data_mod_ma=[dataold(:,1:index_rem_ma-1) dataold(:,index_rem_ma+1:end)];

data_mod=[data_mod_sh,data_mod_me,data_mod_rn,data_mod_pp,data_mod_ts,
data_mod_ma]; index_rem=[index_rem_sh,index_rem_me,index_rem_rn,index_rem_pp,
index_rem_ts,index_rem_ma]; H=[Hsh;Hme;Hrn;Hpp;Hts;Hma];


j=1;

return

%%%%%%%%%%%%%%%%%%%%%%%%

%%example
% d =[0.4600    0.3400    0.1400    0.0300    0.9218    1.0000
%      0.5000    0.3400    0.1500    0.0200    0.7382    1.0000
%      0.4400    0.2900    0.1400    0.0200    0.1763    1.0000
%      0.7600    0.3000    0.6600    0.2100    0.4057    2.0000
%      0.4900    0.2500    0.4500    0.1700    0.9355    2.0000
%      0.7300    0.2900    0.6300    0.1800    0.9169    2.0000]
% [data_mod, index_rem,SIM,SIM1,H]=feat_sel_sim1(d, 'luca', 1)

function [data_mod, index_rem,SIM,SIM1,H]=feat_sel_sim_Frank(data, s)
%

%    p in (0, \infty) as default p=1. if nargin
p=1;
end

l=max(data(:,end)); % #-classes m=size(data,1); % #-samples t=size(data,2)-1;
% #-features dataold=data;
tmp=[];
% forming idealvec using arithmetic mean idealvec=zeros(l,t);
for k=1:l idealvec_s(k,:)=mean(data(find(data(:,end)==k),1:t));
end

%scaling data between [0,1] data_v=data(:,1:t); data_c=data(:,t+1); %labels
mins_v = min(data_v);
Ones = ones(size(data_v));
data_v = data_v+Ones*diag(abs(mins_v)); for k=1:l
tmp=[tmp;abs(mins_v)]; end
tmp;
```

```
idealvec_s = idealvec_s+tmp; maxs_v = max(data_v);
data_v = data_v*diag(maxs_v.^(-1)); idealvec_s=idealvec_s./repmat(maxs_v,
1,1); data = [data_v, data_c];
% sample data datalearn_s=data(:,1:t); land=0.1;
% similarities sim=zeros(t,m,l);

for j=1:m for i=1:t
for k=1:l
%%similarity Luca    with p=1
% a=(1-datalearn_s(j,i))
% b=(idealvec_s(k,i))
% 1-a=(1-(1-datalearn_s(j,i)))
% 1-b=(1-(idealvec_s(k,i)))
% (1-a)^p=((1-(1-datalearn_s(j,i)))^p)
% (1-b)^p=((1-(idealvec_s(k,i)))^p)
% (1-a)(1-b)=((1-(1-datalearn_s(j,i)))*(1-(idealvec_s(k,i))))
% (1-a)^p+(1-b)^p-(1-a)^p*(1-b)^p=(((1-(1-datalearn_s(j,i)))^p)
+((1-(idealvec_s(k,i)))^p)-(((1-(1- datalearn_s(j,i)))^p)*((1-(idealvec_s(k,
i)))^p)))^(1/p)
a=1-datalearn_s(j,i); b=1-idealvec_s(k,i);
sn_1(i,j,k)=1-(((log(1+((((s^(1-a))-1)*((s^(1-b))-1))/(s-1))))/log(s)));
sn_2(i,j,k)=1-(((log(1+((((s^(1-(1-a)))-1)*((s^(1-(1-b)))-1))/(s-1))))/log
(s)));
%sn_1(i,j,k)=1-(((log(1+((s^(1-datalearn_s(j,i)))-1)*((s^(1-(idealvec_s(k,
i))))-1)/(s-1))))/log(s));
%sn_2(i,j,k)=1-(((log(1+((s^(datalearn_s(j,i)))-1)*((s^(1-(1-idealvec_s(k,
i))))-1)/(s-1))))/log(s));
sim(i,j,k)=log(1+(((s^sn_1(i,j,k))-1)*((s^sn_2(i,j,k))-1)/(s-1)))/log(s); end
endend
%class 1,2,3,4,. n
% SIM(:,:,1) = similarity for class 1
% SIM(:,:,2) = similarity for class 2 SIM=sim;
% reduce number of dimensions in sim sim=reshape(sim,t,m*l)';
SIM1=sim;

% possibility for two different entropy measures
% moodifying zero and one values of the similarity values to work with
% De Luca's entropy measure
%%%%%%Hsh(1)
delta=1E-10; sim(find(sim==0))=delta; sim(find(sim==1))=1-delta;
Hsh=sum(-sim.*log(sim)-(1-sim).*log(1-sim));
Hsh=(Hsh-min(Hsh))/(max(Hsh)-min(Hsh));


% %%%%%%Hme(2)
Hme=sum(sin(pi/2*sim)+sin(pi/2*(1-sim))-1); Hme=(Hme-min(Hme))/(max(Hme)-min
(Hme));

%%%%%%%%Hrn(3)
alfa_rn=0.8;
Hrn=log(sum((sim.^alfa_rn)+((1-sim).^alfa_rn)))/(1-alfa_rn); Hrn=(Hrn-
min(Hrn))/(max(Hrn)-min(Hrn));
```

```
% %%%%%%%Hpp(4)
Hpp=sum((sim.*exp(1.-sim))+(((1-sim).*exp(1.-(1-sim))))); Hpp=(Hpp-min(Hpp))/
(max(Hpp)-min(Hpp));

% %%%%%%%%Hts(5)
alfa_ts=0.5;
Hts=1-(sum((sim.^alfa_ts)+((1-sim).^alfa_ts)))/(alfa_ts-1);
Hts=(Hts-min(Hts))/(max(Hts)-min(Hts));

%%%%%%%%%Hma(6)
gama=2; a_ma=0.5; b_ma=3; alfa_ma=0.1; beta_ma=2;
Hma=(sum(((sim.^alfa_ma).*exp(a_ma.*sim.^alfa_ma+b_ma).^beta_ma)+((1-sim).
^alfa_ma).
*exp(a_ma.*(1-sim).^alfa_ma+b_ma).^beta_ma)); Hma=(Hma-min(Hma))/(max(Hma)-
min(Hma));

% find maximum feature
[i_sh, index_rem_sh]=max(Hsh); [i_me, index_rem_me]=max(Hme);
[i_rn, index_rem_rn]=max(Hrn); [i_pp, index_rem_pp]=max(Hpp);
[i_ts, index_rem_ts]=max(Hts); [i_ma, index_rem_ma]=max(Hma);

% removing feature from the data
data_mod_sh=[dataold(:,1:index_rem_sh-1) dataold(:,index_rem_sh+1:end)];
data_mod_me=[dataold(:,1:index_rem_me-1) dataold(:,index_rem_me+1:end)];
data_mod_rn=[dataold(:,1:index_rem_rn-1) dataold(:,index_rem_rn+1:end)];
data_mod_pp=[dataold(:,1:index_rem_pp-1) dataold(:,index_rem_pp+1:end)];
data_mod_ts=[dataold(:,1:index_rem_ts-1) dataold(:,index_rem_ts+1:end)];
data_mod_ma=[dataold(:,1:index_rem_ma-1) dataold(:,index_rem_ma+1:end)];

data_mod=[data_mod_sh,data_mod_me,data_mod_rn,data_mod_pp,data_mod_ts,
data_mod_ma]; index_rem=[index_rem_sh,index_rem_me,index_rem_rn,index_rem_pp,
index_rem_ts,index_rem_ma]; H=[Hsh;Hme;Hrn;Hpp;Hts;Hma];

j=1;

return

%%%%%%%%%%%%%%%%%%%%%%%

%%example
% d =[0.4600    0.3400    0.1400    0.0300    0.9218    1.0000
%     0.5000    0.3400    0.1500    0.0200    0.7382    1.0000
%     0.4400    0.2900    0.1400    0.0200    0.1763    1.0000
%     0.7600    0.3000    0.6600    0.2100    0.4057    2.0000
%     0.4900    0.2500    0.4500    0.1700    0.9355    2.0000
%     0.7300    0.2900    0.6300    0.1800    0.9169    2.0000]
% [data_mod, index_rem,SIM,SIM1,H]=feat_sel_sim1(d, 'luca', 1)

function [data_mod, index_rem,SIM,SIM1,H]=feat_sel_sim_luca(data, p)
%
```

```
%    p in (0, \infty) as default p=1. if nargin
p=1;
end

l=max(data(:,end)); % #-classes m=size(data,1); % #-samples t=size(data,2)-1;
% #-features dataold=data;
tmp=[];
% forming idealvec using arithmetic mean idealvec=zeros(1,t);
for k=1:l idealvec_s(k,:)=mean(data(find(data(:,end)==k),1:t));
end

%scaling data between [0,1] data_v=data(:,1:t); data_c=data(:,t+1); %labels
mins_v = min(data_v);
Ones = ones(size(data_v));
data_v = data_v+Ones*diag(abs(mins_v)); for k=1:l
tmp=[tmp;abs(mins_v)]; end
tmp;
idealvec_s = idealvec_s+tmp; maxs_v = max(data_v);
data_v = data_v*diag(maxs_v.^(-1)); idealvec_s=idealvec_s./repmat(maxs_v,
l,1); data = [data_v, data_c];
% sample data datalearn_s=data(:,1:t); land=0.1;
% similarities sim=zeros(t,m,l);

for j=1:m for i=1:t
for k=1:l
%%similarity Luca    with p=1
sim(i,j,k)=(1-abs(idealvec_s(k,i)^p-datalearn_s(j,i))^p)^(1/p); end
end end
%class 1,2,3,4,.    n
% SIM(:,:,1) = similarity for class 1
% SIM(:,:,2) = similarity for class 2 SIM=sim;
% reduce number of dimensions in sim sim=reshape(sim,t,m*l)';
SIM1=sim;

% possibility for two different entropy measures
% moodifying zero and one values of the similarity values to work with
% De Luca's entropy measure
%%%%%%Hsh(1)
delta=1E-10; sim(find(sim==0))=delta; sim(find(sim==1))=1-delta;
Hsh=sum(-sim.*log(sim)-(1-sim).*log(1-sim));
Hsh=(Hsh-min(Hsh))/(max(Hsh)-min(Hsh));

% %%%%%%Hme(2)
Hme=sum(sin(pi/2*sim)+sin(pi/2*(1-sim))-1); Hme=(Hme-min(Hme))/(max(Hme)-min
(Hme));

%%%%%%%%Hrn(3)
alfa_rn=0.8;
Hrn=log(sum((sim.^alfa_rn)+((1-sim).^alfa_rn)))/(1-alfa_rn); Hrn=(Hrn-
min(Hrn))/(max(Hrn)-min(Hrn));
```

```
% %%%%%%%Hpp(4)
Hpp=sum((sim.*exp(1.-sim))+(((1-sim).*exp(1.-(1-sim))))); Hpp=(Hpp-min(Hpp))/
(max(Hpp)-min(Hpp));

% %%%%%%%%Hts(5)
alfa_ts=0.5;
Hts=1-(sum((sim.^alfa_ts)+((1-sim).^alfa_ts)))/(alfa_ts-1);
Hts=(Hts-min(Hts))/(max(Hts)-min(Hts));

%%%%%%%%%Hma(6)
gama=2; a_ma=0.5; b_ma=3; alfa_ma=0.1; beta_ma=2;
Hma=(sum(((sim.^alfa_ma).*exp(a_ma.*sim.^alfa_ma+b_ma).^beta_ma)+((1-sim).
^alfa_ma).
*exp(a_ma.*(1-sim).^alfa_ma+b_ma).^beta_ma)); Hma=(Hma-min(Hma))/(max(Hma)-
min(Hma));

% find maximum feature
[i_sh, index_rem_sh]=max(Hsh); [i_me, index_rem_me]=max(Hme);
[i_rn, index_rem_rn]=max(Hrn); [i_pp, index_rem_pp]=max(Hpp);
[i_ts, index_rem_ts]=max(Hts); [i_ma, index_rem_ma]=max(Hma);

% removing feature from the data
data_mod_sh=[dataold(:,1:index_rem_sh-1) dataold(:,index_rem_sh+1:end)];
data_mod_me=[dataold(:,1:index_rem_me-1) dataold(:,index_rem_me+1:end)];
data_mod_rn=[dataold(:,1:index_rem_rn-1) dataold(:,index_rem_rn+1:end)];
data_mod_pp=[dataold(:,1:index_rem_pp-1) dataold(:,index_rem_pp+1:end)];
data_mod_ts=[dataold(:,1:index_rem_ts-1) dataold(:,index_rem_ts+1:end)];
data_mod_ma=[dataold(:,1:index_rem_ma-1) dataold(:,index_rem_ma+1:end)];

data_mod=[data_mod_sh,data_mod_me,data_mod_rn,data_mod_pp,data_mod_ts,
data_mod_ma]; index_rem=[index_rem_sh,index_rem_me,index_rem_rn,index_rem_pp,
index_rem_ts,index_rem_ma]; H=[Hsh;Hme;Hrn;Hpp;Hts;Hma];

j=1;

return

%%%%%%%%%%%%%%%%%%%%%%%%

%%example
% d =[0.4600    0.3400    0.1400    0.0300    0.9218    1.0000
%     0.5000    0.3400    0.1500    0.0200    0.7382    1.0000
%     0.4400    0.2900    0.1400    0.0200    0.1763    1.0000
%     0.7600    0.3000    0.6600    0.2100    0.4057    2.0000
%     0.4900    0.2500    0.4500    0.1700    0.9355    2.0000
%     0.7300    0.2900    0.6300    0.1800    0.9169    2.0000]
% [data_mod, index_rem,SIM,SIM1,H]=feat_sel_sim1(d, 'luca', 1)

function [data_mod, index_rem,SIM,SIM1,H]=feat_sel_sim_Weber(data, landa)
%
```

```
%    p in (0, \infty) as default p=1. if nargin
p=1;
end

l=max(data(:,end)); % #-classes m=size(data,1); % #-samples t=size(data,2)-1;
% #-features dataold=data;
tmp=[];
% forming idealvec using arithmetic mean idealvec=zeros(1,t);
for k=1:l idealvec_s(k,:)=mean(data(find(data(:,end)==k),1:t));
end

%scaling data between [0,1] data_v=data(:,1:t); data_c=data(:,t+1); %labels
mins_v = min(data_v);
Ones = ones(size(data_v));
data_v = data_v+Ones*diag(abs(mins_v)); for k=1:l
tmp=[tmp;abs(mins_v)]; end
tmp;
idealvec_s = idealvec_s+tmp; maxs_v = max(data_v);
data_v = data_v*diag(maxs_v.^(-1)); idealvec_s=idealvec_s./repmat(maxs_v,
l,1); data = [data_v, data_c];
% sample data datalearn_s=data(:,1:t);
%land=0.1;
% similarities sim=zeros(t,m,l);

for j=1:m for i=1:t
for k=1:l
%similarity Weber    with landa=
sn_1(i,j,k)=min(1,(1-datalearn_s(j,i))+(idealvec_s(k,i))-((landa/(1-landa))*
((1-datalearn_s(j,i))* (idealvec_s(k,i)))));
sn_2(i,j,k)=min(1,(datalearn_s(j,i))+(1-idealvec_s(k,i))+((landa/(1-landa))*
((datalearn_s(j,i))*(1- idealvec_s(k,i)))));
sim(i,j,k)=max(0,(sn_1(i,j,k)+sn_2(i,j,k)+landa*sn_1(i,j,k)*sn_2(i,j,k)-1)/(1
+landa)); end
endend
%class 1,2,3,4,. n
% SIM(:,:,1) = similarity for class 1
% SIM(:,:,2) = similarity for class 2 SIM=sim;
% reduce number of dimensions in sim sim=reshape(sim,t,m*l)';
SIM1=sim;

% possibility for two different entropy measures
% moodifying zero and one values of the similarity values to work with
% De Luca's entropy measure
%%%%%%Hsh(1)
delta=1E-10; sim(find(sim==0))=delta; sim(find(sim==1))=1-delta;
Hsh=sum(-sim.*log(sim)-(1-sim).*log(1-sim));
Hsh=(Hsh-min(Hsh))/(max(Hsh)-min(Hsh));

% %%%%%%Hme(2)
Hme=sum(sin(pi/2*sim)+sin(pi/2*(1-sim))-1); Hme=(Hme-min(Hme))/(max(Hme)-min
(Hme));
```

```
%%%%%%%%Hrn(3)
alfa_rn=0.8;
Hrn=log(sum((sim.^alfa_rn)+((1-sim).^alfa_rn)))/(1-alfa_rn); Hrn=(Hrn-
min(Hrn))/(max(Hrn)-min(Hrn));

% %%%%%%%Hpp(4)
Hpp=sum((sim.*exp(1.-sim))+(((1-sim).*exp(1.-(1-sim))))); Hpp=(Hpp-min(Hpp))/
(max(Hpp)-min(Hpp));

% %%%%%%%Hts(5)
alfa_ts=0.5;
Hts=1-(sum((sim.^alfa_ts)+((1-sim).^alfa_ts)))/(alfa_ts-1);
Hts=(Hts-min(Hts))/(max(Hts)-min(Hts));

%%%%%%%%Hma(6)
gama=2; a_ma=0.5; b_ma=3; alfa_ma=0.1; beta_ma=2;
Hma=(sum(((sim.^alfa_ma).*exp(a_ma.*sim.^alfa_ma+b_ma).^beta_ma)+((1-sim).
^alfa_ma).
*exp(a_ma.*(1-sim).^alfa_ma+b_ma).^beta_ma)); Hma=(Hma-min(Hma))/(max(Hma)-
min(Hma));

% find maximum feature
[i_sh, index_rem_sh]=max(Hsh); [i_me, index_rem_me]=max(Hme);
[i_rn, index_rem_rn]=max(Hrn); [i_pp, index_rem_pp]=max(Hpp);
[i_ts, index_rem_ts]=max(Hts); [i_ma, index_rem_ma]=max(Hma);

% removing feature from the data
data_mod_sh=[dataold(:,1:index_rem_sh-1) dataold(:,index_rem_sh+1:end)];
data_mod_me=[dataold(:,1:index_rem_me-1) dataold(:,index_rem_me+1:end)];
data_mod_rn=[dataold(:,1:index_rem_rn-1) dataold(:,index_rem_rn+1:end)];
data_mod_pp=[dataold(:,1:index_rem_pp-1) dataold(:,index_rem_pp+1:end)];
data_mod_ts=[dataold(:,1:index_rem_ts-1) dataold(:,index_rem_ts+1:end)];
data_mod_ma=[dataold(:,1:index_rem_ma-1) dataold(:,index_rem_ma+1:end)];

data_mod=[data_mod_sh,data_mod_me,data_mod_rn,data_mod_pp,data_mod_ts,
data_mod_ma]; index_rem=[index_rem_sh,index_rem_me,index_rem_rn,index_rem_pp,
index_rem_ts,index_rem_ma]; H=[Hsh;Hme;Hrn;Hpp;Hts;Hma];

j=1;

return

%%%%%%%%%%%%%%%%%%%%%%%

%%example
% d =[0.4600    0.3400    0.1400    0.0300    0.9218    1.0000
%     0.5000    0.3400    0.1500    0.0200    0.7382    1.0000
%     0.4400    0.2900    0.1400    0.0200    0.1763    1.0000
%     0.7600    0.3000    0.6600    0.2100    0.4057    2.0000
%     0.4900    0.2500    0.4500    0.1700    0.9355    2.0000
```

```
%     0.7300     0.2900     0.6300     0.1800     0.9169     2.0000]
% [data_mod, index_rem,SIM,SIM1,H]=feat_sel_sim1(d, 'luca', 1)

function [data_mod, index_rem,SIM,SIM1,H]=feat_sel_sim_Yager(data, omega)
%

%    p in (0, \infty) as default p=1. if nargin
p=1;
end

l=max(data(:,end)); % #-classes m=size(data,1); % #-samples t=size(data,2)-1;
% #-features dataold=data;
tmp=[];
% forming idealvec using arithmetic mean idealvec=zeros(l,t);
for k=1:l idealvec_s(k,:)=mean(data(find(data(:,end)==k),1:t));
end

%scaling data between [0,1] data_v=data(:,1:t); data_c=data(:,t+1); %labels
mins_v = min(data_v);
Ones = ones(size(data_v));
data_v = data_v+Ones*diag(abs(mins_v)); for k=1:l
tmp=[tmp;abs(mins_v)]; end
tmp;
idealvec_s = idealvec_s+tmp; maxs_v = max(data_v);
data_v = data_v*diag(maxs_v.^(-1)); idealvec_s=idealvec_s./repmat(maxs_v,
l,1); data = [data_v, data_c];
% sample data datalearn_s=data(:,1:t); land=0.1;
% similarities sim=zeros(t,m,l);

for j=1:m for i=1:t
for k=1:l
%%similarity Luca    with p=1
sn_1(i,j,k)=min(1,(1-datalearn_s(j,i))^omega+(idealvec_s(k,i)^omega)^(1/
omega)); sn_2(i,j,k)=min(1,(datalearn_s(j,i))^omega+(1-idealvec_s(k,i)^omega)
^(1/omega)); sim(i,j,k)=1-min(1,((1-sn_1(i,j,k))^omega+(1-sn_1(i,j,k))^omega)
^(1/omega));
end end
end
%class 1,2,3,4,.    n
% SIM(:,:,1) = similarity for class 1
% SIM(:,:,2) = similarity for class 2 SIM=sim;
% reduce number of dimensions in sim sim=reshape(sim,t,m*l)';
SIM1=sim;

% possibility for two different entropy measures
% moodifying zero and one values of the similarity values to work with
% De Luca's entropy measure
%%%%%%Hsh(1)
delta=1E-10; sim(find(sim==0))=delta; sim(find(sim==1))=1-delta;
Hsh=sum(-sim.*log(sim)-(1-sim).*log(1-sim));
Hsh=(Hsh-min(Hsh))/(max(Hsh)-min(Hsh));
```

```
% %%%%%%Hme(2)
Hme=sum(sin(pi/2*sim)+sin(pi/2*(1-sim))-1); Hme=(Hme-min(Hme))/(max(Hme)-min
(Hme));

%%%%%%%%Hrn(3)
alfa_rn=0.8;
Hrn=log(sum((sim.^alfa_rn)+((1-sim).^alfa_rn)))/(1-alfa_rn); Hrn=(Hrn-
min(Hrn))/(max(Hrn)-min(Hrn));

% %%%%%%%Hpp(4)
Hpp=sum((sim.*exp(1.-sim))+(((1-sim).*exp(1.-(1-sim))))); Hpp=(Hpp-min(Hpp))/
(max(Hpp)-min(Hpp));


% %%%%%%%%Hts(5)
alfa_ts=0.5;
Hts=1-(sum((sim.^alfa_ts)+((1-sim).^alfa_ts)))/(alfa_ts-1);
Hts=(Hts-min(Hts))/(max(Hts)-min(Hts));

%%%%%%%%%Hma(6)
gama=2; a_ma=0.5; b_ma=3; alfa_ma=0.1; beta_ma=2;
Hma=(sum(((sim.^alfa_ma).*exp(a_ma.*sim.^alfa_ma+b_ma).^beta_ma)+((1-sim).
^alfa_ma).
*exp(a_ma.*(1-sim).^alfa_ma+b_ma).^beta_ma)); Hma=(Hma-min(Hma))/(max(Hma)-
min(Hma));

% find maximum feature
[i_sh, index_rem_sh]=max(Hsh); [i_me, index_rem_me]=max(Hme); [i_rn,
index_rem_rn]=max(Hrn); [i_pp, index_rem_pp]=max(Hpp); [i_ts, index_rem_ts]
=max(Hts); [i_ma, index_rem_ma]=max(Hma);

% removing feature from the data
data_mod_sh=[dataold(:,1:index_rem_sh-1) dataold(:,index_rem_sh+1:end)];
data_mod_me=[dataold(:,1:index_rem_me-1) dataold(:,index_rem_me+1:end)];
data_mod_rn=[dataold(:,1:index_rem_rn-1) dataold(:,index_rem_rn+1:end)];
data_mod_pp=[dataold(:,1:index_rem_pp-1) dataold(:,index_rem_pp+1:end)];
data_mod_ts=[dataold(:,1:index_rem_ts-1) dataold(:,index_rem_ts+1:end)];
data_mod_ma=[dataold(:,1:index_rem_ma-1) dataold(:,index_rem_ma+1:end)];

data_mod=[data_mod_sh,data_mod_me,data_mod_rn,data_mod_pp,data_mod_ts,
data_mod_ma]; index_rem=[index_rem_sh,index_rem_me,index_rem_rn,index_rem_pp,
index_rem_ts,index_rem_ma]; H=[Hsh;Hme;Hrn;Hpp;Hts;Hma];

j=1;

return

%%%%%%%%%%%%%%%%%%%%%%%%
```

```
%%example
% d =[0.4600     0.3400     0.1400     0.0300     0.9218     1.0000
%       0.5000     0.3400     0.1500     0.0200     0.7382     1.0000
%       0.4400     0.2900     0.1400     0.0200     0.1763     1.0000
%       0.7600     0.3000     0.6600     0.2100     0.4057     2.0000
%       0.4900     0.2500     0.4500     0.1700     0.9355     2.0000
%       0.7300     0.2900     0.6300     0.1800     0.9169     2.0000]
% [data_mod, index_rem,SIM,SIM1,H]=feat_sel_sim1(d, 'luca', 1)

function [data_mod, index_rem,SIM,SIM1,H]=feat_sel_sim_YU(data, land)
%

%    p in (0, \infty) as default p=1. if nargin
p=1;
end

l=max(data(:,end)); % #-classes m=size(data,1); % #-samples t=size(data,2)-1;
% #-features dataold=data;
tmp=[];
% forming idealvec using arithmetic mean idealvec=zeros(l,t);
for k=1:l idealvec_s(k,:)=mean(data(find(data(:,end)==k),1:t));
end

%scaling data between [0,1] data_v=data(:,1:t); data_c=data(:,t+1); %labels
mins_v = min(data_v);
Ones = ones(size(data_v));
data_v = data_v+Ones*diag(abs(mins_v)); for k=1:l
tmp=[tmp;abs(mins_v)]; end
tmp;
idealvec_s = idealvec_s+tmp; maxs_v = max(data_v);
data_v = data_v*diag(maxs_v.^(-1)); idealvec_s=idealvec_s./repmat(maxs_v,
l,1); data = [data_v, data_c];

% sample data datalearn_s=data(:,1:t);
%land=0.1;
% similarities

sim=zeros(t,m,l); for j=1:m
for i=1:t
for k=1:l
%similarity YU's    with landa=0.1
sn_1(i,j,k)=min(1,(1-datalearn_s(j,i))+(idealvec_s(k,i))
+(land*((1-datalearn_s(j,i))*(idealvec_s(k,i)))));

i)))));
sn_2(i,j,k)=min(1,(datalearn_s(j,i))+(1-idealvec_s(k,i))+(land*((datalearn_s
(j,i))*(1- idealvec_s(k, sim(i,j,k)=max(0,(1+land)*(sn_1(i,j,k)+sn_2(i,j,k)-
1)-((land)*(sn_1(i,j,k)*sn_2(i,j,k))));
end
endend
```

```
%class 1,2,3,4,. n
% SIM(:,:,1) = similarity for class 1
% SIM(:,:,2) = similarity for class 2 SIM=sim;
% reduce number of dimensions in sim sim=reshape(sim,t,m*l)';
SIM1=sim;

% possibility for two different entropy measures
% moodifying zero and one values of the similarity values to work with
% De Luca's entropy measure
%%%%%%Hsh(1)
delta=1E-10; sim(find(sim==0))=delta; sim(find(sim==1))=1-delta;
Hsh=sum(-sim.*log(sim)-(1-sim).*log(1-sim));
%Hsh=(Hsh-min(Hsh))/(max(Hsh)-min(Hsh));

% %%%%%%Hme(2)
Hme=sum(sin(pi/2*sim)+sin(pi/2*(1-sim))-1);
%Hme=(Hme-min(Hme))/(max(Hme)-min(Hme));

%%%%%%%%Hrn(3)
alfa_rn=0.8;
Hrn=log(sum((sim.^alfa_rn)+((1-sim).^alfa_rn)))/(1-alfa_rn);
%Hrn=(Hrn-min(Hrn))/(max(Hrn)-min(Hrn));

% %%%%%%%Hpp(4)
Hpp=sum((sim.*exp(1.-sim))+(((1-sim).*exp(1.-(1-sim)))));
%Hpp=(Hpp-min(Hpp))/(max(Hpp)-min(Hpp));

% %%%%%%%%Hts(5)
alfa_ts=0.5;
Hts=1-(sum((sim.^alfa_ts)+((1-sim).^alfa_ts)))/(alfa_ts-1);
%Hts=(Hts-min(Hts))/(max(Hts)-min(Hts));

%%%%%%%%%Hma(6)
gama=2; a_ma=0.5; b_ma=3; alfa_ma=0.1; beta_ma=2;
Hma=(sum(((sim.^alfa_ma).*exp(a_ma.*sim.^alfa_ma+b_ma).^beta_ma)+((1-sim).
^alfa_ma).
*exp(a_ma.*(1-sim).^alfa_ma+b_ma).^beta_ma));
%Hma=(Hma-min(Hma))/(max(Hma)-min(Hma));

% find maximum feature
[i_sh, index_rem_sh]=max(Hsh); [i_me, index_rem_me]=max(Hme);
[i_rn, index_rem_rn]=max(Hrn); [i_pp, index_rem_pp]=max(Hpp);
[i_ts, index_rem_ts]=max(Hts); [i_ma, index_rem_ma]=max(Hma);

% removing feature from the data
data_mod_sh=[dataold(:,1:index_rem_sh-1) dataold(:,index_rem_sh+1:end)];
data_mod_me=[dataold(:,1:index_rem_me-1) dataold(:,index_rem_me+1:end)];
data_mod_rn=[dataold(:,1:index_rem_rn-1) dataold(:,index_rem_rn+1:end)];
data_mod_pp=[dataold(:,1:index_rem_pp-1) dataold(:,index_rem_pp+1:end)];
data_mod_ts=[dataold(:,1:index_rem_ts-1) dataold(:,index_rem_ts+1:end)];
data_mod_ma=[dataold(:,1:index_rem_ma-1) dataold(:,index_rem_ma+1:end)];
```

```
data_mod=[data_mod_sh,data_mod_me,data_mod_rn,data_mod_pp,data_mod_ts,
data_mod_ma]; index_rem=[index_rem_sh,index_rem_me,index_rem_rn,index_rem_pp,
index_rem_ts,index_rem_ma]; H=[Hsh;Hme;Hrn;Hpp;Hts;Hma];

j=1;


return

%%%%%%%%%%%%%%%%%%%%%%%%

%%example
% d =[0.4600    0.3400    0.1400    0.0300    0.9218    1.0000
%     0.5000    0.3400    0.1500    0.0200    0.7382    1.0000
%     0.4400    0.2900    0.1400    0.0200    0.1763    1.0000
%     0.7600    0.3000    0.6600    0.2100    0.4057    2.0000
%     0.4900    0.2500    0.4500    0.1700    0.9355    2.0000
%     0.7300    0.2900    0.6300    0.1800    0.9169    2.0000]
% [data_mod, index_rem,SIM,SIM1,H]=feat_sel_sim1(d, 'luca', 1)

function [data_mod, index_rem,SIM,SIM1,H]=feat_sel_sim1(data, measure, p)
%

%    p in (0, \infty) as default p=1. if nargin
p=1;
end
if nargin<2 measure='luca'
end

if nargin<2 measure='RN'
end

l=max(data(:,end)); % #-classes m=size(data,1); % #-samples t=size(data,2)-1;
% #-features dataold=data;
tmp=[];
% forming idealvec using arithmetic mean idealvec=zeros(l,t);
for k=1:l idealvec_s(k,:)=mean(data(find(data(:,end)==k),1:t));
end

%scaling data between [0,1] data_v=data(:,1:t); data_c=data(:,t+1); %labels
mins_v = min(data_v);
Ones = ones(size(data_v));
data_v = data_v+Ones*diag(abs(mins_v)); for k=1:l
tmp=[tmp;abs(mins_v)]; end
tmp;
idealvec_s = idealvec_s+tmp; maxs_v = max(data_v);
data_v = data_v*diag(maxs_v.^(-1));

idealvec_s=idealvec_s./repmat(maxs_v,l,1); data = [data_v, data_c];
% sample data datalearn_s=data(:,1:t); land=0.1;
% similarities sim=zeros(t,m,l); for j=1:m
```

```
for i=1:t
for k=1:l
%%similarity Luca    with p=1
sim(i,j,k)=(1-abs(idealvec_s(k,i)^p-datalearn_s(j,i))^p)^(1/p);
%%similarity YU's    with landa=0.1
%    sn_1(i,j,k)=min(1,(1-datalearn_s(j,i))+(idealvec_s(k,i))+(land*((1-
datalearn_s(j,i))*(idealvec_s (k,i)))));
%    sn_2(i,j,k)=min(1,(datalearn_s(j,i))+(1-idealvec_s(k,i))
+(land*((datalearn_s(j,i))*(1- idealvec_s (k,i)))));
%        sim(i,j,k)=max(0,(1-land)*(sn_1(i,j,k)+sn_2(i,j,k)-1)-((land)*(sn_1
(i,j,k)*sn_2(i,j,k)))); end
endend
%class 1,2,3,4,. n
% SIM(:,:,1) = similarity for class 1
% SIM(:,:,2) = similarity for class 2 SIM=sim;
% reduce number of dimensions in sim sim=reshape(sim,t,m*l)';
SIM1=sim;

% possibility for two different entropy measures if measure=='luca'
% moodifying zero and one values of the similarity values to work with
% De Luca's entropy measure
%Hsh(1)
delta=1E-10; sim(find(sim==0))=delta; sim(find(sim==1))=1-delta;
H=sum(-sim.*log(sim)-(1-sim).*log(1-sim));
H=(H-min(H))/(max(H)-min(H)); end


if measure=='park'
%    %Hme(2)
%    H=sum(sin(pi/2*sim)+sin(pi/2*(1-sim))-1);
%    %Hrn(3)
%    alfa=0.8;
%    H=log(sum((sim.^alfa)+((1-sim).^alfa)))/(1-alfa);
%    %Hpp(4)
%    H=sum((sim.*exp(1.-sim))+(((1-sim).*exp(1.-(1-sim)))));
%    %Hts(5)
%    alfa=0.5;
%    H=1-(sum((sim.^alfa)+((1-sim).^alfa)))/(alfa-1);
%Hma(6)
gama=2; a=0.5; b=3;
alfa=0.1; beta=2;
H=(sum(((sim.^alfa).*exp(a.*sim.^alfa+b).^beta)+((1-sim).^alfa).*exp(a.*(1-
sim).^alfa+b).
^beta));

H=(H-min(H))/(max(H)-min(H));
end

% find maximum feature [i, index_rem]=max(H);

% removing feature from the data data_mod=[dataold(:,1:index_rem-1) dataold
(:,index_rem+1:end)]; j=1;
```

```
return

%%%%%%%%%%%%%%%%%%%%%%%

%%example
% d =[0.4600    0.3400    0.1400    0.0300    0.9218    1.0000
%     0.5000    0.3400    0.1500    0.0200    0.7382    1.0000
%     0.4400    0.2900    0.1400    0.0200    0.1763    1.0000
```

```
%     0.7600    0.3000    0.6600    0.2100    0.4057    2.0000
%     0.4900    0.2500    0.4500    0.1700    0.9355    2.0000
%     0.7300    0.2900    0.6300    0.1800    0.9169    2.0000]
```

```
% [data_mod, index_rem,SIM,SIM1,H]=feat_sel_sim1(d, 'luca', 1)

d =[0.4600    0.3400    0.1400    0.0300    0.9218    1.0000
    0.5000    0.3400    0.1500    0.0200    0.7382    2.0000
    0.4400    0.2900    0.1400    0.0200    0.1763    1.0000
    0.7600    0.3000    0.6600    0.2100    0.4057    2.0000
    0.4900    0.2500    0.4500    0.1700    0.9355    1.0000
    0.7300    0.2900    0.6300    0.1800    0.9169    2.0000];

xdata=d(1:5,1:2,:);
group=d(1:5,6,:);
xnew=d(4:6,1:2);

figure,
svmStruct_linear_QP = svmtrain(xdata,group,'showplot',
true,'kernel_function','linear','method','QP')
svmStruct_linear_SMO = svmtrain(xdata,group,'showplot',
true,'kernel_function','linear','method','SMO')
svmStruct_linear_LS = svmtrain(xdata,group,'showplot',
true,'kernel_function','linear','method','LS')
newClasses_linear_QP = svmclassify(svmStruct_linear_QP,xnew)
newClasses_linear_SMO = svmclassify(svmStruct_linear_SMO,xnew)
newClasses_linear_LS = svmclassify(svmStruct_linear_LS,xnew)

figure,
svmStruct_quadratic_QP = svmtrain(xdata,group,'showplot',
true,'kernel_function','quadratic','method','QP')
svmStruct_quadratic_SMO = svmtrain(xdata,group,'showplot',
true,'kernel_function','quadratic','method','SMO') svmStruct_quadratic_LS =
svmtrain(xdata,group,'showplot',
true,'kernel_function','quadratic','method','LS')
newClasses_quadratic_QP = svmclassify(svmStruct_quadratic_QP,xnew)
newClasses_quadratic_SMO = svmclassify(svmStruct_quadratic_SMO,xnew)
newClasses_quadratic_LS = svmclassify(svmStruct_quadratic_LS,xnew)
```

```
figure,
svmStruct_polynomial_QP = svmtrain(xdata,group,'showplot',
true,'kernel_function','polynomial','method','QP') svmStruct_polynomial_SMO =
svmtrain(xdata,group,'showplot',
true,'kernel_function','polynomial','method','SMO') svmStruct_polynomial_LS =
svmtrain(xdata,group,'showplot',
true,'kernel_function','polynomial','method','LS') newClasses_polynomial_QP =
svmclassify(svmStruct_polynomial_QP,xnew)
newClasses_polynomial_SMO = svmclassify(svmStruct_polynomial_SMO,xnew)
newClasses_polynomial_LS = svmclassify(svmStruct_polynomial_LS,xnew)

figure,
svmStruct_rbf_QP = svmtrain(xdata,group,'showplot',
true,'kernel_function','rbf','method','QP') svmStruct_rbf_SMO = svmtrain
(xdata,group,'showplot',true,'kernel_function','rbf','method','SMO')
svmStruct_rbf_LS = svmtrain(xdata,group,'showplot',
true,'kernel_function','rbf','method','LS') newClasses_rbf_QP = svmclassify
(svmStruct_rbf_QP,xnew)
newClasses_rbf_SMO = svmclassify(svmStruct_rbf_SMO,xnew) newClasses_rbf_LS =
svmclassify(svmStruct_rbf_LS,xnew)

figure,
svmStruct_mlp_QP = svmtrain(xdata,group,'showplot',
true,'kernel_function','mlp','method','QP') svmStruct_mlp_SMO = svmtrain
(xdata,group,'showplot',true,'kernel_function','mlp','method','SMO')
svmStruct_mlp_LS = svmtrain(xdata,group,'showplot',
true,'kernel_function','mlp','method','LS') newClasses_mlp_QP = svmclassify
(svmStruct_mlp_QP,xnew)
newClasses_mlp_SMO = svmclassify(svmStruct_mlp_SMO,xnew) newClasses_mlp_LS =
svmclassify(svmStruct_mlp_LS,xnew)
%% *************custom kernel rng(1); % For reproducibility
n = 100; % Number of points per quadrant

r1 = sqrt(rand(2*n,1));     % Random radii
t1 = [pi/2*rand(n,1); (pi/2*rand(n,1)+pi)]; % Random angles for Q1 and Q3 X1 =
[r1.*cos(t1) r1.*sin(t1)];     % Polar-to-Cartesian conversion

r2 = sqrt(rand(2*n,1));
t2 = [pi/2*rand(n,1)+pi/2; (pi/2*rand(n,1)-pi/2)]; % Random angles for Q2 and
Q4 X2 = [r2.*cos(t2) r2.*sin(t2)];

X = [X1; X2];     % Predictors Y = ones(4*n,1);
Y(2*n + 1:end) = -1; % Labels
```

Index

Automation and Computational Intelligence for Road Maintenance and Management: Advances and Applications, First Edition. Hamzeh Zakeri, Fereidoon Moghadas Nejad, and Amir H. Gandomi.

l

m

n

o

p

q

r

s

t

u

Printed and bound by CPI Group (UK) Ltd, Croydon, CR0 4YY
16/08/2022
03141872-0001